NUMERICAL LINEAR ALGEBRA

PHYSICAL SCIENCES

SOCIAL AND BEHAVIORAL SCIENCES

STATISTICS

Elementary Linear Algebra

SIXTH EDITION

RON LARSON
The Pennsylvania State University
The Behrend College

DAVID C. FALVO
The Pennsylvania State University
The Behrend College

HOUGHTON MIFFLIN HARCOURT PUBLISHING COMPANY *Boston* *New York*

Publisher: Richard Stratton
Senior Sponsoring Editor: Cathy Cantin
Senior Marketing Manager: Jennifer Jones
Discipline Product Manager: Gretchen Rice King
Associate Editor: Janine Tangney
Associate Editor: Jeannine Lawless
Senior Project Editor: Kerry Falvey
Program Manager: Touraj Zadeh
Senior Media Producer: Douglas Winicki
Senior Content Manager: Maren Kunert
Art and Design Manager: Jill Haber
Cover Design Manager: Anne S. Katzeff
Senior Photo Editor: Jennifer Meyer Dare
Senior Composition Buyer: Chuck Dutton
New Title Project Manager: Susan Peltier
Manager of New Title Project Management: Pat O'Neill
Editorial Assistant: Amy Haines
Marketing Assistant: Michael Moore
Editorial Assistant: Laura Collins

Cover image: © Carl Reader/age fotostock

Printed in the U.S.A.

Library of Congress Control Number: 2007940572

Instructor's examination copy
ISBN-13: 978-0-547-00481-5
ISBN-10: 0-547-00481-8

For orders, use student text ISBNs
ISBN-13: 978-0-618-78376-2
ISBN-10: 0-618-78376-8

123456789-DOC-12 11 10 09 08

Contents

*Available online at **college.hmco.com/pic/larsonELA6e.***

A Word from the Authors

Welcome! We have designed *Elementary Linear Algebra,* Sixth Edition, for the introductory linear algebra course.

Students embarking on a linear algebra course should have a thorough knowledge of algebra, and familiarity with analytic geometry and trigonometry. We do not assume that calculus is a prerequisite for this course, but we do include examples and exercises requiring calculus in the text. These exercises are clearly labeled and can be omitted if desired.

Many students will encounter mathematical formalism for the first time in this course. As a result, our primary goal is to present the major concepts of linear algebra clearly and concisely. To this end, we have carefully selected the examples and exercises to balance theory with applications and geometrical intuition.

The order and coverage of topics were chosen for maximum efficiency, effectiveness, and balance. For example, in Chapter 4 we present the main ideas of vector spaces and bases, beginning with a brief look leading into the vector space concept as a natural extension of these familiar examples. This material is often the most difficult for students, but our approach to linear independence, span, basis, and dimension is carefully explained and illustrated by examples. The eigenvalue problem is developed in detail in Chapter 7, but we lay an intuitive foundation for students earlier in Section 1.2, Section 3.1, and Chapter 4.

Additional online Chapters 8, 9, and 10 cover complex vector spaces, linear programming, and numerical methods. They can be found on the student website for this text at *college.hmco.com/pic/larsonELA6e*.

Please read on to learn more about the features of the Sixth Edition.

We hope you enjoy this new edition of *Elementary Linear Algebra*.

Acknowledgments

We would like to thank the many people who have helped us during various stages of the project. In particular, we appreciate the efforts of the following colleagues who made many helpful suggestions along the way:

Elwyn Davis, *Pittsburg State University, VA*
Gary Hull, *Frederick Community College, MD*
Dwayne Jennings, *Union University, TN*
Karl Reitz, *Chapman University, CA*
Cindia Stewart, *Shenandoah University, VA*
Richard Vaughn, *Paradise Valley Community College, AZ*
Charles Waters, *Minnesota State University–Mankato, MN*
Donna Weglarz, *Westwood College–DuPage, IL*
John Woods, *Southwestern Oklahoma State University, OK*

We would like to thank Bruce H. Edwards, The University of Florida, for his contributions to previous editions of *Elementary Linear Algebra.*

We would also like to thank Helen Medley for her careful accuracy checking of the textbook.

On a personal level, we are grateful to our wives, Deanna Gilbert Larson and Susan Falvo, for their love, patience, and support. Also, special thanks go to R. Scott O'Neil.

Ron Larson
David C. Falvo

Theorems and Proofs

THEOREM 2.9 The Inverse of a Product If A and B are invertible matrices of size n, then AB is invertible and

$$(AB)^{-1} = B^{-1}A^{-1}.$$

Theorems are presented in clear and mathematically precise language.

Key theorems are also available via **PowerPoint® Presentation** on the instructor website. They can be displayed in class using a computer monitor or projector, or printed out for use as class handouts.

Students will gain experience solving **proofs** presented in several different ways:

- Some proofs are presented in **outline form,** omitting the need for burdensome calculations.
- Specialized exercises labeled **Guided Proofs** lead students through the initial steps of constructing proofs and then utilizing the results.
- The proofs of several theorems are left as **exercises,** to give students additional practice.

PROOF To begin, observe that if E is an elementary matrix, then, by Theorem 3.3, the next few statements are true. If E is obtained from I by interchanging two rows, then $|E| = -1$. If E is obtained by multiplying a row of I by a nonzero constant c, then $|E| = c$. If E is obtained by adding a multiple of one row of I to another row of I, then $|E| = 1$. Additionally, by Theorem 2.12, if E results from performing an elementary row operation on I and the same elementary row operation is performed on B, then the matrix EB results. It follows that

$$|EB| = |E|\,|B|.$$

This can be generalized to conclude that $|E_k \cdots E_2E_1B| = |E_k| \cdots |E_2|\,|E_1|\,|B|$, where E_i is an elementary matrix. Now consider the matrix AB. If A is *nonsingular*, then, by Theorem 2.14, it can be written as the product of elementary matrices $A = E_k \cdots E_2E_1$ and you can write

$|A|\,|B|$.

56. Guided Proof Prove Theorem 3.9: If A is a square matrix, then $\det(A) = \det(A^T)$.

Getting Started: To prove that the determinants of A and A^T are equal, you need to show that their cofactor expansions are equal. Because the cofactors are $\pm$ determinants of smaller matrices, you need to use mathematical induction.

(i) Initial step for induction: If A is of order 1, then $A = [a_{11}] = A^T$, so $\det(A) = \det(A^T) = a_{11}$.

(ii) Assume the inductive hypothesis holds for all matrices of order $n - 1$. Let A be a square matrix of order n. Write an expression for the determinant of A by expanding by the first row.

(iii) Write an expression for the determinant of A^T by expanding by the first column.

(iv) Compare the expansions in (i) and (ii). The entries of the first row of A are the same as the entries of the first column of A^T. Compare cofactors (these are the $\pm$ determinants of smaller matrices that are transposes of one another) and use the inductive hypothesis to conclude that they are equal as well.

Real World Applications

REVISED! Each chapter ends with a section on **real-life applications** of linear algebra concepts, covering interesting topics such as:

- Computer graphics
- Cryptography
- Population growth and more!

A full listing of the applications can be found in the **Index of Applications** inside the front cover.

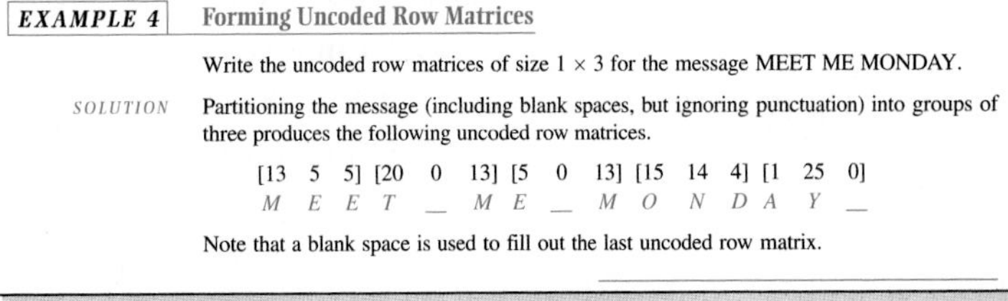
EXAMPLE 4 **Forming Uncoded Row Matrices**

Write the uncoded row matrices of size 1×3 for the message MEET ME MONDAY.

SOLUTION Partitioning the message (including blank spaces, but ignoring punctuation) into groups of three produces the following uncoded row matrices.

$$\begin{matrix}[13 & 5 & 5] & [20 & 0 & 13] & [5 & 0 & 13] & [15 & 14 & 4] & [1 & 25 & 0] \\ M & E & E & T & \_ & M & E & \_ & M & O & N & D & A & Y & \_\end{matrix}$$

Note that a blank space is used to fill out the last uncoded row matrix.

INDEX OF APPLICATIONS

Conceptual Understanding

CHAPTER OBJECTIVES

- Find the determinants of a 2 × 2 matrix and a triangular matrix.
- Find the minors and cofactors of a matrix and use expansion by cofactors to find the determinant of a matrix.
- Use elementary row or column operations to evaluate the determinant of a matrix.
- Recognize conditions that yield zero determinants.
- Find the determinant of an elementary matrix.
- Use the determinant and properties of the determinant to decide whether a matrix is singular or nonsingular, and recognize equivalent conditions for a nonsingular matrix.
- Verify and find an eigenvalue and an eigenvector of a matrix.

NEW! **Chapter Objectives** are now listed on each chapter opener page. These objectives highlight the key concepts covered in the chapter, to serve as a guide to student learning.

The **Discovery** features are designed to help students develop an intuitive understanding of mathematical concepts and relationships.

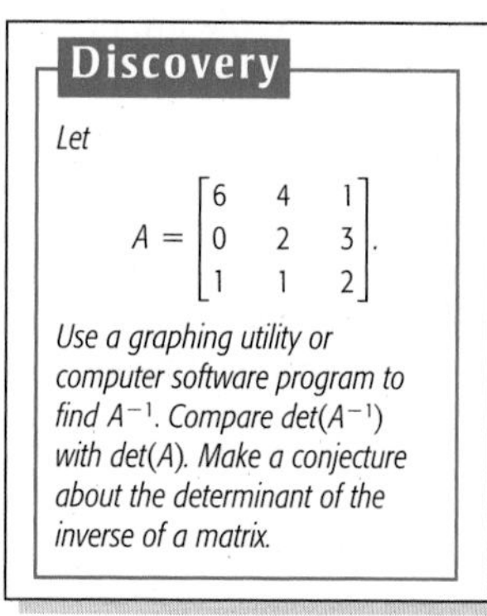

Discovery

Let

$$A = \begin{bmatrix} 6 & 4 & 1 \\ 0 & 2 & 3 \\ 1 & 1 & 2 \end{bmatrix}.$$

Use a graphing utility or computer software program to find A^{-1}. Compare $\det(A^{-1})$ with $\det(A)$. Make a conjecture about the determinant of the inverse of a matrix.

True or False? In Exercises 62–65, determine whether each statement is true or false. If a statement is true, give a reason or cite an appropriate statement from the text. If a statement is false, provide an example that shows the statement is not true in all cases or cite an appropriate statement from the text.

62. (a) The nullspace of A is also called the solution space of A.

(b) The nullspace of A is the solution space of the homogeneous system $A\mathbf{x} = \mathbf{0}$.

63. (a) If an $m \times n$ matrix A is row-equivalent to an $m \times n$ matrix B, then the row space of A is equivalent to the row space of B.

(b) If A is an $m \times n$ matrix of rank r, then the dimension of the solution space of $A\mathbf{x} = \mathbf{0}$ is $m - r$.

True or False? exercises test students' knowledge of core concepts. Students are asked to give examples or justifications to support their conclusions.

Graphics and Geometric Emphasis

Visualization skills are necessary for the understanding of mathematical concepts and theory. The Sixth Edition includes the following resources to help develop these skills:

- **Graphs** accompany examples, particularly when representing vector spaces and inner product spaces.
- **Computer-generated illustrations** offer geometric interpretations of problems.

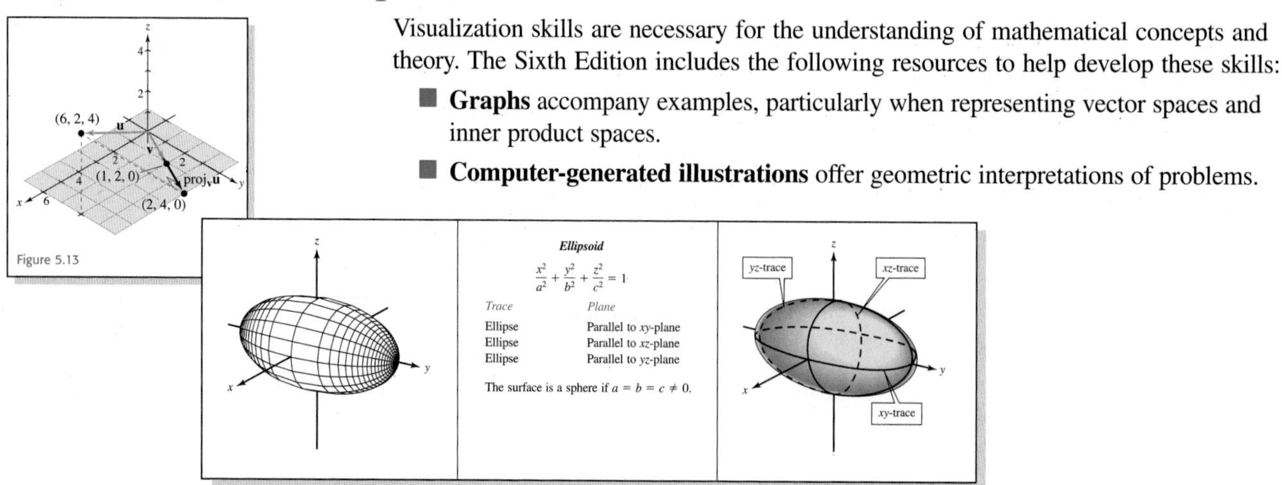

Problem Solving and Review

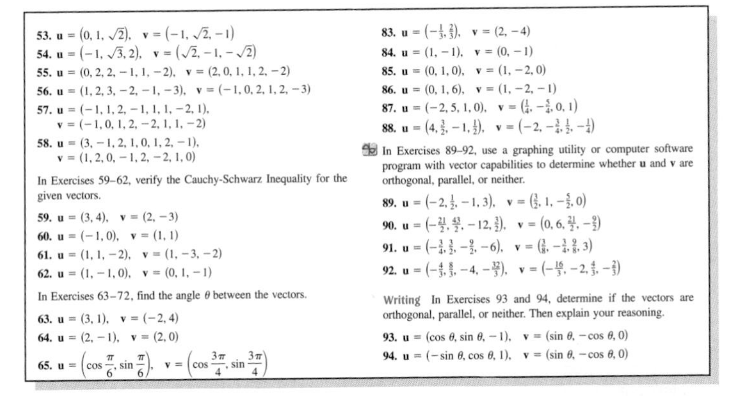

53. $\mathbf{u} = (0, 1, \sqrt{2}), \quad \mathbf{v} = (-1, \sqrt{2}, -1)$
54. $\mathbf{u} = (-1, \sqrt{3}, 2), \quad \mathbf{v} = (\sqrt{2}, -1, -\sqrt{2})$
55. $\mathbf{u} = (0, 2, 2, -1, 1, -2), \quad \mathbf{v} = (2, 0, 1, 1, 2, -2)$
56. $\mathbf{u} = (1, 2, 3, -2, -1, -3), \quad \mathbf{v} = (-1, 0, 2, 1, 2, -3)$
57. $\mathbf{u} = (-1, 1, 2, -1, 1, 1, -2, 1),$ $\mathbf{v} = (-1, 0, 1, 2, -2, 1, 1, -2)$
58. $\mathbf{u} = (3, -1, 2, 1, 0, 1, 2, -1),$ $\mathbf{v} = (1, 2, 0, -1, 2, -2, 1, 0)$

In Exercises 59–62, verify the Cauchy-Schwarz Inequality for the given vectors.

59. $\mathbf{u} = (3, 4), \quad \mathbf{v} = (2, -3)$
60. $\mathbf{u} = (-1, 0), \quad \mathbf{v} = (1, 1)$
61. $\mathbf{u} = (1, 1, -2), \quad \mathbf{v} = (1, -3, -2)$
62. $\mathbf{u} = (1, -1, 0), \quad \mathbf{v} = (0, 1, -1)$

In Exercises 63–72, find the angle θ between the vectors.

63. $\mathbf{u} = (3, 1), \quad \mathbf{v} = (-2, 4)$
64. $\mathbf{u} = (2, -1), \quad \mathbf{v} = (2, 0)$
65. $\mathbf{u} = \left(\cos\frac{\pi}{6}, \sin\frac{\pi}{6}\right), \quad \mathbf{v} = \left(\cos\frac{3\pi}{4}, \sin\frac{3\pi}{4}\right)$

83. $\mathbf{u} = \left(-\frac{1}{3}, \frac{2}{3}\right), \quad \mathbf{v} = (2, -4)$
84. $\mathbf{u} = (1, -1), \quad \mathbf{v} = (0, -1)$
85. $\mathbf{u} = (0, 1, 0), \quad \mathbf{v} = (1, -2, 0)$
86. $\mathbf{u} = (0, 1, 6), \quad \mathbf{v} = (1, -2, -1)$
87. $\mathbf{u} = (-2, 5, 1, 0), \quad \mathbf{v} = \left(\frac{1}{4}, -\frac{5}{4}, 0, 1\right)$
88. $\mathbf{u} = \left(4, \frac{3}{2}, -1, \frac{1}{2}\right), \quad \mathbf{v} = \left(-2, -\frac{3}{4}, \frac{1}{2}, -\frac{1}{4}\right)$

In Exercises 89–92, use a graphing utility or computer software program with vector capabilities to determine whether **u** and **v** are orthogonal, parallel, or neither.

89. $\mathbf{u} = \left(-2, \frac{1}{2}, -1, 3\right), \quad \mathbf{v} = \left(\frac{3}{2}, 1, -\frac{5}{2}, 0\right)$
90. $\mathbf{u} = \left(-\frac{21}{2}, \frac{43}{2}, -12, \frac{3}{2}\right), \quad \mathbf{v} = \left(0, 6, \frac{21}{2}, -\frac{9}{2}\right)$
91. $\mathbf{u} = \left(-\frac{3}{4}, \frac{3}{2}, -\frac{9}{2}, -6\right), \quad \mathbf{v} = \left(\frac{3}{8}, -\frac{3}{4}, \frac{9}{8}, 3\right)$
92. $\mathbf{u} = \left(-\frac{4}{3}, \frac{8}{3}, -4, -\frac{32}{3}\right), \quad \mathbf{v} = \left(-\frac{16}{3}, -2, \frac{4}{3}, -\frac{2}{3}\right)$

Writing In Exercises 93 and 94, determine if the vectors are orthogonal, parallel, or neither. Then explain your reasoning.

93. $\mathbf{u} = (\cos\theta, \sin\theta, -1), \quad \mathbf{v} = (\sin\theta, -\cos\theta, 0)$
94. $\mathbf{u} = (-\sin\theta, \cos\theta, 1), \quad \mathbf{v} = (\sin\theta, -\cos\theta, 0)$

REVISED! Comprehensive section and chapter exercise sets give students practice in problem-solving techniques and test their understanding of mathematical concepts. A wide variety of exercise types are represented, including:

- **Writing exercises**
- **Guided Proof exercises**
- **Technology exercises,** indicated throughout the text with the technology icon.
- **Applications** exercises
- Exercises utilizing **electronic data sets,** indicated by HM and found on the student website at ***college.hmco.com/pic/larsonELA6e***

Each chapter includes two **Chapter Projects,** which offer the opportunity for group activities or more extensive homework assignments.

Chapter Projects are focused on theoretical concepts or applications, and many encourage the use of technology.

CHAPTER 3 Projects

1 Eigenvalues and Stochastic Matrices

In Section 2.5, you studied a consumer preference model for competing cable television companies. The matrix representing the transition probabilities was

$$P = \begin{bmatrix} 0.70 & 0.15 & 0.15 \\ 0.20 & 0.80 & 0.15 \\ 0.10 & 0.05 & 0.70 \end{bmatrix}.$$

When provided with the initial state matrix X, you observed that the number of subscribers after 1 year is the product PX.

$$X = \begin{bmatrix} 15{,}000 \\ 20{,}000 \\ 65{,}000 \end{bmatrix} \longrightarrow PX = \begin{bmatrix} 0.70 & 0.15 & 0.15 \\ 0.20 & 0.80 & 0.15 \\ 0.10 & 0.05 & 0.70 \end{bmatrix} \begin{bmatrix} 15{,}000 \\ 20{,}000 \\ 65{,}000 \end{bmatrix} = \begin{bmatrix} 23{,}250 \\ 28{,}750 \\ 48{,}000 \end{bmatrix}$$

Cumulative Tests follow chapters 3, 5, and 7, and help students synthesize the knowledge they have accumulated throughout the text, as well as prepare for exams and future mathematics courses.

CHAPTERS 4 & 5 Cumulative Test

Take this test as you would take a test in class. After you are done, check your work against the answers in the back of the book.

1. Given the vectors $\mathbf{v} = (1, -2)$ and $\mathbf{w} = (2, -5)$, find and sketch each vector.
   (a) $\mathbf{v} + \mathbf{w}$ (b) $3\mathbf{v}$ (c) $2\mathbf{v} - 4\mathbf{w}$
2. If possible, write $\mathbf{w} = (2, 4, 1)$ as a linear combination of the vectors $\mathbf{v}_1$, $\mathbf{v}_2$, and $\mathbf{v}_3$.
   $\mathbf{v}_1 = (1, 2, 0), \quad \mathbf{v}_2 = (-1, 0, 1), \quad \mathbf{v}_3 = (0, 3, 0)$
3. Prove that the set of all singular 2×2 matrices is not a vector space.

Historical Emphasis

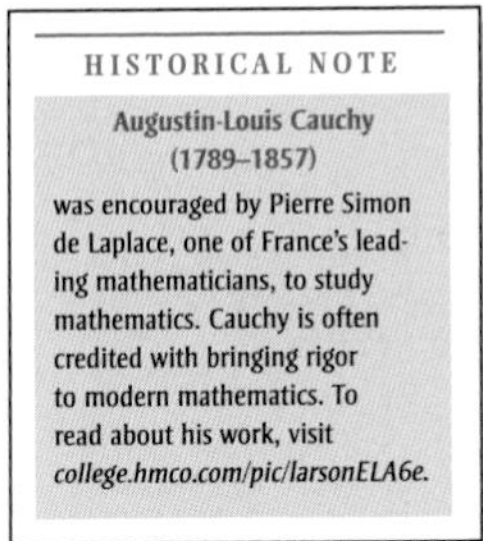

HISTORICAL NOTE

Augustin-Louis Cauchy (1789–1857) was encouraged by Pierre Simon de Laplace, one of France's leading mathematicians, to study mathematics. Cauchy is often credited with bringing rigor to modern mathematics. To read about his work, visit *college.hmco.com/pic/larsonELA6e.*

NEW! **Historical Notes** are included throughout the text and feature brief biographies of prominent mathematicians who contributed to linear algebra.

Students are directed to the Web to read the full biographies, which are available via **PowerPoint® Presentation.**

Computer Algebra Systems and Graphing Calculators

Technology Note

You can use a graphing utility or computer software program to find the unit vector for a given vector. For example, you can use a graphing utility to find the unit vector for **v** = (−3, 4), which may appear as:

```
VECTOR:V        2
 e1=-3
 e2=4
```

```
unitV V
            [-.6 .8]
```

The **Technology Note** feature in the text indicates how students can utilize graphing calculators and computer algebra systems appropriately in the problem-solving process.

NEW! **Online Technology Guide** provides the coverage students need to use computer algebra systems and graphing calculators with this text.

Provided on the accompanying student website, this guide includes **CAS and graphing calculator keystrokes for select examples in the text.** These examples feature an accompanying Technology Note, directing students to the Guide for instruction on using their CAS/graphing calculator to solve the example.

In addition, the Guide provides an **Introduction to MATLAB, Maple, Mathematica, and Graphing Calculators,** as well as a section on **Technology Pitfalls.**

Technology Note

You can use a computer software program or graphing utility with a built-in power regression program to verify the result of Example 10. For example, using the data in Table 5.2 and a graphing utility, a power fit program would result in an answer of (or very similar to) $y \approx 1.00042x^{1.49954}$. Keystrokes and programming syntax for these utilities/programs applicable to Example 10 are provided in the **Online Technology Guide,** available at *college.hmco.com/pic/larsonELA6e*.

EXAMPLE 7 Using Elimination to Rewrite a System in Row-Echelon Form

Solve the system.

$$\begin{aligned} x - 2y + 3z &= 9 \\ -x + 3y \qquad &= -4 \\ 2x - 5y + 5z &= 17 \end{aligned}$$

Keystrokes for TI-83
Enter the system into matrix A.
To rewrite the system in row-echelon form, use the following keystrokes.
MATRX → ALPHA [A] MATRX ENTER ENTER

Keystrokes for TI-83 Plus
Enter the system into matrix A.
To rewrite the system in row-echelon form, use the following keystrokes.
2nd [MATRX] → ALPHA [A] 2nd [MATRX] ENTER ENTER

Keystrokes for TI-84 Plus
Enter the system into matrix A.
To rewrite the system in row-echelon form, use the following keystrokes.
2nd [MATRIX] → ALPHA [A] 2nd [MATRIX] ENTER ENTER

Keystrokes for TI-86
Enter the system into matrix A.
To rewrite the system in row-echelon form, use the following keystrokes.
2nd [MATRX] F4 F4 ALPHA [A] ENTER

Part I: Texas Instruments TI-83, TI-83 Plus, TI-84 Plus Graphing Calculator

I.1 Systems of Linear Equations

I.1.1 Basics: Press the ON key to begin using your TI-83 calculator. If you need to adjust the display contrast, first press 2nd, then press and hold ▲ (the *up* arrow key) to increase the contrast or ▼ (the *down* arrow key) to decrease the contrast. As you press and hold ▲ or ▼, an integer between 0 (lightest) and 9 (darkest) appears in the upper right corner of the display. When you have finished with the calculator, turn it off to conserve battery power by pressing 2nd and then OFF.

Check the TI-83's settings by pressing MODE. If necessary, use the arrow key to move the blinking cursor to a setting you want to change. Press ENTER to select a new setting. To start, select the options along the left side of the MODE menu as illustrated in Figure I.1: normal display, floating display decimals, radian measure, function graphs, connected lines, sequential plotting, real number system, and full screen display. Details on alternative options will be given later in this guide. For now, leave the MODE menu by pressing CLEAR.

The **Graphing Calculator Keystroke Guide** offers commands and instructions for various calculators and includes examples with step-by-step solutions, technology tips, and programs.

The Graphing Calculator Keystroke Guide covers TI-83/TI-83 PLUS, TI-84 PLUS, TI-86, TI-89, TI-92, and Voyage 200.

Also available on the student website:

- **Electronic Data Sets** are designed to be used with select exercises in the text and help students reinforce and broaden their technology skills using graphing calculators and computer algebra systems.
- **MATLAB Exercises** enhance students' understanding of concepts using MATLAB software. These optional exercises correlate to chapters in the text.

Additional Resources ■ Get More from Your Textbook

Instructor Resources	*Student Resources*
Instructor Website This website offers instructors a variety of resources, including: ■ **Instructor's Solutions Manual,** featuring complete solutions to all even-numbered exercises in the text. ■ **Digital Art and Figures,** featuring key theorems from the text.	**Student Website** This website offers comprehensive study resources, including: ■ ***NEW!*** **Online Multimedia eBook** ■ ***NEW!*** **Online Technology Guide** ■ **Electronic Simulations** ■ **MATLAB Exercises** ■ **Graphing Calculator Keystroke Guide** ■ **Chapters 8, 9, and 10** ■ **Electronic Data Sets** ■ **Historical Note Biographies**
NEW! **HM Testing™ (Powered by Diploma®) "Testing the way you want it"** HM Testing provides instructors with a wide array of new algorithmic exercises along with improved functionality and ease of use. Instructors can create, author/edit algorithmic questions, customize, and deliver multiple types of tests.	**Student Solutions Manual** Contains complete solutions to all odd-numbered exercises in the text.

HM Math SPACE with Eduspace®: Houghton Mifflin's Online Learning Tool (powered by Blackboard®) This web-based learning system provides instructors and students with powerful course management tools and text-specific content to support all of their online teaching and learning needs. Eduspace now includes:

- ***NEW!*** **WebAssign®** Developed by teachers, for teachers, WebAssign allows instructors to create assignments from an abundant ready-to-use database of algorithmic questions, or write and customize their own exercises. With WebAssign, instructors can: create, post, and review assignments 24 hours a day, 7 days a week; deliver, collect, grade, and record assignments instantly; offer more practice exercises, quizzes and homework; assess student performance to keep abreast of individual progress; and capture the attention of online or distance-learning students.
- 

SMARTHINKING® Live, Online Tutoring SMARTHINKING provides an easy-to-use and effective online, text-specific tutoring service. A dynamic **Whiteboard** and a **Graphing Calculator** function enable students and e-structors to collaborate easily.

Online Course Content for Blackboard®, WebCT®, and eCollege® Deliver program- or text-specific Houghton Mifflin content online using your institution's local course management system. Houghton Mifflin offers homework and other resources formatted for Blackboard, WebCT, eCollege, and other course management systems. Add to an existing online course or create a new one by selecting from a wide range of powerful learning and instructional materials.

For more information, visit ***college.hmco.com/pic/larson/ELA6e*** or contact your local Houghton Mifflin sales representative.

What Is Linear Algebra?

To answer the question "What is linear algebra?," take a closer look at what you will study in this course. The most fundamental theme of linear algebra, and the first topic covered in this textbook, is the theory of systems of linear equations. You have probably encountered small systems of linear equations in your previous mathematics courses. For example, suppose you travel on an airplane between two cities that are 5000 kilometers apart. If the trip one way against a headwind takes $6\frac{1}{4}$ hours and the return trip the same day in the direction of the wind takes only 5 hours, can you find the ground speed of the plane and the speed of the wind, assuming that both remain constant?

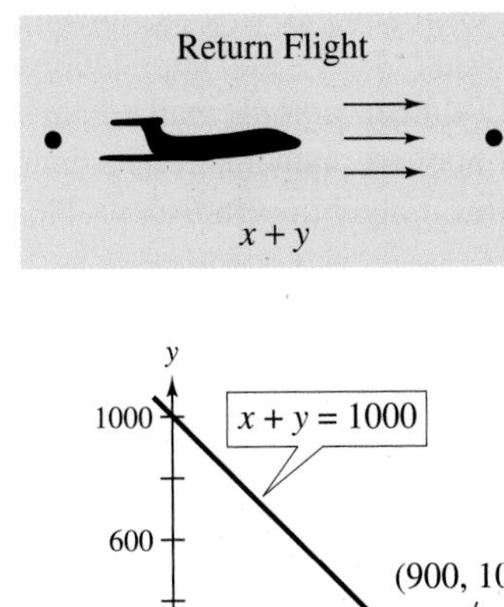

If you let x represent the speed of the plane and y the speed of the wind, then the following system models the problem.

$$\begin{aligned} 6.25(x - y) &= 5000 \\ 5(x + y) &= 5000 \end{aligned}$$

This system of two equations and two unknowns simplifies to

$$\begin{aligned} x - y &= 800 \\ x + y &= 1000, \end{aligned}$$

and the solution is $x = 900$ kilometers per hour and $y = 100$ kilometers per hour. Geometrically, this system represents two lines in the xy-plane. You can see in the figure that these lines intersect at the point (900, 100), which verifies the answer that was obtained.

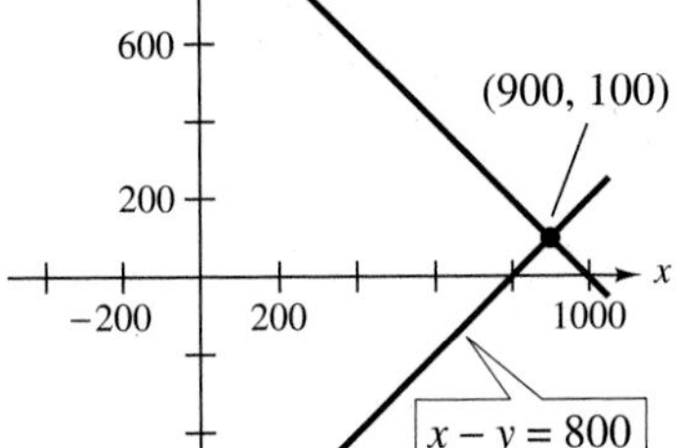

The lines intersect at (900, 100).

Solving systems of linear equations is one of the most important applications of linear algebra. It has been argued that the majority of all mathematical problems encountered in scientific and industrial applications involve solving a linear system at some point. Linear applications arise in such diverse areas as engineering, chemistry, economics, business, ecology, biology, and psychology.

Of course, the small system presented in the airplane example above is very easy to solve. In real-world situations, it is not unusual to have to solve systems of hundreds or even thousands of equations. One of the early goals of this course is to develop an algorithm that helps solve larger systems in an orderly manner and is amenable to computer implementation.

The first three chapters of this textbook cover linear systems and two other computational areas you may have studied before: matrices and determinants. These discussions prepare the way for the central theoretical topic of linear algebra: the concept of a vector space. Vector spaces generalize the familiar properties of vectors in the plane. It is at this point in the text that you will begin to write proofs and learn to verify theoretical properties of vector spaces.

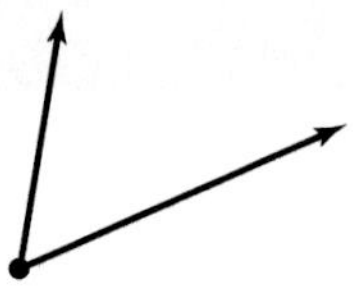

Vectors in the Plane

The concept of a vector space permits you to develop an entire theory of its properties. The theorems you prove will apply to all vector spaces. For example, in Chapter 6 you will study linear transformations, which are special functions between vector spaces. The applications of linear transformations appear almost everywhere—computer graphics, differential equations, and satellite data transmission, to name just a few examples.

Another major focus of linear algebra is the so-called eigenvalue (ī–gən–value) problem. Eigenvalues are certain numbers associated with square matrices and are fundamental in applications as diverse as population dynamics, electrical networks, chemical reactions, differential equations, and economics.

LINEAR ALGEBRA The branch of algebra in which one studies vector (linear) spaces, linear operators (linear mappings), and linear, bilinear, and quadratic functions (functionals and forms) on vector spaces. *(Encyclopedia of Mathematics, Kluwer Academic Press, 1990)*

Linear algebra strikes a wonderful balance between computation and theory. As you proceed, you will become adept at matrix computations and will simultaneously develop abstract reasoning skills. Furthermore, you will see immediately that the applications of linear algebra to other disciplines are plentiful. In fact, you will notice that each chapter of this textbook closes with a section of applications. You might want to peruse some of these sections to see the many diverse areas to which linear algebra can be applied. (An index of these applications is given on the inside front cover.)

Linear algebra has become a central course for mathematics majors as well as students of science, business, and engineering. Its balance of computation, theory, and applications to real life, geometry, and other areas makes linear algebra unique among mathematics courses. For the many people who make use of pure and applied mathematics in their professional careers, an understanding and appreciation of linear algebra is indispensable.

Systems of Linear Equations

CHAPTER OBJECTIVES

- Recognize, graph, and solve a system of linear equations in n variables.
- Use back-substitution to solve a system of linear equations.
- Determine whether a system of linear equations is consistent or inconsistent.
- Determine if a matrix is in row-echelon form or reduced row-echelon form.
- Use elementary row operations with back-substitution to solve a system in row-echelon form.
- Use elimination to rewrite a system in row-echelon form.
- Write an augmented or coefficient matrix from a system of linear equations, or translate a matrix into a system of linear equations.
- Solve a system of linear equations using Gaussian elimination and Gaussian elimination with back-substitution.
- Solve a homogeneous system of linear equations.
- Set up and solve a system of equations to fit a polynomial function to a set of data points, as well as to represent a network.

1.1 Introduction to Systems of Linear Equations

Linear algebra is a branch of mathematics rich in theory and applications. This text strikes a balance between the theoretical and the practical. Because linear algebra arose from the study of systems of linear equations, you shall begin with linear equations. Although some material in this first chapter will be familiar to you, it is suggested that you carefully study the methods presented here. Doing so will cultivate and clarify your intuition for the more abstract material that follows.

The study of linear algebra demands familiarity with algebra, analytic geometry, and trigonometry. Occasionally you will find examples and exercises requiring a knowledge of calculus; these are clearly marked in the text.

Early in your study of linear algebra you will discover that many of the solution methods involve dozens of arithmetic steps, so it is essential to strive to avoid careless errors. A computer or calculator can be very useful in checking your work, as well as in performing many of the routine computations in linear algebra.

HISTORICAL NOTE

Carl Friedrich Gauss (1777–1855)

is often ranked—along with Archimedes and Newton—as one of the greatest mathematicians in history. To read about his contributions to linear algebra, visit *college.hmco.com/pic/larsonELA6e.*

Linear Equations in n Variables

Recall from analytic geometry that the equation of a line in two-dimensional space has the form

$$a_1x + a_2y = b, \quad a_1, a_2, \text{ and } b \text{ are constants.}$$

This is a **linear equation in two variables** x and y. Similarly, the equation of a plane in three-dimensional space has the form

$$a_1x + a_2y + a_3z = b, \quad a_1, a_2, a_3, \text{ and } b \text{ are constants.}$$

Such an equation is called a **linear equation in three variables** x, y, and z. In general, a linear equation in n variables is defined as follows.

Definition of a Linear Equation in *n* Variables

A **linear equation in *n* variables** $x_1, x_2, x_3, \ldots, x_n$ has the form

$$a_1x_1 + a_2x_2 + a_3x_3 + \cdot \cdot \cdot + a_nx_n = b.$$

The **coefficients** $a_1, a_2, a_3, \ldots, a_n$ are real numbers, and the **constant term** b is a real number. The number a_1 is the **leading coefficient,** and x_1 is the **leading variable.**

REMARK: Letters that occur early in the alphabet are used to represent constants, and letters that occur late in the alphabet are used to represent variables.

Linear equations have no products or roots of variables and no variables involved in trigonometric, exponential, or logarithmic functions. Variables appear only to the first power. Example 1 lists some equations that are linear and some that are not linear.

EXAMPLE 1 Examples of Linear Equations and Nonlinear Equations

Each equation is linear.

(a) $3x + 2y = 7$ (b) $\frac{1}{2}x + y - \pi z = \sqrt{2}$

(c) $x_1 - 2x_2 + 10x_3 + x_4 = 0$ (d) $\left(\sin \frac{\pi}{2}\right)x_1 - 4x_2 = e^2$

Each equation is not linear.

(a) $xy + z = 2$ (b) $e^x - 2y = 4$

(c) $\sin x_1 + 2x_2 - 3x_3 = 0$ (d) $\frac{1}{x} + \frac{1}{y} = 4$

A **solution** of a linear equation in n variables is a sequence of n real numbers $s_1, s_2, s_3, \ldots, s_n$ arranged so the equation is satisfied when the values

$$x_1 = s_1, \quad x_2 = s_2, \quad x_3 = s_3, \quad \ldots, \quad x_n = s_n$$

are substituted into the equation. For example, the equation

$$x_1 + 2x_2 = 4$$

is satisfied when $x_1 = 2$ and $x_2 = 1$. Some other solutions are $x_1 = -4$ and $x_2 = 4$, $x_1 = 0$ and $x_2 = 2$, and $x_1 = -2$ and $x_2 = 3$.

The set of *all* solutions of a linear equation is called its **solution set,** and when this set is found, the equation is said to have been **solved.** To describe the entire solution set of a linear equation, a **parametric representation** is often used, as illustrated in Examples 2 and 3.

EXAMPLE 2 Parametric Representation of a Solution Set

Solve the linear equation $x_1 + 2x_2 = 4$.

SOLUTION To find the solution set of an equation involving two variables, solve for one of the variables in terms of the other variable. If you solve for x_1 in terms of x_2, you obtain

$$x_1 = 4 - 2x_2.$$

In this form, the variable x_2 is **free,** which means that it can take on any real value. The variable x_1 is not free because its value depends on the value assigned to x_2. To represent the infinite number of solutions of this equation, it is convenient to introduce a third variable t called a **parameter.** By letting $x_2 = t$, you can represent the solution set as

$$x_1 = 4 - 2t, \qquad x_2 = t, \qquad t \text{ is any real number.}$$

Particular solutions can be obtained by assigning values to the parameter t. For instance, $t = 1$ yields the solution $x_1 = 2$ and $x_2 = 1$, and $t = 4$ yields the solution $x_1 = -4$ and $x_2 = 4$.

The solution set of a linear equation can be represented parametrically in more than one way. In Example 2 you could have chosen x_1 to be the free variable. The parametric representation of the solution set would then have taken the form

$$x_1 = s, \qquad x_2 = 2 - \tfrac{1}{2}s, \qquad s \text{ is any real number.}$$

For convenience, choose the variables that occur last in a given equation to be free variables.

EXAMPLE 3 Parametric Representation of a Solution Set

Solve the linear equation $3x + 2y - z = 3$.

SOLUTION Choosing y and z to be the free variables, begin by solving for x to obtain

$$3x = 3 - 2y + z$$
$$x = 1 - \tfrac{2}{3}y + \tfrac{1}{3}z.$$

Letting $y = s$ and $z = t$, you obtain the parametric representation

$$x = 1 - \tfrac{2}{3}s + \tfrac{1}{3}t, \qquad y = s, \qquad z = t$$

where s and t are any real numbers. Two particular solutions are

$$x = 1, y = 0, z = 0 \quad \text{and} \quad x = 1, y = 1, z = 2.$$

Systems of Linear Equations

A **system of m linear equations in n variables** is a set of m equations, each of which is linear in the same n variables:

$$\begin{aligned}
a_{11}x_1 + a_{12}x_2 + a_{13}x_3 + \cdot \cdot \cdot + a_{1n}x_n &= b_1 \\
a_{21}x_1 + a_{22}x_2 + a_{23}x_3 + \cdot \cdot \cdot + a_{2n}x_n &= b_2 \\
a_{31}x_1 + a_{32}x_2 + a_{33}x_3 + \cdot \cdot \cdot + a_{3n}x_n &= b_3 \\
&\vdots \\
a_{m1}x_1 + a_{m2}x_2 + a_{m3}x_3 + \cdot \cdot \cdot + a_{mn}x_n &= b_m.
\end{aligned}$$

REMARK: The double-subscript notation indicates a_{ij} is the coefficient of x_j in the ith equation.

A **solution** of a system of linear equations is a sequence of numbers $s_1, s_2, s_3, \ldots, s_n$ that is a solution of each of the linear equations in the system. For example, the system

$$\begin{aligned}
3x_1 + 2x_2 &= 3 \\
-x_1 + x_2 &= 4
\end{aligned}$$

has $x_1 = -1$ and $x_2 = 3$ as a solution because *both* equations are satisfied when $x_1 = -1$ and $x_2 = 3$. On the other hand, $x_1 = 1$ and $x_2 = 0$ is not a solution of the system because these values satisfy only the first equation in the system.

Discovery

Graph the two lines

$$\begin{aligned}
3x - y &= 1 \\
2x - y &= 0
\end{aligned}$$

in the xy-plane. Where do they intersect? How many solutions does this system of linear equations have?

Repeat this analysis for the pairs of lines

$$\begin{aligned}
3x - y &= 1 \qquad & 3x - y &= 1 \\
3x - y &= 0 \qquad & 6x - 2y &= 2.
\end{aligned}$$

In general, what basic types of solution sets are possible for a system of two equations in two unknowns?

It is possible for a system of linear equations to have exactly one solution, an infinite number of solutions, or no solution. A system of linear equations is called **consistent** if it has at least one solution and **inconsistent** if it has no solution.

EXAMPLE 4 Systems of Two Equations in Two Variables

Solve each system of linear equations, and graph each system as a pair of straight lines.

(a) $\begin{aligned} x + y &= 3 \\ x - y &= -1 \end{aligned}$ (b) $\begin{aligned} x + y &= 3 \\ 2x + 2y &= 6 \end{aligned}$ (c) $\begin{aligned} x + y &= 3 \\ x + y &= 1 \end{aligned}$

SOLUTION

(a) This system has exactly one solution, $x = 1$ and $y = 2$. The solution can be obtained by adding the two equations to give $2x = 2$, which implies $x = 1$ and so $y = 2$. The graph of this system is represented by two *intersecting* lines, as shown in Figure 1.1(a).

(b) This system has an infinite number of solutions because the second equation is the result of multiplying both sides of the first equation by 2. A parametric representation of the solution set is shown as

$$x = 3 - t, \qquad y = t, \quad t \text{ is any real number.}$$

The graph of this system is represented by two *coincident* lines, as shown in Figure 1.1(b).

(c) This system has no solution because it is impossible for the sum of two numbers to be 3 and 1 simultaneously. The graph of this system is represented by two *parallel* lines, as shown in Figure 1.1(c).

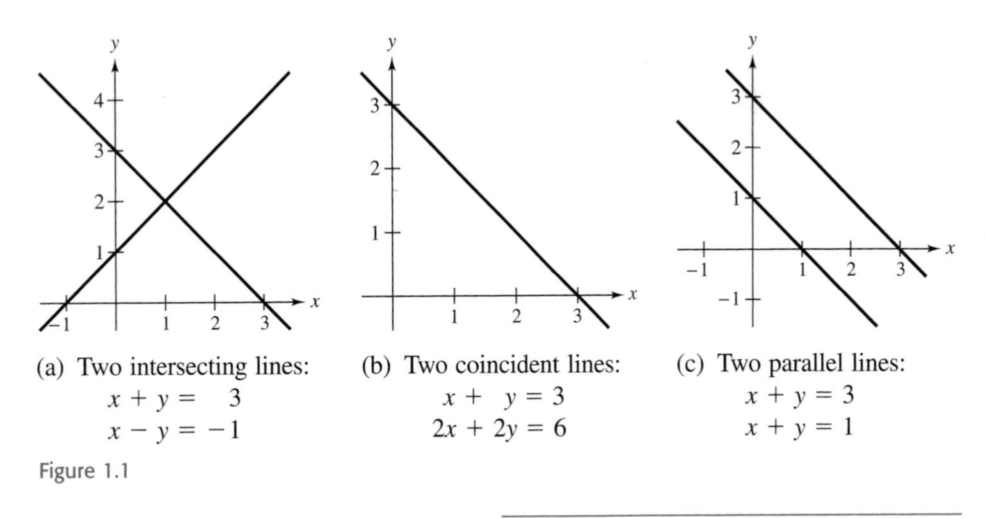

(a) Two intersecting lines: $x + y = 3$, $x - y = -1$

(b) Two coincident lines: $x + y = 3$, $2x + 2y = 6$

(c) Two parallel lines: $x + y = 3$, $x + y = 1$

Figure 1.1

Example 4 illustrates the three basic types of solution sets that are possible for a system of linear equations. This result is stated here without proof. (The proof is provided later in Theorem 2.5.)

Number of Solutions of a System of Linear Equations

For a system of linear equations in n variables, precisely one of the following is true.

1. The system has exactly one solution (consistent system).
2. The system has an infinite number of solutions (consistent system).
3. The system has no solution (inconsistent system).

Solving a System of Linear Equations

Which system is easier to solve algebraically?

$$\begin{aligned} x - 2y + 3z &= 9 \\ -x + 3y &= -4 \\ 2x - 5y + 5z &= 17 \end{aligned} \qquad \begin{aligned} x - 2y + 3z &= 9 \\ y + 3z &= 5 \\ z &= 2 \end{aligned}$$

The system on the right is clearly easier to solve. This system is in **row-echelon form,** which means that it follows a stair-step pattern and has leading coefficients of 1. To solve such a system, use a procedure called **back-substitution.**

EXAMPLE 5 Using Back-Substitution to Solve a System in Row-Echelon Form

Use back-substitution to solve the system.

$$\begin{aligned} x - 2y &= 5 \qquad \text{Equation 1} \\ y &= -2 \qquad \text{Equation 2} \end{aligned}$$

SOLUTION From Equation 2 you know that $y = -2$. By substituting this value of y into Equation 1, you obtain

$$\begin{aligned} x - 2(-2) &= 5 \qquad \text{Substitute } y = -2. \\ x &= 1. \qquad \text{Solve for } x. \end{aligned}$$

The system has exactly one solution: $x = 1$ and $y = -2$.

The term "back-substitution" implies that you work *backward.* For instance, in Example 5, the second equation gave you the value of y. Then you substituted that value into the first equation to solve for x. Example 6 further demonstrates this procedure.

EXAMPLE 6 Using Back-Substitution to Solve a System in Row-Echelon Form

Solve the system.

$$\begin{aligned} x - 2y + 3z &= 9 \qquad \text{Equation 1} \\ y + 3z &= 5 \qquad \text{Equation 2} \\ z &= 2 \qquad \text{Equation 3} \end{aligned}$$

SOLUTION From Equation 3 you already know the value of z. To solve for y, substitute $z = 2$ into Equation 2 to obtain

$$\begin{aligned} y + 3(2) &= 5 && \text{Substitute } z = 2. \\ y &= -1. && \text{Solve for } y. \end{aligned}$$

Finally, substitute $y = -1$ and $z = 2$ in Equation 1 to obtain

$$\begin{aligned} x - 2(-1) + 3(2) &= 9 && \text{Substitute } y = -1,\ z = 2. \\ x &= 1. && \text{Solve for } x. \end{aligned}$$

The solution is $x = 1$, $y = -1$, and $z = 2$.

Two systems of linear equations are called **equivalent** if they have precisely the same solution set. To solve a system that is not in row-echelon form, first change it to an *equivalent* system that is in row-echelon form by using the operations listed below.

Operations That Lead to Equivalent Systems of Equations

Each of the following operations on a system of linear equations produces an *equivalent* system.

1. Interchange two equations.
2. Multiply an equation by a nonzero constant.
3. Add a multiple of an equation to another equation.

Rewriting a system of linear equations in row-echelon form usually involves a *chain* of equivalent systems, each of which is obtained by using one of the three basic operations. This process is called **Gaussian elimination,** after the German mathematician Carl Friedrich Gauss (1777–1855).

EXAMPLE 7 Using Elimination to Rewrite a System in Row-Echelon Form

Solve the system.

$$\begin{aligned} x - 2y + 3z &= 9 \\ -x + 3y &= -4 \\ 2x - 5y + 5z &= 17 \end{aligned}$$

SOLUTION Although there are several ways to begin, you want to use a systematic procedure that can be applied easily to large systems. Work from the upper left corner of the system, saving the x in the upper left position and eliminating the other x's from the first column.

$$\begin{aligned} x - 2y + 3z &= 9 \\ y + 3z &= 5 \\ 2x - 5y + 5z &= 17 \end{aligned}$$

← Adding the first equation to the second equation produces a new second equation.

$$\begin{aligned} x - 2y + 3z &= 9 \\ y + 3z &= 5 \\ -y - z &= -1 \end{aligned}$$

← Adding -2 times the first equation to the third equation produces a new third equation.

Now that everything but the first x has been eliminated from the first column, work on the second column.

$$\begin{aligned} x - 2y + 3z &= 9 \\ y + 3z &= 5 \\ 2z &= 4 \end{aligned}$$

← **Adding the second equation to the third equation produces a new third equation.**

$$\begin{aligned} x - 2y + 3z &= 9 \\ y + 3z &= 5 \\ z &= 2 \end{aligned}$$

← **Multiplying the third equation by $\frac{1}{2}$ produces a new third equation.**

This is the same system you solved in Example 6, and, as in that example, the solution is

$$x = 1, \qquad y = -1, \qquad z = 2.$$

Each of the three equations in Example 7 is represented in a three-dimensional coordinate system by a plane. Because the unique solution of the system is the point

$$(x, y, z) = (1, -1, 2),$$

the three planes intersect at the point represented by these coordinates, as shown in Figure 1.2.

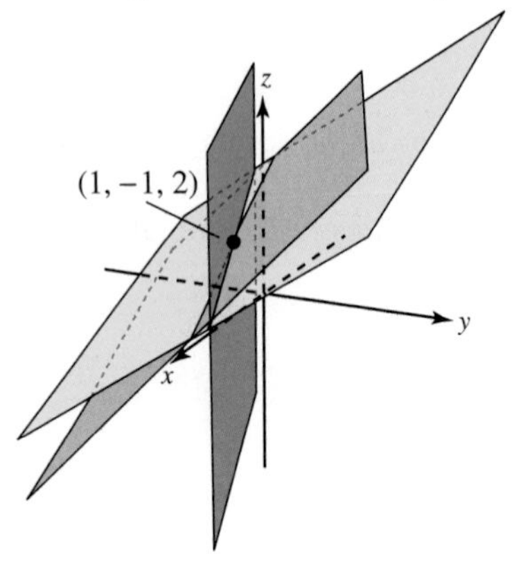

Figure 1.2

Technology Note

Many graphing utilities and computer software programs can solve a system of m linear equations in n variables. Try solving the system in Example 7 using the simultaneous equation solver feature of your graphing utility or computer software program. Keystrokes and programming syntax for these utilities/programs applicable to Example 7 are provided in the **Online Technology Guide,** available at *college.hmco.com/pic/larsonELA6e*.

Because many steps are required to solve a system of linear equations, it is very easy to make errors in arithmetic. It is suggested that you develop the habit of *checking your solution by substituting it into each equation in the original system.* For instance, in Example 7, you can check the solution $x = 1$, $y = -1$, and $z = 2$ as follows.

$$\begin{aligned} &\text{Equation 1:} && (1) - 2(-1) + 3(2) = 9 \\ &\text{Equation 2:} && -(1) + 3(-1) = -4 \\ &\text{Equation 3:} && 2(1) - 5(-1) + 5(2) = 17 \end{aligned}$$

Substitute solution in each equation of the original system.

Each of the systems in Examples 5, 6, and 7 has exactly one solution. You will now look at an inconsistent system—one that has no solution. The key to recognizing an inconsistent system is reaching a false statement such as $0 = 7$ at some stage of the elimination process. This is demonstrated in Example 8.

EXAMPLE 8 An Inconsistent System

Solve the system.

$$\begin{aligned} x_1 - 3x_2 + x_3 &= 1 \\ 2x_1 - x_2 - 2x_3 &= 2 \\ x_1 + 2x_2 - 3x_3 &= -1 \end{aligned}$$

SOLUTION

$$\begin{aligned} x_1 - 3x_2 + x_3 &= 1 \\ 5x_2 - 4x_3 &= 0 \\ x_1 + 2x_2 - 3x_3 &= -1 \end{aligned}$$

Adding -2 times the first equation to the second equation produces a new second equation.

$$\begin{aligned} x_1 - 3x_2 + x_3 &= 1 \\ 5x_2 - 4x_3 &= 0 \\ 5x_2 - 4x_3 &= -2 \end{aligned}$$

Adding -1 times the first equation to the third equation produces a new third equation.

(Another way of describing this operation is to say that you *subtracted* the first equation from the third equation to produce a new third equation.) Now, continuing the elimination process, add -1 times the second equation to the third equation to produce a new third equation.

$$\begin{aligned} x_1 - 3x_2 + x_3 &= 1 \\ 5x_2 - 4x_3 &= 0 \\ 0 &= -2 \end{aligned}$$

Adding -1 times the second equation to the third equation produces a new third equation.

Because the third "equation" is a false statement, this system has no solution. Moreover, because this system is equivalent to the original system, you can conclude that the original system also has no solution.

As in Example 7, the three equations in Example 8 represent planes in a three-dimensional coordinate system. In this example, however, the system is inconsistent. So, the planes do not have a point in common, as shown in Figure 1.3 on the next page.

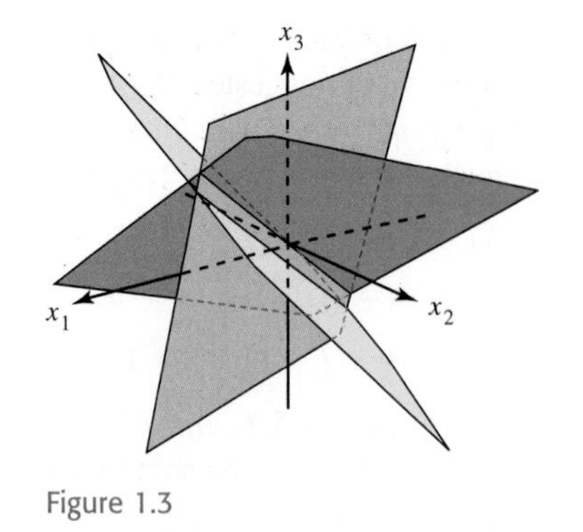

Figure 1.3

This section ends with an example of a system of linear equations that has an infinite number of solutions. You can represent the solution set for such a system in parametric form, as you did in Examples 2 and 3.

EXAMPLE 9 A System with an Infinite Number of Solutions

Solve the system.

$$\begin{aligned} x_2 - x_3 &= 0 \\ x_1 \qquad - 3x_3 &= -1 \\ -x_1 + 3x_2 \qquad &= 1 \end{aligned}$$

SOLUTION Begin by rewriting the system in row-echelon form as follows.

$$\begin{aligned} x_1 \qquad - 3x_3 &= -1 \\ x_2 - x_3 &= 0 \\ -x_1 + 3x_2 \qquad &= 1 \end{aligned}$$

The first two equations are interchanged.

$$\begin{aligned} x_1 \qquad - 3x_3 &= -1 \\ x_2 - x_3 &= 0 \\ 3x_2 - 3x_3 &= 0 \end{aligned}$$

Adding the first equation to the third equation produces a new third equation.

$$\begin{aligned} x_1 \qquad - 3x_3 &= -1 \\ x_2 - x_3 &= 0 \\ 0 &= 0 \end{aligned}$$

Adding -3 times the second equation to the third equation eliminates the third equation.

Because the third equation is unnecessary, omit it to obtain the system shown below.

$$\begin{aligned} x_1 \qquad - 3x_3 &= -1 \\ x_2 - x_3 &= 0 \end{aligned}$$

To represent the solutions, choose x_3 to be the free variable and represent it by the parameter t. Because $x_2 = x_3$ and $x_1 = 3x_3 - 1$, you can describe the solution set as

$$x_1 = 3t - 1, \qquad x_2 = t, \qquad x_3 = t, \quad t \text{ is any real number.}$$

Discovery

Graph the two lines represented by the system of equations.

$$\begin{aligned} x - 2y &= 1 \\ -2x + 3y &= -3 \end{aligned}$$

You can use Gaussian elimination to solve this system as follows.

$$\begin{aligned} x - 2y &= 1 \\ -1y &= -1 \end{aligned} \qquad \begin{aligned} x - 2y &= 1 \\ y &= 1 \end{aligned} \qquad \begin{aligned} x &= 3 \\ y &= 1 \end{aligned}$$

Graph the system of equations you obtain at each step of this process. What do you observe about the lines? You are asked to repeat this graphical analysis for other systems in Exercises 91 and 92.

SECTION 1.1 Exercises

In Exercises 1–6, determine whether the equation is linear in the variables x and y.

1. $2x - 3y = 4$

2. $3x - 4xy = 0$

3. $\frac{3}{y} + \frac{2}{x} - 1 = 0$

4. $x^2 + y^2 = 4$

5. $2 \sin x - y = 14$

6. $(\sin 2)x - y = 14$

In Exercises 7–10, find a parametric representation of the solution set of the linear equation.

7. $2x - 4y = 0$

8. $3x - \frac{1}{2}y = 9$

9. $x + y + z = 1$

10. $13x_1 - 26x_2 + 39x_3 = 13$

In Exercises 11–16, use back-substitution to solve the system.

11. $\begin{aligned} x_1 - x_2 &= 2 \\ x_2 &= 3 \end{aligned}$

12. $\begin{aligned} 2x_1 - 4x_2 &= 6 \\ 3x_2 &= 9 \end{aligned}$

13. $\begin{aligned} -x + y - z &= 0 \\ 2y + z &= 3 \\ \tfrac{1}{2}z &= 0 \end{aligned}$

14. $\begin{aligned} x - y \quad &= 4 \\ 2y + z &= 6 \\ 3z &= 6 \end{aligned}$

15. $\begin{aligned} 5x_1 + 2x_2 + x_3 &= 0 \\ 2x_1 + x_2 \quad &= 0 \end{aligned}$

16. $\begin{aligned} x_1 + x_2 + x_3 &= 0 \\ x_2 \quad &= 0 \end{aligned}$

In Exercises 17–30, graph each system of equations as a pair of lines in the xy-plane. Solve each system and interpret your answer.

17. $\begin{aligned} 2x + y &= 4 \\ x - y &= 2 \end{aligned}$

18. $\begin{aligned} x + 3y &= 2 \\ -x + 2y &= 3 \end{aligned}$

19. $\begin{aligned} x - y &= 1 \\ -2x + 2y &= 5 \end{aligned}$

20. $\begin{aligned} \tfrac{1}{2}x - \tfrac{1}{3}y &= 1 \\ -2x + \tfrac{4}{3}y &= -4 \end{aligned}$

21. $\begin{aligned} 3x - 5y &= 7 \\ 2x + y &= 9 \end{aligned}$

22. $\begin{aligned} -x + 3y &= 17 \\ 4x + 3y &= 7 \end{aligned}$

23. $\begin{aligned} 2x - y &= 5 \\ 5x - y &= 11 \end{aligned}$

24. $\begin{aligned} x - 5y &= 21 \\ 6x + 5y &= 21 \end{aligned}$

25. $\begin{aligned} \frac{x+3}{4} + \frac{y-1}{3} &= 1 \\ 2x - y &= 12 \end{aligned}$

26. $\begin{aligned} \frac{x-1}{2} + \frac{y+2}{3} &= 4 \\ x - 2y &= 5 \end{aligned}$

27. $\begin{aligned} 0.05x - 0.03y &= 0.07 \\ 0.07x + 0.02y &= 0.16 \end{aligned}$

28. $\begin{aligned} 0.2x - 0.5y &= -27.8 \\ 0.3x + 0.4y &= 68.7 \end{aligned}$

29. $\begin{aligned} \frac{x}{4} + \frac{y}{6} &= 1 \\ x - y &= 3 \end{aligned}$

30. $\begin{aligned} \frac{2}{3}x + \frac{1}{6}y &= \frac{2}{3} \\ 4x + y &= 4 \end{aligned}$

In Exercises 31–36, complete the following set of tasks for each system of equations.

(a) Use a graphing utility to graph the equations in the system.
(b) Use the graphs to determine whether the system is consistent or inconsistent.
(c) If the system is consistent, approximate the solution.
(d) Solve the system algebraically.
(e) Compare the solution in part (d) with the approximation in part (c). What can you conclude?

31. $\begin{aligned} -3x - y &= 3 \\ 6x + 2y &= 1 \end{aligned}$

32. $\begin{aligned} 4x - 5y &= 3 \\ -8x + 10y &= 14 \end{aligned}$

33. $\begin{aligned} 2x - 8y &= 3 \\ \tfrac{1}{2}x + y &= 0 \end{aligned}$

34. $\begin{aligned} 9x - 4y &= 5 \\ \tfrac{1}{2}x + \tfrac{1}{3}y &= 0 \end{aligned}$

35. $\begin{aligned} 4x - 8y &= 9 \\ 0.8x - 1.6y &= 1.8 \end{aligned}$

36. $\begin{aligned} -5.3x + 2.1y &= 1.25 \\ 15.9x - 6.3y &= -3.75 \end{aligned}$

The symbol indicates an exercise in which you are instructed to use a graphing utility or a symbolic computer software program.

In Exercises 37–56, solve the system of linear equations.

37. $\begin{aligned} x_1 - x_2 &= 0 \\ 3x_1 - 2x_2 &= -1 \end{aligned}$

38. $\begin{aligned} 3x + 2y &= 2 \\ 6x + 4y &= 14 \end{aligned}$

39. $\begin{aligned} 2u + v &= 120 \\ u + 2v &= 120 \end{aligned}$

40. $\begin{aligned} x_1 - 2x_2 &= 0 \\ 6x_1 + 2x_2 &= 0 \end{aligned}$

41. $\begin{aligned} 9x - 3y &= -1 \\ \tfrac{1}{5}x + \tfrac{2}{5}y &= -\tfrac{1}{3} \end{aligned}$

42. $\begin{aligned} \tfrac{2}{3}x_1 + \tfrac{1}{6}x_2 &= 0 \\ 4x_1 + x_2 &= 0 \end{aligned}$

43. $\begin{aligned} \frac{x-1}{2} + \frac{y+2}{3} &= 4 \\ x - 2y &= 5 \end{aligned}$

44. $\begin{aligned} \frac{x_1+3}{4} + \frac{x_2-1}{3} &= 1 \\ 2x_1 - x_2 &= 12 \end{aligned}$

45. $\begin{aligned} 0.02x_1 - 0.05x_2 &= -0.19 \\ 0.03x_1 + 0.04x_2 &= 0.52 \end{aligned}$

46. $\begin{aligned} 0.05x_1 - 0.03x_2 &= 0.21 \\ 0.07x_1 + 0.02x_2 &= 0.17 \end{aligned}$

47. $\begin{aligned} x + y + z &= 6 \\ 2x - y + z &= 3 \\ 3x - z &= 0 \end{aligned}$

48. $\begin{aligned} x + y + z &= 2 \\ -x + 3y + 2z &= 8 \\ 4x + y &= 4 \end{aligned}$

49. $\begin{aligned} 3x_1 - 2x_2 + 4x_3 &= 1 \\ x_1 + x_2 - 2x_3 &= 3 \\ 2x_1 - 3x_2 + 6x_3 &= 8 \end{aligned}$

50. $\begin{aligned} 5x_1 - 3x_2 + 2x_3 &= 3 \\ 2x_1 + 4x_2 - x_3 &= 7 \\ x_1 - 11x_2 + 4x_3 &= 3 \end{aligned}$

51. $\begin{aligned} 2x_1 + x_2 - 3x_3 &= 4 \\ 4x_1 + 2x_3 &= 10 \\ -2x_1 + 3x_2 - 13x_3 &= -8 \end{aligned}$

52. $\begin{aligned} x_1 + 4x_3 &= 13 \\ 4x_1 - 2x_2 + x_3 &= 7 \\ 2x_1 - 2x_2 - 7x_3 &= -19 \end{aligned}$

53. $\begin{aligned} x - 3y + 2z &= 18 \\ 5x - 15y + 10z &= 18 \end{aligned}$

54. $\begin{aligned} x_1 - 2x_2 + 5x_3 &= 2 \\ 3x_1 + 2x_2 - x_3 &= -2 \end{aligned}$

55. $\begin{aligned} x + y + z + w &= 6 \\ 2x + 3y - w &= 0 \\ -3x + 4y + z + 2w &= 4 \\ x + 2y - z + w &= 0 \end{aligned}$

56. $\begin{aligned} x_1 + 3x_4 &= 4 \\ 2x_2 - x_3 - x_4 &= 0 \\ 3x_2 - 2x_4 &= 1 \\ 2x_1 - x_2 + 4x_3 &= 5 \end{aligned}$

In Exercises 57–64, use a computer software program or graphing utility to solve the system of linear equations.

57. $\begin{aligned} x_1 + 0.5x_2 + 0.33x_3 + 0.25x_4 &= 1.1 \\ 0.5x_1 + 0.33x_2 + 0.25x_3 + 0.21x_4 &= 1.2 \\ 0.33x_1 + 0.25x_2 + 0.2x_3 + 0.17x_4 &= 1.3 \\ 0.25x_1 + 0.2x_2 + 0.17x_3 + 0.14x_4 &= 1.4 \end{aligned}$

The symbol HM indicates that electronic data sets for these exercises are available at *college.hmco.com/pic/larsonELA6e*. These data sets are compatible with each of the following technologies: MATLAB, *Mathematica*, Maple, Derive, TI-83/TI-83 Plus, TI-84/TI-84 Plus, TI-86, TI-89, TI-92, and TI-92 Plus.

58. $\begin{aligned} 0.1x - 2.5y + 1.2z - 0.75w &= 108 \\ 2.4x + 1.5y - 1.8z + 0.25w &= -81 \\ 0.4x - 3.2y + 1.6z - 1.4w &= 148.8 \\ 1.6x + 1.2y - 3.2z + 0.6w &= -143.2 \end{aligned}$

59. $\begin{aligned} 123.5x + 61.3y - 32.4z &= -262.74 \\ 54.7x - 45.6y + 98.2z &= 197.4 \\ 42.4x - 89.3y + 12.9z &= 33.66 \end{aligned}$

60. $\begin{aligned} 120.2x + 62.4y - 36.5z &= 258.64 \\ 56.8x - 42.8y + 27.3z &= -71.44 \\ 88.1x + 72.5y - 28.5z &= 225.88 \end{aligned}$

61. $\begin{aligned} \tfrac{1}{2}x_1 - \tfrac{3}{7}x_2 + \tfrac{2}{9}x_3 &= \tfrac{349}{630} \\ \tfrac{2}{3}x_1 + \tfrac{4}{9}x_2 - \tfrac{2}{5}x_3 &= -\tfrac{19}{45} \\ \tfrac{4}{5}x_1 - \tfrac{1}{8}x_2 + \tfrac{4}{3}x_3 &= \tfrac{139}{150} \end{aligned}$

62. $\begin{aligned} \tfrac{1}{4}x_1 - \tfrac{3}{5}x_2 + \tfrac{1}{3}x_3 &= \tfrac{43}{60} \\ \tfrac{2}{5}x_1 + \tfrac{1}{4}x_2 - \tfrac{5}{6}x_3 &= -\tfrac{331}{600} \\ \tfrac{3}{4}x_1 - \tfrac{2}{5}x_2 + \tfrac{1}{5}x_3 &= \tfrac{81}{100} \end{aligned}$

63. $\begin{aligned} \tfrac{1}{8}x - \tfrac{1}{7}y + \tfrac{1}{6}z - \tfrac{1}{5}w &= 1 \\ \tfrac{1}{7}x + \tfrac{1}{6}y - \tfrac{1}{5}z + \tfrac{1}{4}w &= 1 \\ \tfrac{1}{6}x - \tfrac{1}{5}y + \tfrac{1}{4}z - \tfrac{1}{3}w &= 1 \\ \tfrac{1}{5}x + \tfrac{1}{4}y - \tfrac{1}{3}z + \tfrac{1}{2}w &= 1 \end{aligned}$

64. $\begin{aligned} \tfrac{1}{8}x + \tfrac{1}{7}y - \tfrac{1}{6}z + \tfrac{1}{5}w &= 1 \\ \tfrac{1}{7}x - \tfrac{1}{6}y + \tfrac{1}{5}z - \tfrac{1}{4}w &= 1 \\ \tfrac{1}{6}x + \tfrac{1}{5}y - \tfrac{1}{4}z + \tfrac{1}{3}w &= 1 \\ \tfrac{1}{5}x - \tfrac{1}{4}y + \tfrac{1}{3}z - \tfrac{1}{2}w &= 1 \end{aligned}$

In Exercises 65–68, state why each system of equations must have at least one solution. Then solve the system and determine if it has exactly one solution or an infinite number of solutions.

65. $\begin{aligned} 4x + 3y + 17z &= 0 \\ 5x + 4y + 22z &= 0 \\ 4x + 2y + 19z &= 0 \end{aligned}$

66. $\begin{aligned} 2x + 3y &= 0 \\ 4x + 3y - z &= 0 \\ 8x + 3y + 3z &= 0 \end{aligned}$

67. $\begin{aligned} 5x + 5y - z &= 0 \\ 10x + 5y + 2z &= 0 \\ 5x + 15y - 9z &= 0 \end{aligned}$

68. $\begin{aligned} 12x + 5y + z &= 0 \\ 12x + 4y - z &= 0 \end{aligned}$

True or False? In Exercises 69 and 70, determine whether each statement is true or false. If a statement is true, give a reason or cite an appropriate statement from the text. If a statement is false, provide an example that shows the statement is not true in all cases or cite an appropriate statement from the text.

69. (a) A system of one linear equation in two variables is always consistent.

(b) A system of two linear equations in three variables is always consistent.

(c) If a linear system is consistent, then it has an infinite number of solutions.

70. (a) A system of linear equations can have exactly two solutions.
(b) Two systems of linear equations are equivalent if they have the same solution set.
(c) A system of three linear equations in two variables is always inconsistent.

71. Find a system of two equations in two variables, x_1 and x_2, that has the solution set given by the parametric representation $x_1 = t$ and $x_2 = 3t - 4$, where t is any real number. Then show that the solutions to your system can also be written as

$x_1 = \frac{4}{3} + \frac{t}{3}$ and $x_2 = t$.

72. Find a system of two equations in three variables, x_1, x_2, and x_3, that has the solution set given by the parametric representation

$x_1 = t, \quad x_2 = s,$ and $x_3 = 3 + s - t,$

where s and t are any real numbers. Then show that the solutions to your system can also be written as

$x_1 = 3 + s - t, \; x_2 = s,$ and $x_3 = t.$

In Exercises 73–76, solve the system of equations by letting $A = 1/x$, $B = 1/y$, and $C = 1/z$.

73. $$\begin{aligned} \frac{12}{x} - \frac{12}{y} &= 7 \\ \frac{3}{x} + \frac{4}{y} &= 0 \end{aligned}$$

74. $$\begin{aligned} \frac{2}{x} + \frac{3}{y} &= 0 \\ \frac{3}{x} - \frac{4}{y} &= -\frac{25}{6} \end{aligned}$$

75. $$\begin{aligned} \frac{2}{x} + \frac{1}{y} - \frac{3}{z} &= 4 \\ \frac{4}{x} \qquad + \frac{2}{z} &= 10 \\ -\frac{2}{x} + \frac{3}{y} - \frac{13}{z} &= -8 \end{aligned}$$

76. $$\begin{aligned} \frac{2}{x} + \frac{1}{y} - \frac{2}{z} &= 5 \\ \frac{3}{x} - \frac{4}{y} \qquad &= -1 \\ \frac{2}{x} + \frac{1}{y} + \frac{3}{z} &= 0 \end{aligned}$$

In Exercises 77 and 78, solve the system of linear equations for x and y.

77. $$\begin{aligned} (\cos\theta)x + (\sin\theta)y &= 1 \\ (-\sin\theta)x + (\cos\theta)y &= 0 \end{aligned}$$

78. $$\begin{aligned} (\cos\theta)x + (\sin\theta)y &= 1 \\ (-\sin\theta)x + (\cos\theta)y &= 1 \end{aligned}$$

In Exercises 79–84, determine the value(s) of k such that the system of linear equations has the indicated number of solutions.

79. An infinite number of solutions

$$\begin{aligned} 4x + ky &= 6 \\ kx + y &= -3 \end{aligned}$$

80. An infinite number of solutions

$$\begin{aligned} kx + y &= 4 \\ 2x - 3y &= -12 \end{aligned}$$

81. Exactly one solution

$$\begin{aligned} x + ky &= 0 \\ kx + y &= 0 \end{aligned}$$

82. No solution

$$\begin{aligned} x + ky &= 2 \\ kx + y &= 4 \end{aligned}$$

83. No solution

$$\begin{aligned} x + 2y + kz &= 6 \\ 3x + 6y + 8z &= 4 \end{aligned}$$

84. Exactly one solution

$$\begin{aligned} kx + 2ky + 3kz &= 4k \\ x + y + z &= 0 \\ 2x - y + z &= 1 \end{aligned}$$

85. Determine the values of k such that the system of linear equations does not have a unique solution.

$$\begin{aligned} x + y + kz &= 3 \\ x + ky + z &= 2 \\ kx + y + z &= 1 \end{aligned}$$

86. Find values of a, b, and c such that the system of linear equations has (a) exactly one solution, (b) an infinite number of solutions, and (c) no solution.

$$\begin{aligned} x + 5y + z &= 0 \\ x + 6y - z &= 0 \\ 2x + ay + bz &= c \end{aligned}$$

87. Writing Consider the system of linear equations in x and y.

$$\begin{aligned} a_1x + b_1y &= c_1 \\ a_2x + b_2y &= c_2 \\ a_3x + b_3y &= c_3 \end{aligned}$$

Describe the graphs of these three equations in the xy-plane when the system has (a) exactly one solution, (b) an infinite number of solutions, and (c) no solution.

88. Writing Explain why the system of linear equations in Exercise 87 must be consistent if the constant terms c_1, c_2, and c_3 are all zero.

89. Show that if $ax^2 + bx + c = 0$ for all x, then $a = b = c = 0$.

90. Consider the system of linear equations in x and y.

$$\begin{aligned} ax + by &= e \\ cx + dy &= f \end{aligned}$$

Under what conditions will the system have exactly one solution?

In Exercises 91 and 92, sketch the lines determined by the system of linear equations. Then use Gaussian elimination to solve the system. At each step of the elimination process, sketch the corresponding lines. What do you observe about these lines?

91. $$\begin{aligned} x - 4y &= -3 \\ 5x - 6y &= 13 \end{aligned}$$

92. $$\begin{aligned} 2x - 3y &= 7 \\ -4x + 6y &= -14 \end{aligned}$$

Writing In Exercises 93 and 94, the graphs of two equations are shown and appear to be parallel. Solve the system of equations algebraically. Explain why the graphs are misleading.

93. $100y - x = 200$
$99y - x = -198$

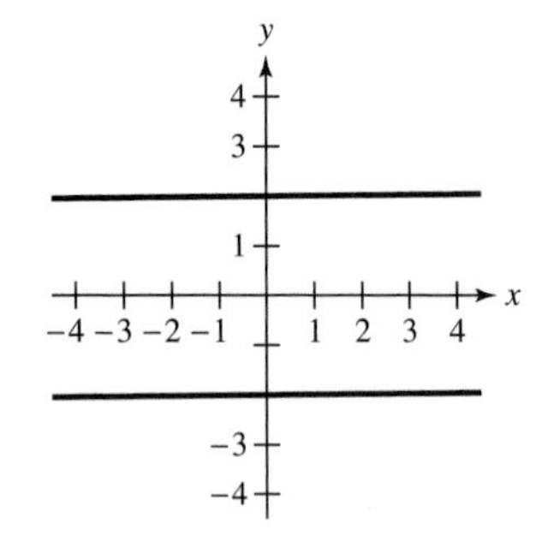

94. $21x - 20y = 0$
$13x - 12y = 120$

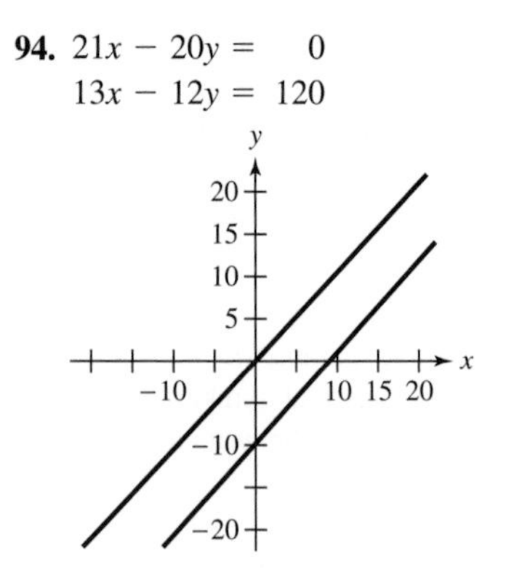

1.2 Gaussian Elimination and Gauss-Jordan Elimination

In Section 1.1, Gaussian elimination was introduced as a procedure for solving a system of linear equations. In this section you will study this procedure more thoroughly, beginning with some definitions. The first is the definition of a **matrix.**

Definition of a Matrix

If m and n are positive integers, then an $m \times n$ **matrix** is a rectangular array

$$\underbrace{\begin{bmatrix} a_{11} & a_{12} & a_{13} & \cdots & a_{1n} \\ a_{21} & a_{22} & a_{23} & \cdots & a_{2n} \\ a_{31} & a_{32} & a_{33} & \cdots & a_{3n} \\ \vdots & \vdots & \vdots & & \vdots \\ a_{m1} & a_{m2} & a_{m3} & \cdots & a_{mn} \end{bmatrix}}_{n \text{ columns}} \Bigg\} \; m \text{ rows}$$

in which each **entry,** a_{ij}, of the matrix is a number. An $m \times n$ matrix (read "m by n") has m **rows** (horizontal lines) and n **columns** (vertical lines).

REMARK: The plural of matrix is *matrices.* If each entry of a matrix is a *real* number, then the matrix is called a **real matrix.** Unless stated otherwise, all matrices in this text are assumed to be real matrices.

The entry a_{ij} is located in the ith row and the jth column. The index i is called the **row subscript** because it identifies the row in which the entry lies, and the index j is called the **column subscript** because it identifies the column in which the entry lies.

A matrix with m rows and n columns (an $m \times n$ matrix) is said to be of **size** $m \times n$. If $m = n$, the matrix is called **square** of **order** n. For a square matrix, the entries a_{11}, a_{22}, $a_{33}, \ldots$ are called the **main diagonal** entries.

EXAMPLE 1 Examples of Matrices

Each matrix has the indicated size.

(a) Size: 1×1

$$[2]$$

(b) Size: 2×2

$$\begin{bmatrix} 0 & 0 \\ 0 & 0 \end{bmatrix}$$

(c) Size: 1×4

$$\begin{bmatrix} 1 & -3 & 0 & \frac{1}{2} \end{bmatrix}$$

(d) Size: 3×2

$$\begin{bmatrix} e & \pi \\ 2 & \sqrt{2} \\ -7 & 4 \end{bmatrix}$$

One very common use of matrices is to represent systems of linear equations. The matrix derived from the coefficients and constant terms of a system of linear equations is called the **augmented matrix** of the system. The matrix containing only the coefficients of the system is called the **coefficient matrix** of the system. Here is an example.

System

$$\begin{aligned} x - 4y + 3z &= 5 \\ -x + 3y - z &= -3 \\ 2x \qquad - 4z &= 6 \end{aligned}$$

Augmented Matrix

$$\begin{bmatrix} 1 & -4 & 3 & 5 \\ -1 & 3 & -1 & -3 \\ 2 & 0 & -4 & 6 \end{bmatrix}$$

Coefficient Matrix

$$\begin{bmatrix} 1 & -4 & 3 \\ -1 & 3 & -1 \\ 2 & 0 & -4 \end{bmatrix}$$

REMARK: Use 0 to indicate coefficients of zero. The coefficient of y in the third equation is zero, so a 0 takes its place in the matrix. Also note the fourth column of constant terms in the augmented matrix.

When forming either the coefficient matrix or the augmented matrix of a system, you should begin by aligning the variables in the equations vertically.

Given System

$$\begin{aligned} x_1 + 3x_2 &= 9 \\ -x_2 + 4x_3 &= -2 \\ x_1 - 5x_3 &= 0 \end{aligned}$$

Align Variables

$$\begin{aligned} x_1 + 3x_2 \qquad\quad &= 9 \\ -x_2 + 4x_3 &= -2 \\ x_1 \qquad\quad - 5x_3 &= 0 \end{aligned}$$

Augmented Matrix

$$\begin{bmatrix} 1 & 3 & 0 & 9 \\ 0 & -1 & 4 & -2 \\ 1 & 0 & -5 & 0 \end{bmatrix}$$

Elementary Row Operations

In the previous section you studied three operations that can be used on a system of linear equations to produce equivalent systems.

1. Interchange two equations.
2. Multiply an equation by a nonzero constant.
3. Add a multiple of an equation to another equation.

In matrix terminology these three operations correspond to **elementary row operations.** An elementary row operation on an augmented matrix produces a new augmented matrix corresponding to a new (but equivalent) system of linear equations. Two matrices are said to be **row-equivalent** if one can be obtained from the other by a finite sequence of elementary row operations.

Elementary Row Operations

1. Interchange two rows.
2. Multiply a row by a nonzero constant.
3. Add a multiple of a row to another row.

Although elementary row operations are simple to perform, they involve a lot of arithmetic. Because it is easy to make a mistake, you should get in the habit of noting the elementary row operation performed in each step so that it is easier to check your work.

Because solving some systems involves several steps, it is helpful to use a shorthand method of notation to keep track of each elementary row operation you perform. This notation is introduced in the next example.

EXAMPLE 2 Elementary Row Operations

(a) Interchange the first and second rows.

Original Matrix / *New Row-Equivalent Matrix* / *Notation*

$$\begin{bmatrix} 0 & 1 & 3 & 4 \\ -1 & 2 & 0 & 3 \\ 2 & -3 & 4 & 1 \end{bmatrix} \qquad \begin{bmatrix} -1 & 2 & 0 & 3 \\ 0 & 1 & 3 & 4 \\ 2 & -3 & 4 & 1 \end{bmatrix} \qquad R_1 \leftrightarrow R_2$$

(b) Multiply the first row by $\frac{1}{2}$ to produce a new first row.

Original Matrix / *New Row-Equivalent Matrix* / *Notation*

$$\begin{bmatrix} 2 & -4 & 6 & -2 \\ 1 & 3 & -3 & 0 \\ 5 & -2 & 1 & 2 \end{bmatrix} \qquad \begin{bmatrix} 1 & -2 & 3 & -1 \\ 1 & 3 & -3 & 0 \\ 5 & -2 & 1 & 2 \end{bmatrix} \qquad \left(\tfrac{1}{2}\right)R_1 \rightarrow R_1$$

(c) Add -2 times the first row to the third row to produce a new third row.

Original Matrix / *New Row-Equivalent Matrix* / *Notation*

$$\begin{bmatrix} 1 & 2 & -4 & 3 \\ 0 & 3 & -2 & -1 \\ 2 & 1 & 5 & -2 \end{bmatrix} \qquad \begin{bmatrix} 1 & 2 & -4 & 3 \\ 0 & 3 & -2 & -1 \\ 0 & -3 & 13 & -8 \end{bmatrix} \qquad R_3 + (-2)R_1 \rightarrow R_3$$

REMARK: Notice in Example 2(c) that adding -2 times row 1 to row 3 does not change row 1.

Technology Note

Many graphing utilities and computer software programs can perform elementary row operations on matrices. If you are using a graphing utility, your screens for Example 2(c) may look like those shown below. Keystrokes and programming syntax for these utilities/programs applicable to Example 2(c) are provided in the **Online Technology Guide,** available at *college.hmco.com/pic/larsonELA6e*.

```
A
        [[1 2 -4 3 ]
         [0 3 -2 -1]
         [2 1 5  -2]]
```

```
mRAdd(-2,A,1,3)
        [[1 2  -4 3 ]
         [0 3  -2 -1]
         [0 -3 13 -8]]
```

In Example 7 in Section 1.1, you used Gaussian elimination with back-substitution to solve a system of linear equations. You will now learn the matrix version of Gaussian elimination. The two methods used in the next example are essentially the same. The basic difference is that with the matrix method there is no need to rewrite the variables over and over again.

EXAMPLE 3 Using Elementary Row Operations to Solve a System

Linear System

$$\begin{aligned} x - 2y + 3z &= 9 \\ -x + 3y &= -4 \\ 2x - 5y + 5z &= 17 \end{aligned}$$

Associated Augmented Matrix

$$\begin{bmatrix} 1 & -2 & 3 & 9 \\ -1 & 3 & 0 & -4 \\ 2 & -5 & 5 & 17 \end{bmatrix}$$

Add the first equation to the second equation.

$$\begin{aligned} x - 2y + 3z &= 9 \\ y + 3z &= 5 \\ 2x - 5y + 5z &= 17 \end{aligned}$$

Add the first row to the second row to produce a new second row.

$$\begin{bmatrix} 1 & -2 & 3 & 9 \\ 0 & 1 & 3 & 5 \\ 2 & -5 & 5 & 17 \end{bmatrix} \quad R_2 + R_1 \rightarrow R_2$$

Add -2 times the first equation to the third equation.

$$\begin{aligned} x - 2y + 3z &= 9 \\ y + 3z &= 5 \\ -y - z &= -1 \end{aligned}$$

Add -2 times the first row to the third row to produce a new third row.

$$\begin{bmatrix} 1 & -2 & 3 & 9 \\ 0 & 1 & 3 & 5 \\ 0 & -1 & -1 & -1 \end{bmatrix} \quad R_3 + (-2)R_1 \rightarrow R_3$$

Add the second equation to the third equation.

$$\begin{aligned} x - 2y + 3z &= 9 \\ y + 3z &= 5 \\ 2z &= 4 \end{aligned}$$

Add the second row to the third row to produce a new third row.

$$\begin{bmatrix} 1 & -2 & 3 & 9 \\ 0 & 1 & 3 & 5 \\ 0 & 0 & 2 & 4 \end{bmatrix} \quad R_3 + R_2 \rightarrow R_3$$

Multiply the third equation by $\frac{1}{2}$.

$$\begin{aligned} x - 2y + 3z &= 9 \\ y + 3z &= 5 \\ z &= 2 \end{aligned}$$

Multiply the third row by $\frac{1}{2}$ to produce a new third row.

$$\begin{bmatrix} 1 & -2 & 3 & 9 \\ 0 & 1 & 3 & 5 \\ 0 & 0 & 1 & 2 \end{bmatrix} \quad \left(\tfrac{1}{2}\right)R_3 \rightarrow R_3$$

Now you can use back-substitution to find the solution, as in Example 6 in Section 1.1. The solution is $x = 1$, $y = -1$, and $z = 2$.

The last matrix in Example 3 is said to be in **row-echelon form.** The term *echelon* refers to the stair-step pattern formed by the nonzero elements of the matrix. To be in row-echelon form, a matrix must have the properties listed below.

Definition of Row-Echelon Form of a Matrix

A matrix in **row-echelon form** has the following properties.

1. All rows consisting entirely of zeros occur at the bottom of the matrix.
2. For each row that does not consist entirely of zeros, the first nonzero entry is 1 (called a **leading 1**).
3. For two successive (nonzero) rows, the leading 1 in the higher row is farther to the left than the leading 1 in the lower row.

REMARK: A matrix in row-echelon form is in **reduced row-echelon form** if every column that has a leading 1 has zeros in every position above and below its leading 1.

EXAMPLE 4 Row-Echelon Form

The matrices below are in row-echelon form.

$$\text{(a)} \begin{bmatrix} 1 & 2 & -1 & 4 \\ 0 & 1 & 0 & 3 \\ 0 & 0 & 1 & -2 \end{bmatrix} \qquad \text{(b)} \begin{bmatrix} 0 & 1 & 0 & 5 \\ 0 & 0 & 1 & 3 \\ 0 & 0 & 0 & 0 \end{bmatrix}$$

$$\text{(c)} \begin{bmatrix} 1 & -5 & 2 & -1 & 3 \\ 0 & 0 & 1 & 3 & -2 \\ 0 & 0 & 0 & 1 & 4 \\ 0 & 0 & 0 & 0 & 1 \end{bmatrix} \qquad \text{(d)} \begin{bmatrix} 1 & 0 & 0 & -1 \\ 0 & 1 & 0 & 2 \\ 0 & 0 & 1 & 3 \\ 0 & 0 & 0 & 0 \end{bmatrix}$$

The matrices shown in parts (b) and (d) are in *reduced* row-echelon form. The matrices listed below are not in row-echelon form.

$$\text{(e)} \begin{bmatrix} 1 & 2 & -3 & 4 \\ 0 & 2 & 1 & -1 \\ 0 & 0 & 1 & -3 \end{bmatrix} \qquad \text{(f)} \begin{bmatrix} 1 & 2 & -1 & 2 \\ 0 & 0 & 0 & 0 \\ 0 & 1 & 2 & -4 \end{bmatrix}$$

Use a graphing utility or a computer software program to find the reduced row-echelon form of the matrix in part (f) of Example 4. Keystrokes and programming syntax for these utilities/programs applicable to Example 4(f) are provided in the **Online Technology Guide,** available at *college.hmco.com/pic/larsonELA6e.*

It can be shown that every matrix is row-equivalent to a matrix in row-echelon form. For instance, in Example 4 you could change the matrix in part (e) to row-echelon form by multiplying the second row in the matrix by $\frac{1}{2}$.

The method of using Gaussian elimination with back-substitution to solve a system is as follows.

REMARK: For keystrokes and programming syntax regarding specific graphing utilities and computer software programs involving Example 4(f), please visit *college.hmco.com/pic/larsonELA6e*. Similar exercises and projects are also available on the website.

Gaussian Elimination with Back-Substitution

1. Write the augmented matrix of the system of linear equations.
2. Use elementary row operations to rewrite the augmented matrix in row-echelon form.
3. Write the system of linear equations corresponding to the matrix in row-echelon form, and use back-substitution to find the solution.

Gaussian elimination with back-substitution works well as an algorithmic method for solving systems of linear equations. For this algorithm, the order in which the elementary row operations are performed is important. Move from *left to right by columns*, changing all entries directly below the leading 1's to zeros.

EXAMPLE 5 Gaussian Elimination with Back-Substitution

Solve the system.

$$\begin{aligned} x_2 + x_3 - 2x_4 &= -3 \\ x_1 + 2x_2 - x_3 &= 2 \\ 2x_1 + 4x_2 + x_3 - 3x_4 &= -2 \\ x_1 - 4x_2 - 7x_3 - x_4 &= -19 \end{aligned}$$

SOLUTION The augmented matrix for this system is

$$\begin{bmatrix} 0 & 1 & 1 & -2 & -3 \\ 1 & 2 & -1 & 0 & 2 \\ 2 & 4 & 1 & -3 & -2 \\ 1 & -4 & -7 & -1 & -19 \end{bmatrix}.$$

Obtain a leading 1 in the upper left corner and zeros elsewhere in the first column.

$$\begin{bmatrix} 1 & 2 & -1 & 0 & 2 \\ 0 & 1 & 1 & -2 & -3 \\ 2 & 4 & 1 & -3 & -2 \\ 1 & -4 & -7 & -1 & -19 \end{bmatrix}$$

← The first two rows are interchanged. $R_1 \leftrightarrow R_2$

$$\begin{bmatrix} 1 & 2 & -1 & 0 & 2 \\ 0 & 1 & 1 & -2 & -3 \\ 0 & 0 & 3 & -3 & -6 \\ 1 & -4 & -7 & -1 & -19 \end{bmatrix}$$

← Adding -2 times the first row to the third row produces a new third row. $R_3 + (-2)R_1 \rightarrow R_3$

$$\begin{bmatrix} 1 & 2 & -1 & 0 & 2 \\ 0 & 1 & 1 & -2 & -3 \\ 0 & 0 & 3 & -3 & -6 \\ 0 & -6 & -6 & -1 & -21 \end{bmatrix}$$

Adding −1 times the first row to the fourth row produces a new fourth row. $R_4 + (-1)R_1 \rightarrow R_4$

Now that the first column is in the desired form, you should change the second column as shown below.

$$\begin{bmatrix} 1 & 2 & -1 & 0 & 2 \\ 0 & 1 & 1 & -2 & -3 \\ 0 & 0 & 3 & -3 & -6 \\ 0 & 0 & 0 & -13 & -39 \end{bmatrix}$$

Adding 6 times the second row to the fourth row produces a new fourth row. $R_4 + (6)R_2 \rightarrow R_4$

To write the third column in proper form, multiply the third row by $\frac{1}{3}$.

$$\begin{bmatrix} 1 & 2 & -1 & 0 & 2 \\ 0 & 1 & 1 & -2 & -3 \\ 0 & 0 & 1 & -1 & -2 \\ 0 & 0 & 0 & -13 & -39 \end{bmatrix}$$

Multiplying the third row by $\frac{1}{3}$ produces a new third row. $\left(\frac{1}{3}\right)R_3 \rightarrow R_3$

Similarly, to write the fourth column in proper form, you should multiply the fourth row by $-\frac{1}{13}$.

$$\begin{bmatrix} 1 & 2 & -1 & 0 & 2 \\ 0 & 1 & 1 & -2 & -3 \\ 0 & 0 & 1 & -1 & -2 \\ 0 & 0 & 0 & 1 & 3 \end{bmatrix}$$

Multiplying the fourth row by $-\frac{1}{13}$ produces a new fourth row. $\left(-\frac{1}{13}\right)R_4 \rightarrow R_4$

The matrix is now in row-echelon form, and the corresponding system of linear equations is as shown below.

$$\begin{aligned} x_1 + 2x_2 - x_3 &= 2 \\ x_2 + x_3 - 2x_4 &= -3 \\ x_3 - x_4 &= -2 \\ x_4 &= 3 \end{aligned}$$

Using back-substitution, you can determine that the solution is

$$x_1 = -1, \quad x_2 = 2, \quad x_3 = 1, \quad x_4 = 3.$$

When solving a system of linear equations, remember that it is possible for the system to have no solution. If during the elimination process you obtain a row with all zeros except for the last entry, it is unnecessary to continue the elimination process. You can simply conclude that the system is inconsistent and has no solution.

EXAMPLE 6 **A System with No Solution**

Solve the system.

$$\begin{aligned} x_1 - x_2 + 2x_3 &= 4 \\ x_1 + x_3 &= 6 \\ 2x_1 - 3x_2 + 5x_3 &= 4 \\ 3x_1 + 2x_2 - x_3 &= 1 \end{aligned}$$

SOLUTION The augmented matrix for this system is

$$\begin{bmatrix} 1 & -1 & 2 & 4 \\ 1 & 0 & 1 & 6 \\ 2 & -3 & 5 & 4 \\ 3 & 2 & -1 & 1 \end{bmatrix}.$$

Apply Gaussian elimination to the augmented matrix.

$$\begin{bmatrix} 1 & -1 & 2 & 4 \\ 0 & 1 & -1 & 2 \\ 2 & -3 & 5 & 4 \\ 3 & 2 & -1 & 1 \end{bmatrix} \qquad R_2 + (-1)R_1 \rightarrow R_2$$

$$\begin{bmatrix} 1 & -1 & 2 & 4 \\ 0 & 1 & -1 & 2 \\ 0 & -1 & 1 & -4 \\ 3 & 2 & -1 & 1 \end{bmatrix} \qquad R_3 + (-2)R_1 \rightarrow R_3$$

$$\begin{bmatrix} 1 & -1 & 2 & 4 \\ 0 & 1 & -1 & 2 \\ 0 & -1 & 1 & -4 \\ 0 & 5 & -7 & -11 \end{bmatrix} \qquad R_4 + (-3)R_1 \rightarrow R_4$$

$$\begin{bmatrix} 1 & -1 & 2 & 4 \\ 0 & 1 & -1 & 2 \\ 0 & 0 & 0 & -2 \\ 0 & 5 & -7 & -11 \end{bmatrix} \qquad R_3 + R_2 \rightarrow R_3$$

Note that the third row of this matrix consists of all zeros except for the last entry. This means that the original system of linear equations is *inconsistent.* You can see why this is true by converting back to a system of linear equations.

$$\begin{aligned} x_1 - x_2 + 2x_3 &= 4 \\ x_2 - x_3 &= 2 \\ 0 &= -2 \\ 5x_2 - 7x_3 &= -11 \end{aligned}$$

Because the third "equation" is a false statement, the system has no solution.

Discovery

Consider the system of linear equations.

$$\begin{aligned} 2x_1 + 3x_2 + 5x_3 &= 0 \\ -5x_1 + 6x_2 - 17x_3 &= 0 \\ 7x_1 - 4x_2 + 3x_3 &= 0 \end{aligned}$$

Without doing any row operations, explain why this system is consistent.

The system below has more variables than equations. Why does it have an infinite number of solutions?

$$\begin{aligned} 2x_1 + 3x_2 + 5x_3 + 2x_4 &= 0 \\ -5x_1 + 6x_2 - 17x_3 - 3x_4 &= 0 \\ 7x_1 - 4x_2 + 3x_3 + 13x_4 &= 0 \end{aligned}$$

Gauss-Jordan Elimination

With Gaussian elimination, you apply elementary row operations to a matrix to obtain a (row-equivalent) row-echelon form. A second method of elimination, called **Gauss-Jordan elimination** after Carl Gauss and Wilhelm Jordan (1842–1899), continues the reduction process until a *reduced* row-echelon form is obtained. This procedure is demonstrated in the next example.

EXAMPLE 7 Gauss-Jordan Elimination

Use Gauss-Jordan elimination to solve the system.

$$\begin{aligned} x - 2y + 3z &= 9 \\ -x + 3y &= -4 \\ 2x - 5y + 5z &= 17 \end{aligned}$$

SOLUTION In Example 3, Gaussian elimination was used to obtain the row-echelon form

$$\begin{bmatrix} 1 & -2 & 3 & 9 \\ 0 & 1 & 3 & 5 \\ 0 & 0 & 1 & 2 \end{bmatrix}.$$

Now, rather than using back-substitution, apply elementary row operations until you obtain a matrix in reduced row-echelon form. To do this, you must produce zeros above each of the leading 1's, as follows.

$$\begin{bmatrix} 1 & 0 & 9 & 19 \\ 0 & 1 & 3 & 5 \\ 0 & 0 & 1 & 2 \end{bmatrix} \quad R_1 + (2)R_2 \rightarrow R_1$$

$$\begin{bmatrix} 1 & 0 & 9 & 19 \\ 0 & 1 & 0 & -1 \\ 0 & 0 & 1 & 2 \end{bmatrix} \quad R_2 + (-3)R_3 \rightarrow R_2$$

$$\begin{bmatrix} 1 & 0 & 0 & 1 \\ 0 & 1 & 0 & -1 \\ 0 & 0 & 1 & 2 \end{bmatrix} \quad R_1 + (-9)R_3 \rightarrow R_1$$

Now, converting back to a system of linear equations, you have

$$\begin{aligned} x \quad\quad\quad &= 1 \\ y \quad\quad &= -1 \\ z &= 2. \end{aligned}$$

The Gaussian and Gauss-Jordan elimination procedures employ an algorithmic approach easily adapted to computer use. These elimination procedures, however, make no effort to avoid fractional coefficients. For instance, if the system in Example 7 had been listed as

$$\begin{aligned} 2x - 5y + 5z &= 17 \\ x - 2y + 3z &= 9 \\ -x + 3y \quad\quad &= -4 \end{aligned}$$

both procedures would have required multiplying the first row by $\frac{1}{2}$, which would have introduced fractions in the first row. For hand computations, fractions can sometimes be avoided by judiciously choosing the order in which elementary row operations are applied.

REMARK: No matter which order you use, the reduced row-echelon form will be the same.

The next example demonstrates how Gauss-Jordan elimination can be used to solve a system with an infinite number of solutions.

EXAMPLE 8 A System with an Infinite Number of Solutions

Solve the system of linear equations.

$$\begin{aligned} 2x_1 + 4x_2 - 2x_3 &= 0 \\ 3x_1 + 5x_2 \quad\quad &= 1 \end{aligned}$$

SOLUTION The augmented matrix of the system of linear equations is

$$\begin{bmatrix} 2 & 4 & -2 & 0 \\ 3 & 5 & 0 & 1 \end{bmatrix}.$$

Using a graphing utility, a computer software program, or Gauss-Jordan elimination, you can verify that the reduced row-echelon form of the matrix is

$$\begin{bmatrix} 1 & 0 & 5 & 2 \\ 0 & 1 & -3 & -1 \end{bmatrix}.$$

The corresponding system of equations is

$$\begin{aligned} x_1 \qquad + 5x_3 &= 2 \\ x_2 - 3x_3 &= -1. \end{aligned}$$

Now, using the parameter t to represent the *nonleading* variable x_3, you have

$$x_1 = 2 - 5t, \qquad x_2 = -1 + 3t, \qquad x_3 = t, \qquad \text{where } t \text{ is any real number.}$$

REMARK: Note that in Example 8 an arbitrary parameter was assigned to the nonleading variable x_3. You subsequently solved for the leading variables x_1 and x_2 as functions of t.

You have looked at two elimination methods for solving a system of linear equations. Which is better? To some degree the answer depends on personal preference. In real-life applications of linear algebra, systems of linear equations are usually solved by computer. Most computer programs use a form of Gaussian elimination, with special emphasis on ways to reduce rounding errors and minimize storage of data. Because the examples and exercises in this text are generally much simpler and focus on the underlying concepts, you will need to know both elimination methods.

Homogeneous Systems of Linear Equations

As the final topic of this section, you will look at systems of linear equations in which each of the constant terms is zero. We call such systems **homogeneous.** For example, a homogeneous system of m equations in n variables has the form

$$\begin{aligned} a_{11}x_1 + a_{12}x_2 + a_{13}x_3 + \cdots + a_{1n}x_n &= 0 \\ a_{21}x_1 + a_{22}x_2 + a_{23}x_3 + \cdots + a_{2n}x_n &= 0 \\ a_{31}x_1 + a_{32}x_2 + a_{33}x_3 + \cdots + a_{3n}x_n &= 0 \\ &\vdots \\ a_{m1}x_1 + a_{m2}x_2 + a_{m3}x_3 + \cdots + a_{mn}x_n &= 0. \end{aligned}$$

It is easy to see that a homogeneous system must have at least one solution. Specifically, if all variables in a homogeneous system have the value zero, then each of the equations must be satisfied. Such a solution is called **trivial** (or **obvious**). For instance, a homogeneous system of three equations in the three variables x_1, x_2, and x_3 must have $x_1 = 0$, $x_2 = 0$, and $x_3 = 0$ as a trivial solution.

EXAMPLE 9 Solving a Homogeneous System of Linear Equations

Solve the system of linear equations.

$$\begin{aligned} x_1 - x_2 + 3x_3 &= 0 \\ 2x_1 + x_2 + 3x_3 &= 0 \end{aligned}$$

SOLUTION Applying Gauss-Jordan elimination to the augmented matrix

$$\begin{bmatrix} 1 & -1 & 3 & 0 \\ 2 & 1 & 3 & 0 \end{bmatrix}$$

yields the matrix shown below.

$$\begin{bmatrix} 1 & -1 & 3 & 0 \\ 0 & 3 & -3 & 0 \end{bmatrix} \qquad R_2 + (-2)R_1 \rightarrow R_2$$

$$\begin{bmatrix} 1 & -1 & 3 & 0 \\ 0 & 1 & -1 & 0 \end{bmatrix} \qquad \left(\tfrac{1}{3}\right)R_2 \rightarrow R_2$$

$$\begin{bmatrix} 1 & 0 & 2 & 0 \\ 0 & 1 & -1 & 0 \end{bmatrix} \qquad R_1 + R_2 \rightarrow R_1$$

The system of equations corresponding to this matrix is

$$\begin{aligned} x_1 \qquad + 2x_3 &= 0 \\ x_2 - x_3 &= 0. \end{aligned}$$

Using the parameter $t = x_3$, the solution set is

$$x_1 = -2t, \qquad x_2 = t, \qquad x_3 = t, \quad t \text{ is any real number.}$$

This system of equations has an infinite number of solutions, one of which is the trivial solution (given by $t = 0$).

Example 9 illustrates an important point about homogeneous systems of linear equations. You began with two equations in three variables and discovered that the system has an infinite number of solutions. In general, a homogeneous system with fewer equations than variables has an infinite number of solutions.

THEOREM 1.1 The Number of Solutions of a Homogeneous System

Every homogeneous system of linear equations is consistent. Moreover, if the system has fewer equations than variables, then it must have an infinite number of solutions.

A proof of Theorem 1.1 can be done using the same procedures as those used in Example 9, but for a general matrix.

SECTION 1.2 Exercises

In Exercises 1–8, determine the size of the matrix.

1. $\begin{bmatrix} 1 & 2 & -4 \\ 3 & -4 & 6 \\ 0 & 1 & 2 \end{bmatrix}$

2. $\begin{bmatrix} 2 & -1 & 4 & 2 \\ 1 & 0 & 2 & -6 \end{bmatrix}$

3. $\begin{bmatrix} 2 & -1 & -1 & 1 \\ -6 & 2 & 0 & 1 \end{bmatrix}$

4. $\begin{bmatrix} 1 & 2 & 3 & 0 & 1 \end{bmatrix}$

5. $\begin{bmatrix} 1 & 2 & 3 & 4 & -10 \end{bmatrix}$

6. $\begin{bmatrix} -1 \end{bmatrix}$

7. $\begin{bmatrix} 8 & 6 & 4 & 1 & 3 \\ 2 & 1 & -7 & 4 & 1 \\ 1 & 1 & -1 & 2 & 1 \\ 1 & -1 & 2 & 0 & 0 \end{bmatrix}$

8. $\begin{bmatrix} 1 \\ 2 \\ -1 \\ -2 \end{bmatrix}$

In Exercises 9–14, determine whether the matrix is in row-echelon form. If it is, determine whether it is also in reduced row-echelon form.

9. $\begin{bmatrix} 1 & 0 & 0 & 0 \\ 0 & 1 & 1 & 2 \\ 0 & 0 & 0 & 0 \end{bmatrix}$

10. $\begin{bmatrix} 0 & 1 & 0 & 0 \\ 1 & 0 & 2 & 1 \end{bmatrix}$

11. $\begin{bmatrix} 2 & 0 & 1 & 3 \\ 0 & -1 & 1 & 4 \\ 0 & 0 & 0 & 1 \end{bmatrix}$

12. $\begin{bmatrix} 1 & 0 & 2 & 1 \\ 0 & 1 & 3 & 4 \\ 0 & 0 & 1 & 0 \end{bmatrix}$

13. $\begin{bmatrix} 0 & 0 & 1 & 0 & 0 \\ 0 & 0 & 0 & 1 & 0 \\ 0 & 0 & 0 & 2 & 0 \end{bmatrix}$

14. $\begin{bmatrix} 1 & 0 & 0 & 0 \\ 0 & 0 & 0 & 1 \\ 0 & 0 & 0 & 0 \end{bmatrix}$

In Exercises 15–22, find the solution set of the system of linear equations represented by the augmented matrix.

15. $\begin{bmatrix} 1 & 0 & 0 \\ 0 & 1 & 2 \end{bmatrix}$

16. $\begin{bmatrix} 1 & 0 & 2 \\ 0 & 1 & 3 \end{bmatrix}$

17. $\begin{bmatrix} 1 & -1 & 0 & 3 \\ 0 & 1 & -2 & 1 \\ 0 & 0 & 1 & -1 \end{bmatrix}$

18. $\begin{bmatrix} 1 & 2 & 1 & 0 \\ 0 & 0 & 1 & -1 \\ 0 & 0 & 0 & 0 \end{bmatrix}$

19. $\begin{bmatrix} 2 & 1 & -1 & 3 \\ 1 & -1 & 1 & 0 \\ 0 & 1 & 2 & 1 \end{bmatrix}$

20. $\begin{bmatrix} 2 & 1 & 1 & 0 \\ 1 & -2 & 1 & -2 \\ 1 & 0 & 1 & 0 \end{bmatrix}$

21. $\begin{bmatrix} 1 & 2 & 0 & 1 & 4 \\ 0 & 1 & 2 & 1 & 3 \\ 0 & 0 & 1 & 2 & 1 \\ 0 & 0 & 0 & 1 & 4 \end{bmatrix}$

22. $\begin{bmatrix} 1 & 2 & 0 & 1 & 3 \\ 0 & 1 & 3 & 0 & 1 \\ 0 & 0 & 1 & 2 & 0 \\ 0 & 0 & 0 & 0 & 2 \end{bmatrix}$

In Exercises 23–36, solve the system using either Gaussian elimination with back-substitution or Gauss-Jordan elimination.

23. $\begin{aligned} x + 2y &= 7 \\ 2x + y &= 8 \end{aligned}$

24. $\begin{aligned} 2x + 6y &= 16 \\ -2x - 6y &= -16 \end{aligned}$

25. $\begin{aligned} -x + 2y &= 1.5 \\ 2x - 4y &= 3 \end{aligned}$

26. $\begin{aligned} 2x - y &= -0.1 \\ 3x + 2y &= 1.6 \end{aligned}$

27. $\begin{aligned} -3x + 5y &= -22 \\ 3x + 4y &= 4 \\ 4x - 8y &= 32 \end{aligned}$

28. $\begin{aligned} x + 2y &= 0 \\ x + y &= 6 \\ 3x - 2y &= 8 \end{aligned}$

29. $\begin{aligned} x_1 - 3x_3 &= -2 \\ 3x_1 + x_2 - 2x_3 &= 5 \\ 2x_1 + 2x_2 + x_3 &= 4 \end{aligned}$

30. $\begin{aligned} 2x_1 - x_2 + 3x_3 &= 24 \\ 2x_2 - x_3 &= 14 \\ 7x_1 - 5x_2 &= 6 \end{aligned}$

31. $\begin{aligned} x_1 + x_2 - 5x_3 &= 3 \\ x_1 - 2x_3 &= 1 \\ 2x_1 - x_2 - x_3 &= 0 \end{aligned}$

32. $\begin{aligned} 2x_1 + 3x_3 &= 3 \\ 4x_1 - 3x_2 + 7x_3 &= 5 \\ 8x_1 - 9x_2 + 15x_3 &= 10 \end{aligned}$

33. $\begin{aligned} 4x + 12y - 7z - 20w &= 22 \\ 3x + 9y - 5z - 28w &= 30 \end{aligned}$

34. $\begin{aligned} x + 2y + z &= 8 \\ -3x - 6y - 3z &= -21 \end{aligned}$

35. $\begin{aligned} 3x + 3y + 12z &= 6 \\ x + y + 4z &= 2 \\ 2x + 5y + 20z &= 10 \\ -x + 2y + 8z &= 4 \end{aligned}$

36. $\begin{aligned} 2x + y - z + 2w &= -6 \\ 3x + 4y + w &= 1 \\ x + 5y + 2z + 6w &= -3 \\ 5x + 2y - z - w &= 3 \end{aligned}$

In Exercises 37–42, use a computer software program or graphing utility to solve the system of linear equations.

37. $\begin{aligned} x_1 - 2x_2 + 5x_3 - 3x_4 &= 23.6 \\ x_1 + 4x_2 - 7x_3 - 2x_4 &= 45.7 \\ 3x_1 - 5x_2 + 7x_3 + 4x_4 &= 29.9 \end{aligned}$

38. $\begin{aligned} 23.4x - 45.8y + 43.7z &= 87.2 \\ 86.4x + 12.3y - 56.9z &= 14.5 \\ 93.6x - 50.7y + 12.6z &= 44.4 \end{aligned}$

39.
$$\begin{aligned} x_1 - x_2 + 2x_3 + 2x_4 + 6x_5 &= 6 \\ 3x_1 - 2x_2 + 4x_3 + 4x_4 + 12x_5 &= 14 \\ x_2 - x_3 - x_4 - 3x_5 &= -3 \\ 2x_1 - 2x_2 + 4x_3 + 5x_4 + 15x_5 &= 10 \\ 2x_1 - 2x_2 + 4x_3 + 4x_4 + 13x_5 &= 13 \end{aligned}$$

40.
$$\begin{aligned} x_1 + x_2 - 2x_3 + 3x_4 + 2x_5 &= 9 \\ 3x_1 + 3x_2 - x_3 + x_4 + x_5 &= 5 \\ 2x_1 + 2x_2 - x_3 + x_4 - 2x_5 &= 1 \\ 4x_1 + 4x_2 + x_3 \qquad - 3x_5 &= 4 \\ 8x_1 + 5x_2 - 2x_3 - x_4 + 2x_5 &= 3 \end{aligned}$$

41.
$$\begin{aligned} 4x_1 - 3x_2 + x_3 - x_4 + 2x_5 - x_6 &= 8 \\ x_1 - 2x_2 + x_3 - 3x_4 + x_5 - 4x_6 &= 4 \\ 2x_1 + x_2 - 3x_3 + x_4 - 2x_5 + 5x_6 &= 2 \\ -2x_1 + 3x_2 - x_3 + x_4 - x_5 + 2x_6 &= -7 \\ x_1 - 3x_2 + x_3 - 2x_4 + x_5 - 2x_6 &= 9 \\ 5x_1 - 4x_2 - x_3 - x_4 + 4x_5 + 5x_6 &= 9 \end{aligned}$$

42.
$$\begin{aligned} x_1 + 2x_2 - 2x_3 + 2x_4 - x_5 + 3x_6 &= 0 \\ 2x_1 - x_2 + 3x_3 + x_4 - 3x_5 + 2x_6 &= 17 \\ x_1 + 3x_2 - 2x_3 + x_4 - 2x_5 - 3x_6 &= -5 \\ 3x_1 - 2x_2 + x_3 - x_4 + 3x_5 - 2x_6 &= -1 \\ -x_1 - 2x_2 + x_3 + 2x_4 - 2x_5 + 3x_6 &= 10 \\ x_1 - 3x_2 + x_3 + 3x_4 - 2x_5 + x_6 &= 11 \end{aligned}$$

In Exercises 43–46, solve the homogeneous linear system corresponding to the coefficient matrix provided.

43. $\begin{bmatrix} 1 & 0 & 0 \\ 0 & 1 & 1 \\ 0 & 0 & 0 \end{bmatrix}$ **44.** $\begin{bmatrix} 1 & 0 & 0 & 0 \\ 0 & 1 & 1 & 0 \end{bmatrix}$

45. $\begin{bmatrix} 1 & 0 & 0 & 1 \\ 0 & 0 & 1 & 0 \\ 0 & 0 & 0 & 0 \end{bmatrix}$ **46.** $\begin{bmatrix} 0 & 0 & 0 \\ 0 & 0 & 0 \\ 0 & 0 & 0 \end{bmatrix}$

47. Consider the matrix $A = \begin{bmatrix} 1 & k & 2 \\ -3 & 4 & 1 \end{bmatrix}$.

(a) If A is the *augmented* matrix of a system of linear equations, determine the number of equations and the number of variables.

(b) If A is the *augmented* matrix of a system of linear equations, find the value(s) of k such that the system is consistent.

(c) If A is the *coefficient* matrix of a *homogeneous* system of linear equations, determine the number of equations and the number of variables.

(d) If A is the *coefficient* matrix of a *homogeneous* system of linear equations, find the value(s) of k such that the system is consistent.

48. Consider the matrix $A = \begin{bmatrix} 2 & -1 & 3 \\ -4 & 2 & k \\ 4 & -2 & 6 \end{bmatrix}$.

(a) If A is the *augmented* matrix of a system of linear equations, determine the number of equations and the number of variables.

(b) If A is the *augmented* matrix of a system of linear equations, find the value(s) of k such that the system is consistent.

(c) If A is the *coefficient* matrix of a *homogeneous* system of linear equations, determine the number of equations and the number of variables.

(d) If A is the *coefficient* matrix of a *homogeneous* system of linear equations, find the value(s) of k such that the system is consistent.

In Exercises 49 and 50, find values of a, b, and c (if possible) such that the system of linear equations has (a) a unique solution, (b) no solution, and (c) an infinite number of solutions.

49.
$$\begin{aligned} x + y \qquad &= 2 \\ y + z &= 2 \\ x \qquad + z &= 2 \\ ax + by + cz &= 0 \end{aligned}$$

50.
$$\begin{aligned} x + y \qquad &= 0 \\ y + z &= 0 \\ x \qquad + z &= 0 \\ ax + by + cz &= 0 \end{aligned}$$

51. The system below has one solution: $x = 1$, $y = -1$, and $z = 2$.

$$\begin{aligned} 4x - 2y + 5z &= 16 && \text{Equation 1} \\ x + y \qquad &= 0 && \text{Equation 2} \\ -x - 3y + 2z &= 6 && \text{Equation 3} \end{aligned}$$

Solve the systems provided by (a) Equations 1 and 2, (b) Equations 1 and 3, and (c) Equations 2 and 3. (d) How many solutions does each of these systems have?

52. Assume the system below has a unique solution.

$$\begin{aligned} a_{11}x_1 + a_{12}x_2 + a_{13}x_3 &= b_1 && \text{Equation 1} \\ a_{21}x_1 + a_{22}x_2 + a_{23}x_3 &= b_2 && \text{Equation 2} \\ a_{31}x_1 + a_{32}x_2 + a_{33}x_3 &= b_3 && \text{Equation 3} \end{aligned}$$

Does the system composed of Equations 1 and 2 have a unique solution, no solution, or an infinite number of solutions?

In Exercises 53 and 54, find the unique reduced row-echelon matrix that is row-equivalent to the matrix provided.

53. $\begin{bmatrix} 1 & 2 \\ -1 & 2 \end{bmatrix}$ **54.** $\begin{bmatrix} 1 & 2 & 3 \\ 4 & 5 & 6 \\ 7 & 8 & 9 \end{bmatrix}$

55. Writing Describe all possible 2×2 reduced row-echelon matrices. Support your answer with examples.

56. Writing Describe all possible 3×3 reduced row-echelon matrices. Support your answer with examples.

True or False? In Exercises 57 and 58, determine whether each statement is true or false. If a statement is true, give a reason or cite an appropriate statement from the text. If a statement is false, provide an example that shows the statement is not true in all cases or cite an appropriate statement from the text.

57. (a) A 6×3 matrix has six rows.

(b) Every matrix is row-equivalent to a matrix in row-echelon form.

(c) If the row-echelon form of the augmented matrix of a system of linear equations contains the row $[1\ 0\ 0\ 0\ 0]$, then the original system is inconsistent.

(d) A homogeneous system of four linear equations in six variables has an infinite number of solutions.

58. (a) A 4×7 matrix has four columns.

(b) Every matrix has a unique reduced row-echelon form.

(c) A homogeneous system of four linear equations in four variables is always consistent.

(d) Multiplying a row of a matrix by a constant is one of the elementary row operations.

In Exercises 59 and 60, determine conditions on a, b, c, and d such that the matrix

$$\begin{bmatrix} a & b \\ c & d \end{bmatrix}$$

will be row-equivalent to the given matrix.

59. $\begin{bmatrix} 1 & 0 \\ 0 & 1 \end{bmatrix}$ **60.** $\begin{bmatrix} 1 & 0 \\ 0 & 0 \end{bmatrix}$

In Exercises 61 and 62, find all values of λ (the Greek letter lambda) such that the homogeneous system of linear equations will have nontrivial solutions.

61. $\begin{aligned} (\lambda - 2)x + y &= 0 \\ x + (\lambda - 2)y &= 0 \end{aligned}$

62. $\begin{aligned} (\lambda - 1)x + 2y &= 0 \\ x + \lambda y &= 0 \end{aligned}$

63. Writing Is it possible for a system of linear equations with fewer equations than variables to have no solution? If so, give an example.

64. Writing Does a matrix have a unique row-echelon form? Illustrate your answer with examples. Is the reduced row-echelon form unique?

65. Writing Consider the 2×2 matrix $\begin{bmatrix} a & b \\ c & d \end{bmatrix}$.

Perform the sequence of row operations.

(a) Add (-1) times the second row to the first row.

(b) Add 1 times the first row to the second row.

(c) Add (-1) times the second row to the first row.

(d) Multiply the first row by (-1).

What happened to the original matrix? Describe, in general, how to interchange two rows of a matrix using only the second and third elementary row operations.

66. The augmented matrix represents a system of linear equations that has been reduced using Gauss-Jordan elimination. Write a system of equations with nonzero coefficients that is represented by the reduced matrix.

$$\begin{bmatrix} 1 & 0 & 3 & -2 \\ 0 & 1 & 4 & 1 \\ 0 & 0 & 0 & 0 \end{bmatrix}$$

There are many correct answers.

67. Writing Describe the row-echelon form of an augmented matrix that corresponds to a system of linear equations that is inconsistent.

68. Writing Describe the row-echelon form of an augmented matrix that corresponds to a system of linear equations that has infinitely many solutions.

69. Writing In your own words, describe the difference between a matrix in row-echelon form and a matrix in reduced row-echelon form.

1.3 Applications of Systems of Linear Equations

Systems of linear equations arise in a wide variety of applications and are one of the central themes in linear algebra. In this section you will look at two such applications, and you will see many more in subsequent chapters. The first application shows how to fit a polynomial function to a set of data points in the plane. The second application focuses on networks and Kirchhoff's Laws for electricity.

Polynomial Curve Fitting

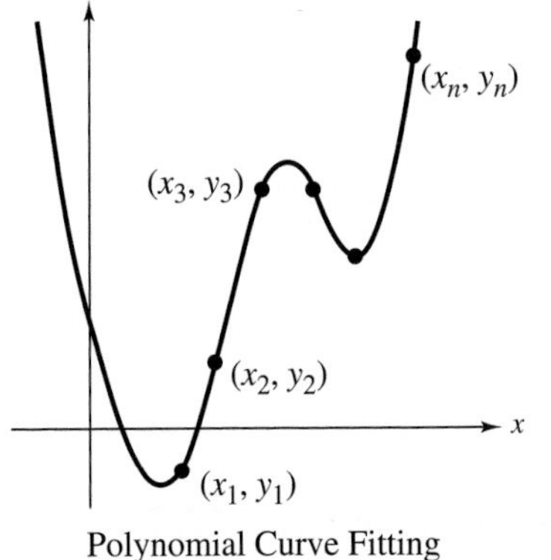

Polynomial Curve Fitting

Figure 1.4

Suppose a collection of data is represented by n points in the xy-plane,

$$(x_1, y_1), (x_2, y_2), \ldots, (x_n, y_n)$$

and you are asked to find a polynomial function of degree $n - 1$

$$p(x) = a_0 + a_1x + a_2x^2 + \cdots + a_{n-1}x^{n-1}$$

whose graph passes through the specified points. This procedure is called **polynomial curve fitting.** If all x-coordinates of the points are distinct, then there is precisely one polynomial function of degree $n - 1$ (or less) that fits the n points, as shown in Figure 1.4.

To solve for the n coefficients of $p(x)$, substitute each of the n points into the polynomial function and obtain n linear equations in n variables $a_0, a_1, a_2, \ldots, a_{n-1}$.

$$\begin{aligned}
a_0 + a_1x_1 + a_2x_1^2 + \cdots + a_{n-1}x_1^{n-1} &= y_1 \\
a_0 + a_1x_2 + a_2x_2^2 + \cdots + a_{n-1}x_2^{n-1} &= y_2 \\
&\vdots \\
a_0 + a_1x_n + a_2x_n^2 + \cdots + a_{n-1}x_n^{n-1} &= y_n
\end{aligned}$$

This procedure is demonstrated with a second-degree polynomial in Example 1.

EXAMPLE 1 Polynomial Curve Fitting

Determine the polynomial $p(x) = a_0 + a_1x + a_2x^2$ whose graph passes through the points $(1, 4)$, $(2, 0)$, and $(3, 12)$.

SOLUTION Substituting $x = 1$, 2, and 3 into $p(x)$ and equating the results to the respective y-values produces the system of linear equations in the variables a_0, a_1, and a_2 shown below.

$$\begin{aligned}
p(1) &= a_0 + a_1(1) + a_2(1)^2 = a_0 + a_1 + a_2 = 4 \\
p(2) &= a_0 + a_1(2) + a_2(2)^2 = a_0 + 2a_1 + 4a_2 = 0 \\
p(3) &= a_0 + a_1(3) + a_2(3)^2 = a_0 + 3a_1 + 9a_2 = 12
\end{aligned}$$

The solution of this system is $a_0 = 24$, $a_1 = -28$, and $a_2 = 8$, so the polynomial function is

$$p(x) = 24 - 28x + 8x^2.$$

HM

Simulation
Explore this concept further with an electronic simulation available on the website *college.hmco.com/pic/larsonELA6e.*

The graph of p is shown in Figure 1.5.

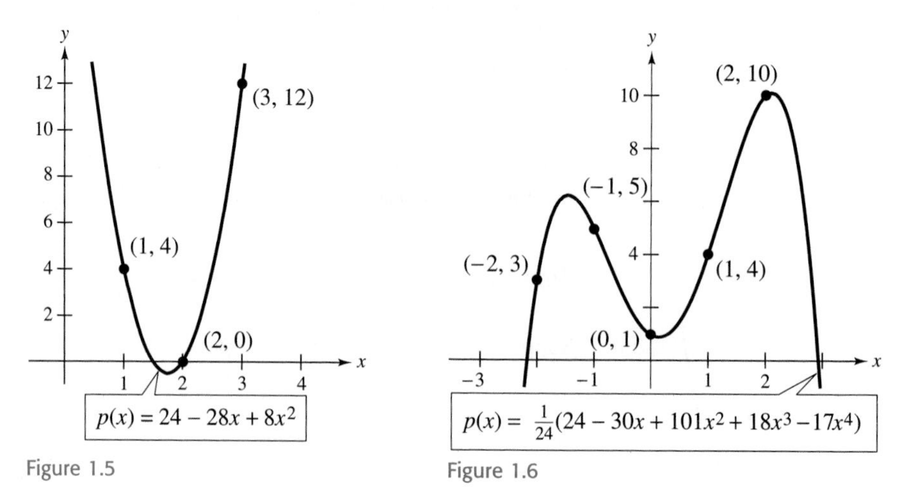

Figure 1.5

Figure 1.6

EXAMPLE 2 **Polynomial Curve Fitting**

Find a polynomial that fits the points $(-2, 3)$, $(-1, 5)$, $(0, 1)$, $(1, 4)$, and $(2, 10)$.

SOLUTION Because you are provided with five points, choose a fourth-degree polynomial function

$$p(x) = a_0 + a_1x + a_2x^2 + a_3x^3 + a_4x^4.$$

Substituting the given points into $p(x)$ produces the system of linear equations listed below.

$$\begin{aligned}
a_0 - 2a_1 + 4a_2 - 8a_3 + 16a_4 &= 3 \\
a_0 - a_1 + a_2 - a_3 + a_4 &= 5 \\
a_0 &= 1 \\
a_0 + a_1 + a_2 + a_3 + a_4 &= 4 \\
a_0 + 2a_1 + 4a_2 + 8a_3 + 16a_4 &= 10
\end{aligned}$$

The solution of these equations is

$$a_0 = 1, \quad a_1 = -\tfrac{30}{24}, \quad a_2 = \tfrac{101}{24}, \quad a_3 = \tfrac{18}{24}, \quad a_4 = -\tfrac{17}{24}$$

which means the polynomial function is

$$\begin{aligned}
p(x) &= 1 - \tfrac{30}{24}x + \tfrac{101}{24}x^2 + \tfrac{18}{24}x^3 - \tfrac{17}{24}x^4 \\
&= \tfrac{1}{24}(24 - 30x + 101x^2 + 18x^3 - 17x^4).
\end{aligned}$$

The graph of p is shown in Figure 1.6.

The system of linear equations in Example 2 is relatively easy to solve because the x-values are small. With a set of points with large x-values, it is usually best to *translate* the values before attempting the curve-fitting procedure. This approach is demonstrated in the next example.

EXAMPLE 3 Translating Large x-Values Before Curve Fitting

Find a polynomial that fits the points

$$\underbrace{(2006, 3)}_{(x_1, y_1)}, \quad \underbrace{(2007, 5)}_{(x_2, y_2)}, \quad \underbrace{(2008, 1)}_{(x_3, y_3)}, \quad \underbrace{(2009, 4)}_{(x_4, y_4)}, \quad \underbrace{(2010, 10)}_{(x_5, y_5)}.$$

SOLUTION Because the given x-values are large, use the translation $z = x - 2008$ to obtain

$$\underbrace{(-2, 3)}_{(z_1, y_1)}, \quad \underbrace{(-1, 5)}_{(z_2, y_2)}, \quad \underbrace{(0, 1)}_{(z_3, y_3)}, \quad \underbrace{(1, 4)}_{(z_4, y_4)}, \quad \underbrace{(2, 10)}_{(z_5, y_5)}.$$

This is the same set of points as in Example 2. So, the polynomial that fits these points is

$$\begin{aligned} p(z) &= \tfrac{1}{24}(24 - 30z + 101z^2 + 18z^3 - 17z^4) \\ &= 1 - \tfrac{5}{4}z + \tfrac{101}{24}z^2 + \tfrac{3}{4}z^3 - \tfrac{17}{24}z^4. \end{aligned}$$

Letting $z = x - 2008$, you have

$$p(x) = 1 - \tfrac{5}{4}(x - 2008) + \tfrac{101}{24}(x - 2008)^2 + \tfrac{3}{4}(x - 2008)^3 - \tfrac{17}{24}(x - 2008)^4.$$

EXAMPLE 4 An Application of Curve Fitting

Find a polynomial that relates the periods of the first three planets to their mean distances from the sun, as shown in Table 1.1. Then test the accuracy of the fit by using the polynomial to calculate the period of Mars. (Distance is measured in astronomical units, and period is measured in years.) (Source: *CRC Handbook of Chemistry and Physics*)

TABLE 1.1

Planet	*Mercury*	*Venus*	*Earth*	*Mars*	*Jupiter*	*Saturn*
Mean Distance	0.387	0.723	1.0	1.523	5.203	9.541
Period	0.241	0.615	1.0	1.881	11.861	29.457

SOLUTION Begin by fitting a quadratic polynomial function

$$p(x) = a_0 + a_1x + a_2x^2$$

to the points (0.387, 0.241), (0.723, 0.615), and (1, 1). The system of linear equations obtained by substituting these points into $p(x)$ is

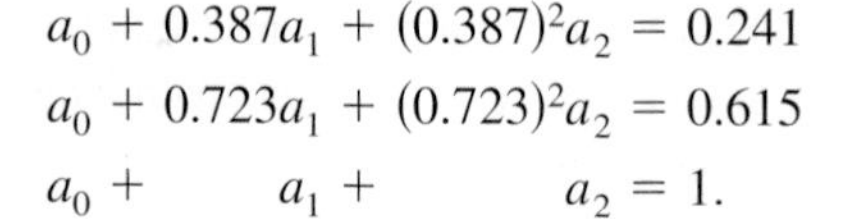

$$a_0 + 0.387a_1 + (0.387)^2a_2 = 0.241$$
$$a_0 + 0.723a_1 + (0.723)^2a_2 = 0.615$$
$$a_0 + a_1 + a_2 = 1.$$

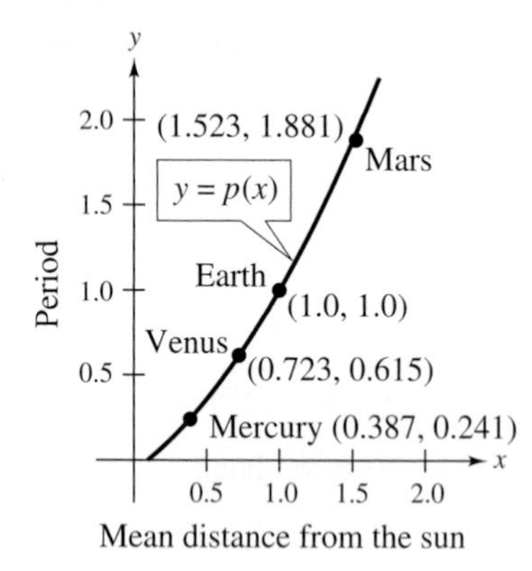

Figure 1.7

The approximate solution of the system is

$$a_0 \approx -0.0634, \quad a_1 \approx 0.6119, \quad a_2 \approx 0.4515$$

which means that the polynomial function can be approximated by

$$p(x) = -0.0634 + 0.6119x + 0.4515x^2.$$

Using $p(x)$ to evaluate the period of Mars produces

$$p(1.523) \approx 1.916 \text{ years.}$$

This estimate is compared graphically with the actual period of Mars in Figure 1.7. Note that the actual period (from Table 1.1) is 1.881 years.

An important lesson may be learned from the application shown in Example 4: The polynomial that fits the given data points is not necessarily an accurate model for the relationship between x and y for x-values other than those corresponding to the given points. Generally, the farther the additional points are from the given points, the worse the fit. For instance, in Example 4 the mean distance of Jupiter is 5.203. The corresponding polynomial approximation for the period is 15.343 years—a poor estimate of Jupiter's actual period of 11.861 years.

The problem of curve fitting can be difficult. Types of functions other than polynomial functions often provide better fits. To see this, look again at the curve-fitting problem in Example 4. Taking the natural logarithms of the distances and periods of the first six planets produces the results shown in Table 1.2 and Figure 1.8.

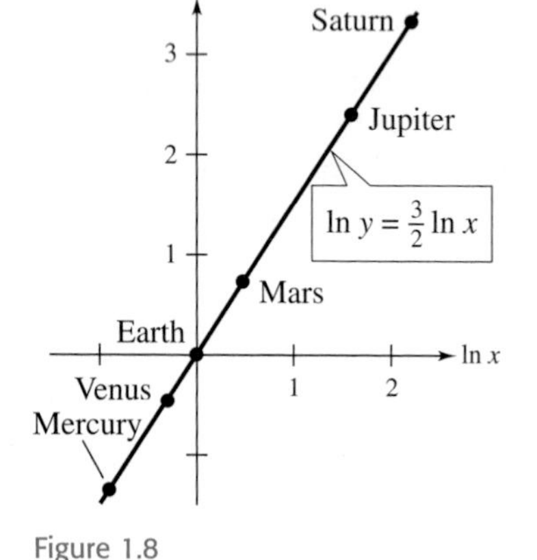

Figure 1.8

TABLE 1.2

Planet	*Mercury*	*Venus*	*Earth*	*Mars*	*Jupiter*	*Saturn*
Mean Distance (x)	0.387	0.723	1.0	1.523	5.203	9.541
Natural Log of Mean	−0.949	−0.324	0.0	0.421	1.649	2.256
Period (y)	0.241	0.615	1.0	1.881	11.861	29.457
Natural Log of Period	−1.423	−0.486	0.0	0.632	2.473	3.383

Now, fitting a polynomial to the logarithms of the distances and periods produces the *linear relationship* between $\ln x$ and $\ln y$ shown below.

$$\ln y = \frac{3}{2} \ln x$$

From this equation it follows that $y = x^{3/2}$, or $y^2 = x^3$.

In other words, the square of the period (in years) of each planet is equal to the cube of its mean distance (in astronomical units) from the sun. This relationship was first discovered by Johannes Kepler in 1619.

Network Analysis

Networks composed of branches and junctions are used as models in many diverse fields such as economics, traffic analysis, and electrical engineering.

In such models it is assumed that the total flow into a junction is equal to the total flow out of the junction. For example, because the junction shown in Figure 1.9 has 25 units flowing into it, there must be 25 units flowing out of it. This is represented by the linear equation

$$x_1 + x_2 = 25.$$

Because each junction in a network gives rise to a linear equation, you can analyze the flow through a network composed of several junctions by solving a system of linear equations. This procedure is illustrated in Example 5.

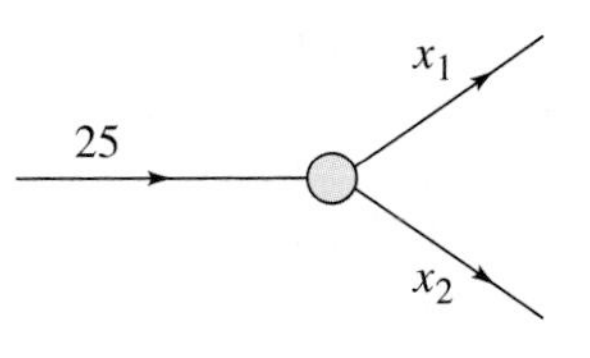

Figure 1.9

EXAMPLE 5 **Analysis of a Network**

Set up a system of linear equations to represent the network shown in Figure 1.10, and solve the system.

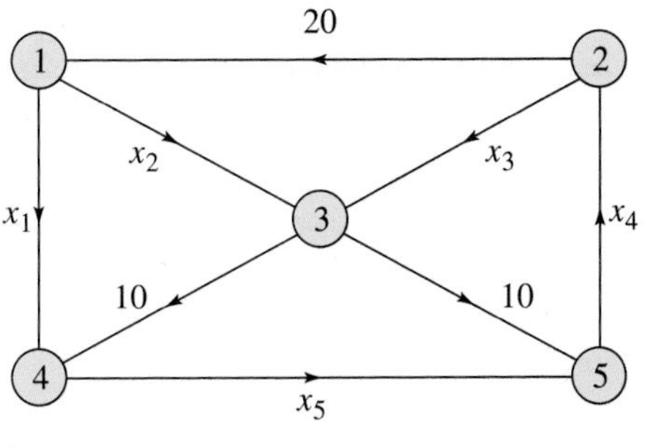

Figure 1.10

SOLUTION Each of the network's five junctions gives rise to a linear equation, as shown below.

$$\begin{aligned}
x_1 + x_2 \qquad\qquad\qquad &= 20 \qquad \text{Junction 1}\\
x_3 - x_4 \qquad &= -20 \qquad \text{Junction 2}\\
x_2 + x_3 \qquad\qquad &= 20 \qquad \text{Junction 3}\\
x_1 \qquad\qquad\qquad - x_5 &= -10 \qquad \text{Junction 4}\\
-x_4 + x_5 &= -10 \qquad \text{Junction 5}
\end{aligned}$$

The augmented matrix for this system is

$$\begin{bmatrix} 1 & 1 & 0 & 0 & 0 & 20 \\ 0 & 0 & 1 & -1 & 0 & -20 \\ 0 & 1 & 1 & 0 & 0 & 20 \\ 1 & 0 & 0 & 0 & -1 & -10 \\ 0 & 0 & 0 & -1 & 1 & -10 \end{bmatrix}.$$

Gauss-Jordan elimination produces the matrix

$$\begin{bmatrix} 1 & 0 & 0 & 0 & -1 & -10 \\ 0 & 1 & 0 & 0 & 1 & 30 \\ 0 & 0 & 1 & 0 & -1 & -10 \\ 0 & 0 & 0 & 1 & -1 & 10 \\ 0 & 0 & 0 & 0 & 0 & 0 \end{bmatrix}.$$

From the matrix above, you can see that

$$x_1 - x_5 = -10, \quad x_2 + x_5 = 30, \quad x_3 - x_5 = -10, \quad \text{and} \quad x_4 - x_5 = 10.$$

Letting $t = x_5$, you have

$$x_1 = t - 10, \quad x_2 = -t + 30, \quad x_3 = t - 10, \quad x_4 = t + 10, \quad x_5 = t$$

where t is a real number, so this system has an infinite number of solutions.

In Example 5, suppose you could control the amount of flow along the branch labeled x_5. Using the solution from Example 5, you could then control the flow represented by each of the other variables. For instance, letting $t = 10$ would reduce the flow of x_1 and x_3 to zero, as shown in Figure 1.11. Similarly, letting $t = 20$ would produce the network shown in Figure 1.12.

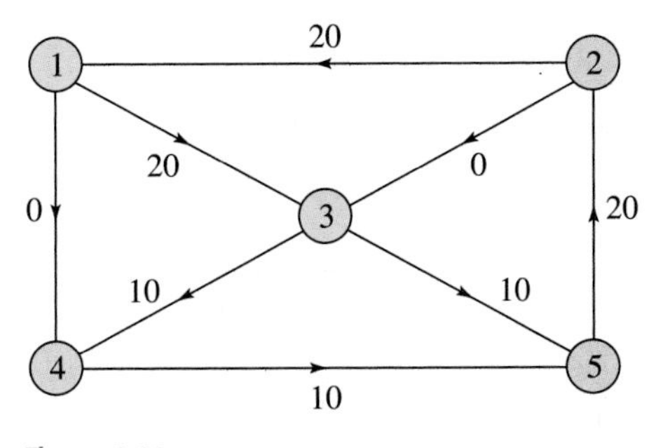

Figure 1.11

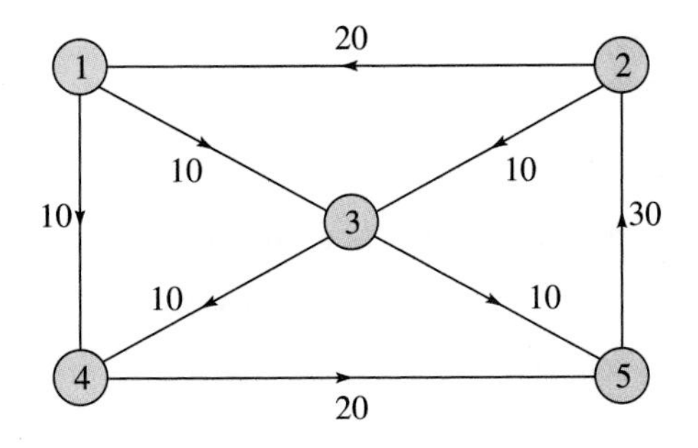

Figure 1.12

You can see how the type of network analysis demonstrated in Example 5 could be used in problems dealing with the flow of traffic through the streets of a city or the flow of water through an irrigation system.

An electrical network is another type of network where analysis is commonly applied. An analysis of such a system uses two properties of electrical networks known as **Kirchhoff's Laws.**

1. All the current flowing into a junction must flow out of it.
2. The sum of the products IR (I is current and R is resistance) around a closed path is equal to the total voltage in the path.

In an electrical network, current is measured in amps, resistance in ohms, and the product of current and resistance in volts. Batteries are represented by the symbol —|⊢— . The larger vertical bar denotes where the current flows out of the terminal. Resistance is denoted by the symbol -/\/\/\- . The direction of the current is indicated by an arrow in the branch.

REMARK: A *closed* path is a sequence of branches such that the beginning point of the first branch coincides with the end point of the last branch.

EXAMPLE 6 Analysis of an Electrical Network

Determine the currents I_1, I_2, and I_3 for the electrical network shown in Figure 1.13.

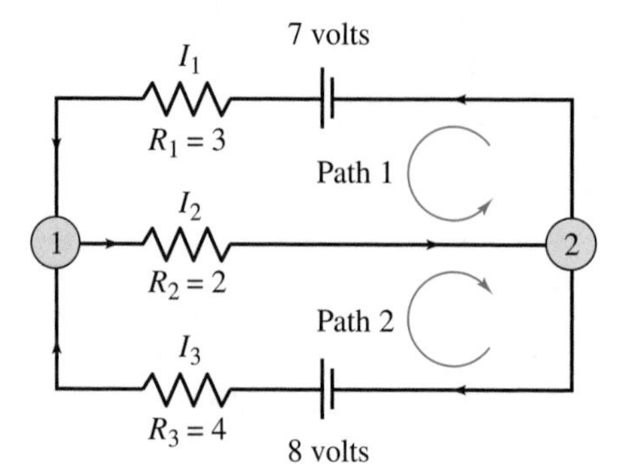

Figure 1.13

SOLUTION Applying Kirchhoff's first law to either junction produces

$$I_1 + I_3 = I_2 \qquad \textbf{Junction 1 or Junction 2}$$

and applying Kirchhoff's second law to the two paths produces

$$R_1I_1 + R_2I_2 = 3I_1 + 2I_2 = 7 \qquad \textbf{Path 1}$$
$$R_2I_2 + R_3I_3 = 2I_2 + 4I_3 = 8. \qquad \textbf{Path 2}$$

So, you have the following system of three linear equations in the variables I_1, I_2, and I_3.

$$\begin{aligned} I_1 - I_2 + I_3 &= 0 \\ 3I_1 + 2I_2 &= 7 \\ 2I_2 + 4I_3 &= 8 \end{aligned}$$

Applying Gauss-Jordan elimination to the augmented matrix

$$\begin{bmatrix} 1 & -1 & 1 & 0 \\ 3 & 2 & 0 & 7 \\ 0 & 2 & 4 & 8 \end{bmatrix}$$

produces the reduced row-echelon form

$$\begin{bmatrix} 1 & 0 & 0 & 1 \\ 0 & 1 & 0 & 2 \\ 0 & 0 & 1 & 1 \end{bmatrix}$$

which means $I_1 = 1$ amp, $I_2 = 2$ amps, and $I_3 = 1$ amp.

EXAMPLE 7 Analysis of an Electrical Network

Determine the currents I_1, I_2, I_3, I_4, I_5, and I_6 for the electrical network shown in Figure 1.14.

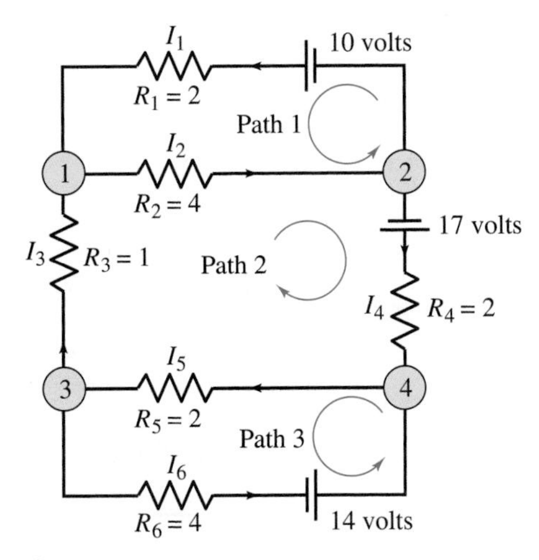

Figure 1.14

SOLUTION Applying Kirchhoff's first law to the four junctions produces

$$I_1 + I_3 = I_2 \quad \textbf{Junction 1}$$
$$I_1 + I_4 = I_2 \quad \textbf{Junction 2}$$
$$I_3 + I_6 = I_5 \quad \textbf{Junction 3}$$
$$I_4 + I_6 = I_5 \quad \textbf{Junction 4}$$

and applying Kirchhoff's second law to the three paths produces

$$\begin{aligned} 2I_1 + 4I_2 \qquad\qquad\qquad\qquad &= 10 \quad \textbf{Path 1} \\ 4I_2 + I_3 + 2I_4 + 2I_5 \qquad &= 17 \quad \textbf{Path 2} \\ 2I_5 + 4I_6 &= 14. \quad \textbf{Path 3} \end{aligned}$$

You now have the following system of seven linear equations in the variables I_1, I_2, I_3, I_4, I_5, and I_6.

$$\begin{aligned} I_1 - I_2 + I_3 \qquad\qquad\qquad &= 0 \\ I_1 - I_2 \qquad + I_4 \qquad\qquad &= 0 \\ I_3 \qquad - I_5 + I_6 &= 0 \\ I_4 - I_5 + I_6 &= 0 \\ 2I_1 + 4I_2 \qquad\qquad\qquad &= 10 \\ 4I_2 + I_3 + 2I_4 + 2I_5 \qquad &= 17 \\ 2I_5 + 4I_6 &= 14 \end{aligned}$$

Using Gauss-Jordan elimination, a graphing utility, or a computer software program, you can solve this system to obtain

$$I_1 = 1, \quad I_2 = 2, \quad I_3 = 1, \quad I_4 = 1, \quad I_5 = 3, \quad \text{and} \quad I_6 = 2$$

meaning $I_1 = 1$ amp, $I_2 = 2$ amps, $I_3 = 1$ amp, $I_4 = 1$ amp, $I_5 = 3$ amps, and $I_6 = 2$ amps.

SECTION 1.3 Exercises

Polynomial Curve Fitting

In Exercises 1–6, (a) determine the polynomial function whose graph passes through the given points, and (b) sketch the graph of the polynomial function, showing the given points.

1. $(2, 5), (3, 2), (4, 5)$

2. $(2, 4), (3, 4), (4, 4)$

3. $(2, 4), (3, 6), (5, 10)$

4. $(-1, 3), (0, 0), (1, 1), (4, 58)$

5. $(2006, 5), (2007, 7), (2008, 12)$ $(z = x - 2007)$

6. $(2005, 150), (2006, 180), (2007, 240), (2008, 360)$ $(z = x - 2005)$

7. Writing Try to fit the graph of a polynomial function to the values shown in the table. What happens, and why?

x	1	2	3	3	4
y	1	1	2	3	4

8. The graph of a function f passes through the points $(0, 1)$, $\left(2, \frac{1}{3}\right)$, and $\left(4, \frac{1}{5}\right)$. Find a quadratic function whose graph passes through these points.

9. Find a polynomial function p of degree 2 or less that passes through the points $(0, 1)$, $(2, 3)$, and $(4, 5)$. Then sketch the graph of $y = 1/p(x)$ and compare this graph with the graph of the polynomial function found in Exercise 8.

10. Calculus The graph of a parabola passes through the points $(0, 1)$ and $\left(\frac{1}{2}, \frac{1}{2}\right)$ and has a horizontal tangent at $\left(\frac{1}{2}, \frac{1}{2}\right)$. Find an equation for the parabola and sketch its graph.

11. Calculus The graph of a cubic polynomial function has horizontal tangents at $(1, -2)$ and $(-1, 2)$. Find an equation for the cubic and sketch its graph.

12. Find an equation of the circle passing through the points $(1, 3)$, $(-2, 6)$, and $(4, 2)$.

13. The U.S. census lists the population of the United States as 227 million in 1980, 249 million in 1990, and 281 million in 2000. Fit a second-degree polynomial passing through these three points and use it to predict the population in 2010 and in 2020. (Source: U.S. Census Bureau)

14. The U.S. population figures for the years 1920, 1930, 1940, and 1950 are shown in the table. (Source: U.S. Census Bureau)

Year	1920	1930	1940	1950
Population (in millions)	106	123	132	151

(a) Find a cubic polynomial that fits these data and use it to estimate the population in 1960.

(b) The actual population in 1960 was 179 million. How does your estimate compare?

15. The net profits (in millions of dollars) for Microsoft from 2000 to 2007 are shown in the table. (Source: Microsoft Corporation)

Year	2000	2001	2002	2003
Net Profit	9421	10,003	10,384	10,526

Year	2004	2005	2006	2007
Net Profit	11,330	12,715	12,599	14,410

(a) Set up a system of equations to fit the data for the years 2001, 2003, 2005, and 2007 to a cubic model.

(b) Solve the system. Does the solution produce a reasonable model for predicting future net profits? Explain.

16. The sales (in billions of dollars) for Wal-Mart stores from 2000 to 2007 are shown in the table. (Source: Wal-Mart)

Year	2000	2001	2002	2003
Sales	191.3	217.8	244.5	256.3

Year	2004	2005	2006	2007
Sales	285.2	312.4	346.5	377.0

(a) Set up a system of equations to fit the data for the years 2001, 2003, 2005, and 2007 to a cubic model.
(b) Solve the system. Does the solution produce a reasonable model for predicting future sales? Explain.

17. Use $\sin 0 = 0$, $\sin(\pi/2) = 1$, and $\sin \pi = 0$ to estimate $\sin(\pi/3)$.

18. Use $\log_2 1 = 0$, $\log_2 2 = 1$, and $\log_2 4 = 2$ to estimate $\log_2 3$.

19. Guided Proof Prove that if a polynomial function $p(x) = a_0 + a_1x + a_2x^2$ is zero for $x = -1$, $x = 0$, and $x = 1$, then $a_0 = a_1 = a_2 = 0$.

Getting Started: Write a system of linear equations and solve the system for a_0, a_1, and a_2.

(i) Substitute $x = -1, 0,$ and 1 into $p(x)$.
(ii) Set the result equal to 0.
(iii) Solve the resulting system of linear equations in the variables a_0, a_1, and a_2.

20. The statement in Exercise 19 can be generalized: If a polynomial function $p(x) = a_0 + a_1x + \cdots + a_{n-1}x^{n-1}$ is zero for more than $n - 1$ x-values, then $a_0 = a_1 = \cdots = a_{n-1} = 0$. Use this result to prove that there is at most one polynomial function of degree $n - 1$ (or less) whose graph passes through n points in the plane with distinct x-coordinates.

Network Analysis

21. Water is flowing through a network of pipes (in thousands of cubic meters per hour), as shown in Figure 1.15.

(a) Solve this system for the water flow represented by x_i, $i = 1, 2, \ldots, 7$.
(b) Find the water flow when $x_6 = x_7 = 0$.
(c) Find the water flow when $x_5 = 1000$ and $x_6 = 0$.

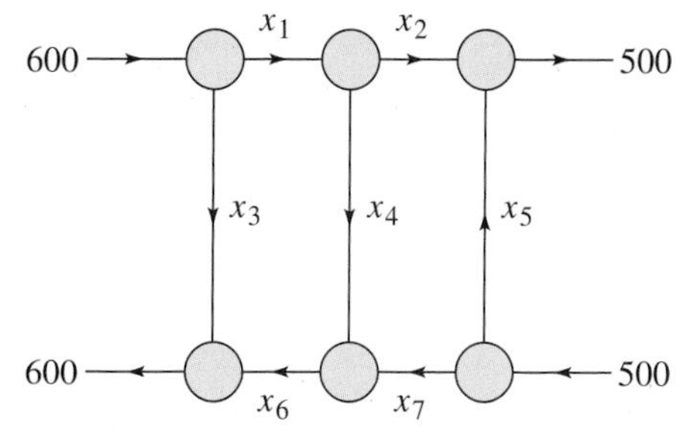

Figure 1.15

22. The flow of traffic (in vehicles per hour) through a network of streets is shown in Figure 1.16.

(a) Solve this system for x_i, $i = 1, 2, \ldots, 5$.
(b) Find the traffic flow when $x_2 = 200$ and $x_3 = 50$.
(c) Find the traffic flow when $x_2 = 150$ and $x_3 = 0$.

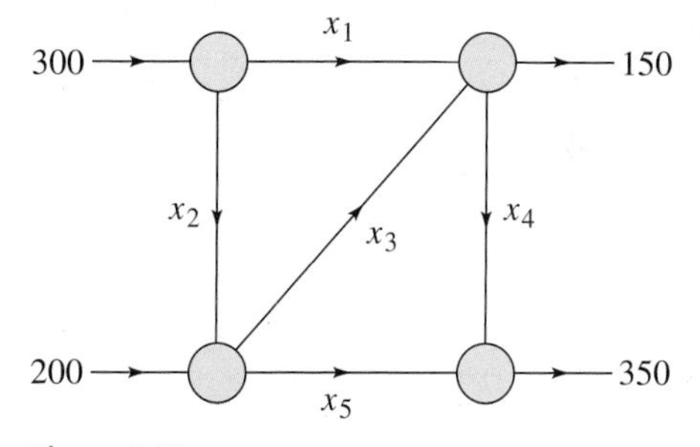

Figure 1.16

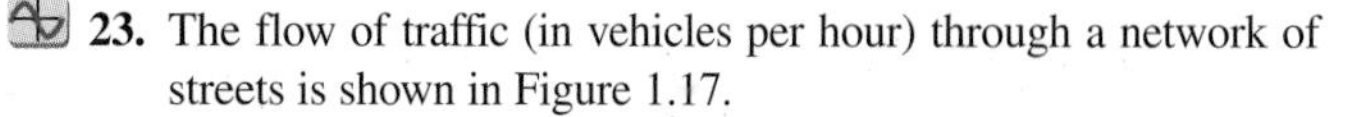

23. The flow of traffic (in vehicles per hour) through a network of streets is shown in Figure 1.17.

(a) Solve this system for x_i, $i = 1, 2, 3, 4$.
(b) Find the traffic flow when $x_4 = 0$.
(c) Find the traffic flow when $x_4 = 100$.

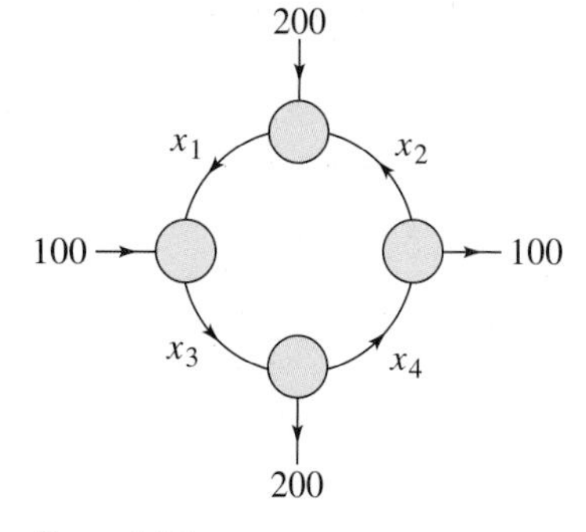

Figure 1.17

24. The flow of traffic (in vehicles per hour) through a network of streets is shown in Figure 1.18.

(a) Solve this system for x_i, $i = 1, 2, \ldots, 5$.
(b) Find the traffic flow when $x_3 = 0$ and $x_5 = 100$.
(c) Find the traffic flow when $x_3 = x_5 = 100$.

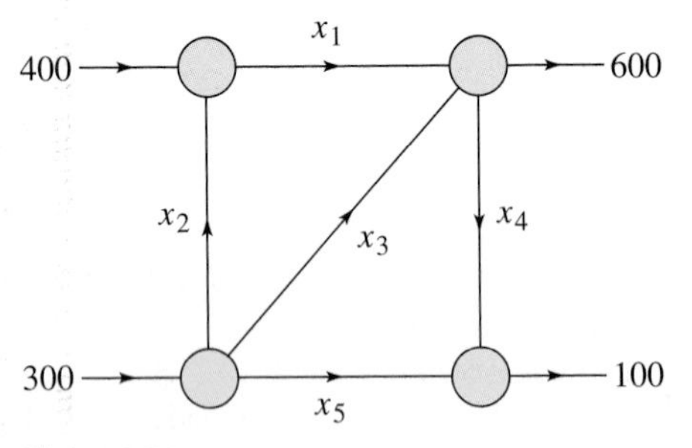

Figure 1.18

25. Determine the currents I_1, I_2, and I_3 for the electrical network shown in Figure 1.19.

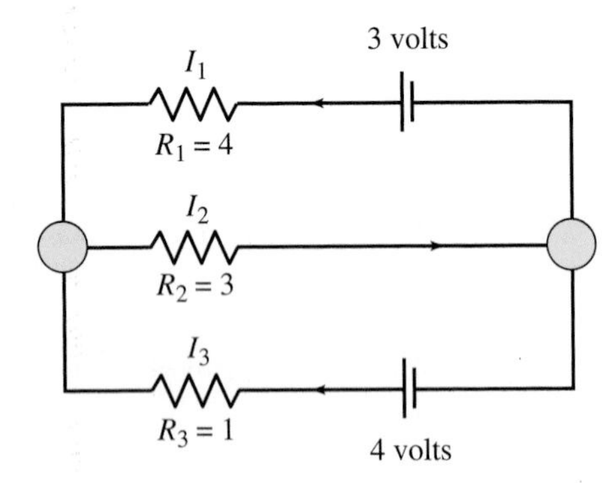

Figure 1.19

26. Determine the currents I_1, I_2, and I_3 for the electrical network shown in Figure 1.20.

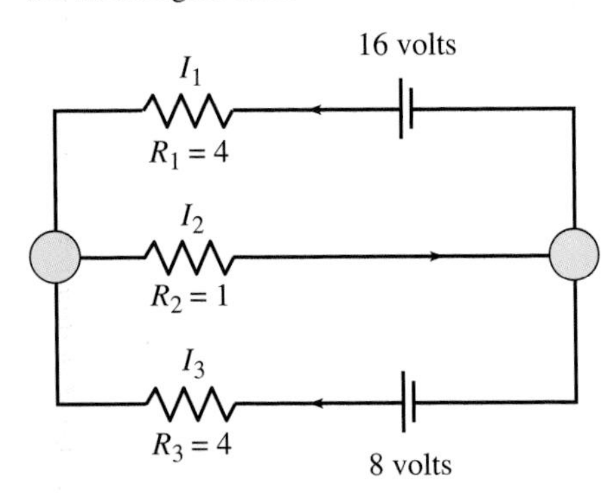

Figure 1.20

27. (a) Determine the currents I_1, I_2, and I_3 for the electrical network shown in Figure 1.21.

(b) How is the result affected when A is changed to 2 volts and B is changed to 6 volts?

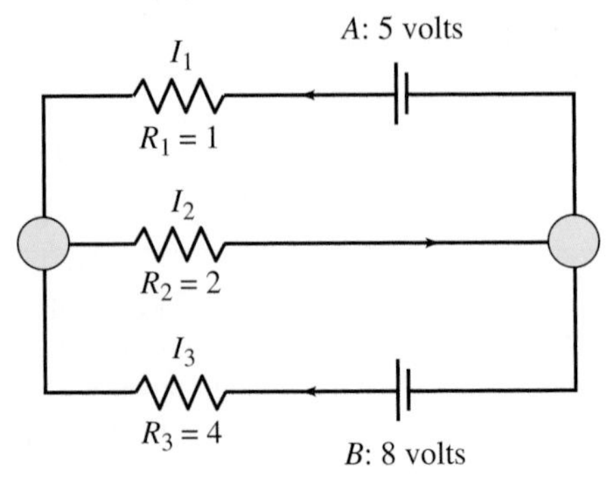

Figure 1.21

28. Determine the currents I_1, I_2, I_3, I_4, I_5, and I_6 for the electrical network shown in Figure 1.22.

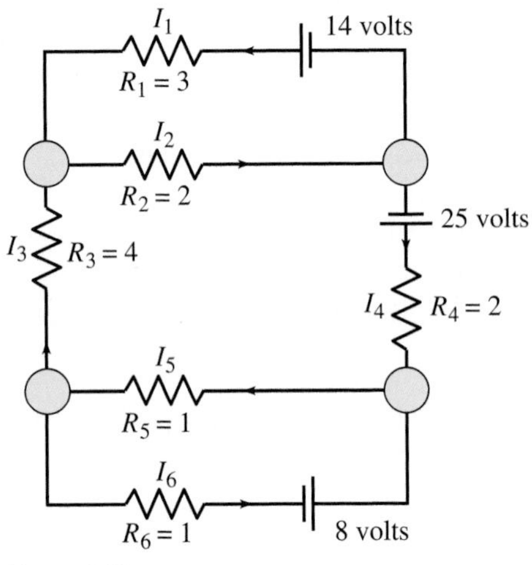

Figure 1.22

In Exercises 29–32, use a system of equations to write the partial fraction decomposition of the rational expression. Then solve the system using matrices.

29. $\dfrac{4x^2}{(x+1)^2(x-1)} = \dfrac{A}{x-1} + \dfrac{B}{x+1} + \dfrac{C}{(x+1)^2}$

30. $\dfrac{8x^2}{(x-1)^2(x+1)} = \dfrac{A}{x+1} + \dfrac{B}{x-1} + \dfrac{C}{(x-1)^2}$

31. $\dfrac{20 - x^2}{(x+2)(x-2)^2} = \dfrac{A}{x+2} + \dfrac{B}{x-2} + \dfrac{C}{(x-2)^2}$

32. $\dfrac{3x^2 - 7x - 12}{(x+4)(x-4)^2} = \dfrac{A}{x+4} + \dfrac{B}{x-4} + \dfrac{C}{(x-4)^2}$

In Exercises 33 and 34, find the values of x, y, and λ that satisfy the system of equations. Such systems arise in certain problems of calculus, and λ is called the Lagrange multiplier.

33. $\begin{aligned} 2x \quad\; + \lambda \quad &= 0 \\ 2y + \lambda \quad &= 0 \\ x + y \quad - 4 &= 0 \end{aligned}$

34. $\begin{aligned} 2y + 2\lambda + \quad 2 &= 0 \\ 2x \qquad \lambda + \quad 1 &= 0 \\ 2x + y \qquad - 100 &= 0 \end{aligned}$

35. In Super Bowl XLI on February 4, 2007, the Indianapolis Colts beat the Chicago Bears by a score of 29 to 17. The total points scored came from 13 scoring plays, which were a combination of touchdowns, extra-point kicks, and field goals, worth 6, 1, and 3 points, respectively. The numbers of field goals and extra-point kicks were equal. Write a system of equations to represent this event. Then determine the number of each type of scoring play. (Source: National Football League)

36. In the 2007 Fiesta Bowl Championship Series on January 8, 2007, the University of Florida Gators defeated the Ohio State University Buckeyes by a score of 41 to 14. The total points scored came from a combination of touchdowns, extra-point kicks, and field goals, worth 6, 1, and 3 points, respectively. The numbers of touchdowns and extra-point kicks were equal. The number of touchdowns was one more than three times the number of field goals. Write a system of equations to represent this event. Then determine the number of each type of scoring play. (Source: www.fiestabowl.org)

CHAPTER 1 Review Exercises

In Exercises 1–8, determine whether the equation is linear in the variables x and y.

1. $2x - y^2 = 4$

2. $2xy - 6y = 0$

3. $(\sin \pi)x + y = 2$

4. $e^{-2}x + 5y = 8$

5. $\dfrac{2}{x} + 4y = 3$

6. $\dfrac{4}{y} - x = 10$

7. $\frac{1}{2}x - \frac{1}{4}y = 0$

8. $\frac{3}{5}x + \frac{7}{10}y = 2$

In Exercises 9 and 10, find a parametric representation of the solution set of the linear equation.

9. $-4x + 2y - 6z = 1$

10. $3x_1 + 2x_2 - 4x_3 = 0$

In Exercises 11–22, solve the system of linear equations.

11. $\begin{aligned} x + y &= 2 \\ 3x - y &= 0 \end{aligned}$

12. $\begin{aligned} x + y &= -1 \\ 3x + 2y &= 0 \end{aligned}$

13. $\begin{aligned} 3y &= 2x \\ y &= x + 4 \end{aligned}$

14. $\begin{aligned} x &= y + 3 \\ 4x &= y + 10 \end{aligned}$

15. $\begin{aligned} y + x &= 0 \\ 2x + y &= 0 \end{aligned}$

16. $\begin{aligned} y &= -4x \\ y &= x \end{aligned}$

17. $\begin{aligned} x - y &= 9 \\ -x + y &= 1 \end{aligned}$

18. $\begin{aligned} 40x_1 + 30x_2 &= 24 \\ 20x_1 + 15x_2 &= -14 \end{aligned}$

19. $\begin{aligned} 0.2x_1 + 0.3x_2 &= 0.14 \\ 0.4x_1 + 0.5x_2 &= 0.20 \end{aligned}$

20. $\begin{aligned} 0.2x - 0.1y &= 0.07 \\ 0.4x - 0.5y &= -0.01 \end{aligned}$

21. $\begin{aligned} \tfrac{1}{2}x - \tfrac{1}{3}y &= 0 \\ 3x + 2(y + 5) &= 10 \end{aligned}$

22. $\begin{aligned} \tfrac{1}{3}x + \tfrac{4}{7}y &= 3 \\ 2x + 3y &= 15 \end{aligned}$

In Exercises 23 and 24, determine the size of the matrix.

23. $\begin{bmatrix} 2 & 3 & -1 \\ 0 & 5 & 1 \end{bmatrix}$

24. $\begin{bmatrix} 2 & 1 \\ -4 & -1 \\ 0 & 5 \end{bmatrix}$

In Exercises 25–28, determine whether the matrix is in row-echelon form. If it is, determine whether it is also in reduced row-echelon form.

25. $\begin{bmatrix} 1 & 0 & 1 & 1 \\ 0 & 1 & 2 & 1 \\ 0 & 0 & 0 & 1 \end{bmatrix}$

26. $\begin{bmatrix} 1 & 2 & -3 & 0 \\ 0 & 0 & 0 & 1 \\ 0 & 0 & 0 & 0 \end{bmatrix}$

27. $\begin{bmatrix} -1 & 2 & 1 \\ 0 & 1 & 0 \\ 0 & 0 & 1 \end{bmatrix}$

28. $\begin{bmatrix} 0 & 1 & 0 & 0 \\ 0 & 0 & 1 & 2 \\ 0 & 0 & 0 & 0 \end{bmatrix}$

In Exercises 29 and 30, find the solution set of the system of linear equations represented by the augmented matrix.

29. $\begin{bmatrix} 1 & 2 & 0 & 0 \\ 0 & 0 & 1 & 0 \\ 0 & 0 & 0 & 0 \end{bmatrix}$

30. $\begin{bmatrix} 1 & 2 & 3 & 0 \\ 0 & 0 & 0 & 1 \\ 0 & 0 & 0 & 0 \end{bmatrix}$

In Exercises 31–40, solve the system using either Gaussian elimination with back-substitution or Gauss-Jordan elimination.

31. $\begin{aligned} -x + y + 2z &= 1 \\ 2x + 3y + z &= -2 \\ 5x + 4y + 2z &= 4 \end{aligned}$

32. $\begin{aligned} 2x + 3y + z &= 10 \\ 2x - 3y - 3z &= 22 \\ 4x - 2y + 3z &= -2 \end{aligned}$

33. $\begin{aligned} 2x + 3y + 3z &= 3 \\ 6x + 6y + 12z &= 13 \\ 12x + 9y - z &= 2 \end{aligned}$

34. $\begin{aligned} 2x \qquad + 6z &= -9 \\ 3x - 2y + 11z &= -16 \\ 3x - y + 7z &= -11 \end{aligned}$

35. $$\begin{aligned} x - 2y + z &= -6 \\ 2x - 3y &= -7 \\ -x + 3y - 3z &= 11 \end{aligned}$$

36. $$\begin{aligned} x + 2y + 6z &= 1 \\ 2x + 5y + 15z &= 4 \\ 3x + y + 3z &= -6 \end{aligned}$$

37. $$\begin{aligned} 2x + y + 2z &= 4 \\ 2x + 2y &= 5 \\ 2x - y + 6z &= 2 \end{aligned}$$

38. $$\begin{aligned} 2x_1 + 5x_2 - 19x_3 &= 34 \\ 3x_1 + 8x_2 - 31x_3 &= 54 \end{aligned}$$

39. $$\begin{aligned} 2x_1 + x_2 + x_3 + 2x_4 &= -1 \\ 5x_1 - 2x_2 + x_3 - 3x_4 &= 0 \\ -x_1 + 3x_2 + 2x_3 + 2x_4 &= 1 \\ 3x_1 + 2x_2 + 3x_3 - 5x_4 &= 12 \end{aligned}$$

40. $$\begin{aligned} x_1 + 5x_2 + 3x_3 &= 14 \\ 4x_2 + 2x_3 + 5x_4 &= 3 \\ 3x_3 + 8x_4 + 6x_5 &= 16 \\ 2x_1 + 4x_2 - 2x_5 &= 0 \\ 2x_1 - x_3 &= 0 \end{aligned}$$

In Exercises 41–46, use the matrix capabilities of a graphing utility to reduce the augmented matrix corresponding to the system of equations to solve the system.

41. $$\begin{aligned} 3x + 3y + 12z &= 6 \\ x + y + 4z &= 2 \\ 2x + 5y + 20z &= 10 \\ -x + 2y + 8z &= 4 \end{aligned}$$

42. $$\begin{aligned} 2x + 10y + 2z &= 6 \\ x + 5y + 2z &= 6 \\ x + 5y + z &= 3 \\ -3x - 15y + 3z &= -9 \end{aligned}$$

43. $$\begin{aligned} 2x + y - z + 2w &= -6 \\ 3x + 4y + w &= 1 \\ x + 5y + 2z + 6w &= -3 \\ 5x + 2y - z - w &= 3 \end{aligned}$$

44. $$\begin{aligned} x + 2y + z + 4w &= 11 \\ 3x + 6y + 5z + 12w &= 30 \\ x + 3y - 3z + 2w &= -5 \\ 6x - y - z + w &= -9 \end{aligned}$$

45. $$\begin{aligned} x + y + z + w &= 0 \\ 2x + 3y + z - 2w &= 0 \\ 3x + 5y + z &= 0 \end{aligned}$$

46. $$\begin{aligned} x + 2y + z + 3w &= 0 \\ x - y + w &= 0 \\ 5y - z + 2w &= 0 \end{aligned}$$

In Exercises 47–50, solve the homogeneous system of linear equations.

47. $$\begin{aligned} x_1 - 2x_2 - 8x_3 &= 0 \\ 3x_1 + 2x_2 &= 0 \\ -x_1 + x_2 + 7x_3 &= 0 \end{aligned}$$

48. $$\begin{aligned} 2x_1 + 4x_2 - 7x_3 &= 0 \\ x_1 - 3x_2 + 9x_3 &= 0 \\ 6x_1 + 9x_3 &= 0 \end{aligned}$$

49. $$\begin{aligned} 2x_1 - 8x_2 + 4x_3 &= 0 \\ 3x_1 - 10x_2 + 7x_3 &= 0 \\ 10x_2 + 5x_3 &= 0 \end{aligned}$$

50. $$\begin{aligned} x_1 + 3x_2 + 5x_3 &= 0 \\ x_1 + 4x_2 + \tfrac{1}{2}x_3 &= 0 \end{aligned}$$

51. Determine the value of k such that the system of linear equations is inconsistent.

$$\begin{aligned} kx + y &= 0 \\ x + ky &= 1 \end{aligned}$$

52. Determine the value of k such that the system of linear equations has exactly one solution.

$$\begin{aligned} x - y + 2z &= 0 \\ -x + y - z &= 0 \\ x + ky + z &= 0 \end{aligned}$$

53. Find conditions on a and b such that the system of linear equations has (a) no solution, (b) exactly one solution, and (c) an infinite number of solutions.

$$\begin{aligned} x + 2y &= 3 \\ ax + by &= -9 \end{aligned}$$

54. Find (if possible) conditions on a, b, and c such that the system of linear equations has (a) no solution, (b) exactly one solution, and (c) an infinite number of solutions.

$$\begin{aligned} 2x - y + z &= a \\ x + y + 2z &= b \\ 3y + 3z &= c \end{aligned}$$

55. Writing Describe a method for showing that two matrices are row-equivalent. Are the two matrices below row-equivalent?

$$\begin{bmatrix} 1 & 1 & 2 \\ 0 & -1 & 2 \\ 3 & 1 & 2 \end{bmatrix} \quad \text{and} \quad \begin{bmatrix} 1 & 2 & 3 \\ 4 & 3 & 6 \\ 5 & 5 & 10 \end{bmatrix}$$

56. Writing Describe all possible 2×3 reduced row-echelon matrices. Support your answer with examples.

57. Let $n \geq 3$. Find the reduced row-echelon form of the $n \times n$ matrix.

$$\begin{bmatrix} 1 & 2 & 3 & \cdots & n \\ n+1 & n+2 & n+3 & \cdots & 2n \\ 2n+1 & 2n+2 & 2n+3 & \cdots & 3n \\ \vdots & \vdots & \vdots & & \vdots \\ n^2-n+1 & n^2-n+2 & n^2-n+3 & \cdots & n^2 \end{bmatrix}$$

58. Find all values of λ for which the homogeneous system of linear equations has nontrivial solutions.

$$\begin{aligned} (\lambda + 2)x_1 - 2x_2 + 3x_3 &= 0 \\ -2x_1 + (\lambda - 1)x_2 + 6x_3 &= 0 \\ x_1 + 2x_2 + \lambda x_3 &= 0 \end{aligned}$$

True or False? In Exercises 59 and 60, determine whether each statement is true or false. If a statement is true, give a reason or cite an appropriate statement from the text. If a statement is false, provide an example that shows the statement is not true in all cases or cite an appropriate statement from the text.

59. (a) The solution set of a linear equation can be parametrically represented in only one way.

(b) A consistent system of linear equations can have an infinite number of solutions.

60. (a) A homogeneous system of linear equations must have at least one solution.

(b) A system of linear equations with fewer equations than variables always has at least one solution.

61. The University of Tennessee Lady Volunteers defeated the Rutgers University Scarlet Knights 59 to 46. The Lady Volunteers' scoring resulted from a combination of three-point baskets, two-point baskets, and one-point free throws. There were three times as many two-point baskets as three-point baskets. The number of free throws was one less than the number of two-point baskets. (Source: National Collegiate Athletic Association)

(a) Set up a system of linear equations to find the numbers of three-point baskets, two-point baskets, and one-point free throws scored by the Lady Volunteers.

(b) Solve your system.

62. In Super Bowl I, on January 15, 1967, the Green Bay Packers defeated the Kansas City Chiefs by a score of 35 to 10. The total points scored came from a combination of touchdowns, extra-point kicks, and field goals, worth 6, 1, and 3 points, respectively. The numbers of touchdowns and extra-point kicks were equal. There were six times as many touchdowns as field goals. (Source: National Football League)

(a) Set up a system of linear equations to find the numbers of touchdowns, extra-point kicks, and field goals that were scored.

(b) Solve your system.

In Exercises 63 and 64, use a system of equations to write the partial fraction decomposition of the rational expression. Then solve the system using matrices.

63. $\dfrac{3x^2 - 3x - 2}{(x + 2)(x - 2)^2} = \dfrac{A}{x + 2} + \dfrac{B}{x - 2} + \dfrac{C}{(x - 2)^2}$

64. $\dfrac{3x^2 + 3x - 2}{(x + 1)^2(x - 1)} = \dfrac{A}{x + 1} + \dfrac{B}{x - 1} + \dfrac{C}{(x + 1)^2}$

Polynomial Curve Fitting

In Exercises 65 and 66, (a) determine the polynomial whose graph passes through the given points, and (b) sketch the graph of the polynomial, showing the given points.

65. $(2, 5), (3, 0), (4, 20)$

66. $(-1, -1), (0, 0), (1, 1), (2, 4)$

67. A company has sales (measured in millions) of \$50, \$60, and \$75 during three consecutive years. Find a quadratic function that fits these data, and use it to predict the sales during the fourth year.

68. The polynomial function $p(x) = a_0 + a_1x + a_2x^2 + a_3x^3$ is zero when $x = 1, 2, 3$, and 4. What are the values of a_0, a_1, a_2, and a_3?

69. A wildlife management team studied the population of deer in one small tract of a wildlife preserve. The population and the number of years since the study began are shown in the table.

Year	0	4	80
Population	80	68	30

(a) Set up a system of equations to fit the data to a quadratic polynomial function.

(b) Solve your system.

(c) Use a graphing utility to fit a quadratic model to the data.

(d) Compare the quadratic polynomial function in part (b) with the model in part (c).

(e) Cite the statement from the text that verifies your results.

70. A research team studied the average monthly temperatures of a small lake over a period of about one year. The temperatures and the numbers of months since the study began are shown in the table.

Month	0	6	12
Temperature	40	73	52

(a) Set up a system of equations to fit the data to a quadratic polynomial function.

(b) Solve your system.

(c) Use a graphing utility to fit a quadratic model to the data.

(d) Compare the quadratic polynomial function in part (b) with the model in part (c).

(e) Cite the statement from the text that verifies your results.

Network Analysis

71. Determine the currents I_1, I_2, and I_3 for the electrical network shown in Figure 1.23.

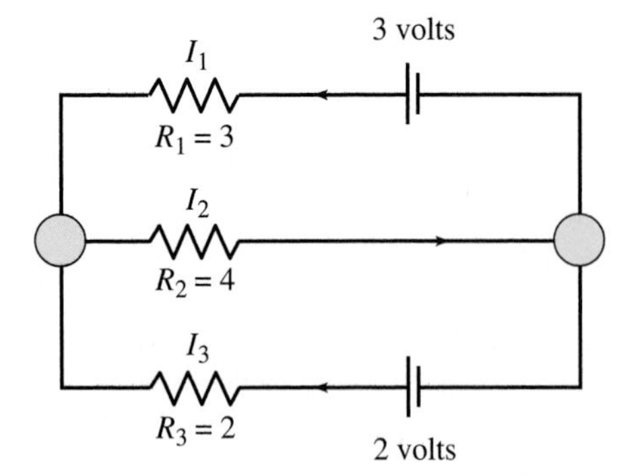

Figure 1.23

72. The flow through a network is shown in Figure 1.24.

(a) Solve the system for x_i, $i = 1, 2, \ldots, 6$.

(b) Find the flow when $x_3 = 100$, $x_5 = 50$, and $x_6 = 50$.

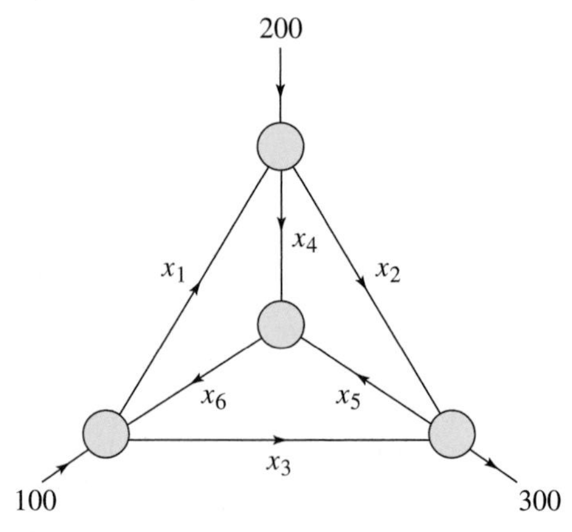

Figure 1.24

CHAPTER 1 Projects

1 Graphing Linear Equations

You saw in Section 1.1 that a system of two linear equations in two variables x and y can be represented geometrically as two lines in the plane. These lines can intersect at a point, coincide, or be parallel, as indicated in Figure 1.25.

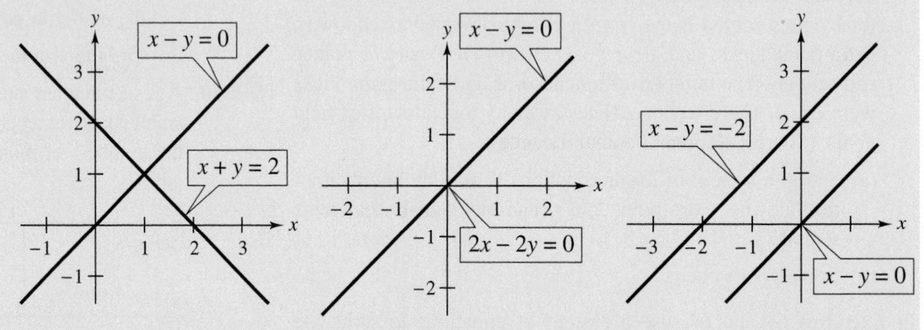

Figure 1.25

1. Consider the system below, where a and b are constants. Answer the questions that follow. For Questions (a)–(c), if an answer is yes, give an example. Otherwise, explain why the answer is no.

$$\begin{aligned} 2x - y &= 3 \\ ax + by &= 6 \end{aligned}$$

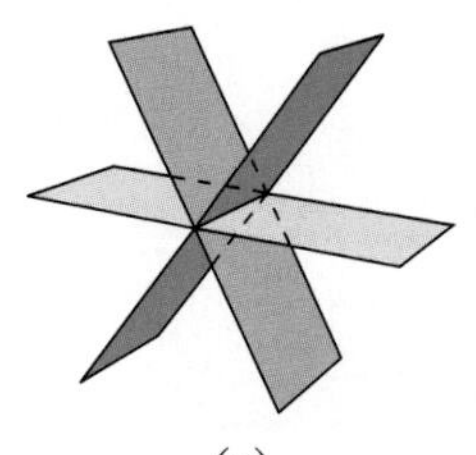
(a)

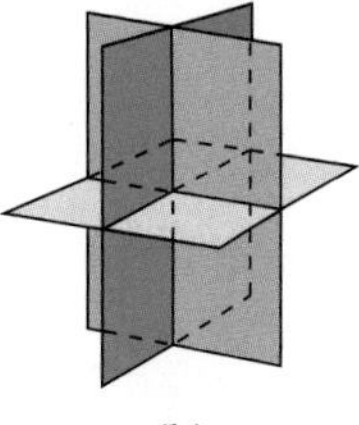
(b)

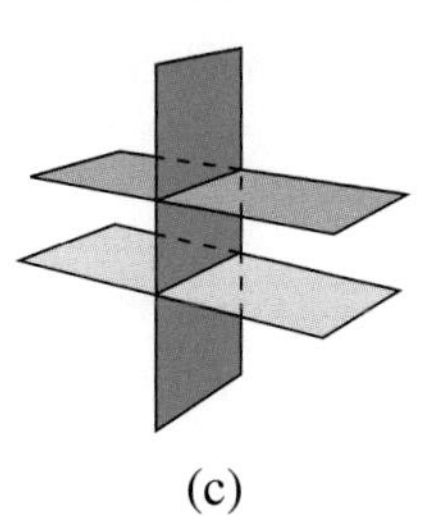
(c)

Figure 1.26

(a) Can you find values of a and b for which the resulting system has a unique solution?
(b) Can you find values of a and b for which the resulting system has an infinite number of solutions?
(c) Can you find values of a and b for which the resulting system has no solution?
(d) Graph the resulting lines for each of the systems in parts (a), (b), and (c).

2. Now consider a system of three linear equations in x, y, and z. Each equation represents a plane in the three-dimensional coordinate system.
(a) Find an example of a system represented by three planes intersecting in a line, as shown in Figure 1.26(a).
(b) Find an example of a system represented by three planes intersecting at a point, as shown in Figure 1.26(b).
(c) Find an example of a system represented by three planes with no common intersection, as shown in Figure 1.26(c).
(d) Are there other configurations of three planes not covered by the three examples in parts (a), (b), and (c)? Explain.

2 Underdetermined and Overdetermined Systems of Equations

The next system of linear equations is said to be **underdetermined** because there are more variables than equations.

$$\begin{aligned} x_1 + 2x_2 - 3x_3 &= 4 \\ 2x_1 - x_2 + 4x_3 &= -3 \end{aligned}$$

Similarly, the following system is **overdetermined** because there are more equations than variables.

$$\begin{aligned} x_1 + 3x_2 &= 5 \\ 2x_1 - 2x_2 &= -3 \\ -x_1 + 7x_2 &= 0 \end{aligned}$$

You can explore whether the number of variables and the number of equations have any bearing on the consistency of a system of linear equations. For Exercises 1–4, if an answer is yes, give an example. Otherwise, explain why the answer is no.

1. Can you find a consistent underdetermined linear system?
2. Can you find a consistent overdetermined linear system?
3. Can you find an inconsistent underdetermined linear system?
4. Can you find an inconsistent overdetermined linear system?
5. Explain why you would expect an overdetermined linear system to be inconsistent. Must this always be the case?
6. Explain why you would expect an underdetermined linear system to have an infinite number of solutions. Must this always be the case?

Matrices

CHAPTER OBJECTIVES

- Write a system of linear equations represented by a matrix, as well as write the matrix form of a system of linear equations.
- Write and solve a system of linear equations in the form $A\mathbf{x} = \mathbf{b}$.
- Use properties of matrix operations to solve matrix equations.
- Find the transpose of a matrix, the inverse of a matrix, and the inverse of a matrix product (if they exist).
- Factor a matrix into a product of elementary matrices, and determine when they are invertible.
- Find and use the *LU*-factorization of a matrix to solve a system of linear equations.
- Use a stochastic matrix to measure consumer preference.
- Use matrix multiplication to encode and decode messages.
- Use matrix algebra to analyze economic systems (Leontief input-output models).
- Use the method of least squares to find the least squares regression line for a set of data.

2.1 Operations with Matrices

In Section 1.2 you used matrices to solve systems of linear equations. Matrices, however, can be used to do much more than that. There is a rich mathematical theory of matrices, and its applications are numerous. This section and the next introduce some fundamentals of matrix theory.

It is standard mathematical convention to represent matrices in any one of the following three ways.

1. A matrix can be denoted by an uppercase letter such as

$$A, B, C, \ldots .$$

2. A matrix can be denoted by a representative element enclosed in brackets, such as

$$[a_{ij}], [b_{ij}], [c_{ij}], \ldots .$$

3. A matrix can be denoted by a rectangular array of numbers

$$\begin{bmatrix} a_{11} & a_{12} & a_{13} & \cdots & a_{1n} \\ a_{21} & a_{22} & a_{23} & \cdots & a_{2n} \\ a_{31} & a_{32} & a_{33} & \cdots & a_{3n} \\ \vdots & \vdots & \vdots & & \vdots \\ a_{m1} & a_{m2} & a_{m3} & \cdots & a_{mn} \end{bmatrix}.$$

As mentioned in Chapter 1, the matrices in this text are primarily *real matrices.* That is, their entries contain real numbers.

Two matrices are said to be **equal** if their corresponding entries are equal.

Definition of Equality of Matrices

Two matrices $A = [a_{ij}]$ and $B = [b_{ij}]$ are **equal** if they have the same size $(m \times n)$ and

$$a_{ij} = b_{ij}$$

for $1 \leq i \leq m$ and $1 \leq j \leq n$.

EXAMPLE 1 **Equality of Matrices**

Consider the four matrices

$$A = \begin{bmatrix} 1 & 2 \\ 3 & 4 \end{bmatrix}, \qquad B = \begin{bmatrix} 1 \\ 3 \end{bmatrix},$$

$$C = \begin{bmatrix} 1 & 3 \end{bmatrix}, \qquad \text{and} \qquad D = \begin{bmatrix} 1 & 2 \\ x & 4 \end{bmatrix}.$$

Matrices A and B are **not** equal because they are of different sizes. Similarly, B and C are not equal. Matrices A and D are equal if and only if $x = 3$.

REMARK: The phrase "if and only if" means the statement is true in both directions. For example, "p if and only if q" means that p implies q and q implies p.

A matrix that has only one column, such as matrix B in Example 1, is called a **column matrix** or **column vector.** Similarly, a matrix that has only one row, such as matrix C in Example 1, is called a **row matrix** or **row vector.** Boldface lowercase letters are often used to designate column matrices and row matrices. For instance, matrix A in Example 1 can be partitioned into the two column matrices $\mathbf{a}_1 = \begin{bmatrix} 1 \\ 3 \end{bmatrix}$ and $\mathbf{a}_2 = \begin{bmatrix} 2 \\ 4 \end{bmatrix}$, as follows.

$$A = \begin{bmatrix} 1 & 2 \\ 3 & 4 \end{bmatrix} = \left[\begin{array}{c:c} 1 & 2 \\ 3 & 4 \end{array}\right] = [\mathbf{a}_1 \ \vdots \ \mathbf{a}_2]$$

Matrix Addition

You can **add** two matrices (of the same size) by adding their corresponding entries.

Definition of Matrix Addition

If $A = [a_{ij}]$ and $B = [b_{ij}]$ are matrices of size $m \times n$, then their **sum** is the $m \times n$ matrix given by

$$A + B = [a_{ij} + b_{ij}].$$

The sum of two matrices of different sizes is undefined.

EXAMPLE 2 Addition of Matrices

(a) $\begin{bmatrix} -1 & 2 \\ 0 & 1 \end{bmatrix} + \begin{bmatrix} 1 & 3 \\ -1 & 2 \end{bmatrix} = \begin{bmatrix} -1+1 & 2+3 \\ 0-1 & 1+2 \end{bmatrix} = \begin{bmatrix} 0 & 5 \\ -1 & 3 \end{bmatrix}$

(b) $\begin{bmatrix} 0 & 1 & -2 \\ 1 & 2 & 3 \end{bmatrix} + \begin{bmatrix} 0 & 0 & 0 \\ 0 & 0 & 0 \end{bmatrix} = \begin{bmatrix} 0 & 1 & -2 \\ 1 & 2 & 3 \end{bmatrix}$

(c) $\begin{bmatrix} 1 \\ -3 \\ -2 \end{bmatrix} + \begin{bmatrix} -1 \\ 3 \\ 2 \end{bmatrix} = \begin{bmatrix} 0 \\ 0 \\ 0 \end{bmatrix}$

(d) The sum of

$$A = \begin{bmatrix} 2 & 1 & 0 \\ 4 & 0 & -1 \\ 3 & -2 & 2 \end{bmatrix} \quad \text{and} \quad B = \begin{bmatrix} 0 & 1 \\ -1 & 3 \\ 2 & 4 \end{bmatrix}$$

is undefined.

Scalar Multiplication

When working with matrices, real numbers are referred to as **scalars.** You can multiply a matrix A by a scalar c by multiplying each entry in A by c.

Definition of Scalar Multiplication

If $A = [a_{ij}]$ is an $m \times n$ matrix and c is a scalar, then the **scalar multiple** of A by c is the $m \times n$ matrix given by

$$cA = [ca_{ij}].$$

You can use $-A$ to represent the scalar product $(-1)A$. If A and B are of the same size, $A - B$ represents the sum of A and $(-1)B$. That is,

$$A - B = A + (-1)B. \qquad \text{Subtraction of matrices}$$

EXAMPLE 3 **Scalar Multiplication and Matrix Subtraction**

For the matrices

$$A = \begin{bmatrix} 1 & 2 & 4 \\ -3 & 0 & -1 \\ 2 & 1 & 2 \end{bmatrix} \quad \text{and} \quad B = \begin{bmatrix} 2 & 0 & 0 \\ 1 & -4 & 3 \\ -1 & 3 & 2 \end{bmatrix}$$

find (a) $3A$, (b) $-B$, and (c) $3A - B$.

SOLUTION (a) $3A = 3\begin{bmatrix} 1 & 2 & 4 \\ -3 & 0 & -1 \\ 2 & 1 & 2 \end{bmatrix} = \begin{bmatrix} 3(1) & 3(2) & 3(4) \\ 3(-3) & 3(0) & 3(-1) \\ 3(2) & 3(1) & 3(2) \end{bmatrix} = \begin{bmatrix} 3 & 6 & 12 \\ -9 & 0 & -3 \\ 6 & 3 & 6 \end{bmatrix}$

(b) $-B = (-1)\begin{bmatrix} 2 & 0 & 0 \\ 1 & -4 & 3 \\ -1 & 3 & 2 \end{bmatrix} = \begin{bmatrix} -2 & 0 & 0 \\ -1 & 4 & -3 \\ 1 & -3 & -2 \end{bmatrix}$

(c) $3A - B = \begin{bmatrix} 3 & 6 & 12 \\ -9 & 0 & -3 \\ 6 & 3 & 6 \end{bmatrix} - \begin{bmatrix} 2 & 0 & 0 \\ 1 & -4 & 3 \\ -1 & 3 & 2 \end{bmatrix} = \begin{bmatrix} 1 & 6 & 12 \\ -10 & 4 & -6 \\ 7 & 0 & 4 \end{bmatrix}$

Matrix Multiplication

REMARK: It is often convenient to rewrite a matrix B as cA by factoring c out of every entry in matrix B. For instance, the scalar $\frac{1}{2}$ has been factored out of the matrix below.

$$\begin{bmatrix} \frac{1}{2} & -\frac{3}{2} \\ \frac{5}{2} & \frac{1}{2} \end{bmatrix} = \frac{1}{2}\begin{bmatrix} 1 & -3 \\ 5 & 1 \end{bmatrix}$$
$$B = cA$$

The third basic matrix operation is **matrix multiplication.** To see the usefulness of this operation, consider the following application in which matrices are helpful for organizing information.

A football stadium has three concession areas, located in the south, north, and west stands. The top-selling items are peanuts, hot dogs, and soda. Sales for a certain day are recorded in the first matrix below, and the prices (in dollars) of the three items are given in the second matrix.

	Number of Items Sold				
	Peanuts	*Hot Dogs*	*Soda*	*Selling Price*	
South stand	120	250	305	2.00	*Peanuts*
North stand	207	140	419	3.00	*Hot Dogs*
West stand	29	120	190	2.75	*Soda*

To calculate the total sales of the three top-selling items at the south stand, you can multiply each entry in the first row of the matrix on the left by the corresponding entry in the price column matrix on the right and add the results. The south stand sales are

$$(120)(2.00) + (250)(3.00) + (305)(2.75) = \$1828.75. \qquad \text{South stand sales}$$

Similarly, you can calculate the sales for the other two stands as follows.

$$(207)(2.00) + (140)(3.00) + (419)(2.75) = \$1986.25 \qquad \text{North stand sales}$$
$$(29)(2.00) + (120)(3.00) + (190)(2.75) = \$940.50 \qquad \text{West stand sales}$$

The preceding computations are examples of matrix multiplication. You can write the product of the 3×3 matrix indicating the number of items sold and the 3×1 matrix indicating the selling prices as follows.

$$\begin{bmatrix} 120 & 250 & 305 \\ 207 & 140 & 419 \\ 29 & 120 & 190 \end{bmatrix} \begin{bmatrix} 2.00 \\ 3.00 \\ 2.75 \end{bmatrix} = \begin{bmatrix} 1828.75 \\ 1986.25 \\ 940.50 \end{bmatrix}$$

The product of these matrices is the 3×1 matrix giving the total sales for each of the three stands.

The general definition of the product of two matrices shown below is based on the ideas just developed. Although at first glance this definition may seem unusual, you will see that it has many practical applications.

Definition of Matrix Multiplication

If $A = [a_{ij}]$ is an $m \times n$ matrix and $B = [b_{ij}]$ is an $n \times p$ matrix, then the **product** AB is an $m \times p$ matrix

$$AB = [c_{ij}]$$

where

$$c_{ij} = \sum_{k=1}^{n} a_{ik}b_{kj} = a_{i1}b_{1j} + a_{i2}b_{2j} + a_{i3}b_{3j} + \cdot \cdot \cdot + a_{in}b_{nj}.$$

This definition means that the entry in the ith row and the jth column of the product AB is obtained by multiplying the entries in the ith row of A by the corresponding entries in the jth column of B and then adding the results. The next example illustrates this process.

EXAMPLE 4 Finding the Product of Two Matrices

Find the product AB, where

$$A = \begin{bmatrix} -1 & 3 \\ 4 & -2 \\ 5 & 0 \end{bmatrix} \quad \text{and} \quad B = \begin{bmatrix} -3 & 2 \\ -4 & 1 \end{bmatrix}.$$

SOLUTION First note that the product AB is defined because A has size 3×2 and B has size 2×2. Moreover, the product AB has size 3×2 and will take the form

$$\begin{bmatrix} -1 & 3 \\ 4 & -2 \\ 5 & 0 \end{bmatrix} \begin{bmatrix} -3 & 2 \\ -4 & 1 \end{bmatrix} = \begin{bmatrix} c_{11} & c_{12} \\ c_{21} & c_{22} \\ c_{31} & c_{32} \end{bmatrix}.$$

To find c_{11} (the entry in the first row and first column of the product), multiply corresponding entries in the first row of A and the first column of B. That is,

$$c_{11} = (-1)(-3) + (3)(-4) = -9$$

$$\begin{bmatrix} -1 & 3 \\ 4 & -2 \\ 5 & 0 \end{bmatrix} \begin{bmatrix} -3 & 2 \\ -4 & 1 \end{bmatrix} = \begin{bmatrix} -9 & c_{12} \\ c_{21} & c_{22} \\ c_{31} & c_{32} \end{bmatrix}.$$

HISTORICAL NOTE

Arthur Cayley
(1821–1895)
showed signs of mathematical genius at an early age, but ironically wasn't able to find a position as a mathematician upon graduating from college. Ultimately, however, Cayley made major contributions to linear algebra. To read about his work, visit *college.hmco.com/pic/larsonELA6e.*

Similarly, to find c_{12}, multiply corresponding entries in the first row of A and the second column of B to obtain

$$c_{12} = (-1)(2) + (3)(1) = 1$$

$$\begin{bmatrix} -1 & 3 \\ 4 & -2 \\ 5 & 0 \end{bmatrix} \begin{bmatrix} -3 & 2 \\ -4 & 1 \end{bmatrix} = \begin{bmatrix} -9 & 1 \\ c_{21} & c_{22} \\ c_{31} & c_{32} \end{bmatrix}.$$

Continuing this pattern produces the results shown below.

$$\begin{aligned} c_{21} &= (4)(-3) + (-2)(-4) = -4 \\ c_{22} &= (4)(2) + (-2)(1) = 6 \\ c_{31} &= (5)(-3) + (0)(-4) = -15 \\ c_{32} &= (5)(2) + (0)(1) = 10 \end{aligned}$$

The product is

$$AB = \begin{bmatrix} -1 & 3 \\ 4 & -2 \\ 5 & 0 \end{bmatrix} \begin{bmatrix} -3 & 2 \\ -4 & 1 \end{bmatrix} = \begin{bmatrix} -9 & 1 \\ -4 & 6 \\ -15 & 10 \end{bmatrix}.$$

Be sure you understand that for the product of two matrices to be defined, the number of columns of the first matrix must equal the number of rows of the second matrix. That is,

$$\underset{m \times n}{A} \quad \underset{n \times p}{B} \quad = \quad \underset{m \times p}{AB.}$$

equal

size of AB

So, the product BA is not defined for matrices such as A and B in Example 4.

The general pattern for matrix multiplication is as follows. To obtain the element in the ith row and the jth column of the product AB, use the ith row of A and the jth column of B.

$$\begin{bmatrix} a_{11} & a_{12} & a_{13} & \cdots & a_{1n} \\ a_{21} & a_{22} & a_{23} & \cdots & a_{2n} \\ \vdots & \vdots & \vdots & & \vdots \\ a_{i1} & a_{i2} & a_{i3} & \cdots & a_{in} \\ \vdots & \vdots & \vdots & & \vdots \\ a_{m1} & a_{m2} & a_{m3} & \cdots & a_{mn} \end{bmatrix} \begin{bmatrix} b_{11} & b_{12} & \cdots & b_{1j} & \cdots & b_{1p} \\ b_{21} & b_{22} & \cdots & b_{2j} & \cdots & b_{2p} \\ b_{31} & b_{32} & \cdots & b_{3j} & \cdots & b_{3p} \\ \vdots & \vdots & & \vdots & & \vdots \\ b_{n1} & b_{n2} & \cdots & b_{nj} & \cdots & b_{np} \end{bmatrix} = \begin{bmatrix} c_{11} & c_{12} & \cdots & c_{1j} & \cdots & c_{1p} \\ c_{21} & c_{22} & \cdots & c_{2j} & \cdots & c_{2p} \\ \vdots & \vdots & & \vdots & & \vdots \\ c_{i1} & c_{i2} & \cdots & c_{ij} & \cdots & c_{ip} \\ \vdots & \vdots & & \vdots & & \vdots \\ c_{m1} & c_{m2} & \cdots & c_{mj} & \cdots & c_{mp} \end{bmatrix}$$

$$a_{i1}b_{1j} + a_{i2}b_{2j} + a_{i3}b_{3j} + \cdots + a_{in}b_{nj} = c_{ij}$$

Discovery

Let

$$A = \begin{bmatrix} 1 & 2 \\ 3 & 4 \end{bmatrix} \quad \text{and} \quad B = \begin{bmatrix} 0 & 1 \\ 1 & 2 \end{bmatrix}.$$

Calculate $A + B$ and $B + A$.

In general, is the operation of matrix addition commutative? Now calculate AB and BA. Is matrix multiplication commutative?

EXAMPLE 5 **Matrix Multiplication**

(a) $$\begin{bmatrix} 1 & 0 & 3 \\ 2 & -1 & -2 \end{bmatrix} \begin{bmatrix} -2 & 4 & 2 \\ 1 & 0 & 0 \\ -1 & 1 & -1 \end{bmatrix} = \begin{bmatrix} -5 & 7 & -1 \\ -3 & 6 & 6 \end{bmatrix}$$
$$2 \times 3 \qquad 3 \times 3 \qquad 2 \times 3$$

(b) $$\begin{bmatrix} 3 & 4 \\ -2 & 5 \end{bmatrix} \begin{bmatrix} 1 & 0 \\ 0 & 1 \end{bmatrix} = \begin{bmatrix} 3 & 4 \\ -2 & 5 \end{bmatrix}$$
$$2 \times 2 \qquad 2 \times 2 \qquad 2 \times 2$$

(c) $$\begin{bmatrix} 1 & 2 \\ 1 & 1 \end{bmatrix} \begin{bmatrix} -1 & 2 \\ 1 & -1 \end{bmatrix} = \begin{bmatrix} 1 & 0 \\ 0 & 1 \end{bmatrix}$$
$$2 \times 2 \qquad 2 \times 2 \qquad 2 \times 2$$

(d) $$\begin{bmatrix} 1 & -2 & -3 \end{bmatrix} \begin{bmatrix} 2 \\ -1 \\ 1 \end{bmatrix} = \begin{bmatrix} 1 \end{bmatrix}$$
$$1 \times 3 \qquad 3 \times 1 \qquad 1 \times 1$$

$$\text{(e)}\ \begin{bmatrix} 2 \\ -1 \\ 1 \end{bmatrix} \begin{bmatrix} 1 & -2 & -3 \end{bmatrix} = \begin{bmatrix} 2 & -4 & -6 \\ -1 & 2 & 3 \\ 1 & -2 & -3 \end{bmatrix}$$

$$3 \times 1 \qquad 1 \times 3 \qquad 3 \times 3$$

REMARK: Note the difference between the two products in parts (d) and (e) of Example 5. In general, matrix multiplication is not commutative. It is usually not true that the product AB is equal to the product BA. (See Section 2.2 for further discussion of the noncommutativity of matrix multiplication.)

Technology Note

Most graphing utilities and computer software programs can perform matrix addition, scalar multiplication, and matrix multiplication. If you are using a graphing utility, your screens for Example 5(c) may look like:

A
[[1 2]
[1 1]]
B
[[-1 2]
[1 -1]]

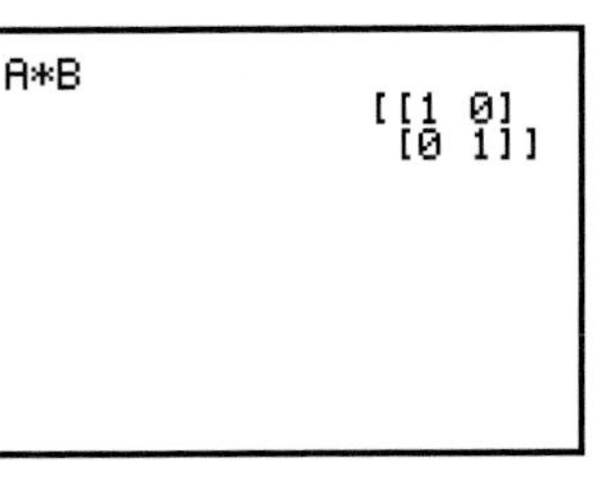

Keystrokes and programming syntax for these utilities/programs applicable to Example 5(c) are provided in the **Online Technology Guide,** available at *college.hmco.com/pic/larsonELA6e*.

Systems of Linear Equations

One practical application of matrix multiplication is representing a system of linear equations. Note how the system

$$\begin{aligned} a_{11}x_1 + a_{12}x_2 + a_{13}x_3 &= b_1 \\ a_{21}x_1 + a_{22}x_2 + a_{23}x_3 &= b_2 \\ a_{31}x_1 + a_{32}x_2 + a_{33}x_3 &= b_3 \end{aligned}$$

can be written as the matrix equation $A\mathbf{x} = \mathbf{b}$, where A is the coefficient matrix of the system, and $\mathbf{x}$ and $\mathbf{b}$ are column matrices. You can write the system as

$$\underbrace{\begin{bmatrix} a_{11} & a_{12} & a_{13} \\ a_{21} & a_{22} & a_{23} \\ a_{31} & a_{32} & a_{33} \end{bmatrix}}_{A} \underbrace{\begin{bmatrix} x_1 \\ x_2 \\ x_3 \end{bmatrix}}_{\mathbf{x}} = \underbrace{\begin{bmatrix} b_1 \\ b_2 \\ b_3 \end{bmatrix}}_{\mathbf{b}}.$$

EXAMPLE 6 **Solving a System of Linear Equations**

Solve the matrix equation $A\mathbf{x} = \mathbf{0}$, where

$$A = \begin{bmatrix} 1 & -2 & 1 \\ 2 & 3 & -2 \end{bmatrix}, \quad \mathbf{x} = \begin{bmatrix} x_1 \\ x_2 \\ x_3 \end{bmatrix}, \quad \text{and} \quad \mathbf{0} = \begin{bmatrix} 0 \\ 0 \end{bmatrix}.$$

SOLUTION As a system of linear equations, $A\mathbf{x} = \mathbf{0}$ looks like

$$\begin{aligned} x_1 - 2x_2 + x_3 &= 0 \\ 2x_1 + 3x_2 - 2x_3 &= 0. \end{aligned}$$

Using Gauss-Jordan elimination on the augmented matrix of this system, you obtain

$$\begin{bmatrix} 1 & 0 & -\frac{1}{7} & 0 \\ 0 & 1 & -\frac{4}{7} & 0 \end{bmatrix}.$$

So, the system has an infinite number of solutions. Here a convenient choice of a parameter is $x_3 = 7t$, and you can write the solution set as

$$x_1 = t, \qquad x_2 = 4t, \qquad x_3 = 7t, \quad t \text{ is any real number.}$$

In matrix terminology, you have found that the matrix equation

$$\begin{bmatrix} 1 & -2 & 1 \\ 2 & 3 & -2 \end{bmatrix} \begin{bmatrix} x_1 \\ x_2 \\ x_3 \end{bmatrix} = \begin{bmatrix} 0 \\ 0 \end{bmatrix}$$

has an infinite number of solutions represented by

$$\mathbf{x} = \begin{bmatrix} x_1 \\ x_2 \\ x_3 \end{bmatrix} = \begin{bmatrix} t \\ 4t \\ 7t \end{bmatrix} = t \begin{bmatrix} 1 \\ 4 \\ 7 \end{bmatrix}, \quad t \text{ is any scalar.}$$

That is, any scalar multiple of the column matrix on the right is a solution.

Partitioned Matrices

The system $A\mathbf{x} = \mathbf{b}$ can be represented in a more convenient way by partitioning the matrices A and $\mathbf{x}$ in the following manner. If

$$A = \begin{bmatrix} a_{11} & a_{12} & \cdots & a_{1n} \\ a_{21} & a_{22} & \cdots & a_{2n} \\ \vdots & \vdots & & \vdots \\ a_{m1} & a_{m2} & \cdots & a_{mn} \end{bmatrix}, \quad \mathbf{x} = \begin{bmatrix} x_1 \\ x_2 \\ \vdots \\ x_n \end{bmatrix}, \quad \text{and} \quad \mathbf{b} = \begin{bmatrix} b_1 \\ b_2 \\ \vdots \\ b_m \end{bmatrix}$$

are the coefficient matrix, the column matrix of unknowns, and the right-hand side, respectively, of the $m \times n$ linear system $A\mathbf{x} = \mathbf{b}$, then you can write

$$A\mathbf{x} = \mathbf{b}$$

$$\begin{bmatrix} a_{11} & a_{12} & \cdots & a_{1n} \\ a_{21} & a_{22} & \cdots & a_{2n} \\ \vdots & \vdots & & \vdots \\ a_{m1} & a_{m2} & \cdots & a_{mn} \end{bmatrix} \begin{bmatrix} x_1 \\ x_2 \\ \vdots \\ x_n \end{bmatrix} = \mathbf{b}$$

$$\begin{bmatrix} a_{11}x_1 + a_{12}x_2 + \cdots + a_{1n}x_n \\ a_{21}x_1 + a_{22}x_2 + \cdots + a_{2n}x_n \\ \vdots \\ a_{m1}x_1 + a_{m2}x_2 + \cdots + a_{mn}x_n \end{bmatrix} = \mathbf{b}$$

$$x_1 \begin{bmatrix} a_{11} \\ a_{21} \\ \vdots \\ a_{m1} \end{bmatrix} + x_2 \begin{bmatrix} a_{12} \\ a_{22} \\ \vdots \\ a_{m2} \end{bmatrix} + \cdots + x_n \begin{bmatrix} a_{1n} \\ a_{2n} \\ \vdots \\ a_{mn} \end{bmatrix} = \mathbf{b}.$$

In other words,

$$A\mathbf{x} = x_1\mathbf{a}_1 + x_2\mathbf{a}_2 + \cdots + x_n\mathbf{a}_n = \mathbf{b},$$

where $\mathbf{a}_1, \mathbf{a}_2, \ldots, \mathbf{a}_n$ are the columns of the matrix A. The expression

$$x_1 \begin{bmatrix} a_{11} \\ a_{21} \\ \vdots \\ a_{m1} \end{bmatrix} + x_2 \begin{bmatrix} a_{12} \\ a_{22} \\ \vdots \\ a_{m2} \end{bmatrix} + \cdots + x_n \begin{bmatrix} a_{1n} \\ a_{2n} \\ \vdots \\ a_{mn} \end{bmatrix}$$

is called a **linear combination** of the column matrices $\mathbf{a}_1, \mathbf{a}_2, \ldots, \mathbf{a}_n$ with **coefficients** $x_1, x_2, \ldots, x_n$.

In general, the matrix product $A\mathbf{x}$ is a linear combination of the column vectors $\mathbf{a}_1, \mathbf{a}_2, \ldots, \mathbf{a}_n$ that form the coefficient matrix A. Furthermore, the system $A\mathbf{x} = \mathbf{b}$ is consistent if and only if $\mathbf{b}$ can be expressed as such a linear combination, where the coefficients of the linear combination are a solution of the system.

EXAMPLE 7 Solving a System of Linear Equations

The linear system

$$\begin{aligned} x_1 + 2x_2 + 3x_3 &= 0 \\ 4x_1 + 5x_2 + 6x_3 &= 3 \\ 7x_1 + 8x_2 + 9x_3 &= 6 \end{aligned}$$

Technology Note

Many real-life applications of linear systems involve enormous numbers of equations and variables. For example, a flight crew scheduling problem for American Airlines required the manipulation of matrices with 837 rows and more than 12,750,000 columns. This application of *linear programming* required that the problem be partitioned into smaller pieces and then solved on a CRAY Y-MP supercomputer.

(Source: Very-Large Scale Linear Programming, A Case Study in Combining Interior Point and Simplex Methods, Bixby, Robert E., et al., *Operations Research,* 40, no. 5, 1992.)

can be rewritten as a matrix equation $A\mathbf{x} = \mathbf{b}$, as follows.

$$x_1\begin{bmatrix}1\\4\\7\end{bmatrix} + x_2\begin{bmatrix}2\\5\\8\end{bmatrix} + x_3\begin{bmatrix}3\\6\\9\end{bmatrix} = \begin{bmatrix}0\\3\\6\end{bmatrix}$$

Using Gaussian elimination, you can show that this system has an infinite number of solutions, one of which is $x_1 = 1, x_2 = 1, x_3 = -1$.

$$1\begin{bmatrix}1\\4\\7\end{bmatrix} + 1\begin{bmatrix}2\\5\\8\end{bmatrix} + (-1)\begin{bmatrix}3\\6\\9\end{bmatrix} = \begin{bmatrix}0\\3\\6\end{bmatrix}$$

That is, $\mathbf{b}$ can be expressed as a linear combination of the columns of A. This representation of one column vector in terms of others is a fundamental theme of linear algebra.

Just as you partitioned A into columns and $\mathbf{x}$ into rows, it is often useful to consider an $m \times n$ matrix partitioned into smaller matrices. For example, the matrix on the left below can be partitioned as shown below at the right.

$$\begin{bmatrix}1 & 2 & 0 & 0\\3 & 4 & 0 & 0\\-1 & -2 & 2 & 1\end{bmatrix} \qquad \left[\begin{array}{cc|cc}1 & 2 & 0 & 0\\3 & 4 & 0 & 0\\\hline -1 & -2 & 2 & 1\end{array}\right]$$

The matrix could also be partitioned into column matrices

$$\left[\begin{array}{c|c|c|c}1 & 2 & 0 & 0\\3 & 4 & 0 & 0\\-1 & -2 & 2 & 1\end{array}\right] = [\mathbf{c}_1 \quad \mathbf{c}_2 \quad \mathbf{c}_3 \quad \mathbf{c}_4]$$

or row matrices

$$\left[\begin{array}{cccc}1 & 2 & 0 & 0\\\hline 3 & 4 & 0 & 0\\\hline -1 & -2 & 2 & 1\end{array}\right] = \begin{bmatrix}\mathbf{r}_1\\\mathbf{r}_2\\\mathbf{r}_3\end{bmatrix}.$$

SECTION 2.1 Exercises

In Exercises 1–6, find (a) $A + B$, (b) $A - B$, (c) $2A$, (d) $2A - B$, and (e) $B + \frac{1}{2}A$.

1. $A = \begin{bmatrix}1 & -1\\2 & -1\end{bmatrix}, \quad B = \begin{bmatrix}2 & -1\\-1 & 8\end{bmatrix}$

2. $A = \begin{bmatrix}1 & 2\\2 & 1\end{bmatrix}, \quad B = \begin{bmatrix}-3 & -2\\4 & 2\end{bmatrix}$

3. $A = \begin{bmatrix}6 & -1\\2 & 4\\-3 & 5\end{bmatrix}, \quad B = \begin{bmatrix}1 & 4\\-1 & 5\\1 & 10\end{bmatrix}$

4. $A = \begin{bmatrix}2 & 1 & 1\\-1 & -1 & 4\end{bmatrix}, \quad B = \begin{bmatrix}2 & -3 & 4\\-3 & 1 & -2\end{bmatrix}$

5. $A = \begin{bmatrix}3 & 2 & -1\\2 & 4 & 5\\0 & 1 & 2\end{bmatrix}, \quad B = \begin{bmatrix}0 & 2 & 1\\5 & 4 & 2\\2 & 1 & 0\end{bmatrix}$

6. $A = \begin{bmatrix}2 & 3 & 4\\0 & 1 & -1\\2 & 0 & 1\end{bmatrix}, \quad B = \begin{bmatrix}0 & 6 & 2\\4 & 1 & 0\\-1 & 2 & 4\end{bmatrix}$

7. Find (a) c_{21} and (b) c_{13}, where $C = 2A - 3B$,

$$A = \begin{bmatrix} 5 & 4 & 4 \\ -3 & 1 & 2 \end{bmatrix}, \text{ and } B = \begin{bmatrix} 1 & 2 & -7 \\ 0 & -5 & 1 \end{bmatrix}.$$

8. Find (a) c_{23} and (b) c_{32}, where $C = 5A + 2B$,

$$A = \begin{bmatrix} 4 & 11 & -9 \\ 0 & 3 & 2 \\ -3 & 1 & 1 \end{bmatrix}, \text{ and } B = \begin{bmatrix} 1 & 0 & 5 \\ -4 & 6 & 11 \\ -6 & 4 & 9 \end{bmatrix}.$$

9. Solve for x, y, and z in the matrix equation

$$4\begin{bmatrix} x & y \\ z & -1 \end{bmatrix} = 2\begin{bmatrix} y & z \\ -x & 1 \end{bmatrix} + 2\begin{bmatrix} 4 & x \\ 5 & -x \end{bmatrix}.$$

10. Solve for x, y, z, and w in the matrix equation

$$\begin{bmatrix} w & x \\ y & x \end{bmatrix} = \begin{bmatrix} -4 & 3 \\ 2 & -1 \end{bmatrix} + 2\begin{bmatrix} y & w \\ z & x \end{bmatrix}.$$

In Exercises 11–18, find (a) AB and (b) BA (if they are defined).

11. $A = \begin{bmatrix} 1 & 2 \\ 4 & 2 \end{bmatrix}, \quad B = \begin{bmatrix} 2 & -1 \\ -1 & 8 \end{bmatrix}$

12. $A = \begin{bmatrix} 1 & -1 & 7 \\ 2 & -1 & 8 \\ 3 & 1 & -1 \end{bmatrix}, \quad B = \begin{bmatrix} 1 & 1 & 2 \\ 2 & 1 & 1 \\ 1 & -3 & 2 \end{bmatrix}$

13. $A = \begin{bmatrix} 2 & 1 \\ -3 & 4 \\ 1 & 6 \end{bmatrix}, \quad B = \begin{bmatrix} 0 & -1 & 0 \\ 4 & 0 & 2 \\ 8 & -1 & 7 \end{bmatrix}$

14. $A = \begin{bmatrix} 3 & 2 & 1 \end{bmatrix}, \quad B = \begin{bmatrix} 2 \\ 3 \\ 0 \end{bmatrix}$

15. $A = \begin{bmatrix} -1 & 3 \\ 4 & -5 \\ 0 & 2 \end{bmatrix}, \quad B = \begin{bmatrix} 1 & 2 \\ 0 & 7 \end{bmatrix}$

16. $A = \begin{bmatrix} 0 & -1 & 0 \\ 4 & 0 & 2 \\ 8 & -1 & 7 \end{bmatrix}, \quad B = \begin{bmatrix} 2 \\ -3 \\ 1 \end{bmatrix}$

17. $A = \begin{bmatrix} 6 \\ -2 \\ 1 \\ 6 \end{bmatrix}, \quad B = \begin{bmatrix} 10 & 12 \end{bmatrix}$

18. $A = \begin{bmatrix} 1 & 0 & 3 & -2 & 4 \\ 6 & 13 & 8 & -17 & 20 \end{bmatrix}, \quad B = \begin{bmatrix} 1 & 6 \\ 4 & 2 \end{bmatrix}$

In Exercises 19 and 20, find (a) $2A + B$, (b) $3B - A$, (c) AB, and (d) BA (if they are defined).

19. $A = \begin{bmatrix} 2 & -2 & 4 & 1 & 0 & 3 \\ -1 & 4 & 2 & -2 & -1 & 3 \\ 3 & -3 & 1 & 2 & 3 & -4 \\ 2 & -1 & 3 & 0 & 1 & 2 \\ 5 & 1 & -2 & -4 & 1 & 3 \\ 2 & 2 & 3 & -4 & -1 & -2 \end{bmatrix}$

$B = \begin{bmatrix} 1 & 2 & -3 & 4 & 1 & 2 \\ 2 & -3 & 1 & 3 & -1 & 2 \\ 0 & -2 & -3 & 0 & 1 & -1 \\ 1 & 2 & 3 & 2 & 1 & -1 \\ 2 & -1 & -3 & 0 & 4 & 2 \\ 1 & -2 & 4 & -2 & -4 & -1 \end{bmatrix}$

20. $A = \begin{bmatrix} 2 & 1 & 3 & 2 & -1 \\ 3 & -1 & 0 & 1 & 2 \\ 2 & 1 & -3 & 3 & -2 \\ -4 & 0 & 2 & -3 & 1 \\ 1 & 0 & -1 & 2 & 4 \\ 2 & -3 & 2 & 1 & -4 \end{bmatrix}$

$B = \begin{bmatrix} 5 & 2 & 1 & 3 & 2 & 1 \\ -4 & -2 & 2 & -1 & 3 & -1 \\ 4 & 0 & 1 & 3 & -2 & 1 \\ -1 & 2 & -3 & -1 & 2 & 3 \\ -2 & 1 & 4 & 3 & -2 & 2 \\ 1 & -2 & 3 & 4 & -2 & -1 \end{bmatrix}$

In Exercises 21–28, let A, B, C, D, and E be matrices with the provided orders.

A: 3×4 B: 3×4 C: 4×2 D: 4×2 E: 4×3

If defined, determine the size of the matrix. If not defined, provide an explanation.

21. $A + B$ **22.** $C + E$ **23.** $\frac{1}{2}D$ **24.** $-4A$

25. AC **26.** BE **27.** $E - 2A$ **28.** $2D + C$

In Exercises 29–36, write the system of linear equations in the form $A\mathbf{x} = \mathbf{b}$ and solve this matrix equation for $\mathbf{x}$.

29. $\begin{aligned} -x_1 + x_2 &= 4 \\ -2x_1 + x_2 &= 0 \end{aligned}$

30. $\begin{aligned} 2x_1 + 3x_2 &= 5 \\ x_1 + 4x_2 &= 10 \end{aligned}$

31. $\begin{aligned} -2x_1 - 3x_2 &= -4 \\ 6x_1 + x_2 &= -36 \end{aligned}$

32. $\begin{aligned} -4x_1 + 9x_2 &= -13 \\ x_1 - 3x_2 &= 12 \end{aligned}$

33.
$$\begin{aligned} x_1 - 2x_2 + 3x_3 &= 9 \\ -x_1 + 3x_2 - x_3 &= -6 \\ 2x_1 - 5x_2 + 5x_3 &= 17 \end{aligned}$$

34.
$$\begin{aligned} x_1 + x_2 - 3x_3 &= -1 \\ -x_1 + 2x_2 &= 1 \\ x_1 - x_2 + x_3 &= 2 \end{aligned}$$

35.
$$\begin{aligned} x_1 - 5x_2 + 2x_3 &= -20 \\ -3x_1 + x_2 - x_3 &= 8 \\ -2x_2 + 5x_3 &= -16 \end{aligned}$$

36.
$$\begin{aligned} x_1 - x_2 + 4x_3 &= 17 \\ x_1 + 3x_2 &= -11 \\ -6x_2 + 5x_3 &= 40 \end{aligned}$$

In Exercises 37 and 38, solve the matrix equation for A.

37. $\begin{bmatrix} 1 & 2 \\ 3 & 5 \end{bmatrix} A = \begin{bmatrix} 1 & 0 \\ 0 & 1 \end{bmatrix}$ **38.** $\begin{bmatrix} 2 & -1 \\ 3 & -2 \end{bmatrix} A = \begin{bmatrix} 1 & 0 \\ 0 & 1 \end{bmatrix}$

In Exercises 39 and 40, solve the matrix equation for a, b, c, and d.

39. $\begin{bmatrix} 1 & 2 \\ 3 & 4 \end{bmatrix} \begin{bmatrix} a & b \\ c & d \end{bmatrix} = \begin{bmatrix} 6 & 3 \\ 19 & 2 \end{bmatrix}$

40. $\begin{bmatrix} a & b \\ c & d \end{bmatrix} \begin{bmatrix} 2 & 1 \\ 3 & 1 \end{bmatrix} = \begin{bmatrix} 3 & 17 \\ 4 & -1 \end{bmatrix}$

41. Find conditions on w, x, y, and z such that $AB = BA$ for the matrices below.

$$A = \begin{bmatrix} w & x \\ y & z \end{bmatrix} \quad \text{and} \quad B = \begin{bmatrix} 1 & 1 \\ -1 & 1 \end{bmatrix}$$

42. Verify $AB = BA$ for the following matrices.

$$A = \begin{bmatrix} \cos\alpha & -\sin\alpha \\ \sin\alpha & \cos\alpha \end{bmatrix} \quad \text{and} \quad B = \begin{bmatrix} \cos\beta & -\sin\beta \\ \sin\beta & \cos\beta \end{bmatrix}$$

A square matrix

$$A = \begin{bmatrix} a_{11} & 0 & 0 & \cdots & 0 \\ 0 & a_{22} & 0 & \cdots & 0 \\ 0 & 0 & a_{33} & \cdots & 0 \\ \vdots & \vdots & \vdots & & \vdots \\ 0 & 0 & \cdots & 0 & a_{nn} \end{bmatrix}$$

is called a **diagonal matrix** if all entries that are not on the main diagonal are zero. In Exercises 43 and 44, find the product AA for the diagonal matrix.

43. $A = \begin{bmatrix} -1 & 0 & 0 \\ 0 & 2 & 0 \\ 0 & 0 & 3 \end{bmatrix}$ **44.** $A = \begin{bmatrix} 2 & 0 & 0 \\ 0 & -3 & 0 \\ 0 & 0 & 0 \end{bmatrix}$

In Exercises 45 and 46, find the products AB and BA for the diagonal matrices.

45. $A = \begin{bmatrix} 2 & 0 \\ 0 & -3 \end{bmatrix}$, $B = \begin{bmatrix} -5 & 0 \\ 0 & 4 \end{bmatrix}$

46. $A = \begin{bmatrix} 3 & 0 & 0 \\ 0 & -5 & 0 \\ 0 & 0 & 0 \end{bmatrix}$, $B = \begin{bmatrix} -7 & 0 & 0 \\ 0 & 4 & 0 \\ 0 & 0 & 12 \end{bmatrix}$

47. Guided Proof Prove that if A and B are diagonal matrices (of the same size), then $AB = BA$.

Getting Started: To prove that the matrices AB and BA are equal, you need to show that their corresponding entries are equal.

(i) Begin your proof by letting $A = [a_{ij}]$ and $B = [b_{ij}]$ be two diagonal $n \times n$ matrices.
(ii) The ijth entry of the product AB is $c_{ij} = \sum_{k=1}^{n} a_{ik}b_{kj}$.
(iii) Evaluate the entries c_{ij} for the two cases $i \neq j$ and $i = j$.
(iv) Repeat this analysis for the product BA.

48. Writing Let A and B be 3×3 matrices, where A is diagonal.

(a) Describe the product AB. Illustrate your answer with examples.
(b) Describe the product BA. Illustrate your answer with examples.
(c) How do the results in parts (a) and (b) change if the diagonal entries of A are all equal?

In Exercises 49–52, find the trace of the matrix. The **trace** of an $n \times n$ matrix A is the sum of the main diagonal entries. That is, $\text{Tr}(A) = a_{11} + a_{22} + \cdots + a_{nn}$.

49. $\begin{bmatrix} 1 & 2 & 3 \\ 0 & -2 & 4 \\ 3 & 1 & 3 \end{bmatrix}$ **50.** $\begin{bmatrix} 1 & 0 & 0 \\ 0 & 1 & 0 \\ 0 & 0 & 1 \end{bmatrix}$

51. $\begin{bmatrix} 1 & 0 & 2 & 1 \\ 0 & 1 & -1 & 2 \\ 4 & 2 & 1 & 0 \\ 0 & 0 & 5 & 1 \end{bmatrix}$ **52.** $\begin{bmatrix} 1 & 4 & 3 & 2 \\ 4 & 0 & 6 & 1 \\ 3 & 6 & 2 & 1 \\ 2 & 1 & 1 & -3 \end{bmatrix}$

53. Prove that each statement is true if A and B are square matrices of order n and c is a scalar.

(a) $\text{Tr}(A + B) = \text{Tr}(A) + \text{Tr}(B)$
(b) $\text{Tr}(cA) = c\text{Tr}(A)$

54. Prove that if A and B are square matrices of order n, then $\text{Tr}(AB) = \text{Tr}(BA)$.

55. Show that the matrix equation has no solution.

$$\begin{bmatrix} 1 & 1 \\ 1 & 1 \end{bmatrix} A = \begin{bmatrix} 1 & 0 \\ 0 & 1 \end{bmatrix}$$

56. Show that no 2×2 matrices A and B exist that satisfy the matrix equation

$$AB - BA = \begin{bmatrix} 1 & 0 \\ 0 & 1 \end{bmatrix}.$$

57. Let $i = \sqrt{-1}$ and let $A = \begin{bmatrix} i & 0 \\ 0 & i \end{bmatrix}$ and $B = \begin{bmatrix} 0 & -i \\ i & 0 \end{bmatrix}$.

(a) Find A^2, A^3, and A^4. Identify any similarities among i^2, i^3, and i^4.

(b) Find and identify B^2.

58. Guided Proof Prove that if the product AB is a square matrix, then the product BA is defined.

Getting Started: To prove that the product BA is defined, you need to show that the number of columns of B equals the number of rows of A.

(i) Begin your proof by noting that the number of columns of A equals the number of rows of B.

(ii) You can then assume that A has size $m \times n$ and B has size $n \times p$.

(iii) Use the hypothesis that the product AB is a square matrix.

59. Prove that if both products AB and BA are defined, then AB and BA are square matrices.

60. Let A and B be two matrices such that the product AB is defined. Show that if A has two identical rows, then the corresponding two rows of AB are also identical.

61. Let A and B be $n \times n$ matrices. Show that if the ith row of A has all zero entries, then the ith row of AB will have all zero entries. Give an example using 2×2 matrices to show that the converse is not true.

62. The columns of matrix T show the coordinates of the vertices of a triangle. Matrix A is a transformation matrix.

$$A = \begin{bmatrix} 0 & -1 \\ 1 & 0 \end{bmatrix}, \quad T = \begin{bmatrix} 1 & 2 & 3 \\ 1 & 4 & 2 \end{bmatrix}$$

(a) Find AT and AAT. Then sketch the original triangle and the two transformed triangles. What transformation does A represent?

(b) A triangle is determined by AAT. Describe the transformation process that produces the triangle determined by AT and then the triangle determined by T.

63. A corporation has three factories, each of which manufactures acoustic guitars and electric guitars. The number of guitars of type i produced at factory j in one day is represented by a_{ij} in the matrix

$$A = \begin{bmatrix} 70 & 50 & 25 \\ 35 & 100 & 70 \end{bmatrix}.$$

Find the production levels if production is increased by 20%.

64. A corporation has four factories, each of which manufactures sport utility vehicles and pickup trucks. The number of vehicles of type i produced at factory j in one day is represented by a_{ij} in the matrix

$$A = \begin{bmatrix} 100 & 90 & 70 & 30 \\ 40 & 20 & 60 & 60 \end{bmatrix}.$$

Find the production levels if production is increased by 10%.

65. A fruit grower raises two crops, apples and peaches. Each of these crops is shipped to three different outlets. The number of units of crop i that are shipped to outlet j is represented by a_{ij} in the matrix

$$A = \begin{bmatrix} 125 & 100 & 75 \\ 100 & 175 & 125 \end{bmatrix}.$$

The profit per unit is represented by the matrix

$B = [\$3.50 \quad \$6.00]$.

Find the product BA and state what each entry of the product represents.

66. A company manufactures tables and chairs at two locations. Matrix C gives the total cost of manufacturing each product at each location.

$$C = \begin{matrix} \textit{Tables} \\ \textit{Chairs} \end{matrix} \begin{bmatrix} \overset{\textit{Location 1}}{627} & \overset{\textit{Location 2}}{681} \\ 135 & 150 \end{bmatrix}$$

(a) If labor accounts for about $\frac{2}{3}$ of the total cost, determine the matrix L that gives the labor cost for each product at each location. What matrix operation did you use?

(b) Find the matrix M that gives material costs for each product at each location. (Assume there are only labor and material costs.)

True or False? In Exercises 67 and 68, determine whether each statement is true or false. If a statement is true, give a reason or cite an appropriate statement from the text. If a statement is false, provide an example that shows the statement is not true in all cases or cite an appropriate statement from the text.

67. (a) For the product of two matrices to be defined, the number of columns of the first matrix must equal the number of rows of the second matrix.

(b) The system $A\mathbf{x} = \mathbf{b}$ is consistent if and only if $\mathbf{b}$ can be expressed as a linear combination, where the coefficients of the linear combination are a solution of the system.

68. (a) If A is an $m \times n$ matrix and B is an $n \times r$ matrix, then the product AB is an $m \times r$ matrix.

(b) The matrix equation $A\mathbf{x} = \mathbf{b}$, where A is the coefficient matrix and $\mathbf{x}$ and $\mathbf{b}$ are column matrices, can be used to represent a system of linear equations.

69. Writing The matrix

$$P = \begin{array}{r} \\ \textit{From } R \\ \textit{From } D \\ \textit{From } I \end{array} \begin{array}{c} \begin{array}{ccc} \textit{To } R & \textit{To } D & \textit{To } I \end{array} \\ \begin{bmatrix} 0.75 & 0.15 & 0.10 \\ 0.20 & 0.60 & 0.20 \\ 0.30 & 0.40 & 0.30 \end{bmatrix} \end{array}$$

represents the proportions of a voting population that change from party i to party j in a given election. That is, p_{ij} $(i \neq j)$ represents the proportion of the voting population that changes from party i to party j, and p_{ii} represents the proportion that remains loyal to party i from one election to the next. Find the product of P with itself. What does this product represent?

70. The matrices show the numbers of people (in thousands) who lived in various regions of the United States in 2005 and the numbers of people (in thousands) projected to live in those regions in 2015. The regional populations are separated into three age categories. (Source: U.S. Census Bureau)

		2005	
	0–17	18–64	65+
Northeast	12,607	34,418	6286
Midwest	16,131	41,395	7177
South	26,728	63,911	11,689
Mountain	5306	12,679	2020
Pacific	12,524	30,741	4519

		2015	
	0–17	18–64	65+
Northeast	12,441	35,289	8835
Midwest	16,363	42,250	9955
South	29,373	73,496	17,572
Mountain	5263	14,231	3337
Pacific	12,826	33,292	7086

(a) The total population in 2005 was 288,131,000 and the projected total population in 2015 is 321,609,000. Rewrite the matrices to give the information as percents of the total population.

(b) Write a matrix that gives the projected changes in the percents of the population in the various regions and age groups from 2005 to 2015.

(c) Based on the result of part (b), which age group(s) is (are) projected to show relative growth from 2005 to 2015?

In Exercises 71 and 72, perform the indicated block multiplication of matrices A and B. If matrices A and B have each been partitioned into four submatrices

$$A = \begin{bmatrix} A_{11} & A_{12} \\ A_{21} & A_{22} \end{bmatrix} \quad \text{and} \quad B = \begin{bmatrix} B_{11} & B_{12} \\ B_{21} & B_{22} \end{bmatrix},$$

then you can **block multiply** A and B, provided the sizes of the submatrices are such that the matrix multiplications and additions are defined.

$$AB = \begin{bmatrix} A_{11} & A_{12} \\ A_{21} & A_{22} \end{bmatrix} \begin{bmatrix} B_{11} & B_{12} \\ B_{21} & B_{22} \end{bmatrix}$$
$$= \begin{bmatrix} A_{11}B_{11} + A_{12}B_{21} & A_{11}B_{12} + A_{12}B_{22} \\ A_{21}B_{11} + A_{22}B_{21} & A_{21}B_{12} + A_{22}B_{22} \end{bmatrix}$$

71. $A = \left[\begin{array}{cc|cc} 1 & 2 & 0 & 0 \\ 0 & 1 & 0 & 0 \\ \hline 0 & 0 & 2 & 1 \end{array}\right], \quad B = \left[\begin{array}{cc|c} 1 & 2 & 0 \\ -1 & 1 & 0 \\ \hline 0 & 0 & 1 \\ 0 & 0 & 3 \end{array}\right]$

72. $A = \left[\begin{array}{cc|cc} 0 & 0 & 1 & 0 \\ 0 & 0 & 0 & 1 \\ \hline -1 & 0 & 0 & 0 \\ 0 & -1 & 0 & 0 \end{array}\right], \quad B = \left[\begin{array}{cc|cc} 1 & 2 & 3 & 4 \\ 5 & 6 & 7 & 8 \\ \hline 1 & 2 & 3 & 4 \\ 5 & 6 & 7 & 8 \end{array}\right]$

In Exercises 73–76, express the column matrix $\mathbf{b}$ as a linear combination of the columns of A.

73. $A = \begin{bmatrix} 1 & -1 & 2 \\ 3 & -3 & 1 \end{bmatrix}$, $\mathbf{b} = \begin{bmatrix} -1 \\ 7 \end{bmatrix}$

74. $A = \begin{bmatrix} 1 & 2 & 4 \\ -1 & 0 & 2 \\ 0 & 1 & 3 \end{bmatrix}$, $\mathbf{b} = \begin{bmatrix} 1 \\ 3 \\ 2 \end{bmatrix}$

75. $A = \begin{bmatrix} 1 & 1 & -5 \\ 1 & 0 & -1 \\ 2 & -1 & -1 \end{bmatrix}$, $\mathbf{b} = \begin{bmatrix} 3 \\ 1 \\ 0 \end{bmatrix}$

76. $A = \begin{bmatrix} -3 & 5 \\ 3 & 4 \\ 4 & -8 \end{bmatrix}$, $\mathbf{b} = \begin{bmatrix} -22 \\ 4 \\ 32 \end{bmatrix}$

2.2 Properties of Matrix Operations

In Section 2.1 you concentrated on the mechanics of the three basic matrix operations: matrix addition, scalar multiplication, and matrix multiplication. This section begins to develop the **algebra of matrices.** You will see that this algebra shares many (but not all) of the properties of the algebra of real numbers. Several properties of matrix addition and scalar multiplication are listed below.

THEOREM 2.1
Properties of Matrix Addition and Scalar Multiplication

If A, B, and C are $m \times n$ matrices and c and d are scalars, then the following properties are true.

1. $A + B = B + A$ — **Commutative property of addition**
2. $A + (B + C) = (A + B) + C$ — **Associative property of addition**
3. $(cd)A = c(dA)$ — **Associative property of multiplication**
4. $1A = A$ — **Multiplicative identity**
5. $c(A + B) = cA + cB$ — **Distributive property**
6. $(c + d)A = cA + dA$ — **Distributive property**

PROOF

The proofs of these six properties follow directly from the definitions of matrix addition, scalar multiplication, and the corresponding properties of real numbers. For example, to prove the commutative property of *matrix addition,* let $A = [a_{ij}]$ and $B = [b_{ij}]$. Then, using the commutative property of *addition of real numbers,* write

$$A + B = [a_{ij} + b_{ij}] = [b_{ij} + a_{ij}] = B + A.$$

Similarly, to prove Property 5, use the distributive property (for real numbers) of multiplication over addition to write

$$c(A + B) = [c(a_{ij} + b_{ij})] = [ca_{ij} + cb_{ij}] = cA + cB.$$

The proofs of the remaining four properties are left as exercises. (See Exercises 47–50.)

In the preceding section, matrix addition was defined as the sum of *two* matrices, making it a binary operation. The associative property of matrix addition now allows you to write expressions such as $A + B + C$ as $(A + B) + C$ or as $A + (B + C)$. This same reasoning applies to sums of four or more matrices.

EXAMPLE 1 **Addition of More than Two Matrices**

By adding corresponding entries, you can obtain the sum of four matrices shown below.

$$\begin{bmatrix} 1 \\ 2 \\ -3 \end{bmatrix} + \begin{bmatrix} -1 \\ -1 \\ 2 \end{bmatrix} + \begin{bmatrix} 0 \\ 1 \\ 4 \end{bmatrix} + \begin{bmatrix} 2 \\ -3 \\ -2 \end{bmatrix} = \begin{bmatrix} 2 \\ -1 \\ 1 \end{bmatrix}$$

One important property of the addition of real numbers is that the number 0 serves as the additive identity. That is, $c + 0 = c$ for any real number c. For matrices, a similar property holds. Specifically, if A is an $m \times n$ matrix and O_{mn} is the $m \times n$ matrix consisting entirely of zeros, then $A + O_{mn} = A$. The matrix O_{mn} is called a **zero matrix,** and it serves as the **additive identity** for the set of all $m \times n$ matrices. For example, the following matrix serves as the additive identity for the set of all 2×3 matrices.

$$O_{23} = \begin{bmatrix} 0 & 0 & 0 \\ 0 & 0 & 0 \end{bmatrix}$$

When the size of the matrix is understood, you may denote a zero matrix simply by 0.

The following properties of zero matrices are easy to prove, and their proofs are left as an exercise. (See Exercise 51.)

THEOREM 2.2 Properties of Zero Matrices

If A is an $m \times n$ matrix and c is a scalar, then the following properties are true.

1. $A + O_{mn} = A$
2. $A + (-A) = O_{mn}$
3. If $cA = O_{mn}$, then $c = 0$ or $A = O_{mn}$.

REMARK: Property 2 can be described by saying that matrix $-A$ is the **additive inverse** of A.

The algebra of real numbers and the algebra of matrices have many similarities. For example, compare the solutions below.

Real Numbers (Solve for x.)	$m \times n$ Matrices (Solve for X.)
$x + a = b$	$X + A = B$
$x + a + (-a) = b + (-a)$	$X + A + (-A) = B + (-A)$
$x + 0 = b - a$	$X + O = B - A$
$x = b - a$	$X = B - A$

The process of solving a matrix equation is demonstrated in Example 2.

EXAMPLE 2 **Solving a Matrix Equation**

Solve for X in the equation $3X + A = B$, where

$$A = \begin{bmatrix} 1 & -2 \\ 0 & 3 \end{bmatrix} \quad \text{and} \quad B = \begin{bmatrix} -3 & 4 \\ 2 & 1 \end{bmatrix}.$$

SOLUTION Begin by solving the equation for X to obtain

$$3X = B - A \quad \Longrightarrow \quad X = \tfrac{1}{3}(B - A).$$

Now, using the given matrices A and B, you have

$$X = \frac{1}{3}\left(\begin{bmatrix} -3 & 4 \\ 2 & 1 \end{bmatrix} - \begin{bmatrix} 1 & -2 \\ 0 & 3 \end{bmatrix}\right) = \frac{1}{3}\begin{bmatrix} -4 & 6 \\ 2 & -2 \end{bmatrix} = \begin{bmatrix} -\frac{4}{3} & 2 \\ \frac{2}{3} & -\frac{2}{3} \end{bmatrix}.$$

Properties of Matrix Multiplication

In the next theorem, the algebra of matrices is extended to include some useful properties of matrix multiplication. The proof of Property 2 is presented below. The proofs of the remaining properties are left as an exercise. (See Exercise 52.)

THEOREM 2.3 Properties of Matrix Multiplication

If A, B, and C are matrices (with sizes such that the given matrix products are defined) and c is a scalar, then the following properties are true.

1. $A(BC) = (AB)C$ **Associative property of multiplication**
2. $A(B + C) = AB + AC$ **Distributive property**
3. $(A + B)C = AC + BC$ **Distributive property**
4. $c(AB) = (cA)B = A(cB)$

PROOF To prove Property 2, show that the matrices $A(B + C)$ and $AB + AC$ are equal by showing that their corresponding entries are equal. Assume A has size $m \times n$, B has size $n \times p$, and C has size $n \times p$. Using the definition of matrix multiplication, the entry in the ith row and jth column of $A(B + C)$ is $a_{i1}(b_{1j} + c_{1j}) + \cdots + a_{in}(b_{nj} + c_{nj})$. Moreover, the entry in the ith row and jth column of $AB + AC$ is

$$(a_{i1}b_{1j} + \cdots + a_{in}b_{nj}) + (a_{i1}c_{1j} + \cdots + a_{in}c_{nj}).$$

By distributing and regrouping, you can see that these two ijth entries are equal. So,

$$A(B + C) = AB + AC.$$

The associative property of matrix multiplication permits you to write such matrix products as ABC without ambiguity, as demonstrated in Example 3.

EXAMPLE 3 **Matrix Multiplication Is Associative**

Find the matrix product ABC by grouping the factors first as $(AB)C$ and then as $A(BC)$. Show that the same result is obtained from both processes.

$$A = \begin{bmatrix} 1 & -2 \\ 2 & -1 \end{bmatrix}, \quad B = \begin{bmatrix} 1 & 0 & 2 \\ 3 & -2 & 1 \end{bmatrix}, \quad C = \begin{bmatrix} -1 & 0 \\ 3 & 1 \\ 2 & 4 \end{bmatrix}$$

SOLUTION Grouping the factors as $(AB)C$, you have

$$(AB)C = \left(\begin{bmatrix} 1 & -2 \\ 2 & -1 \end{bmatrix}\begin{bmatrix} 1 & 0 & 2 \\ 3 & -2 & 1 \end{bmatrix}\right)\begin{bmatrix} -1 & 0 \\ 3 & 1 \\ 2 & 4 \end{bmatrix}$$

$$= \begin{bmatrix} -5 & 4 & 0 \\ -1 & 2 & 3 \end{bmatrix}\begin{bmatrix} -1 & 0 \\ 3 & 1 \\ 2 & 4 \end{bmatrix} = \begin{bmatrix} 17 & 4 \\ 13 & 14 \end{bmatrix}.$$

Grouping the factors as $A(BC)$, you obtain the same result.

$$A(BC) = \begin{bmatrix} 1 & -2 \\ 2 & -1 \end{bmatrix}\left(\begin{bmatrix} 1 & 0 & 2 \\ 3 & -2 & 1 \end{bmatrix}\begin{bmatrix} -1 & 0 \\ 3 & 1 \\ 2 & 4 \end{bmatrix}\right)$$

$$= \begin{bmatrix} 1 & -2 \\ 2 & -1 \end{bmatrix}\begin{bmatrix} 3 & 8 \\ -7 & 2 \end{bmatrix} = \begin{bmatrix} 17 & 4 \\ 13 & 14 \end{bmatrix}$$

Note that no commutative property for matrix multiplication is listed in Theorem 2.3. Although the product AB is defined, it can easily happen that A and B are not of the proper sizes to define the product BA. For instance, if A is of size 2×3 and B is of size 3×3, then the product AB is defined but the product BA is not. The next example shows that even if both products AB and BA are defined, they may not be equal.

EXAMPLE 4 **Noncommutativity of Matrix Multiplication**

Show that AB and BA are not equal for the matrices

$$A = \begin{bmatrix} 1 & 3 \\ 2 & -1 \end{bmatrix} \quad \text{and} \quad B = \begin{bmatrix} 2 & -1 \\ 0 & 2 \end{bmatrix}.$$

SOLUTION

$$AB = \begin{bmatrix} 1 & 3 \\ 2 & -1 \end{bmatrix}\begin{bmatrix} 2 & -1 \\ 0 & 2 \end{bmatrix} = \begin{bmatrix} 2 & 5 \\ 4 & -4 \end{bmatrix}$$

$$BA = \begin{bmatrix} 2 & -1 \\ 0 & 2 \end{bmatrix}\begin{bmatrix} 1 & 3 \\ 2 & -1 \end{bmatrix} = \begin{bmatrix} 0 & 7 \\ 4 & -2 \end{bmatrix}$$

$$AB \neq BA$$

Do not conclude from Example 4 that the matrix products AB and BA are *never* the same. Sometimes they are the same. For example, try multiplying the following matrices, first in the order AB and then in the order BA.

$$A = \begin{bmatrix} 1 & 2 \\ 1 & 1 \end{bmatrix} \quad \text{and} \quad B = \begin{bmatrix} -2 & 4 \\ 2 & -2 \end{bmatrix}$$

You will see that the two products are equal. The point is this: Although AB and BA are sometimes equal, AB and BA are usually not equal.

Another important quality of matrix algebra is that it does not have a general cancellation property for matrix multiplication. That is, if $AC = BC$, it is not necessarily true that $A = B$. This is demonstrated in Example 5. (In the next section you will see that, for some special types of matrices, cancellation is valid.)

EXAMPLE 5 An Example in Which Cancellation Is Not Valid

Show that $AC = BC$.

$$A = \begin{bmatrix} 1 & 3 \\ 0 & 1 \end{bmatrix}, \quad B = \begin{bmatrix} 2 & 4 \\ 2 & 3 \end{bmatrix}, \quad C = \begin{bmatrix} 1 & -2 \\ -1 & 2 \end{bmatrix}$$

SOLUTION

$$AC = \begin{bmatrix} 1 & 3 \\ 0 & 1 \end{bmatrix}\begin{bmatrix} 1 & -2 \\ -1 & 2 \end{bmatrix} = \begin{bmatrix} -2 & 4 \\ -1 & 2 \end{bmatrix}$$

$$BC = \begin{bmatrix} 2 & 4 \\ 2 & 3 \end{bmatrix}\begin{bmatrix} 1 & -2 \\ -1 & 2 \end{bmatrix} = \begin{bmatrix} -2 & 4 \\ -1 & 2 \end{bmatrix}$$

$AC = BC$, even though $A \neq B$.

You will now look at a special type of *square* matrix that has 1's on the main diagonal and 0's elsewhere.

$$I_n = \begin{bmatrix} 1 & 0 & 0 & \cdots & 0 \\ 0 & 1 & 0 & \cdots & 0 \\ 0 & 0 & 1 & \cdots & 0 \\ \vdots & \vdots & \vdots & & \vdots \\ 0 & 0 & 0 & \cdots & 1 \end{bmatrix}$$

For instance, if $n = 1, 2,$ or 3, we have

$$\underset{1 \times 1}{I_1 = [1]}, \quad \underset{2 \times 2}{I_2 = \begin{bmatrix} 1 & 0 \\ 0 & 1 \end{bmatrix}}, \quad \underset{3 \times 3}{I_3 = \begin{bmatrix} 1 & 0 & 0 \\ 0 & 1 & 0 \\ 0 & 0 & 1 \end{bmatrix}}.$$

When the order of the matrix is understood to be n, you may denote I_n simply as I.

As stated in Theorem 2.4 on the next page, the matrix I_n serves as the **identity** for matrix multiplication; it is called the **identity matrix of order *n*.** The proof of this theorem is left as an exercise. (See Exercise 53.)

THEOREM 2.4 Properties of the Identity Matrix

If A is a matrix of size $m \times n$, then the following properties are true.

1. $AI_n = A$
2. $I_mA = A$

As a special case of this theorem, note that if A is a *square* matrix of order n, then

$$AI_n = I_nA = A.$$

EXAMPLE 6 Multiplication by an Identity Matrix

(a) $$\begin{bmatrix} 3 & -2 \\ 4 & 0 \\ -1 & 1 \end{bmatrix}\begin{bmatrix} 1 & 0 \\ 0 & 1 \end{bmatrix} = \begin{bmatrix} 3 & -2 \\ 4 & 0 \\ -1 & 1 \end{bmatrix}$$

(b) $$\begin{bmatrix} 1 & 0 & 0 \\ 0 & 1 & 0 \\ 0 & 0 & 1 \end{bmatrix}\begin{bmatrix} -2 \\ 1 \\ 4 \end{bmatrix} = \begin{bmatrix} -2 \\ 1 \\ 4 \end{bmatrix}$$

For repeated multiplication of *square* matrices, you can use the same exponential notation used with real numbers. That is, $A^1 = A$, $A^2 = AA$, and for a positive integer k, A^k is

$$A^k = \underbrace{AA \cdot \cdot \cdot A}_{k \text{ factors}}.$$

It is convenient also to define $A^0 = I_n$ (where A is a square matrix of order n). These definitions allow you to establish the properties

1. $A^jA^k = A^{j+k}$ and 2. $(A^j)^k = A^{jk}$

where j and k are nonnegative integers.

EXAMPLE 7 Repeated Multiplication of a Square Matrix

Find A^3 for the matrix $A = \begin{bmatrix} 2 & -1 \\ 3 & 0 \end{bmatrix}$.

SOLUTION $$A^3 = \left(\begin{bmatrix} 2 & -1 \\ 3 & 0 \end{bmatrix}\begin{bmatrix} 2 & -1 \\ 3 & 0 \end{bmatrix}\right)\begin{bmatrix} 2 & -1 \\ 3 & 0 \end{bmatrix} = \begin{bmatrix} 1 & -2 \\ 6 & -3 \end{bmatrix}\begin{bmatrix} 2 & -1 \\ 3 & 0 \end{bmatrix} = \begin{bmatrix} -4 & -1 \\ 3 & -6 \end{bmatrix}$$

In Section 1.1 you saw that a system of linear equations must have exactly one solution, an infinite number of solutions, or no solution. Using the matrix algebra developed so far, you can now prove that this is true.

THEOREM 2.5
Number of Solutions of a System of Linear Equations

For a system of linear equations in n variables, precisely one of the following is true.

1. The system has exactly one solution.
2. The system has an infinite number of solutions.
3. The system has no solution.

PROOF

Represent the system by the matrix equation $A\mathbf{x} = \mathbf{b}$. If the system has exactly one solution or no solution, then there is nothing to prove. So, you can assume that the system has at least two distinct solutions $\mathbf{x}_1$ and $\mathbf{x}_2$. The proof will be complete if you can show that this assumption implies that the system has an infinite number of solutions. Because $\mathbf{x}_1$ and $\mathbf{x}_2$ are solutions, you have $A\mathbf{x}_1 = A\mathbf{x}_2 = \mathbf{b}$ and $A(\mathbf{x}_1 - \mathbf{x}_2) = O$. This implies that the (nonzero) column matrix $\mathbf{x}_h = \mathbf{x}_1 - \mathbf{x}_2$ is a solution of the homogeneous system of linear equations $A\mathbf{x} = O$. It can now be said that for any scalar c,

$$A(\mathbf{x}_1 + c\mathbf{x}_h) = A\mathbf{x}_1 + A(c\mathbf{x}_h) = \mathbf{b} + c(A\mathbf{x}_h) = \mathbf{b} + cO = \mathbf{b}.$$

So $\mathbf{x}_1 + c\mathbf{x}_h$ is a solution of $A\mathbf{x} = \mathbf{b}$ for any scalar c. Because there are an infinite number of possible values of c and each value produces a different solution, you can conclude that the system has an infinite number of solutions.

The Transpose of a Matrix

The **transpose** of a matrix is formed by writing its columns as rows. For instance, if A is the $m \times n$ matrix shown by

$$A = \begin{bmatrix} a_{11} & a_{12} & a_{13} & \cdots & a_{1n} \\ a_{21} & a_{22} & a_{23} & \cdots & a_{2n} \\ a_{31} & a_{32} & a_{33} & \cdots & a_{3n} \\ \vdots & \vdots & \vdots & & \vdots \\ a_{m1} & a_{m2} & a_{m3} & \cdots & a_{mn} \end{bmatrix},$$

Size: $m \times n$

then the transpose, denoted by A^T, is the $n \times m$ matrix below

$$A^T = \begin{bmatrix} a_{11} & a_{21} & a_{31} & \cdots & a_{m1} \\ a_{12} & a_{22} & a_{32} & \cdots & a_{m2} \\ a_{13} & a_{23} & a_{33} & \cdots & a_{m3} \\ \vdots & \vdots & \vdots & & \vdots \\ a_{1n} & a_{2n} & a_{3n} & \cdots & a_{mn} \end{bmatrix}.$$

Size: $n \times m$

EXAMPLE 8 **The Transpose of a Matrix**

Find the transpose of each matrix.

(a) $A = \begin{bmatrix} 2 \\ 8 \end{bmatrix}$ (b) $B = \begin{bmatrix} 1 & 2 & 3 \\ 4 & 5 & 6 \\ 7 & 8 & 9 \end{bmatrix}$ (c) $C = \begin{bmatrix} 1 & 2 & 0 \\ 2 & 1 & 0 \\ 0 & 0 & 1 \end{bmatrix}$ (d) $D = \begin{bmatrix} 0 & 1 \\ 2 & 4 \\ 1 & -1 \end{bmatrix}$

SOLUTION (a) $A^T = \begin{bmatrix} 2 & 8 \end{bmatrix}$ (b) $B^T = \begin{bmatrix} 1 & 4 & 7 \\ 2 & 5 & 8 \\ 3 & 6 & 9 \end{bmatrix}$ (c) $C^T = \begin{bmatrix} 1 & 2 & 0 \\ 2 & 1 & 0 \\ 0 & 0 & 1 \end{bmatrix}$

(d) $D^T = \begin{bmatrix} 0 & 2 & 1 \\ 1 & 4 & -1 \end{bmatrix}$

Discovery

Let $A = \begin{bmatrix} 1 & 2 \\ 3 & 4 \end{bmatrix}$ *and* $B = \begin{bmatrix} 3 & 5 \\ 1 & -1 \end{bmatrix}$.

Calculate $(AB)^T$, A^TB^T, *and* B^TA^T. *Make a conjecture about the transpose of a product of two square matrices. Select two other square matrices to check your conjecture.*

REMARK: Note that the square matrix in part (c) of Example 8 is equal to its transpose. Such a matrix is called **symmetric.** A matrix A is symmetric if $A = A^T$. From this definition it is clear that a symmetric matrix must be square. Also, if $A = [a_{ij}]$ is a symmetric matrix, then $a_{ij} = a_{ji}$ for all $i \neq j$.

THEOREM 2.6 Properties of Transposes

If A and B are matrices (with sizes such that the given matrix operations are defined) and c is a scalar, then the following properties are true.

1. $(A^T)^T = A$ — Transpose of a transpose
2. $(A + B)^T = A^T + B^T$ — Transpose of a sum
3. $(cA)^T = c(A^T)$ — Transpose of a scalar multiple
4. $(AB)^T = B^TA^T$ — Transpose of a product

PROOF Because the transpose operation interchanges rows and columns, Property 1 seems to make sense. To prove Property 1, let A be an $m \times n$ matrix. Observe that A^T has size $n \times m$ and $(A^T)^T$ has size $m \times n$, the same as A. To show that $(A^T)^T = A$ you must show that the ijth entries are the same. Let a_{ij} be the ijth entry of A. Then a_{ij} is the jith entry of A^T, and the ijth entry of $(A^T)^T$. This proves Property 1. The proofs of the remaining properties are left as an exercise. (See Exercise 54.)

REMARK: Remember that you *reverse the order* of multiplication when forming the transpose of a product. That is, the transpose of AB is $(AB)^T = B^TA^T$ and is *not* usually equal to A^TB^T.

Properties 2 and 4 can be generalized to cover sums or products of any finite number of matrices. For instance, the transpose of the sum of three matrices is

$$(A + B + C)^T = A^T + B^T + C^T,$$

and the transpose of the product of three matrices is

$$(ABC)^T = C^TB^TA^T.$$

EXAMPLE 9 **Finding the Transpose of a Product**

Show that $(AB)^T$ and B^TA^T are equal.

$$A = \begin{bmatrix} 2 & 1 & -2 \\ -1 & 0 & 3 \\ 0 & -2 & 1 \end{bmatrix} \quad \text{and} \quad B = \begin{bmatrix} 3 & 1 \\ 2 & -1 \\ 3 & 0 \end{bmatrix}$$

SOLUTION

$$AB = \begin{bmatrix} 2 & 1 & -2 \\ -1 & 0 & 3 \\ 0 & -2 & 1 \end{bmatrix} \begin{bmatrix} 3 & 1 \\ 2 & -1 \\ 3 & 0 \end{bmatrix} = \begin{bmatrix} 2 & 1 \\ 6 & -1 \\ -1 & 2 \end{bmatrix}$$

$$(AB)^T = \begin{bmatrix} 2 & 6 & -1 \\ 1 & -1 & 2 \end{bmatrix}$$

$$B^TA^T = \begin{bmatrix} 3 & 2 & 3 \\ 1 & -1 & 0 \end{bmatrix} \begin{bmatrix} 2 & -1 & 0 \\ 1 & 0 & -2 \\ -2 & 3 & 1 \end{bmatrix} = \begin{bmatrix} 2 & 6 & -1 \\ 1 & -1 & 2 \end{bmatrix}$$

$$(AB)^T = B^TA^T$$

EXAMPLE 10 **The Product of a Matrix and Its Transpose**

For the matrix

$$A = \begin{bmatrix} 1 & 3 \\ 0 & -2 \\ -2 & -1 \end{bmatrix}$$

find the product AA^T and show that it is symmetric.

SOLUTION Because

$$AA^T = \begin{bmatrix} 1 & 3 \\ 0 & -2 \\ -2 & -1 \end{bmatrix} \begin{bmatrix} 1 & 0 & -2 \\ 3 & -2 & -1 \end{bmatrix} = \begin{bmatrix} 10 & -6 & -5 \\ -6 & 4 & 2 \\ -5 & 2 & 5 \end{bmatrix}$$

it follows that $AA^T = (AA^T)^T$, so AA^T is symmetric.

REMARK: The property demonstrated in Example 10 is true in general. That is, for any matrix A, the matrix given by $B = AA^T$ is symmetric. You are asked to prove this result in Exercise 55.

SECTION 2.2 Exercises

In Exercises 1–6, perform the indicated operations when $a = 3$, $b = -4$, and

$$A = \begin{bmatrix} 1 & 2 \\ 3 & 4 \end{bmatrix}, \quad B = \begin{bmatrix} 0 & 1 \\ -1 & 2 \end{bmatrix}, \quad O = \begin{bmatrix} 0 & 0 \\ 0 & 0 \end{bmatrix}.$$

1. $aA + bB$ **2.** $A + B$ **3.** $ab(B)$

4. $(a + b)B$ **5.** $(a - b)(A - B)$ **6.** $(ab)O$

7. Solve for X when

$$A = \begin{bmatrix} -4 & 0 \\ 1 & -5 \\ -3 & 2 \end{bmatrix} \text{ and } B = \begin{bmatrix} 1 & 2 \\ -2 & 1 \\ 4 & 4 \end{bmatrix}.$$

(a) $3X + 2A = B$ (b) $2A - 5B = 3X$

(c) $X - 3A + 2B = O$ (d) $6X - 4A - 3B = O$

8. Solve for X when

$$A = \begin{bmatrix} -2 & -1 \\ 1 & 0 \\ 3 & -4 \end{bmatrix} \text{ and } B = \begin{bmatrix} 0 & 3 \\ 2 & 0 \\ -4 & -1 \end{bmatrix}.$$

(a) $X = 3A - 2B$ (b) $2X = 2A - B$

(c) $2X + 3A = B$ (d) $2A + 4B = -2X$

In Exercises 9–14, perform the indicated operations, provided that $c = -2$ and

$$A = \begin{bmatrix} 1 & 2 & 3 \\ 0 & 1 & -1 \end{bmatrix}, \quad B = \begin{bmatrix} 1 & 3 \\ -1 & 2 \end{bmatrix}, \quad C = \begin{bmatrix} 0 & 1 \\ -1 & 0 \end{bmatrix},$$

$$O = \begin{bmatrix} 0 & 0 \\ 0 & 0 \end{bmatrix}.$$

9. $B(CA)$ **10.** $C(BC)$ **11.** $(B + C)A$

12. $B(C + O)$ **13.** $(cB)(C + C)$ **14.** $B(cA)$

In Exercises 15 and 16, demonstrate that if $AC = BC$, then A is *not* necessarily equal to B for the following matrices.

15. $A = \begin{bmatrix} 0 & 1 \\ 0 & 1 \end{bmatrix}, B = \begin{bmatrix} 1 & 0 \\ 1 & 0 \end{bmatrix}, C = \begin{bmatrix} 2 & 3 \\ 2 & 3 \end{bmatrix}$

16. $A = \begin{bmatrix} 1 & 2 & 3 \\ 0 & 5 & 4 \\ 3 & -2 & 1 \end{bmatrix}, B = \begin{bmatrix} 4 & -6 & 3 \\ 5 & 4 & 4 \\ -1 & 0 & 1 \end{bmatrix},$

$C = \begin{bmatrix} 0 & 0 & 0 \\ 0 & 0 & 0 \\ 4 & -2 & 3 \end{bmatrix}$

In Exercises 17 and 18, demonstrate that if $AB = O$, then it is *not* necessarily true that $A = O$ or $B = O$ for the following matrices.

17. $A = \begin{bmatrix} 3 & 3 \\ 4 & 4 \end{bmatrix}$ and $B = \begin{bmatrix} 1 & -1 \\ -1 & 1 \end{bmatrix}$

18. $A = \begin{bmatrix} 2 & 4 \\ 2 & 4 \end{bmatrix}$ and $B = \begin{bmatrix} 1 & -2 \\ -\frac{1}{2} & 1 \end{bmatrix}$

In Exercises 19–22, perform the indicated operations when

$$A = \begin{bmatrix} 1 & 2 \\ 0 & -1 \end{bmatrix} \text{ and } I = \begin{bmatrix} 1 & 0 \\ 0 & 1 \end{bmatrix}.$$

19. A^2 **20.** A^4

21. $A(I + A)$ **22.** $A + IA$

In Exercises 23–28, find (a) A^T, (b) A^TA, and (c) AA^T.

23. $A = \begin{bmatrix} 4 & 2 & 1 \\ 0 & 2 & -1 \end{bmatrix}$ **24.** $A = \begin{bmatrix} 1 & -1 \\ 3 & 4 \\ 0 & -2 \end{bmatrix}$

25. $A = \begin{bmatrix} 2 & 1 & -3 \\ 1 & 4 & 1 \\ 0 & 2 & 1 \end{bmatrix}$ **26.** $A = \begin{bmatrix} -7 & 11 & 12 \\ 4 & -3 & 1 \\ 6 & -1 & 3 \end{bmatrix}$

27. $A = \begin{bmatrix} 0 & -4 & 3 & 2 \\ 8 & 4 & 0 & 1 \\ -2 & 3 & 5 & 1 \\ 0 & 0 & -3 & 2 \end{bmatrix}$

28. $A = \begin{bmatrix} 4 & -3 & 2 & 0 \\ 2 & 0 & 11 & -1 \\ -1 & -2 & 0 & 3 \\ 14 & -2 & 12 & -9 \\ 6 & 8 & -5 & 4 \end{bmatrix}$

Writing In Exercises 29 and 30, explain why the formula is *not* valid for matrices. Illustrate your argument with examples.

29. $(A + B)(A - B) = A^2 - B^2$

30. $(A + B)(A + B) = A^2 + 2AB + B^2$

In Exercises 31–34, verify that $(AB)^T = B^TA^T$.

31. $A = \begin{bmatrix} -1 & 1 & -2 \\ 2 & 0 & 1 \end{bmatrix}$ and $B = \begin{bmatrix} -3 & 0 \\ 1 & 2 \\ 1 & -1 \end{bmatrix}$

32. $A = \begin{bmatrix} 1 & 2 \\ 0 & -2 \end{bmatrix}$ and $B = \begin{bmatrix} -3 & -1 \\ 2 & 1 \end{bmatrix}$

33. $A = \begin{bmatrix} 2 & 1 \\ 0 & 1 \\ -2 & 1 \end{bmatrix}$ and $B = \begin{bmatrix} 2 & 3 & 1 \\ 0 & 4 & -1 \end{bmatrix}$

34. $A = \begin{bmatrix} 2 & 1 & -1 \\ 0 & 1 & 3 \\ 4 & 0 & 2 \end{bmatrix}$ and $B = \begin{bmatrix} 1 & 0 & -1 \\ 2 & 1 & -2 \\ 0 & 1 & 3 \end{bmatrix}$

True or False? In Exercises 35 and 36, determine whether each statement is true or false. If a statement is true, give a reason or cite an appropriate statement from the text. If a statement is false, provide an example that shows the statement is not true in all cases or cite an appropriate statement from the text.

35. (a) Matrix addition is commutative.

(b) Matrix multiplication is associative.

(c) The transpose of the product of two matrices equals the product of their transposes; that is, $(AB)^T = A^TB^T$.

(d) For any matrix C, the matrix CC^T is symmetric.

36. (a) Matrix multiplication is commutative.

(b) Every matrix A has an additive inverse.

(c) If the matrices A, B, and C satisfy $AB = AC$, then $B = C$.

(d) The transpose of the sum of two matrices equals the sum of their transposes.

37. Consider the matrices shown below.

$$X = \begin{bmatrix} 1 \\ 0 \\ 1 \end{bmatrix}, \quad Y = \begin{bmatrix} 1 \\ 1 \\ 0 \end{bmatrix}, \quad Z = \begin{bmatrix} 2 \\ -1 \\ 3 \end{bmatrix}, \quad W = \begin{bmatrix} 1 \\ 1 \\ 1 \end{bmatrix}, \quad O = \begin{bmatrix} 0 \\ 0 \\ 0 \end{bmatrix}$$

(a) Find scalars a and b such that $Z = aX + bY$.

(b) Show that there do not exist scalars a and b such that $W = aX + bY$.

(c) Show that if $aX + bY + cW = O$, then $a = b = c = 0$.

(d) Find scalars a, b, and c, not all equal to zero, such that $aX + bY + cZ = O$.

38. Consider the matrices shown below.

$$X = \begin{bmatrix} 1 \\ 2 \\ 3 \end{bmatrix}, \quad Y = \begin{bmatrix} 1 \\ 0 \\ 2 \end{bmatrix}, \quad Z = \begin{bmatrix} 1 \\ 4 \\ 4 \end{bmatrix}, \quad W = \begin{bmatrix} 0 \\ 0 \\ 1 \end{bmatrix}, \quad O = \begin{bmatrix} 0 \\ 0 \\ 0 \end{bmatrix}$$

(a) Find scalars a and b such that $Z = aX + bY$.

(b) Show that there do not exist scalars a and b such that $W = aX + bY$.

(c) Show that if $aX + bY + cW = O$, then $a = b = c = 0$.

(d) Find scalars a, b, and c, not all equal to zero, such that $aX + bY + cZ = O$.

In Exercises 39 and 40, compute the power of A for the matrix

$$A = \begin{bmatrix} 1 & 0 & 0 \\ 0 & -1 & 0 \\ 0 & 0 & 1 \end{bmatrix}.$$

39. A^{19} **40.** A^{20}

An nth root of a matrix B is a matrix A such that $A^n = B$. In Exercises 41 and 42, find the nth root of the matrix B.

41. $B = \begin{bmatrix} 9 & 0 \\ 0 & 4 \end{bmatrix}, \quad n = 2$ **42.** $B = \begin{bmatrix} 8 & 0 & 0 \\ 0 & -1 & 0 \\ 0 & 0 & 27 \end{bmatrix}, \quad n = 3$

In Exercises 43–46, use the given definition to find $f(A)$: If f is the polynomial function,

$f(x) = a_0 + a_1x + a_2x^2 + \cdots + a_nx^n$,

then for an $n \times n$ matrix A, $f(A)$ is defined to be

$f(A) = a_0I_n + a_1A + a_2A^2 + \cdots + a_nA^n$.

43. $f(x) = x^2 - 5x + 2, \quad A = \begin{bmatrix} 2 & 0 \\ 4 & 5 \end{bmatrix}$

44. $f(x) = x^2 - 7x + 6, \quad A = \begin{bmatrix} 5 & 4 \\ 1 & 2 \end{bmatrix}$

45. $f(x) = x^2 - 3x + 2, \quad A = \begin{bmatrix} 2 & 1 \\ -1 & 0 \end{bmatrix}$

46. $f(x) = x^3 - 2x^2 + 5x - 10, \quad A = \begin{bmatrix} 2 & 1 & -1 \\ 1 & 0 & 2 \\ -1 & 1 & 3 \end{bmatrix}$

47. Guided Proof Prove the associative property of matrix addition: $A + (B + C) = (A + B) + C$.

Getting Started: To prove that $A + (B + C)$ and $(A + B) + C$ are equal, show that their corresponding entries are the same.

(i) Begin your proof by letting A, B, and C be $m \times n$ matrices.

(ii) Observe that the ijth entry of $B + C$ is $b_{ij} + c_{ij}$.

(iii) Furthermore, the ijth entry of $A + (B + C)$ is $a_{ij} + (b_{ij} + c_{ij})$.

(iv) Determine the ijth entry of $(A + B) + C$.

48. Prove the associative property of scalar multiplication: $(cd)A = c(dA)$.

49. Prove that the scalar 1 is the identity for scalar multiplication: $1A = A$.

50. Prove the following distributive property: $(c + d)A = cA + dA$.

51. Prove Theorem 2.2.

52. Complete the proof of Theorem 2.3.

(a) Prove the associative property of multiplication: $A(BC) = (AB)C$.

(b) Prove the distributive property: $(A + B)C = AC + BC$.

(c) Prove the property: $c(AB) = (cA)B = A(cB)$.

53. Prove Theorem 2.4.

54. Prove Properties 2, 3, and 4 of Theorem 2.6.

55. Guided Proof Prove that if A is an $m \times n$ matrix, then AA^T and A^TA are symmetric matrices.

Getting Started: To prove that AA^T is symmetric, you need to show that it is equal to its transpose, $(AA^T)^T = AA^T$.

(i) Begin your proof with the left-hand matrix expression $(AA^T)^T$.

(ii) Use the properties of the transpose operation to show that it can be simplified to equal the right-hand expression, AA^T.

(iii) Repeat this analysis for the product A^TA.

56. Give an example of two 2×2 matrices A and B such that $(AB)^T \neq A^TB^T$.

In Exercises 57–60, determine whether the matrix is symmetric, skew-symmetric, or neither. A square matrix is called **skew-symmetric** if $A^T = -A$.

57. $A = \begin{bmatrix} 0 & 2 \\ -2 & 0 \end{bmatrix}$ **58.** $A = \begin{bmatrix} 2 & 1 \\ 1 & 3 \end{bmatrix}$

59. $A = \begin{bmatrix} 0 & 2 & 1 \\ 2 & 0 & 3 \\ 1 & 3 & 0 \end{bmatrix}$ **60.** $A = \begin{bmatrix} 0 & 2 & -1 \\ -2 & 0 & -3 \\ 1 & 3 & 0 \end{bmatrix}$

61. Prove that the main diagonal of a skew-symmetric matrix consists entirely of zeros.

62. Prove that if A and B are $n \times n$ skew-symmetric matrices, then $A + B$ is skew-symmetric.

63. Let A be a square matrix of order n.

(a) Show that $\frac{1}{2}(A + A^T)$ is symmetric.

(b) Show that $\frac{1}{2}(A - A^T)$ is skew-symmetric.

(c) Prove that A can be written as the sum of a symmetric matrix B and a skew-symmetric matrix C, $A = B + C$.

(d) Write the matrix

$$A = \begin{bmatrix} 2 & 5 & 3 \\ -3 & 6 & 0 \\ 4 & 1 & 1 \end{bmatrix}$$

as the sum of a skew-symmetric matrix and a symmetric matrix.

64. Prove that if A is an $n \times n$ matrix, then $A - A^T$ is skew-symmetric.

65. Let A and B be two $n \times n$ symmetric matrices.

(a) Give an example to show that the product AB is not necessarily symmetric.

(b) Prove that AB is symmetric if and only if $AB = BA$.

66. Consider matrices of the form

$$A = \begin{bmatrix} 0 & a_{12} & a_{13} & a_{14} & \cdots & a_{1n} \\ 0 & 0 & a_{23} & a_{24} & \cdots & a_{2n} \\ 0 & 0 & 0 & a_{34} & \cdots & a_{3n} \\ \vdots & \vdots & \vdots & \vdots & & \vdots \\ 0 & 0 & 0 & 0 & \cdots & a_{(n-1)n} \\ 0 & 0 & 0 & 0 & \cdots & 0 \end{bmatrix}.$$

(a) Write a 2×2 matrix and a 3×3 matrix in the form of A.

(b) Use a graphing utility or computer software program to raise each of the matrices to higher powers. Describe the result.

(c) Use the result of part (b) to make a conjecture about powers of A if A is a 4×4 matrix. Use a graphing utility to test your conjecture.

(d) Use the results of parts (b) and (c) to make a conjecture about powers of A if A is an $n \times n$ matrix.

2.3 The Inverse of a Matrix

Section 2.2 discussed some of the similarities between the algebra of real numbers and the algebra of matrices. This section further develops the algebra of matrices to include the solutions of matrix equations involving matrix multiplication. To begin, consider the real number equation $ax = b$. To solve this equation for x, multiply both sides of the equation by a^{-1} (provided $a \neq 0$).

$$\begin{aligned} ax &= b \\ (a^{-1}a)x &= a^{-1}b \\ (1)x &= a^{-1}b \\ x &= a^{-1}b \end{aligned}$$

The number a^{-1} is called the *multiplicative inverse* of a because $a^{-1}a$ yields 1 (the identity element for multiplication). The definition of a multiplicative inverse of a matrix is similar.

Definition of the Inverse of a Matrix

An $n \times n$ matrix A is **invertible** (or **nonsingular**) if there exists an $n \times n$ matrix B such that

$$AB = BA = I_n$$

where I_n is the identity matrix of order n. The matrix B is called the (multiplicative) **inverse** of A. A matrix that does not have an inverse is called **noninvertible** (or **singular**).

Nonsquare matrices do not have inverses. To see this, note that if A is of size $m \times n$ and B is of size $n \times m$ (where $m \neq n$), then the products AB and BA are of different sizes and cannot be equal to each other. Indeed, not all square matrices possess inverses. (See Example 4.) The next theorem, however, tells you that if a matrix *does* possess an inverse, then that inverse is unique.

THEOREM 2.7
Uniqueness of an Inverse Matrix

If A is an invertible matrix, then its inverse is unique. The inverse of A is denoted by A^{-1}.

PROOF Because A is invertible, you know it has at least one inverse B such that

$$AB = I = BA.$$

Suppose A has another inverse C such that

$$AC = I = CA.$$

Then you can show that B and C are equal, as follows.

$$\begin{aligned} AB &= I \\ C(AB) &= CI \\ (CA)B &= C \\ IB &= C \\ B &= C \end{aligned}$$

Consequently $B = C$, and it follows that the inverse of a matrix is unique.

Because the inverse A^{-1} of an invertible matrix A is unique, you can call it *the* inverse of A and write

$$AA^{-1} = A^{-1}A = I.$$

EXAMPLE 1 The Inverse of a Matrix

Show that B is the inverse of A, where

$$A = \begin{bmatrix} -1 & 2 \\ -1 & 1 \end{bmatrix} \quad \text{and} \quad B = \begin{bmatrix} 1 & -2 \\ 1 & -1 \end{bmatrix}.$$

SOLUTION Using the definition of an inverse matrix, you can show that B is the inverse of A by showing that $AB = I = BA$, as follows.

$$AB = \begin{bmatrix} -1 & 2 \\ -1 & 1 \end{bmatrix}\begin{bmatrix} 1 & -2 \\ 1 & -1 \end{bmatrix} = \begin{bmatrix} -1+2 & 2-2 \\ -1+1 & 2-1 \end{bmatrix} = \begin{bmatrix} 1 & 0 \\ 0 & 1 \end{bmatrix}$$

$$BA = \begin{bmatrix} 1 & -2 \\ 1 & -1 \end{bmatrix}\begin{bmatrix} -1 & 2 \\ -1 & 1 \end{bmatrix} = \begin{bmatrix} -1+2 & 2-2 \\ -1+1 & 2-1 \end{bmatrix} = \begin{bmatrix} 1 & 0 \\ 0 & 1 \end{bmatrix}$$

REMARK: Recall that it is not always true that $AB = BA$, even if both products are defined. If A and B are both square matrices and $AB = I_n$, however, then it can be shown that $BA = I_n$. Although the proof of this fact is omitted, it implies that in Example 1 you needed only to check that $AB = I_2$.

The next example shows how to use a system of equations to find the inverse of a matrix.

EXAMPLE 2 Finding the Inverse of a Matrix

Find the inverse of the matrix

$$A = \begin{bmatrix} 1 & 4 \\ -1 & -3 \end{bmatrix}.$$

SOLUTION To find the inverse of A, try to solve the matrix equation $AX = I$ for X.

$$\begin{bmatrix} 1 & 4 \\ -1 & -3 \end{bmatrix} \begin{bmatrix} x_{11} & x_{12} \\ x_{21} & x_{22} \end{bmatrix} = \begin{bmatrix} 1 & 0 \\ 0 & 1 \end{bmatrix}$$

$$\begin{bmatrix} x_{11} + 4x_{21} & x_{12} + 4x_{22} \\ -x_{11} - 3x_{21} & -x_{12} - 3x_{22} \end{bmatrix} = \begin{bmatrix} 1 & 0 \\ 0 & 1 \end{bmatrix}$$

Now, by equating corresponding entries, you obtain the two systems of linear equations shown below.

$$\begin{aligned} x_{11} + 4x_{21} &= 1 \\ -x_{11} - 3x_{21} &= 0 \end{aligned} \qquad \begin{aligned} x_{12} + 4x_{22} &= 0 \\ -x_{12} - 3x_{22} &= 1 \end{aligned}$$

Solving the first system, you find that the first column of X is $x_{11} = -3$ and $x_{21} = 1$. Similarly, solving the second system, you find that the second column of X is $x_{12} = -4$ and $x_{22} = 1$. The inverse of A is

$$X = A^{-1} = \begin{bmatrix} -3 & -4 \\ 1 & 1 \end{bmatrix}.$$

Try using matrix multiplication to check this result.

Generalizing the method used to solve Example 2 provides a convenient method for finding an inverse. Notice first that the two systems of linear equations

$$\begin{aligned} x_{11} + 4x_{21} &= 1 \\ -x_{11} - 3x_{21} &= 0 \end{aligned} \qquad \begin{aligned} x_{12} + 4x_{22} &= 0 \\ -x_{12} - 3x_{22} &= 1 \end{aligned}$$

have the *same coefficient matrix.* Rather than solve the two systems represented by

$$\left[\begin{array}{rr|r} 1 & 4 & 1 \\ -1 & -3 & 0 \end{array}\right] \quad \text{and} \quad \left[\begin{array}{rr|r} 1 & 4 & 0 \\ -1 & -3 & 1 \end{array}\right]$$

separately, you can solve them simultaneously. You can do this by **adjoining** the identity matrix to the coefficient matrix to obtain

$$\left[\begin{array}{rr|rr} 1 & 4 & 1 & 0 \\ -1 & -3 & 0 & 1 \end{array}\right].$$

By applying Gauss-Jordan elimination to this matrix, you can solve *both* systems with a single elimination process, as follows.

$$\left[\begin{array}{rr|rr} 1 & 4 & 1 & 0 \\ 0 & 1 & 1 & 1 \end{array}\right] \qquad R_2 + R_1 \rightarrow R_2$$

$$\left[\begin{array}{rr|rr} 1 & 0 & -3 & -4 \\ 0 & 1 & 1 & 1 \end{array}\right] \qquad R_1 + (-4)R_2 \rightarrow R_1$$

Applying Gauss-Jordan elimination to the "doubly augmented" matrix $[A \vdots I]$, you obtain the matrix $[I \vdots A^{-1}]$.

$$\underset{A}{\begin{bmatrix} 1 & 4 \\ -1 & -3 \end{bmatrix}} \; \vdots \; \underset{I}{\begin{bmatrix} 1 & 0 \\ 0 & 1 \end{bmatrix}} \rightarrow \underset{I}{\begin{bmatrix} 1 & 0 \\ 0 & 1 \end{bmatrix}} \; \vdots \; \underset{A^{-1}}{\begin{bmatrix} -3 & -4 \\ 1 & 1 \end{bmatrix}}$$

This procedure (or algorithm) works for an arbitrary $n \times n$ matrix. If A cannot be row reduced to I_n, then A is noninvertible (or singular). This procedure will be formally justified in the next section, after the concept of an elementary matrix is introduced. For now the algorithm is summarized as follows.

Finding the Inverse of a Matrix by Gauss-Jordan Elimination

Let A be a square matrix of order n.

1. Write the $n \times 2n$ matrix that consists of the given matrix A on the left and the $n \times n$ identity matrix I on the right to obtain $[A \; \vdots \; I]$. Note that you separate the matrices A and I by a dotted line. This process is called **adjoining** matrix I to matrix A.
2. If possible, row reduce A to I using elementary row operations on the *entire* matrix $[A \; \vdots \; I]$. The result will be the matrix $[I \; \vdots \; A^{-1}]$. If this is not possible, then A is noninvertible (or singular).
3. Check your work by multiplying AA^{-1} and $A^{-1}A$ to see that $AA^{-1} = I = A^{-1}A$.

EXAMPLE 3 Finding the Inverse of a Matrix

Find the inverse of the matrix.

$$A = \begin{bmatrix} 1 & -1 & 0 \\ 1 & 0 & -1 \\ -6 & 2 & 3 \end{bmatrix}$$

SOLUTION Begin by adjoining the identity matrix to A to form the matrix

$$[A \; \vdots \; I] = \left[\begin{array}{rrr:rrr} 1 & -1 & 0 & 1 & 0 & 0 \\ 1 & 0 & -1 & 0 & 1 & 0 \\ -6 & 2 & 3 & 0 & 0 & 1 \end{array}\right].$$

Now, using elementary row operations, rewrite this matrix in the form $[I \; \vdots \; A^{-1}]$, as follows.

$$\left[\begin{array}{rrr:rrr} 1 & -1 & 0 & 1 & 0 & 0 \\ 0 & 1 & -1 & -1 & 1 & 0 \\ -6 & 2 & 3 & 0 & 0 & 1 \end{array}\right] \qquad R_2 + (-1)R_1 \rightarrow R_2$$

$$\left[\begin{array}{rrr:rrr} 1 & -1 & 0 & 1 & 0 & 0 \\ 0 & 1 & -1 & -1 & 1 & 0 \\ 0 & -4 & 3 & 6 & 0 & 1 \end{array}\right] \qquad R_3 + (6)R_1 \rightarrow R_3$$

$$\left[\begin{array}{rrr:rrr} 1 & -1 & 0 & 1 & 0 & 0 \\ 0 & 1 & -1 & -1 & 1 & 0 \\ 0 & 0 & -1 & 2 & 4 & 1 \end{array}\right] \qquad R_3 + (4)R_2 \rightarrow R_3$$

$$\left[\begin{array}{rrr:rrr} 1 & -1 & 0 & 1 & 0 & 0 \\ 0 & 1 & -1 & -1 & 1 & 0 \\ 0 & 0 & 1 & -2 & -4 & -1 \end{array}\right] \qquad (-1)R_3 \rightarrow R_3$$

$$\left[\begin{array}{rrr:rrr} 1 & -1 & 0 & 1 & 0 & 0 \\ 0 & 1 & 0 & -3 & -3 & -1 \\ 0 & 0 & 1 & -2 & -4 & -1 \end{array}\right] \qquad R_2 + R_3 \rightarrow R_2$$

$$\left[\begin{array}{rrr:rrr} 1 & 0 & 0 & -2 & -3 & -1 \\ 0 & 1 & 0 & -3 & -3 & -1 \\ 0 & 0 & 1 & -2 & -4 & -1 \end{array}\right] \qquad R_1 + R_2 \rightarrow R_1$$

The matrix A is invertible, and its inverse is

$$A^{-1} = \begin{bmatrix} -2 & -3 & -1 \\ -3 & -3 & -1 \\ -2 & -4 & -1 \end{bmatrix}.$$

Try confirming this by showing that $AA^{-1} = I = A^{-1}A$.

Technology Note

Most graphing utilities and computer software programs can calculate the inverse of a square matrix. If you are using a graphing utility, your screens for Example 3 may look like the images below. Keystrokes and programming syntax for these utilities/programs applicable to Example 3 are provided in the **Online Technology Guide,** available at *college.hmco.com/pic/larsonELA6e*.

```
A
        [[1  -1 0 ]
         [1  0  -1]
         [-6 2  3 ]]
```

```
A-1
        [[-2 -3 -1]
         [-3 -3 -1]
         [-2 -4 -1]]
```

The process shown in Example 3 applies to any $n \times n$ matrix and will find the inverse of matrix A, if possible. If matrix A has no inverse, the process will also tell you that. The next example applies the process to a singular matrix (one that has no inverse).

EXAMPLE 4 A Singular Matrix

Show that the matrix has no inverse.

$$A = \begin{bmatrix} 1 & 2 & 0 \\ 3 & -1 & 2 \\ -2 & 3 & -2 \end{bmatrix}$$

SOLUTION Adjoin the identity matrix to A to form

$$[A \,\vdots\, I] = \left[\begin{array}{rrr:rrr} 1 & 2 & 0 & 1 & 0 & 0 \\ 3 & -1 & 2 & 0 & 1 & 0 \\ -2 & 3 & -2 & 0 & 0 & 1 \end{array}\right]$$

and apply Gauss-Jordan elimination as follows.

$$\left[\begin{array}{rrr:rrr} 1 & 2 & 0 & 1 & 0 & 0 \\ 0 & -7 & 2 & -3 & 1 & 0 \\ -2 & 3 & -2 & 0 & 0 & 1 \end{array}\right] \qquad R_2 + (-3)R_1 \rightarrow R_2$$

$$\left[\begin{array}{rrr:rrr} 1 & 2 & 0 & 1 & 0 & 0 \\ 0 & -7 & 2 & -3 & 1 & 0 \\ 0 & 7 & -2 & 2 & 0 & 1 \end{array}\right] \qquad R_3 + (2)R_1 \rightarrow R_3$$

Now, notice that adding the second row to the third row produces a row of zeros on the left side of the matrix.

$$\left[\begin{array}{rrr:rrr} 1 & 2 & 0 & 1 & 0 & 0 \\ 0 & -7 & 2 & -3 & 1 & 0 \\ 0 & 0 & 0 & -1 & 1 & 1 \end{array}\right] \qquad R_3 + R_2 \rightarrow R_3$$

Because the "A portion" of the matrix has a row of zeros, you can conclude that it is not possible to rewrite the matrix $[A \,\vdots\, I]$ in the form $[I \,\vdots\, A^{-1}]$. This means that A has no inverse, or is noninvertible (or singular).

Using Gauss-Jordan elimination to find the inverse of a matrix works well (even as a computer technique) for matrices of size 3×3 or greater. For 2×2 matrices, however, you can use a formula to find the inverse instead of using Gauss-Jordan elimination. This simple formula is explained as follows.

If A is a 2×2 matrix represented by

$$A = \begin{bmatrix} a & b \\ c & d \end{bmatrix},$$

then A is invertible if and only if $ad - bc \neq 0$. Moreover, if $ad - bc \neq 0$, then the inverse is represented by

$$A^{-1} = \frac{1}{ad - bc}\begin{bmatrix} d & -b \\ -c & a \end{bmatrix}.$$

Try verifying this inverse by finding the product AA^{-1}.

REMARK: The denominator $ad - bc$ is called the **determinant** of A. You will study determinants in detail in Chapter 3.

EXAMPLE 5 Finding the Inverse of a 2 x 2 Matrix

If possible, find the inverse of each matrix.

(a) $A = \begin{bmatrix} 3 & -1 \\ -2 & 2 \end{bmatrix}$ (b) $B = \begin{bmatrix} 3 & -1 \\ -6 & 2 \end{bmatrix}$

SOLUTION (a) For the matrix A, apply the formula for the inverse of a 2×2 matrix to obtain $ad - bc = (3)(2) - (-1)(-2) = 4$. Because this quantity is not zero, the inverse is formed by interchanging the entries on the main diagonal and changing the signs of the other two entries, as follows.

$$A^{-1} = \frac{1}{4}\begin{bmatrix} 2 & 1 \\ 2 & 3 \end{bmatrix} = \begin{bmatrix} \frac{1}{2} & \frac{1}{4} \\ \frac{1}{2} & \frac{3}{4} \end{bmatrix}$$

(b) For the matrix B, you have $ad - bc = (3)(2) - (-1)(-6) = 0$, which means that B is noninvertible.

Properties of Inverses

Some important properties of inverse matrices are listed below.

THEOREM 2.8 Properties of Inverse Matrices

If A is an invertible matrix, k is a positive integer, and c is a scalar not equal to zero, then A^{-1}, A^k, cA, and A^T are invertible and the following are true.

1. $(A^{-1})^{-1} = A$
2. $(A^k)^{-1} = \underbrace{A^{-1}A^{-1} \cdot \cdot \cdot A^{-1}}_{k \text{ factors}} = (A^{-1})^k$
3. $(cA)^{-1} = \dfrac{1}{c}A^{-1}, \; c \neq 0$
4. $(A^T)^{-1} = (A^{-1})^T$

PROOF The key to the proofs of Properties 1, 3, and 4 is the fact that the inverse of a matrix is unique (Theorem 2.7). That is, if $BC = CB = I$, then C is the inverse of B.

Property 1 states that the inverse of A^{-1} is A itself. To prove this, observe that $A^{-1}A = AA^{-1} = I$, which means that A is the inverse of A^{-1}. Thus, $A = (A^{-1})^{-1}$.

Similarly, Property 3 states that $\frac{1}{c}A^{-1}$ is the inverse of (cA), $c \neq 0$. To prove this, use the properties of scalar multiplication given in Theorems 2.1 and 2.3, as follows.

$$(cA)\left(\frac{1}{c}A^{-1}\right) = \left(c\frac{1}{c}\right)AA^{-1} = (1)I = I$$

and

$$\left(\frac{1}{c}A^{-1}\right)(cA) = \left(\frac{1}{c}c\right)A^{-1}A = (1)I = I$$

So $\frac{1}{c}A^{-1}$ is the inverse of (cA), which implies that

$$\frac{1}{c}A^{-1} = (cA)^{-1}.$$

Properties 2 and 4 are left for you to prove. (See Exercises 47 and 48.)

For nonsingular matrices, the exponential notation used for repeated multiplication of *square* matrices can be extended to include exponents that are negative integers. This may be done by defining A^{-k} to be

$$A^{-k} = \underbrace{A^{-1}A^{-1} \cdot \cdot \cdot A^{-1}}_{k \text{ factors}} = (A^{-1})^k.$$

Using this convention you can show that the properties $A^jA^k = A^{j+k}$ and $(A^j)^k = A^{jk}$ hold true for any integers j and k.

EXAMPLE 6 The Inverse of the Square of a Matrix

Compute A^{-2} in two different ways and show that the results are equal.

$$A = \begin{bmatrix} 1 & 1 \\ 2 & 4 \end{bmatrix}$$

SOLUTION One way to find A^{-2} is to find $(A^2)^{-1}$ by squaring the matrix A to obtain

$$A^2 = \begin{bmatrix} 3 & 5 \\ 10 & 18 \end{bmatrix}$$

and using the formula for the inverse of a 2×2 matrix to obtain

$$(A^2)^{-1} = \frac{1}{4}\begin{bmatrix} 18 & -5 \\ -10 & 3 \end{bmatrix}$$

$$= \begin{bmatrix} \frac{9}{2} & -\frac{5}{4} \\ -\frac{5}{2} & \frac{3}{4} \end{bmatrix}.$$

Another way to find A^{-2} is to find $(A^{-1})^2$ by finding A^{-1}

$$A^{-1} = \frac{1}{2}\begin{bmatrix} 4 & -1 \\ -2 & 1 \end{bmatrix} = \begin{bmatrix} 2 & -\frac{1}{2} \\ -1 & \frac{1}{2} \end{bmatrix}$$

and then squaring this matrix to obtain

$$(A^{-1})^2 = \begin{bmatrix} \frac{9}{2} & -\frac{5}{4} \\ -\frac{5}{2} & \frac{3}{4} \end{bmatrix}.$$

Note that each method produces the same result.

Discovery

Let $A = \begin{bmatrix} 1 & 2 \\ 1 & 3 \end{bmatrix}$ *and* $B = \begin{bmatrix} 2 & -1 \\ 1 & -1 \end{bmatrix}$.

Calculate $(AB)^{-1}$, $A^{-1}B^{-1}$, *and* $B^{-1}A^{-1}$. *Make a conjecture about the inverse of a product of two nonsingular matrices. Select two other nonsingular matrices and see whether your conjecture holds.*

The next theorem gives a formula for computing the inverse of a product of two matrices.

THEOREM 2.9 The Inverse of a Product

If A and B are invertible matrices of size n, then AB is invertible and

$$(AB)^{-1} = B^{-1}A^{-1}.$$

PROOF To show that $B^{-1}A^{-1}$ is the inverse of AB, you need only show that it conforms to the definition of an inverse matrix. That is,

$$(AB)(B^{-1}A^{-1}) = A(BB^{-1})A^{-1} = A(I)A^{-1} = (AI)A^{-1} = AA^{-1} = I.$$

In a similar way you can show that $(B^{-1}A^{-1})(AB) = I$ and conclude that AB is invertible and has the indicated inverse.

Theorem 2.9 states that the inverse of a product of two invertible matrices is the product of their inverses taken in the *reverse* order. This can be generalized to include the product of several invertible matrices:

$$(A_1A_2A_3 \cdot \cdot \cdot A_n)^{-1} = A_n^{-1} \cdot \cdot \cdot A_3^{-1}A_2^{-1}A_1^{-1}.$$

(See Example 4 in Appendix A.)

EXAMPLE 7 Finding the Inverse of a Matrix Product

Find $(AB)^{-1}$ for the matrices

$$A = \begin{bmatrix} 1 & 3 & 3 \\ 1 & 4 & 3 \\ 1 & 3 & 4 \end{bmatrix} \quad \text{and} \quad B = \begin{bmatrix} 1 & 2 & 3 \\ 1 & 3 & 3 \\ 2 & 4 & 3 \end{bmatrix}$$

using the fact that A^{-1} and B^{-1} are represented by

$$A^{-1} = \begin{bmatrix} 7 & -3 & -3 \\ -1 & 1 & 0 \\ -1 & 0 & 1 \end{bmatrix} \quad \text{and} \quad B^{-1} = \begin{bmatrix} 1 & -2 & 1 \\ -1 & 1 & 0 \\ \frac{2}{3} & 0 & -\frac{1}{3} \end{bmatrix}.$$

SOLUTION Using Theorem 2.9 produces

$$(AB)^{-1} = B^{-1}A^{-1} = \overbrace{\begin{bmatrix} 1 & -2 & 1 \\ -1 & 1 & 0 \\ \frac{2}{3} & 0 & -\frac{1}{3} \end{bmatrix}}^{B^{-1}} \overbrace{\begin{bmatrix} 7 & -3 & -3 \\ -1 & 1 & 0 \\ -1 & 0 & 1 \end{bmatrix}}^{A^{-1}} = \begin{bmatrix} 8 & -5 & -2 \\ -8 & 4 & 3 \\ 5 & -2 & -\frac{7}{3} \end{bmatrix}.$$

REMARK: Note that you *reverse the order* of multiplication to find the inverse of AB. That is, $(AB)^{-1} = B^{-1}A^{-1}$, and the inverse of AB is usually *not* equal to $A^{-1}B^{-1}$.

One important property in the algebra of real numbers is the cancellation property. That is, if $ac = bc$ $(c \neq 0)$, then $a = b$. *Invertible* matrices have similar cancellation properties.

THEOREM 2.10
Cancellation Properties

If C is an invertible matrix, then the following properties hold.

1. If $AC = BC$, then $A = B$. **Right cancellation property**
2. If $CA = CB$, then $A = B$. **Left cancellation property**

PROOF To prove Property 1, use the fact that C is invertible and write

$$\begin{aligned} AC &= BC \\ (AC)C^{-1} &= (BC)C^{-1} \\ A(CC^{-1}) &= B(CC^{-1}) \\ AI &= BI \\ A &= B. \end{aligned}$$

The second property can be proved in a similar way; this is left to you. (See Exercise 50.)

Be sure to remember that Theorem 2.10 can be applied only if C is an *invertible* matrix. If C is not invertible, then cancellation is not usually valid. For instance, Example 5 in Section 2.2 gives an example of a matrix equation $AC = BC$ in which $A \neq B$, because C is not invertible in the example.

Systems of Equations

In Theorem 2.5 you were able to prove that a system of linear equations can have exactly one solution, an infinite number of solutions, or no solution. For *square* systems (those having the same number of equations as variables), you can use the theorem below to determine whether the system has a unique solution.

THEOREM 2.11 Systems of Equations with Unique Solutions

If A is an invertible matrix, then the system of linear equations $A\mathbf{x} = \mathbf{b}$ has a unique solution given by

$$\mathbf{x} = A^{-1}\mathbf{b}.$$

PROOF Because A is nonsingular, the steps shown below are valid.

$$\begin{aligned} A\mathbf{x} &= \mathbf{b} \\ A^{-1}A\mathbf{x} &= A^{-1}\mathbf{b} \\ I\mathbf{x} &= A^{-1}\mathbf{b} \\ \mathbf{x} &= A^{-1}\mathbf{b} \end{aligned}$$

This solution is unique because if $\mathbf{x}_1$ and $\mathbf{x}_2$ were two solutions, you could apply the cancellation property to the equation $A\mathbf{x}_1 = \mathbf{b} = A\mathbf{x}_2$ to conclude that $\mathbf{x}_1 = \mathbf{x}_2$.

Theorem 2.11 is theoretically important, but it is not very practical for solving a system of linear equations. It would require more work to find A^{-1} and then multiply by $\mathbf{b}$ than simply to solve the system using Gaussian elimination with back-substitution. A situation in which you might consider using Theorem 2.11 as a computational technique would be one in which you have *several* systems of linear equations, all of which have the same coefficient matrix A. In such a case, you could find the inverse matrix once and then solve each system by computing the product $A^{-1}\mathbf{b}$. This is demonstrated in Example 8.

EXAMPLE 8 **Solving a System of Equations Using an Inverse Matrix**

Use an inverse matrix to solve each system.

(a) $\begin{aligned} 2x + 3y + z &= -1 \\ 3x + 3y + z &= 1 \\ 2x + 4y + z &= -2 \end{aligned}$

(b) $\begin{aligned} 2x + 3y + z &= 4 \\ 3x + 3y + z &= 8 \\ 2x + 4y + z &= 5 \end{aligned}$

(c) $\begin{aligned} 2x + 3y + z &= 0 \\ 3x + 3y + z &= 0 \\ 2x + 4y + z &= 0 \end{aligned}$

SOLUTION First note that the coefficient matrix for each system is

$$A = \begin{bmatrix} 2 & 3 & 1 \\ 3 & 3 & 1 \\ 2 & 4 & 1 \end{bmatrix}.$$

Using Gauss-Jordan elimination, you can find A^{-1} to be

$$A^{-1} = \begin{bmatrix} -1 & 1 & 0 \\ -1 & 0 & 1 \\ 6 & -2 & -3 \end{bmatrix}.$$

To solve each system, use matrix multiplication, as follows.

(a) $\mathbf{x} = A^{-1}\mathbf{b} = \begin{bmatrix} -1 & 1 & 0 \\ -1 & 0 & 1 \\ 6 & -2 & -3 \end{bmatrix}\begin{bmatrix} -1 \\ 1 \\ -2 \end{bmatrix} = \begin{bmatrix} 2 \\ -1 \\ -2 \end{bmatrix}$

The solution is $x = 2$, $y = -1$, and $z = -2$.

(b) $\mathbf{x} = A^{-1}\mathbf{b} = \begin{bmatrix} -1 & 1 & 0 \\ -1 & 0 & 1 \\ 6 & -2 & -3 \end{bmatrix}\begin{bmatrix} 4 \\ 8 \\ 5 \end{bmatrix} = \begin{bmatrix} 4 \\ 1 \\ -7 \end{bmatrix}$

The solution is $x = 4$, $y = 1$, and $z = -7$.

(c) $\mathbf{x} = A^{-1}\mathbf{b} = \begin{bmatrix} -1 & 1 & 0 \\ -1 & 0 & 1 \\ 6 & -2 & -3 \end{bmatrix}\begin{bmatrix} 0 \\ 0 \\ 0 \end{bmatrix} = \begin{bmatrix} 0 \\ 0 \\ 0 \end{bmatrix}$

The solution is trivial: $x = 0$, $y = 0$, and $z = 0$.

SECTION 2.3 Exercises

In Exercises 1–4, show that B is the inverse of A.

1. $A = \begin{bmatrix} 1 & 2 \\ 3 & 4 \end{bmatrix}$, $B = \begin{bmatrix} -2 & 1 \\ \frac{3}{2} & -\frac{1}{2} \end{bmatrix}$

2. $A = \begin{bmatrix} 1 & -1 \\ 2 & 3 \end{bmatrix}$, $B = \begin{bmatrix} \frac{3}{5} & \frac{1}{5} \\ -\frac{2}{5} & \frac{1}{5} \end{bmatrix}$

3. $A = \begin{bmatrix} -2 & 2 & 3 \\ 1 & -1 & 0 \\ 0 & 1 & 4 \end{bmatrix}$, $B = \frac{1}{3}\begin{bmatrix} -4 & -5 & 3 \\ -4 & -8 & 3 \\ 1 & 2 & 0 \end{bmatrix}$

4. $A = \begin{bmatrix} 2 & -17 & 11 \\ -1 & 11 & -7 \\ 0 & 3 & -2 \end{bmatrix}$, $B = \begin{bmatrix} 1 & 1 & 2 \\ 2 & 4 & -3 \\ 3 & 6 & -5 \end{bmatrix}$

In Exercises 5–24, find the inverse of the matrix (if it exists).

5. $\begin{bmatrix} 1 & 2 \\ 3 & 7 \end{bmatrix}$

6. $\begin{bmatrix} 1 & -2 \\ 2 & -3 \end{bmatrix}$

7. $\begin{bmatrix} -7 & 33 \\ 4 & -19 \end{bmatrix}$

8. $\begin{bmatrix} -1 & 1 \\ 3 & -3 \end{bmatrix}$

9. $\begin{bmatrix} 1 & 1 & 1 \\ 3 & 5 & 4 \\ 3 & 6 & 5 \end{bmatrix}$

10. $\begin{bmatrix} 1 & 2 & 2 \\ 3 & 7 & 9 \\ -1 & -4 & -7 \end{bmatrix}$

11. $\begin{bmatrix} 1 & 2 & -1 \\ 3 & 7 & -10 \\ 7 & 16 & -21 \end{bmatrix}$

12. $\begin{bmatrix} 10 & 5 & -7 \\ -5 & 1 & 4 \\ 3 & 2 & -2 \end{bmatrix}$

13. $\begin{bmatrix} 1 & 1 & 2 \\ 3 & 1 & 0 \\ -2 & 0 & 3 \end{bmatrix}$ **14.** $\begin{bmatrix} 3 & 2 & 5 \\ 2 & 2 & 4 \\ -4 & 4 & 0 \end{bmatrix}$

15. $\begin{bmatrix} 0.1 & 0.2 & 0.3 \\ -0.3 & 0.2 & 0.2 \\ 0.5 & 0.5 & 0.5 \end{bmatrix}$ **16.** $\begin{bmatrix} 2 & 0 & 0 \\ 0 & 3 & 0 \\ 0 & 0 & 5 \end{bmatrix}$

17. $\begin{bmatrix} 1 & 0 & 0 \\ 3 & 4 & 0 \\ 2 & 5 & 5 \end{bmatrix}$ **18.** $\begin{bmatrix} 1 & 0 & 0 \\ 3 & 0 & 0 \\ 2 & 5 & 5 \end{bmatrix}$

19. $\begin{bmatrix} -8 & 0 & 0 & 0 \\ 0 & 1 & 0 & 0 \\ 0 & 0 & 0 & 0 \\ 0 & 0 & 0 & -5 \end{bmatrix}$ **20.** $\begin{bmatrix} 1 & 0 & 0 & 0 \\ 0 & 2 & 0 & 0 \\ 0 & 0 & -2 & 0 \\ 0 & 0 & 0 & 3 \end{bmatrix}$

21. $\begin{bmatrix} 1 & -2 & -1 & -2 \\ 3 & -5 & -2 & -3 \\ 2 & -5 & -2 & -5 \\ -1 & 4 & 4 & 11 \end{bmatrix}$ **22.** $\begin{bmatrix} 4 & 8 & -7 & 14 \\ 2 & 5 & -4 & 6 \\ 0 & 2 & 1 & -7 \\ 3 & 6 & -5 & 10 \end{bmatrix}$

23. $\begin{bmatrix} 1 & 0 & 3 & 0 \\ 0 & 2 & 0 & 4 \\ 1 & 0 & 3 & 0 \\ 0 & 2 & 0 & 4 \end{bmatrix}$ **24.** $\begin{bmatrix} 1 & 3 & -2 & 0 \\ 0 & 2 & 4 & 6 \\ 0 & 0 & -2 & 1 \\ 0 & 0 & 0 & 5 \end{bmatrix}$

In Exercises 25–28, use an inverse matrix to solve each system of linear equations.

25. (a) $\begin{aligned} x + 2y &= -1 \\ x - 2y &= 3 \end{aligned}$
(b) $\begin{aligned} x + 2y &= 10 \\ x - 2y &= -6 \end{aligned}$
(c) $\begin{aligned} x + 2y &= -3 \\ x - 2y &= 0 \end{aligned}$

26. (a) $\begin{aligned} 2x - y &= -3 \\ 2x + y &= 7 \end{aligned}$
(b) $\begin{aligned} 2x - y &= -1 \\ 2x + y &= -3 \end{aligned}$
(c) $\begin{aligned} 2x - y &= 6 \\ 2x + y &= 10 \end{aligned}$

27. (a) $\begin{aligned} x_1 + 2x_2 + x_3 &= 2 \\ x_1 + 2x_2 - x_3 &= 4 \\ x_1 - 2x_2 + x_3 &= -2 \end{aligned}$
(b) $\begin{aligned} x_1 + 2x_2 + x_3 &= 1 \\ x_1 + 2x_2 - x_3 &= 3 \\ x_1 - 2x_2 + x_3 &= -3 \end{aligned}$

28. (a) $\begin{aligned} x_1 + x_2 - 2x_3 &= 0 \\ x_1 - 2x_2 + x_3 &= 0 \\ x_1 - x_2 - x_3 &= -1 \end{aligned}$
(b) $\begin{aligned} x_1 + x_2 - 2x_3 &= -1 \\ x_1 - 2x_2 + x_3 &= 2 \\ x_1 - x_2 - x_3 &= 0 \end{aligned}$

In Exercises 29–32, use a graphing utility or computer software program with matrix capabilities to solve the system of linear equations using an inverse matrix.

29. $\begin{aligned} x_1 + 2x_2 - x_3 + 3x_4 - x_5 &= -3 \\ x_1 - 3x_2 + x_3 + 2x_4 - x_5 &= -3 \\ 2x_1 + x_2 + x_3 - 3x_4 + x_5 &= 6 \\ x_1 - x_2 + 2x_3 + x_4 - x_5 &= 2 \\ 2x_1 + x_2 - x_3 + 2x_4 + x_5 &= -3 \end{aligned}$

30. $\begin{aligned} x_1 + x_2 - x_3 + 3x_4 - x_5 &= 3 \\ 2x_1 + x_2 + x_3 + x_4 + x_5 &= 4 \\ x_1 + x_2 - x_3 + 2x_4 - x_5 &= 3 \\ 2x_1 + x_2 + 4x_3 + x_4 - x_5 &= -1 \\ 3x_1 + x_2 + x_3 - 2x_4 + x_5 &= 5 \end{aligned}$

31. $\begin{aligned} 2x_1 - 3x_2 + x_3 - 2x_4 + x_5 - 4x_6 &= 20 \\ 3x_1 + x_2 - 4x_3 + x_4 - x_5 + 2x_6 &= -16 \\ 4x_1 + x_2 - 3x_3 + 4x_4 - x_5 + 2x_6 &= -12 \\ -5x_1 - x_2 + 4x_3 + 2x_4 - 5x_5 + 3x_6 &= -2 \\ x_1 + x_2 - 3x_3 + 4x_4 - 3x_5 + x_6 &= -15 \\ 3x_1 - x_2 + 2x_3 - 3x_4 + 2x_5 - 6x_6 &= 25 \end{aligned}$

32. $\begin{aligned} 4x_1 - 2x_2 + 4x_3 + 2x_4 - 5x_5 - x_6 &= 1 \\ 3x_1 + 6x_2 - 5x_3 - 6x_4 + 3x_5 + 3x_6 &= -11 \\ 2x_1 - 3x_2 + x_3 + 3x_4 - x_5 - 2x_6 &= 0 \\ -x_1 + 4x_2 - 4x_3 - 6x_4 + 2x_5 + 4x_6 &= -9 \\ 3x_1 - x_2 + 5x_3 + 2x_4 - 3x_5 - 5x_6 &= 1 \\ -2x_1 + 3x_2 - 4x_3 - 6x_4 + x_5 + 2x_6 &= -12 \end{aligned}$

In Exercises 33–36, use the inverse matrices to find (a) $(AB)^{-1}$, (b) $(A^T)^{-1}$, (c) A^{-2}, and (d) $(2A)^{-1}$.

33. $A^{-1} = \begin{bmatrix} 2 & 5 \\ -7 & 6 \end{bmatrix}$, $B^{-1} = \begin{bmatrix} 7 & -3 \\ 2 & 0 \end{bmatrix}$

34. $A^{-1} = \begin{bmatrix} -\frac{2}{7} & \frac{1}{7} \\ \frac{3}{7} & \frac{2}{7} \end{bmatrix}$, $B^{-1} = \begin{bmatrix} \frac{5}{11} & \frac{2}{11} \\ \frac{3}{11} & -\frac{1}{11} \end{bmatrix}$

35. $A^{-1} = \begin{bmatrix} 1 & -\frac{1}{2} & \frac{3}{4} \\ \frac{3}{2} & \frac{1}{2} & -2 \\ \frac{1}{4} & 1 & \frac{1}{2} \end{bmatrix}$, $B^{-1} = \begin{bmatrix} 2 & 4 & \frac{5}{2} \\ -\frac{3}{4} & 2 & \frac{1}{4} \\ \frac{1}{4} & \frac{1}{2} & 2 \end{bmatrix}$

36. $A^{-1} = \begin{bmatrix} 1 & -4 & 2 \\ 0 & 1 & 3 \\ 4 & 2 & 1 \end{bmatrix}$, $B^{-1} = \begin{bmatrix} 6 & 5 & -3 \\ -2 & 4 & -1 \\ 1 & 3 & 4 \end{bmatrix}$

In Exercises 37 and 38, find x such that the matrix is equal to its own inverse.

37. $A = \begin{bmatrix} 3 & x \\ -2 & -3 \end{bmatrix}$ **38.** $A = \begin{bmatrix} 2 & x \\ -1 & -2 \end{bmatrix}$

In Exercises 39 and 40, find x such that the matrix is singular.

39. $A = \begin{bmatrix} 4 & x \\ -2 & -3 \end{bmatrix}$ **40.** $A = \begin{bmatrix} x & 2 \\ -3 & 4 \end{bmatrix}$

In Exercises 41 and 42, find A provided that

41. $(2A)^{-1} = \begin{bmatrix} 1 & 2 \\ 3 & 4 \end{bmatrix}$. **42.** $(4A)^{-1} = \begin{bmatrix} 2 & 4 \\ -3 & 2 \end{bmatrix}$.

In Exercises 43 and 44, show that the matrix is invertible and find its inverse.

43. $A = \begin{bmatrix} \sin\theta & \cos\theta \\ -\cos\theta & \sin\theta \end{bmatrix}$ **44.** $A = \begin{bmatrix} \sec\theta & \tan\theta \\ \tan\theta & \sec\theta \end{bmatrix}$

True or False? In Exercises 45 and 46, determine whether each statement is true or false. If a statement is true, give a reason or cite an appropriate statement from the text. If a statement is false, provide an example that shows the statement is not true in all cases or cite an appropriate statement from the text.

45. (a) The inverse of a nonsingular matrix is unique.

(b) If the matrices A, B, and C satisfy $BA = CA$ and A is invertible, then $B = C$.

(c) The inverse of the product of two matrices is the product of their inverses; that is, $(AB)^{-1} = A^{-1}B^{-1}$.

(d) If A can be row reduced to the identity matrix, then A is nonsingular.

46. (a) The product of four invertible 7×7 matrices is invertible.

(b) The transpose of the inverse of a nonsingular matrix is equal to the inverse of the transpose.

(c) The matrix $\begin{bmatrix} a & b \\ c & d \end{bmatrix}$ is invertible if $ab - dc \neq 0$.

(d) If A is a square matrix, then the system of linear equations $A\mathbf{x} = \mathbf{b}$ has a unique solution.

47. Prove Property 2 of Theorem 2.8: If A is an invertible matrix and k is a positive integer, then

$$(A^k)^{-1} = \underbrace{A^{-1}A^{-1} \cdot \cdot \cdot A^{-1}}_{k \text{ factors}} = (A^{-1})^k$$

48. Prove Property 4 of Theorem 2.8: If A is an invertible matrix, then $(A^T)^{-1} = (A^{-1})^T$.

49. Guided Proof Prove that the inverse of a symmetric nonsingular matrix is symmetric.

Getting Started: To prove that the inverse of A is symmetric, you need to show that $(A^{-1})^T = A^{-1}$.

(i) Let A be a symmetric, nonsingular matrix.

(ii) This means that $A^T = A$ and A^{-1} exists.

(iii) Use the properties of the transpose to show that $(A^{-1})^T$ is equal to A^{-1}.

50. Prove Property 2 of Theorem 2.10: If C is an invertible matrix such that $CA = CB$, then $A = B$.

51. Prove that if $A^2 = A$, then $I - 2A = (I - 2A)^{-1}$.

52. Prove that if A, B, and C are square matrices and $ABC = I$, then B is invertible and $B^{-1} = CA$.

53. Prove that if A is invertible and $AB = O$, then $B = O$.

54. Guided Proof Prove that if $A^2 = A$, then either $A = I$ or A is singular.

Getting Started: You must show that either A is singular or A equals the identity matrix.

(i) Begin your proof by observing that A is either singular or nonsingular.

(ii) If A is singular, then you are done.

(iii) If A is nonsingular, then use the inverse matrix A^{-1} and the hypothesis $A^2 = A$ to show that $A = I$.

55. Writing Is the sum of two invertible matrices invertible? Explain why or why not. Illustrate your conclusion with appropriate examples.

56. Writing Under what conditions will the diagonal matrix

$$A = \begin{bmatrix} a_{11} & 0 & 0 & \cdots & 0 \\ 0 & a_{22} & 0 & \cdots & 0 \\ \vdots & \vdots & \vdots & & \vdots \\ 0 & 0 & 0 & \cdots & a_{nn} \end{bmatrix}$$

be invertible? If A is invertible, find its inverse.

57. Use the result of Exercise 56 to find A^{-1} for each matrix.

(a) $A = \begin{bmatrix} -1 & 0 & 0 \\ 0 & 3 & 0 \\ 0 & 0 & 2 \end{bmatrix}$

(b) $A = \begin{bmatrix} \frac{1}{2} & 0 & 0 \\ 0 & \frac{1}{3} & 0 \\ 0 & 0 & \frac{1}{4} \end{bmatrix}$

58. Let $A = \begin{bmatrix} 1 & 2 \\ -2 & 1 \end{bmatrix}$.

(a) Show that $A^2 - 2A + 5I = O$, where I is the identity matrix of order 2.

(b) Show that $A^{-1} = \frac{1}{5}(2I - A)$.

(c) Show that, in general, for any square matrix satisfying $A^2 - 2A + 5I = 0$, the inverse of A is given by $A^{-1} = \frac{1}{5}(2I - A)$.

59. Let $\mathbf{u}$ be an $n \times 1$ column matrix satisfying $\mathbf{u}^T\mathbf{u} = 1$. The $n \times n$ matrix $H = I_n - 2\mathbf{u}\mathbf{u}^T$ is called a **Householder matrix.**

(a) Prove that H is symmetric and nonsingular.

(b) Let $\mathbf{u} = \begin{bmatrix} \sqrt{2}/2 \\ \sqrt{2}/2 \\ 0 \end{bmatrix}$. Show that $\mathbf{u}^T\mathbf{u} = 1$ and calculate the Householder matrix H.

60. Prove that if the matrix $I - AB$ is nonsingular, then so is $I - BA$.

61. Let A, D, and P be $n \times n$ matrices satisfying $AP = PD$. If P is nonsingular, solve this equation for A. Must it be true that $A = D$?

62. Let A, D, and P be $n \times n$ matrices satisfying $P^{-1}AP = D$. Solve this equation for A. Must it be true that $A = D$?

63. Find an example of a singular 2×2 matrix satisfying $A^2 = A$.

2.4 Elementary Matrices

In Section 1.2, the three elementary row operations for matrices listed below were introduced.

1. Interchange two rows.
2. Multiply a row by a nonzero constant.
3. Add a multiple of a row to another row.

In this section, you will see how matrix multiplication can be used to perform these operations.

Definition of an Elementary Matrix

An $n \times n$ matrix is called an **elementary matrix** if it can be obtained from the identity matrix I_n by a single elementary row operation.

REMARK: The identity matrix I_n is elementary by this definition because it can be obtained from itself by multiplying any one of its rows by 1.

EXAMPLE 1 Elementary Matrices and Nonelementary Matrices

Which of the following matrices are elementary? For those that are, describe the corresponding elementary row operation.

(a) $\begin{bmatrix} 1 & 0 & 0 \\ 0 & 3 & 0 \\ 0 & 0 & 1 \end{bmatrix}$ (b) $\begin{bmatrix} 1 & 0 & 0 \\ 0 & 1 & 0 \end{bmatrix}$ (c) $\begin{bmatrix} 1 & 0 & 0 \\ 0 & 1 & 0 \\ 0 & 0 & 0 \end{bmatrix}$

(d) $\begin{bmatrix} 1 & 0 & 0 \\ 0 & 0 & 1 \\ 0 & 1 & 0 \end{bmatrix}$ (e) $\begin{bmatrix} 1 & 0 \\ 2 & 1 \end{bmatrix}$ (f) $\begin{bmatrix} 1 & 0 & 0 \\ 0 & 2 & 0 \\ 0 & 0 & -1 \end{bmatrix}$

SOLUTION (a) This matrix *is* elementary. It can be obtained by multiplying the second row of I_3 by 3.

(b) This matrix is *not* elementary because it is not square.

(c) This matrix is *not* elementary because it was obtained by multiplying the third row of I_3 by 0 (row multiplication must be by a *nonzero* constant).

(d) This matrix *is* elementary. It can be obtained by interchanging the second and third rows of I_3.
(e) This matrix *is* elementary. It can be obtained by multiplying the first row of I_2 by 2 and adding the result to the second row.
(f) This matrix is *not* elementary because two elementary row operations are required to obtain it from I_3.

Elementary matrices are useful because they enable you to use matrix multiplication to perform elementary row operations, as demonstrated in Example 2.

EXAMPLE 2 Elementary Matrices and Elementary Row Operations

(a) In the matrix product below, E is the elementary matrix in which the first two rows of I_3 have been interchanged.

$$\overset{E}{\begin{bmatrix} 0 & 1 & 0 \\ 1 & 0 & 0 \\ 0 & 0 & 1 \end{bmatrix}} \overset{A}{\begin{bmatrix} 0 & 2 & 1 \\ 1 & -3 & 6 \\ 3 & 2 & -1 \end{bmatrix}} = \begin{bmatrix} 1 & -3 & 6 \\ 0 & 2 & 1 \\ 3 & 2 & -1 \end{bmatrix}$$

Note that the first two rows of A have been interchanged by multiplying *on the left* by E.

(b) In the next matrix product, E is the elementary matrix in which the second row of I_3 has been multiplied by $\frac{1}{2}$.

$$\overset{E}{\begin{bmatrix} 1 & 0 & 0 \\ 0 & \frac{1}{2} & 0 \\ 0 & 0 & 1 \end{bmatrix}} \overset{A}{\begin{bmatrix} 1 & 0 & -4 & 1 \\ 0 & 2 & 6 & -4 \\ 0 & 1 & 3 & 1 \end{bmatrix}} = \begin{bmatrix} 1 & 0 & -4 & 1 \\ 0 & 1 & 3 & -2 \\ 0 & 1 & 3 & 1 \end{bmatrix}$$

Here the size of A is 3×4. A could, however, be any $3 \times n$ matrix and multiplication on the left by E would still result in multiplying the second row of A by $\frac{1}{2}$.

(c) In the product shown below, E is the elementary matrix in which 2 times the first row of I_3 has been added to the second row.

$$\overset{E}{\begin{bmatrix} 1 & 0 & 0 \\ 2 & 1 & 0 \\ 0 & 0 & 1 \end{bmatrix}} \overset{A}{\begin{bmatrix} 1 & 0 & -1 \\ -2 & -2 & 3 \\ 0 & 4 & 5 \end{bmatrix}} = \begin{bmatrix} 1 & 0 & -1 \\ 0 & -2 & 1 \\ 0 & 4 & 5 \end{bmatrix}$$

Note that in the product EA, 2 times the first row of A has been added to the second row.

In each of the three products in Example 2, you were able to perform elementary row operations by multiplying *on the left* by an elementary matrix. This property of elementary matrices is generalized in the next theorem, which is stated without proof.

THEOREM 2.12 Representing Elementary Row Operations

Let E be the elementary matrix obtained by performing an elementary row operation on I_m. If that same elementary row operation is performed on an $m \times n$ matrix A, then the resulting matrix is given by the product EA.

REMARK: Be sure to remember that in Theorem 2.12, A is multiplied *on the left* by the elementary matrix E. Right multiplication by elementary matrices, which involves column operations, will not be considered in this text.

Most applications of elementary row operations require a sequence of operations. For instance, Gaussian elimination usually requires several elementary row operations to row reduce a matrix A. For elementary matrices, this sequence translates into multiplication (on the left) by several elementary matrices. The order of multiplication is important; the elementary matrix immediately to the left of A corresponds to the row operation performed first. This process is demonstrated in Example 3.

EXAMPLE 3 Using Elementary Matrices

Find a sequence of elementary matrices that can be used to write the matrix A in row-echelon form.

$$A = \begin{bmatrix} 0 & 1 & 3 & 5 \\ 1 & -3 & 0 & 2 \\ 2 & -6 & 2 & 0 \end{bmatrix}$$

SOLUTION

Matrix	*Elementary Row Operation*	*Elementary Matrix*
$\begin{bmatrix} 1 & -3 & 0 & 2 \\ 0 & 1 & 3 & 5 \\ 2 & -6 & 2 & 0 \end{bmatrix}$	$R_1 \leftrightarrow R_2$	$E_1 = \begin{bmatrix} 0 & 1 & 0 \\ 1 & 0 & 0 \\ 0 & 0 & 1 \end{bmatrix}$
$\begin{bmatrix} 1 & -3 & 0 & 2 \\ 0 & 1 & 3 & 5 \\ 0 & 0 & 2 & -4 \end{bmatrix}$	$R_3 + (-2)R_1 \rightarrow R_3$	$E_2 = \begin{bmatrix} 1 & 0 & 0 \\ 0 & 1 & 0 \\ -2 & 0 & 1 \end{bmatrix}$
$\begin{bmatrix} 1 & -3 & 0 & 2 \\ 0 & 1 & 3 & 5 \\ 0 & 0 & 1 & -2 \end{bmatrix}$	$(\frac{1}{2})R_3 \rightarrow R_3$	$E_3 = \begin{bmatrix} 1 & 0 & 0 \\ 0 & 1 & 0 \\ 0 & 0 & \frac{1}{2} \end{bmatrix}$

The three elementary matrices E_1, E_2, and E_3 can be used to perform the same elimination.

$$B = E_3E_2E_1A = \begin{bmatrix} 1 & 0 & 0 \\ 0 & 1 & 0 \\ 0 & 0 & \frac{1}{2} \end{bmatrix}\begin{bmatrix} 1 & 0 & 0 \\ 0 & 1 & 0 \\ -2 & 0 & 1 \end{bmatrix}\begin{bmatrix} 0 & 1 & 0 \\ 1 & 0 & 0 \\ 0 & 0 & 1 \end{bmatrix}\begin{bmatrix} 0 & 1 & 3 & 5 \\ 1 & -3 & 0 & 2 \\ 2 & -6 & 2 & 0 \end{bmatrix}$$

$$= \begin{bmatrix} 1 & 0 & 0 \\ 0 & 1 & 0 \\ 0 & 0 & \frac{1}{2} \end{bmatrix}\begin{bmatrix} 1 & 0 & 0 \\ 0 & 1 & 0 \\ -2 & 0 & 1 \end{bmatrix}\begin{bmatrix} 1 & -3 & 0 & 2 \\ 0 & 1 & 3 & 5 \\ 2 & -6 & 2 & 0 \end{bmatrix}$$

$$= \begin{bmatrix} 1 & 0 & 0 \\ 0 & 1 & 0 \\ 0 & 0 & \frac{1}{2} \end{bmatrix}\begin{bmatrix} 1 & -3 & 0 & 2 \\ 0 & 1 & 3 & 5 \\ 0 & 0 & 2 & -4 \end{bmatrix} = \begin{bmatrix} 1 & -3 & 0 & 2 \\ 0 & 1 & 3 & 5 \\ 0 & 0 & 1 & -2 \end{bmatrix}$$

REMARK: The procedure demonstrated in Example 3 is primarily of theoretical interest. In other words, this procedure is not suggested as a practical method for performing Gaussian elimination.

The two matrices in Example 3

$$A = \begin{bmatrix} 0 & 1 & 3 & 5 \\ 1 & -3 & 0 & 2 \\ 2 & -6 & 2 & 0 \end{bmatrix} \quad \text{and} \quad B = \begin{bmatrix} 1 & -3 & 0 & 2 \\ 0 & 1 & 3 & 5 \\ 0 & 0 & 1 & -2 \end{bmatrix}$$

are row-equivalent because you can obtain B by performing a sequence of row operations on A. That is, $B = E_3E_2E_1A$.

The definition of row-equivalent matrices can be restated using elementary matrices, as follows.

Definition of Row Equivalence

Let A and B be $m \times n$ matrices. Matrix B is **row-equivalent** to A if there exists a finite number of elementary matrices $E_1, E_2, \ldots, E_k$ such that

$$B = E_kE_{k-1} \cdots E_2E_1A.$$

You know from Section 2.3 that not all square matrices are invertible. Every elementary matrix, however, is invertible. Moreover, the inverse of an elementary matrix is itself an elementary matrix.

THEOREM 2.13 Elementary Matrices Are Invertible

If E is an elementary matrix, then E^{-1} exists and is an elementary matrix.

To find the inverse of an elementary matrix E, simply reverse the elementary row operation used to obtain E. For instance, you can find the inverse of each of the three elementary matrices shown in Example 3 as follows.

Elementary Matrix		*Inverse Matrix*	
$E_1 = \begin{bmatrix} 0 & 1 & 0 \\ 1 & 0 & 0 \\ 0 & 0 & 1 \end{bmatrix}$	$R_1 \leftrightarrow R_2$	$E_1^{-1} = \begin{bmatrix} 0 & 1 & 0 \\ 1 & 0 & 0 \\ 0 & 0 & 1 \end{bmatrix}$	$R_1 \leftrightarrow R_2$
$E_2 = \begin{bmatrix} 1 & 0 & 0 \\ 0 & 1 & 0 \\ -2 & 0 & 1 \end{bmatrix}$	$R_3 + (-2)R_1 \rightarrow R_3$	$E_2^{-1} = \begin{bmatrix} 1 & 0 & 0 \\ 0 & 1 & 0 \\ 2 & 0 & 1 \end{bmatrix}$	$R_3 + (2)R_1 \rightarrow R_3$
$E_3 = \begin{bmatrix} 1 & 0 & 0 \\ 0 & 1 & 0 \\ 0 & 0 & \frac{1}{2} \end{bmatrix}$	$(\frac{1}{2})R_3 \rightarrow R_3$	$E_3^{-1} = \begin{bmatrix} 1 & 0 & 0 \\ 0 & 1 & 0 \\ 0 & 0 & 2 \end{bmatrix}$	$(2)R_3 \rightarrow R_3$

The following theorem states that every invertible matrix can be written as the product of elementary matrices.

THEOREM 2.14 A Property of Invertible Matrices

A square matrix A is invertible if and only if it can be written as the product of elementary matrices.

PROOF The phrase "if and only if" means that there are actually two parts to the theorem. On the one hand, you have to show that *if* A is invertible, *then* it can be written as the product of elementary matrices. Then you have to show that *if* A can be written as the product of elementary matrices, *then* A is invertible.

To prove the theorem in one direction, assume A is the product of elementary matrices. Then, because every elementary matrix is invertible and the product of invertible matrices is invertible, it follows that A is invertible.

To prove the theorem in the other direction, assume A is invertible. From Theorem 2.11 you know that the system of linear equations represented by $A\mathbf{x} = O$ has only the trivial solution. But this implies that the augmented matrix $[A \;\vdots\; O]$ can be rewritten in the form $[I \;\vdots\; O]$ (using elementary row operations corresponding to $E_1, E_2, \ldots,$ and E_k). We now have $E_k \cdots E_3E_2E_1A = I$ and it follows that $A = E_1^{-1}E_2^{-1}E_3^{-1} \cdots E_k^{-1}$. A can be written as the product of elementary matrices, and the proof is complete.

The first part of this proof is illustrated in Example 4.

EXAMPLE 4 Writing a Matrix as the Product of Elementary Matrices

Find a sequence of elementary matrices whose product is

$$A = \begin{bmatrix} -1 & -2 \\ 3 & 8 \end{bmatrix}.$$

SOLUTION Begin by finding a sequence of elementary row operations that can be used to rewrite A in reduced row-echelon form.

Matrix	*Elementary Row Operation*	*Elementary Matrix*
$\begin{bmatrix} 1 & 2 \\ 3 & 8 \end{bmatrix}$	$(-1)R_1 \rightarrow R_1$	$E_1 = \begin{bmatrix} -1 & 0 \\ 0 & 1 \end{bmatrix}$
$\begin{bmatrix} 1 & 2 \\ 0 & 2 \end{bmatrix}$	$R_2 + (-3)R_1 \rightarrow R_2$	$E_2 = \begin{bmatrix} 1 & 0 \\ -3 & 1 \end{bmatrix}$
$\begin{bmatrix} 1 & 2 \\ 0 & 1 \end{bmatrix}$	$(\frac{1}{2})R_2 \rightarrow R_2$	$E_3 = \begin{bmatrix} 1 & 0 \\ 0 & \frac{1}{2} \end{bmatrix}$
$\begin{bmatrix} 1 & 0 \\ 0 & 1 \end{bmatrix}$	$R_1 + (-2)R_2 \rightarrow R_1$	$E_4 = \begin{bmatrix} 1 & -2 \\ 0 & 1 \end{bmatrix}$

Now, from the matrix product $E_4E_3E_2E_1A = I$, solve for A to obtain $A = E_1^{-1}E_2^{-1}E_3^{-1}E_4^{-1}$. This implies that A is a product of elementary matrices.

$$A = \underset{E_1^{-1}}{\begin{bmatrix} -1 & 0 \\ 0 & 1 \end{bmatrix}} \underset{E_2^{-1}}{\begin{bmatrix} 1 & 0 \\ 3 & 1 \end{bmatrix}} \underset{E_3^{-1}}{\begin{bmatrix} 1 & 0 \\ 0 & 2 \end{bmatrix}} \underset{E_4^{-1}}{\begin{bmatrix} 1 & 2 \\ 0 & 1 \end{bmatrix}} = \begin{bmatrix} -1 & -2 \\ 3 & 8 \end{bmatrix}$$

In Section 2.3 you learned a process for finding the inverse of a nonsingular matrix A. There, you used Gauss-Jordan elimination to reduce the augmented matrix $[A \vdots I]$ to $[I \vdots A^{-1}]$. You can now use Theorem 2.14 to justify this procedure. Specifically, the proof of Theorem 2.14 allows you to write the product

$$I = E_k \cdot \cdot \cdot E_3E_2E_1A.$$

Multiplying both sides of this equation (on the right) by A^{-1}, we can write

$$A^{-1} = E_k \cdot \cdot \cdot E_3E_2E_1I.$$

In other words, a sequence of elementary matrices that reduces A to the identity also can be used to reduce the identity I to A^{-1}. Applying the corresponding sequence of elementary row operations to the matrices A and I simultaneously, you have

$$E_k \cdot \cdot \cdot E_3E_2E_1[A \vdots I] = [I \vdots A^{-1}].$$

Of course, if A is singular, then no such sequence can be found.

The next theorem ties together some important relationships between $n \times n$ matrices and systems of linear equations. The essential parts of this theorem have already been proved (see Theorems 2.11 and 2.14); it is left to you to fill in the other parts of the proof.

THEOREM 2.15
Equivalent Conditions

If A is an $n \times n$ matrix, then the following statements are equivalent.

1. A is invertible.
2. $A\mathbf{x} = \mathbf{b}$ has a unique solution for every $n \times 1$ column matrix $\mathbf{b}$.
3. $A\mathbf{x} = O$ has only the trivial solution.
4. A is row-equivalent to I_n.
5. A can be written as the product of elementary matrices.

The LU-Factorization

Solving systems of linear equations is the most important application of linear algebra. At the heart of the most efficient and modern algorithms for solving linear systems, $A\mathbf{x} = \mathbf{b}$ is the so-called LU-factorization, in which the square matrix A is expressed as a product, $A = LU$. In this product, the square matrix L is **lower triangular,** which means all the entries above the main diagonal are zero. The square matrix U is **upper triangular,** which means all the entries below the main diagonal are zero.

$$\begin{bmatrix} a_{11} & 0 & 0 \\ a_{21} & a_{22} & 0 \\ a_{31} & a_{32} & a_{33} \end{bmatrix} \qquad \begin{bmatrix} a_{11} & a_{12} & a_{13} \\ 0 & a_{22} & a_{23} \\ 0 & 0 & a_{33} \end{bmatrix}$$

3×3 lower triangular matrix $\qquad$ 3×3 upper triangular matrix

By writing $A\mathbf{x} = LU\mathbf{x}$ and letting $U\mathbf{x} = \mathbf{y}$, you can solve for $\mathbf{x}$ in two stages. First solve $L\mathbf{y} = \mathbf{b}$ for $\mathbf{y}$; then solve $U\mathbf{x} = \mathbf{y}$ for $\mathbf{x}$. Each system is easy to solve because the coefficient matrices are triangular. In particular, neither system requires any row operations.

Definition of *LU*-Factorization

If the $n \times n$ matrix A can be written as the product of a lower triangular matrix L and an upper triangular matrix U, then $A = LU$ is an ***LU*-factorization** of A.

EXAMPLE 5 *LU*-Factorizations

(a) $$\begin{bmatrix} 1 & 2 \\ 1 & 0 \end{bmatrix} = \begin{bmatrix} 1 & 0 \\ 1 & 1 \end{bmatrix} \begin{bmatrix} 1 & 2 \\ 0 & -2 \end{bmatrix} = LU$$

is an LU-factorization of the matrix $A = \begin{bmatrix} 1 & 2 \\ 1 & 0 \end{bmatrix}$ as the product of the lower triangular matrix $L = \begin{bmatrix} 1 & 0 \\ 1 & 1 \end{bmatrix}$ and the upper triangular matrix $U = \begin{bmatrix} 1 & 2 \\ 0 & -2 \end{bmatrix}$.

$$\text{(b) } A = \begin{bmatrix} 1 & -3 & 0 \\ 0 & 1 & 3 \\ 2 & -10 & 2 \end{bmatrix} = \begin{bmatrix} 1 & 0 & 0 \\ 0 & 1 & 0 \\ 2 & -4 & 1 \end{bmatrix} \begin{bmatrix} 1 & -3 & 0 \\ 0 & 1 & 3 \\ 0 & 0 & 14 \end{bmatrix} = LU$$

is an LU-factorization of the matrix A.

If a square matrix A can be row reduced to an upper triangular matrix U using only the row operation of adding a multiple of one row to another row below it, then it is easy to find an LU-factorization of the matrix A. All you need to do is keep track of the individual row operations, as indicated in the example below.

EXAMPLE 6 Finding the *LU*-Factorizations of a Matrix

Find the LU-factorization of the matrix $A = \begin{bmatrix} 1 & -3 & 0 \\ 0 & 1 & 3 \\ 2 & -10 & 2 \end{bmatrix}$.

SOLUTION Begin by row reducing A to upper triangular form while keeping track of the elementary matrices used for each row operation.

Matrix	*Elementary Row Operation*	*Elementary Matrix*
$\begin{bmatrix} 1 & -3 & 0 \\ 0 & 1 & 3 \\ 0 & -4 & 2 \end{bmatrix}$	$R_3 + (-2)R_1 \rightarrow R_3$	$E_1 = \begin{bmatrix} 1 & 0 & 0 \\ 0 & 1 & 0 \\ -2 & 0 & 1 \end{bmatrix}$
$\begin{bmatrix} 1 & -3 & 0 \\ 0 & 1 & 3 \\ 0 & 0 & 14 \end{bmatrix}$	$R_3 + (4)R_2 \rightarrow R_3$	$E_2 = \begin{bmatrix} 1 & 0 & 0 \\ 0 & 1 & 0 \\ 0 & 4 & 1 \end{bmatrix}$

The matrix U on the left is upper triangular, and it follows that $E_2E_1A = U$, or $A = E_1^{-1}E_2^{-1}U$. Because the product of the lower triangular matrices

$$E_1^{-1}E_2^{-1} = \begin{bmatrix} 1 & 0 & 0 \\ 0 & 1 & 0 \\ 2 & 0 & 1 \end{bmatrix} \begin{bmatrix} 1 & 0 & 0 \\ 0 & 1 & 0 \\ 0 & -4 & 1 \end{bmatrix} = \begin{bmatrix} 1 & 0 & 0 \\ 0 & 1 & 0 \\ 2 & -4 & 1 \end{bmatrix}$$

is again a lower triangular matrix L, the factorization $A = LU$ is complete. Notice that this is the same LU-factorization that is shown in Example 5(b) at the top of this page.

In general, if A can be row reduced to an upper triangular matrix U using only the row operation of adding a multiple of one row to another row, then A has an LU-factorization.

$$E_k \cdots E_2E_1A = U$$

$$A = E_1^{-1}E_2^{-1} \cdots E_k^{-1}U$$

$$A = LU$$

Here L is the product of the inverses of the elementary matrices used in the row reduction.

Note that the multipliers in Example 6 are -2 and 4, which are the negatives of the corresponding entries in L. This is true in general. If U can be obtained from A using only the row operation of adding a multiple of one row to another row below, then the matrix L is lower triangular with 1's along the diagonal. Furthermore, the negative of each multiplier is in the same position as that of the corresponding zero in U.

Once you have obtained an LU–factorization of a matrix A, you can then solve the system of n linear equations in n variables $A\mathbf{x} = \mathbf{b}$ very efficiently in two steps.

1. Write $\mathbf{y} = U\mathbf{x}$ and solve $L\mathbf{y} = \mathbf{b}$ for $\mathbf{y}$.
2. Solve $U\mathbf{x} = \mathbf{y}$ for $\mathbf{x}$.

The column matrix $\mathbf{x}$ is the solution of the original system because

$$A\mathbf{x} = LU\mathbf{x} = L\mathbf{y} = \mathbf{b}.$$

The second step in this algorithm is just back-substitution, because the matrix U is upper triangular. The first step is similar, except that it starts at the top of the matrix, because L is lower triangular. For this reason, the first step is often called **forward substitution.**

EXAMPLE 7 Solving a Linear System Using *LU*-Factorization

Solve the linear system.

$$\begin{aligned} x_1 - 3x_2 &= -5 \\ x_2 + 3x_3 &= -1 \\ 2x_1 - 10x_2 + 2x_3 &= -20 \end{aligned}$$

SOLUTION You obtained the LU-factorization of the coefficient matrix A in Example 6.

$$A = \begin{bmatrix} 1 & -3 & 0 \\ 0 & 1 & 3 \\ 2 & -10 & 2 \end{bmatrix} = \begin{bmatrix} 1 & 0 & 0 \\ 0 & 1 & 0 \\ 2 & -4 & 1 \end{bmatrix} \begin{bmatrix} 1 & -3 & 0 \\ 0 & 1 & 3 \\ 0 & 0 & 14 \end{bmatrix}$$

First, let $\mathbf{y} = U\mathbf{x}$ and solve the system $L\mathbf{y} = \mathbf{b}$ for $\mathbf{y}$.

$$\begin{bmatrix} 1 & 0 & 0 \\ 0 & 1 & 0 \\ 2 & -4 & 1 \end{bmatrix} \begin{bmatrix} y_1 \\ y_2 \\ y_3 \end{bmatrix} = \begin{bmatrix} -5 \\ -1 \\ -20 \end{bmatrix}$$

This system is easy to solve using forward substitution. Starting with the first equation, you have $y_1 = -5$. The second equation gives $y_2 = -1$. Finally, from the third equation,

$$\begin{aligned} 2y_1 - 4y_2 + y_3 &= -20 \\ y_3 &= -20 - 2y_1 + 4y_2 \\ y_3 &= -20 - 2(-5) + 4(-1) \\ y_3 &= -14. \end{aligned}$$

The solution of $L\mathbf{y} = \mathbf{b}$ is

$$\mathbf{y} = \begin{bmatrix} -5 \\ -1 \\ -14 \end{bmatrix}.$$

Now solve the system $U\mathbf{x} = \mathbf{y}$ for $\mathbf{x}$ using back-substitution.

$$\begin{bmatrix} 1 & -3 & 0 \\ 0 & 1 & 3 \\ 0 & 0 & 14 \end{bmatrix}\begin{bmatrix} x_1 \\ x_2 \\ x_3 \end{bmatrix} = \begin{bmatrix} -5 \\ -1 \\ -14 \end{bmatrix}$$

From the bottom equation, $x_3 = -1$. Then, the second equation gives $x_2 + 3(-1) = -1$, or $x_2 = 2$. Finally, the first equation is $x_1 - 3(2) = -5$, or $x_1 = 1$. So, the solution of the original system of equations is

$$\mathbf{x} = \begin{bmatrix} 1 \\ 2 \\ -1 \end{bmatrix}.$$

SECTION 2.4 Exercises

In Exercises 1–8, determine whether the matrix is elementary. If it is, state the elementary row operation used to produce it.

1. $\begin{bmatrix} 1 & 0 \\ 0 & 2 \end{bmatrix}$

2. $\begin{bmatrix} 1 & 0 & 0 \\ 0 & 0 & 1 \end{bmatrix}$

3. $\begin{bmatrix} 1 & 0 \\ 2 & 1 \end{bmatrix}$

4. $\begin{bmatrix} 0 & 1 \\ 1 & 0 \end{bmatrix}$

5. $\begin{bmatrix} 2 & 0 & 0 \\ 0 & 0 & 1 \\ 0 & 1 & 0 \end{bmatrix}$

6. $\begin{bmatrix} 1 & 0 & 0 \\ 0 & 1 & 0 \\ 2 & 0 & 1 \end{bmatrix}$

7. $\begin{bmatrix} 1 & 0 & 0 & 0 \\ 0 & 1 & 0 & 0 \\ 0 & -5 & 1 & 0 \\ 0 & 0 & 0 & 1 \end{bmatrix}$

8. $\begin{bmatrix} 1 & 0 & 0 & 0 \\ 2 & 1 & 0 & 0 \\ 0 & 0 & 1 & 0 \\ 0 & 0 & -3 & 1 \end{bmatrix}$

In Exercises 9–12, let A, B, and C be

$$A = \begin{bmatrix} 1 & 2 & -3 \\ 0 & 1 & 2 \\ -1 & 2 & 0 \end{bmatrix}, \quad B = \begin{bmatrix} -1 & 2 & 0 \\ 0 & 1 & 2 \\ 1 & 2 & -3 \end{bmatrix},$$

$$C = \begin{bmatrix} 0 & 4 & -3 \\ 0 & 1 & 2 \\ -1 & 2 & 0 \end{bmatrix}.$$

9. Find an elementary matrix E such that $EA = B$.

10. Find an elementary matrix E such that $EA = C$.

11. Find an elementary matrix E such that $EB = A$.

12. Find an elementary matrix E such that $EC = A$.

In Exercises 13–20, find the inverse of the elementary matrix.

13. $\begin{bmatrix} 0 & 1 \\ 1 & 0 \end{bmatrix}$

14. $\begin{bmatrix} 5 & 0 \\ 0 & 1 \end{bmatrix}$

15. $\begin{bmatrix} 0 & 0 & 1 \\ 0 & 1 & 0 \\ 1 & 0 & 0 \end{bmatrix}$

16. $\begin{bmatrix} 1 & 0 & 0 \\ 0 & 1 & 0 \\ 0 & -3 & 1 \end{bmatrix}$

17. $\begin{bmatrix} k & 0 \\ 0 & 1 \end{bmatrix}, \quad k \neq 0$

18. $\begin{bmatrix} k & 0 & 0 \\ 0 & 1 & 0 \\ 0 & 0 & 1 \end{bmatrix}, \quad k \neq 0$

19. $\begin{bmatrix} 1 & 0 & 0 \\ 0 & 0 & 1 \\ 0 & 1 & 0 \end{bmatrix}$

20. $\begin{bmatrix} 1 & 0 & 0 & 0 \\ 0 & 1 & k & 0 \\ 0 & 0 & 1 & 0 \\ 0 & 0 & 0 & 1 \end{bmatrix}$

In Exercises 21–24, find the inverse of the matrix using elementary matrices.

21. $\begin{bmatrix} 3 & -2 \\ 1 & 0 \end{bmatrix}$ **22.** $\begin{bmatrix} 2 & 0 \\ 1 & 1 \end{bmatrix}$

23. $\begin{bmatrix} 1 & 0 & -1 \\ 0 & 6 & -1 \\ 0 & 0 & 4 \end{bmatrix}$ **24.** $\begin{bmatrix} 1 & 0 & -2 \\ 0 & 2 & 1 \\ 0 & 0 & 1 \end{bmatrix}$

In Exercises 25–32, factor the matrix A into a product of elementary matrices.

25. $A = \begin{bmatrix} 1 & 2 \\ 1 & 0 \end{bmatrix}$ **26.** $A = \begin{bmatrix} 0 & 1 \\ 1 & 0 \end{bmatrix}$

27. $A = \begin{bmatrix} 4 & -1 \\ 3 & -1 \end{bmatrix}$ **28.** $A = \begin{bmatrix} 1 & 1 \\ 2 & 1 \end{bmatrix}$

29. $A = \begin{bmatrix} 1 & -2 & 0 \\ -1 & 3 & 0 \\ 0 & 0 & 1 \end{bmatrix}$ **30.** $A = \begin{bmatrix} 1 & 2 & 3 \\ 2 & 5 & 6 \\ 1 & 3 & 4 \end{bmatrix}$

31. $A = \begin{bmatrix} 1 & 0 & 0 & 1 \\ 0 & -1 & 3 & 0 \\ 0 & 0 & 2 & 0 \\ 0 & 0 & 1 & -1 \end{bmatrix}$ **32.** $A = \begin{bmatrix} 4 & 0 & 0 & 2 \\ 0 & 1 & 0 & 1 \\ 0 & 0 & -1 & 2 \\ 1 & 0 & 0 & -2 \end{bmatrix}$

True or False? In Exercises 33 and 34, determine whether each statement is true or false. If a statement is true, give a reason or cite an appropriate statement from the text. If a statement is false, provide an example that shows the statement is not true in all cases or cite an appropriate statement from the text.

33. (a) The identity matrix is an elementary matrix.

(b) If E is an elementary matrix, then $2E$ is an elementary matrix.

(c) The matrix A is row-equivalent to the matrix B if there exists a finite number of elementary matrices $E_1, E_2, \ldots, E_k$ such that $A = E_k E_{k-1} \cdots E_2 E_1 B$.

(d) The inverse of an elementary matrix is an elementary matrix.

34. (a) The zero matrix is an elementary matrix.

(b) A square matrix is nonsingular if it can be written as the product of elementary matrices.

(c) $A\mathbf{x} = O$ has only the trivial solution if and only if $A\mathbf{x} = \mathbf{b}$ has a unique solution for every $n \times 1$ column matrix $\mathbf{b}$.

35. Writing E is the elementary matrix obtained by interchanging two rows in I_n. A is an $n \times n$ matrix.

(a) How will EA compare with A?

(b) Find E^2.

36. Writing E is the elementary matrix obtained by multiplying a row in I_n by a nonzero constant c. A is an $n \times n$ matrix.

(a) How will EA compare with A?

(b) Find E^2.

37. Use elementary matrices to find the inverse of

$$A = \begin{bmatrix} 1 & a & 0 \\ 0 & 1 & 0 \\ 0 & 0 & 1 \end{bmatrix} \begin{bmatrix} 1 & 0 & 0 \\ b & 1 & 0 \\ 0 & 0 & 1 \end{bmatrix} \begin{bmatrix} 1 & 0 & 0 \\ 0 & 1 & 0 \\ 0 & 0 & c \end{bmatrix}, \quad c \neq 0.$$

38. Use elementary matrices to find the inverse of

$$A = \begin{bmatrix} 1 & 0 & 0 \\ 0 & 1 & 0 \\ a & b & c \end{bmatrix}, \quad c \neq 0.$$

39. Writing Is the product of two elementary matrices always elementary? Explain why or why not and provide appropriate examples to illustrate your conclusion.

40. Writing Is the sum of two elementary matrices always elementary? Explain why or why not and provide appropriate examples to illustrate your conclusion.

In Exercises 41–44, find the LU-factorization of the matrix.

41. $\begin{bmatrix} 1 & 0 \\ -2 & 1 \end{bmatrix}$ **42.** $\begin{bmatrix} -2 & 1 \\ -6 & 4 \end{bmatrix}$

43. $\begin{bmatrix} 3 & 0 & 1 \\ 6 & 1 & 1 \\ -3 & 1 & 0 \end{bmatrix}$ **44.** $\begin{bmatrix} 2 & 0 & 0 \\ 0 & -3 & 1 \\ 10 & 12 & 3 \end{bmatrix}$

In Exercises 45 and 46, solve the linear system $A\mathbf{x} = \mathbf{b}$ by
(a) finding the LU-factorization of the coefficient matrix A,
(b) solving the lower triangular system $L\mathbf{y} = \mathbf{b}$, and
(c) solving the upper triangular system $U\mathbf{x} = \mathbf{y}$.

45.
$$\begin{aligned} 2x + y \phantom{{}+z} &= 1 \\ y - z &= 2 \\ -2x + y + z &= -2 \end{aligned}$$

46.
$$\begin{aligned} 2x_1 \phantom{{}+x_2-x_3} &= 4 \\ -2x_1 + x_2 - x_3 &= -4 \\ 6x_1 + 2x_2 + x_3 &= 15 \\ -x_4 &= -1 \end{aligned}$$

47. Writing Suppose you needed to solve many systems of linear equations $A\mathbf{x} = \mathbf{b}_i$, each having the same coefficient matrix A. Explain how you could use the LU-factorization technique to make the task easier, rather than solving each system individually using Gaussian elimination.

48. (a) Show that the matrix

$$A = \begin{bmatrix} 0 & 1 \\ 1 & 0 \end{bmatrix}$$

does not have an LU-factorization.

(b) Find the LU-factorization of the matrix

$$A = \begin{bmatrix} a & b \\ c & d \end{bmatrix}$$

that has 1's along the main diagonal of L. Are there any restrictions on the matrix A?

In Exercises 49–54, determine whether the matrix is idempotent. A square matrix A is **idempotent** if $A^2 = A$.

49. $\begin{bmatrix} 1 & 0 \\ 0 & 0 \end{bmatrix}$

50. $\begin{bmatrix} 0 & 1 \\ 1 & 0 \end{bmatrix}$

51. $\begin{bmatrix} 2 & 3 \\ -1 & -2 \end{bmatrix}$

52. $\begin{bmatrix} 2 & 3 \\ 1 & 2 \end{bmatrix}$

53. $\begin{bmatrix} 0 & 0 & 1 \\ 0 & 1 & 0 \\ 1 & 0 & 0 \end{bmatrix}$

54. $\begin{bmatrix} 0 & 1 & 0 \\ 1 & 0 & 0 \\ 0 & 0 & 1 \end{bmatrix}$

55. Determine a and b such that A is idempotent.

$$A = \begin{bmatrix} 1 & 0 \\ a & b \end{bmatrix}$$

56. Determine conditions on a, b, and c such that A is idempotent.

$$A = \begin{bmatrix} a & 0 \\ b & c \end{bmatrix}$$

57. Prove that if A is an $n \times n$ matrix that is idempotent and invertible, then $A = I_n$.

58. Guided Proof Prove that A is idempotent if and only if A^T is idempotent.

Getting Started: The phrase "if and only if" means that you have to prove two statements:

1. If A is idempotent, then A^T is idempotent.
2. If A^T is idempotent, then A is idempotent.

(i) Begin your proof of the first statement by assuming that A is idempotent.
(ii) This means that $A^2 = A$.
(iii) Use the properties of the transpose to show that A^T is idempotent.
(iv) Begin your proof of the second statement by assuming that A^T is idempotent.

59. Prove that if A and B are idempotent and $AB = BA$, then AB is idempotent.

60. Prove that if A is row-equivalent to B, then B is row-equivalent to A.

61. Guided Proof Prove that if A is row-equivalent to B and B is row-equivalent to C, then A is row-equivalent to C.

Getting Started: To prove that A is row-equivalent to C, you have to find elementary matrices $E_1, \ldots, E_k$ such that $A = E_k \cdots E_1 C$.

(i) Begin your proof by observing that A is row-equivalent to B.
(ii) Meaning, there exist elementary matrices $F_1, \ldots, F_n$ such that $A = F_n \cdots F_1 B$.
(iii) There exist elementary matrices $G_1, \ldots, G_m$ such that $B = G_1 \cdots G_m C$.
(iv) Combine the matrix equations from steps (ii) and (iii).

62. Let A be a nonsingular matrix. Prove that if B is row-equivalent to A, then B is also nonsingular.

2.5 Applications of Matrix Operations

Stochastic Matrices

Many types of applications involve a finite set of *states* $\{S_1, S_2, \ldots, S_n\}$ of a given population. For instance, residents of a city may live downtown or in the suburbs. Voters may vote Democrat, Republican, or for a third party. Soft drink consumers may buy Coca-Cola, Pepsi Cola, or another brand.

The probability that a member of a population will change from the jth state to the ith state is represented by a number p_{ij}, where $0 \le p_{ij} \le 1$. A probability of $p_{ij} = 0$ means that the member is certain *not* to change from the jth state to the ith state, whereas a probability of $p_{ij} = 1$ means that the member is certain to change from the jth state to the ith state.

$$P = \overset{\text{From}}{\overbrace{\begin{matrix} S_1 & S_2 & \cdots & S_n \end{matrix}}} \atop \begin{bmatrix} p_{11} & p_{12} & \cdots & p_{1n} \\ p_{21} & p_{22} & \cdots & p_{2n} \\ \vdots & \vdots & & \vdots \\ p_{n1} & p_{n2} & \cdots & p_{nn} \end{bmatrix} \begin{matrix} S_1 \\ S_2 \\ \vdots \\ S_n \end{matrix} \quad \text{To}$$

P is called the **matrix of transition probabilities** because it gives the probabilities of each possible type of transition (or change) within the population.

At each transition, each member in a given state must either stay in that state or change to another state. For probabilities, this means that the sum of the entries in any column of P is 1. For instance, in the first column we have

$$p_{11} + p_{21} + \cdots + p_{n1} = 1.$$

In general, such a matrix is called **stochastic** (the term "stochastic" means "regarding conjecture"). That is, an $n \times n$ matrix P is called a **stochastic matrix** if each entry is a number between 0 and 1 and each column of P adds up to 1.

EXAMPLE 1 **Examples of Stochastic Matrices and Nonstochastic Matrices**

The matrices in parts (a) and (b) are stochastic, but the matrix in part (c) is not.

$$\text{(a)} \begin{bmatrix} 1 & 0 & 0 \\ 0 & 1 & 0 \\ 0 & 0 & 1 \end{bmatrix} \qquad \text{(b)} \begin{bmatrix} \frac{1}{2} & \frac{1}{3} & \frac{1}{4} \\ \frac{1}{4} & 0 & \frac{3}{4} \\ \frac{1}{4} & \frac{2}{3} & 0 \end{bmatrix} \qquad \text{(c)} \begin{bmatrix} 0.1 & 0.2 & 0.3 \\ 0.2 & 0.3 & 0.4 \\ 0.3 & 0.4 & 0.5 \end{bmatrix}$$

Example 2 describes the use of a stochastic matrix to measure consumer preferences.

EXAMPLE 2 **A Consumer Preference Model**

Two competing companies offer cable television service to a city of 100,000 households. The changes in cable subscriptions each year are shown in Figure 2.1. Company A now has 15,000 subscribers and Company B has 20,000 subscribers. How many subscribers will each company have 1 year from now?

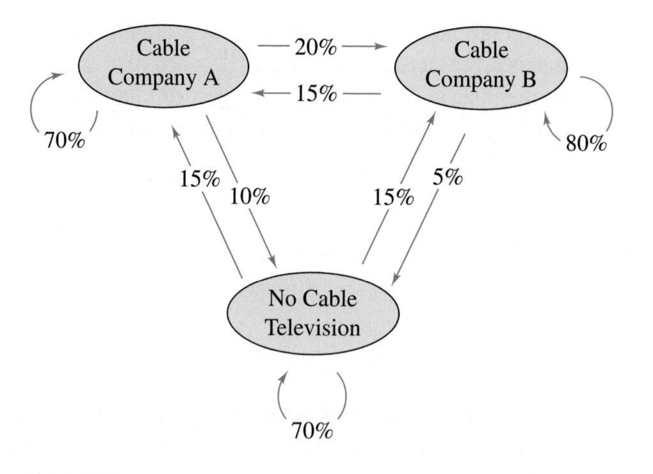

Figure 2.1

SOLUTION The matrix representing the given transition probabilities is

$$P = \overset{\textit{From}}{\overbrace{\begin{matrix} \text{A} & \text{B} & \text{None} \end{matrix}}} \quad \begin{bmatrix} 0.70 & 0.15 & 0.15 \\ 0.20 & 0.80 & 0.15 \\ 0.10 & 0.05 & 0.70 \end{bmatrix} \begin{matrix} \text{A} \\ \text{B} \\ \text{None} \end{matrix} \Bigg\} \textit{To}$$

and the **state matrix** representing the current populations in the three states is

$$X = \begin{bmatrix} 15{,}000 \\ 20{,}000 \\ 65{,}000 \end{bmatrix}. \quad \begin{matrix} \textbf{A} \\ \textbf{B} \\ \textbf{None} \end{matrix}$$

To find the state matrix representing the populations in the three states after one year, multiply P by X to obtain

$$PX = \begin{bmatrix} 0.70 & 0.15 & 0.15 \\ 0.20 & 0.80 & 0.15 \\ 0.10 & 0.05 & 0.70 \end{bmatrix} \begin{bmatrix} 15{,}000 \\ 20{,}000 \\ 65{,}000 \end{bmatrix} = \begin{bmatrix} 23{,}250 \\ 28{,}750 \\ 48{,}000 \end{bmatrix}.$$

After one year, Company A will have 23,250 subscribers and Company B will have 28,750 subscribers.

One of the appeals of the matrix solution in Example 2 is that once the model has been created, it becomes easy to find the state matrices representing future years by repeatedly multiplying by the matrix P. This process is demonstrated in Example 3.

EXAMPLE 3 A Consumer Preference Model

Assuming the matrix of transition probabilities from Example 2 remains the same year after year, find the number of subscribers each cable television company will have after (a) 3 years, (b) 5 years, and (c) 10 years. (The answers in this example have been rounded to the nearest person.)

SOLUTION (a) From Example 2 you know that the number of subscribers after 1 year is

$$PX = \begin{bmatrix} 23{,}250 \\ 28{,}750 \\ 48{,}000 \end{bmatrix}. \quad \begin{matrix} \text{A} \\ \text{B} \\ \text{None} \end{matrix} \qquad \text{After 1 year}$$

Because the matrix of transition probabilities is the same from the first to the third year, the number of subscribers after 3 years is

$$P^3X = \begin{bmatrix} 30{,}283 \\ 39{,}042 \\ 30{,}675 \end{bmatrix}. \quad \begin{matrix} \text{A} \\ \text{B} \\ \text{None} \end{matrix} \qquad \text{After 3 years}$$

After 3 years, Company A will have 30,283 subscribers and Company B will have 39,042 subscribers.

(b) The number of subscribers after 5 years is

$$P^5X = \begin{bmatrix} 32{,}411 \\ 43{,}812 \\ 23{,}777 \end{bmatrix}. \quad \begin{matrix} \text{A} \\ \text{B} \\ \text{None} \end{matrix} \qquad \text{After 5 years}$$

After 5 years, Company A will have 32,411 subscribers and Company B will have 43,812 subscribers.

(c) The number of subscribers after 10 years is

$$P^{10}X = \begin{bmatrix} 33{,}287 \\ 47{,}147 \\ 19{,}566 \end{bmatrix}. \quad \begin{matrix} \text{A} \\ \text{B} \\ \text{None} \end{matrix} \qquad \text{After 10 years}$$

After 10 years, Company A will have 33,287 subscribers and Company B will have 47,147 subscribers.

In Example 3, notice that there is little difference between the numbers of subscribers after 5 years and after 10 years. If the process shown in this example is continued, the numbers of subscribers eventually reach a **steady state.** That is, as long as the matrix P doesn't change, the matrix product P^nX approaches a limit $\overline{X}$. In this particular example, the limit is the steady state matrix

$$\overline{X} = \begin{bmatrix} 33{,}333 \\ 47{,}619 \\ 19{,}048 \end{bmatrix}. \quad \begin{matrix} \text{A} \\ \text{B} \\ \text{None} \end{matrix} \qquad \text{Steady state}$$

You can check to see that $P\overline{X} = \overline{X}$.

Cryptography

A **cryptogram** is a message written according to a secret code (the Greek word *kryptos* means "hidden"). This section describes a method of using matrix multiplication to **encode** and **decode** messages.

Begin by assigning a number to each letter in the alphabet (with 0 assigned to a blank space), as follows.

0 = _	14 = N
1 = A	15 = O
2 = B	16 = P
3 = C	17 = Q
4 = D	18 = R
5 = E	19 = S
6 = F	20 = T
7 = G	21 = U
8 = H	22 = V
9 = I	23 = W
10 = J	24 = X
11 = K	25 = Y
12 = L	26 = Z
13 = M	

Then the message is converted to numbers and partitioned into **uncoded row matrices,** each having n entries, as demonstrated in Example 4.

EXAMPLE 4 Forming Uncoded Row Matrices

Write the uncoded row matrices of size 1×3 for the message MEET ME MONDAY.

SOLUTION Partitioning the message (including blank spaces, but ignoring punctuation) into groups of three produces the following uncoded row matrices.

$$\begin{matrix} [13 & 5 & 5] & [20 & 0 & 13] & [5 & 0 & 13] & [15 & 14 & 4] & [1 & 25 & 0] \\ M & E & E & T & \_ & M & E & \_ & M & O & N & D & A & Y & \_ \end{matrix}$$

Note that a blank space is used to fill out the last uncoded row matrix.

To **encode** a message, choose an $n \times n$ invertible matrix A and multiply the uncoded row matrices (on the right) by A to obtain **coded row matrices.** This process is demonstrated in Example 5.

EXAMPLE 5 Encoding a Message

Use the matrix

$$A = \begin{bmatrix} 1 & -2 & 2 \\ -1 & 1 & 3 \\ 1 & -1 & -4 \end{bmatrix}$$

to encode the message MEET ME MONDAY.

SOLUTION The coded row matrices are obtained by multiplying each of the uncoded row matrices found in Example 4 by the matrix A, as follows.

Uncoded Row Matrix	Encoding Matrix A	Coded Row Matrix

$$\begin{bmatrix} 13 & 5 & 5 \end{bmatrix} \begin{bmatrix} 1 & -2 & 2 \\ -1 & 1 & 3 \\ 1 & -1 & -4 \end{bmatrix} = \begin{bmatrix} 13 & -26 & 21 \end{bmatrix}$$

$$\begin{bmatrix} 20 & 0 & 13 \end{bmatrix} \begin{bmatrix} 1 & -2 & 2 \\ -1 & 1 & 3 \\ 1 & -1 & -4 \end{bmatrix} = \begin{bmatrix} 33 & -53 & -12 \end{bmatrix}$$

$$\begin{bmatrix} 5 & 0 & 13 \end{bmatrix} \begin{bmatrix} 1 & -2 & 2 \\ -1 & 1 & 3 \\ 1 & -1 & -4 \end{bmatrix} = \begin{bmatrix} 18 & -23 & -42 \end{bmatrix}$$

$$\begin{bmatrix} 15 & 14 & 4 \end{bmatrix} \begin{bmatrix} 1 & -2 & 2 \\ -1 & 1 & 3 \\ 1 & -1 & -4 \end{bmatrix} = \begin{bmatrix} 5 & -20 & 56 \end{bmatrix}$$

$$\begin{bmatrix} 1 & 25 & 0 \end{bmatrix} \begin{bmatrix} 1 & -2 & 2 \\ -1 & 1 & 3 \\ 1 & -1 & -4 \end{bmatrix} = \begin{bmatrix} -24 & 23 & 77 \end{bmatrix}$$

The sequence of coded row matrices is

$$[13\ \ -26\ \ 21][33\ \ -53\ \ -12][18\ \ -23\ \ -42][5\ \ -20\ \ 56][-24\ \ 23\ \ 77].$$

Finally, removing the brackets produces the cryptogram below.

$$13\ \ -26\ \ 21\ \ 33\ \ -53\ \ -12\ \ 18\ \ -23\ \ -42\ \ 5\ \ -20\ \ 56\ \ -24\ \ 23\ \ 77$$

For those who do not know the matrix A, decoding the cryptogram found in Example 5 is difficult. But for an authorized receiver who knows the matrix A, decoding is simple. The receiver need only multiply the coded row matrices by A^{-1} to retrieve the uncoded row matrices. In other words, if

$$X = [x_1\ \ x_2\ \cdots\ x_n]$$

is an uncoded $1 \times n$ matrix, then $Y = XA$ is the corresponding encoded matrix. The receiver of the encoded matrix can decode Y by multiplying on the right by A^{-1} to obtain

$$YA^{-1} = (XA)A^{-1} = X.$$

This procedure is demonstrated in Example 6.

Simulation
Explore this concept further with an electronic simulation available on the website *college.hmco.com/pic/larsonELA6e*.

EXAMPLE 6 Decoding a Message

Use the inverse of the matrix

$$A = \begin{bmatrix} 1 & -2 & 2 \\ -1 & 1 & 3 \\ 1 & -1 & -4 \end{bmatrix}$$

to decode the cryptogram

$$13 \quad -26 \quad 21 \quad 33 \quad -53 \quad -12 \quad 18 \quad -23 \quad -42 \quad 5 \quad -20 \quad 56 \quad -24 \quad 23 \quad 77.$$

SOLUTION Begin by using Gauss-Jordan elimination to find A^{-1}.

$$\underset{[A \;\vdots\; I]}{\left[\begin{array}{rrr|rrr} 1 & -2 & 2 & 1 & 0 & 0 \\ -1 & 1 & 3 & 0 & 1 & 0 \\ 1 & -1 & -4 & 0 & 0 & 1 \end{array}\right]} \longrightarrow \underset{[I \;\vdots\; A^{-1}]}{\left[\begin{array}{rrr|rrr} 1 & 0 & 0 & -1 & -10 & -8 \\ 0 & 1 & 0 & -1 & -6 & -5 \\ 0 & 0 & 1 & 0 & -1 & -1 \end{array}\right]}$$

Now, to decode the message, partition the message into groups of three to form the coded row matrices

$$[13 \;\; -26 \;\; 21][33 \;\; -53 \;\; -12][18 \;\; -23 \;\; -42][5 \;\; -20 \;\; 56][-24 \;\; 23 \;\; 77].$$

To obtain the decoded row matrices, multiply each coded row matrix by A^{-1} (on the right).

Coded Row Matrix	*Decoding Matrix A^{-1}*	*Decoded Row Matrix*

$$[13 \;\; -26 \;\; 21]\begin{bmatrix} -1 & -10 & -8 \\ -1 & -6 & -5 \\ 0 & -1 & -1 \end{bmatrix} = [13 \;\; 5 \;\; 5]$$

$$[33 \;\; -53 \;\; -12]\begin{bmatrix} -1 & -10 & -8 \\ -1 & -6 & -5 \\ 0 & -1 & -1 \end{bmatrix} = [20 \;\; 0 \;\; 13]$$

$$[18 \;\; -23 \;\; -42]\begin{bmatrix} -1 & -10 & -8 \\ -1 & -6 & -5 \\ 0 & -1 & -1 \end{bmatrix} = [5 \;\; 0 \;\; 13]$$

$$[5 \;\; -20 \;\; 56]\begin{bmatrix} -1 & -10 & -8 \\ -1 & -6 & -5 \\ 0 & -1 & -1 \end{bmatrix} = [15 \;\; 14 \;\; 4]$$

$$\underset{\text{Coded Row Matrix}}{\begin{bmatrix} -24 & 23 & 77 \end{bmatrix}} \underset{\text{Decoding Matrix } A^{-1}}{\begin{bmatrix} -1 & -10 & -8 \\ -1 & -6 & -5 \\ 0 & -1 & -1 \end{bmatrix}} = \underset{\text{Decoded Row Matrix}}{\begin{bmatrix} 1 & 25 & 0 \end{bmatrix}}$$

The sequence of decoded row matrices is

$$[13 \quad 5 \quad 5]\,[20 \quad 0 \quad 13]\,[5 \quad 0 \quad 13]\,[15 \quad 14 \quad 4]\,[1 \quad 25 \quad 0]$$

and the message is

13	5	5	20	0	13	5	0	13	15	14	4	1	25	0.
M	*E*	*E*	*T*	_	*M*	*E*	_	*M*	*O*	*N*	*D*	*A*	*Y*	_

Leontief Input-Output Models

Matrix algebra has proved effective in analyzing problems concerning the input and output of an economic system. The model discussed here, developed by the American economist Wassily W. Leontief (1906–1999), was first published in 1936. In 1973, Leontief was awarded a Nobel prize for his work in economics.

Suppose that an economic system has n different industries $I_1, I_2, \ldots, I_n$, each of which has **input** needs (raw materials, utilities, etc.) and an **output** (finished product). In producing each unit of output, an industry may use the outputs of other industries, including itself. For example, an electric utility uses outputs from other industries, such as coal and water, and even uses its own electricity.

Let d_{ij} be the amount of output the jth industry needs from the ith industry to produce one unit of output per year. The matrix of these coefficients is called the **input-output matrix.**

$$D = \overset{\text{User (Output)}}{\overset{\begin{matrix} I_1 & I_2 & \cdots & I_n \end{matrix}}{\begin{bmatrix} d_{11} & d_{12} & \cdots & d_{1n} \\ d_{21} & d_{22} & \cdots & d_{2n} \\ \vdots & \vdots & & \vdots \\ d_{n1} & d_{n2} & \cdots & d_{nn} \end{bmatrix}}} \begin{matrix} I_1 \\ I_2 \\ \vdots \\ I_n \end{matrix} \Bigg\} \text{ Supplier (Input)}$$

To understand how to use this matrix, imagine $d_{12} = 0.4$. This means that 0.4 unit of Industry 1's product must be used to produce one unit of Industry 2's product. If $d_{33} = 0.2$, then 0.2 unit of Industry 3's product is needed to produce one unit of its own product. For this model to work, the values of d_{ij} must satisfy $0 \leq d_{ij} \leq 1$ and the sum of the entries in any column must be less than or equal to 1.

EXAMPLE 7 **Forming an Input-Output Matrix**

Consider a simple economic system consisting of three industries: electricity, water, and coal. Production, or output, of one unit of electricity requires 0.5 unit of itself, 0.25 unit of water, and 0.25 unit of coal. Production, or output, of one unit of water requires 0.1 unit of electricity, 0.6 unit of itself, and 0 units of coal. Production, or output, of one unit of coal requires 0.2 unit of electricity, 0.15 unit of water, and 0.5 unit of itself. Find the input-output matrix for this system.

SOLUTION The column entries show the amounts each industry requires from the others, as well as from itself, in order to produce one unit of output.

$$\overbrace{\begin{matrix} E & \quad W & \quad C \end{matrix}}^{\textit{User (Output)}}$$
$$\begin{bmatrix} 0.5 & 0.1 & 0.2 \\ 0.25 & 0.6 & 0.15 \\ 0.25 & 0 & 0.5 \end{bmatrix} \begin{matrix} E \\ W \\ C \end{matrix} \Bigg\} \quad \textit{Supplier (Input)}$$

The row entries show the amounts each industry supplies to the other industries, as well as to itself, in order for that particular industry to produce one unit of output. For instance, the electricity industry supplies 0.5 unit to itself, 0.1 unit to water, and 0.2 unit to coal.

To develop the Leontief input-output model further, let the total output of the ith industry be denoted by x_i. If the economic system is **closed** (meaning that it sells its products only to industries within the system, as in the example above), then the total output of the ith industry is given by the linear equation

$$x_i = d_{i1}x_1 + d_{i2}x_2 + \cdot \cdot \cdot + d_{in}x_n. \qquad \textbf{Closed system}$$

On the other hand, if the industries within the system sell products to nonproducing groups (such as governments or charitable organizations) outside the system, then the system is called **open** and the total output of the ith industry is given by

$$x_i = d_{i1}x_1 + d_{i2}x_2 + \cdot \cdot \cdot + d_{in}x_n + e_i, \qquad \textbf{Open system}$$

where e_i represents the external demand for the ith industry's product. The collection of total outputs for an open system is represented by the following system of n linear equations.

$$\begin{aligned} x_1 &= d_{11}x_1 + d_{12}x_2 + \cdot \cdot \cdot + d_{1n}x_n + e_1 \\ x_2 &= d_{21}x_1 + d_{22}x_2 + \cdot \cdot \cdot + d_{2n}x_n + e_2 \\ &\vdots \\ x_n &= d_{n1}x_1 + d_{n2}x_2 + \cdot \cdot \cdot + d_{nn}x_n + e_n \end{aligned}$$

The matrix form of this system is

$$X = DX + E,$$

where X is called the **output matrix** and E is called the **external demand matrix.**

EXAMPLE 8 **Solving for the Output Matrix of an Open System**

An economic system composed of three industries has the input-output matrix shown below.

$$D = \overset{\text{User (Output)}}{\overbrace{\begin{matrix} \text{A} & \text{B} & \text{C} \end{matrix}}}\ \begin{bmatrix} 0.1 & 0.43 & 0 \\ 0.15 & 0 & 0.37 \\ 0.23 & 0.03 & 0.02 \end{bmatrix} \begin{matrix} \text{A} \\ \text{B} \\ \text{C} \end{matrix} \left.\vphantom{\begin{matrix} 0 \\ 0 \\ 0 \end{matrix}}\right\} \text{Supplier (Input)}$$

Find the output matrix X if the external demands are

$$E = \begin{bmatrix} 20{,}000 \\ 30{,}000 \\ 25{,}000 \end{bmatrix}. \quad \begin{matrix} \text{A} \\ \text{B} \\ \text{C} \end{matrix}$$

(The answers in this example have been rounded to the nearest unit.)

SOLUTION Letting I be the identity matrix, write the equation $X = DX + E$ as $IX - DX = E$, which means

$$(I - D)X = E.$$

Using the matrix D above produces

$$I - D = \begin{bmatrix} 0.9 & -0.43 & 0 \\ -0.15 & 1 & -0.37 \\ -0.23 & -0.03 & 0.98 \end{bmatrix}.$$

Finally, applying Gauss-Jordan elimination to the system of linear equations represented by $(I - D)X = E$ produces

$$\begin{bmatrix} 0.9 & -0.43 & 0 & 20{,}000 \\ -0.15 & 1 & -0.37 & 30{,}000 \\ -0.23 & -0.03 & 0.98 & 25{,}000 \end{bmatrix} \longrightarrow \begin{bmatrix} 1 & 0 & 0 & 46{,}616 \\ 0 & 1 & 0 & 51{,}058 \\ 0 & 0 & 1 & 38{,}014 \end{bmatrix}.$$

So, the output matrix is

$$X = \begin{bmatrix} 46{,}616 \\ 51{,}058 \\ 38{,}014 \end{bmatrix}. \quad \begin{matrix} \text{A} \\ \text{B} \\ \text{C} \end{matrix}$$

To produce the given external demands, the outputs of the three industries must be as follows.

Output for Industry A: 46,616 units

Output for Industry B: 51,058 units

Output for Industry C: 38,014 units

The economic systems described in Examples 7 and 8 are, of course, simple ones. In the real world, an economic system would include many industries or industrial groups. For example, an economic analysis of some of the producing groups in the United States would include the products listed below (taken from the *Statistical Abstract of the United States*).

1. Farm products (grains, livestock, poultry, bulk milk)
2. Processed foods and feeds (beverages, dairy products)
3. Textile products and apparel (yarns, threads, clothing)
4. Hides, skins, and leather (shoes, upholstery)
5. Fuels and power (coal, gasoline, electricity)
6. Chemicals and allied products (drugs, plastic resins)
7. Rubber and plastic products (tires, plastic containers)
8. Lumber and wood products (plywood, pencils)
9. Pulp, paper, and allied products (cardboard, newsprint)
10. Metals and metal products (plumbing fixtures, cans)
11. Machinery and equipment (tractors, drills, computers)
12. Furniture and household durables (carpets, appliances)
13. Nonmetallic mineral products (glass, concrete, bricks)
14. Transportation equipment (automobiles, trucks, planes)
15. Miscellaneous products (toys, cameras, linear algebra texts)

A matrix of order 15×15 would be required to represent even these broad industrial groupings using the Leontief input-output model. A more detailed analysis could easily require an input-output matrix of order greater than 100×100. Clearly, this type of analysis could be done only with the aid of a computer.

Least Squares Regression Analysis

You will now look at a procedure that is used in statistics to develop linear models. The next example demonstrates a visual method for approximating a line of best fit for a given set of data points.

EXAMPLE 9 **A Visual Straight-Line Approximation**

Determine the straight line that best fits the points.

$$(1, 1), (2, 2), (3, 4), (4, 4), \text{ and } (5, 6)$$

SOLUTION Plot the points, as shown in Figure 2.2. It appears that a good choice would be the line whose slope is 1 and whose y-intercept is 0.5. The equation of this line is

$$y = 0.5 + x.$$

An examination of the line shown in Figure 2.2 reveals that you can improve the fit by rotating the line counterclockwise slightly, as shown in Figure 2.3. It seems clear that this new line, the equation of which is $y = 1.2x$, fits the given points better than the original line.

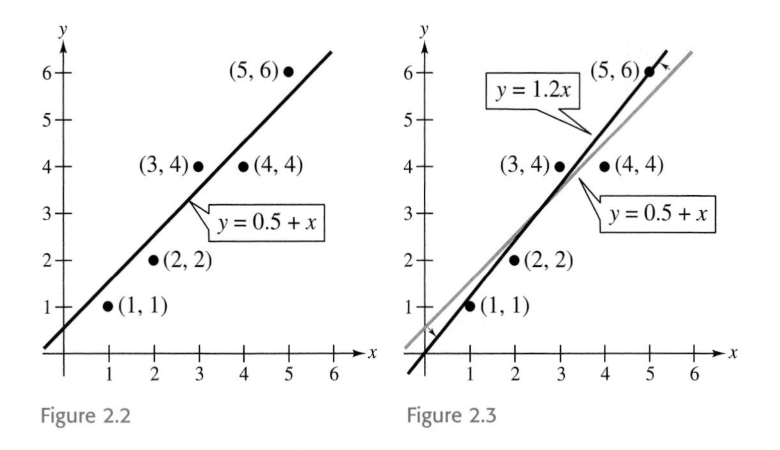

Figure 2.2

Figure 2.3

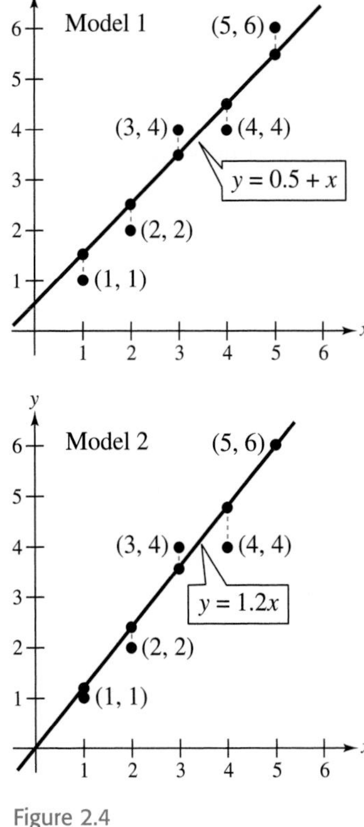

Figure 2.4

One way of measuring how well a function $y = f(x)$ fits a set of points

$$(x_1, y_1), (x_2, y_2), \ldots, (x_n, y_n)$$

is to compute the differences between the values from the function $f(x_i)$ and the actual values y_i, as shown in Figure 2.4. By squaring these differences and summing the results, you obtain a measure of error that is called the **sum of squared error.** The sums of squared errors for our two linear models are shown in Table 2.1 below.

TABLE 2.1

Model 1: $f(x) = 0.5 + x$				*Model 2:* $f(x) = 1.2x$			
x_i	y_i	$f(x_i)$	$[y_i - f(x_i)]^2$	x_i	y_i	$f(x_i)$	$[y_i - f(x_i)]^2$
1	1	1.5	$(-0.5)^2$	1	1	1.2	$(-0.2)^2$
2	2	2.5	$(-0.5)^2$	2	2	2.4	$(-0.4)^2$
3	4	3.5	$(+0.5)^2$	3	4	3.6	$(+0.4)^2$
4	4	4.5	$(-0.5)^2$	4	4	4.8	$(-0.8)^2$
5	6	5.5	$(+0.5)^2$	5	6	6.0	$(+0.0)^2$
Total			1.25	Total			1.00

The sums of squared errors confirm that the second model fits the given points better than the first.

Of all possible linear models for a given set of points, the model that has the best fit is defined to be the one that minimizes the sum of squared error. This model is called the **least squares regression line,** and the procedure for finding it is called the **method of least squares.**

Definition of Least Squares Regression Line

For a set of points $(x_1, y_1), (x_2, y_2), \ldots, (x_n, y_n)$, the **least squares regression line** is given by the linear function

$$f(x) = a_0 + a_1 x$$

that minimizes the sum of squared error

$$[y_1 - f(x_1)]^2 + [y_2 - f(x_2)]^2 + \cdots + [y_n - f(x_n)]^2.$$

To find the least squares regression line for a set of points, begin by forming the system of linear equations

$$\begin{aligned} y_1 &= f(x_1) + [y_1 - f(x_1)] \\ y_2 &= f(x_2) + [y_2 - f(x_2)] \\ &\vdots \\ y_n &= f(x_n) + [y_n - f(x_n)] \end{aligned}$$

where the right-hand term, $[y_i - f(x_i)]$, of each equation is thought of as the error in the approximation of y_i by $f(x_i)$. Then write this error as

$$e_i = y_i - f(x_i)$$

so that the system of equations takes the form

$$\begin{aligned} y_1 &= (a_0 + a_1 x_1) + e_1 \\ y_2 &= (a_0 + a_1 x_2) + e_2 \\ &\vdots \\ y_n &= (a_0 + a_1 x_n) + e_n. \end{aligned}$$

Now, if you define Y, X, A, and E as

$$Y = \begin{bmatrix} y_1 \\ y_2 \\ \vdots \\ y_n \end{bmatrix}, \quad X = \begin{bmatrix} 1 & x_1 \\ 1 & x_2 \\ \vdots & \vdots \\ 1 & x_n \end{bmatrix}, \quad A = \begin{bmatrix} a_0 \\ a_1 \end{bmatrix}, \quad E = \begin{bmatrix} e_1 \\ e_2 \\ \vdots \\ e_n \end{bmatrix}$$

the n linear equations may be replaced by the matrix equation

$$Y = XA + E.$$

Note that the matrix X has two columns, a column of 1's (corresponding to a_0) and a column containing the x_i's. This matrix equation can be used to determine the coefficients of the least squares regression line, as follows.

Matrix Form for Linear Regression

For the regression model $Y = XA + E$, the coefficients of the least squares regression line are given by the matrix equation

$$A = (X^T X)^{-1} X^T Y$$

and the sum of squared error is

$$E^T E.$$

REMARK: You will learn more about this procedure in Section 5.4.

Example 10 demonstrates the use of this procedure to find the least squares regression line for the set of points from Example 9.

EXAMPLE 10 Finding the Least Squares Regression Line

Find the least squares regression line for the points (1, 1), (2, 2), (3, 4), (4, 4), and (5, 6) (see Figure 2.5). Then find the sum of squared error for this regression line.

SOLUTION Using the five points below, the matrices X and Y are

$$X = \begin{bmatrix} 1 & 1 \\ 1 & 2 \\ 1 & 3 \\ 1 & 4 \\ 1 & 5 \end{bmatrix} \quad \text{and} \quad Y = \begin{bmatrix} 1 \\ 2 \\ 4 \\ 4 \\ 6 \end{bmatrix}.$$

This means that

$$X^T X = \begin{bmatrix} 1 & 1 & 1 & 1 & 1 \\ 1 & 2 & 3 & 4 & 5 \end{bmatrix} \begin{bmatrix} 1 & 1 \\ 1 & 2 \\ 1 & 3 \\ 1 & 4 \\ 1 & 5 \end{bmatrix} = \begin{bmatrix} 5 & 15 \\ 15 & 55 \end{bmatrix}$$

and

$$X^T Y = \begin{bmatrix} 1 & 1 & 1 & 1 & 1 \\ 1 & 2 & 3 & 4 & 5 \end{bmatrix} \begin{bmatrix} 1 \\ 2 \\ 4 \\ 4 \\ 6 \end{bmatrix} = \begin{bmatrix} 17 \\ 63 \end{bmatrix}.$$

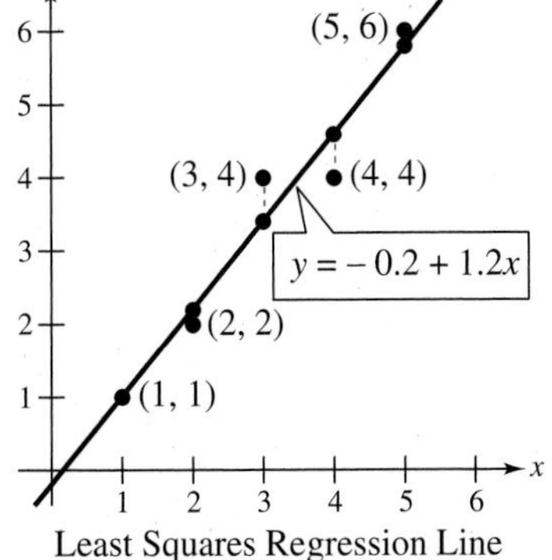

Figure 2.5

Now, using $(X^TX)^{-1}$ to find the coefficient matrix A, you have

$$A = (X^TX)^{-1}X^TY = \frac{1}{50}\begin{bmatrix} 55 & -15 \\ -15 & 5 \end{bmatrix}\begin{bmatrix} 17 \\ 63 \end{bmatrix} = \begin{bmatrix} -0.2 \\ 1.2 \end{bmatrix}.$$

The least squares regression line is

$$y = -0.2 + 1.2x.$$

(See Figure 2.5.) The sum of squared error for this line can be shown to be 0.8, which means that this line fits the data better than either of the two experimental linear models determined earlier.

SECTION 2.5 Exercises

Stochastic Matrices

In Exercises 1–6, determine whether the matrix is stochastic.

1. $\begin{bmatrix} \frac{2}{5} & -\frac{2}{5} \\ \frac{3}{5} & \frac{7}{5} \end{bmatrix}$

2. $\begin{bmatrix} \frac{\sqrt{2}}{2} & \frac{\sqrt{2}}{2} \\ -\frac{\sqrt{2}}{2} & \frac{\sqrt{2}}{2} \end{bmatrix}$

3. $\begin{bmatrix} 0 & 1 & 0 \\ 0 & 0 & 1 \\ 1 & 0 & 0 \end{bmatrix}$

4. $\begin{bmatrix} 0.3 & 0.1 & 0.8 \\ 0.5 & 0.2 & 0.1 \\ 0.2 & 0.7 & 0.1 \end{bmatrix}$

5. $\begin{bmatrix} \frac{1}{3} & \frac{1}{6} & \frac{1}{4} \\ \frac{1}{3} & \frac{2}{3} & \frac{1}{4} \\ \frac{1}{3} & \frac{1}{6} & \frac{1}{2} \end{bmatrix}$

6. $\begin{bmatrix} 1 & 0 & 0 & 0 \\ 0 & 1 & 0 & 0 \\ 0 & 0 & 1 & 0 \\ 0 & 0 & 0 & 1 \end{bmatrix}$

7. The market research department at a manufacturing plant determines that 20% of the people who purchase the plant's product during any month will not purchase it the next month. On the other hand, 30% of the people who do not purchase the product during any month will purchase it the next month. In a population of 1000 people, 100 people purchased the product this month. How many will purchase the product next month? In 2 months?

8. A medical researcher is studying the spread of a virus in a population of 1000 laboratory mice. During any week there is an 80% probability that an infected mouse will overcome the virus, and during the same week there is a 10% probability that a noninfected mouse will become infected. One hundred mice are currently infected with the virus. How many will be infected next week? In 2 weeks?

9. A population of 10,000 is grouped as follows: 5000 nonsmokers, 2500 smokers of one pack or less per day, and 2500 smokers of more than one pack per day. During any month there is a 5% probability that a nonsmoker will begin smoking a pack or less per day, and a 2% probability that a nonsmoker will begin smoking more than a pack per day. For smokers who smoke a pack or less per day, there is a 10% probability of quitting and a 10% probability of increasing to more than a pack per day. For smokers who smoke more than a pack per day, there is a 5% probability of quitting and a 10% probability of dropping to a pack or less per day. How many people will be in each of the 3 groups in 1 month? In 2 months?

10. A population of 100,000 consumers is grouped as follows: 20,000 users of Brand A, 30,000 users of Brand B, and 50,000 who use neither brand. During any month a Brand A user has a 20% probability of switching to Brand B and a 5% probability of not using either brand. A Brand B user has a 15% probability of switching to Brand A and a 10% probability of not using either brand. A nonuser has a 10% probability of purchasing Brand A and a 15% probability of purchasing Brand B. How many people will be in each group in 1 month? In 2 months? In 3 months?

11. A college dormitory houses 200 students. Those who watch an hour or more of television on any day always watch for less than an hour the next day. One-fourth of those who watch television for less than an hour one day will watch an hour or more the next day. Half of the students watched television for an hour or more today. How many will watch television for an hour or more tomorrow? In 2 days? In 30 days?

12. For the matrix of transition probabilities

$$P = \begin{bmatrix} 0.6 & 0.1 & 0.1 \\ 0.2 & 0.7 & 0.1 \\ 0.2 & 0.2 & 0.8 \end{bmatrix},$$

find P^2X and P^3X for the state matrix

$$X = \begin{bmatrix} 100 \\ 100 \\ 800 \end{bmatrix}.$$

Then find the steady state matrix for P.

13. Prove that the product of two 2×2 stochastic matrices is stochastic.

14. Let P be a 2×2 stochastic matrix. Prove that there exists a 2×1 state matrix X with nonnegative entries such that $PX = X$.

Cryptography

In Exercises 15–18, find the uncoded row matrices of the indicated size for the given messages. Then encode the message using the matrix A.

15. *Message:* SELL CONSOLIDATED

Row Matrix Size: 1×3

Encoding Matrix: $A = \begin{bmatrix} 1 & -1 & 0 \\ 1 & 0 & -1 \\ -6 & 2 & 3 \end{bmatrix}$

16. *Message:* PLEASE SEND MONEY

Row Matrix Size: 1×3

Encoding Matrix: $A = \begin{bmatrix} 4 & 2 & 1 \\ -3 & -3 & -1 \\ 3 & 2 & 1 \end{bmatrix}$

17. *Message:* COME HOME SOON

Row Matrix Size: 1×2

Encoding Matrix: $A = \begin{bmatrix} 1 & 2 \\ 3 & 5 \end{bmatrix}$

18. *Message:* HELP IS COMING

Row Matrix Size: 1×4

Encoding Matrix: $A = \begin{bmatrix} -2 & 3 & -1 & -1 \\ -1 & 1 & 1 & 1 \\ -1 & -1 & 1 & 2 \\ 3 & 1 & -2 & -4 \end{bmatrix}$

In Exercises 19–22, find A^{-1} and use it to decode the cryptogram.

19. $A = \begin{bmatrix} 1 & 2 \\ 3 & 5 \end{bmatrix}$,

11 21 64 112 25 50 29 53 23 46 40 75 55 92

20. $A = \begin{bmatrix} 2 & 3 \\ 3 & 4 \end{bmatrix}$,

85 120 6 8 10 15 84 117 42 56 90 125 60 80 30 45 19 26

21. $A = \begin{bmatrix} 1 & 2 & 2 \\ 3 & 7 & 9 \\ -1 & -4 & -7 \end{bmatrix}$,

13 19 10 −1 −33 −77 3 −2 −14 4 1 −9 −5 −25 −47 4 1 −9

22. $A = \begin{bmatrix} 3 & -4 & 2 \\ 0 & 2 & 1 \\ 4 & -5 & 3 \end{bmatrix}$,

112 −140 83 19 −25 13 72 −76 61 95 −118 71 20 21 38 35 −23 36 42 −48 32

23. The cryptogram below was encoded with a 2×2 matrix.

8 21 −15 −10 −13 −13 5 10 5 25 5 19 −1 6 20 40 −18 −18 1 16

The last word of the message is __RON. What is the message?

24. The cryptogram below was encoded with a 2×2 matrix.

5 2 25 11 −2 −7 −15 −15 32 14 −8 −13 38 19 −19 −19 37 16

The last word of the message is __SUE. What is the message?

25. Use a graphing utility or computer software program with matrix capabilities to find A^{-1}. Then decode the cryptogram.

$$A = \begin{bmatrix} 1 & 0 & 2 \\ 2 & -1 & 1 \\ 0 & 1 & 2 \end{bmatrix}$$

38 −14 29 56 −15 62 17 3 38 18 20 76 18 −5 21 29 −7 32 32 9 77 36 −8 48 33 −5 51 41 3 79 12 1 26 58 −22 49 63 −19 69 28 8 67 31 −11 27 41 −18 28

26. A code breaker intercepted the encoded message below.

45 −35 38 −30 18 −18 35 −30 81 −60 42 −28 75 −55 2 −2 22 −21 15 −10

Let $A^{-1} = \begin{bmatrix} w & x \\ y & z \end{bmatrix}$.

(a) You know that $[45 \ \ -35]A^{-1} = [10 \ \ 15]$ and $[38 \ \ -30]A^{-1} = [8 \ \ 14]$, where A^{-1} is the inverse of the encoding matrix A. Write and solve two systems of equations to find w, x, y, and z.

(b) Decode the message.

Leontief Input-Output Models

27. A system composed of two industries, coal and steel, has the following input requirements.

(a) To produce \$1.00 worth of output, the coal industry requires \$0.10 of its own product and \$0.80 of steel.

(b) To produce \$1.00 worth of output, the steel industry requires \$0.10 of its own product and \$0.20 of coal.

Find D, the input-output matrix for this system. Then solve for the output matrix X in the equation $X = DX + E$, where the external demand is

$$E = \begin{bmatrix} 10{,}000 \\ 20{,}000 \end{bmatrix}.$$

28. An industrial system has two industries with the following input requirements.

(a) To produce \$1.00 worth of output, Industry A requires \$0.30 of its own product and \$0.40 of Industry B's product.

(b) To produce \$1.00 worth of output, Industry B requires \$0.20 of its own product and \$0.40 of Industry A's product.

Find D, the input-output matrix for this system. Then solve for the output matrix X in the equation $X = DX + E$, where the external demand is

$$E = \begin{bmatrix} 50{,}000 \\ 30{,}000 \end{bmatrix}.$$

29. A small community includes a farmer, a baker, and a grocer and has the input-output matrix D and external demand matrix E shown below.

$$D = \begin{array}{c} \begin{matrix} \text{Farmer} & \text{Baker} & \text{Grocer} \end{matrix} \\ \begin{bmatrix} 0.40 & 0.50 & 0.50 \\ 0.30 & 0.00 & 0.30 \\ 0.20 & 0.20 & 0.00 \end{bmatrix} \end{array} \begin{matrix} \text{Farmer} \\ \text{Baker} \\ \text{Grocer} \end{matrix} \quad \text{and} \quad E = \begin{bmatrix} 1000 \\ 1000 \\ 1000 \end{bmatrix}$$

Solve for the output matrix X in the equation $X = DX + E$.

30. An industrial system has three industries and the input-output matrix D and external demand matrix E shown below.

$$D = \begin{bmatrix} 0.2 & 0.4 & 0.4 \\ 0.4 & 0.2 & 0.2 \\ 0.0 & 0.2 & 0.2 \end{bmatrix} \quad \text{and} \quad E = \begin{bmatrix} 5000 \\ 2000 \\ 8000 \end{bmatrix}$$

Solve for the output matrix X in the equation $X = DX + E$.

Least Squares Regression Analysis

In Exercises 31–34, (a) sketch the line that appears to be the best fit for the given points, (b) use the method of least squares to find the least squares regression line, and (c) calculate the sum of the squared error.

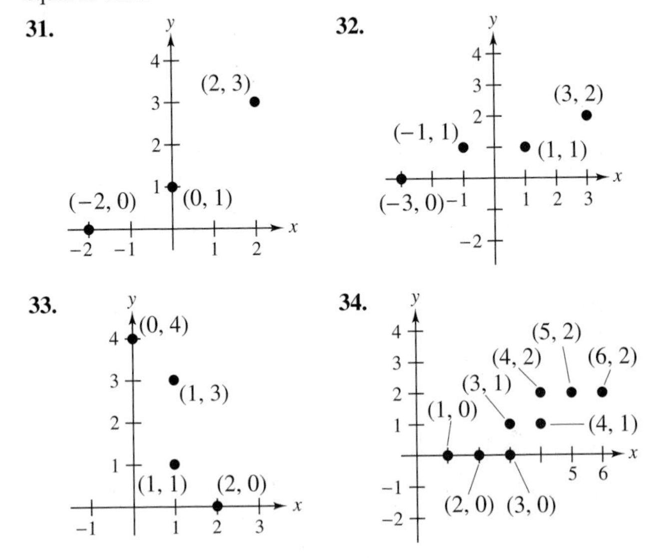

In Exercises 35–42, find the least squares regression line.

35. (0, 0), (1, 1), (2, 4)

36. (1, 0), (3, 3), (5, 6)

37. (−2, 0), (−1, 1), (0, 1), (1, 2)

38. (−4, −1), (−2, 0), (2, 4), (4, 5)

39. (−5, 1), (1, 3), (2, 3), (2, 5)

40. (−3, 4), (−1, 2), (1, 1), (3, 0)

41. (−5, 10), (−1, 8), (3, 6), (7, 4), (5, 5)

42. (0, 6), (4, 3), (5, 0), (8, −4), (10, −5)

43. A fuel refiner wants to know the demand for a certain grade of gasoline as a function of the price. The daily sales y (in gallons) for three different prices of the product are shown in the table.

Price (x)	\$3.00	\$3.25	\$3.50
Demand (y)	4500	3750	3300

(a) Find the least squares regression line for these data.
(b) Estimate the demand when the price is \$3.40.

44. A hardware retailer wants to know the demand for a rechargeable power drill as a function of the price. The monthly sales for four different prices of the drill are shown in the table.

Price (x)	\$25	\$30	\$35	\$40
Demand (y)	82	75	67	55

(a) Find the least squares regression line for these data.
(b) Estimate the demand when the price is \$32.95.

45. The table shows the numbers y of motor vehicle registrations (in millions) in the United States for the years 2000 through 2004. (Source: U.S. Federal Highway Administration)

Year	2000	2001	2002	2003	2004
Number (y)	221.5	230.4	229.6	231.4	237.2

(a) Use the method of least squares to find the least squares regression line for the data. Let t represent the year, with $t = 0$ corresponding to 2000.
(b) Use the linear regression capabilities of a graphing utility to find a linear model for the data. Let t represent the year, with $t = 0$ corresponding to 2000.

46. A wildlife management team studied the reproduction rates of deer in 3 tracts of a wildlife preserve. Each tract contained 5 acres. In each tract the number of females x and the percent of females y that had offspring the following year were recorded. The results are shown in the table.

Number (x)	100	120	140
Percent (y)	75	68	55

(a) Use the method of least squares to find the least squares regression line that models the data.
(b) Use a graphing utility to graph the model and the data in the same viewing window.
(c) Use the model to create a table of estimated values of y. Compare the estimated values with the actual data.
(d) Use the model to estimate the percent of females that had offspring when there were 170 females.
(e) Use the model to estimate the number of females when 40% of the females had offspring.

CHAPTER 2 Review Exercises

In Exercises 1–6, perform the indicated matrix operations.

1. $\begin{bmatrix} 2 & 1 & 0 \\ 0 & 5 & -4 \end{bmatrix} - 3\begin{bmatrix} 5 & 3 & -6 \\ 0 & -2 & 5 \end{bmatrix}$

2. $-2\begin{bmatrix} 1 & 2 \\ 5 & -4 \\ 6 & 0 \end{bmatrix} + 8\begin{bmatrix} 7 & 1 \\ 1 & 2 \\ 1 & 4 \end{bmatrix}$

3. $\begin{bmatrix} 1 & 2 \\ 5 & -4 \\ 6 & 0 \end{bmatrix}\begin{bmatrix} 6 & -2 & 8 \\ 4 & 0 & 0 \end{bmatrix}$

4. $\begin{bmatrix} 1 & 5 \\ 2 & -4 \end{bmatrix}\begin{bmatrix} 6 & -2 & 8 \\ 4 & 0 & 0 \end{bmatrix}$

5. $\begin{bmatrix} 1 & 3 & 2 \\ 0 & 2 & -4 \\ 0 & 0 & 3 \end{bmatrix}\begin{bmatrix} 4 & -3 & 2 \\ 0 & 3 & -1 \\ 0 & 0 & 2 \end{bmatrix}$

6. $\begin{bmatrix} 2 & 1 \\ 6 & 0 \end{bmatrix}\begin{bmatrix} 4 & 2 \\ -3 & 1 \end{bmatrix} + \begin{bmatrix} -2 & 4 \\ 0 & 4 \end{bmatrix}$

In Exercises 7–10, write out the system of linear equations represented by the matrix equation.

7. $\begin{bmatrix} 5 & 4 \\ -1 & 1 \end{bmatrix}\begin{bmatrix} x \\ y \end{bmatrix} = \begin{bmatrix} 2 \\ -22 \end{bmatrix}$

8. $\begin{bmatrix} 2 & -1 \\ 3 & 4 \end{bmatrix}\begin{bmatrix} x \\ y \end{bmatrix} = \begin{bmatrix} 5 \\ -2 \end{bmatrix}$

9. $\begin{bmatrix} 0 & 1 & -2 \\ -1 & 3 & 1 \\ 2 & -2 & 4 \end{bmatrix}\begin{bmatrix} x_1 \\ x_2 \\ x_3 \end{bmatrix} = \begin{bmatrix} -1 \\ 0 \\ 2 \end{bmatrix}$

10. $\begin{bmatrix} 0 & 1 & 2 \\ 3 & 2 & 1 \\ 4 & -3 & -4 \end{bmatrix}\begin{bmatrix} x \\ y \\ z \end{bmatrix} = \begin{bmatrix} 0 \\ -1 \\ -7 \end{bmatrix}$

In Exercises 11–14, write the system of linear equations in matrix form.

11. $\begin{aligned} 2x - y &= 5 \\ 3x + 2y &= -4 \end{aligned}$

12. $\begin{aligned} 2x + y &= -8 \\ x + 4y &= -4 \end{aligned}$

13. $\begin{aligned} 2x_1 + 3x_2 + x_3 &= 10 \\ 2x_1 - 3x_2 - 3x_3 &= 22 \\ 4x_1 - 2x_2 + 3x_3 &= -2 \end{aligned}$

14. $\begin{aligned} -3x_1 - x_2 + x_3 &= 0 \\ 2x_1 + 4x_2 - 5x_3 &= -3 \\ x_1 - 2x_2 + 3x_3 &= 1 \end{aligned}$

In Exercises 15–18, find A^T, A^TA, and AA^T.

15. $\begin{bmatrix} 1 & 2 & -3 \\ 0 & 1 & 2 \end{bmatrix}$

16. $\begin{bmatrix} 3 & -1 \\ 2 & 0 \end{bmatrix}$

17. $\begin{bmatrix} 1 \\ 3 \\ -1 \end{bmatrix}$

18. $\begin{bmatrix} 1 & -2 & -3 \end{bmatrix}$

In Exercises 19–22, find the inverse of the matrix (if it exists).

19. $\begin{bmatrix} 3 & -1 \\ 2 & -1 \end{bmatrix}$

20. $\begin{bmatrix} 4 & -1 \\ -8 & 2 \end{bmatrix}$

21. $\begin{bmatrix} 2 & 3 & 1 \\ 2 & -3 & -3 \\ 4 & 0 & 3 \end{bmatrix}$

22. $\begin{bmatrix} 1 & 1 & 1 & 1 \\ 0 & 1 & 1 & 1 \\ 0 & 0 & 1 & 1 \\ 0 & 0 & 0 & 1 \end{bmatrix}$

In Exercises 23– 26, write the system of linear equations in the form $A\mathbf{x} = \mathbf{b}$. Then find A^{-1} and use it to solve for $\mathbf{x}$.

23. $\begin{aligned} 5x_1 + 4x_2 &= 2 \\ -x_1 + x_2 &= -22 \end{aligned}$

24. $\begin{aligned} 3x_1 + 2x_2 &= 1 \\ x_1 + 4x_2 &= -3 \end{aligned}$

25. $\begin{aligned} -x_1 + x_2 + 2x_3 &= 1 \\ 2x_1 + 3x_2 + x_3 &= -2 \\ 5x_1 + 4x_2 + 2x_3 &= 4 \end{aligned}$

26. $\begin{aligned} x_1 + x_2 + 2x_3 &= 0 \\ x_1 - x_2 + x_3 &= -1 \\ 2x_1 + x_2 + x_3 &= 2 \end{aligned}$

In Exercises 27 and 28, find A.

27. $(3A)^{-1} = \begin{bmatrix} 4 & -1 \\ 2 & 3 \end{bmatrix}$

28. $(2A)^{-1} = \begin{bmatrix} 2 & 4 \\ 0 & 1 \end{bmatrix}$

In Exercises 29 and 30, find x such that the matrix A is nonsingular.

29. $A = \begin{bmatrix} 3 & 1 \\ x & -1 \end{bmatrix}$

30. $A = \begin{bmatrix} 2 & x \\ 1 & 4 \end{bmatrix}$

In Exercises 31 and 32, find the inverse of the elementary matrix.

31. $\begin{bmatrix} 1 & 0 & 4 \\ 0 & 1 & 0 \\ 0 & 0 & 1 \end{bmatrix}$

32. $\begin{bmatrix} 1 & 0 & 0 \\ 0 & 6 & 0 \\ 0 & 0 & 1 \end{bmatrix}$

In Exercises 33–36, factor A into a product of elementary matrices.

33. $A = \begin{bmatrix} 2 & 3 \\ 0 & 1 \end{bmatrix}$

34. $A = \begin{bmatrix} -3 & 13 \\ 1 & -4 \end{bmatrix}$

35. $A = \begin{bmatrix} 1 & 0 & 1 \\ 0 & 1 & -2 \\ 0 & 0 & 4 \end{bmatrix}$

36. $A = \begin{bmatrix} 3 & 0 & 6 \\ 0 & 2 & 0 \\ 1 & 0 & 3 \end{bmatrix}$

37. Find two 2×2 matrices A such that $A^2 = I$.

38. Find two 2×2 matrices A such that $A^2 = O$.

39. Find three 2×2 matrices A such that $A^2 = A$.

40. Find 2×2 matrices A and B such that $AB = O$ but $BA \neq O$.

In Exercises 41 and 42, let the matrices X, Y, Z, and W be

$$X = \begin{bmatrix} 1 \\ 2 \\ 0 \\ 1 \end{bmatrix}, \quad Y = \begin{bmatrix} -1 \\ 0 \\ 3 \\ 2 \end{bmatrix}, \quad Z = \begin{bmatrix} 3 \\ 4 \\ -1 \\ 2 \end{bmatrix}, \quad W = \begin{bmatrix} 3 \\ 2 \\ -4 \\ -1 \end{bmatrix}.$$

41. (a) Find scalars a, b, and c such that $W = aX + bY + cZ$.
(b) Show that there do not exist scalars a and b such that $Z = aX + bY$.

42. Show that if $aX + bY + cZ = O$, then $a = b = c = 0$.

43. Let A, B, and $A + B$ be nonsingular matrices. Prove that $A^{-1} + B^{-1}$ is nonsingular by showing that

$$(A^{-1} + B^{-1})^{-1} = A(A + B)^{-1}B.$$

44. **Writing** Let A, B, and C be $n \times n$ matrices and let C be nonsingular. If $AC = CB$, is it true that $A = B$? If so, prove it. If not, explain why and find an example for which the hypothesis is false.

In Exercises 45 and 46, find the LU-factorization of the matrix.

45. $\begin{bmatrix} 2 & 5 \\ 6 & 14 \end{bmatrix}$

46. $\begin{bmatrix} 1 & 1 & 1 \\ 1 & 2 & 2 \\ 1 & 2 & 3 \end{bmatrix}$

In Exercises 47 and 48, use the LU-factorization of the coefficient matrix to solve the linear system.

47.
$$\begin{aligned} x \quad\quad + z &= 3 \\ 2x + y + 2z &= 7 \\ 3x + 2y + 6z &= 8 \end{aligned}$$

48.
$$\begin{aligned} 2x_1 + x_2 + x_3 - x_4 &= 7 \\ 3x_2 + x_3 - x_4 &= -3 \\ -2x_3 &= 2 \\ 2x_1 + x_2 + x_3 - 2x_4 &= 8 \end{aligned}$$

True or False? In Exercises 49–52, determine whether each statement is true or false. If a statement is true, give a reason or cite an appropriate statement from the text. If a statement is false, provide an example that shows the statement is not true in all cases or cite an appropriate statement from the text.

49. (a) Addition of matrices is not commutative.

(b) The transpose of the sum of matrices is equal to the sum of the transposes of the matrices.

50. (a) The product of a 2×3 matrix and a 3×5 matrix is a matrix that is 5×2.

(b) The transpose of a product is equal to the product of transposes in reverse order.

51. (a) All $n \times n$ matrices are invertible.

(b) If an $n \times n$ matrix A is not symmetric, then A^TA is not symmetric.

52. (a) If A and B are $n \times n$ matrices and A is invertible, then $(ABA^{-1})^2 = AB^2A^{-1}$.

(b) If A and B are nonsingular $n \times n$ matrices, then $A + B$ is a nonsingular matrix.

53. At a convenience store, the numbers of gallons of 87 octane, 89 octane, and 93 octane gasoline sold on Friday, Saturday, and Sunday of a particular holiday weekend are shown by the matrix.

$$\begin{array}{l} \\ \textit{Friday} \\ \textit{Saturday} \\ \textit{Sunday} \end{array} \begin{array}{c} \begin{array}{ccc} 87 & 89 & 93 \end{array} \\ \begin{bmatrix} 580 & 840 & 320 \\ 560 & 420 & 160 \\ 860 & 1020 & 540 \end{bmatrix} \end{array} = A$$

A second matrix gives the selling prices per gallon and the profits per gallon for the three grades of gasoline sold by the convenience store.

$$\begin{array}{l} \\ 87 \\ 89 \\ 93 \end{array} \begin{array}{c} \begin{array}{cc} \textit{Selling Price (per gallon)} & \textit{Profit (per gallon)} \end{array} \\ \begin{bmatrix} 3.05 & 0.05 \\ 3.15 & 0.08 \\ 3.25 & 0.10 \end{bmatrix} \end{array} = B$$

(a) Find AB. What is the meaning of AB in the context of the situation?

(b) Find the convenience store's total gasoline sales profit for Friday through Sunday.

54. At a certain dairy mart, the numbers of gallons of skim, 2%, and whole milk that are sold on Friday, Saturday, and Sunday of a particular week are shown by the matrix.

$$\begin{array}{l} \\ \textit{Friday} \\ \textit{Saturday} \\ \textit{Sunday} \end{array} \begin{array}{c} \begin{array}{ccc} \textit{Skim} & \textit{2\%} & \textit{Whole} \end{array} \\ \begin{bmatrix} 40 & 64 & 52 \\ 60 & 82 & 76 \\ 76 & 96 & 84 \end{bmatrix} \end{array} = A$$

A second matrix gives the selling prices per gallon and the profits per gallon for the three types of milk sold by the dairy mart.

$$\begin{array}{l} \\ \textit{Skim} \\ \textit{2\%} \\ \textit{Whole} \end{array} \begin{array}{c} \begin{array}{cc} \textit{Selling Price (per gallon)} & \textit{Profit (per gallon)} \end{array} \\ \begin{bmatrix} 3.32 & 1.32 \\ 3.22 & 1.07 \\ 3.12 & 0.92 \end{bmatrix} \end{array} = B$$

(a) Find AB. What is the meaning of AB in the context of the situation?

(b) Find the dairy mart's total profit for Friday through Sunday.

55. The numbers of calories burned by individuals of different body weights performing different types of aerobic exercises for a 20-minute time period are shown in the matrix.

$$\begin{array}{l} \\ \textit{Bicycling} \\ \textit{Jogging} \\ \textit{Walking} \end{array} \begin{array}{c} \begin{array}{cc} \textit{120-lb Person} & \textit{150-lb Person} \end{array} \\ \begin{bmatrix} 109 & 136 \\ 127 & 159 \\ 64 & 79 \end{bmatrix} \end{array} = A$$

A 120-pound person and a 150-pound person bicycled for 40 minutes, jogged for 10 minutes, and walked for 60 minutes.

(a) Organize the amounts of time spent exercising in a matrix B.

(b) Find the product BA.

(c) Explain the meaning of the matrix product BA as it applies to this situation.

56. The final grades in a particular linear algebra course at a liberal arts college are determined by grades on two midterms and a final exam. The grades for six students and two possible grading systems are shown in the matrices below.

$$\begin{array}{l} \\ \\ \textit{Student 1} \\ \textit{Student 2} \\ \textit{Student 3} \\ \textit{Student 4} \\ \textit{Student 5} \\ \textit{Student 6} \end{array} \begin{array}{c} \begin{array}{ccc} \textit{Midterm} & \textit{Midterm} & \textit{Final} \\ \textit{1} & \textit{2} & \textit{Exam} \end{array} \\ \begin{bmatrix} 78 & 82 & 80 \\ 84 & 88 & 85 \\ 92 & 93 & 90 \\ 88 & 86 & 90 \\ 74 & 78 & 80 \\ 96 & 95 & 98 \end{bmatrix} \end{array} = A$$

$$\begin{array}{l} \\ \\ \textit{Midterm 1} \\ \textit{Midterm 2} \\ \textit{Final Exam} \end{array} \begin{array}{c} \begin{array}{cc} \textit{Grading} & \textit{Grading} \\ \textit{System 1} & \textit{System 2} \end{array} \\ \begin{bmatrix} 0.25 & 0.20 \\ 0.25 & 0.20 \\ 0.50 & 0.60 \end{bmatrix} \end{array} = B$$

(a) Describe each grading system in matrix B and how it is to be applied.

(b) Compute the numerical grades for the six students using the two grading systems.

(c) Assign each student a letter grade for each grading system using the letter grade scale shown in the table below.

Numerical Grade Range	*Letter Grade*
90–100	A
80–89	B
70–79	C
60–69	D
0–59	F

Note: If necessary, round up to the nearest integer.

Stochastic Matrices

In Exercises 57 and 58, determine whether the matrix is stochastic.

57. $\begin{bmatrix} 1 & 0 & 0 \\ 0 & 0.5 & 0.1 \\ 0 & 0.1 & 0.5 \end{bmatrix}$

58. $\begin{bmatrix} 0.3 & 0.4 & 0.1 \\ 0.2 & 0.4 & 0.5 \\ 0.5 & 0.2 & 0.4 \end{bmatrix}$

In Exercises 59 and 60, use the given matrix of transition probabilities P and state matrix X to find the state matrices PX, P^2X, and P^3X.

59. $P = \begin{bmatrix} \frac{1}{2} & \frac{1}{4} \\ \frac{1}{2} & \frac{3}{4} \end{bmatrix}, \quad X = \begin{bmatrix} 128 \\ 64 \end{bmatrix}$

60. $P = \begin{bmatrix} 0.6 & 0.2 & 0.0 \\ 0.2 & 0.7 & 0.1 \\ 0.2 & 0.1 & 0.9 \end{bmatrix}, \quad X = \begin{bmatrix} 1000 \\ 1000 \\ 1000 \end{bmatrix}$

61. A country is divided into 3 regions. Each year, 10% of the residents of Region 1 move to Region 2 and 5% move to Region 3; 15% of the residents of Region 2 move to Region 1 and 5% move to Region 3; and 10% of the residents of Region 3 move to Region 1 and 10% move to Region 2. This year each region has a population of 100,000. Find the population of each region (a) in 1 year and (b) in 3 years.

62. Find the steady state matrix for the populations described in Exercise 61.

Cryptography

In Exercises 63 and 64, find the uncoded row matrices of the indicated size for the given message. Then encode the message using the matrix A.

63. *Message:* ONE IF BY LAND

Row Matrix Size: 1×2

Encoding Matrix: $A = \begin{bmatrix} 5 & 2 \\ 2 & 1 \end{bmatrix}$

64. *Message:* BEAM ME UP SCOTTY

Row Matrix Size: 1×3

Encoding Matrix: $A = \begin{bmatrix} 2 & 1 & 4 \\ 3 & 1 & 3 \\ -2 & -1 & -3 \end{bmatrix}$

In Exercises 65–68, find A^{-1} to decode the cryptogram. Then decode the message.

65. $A = \begin{bmatrix} 3 & -2 \\ -4 & 3 \end{bmatrix}$,

-45 34 36 -24 -43 37 -23 22 -37 29 57 -38 -39 31

66. $A = \begin{bmatrix} 1 & 4 \\ -1 & -3 \end{bmatrix}$,

11 52 -8 -9 -13 -39 5 20 12 56 5 20 -2 7 9 41 25 100

67. $A = \begin{bmatrix} 2 & -1 & -1 \\ -5 & 2 & 2 \\ 5 & -1 & -2 \end{bmatrix}$,

58 −3 −25 −48 28 19 −40 13 13 −98 39 39 118 −25 −48 28 −14 −14

68. $A = \begin{bmatrix} 1 & 2 & 3 \\ 2 & 5 & 3 \\ 1 & 0 & 8 \end{bmatrix}$,

23 20 132 54 128 102 32 21 203 6 10 23 21 15 129 36 46 173 29 72 45

In Exercises 69 and 70, use a graphing utility or computer software program with matrix capabilities to find A^{-1}, then decode the cryptogram.

69. $\begin{bmatrix} 1 & -2 & 2 \\ -1 & 1 & 3 \\ 1 & -1 & -4 \end{bmatrix}$

−2 2 5 39 −53 −72 −6 −9 93 4 −12 27 31 −49 −16 19 −24 −46 −8 −7 99

70. $\begin{bmatrix} 2 & 0 & 1 \\ 2 & -1 & 0 \\ 1 & 2 & -4 \end{bmatrix}$

66 27 −31 37 5 −9 61 46 −73 46 − 14 9 94 21 −49 32 −4 12 66 31 −53 47 33 −67

Leontief Input-Output Models

71. An industrial system has two industries with the input requirements shown below.

(a) To produce \$1.00 worth of output, Industry A requires \$0.20 of its own product and \$0.30 of Industry B's product.
(b) To produce \$1.00 worth of output, Industry B requires \$0.10 of its own product and \$0.50 of Industry A's product.

Find D, the input-output matrix for this system. Then solve for the output matrix X in the equation $X = DX + E$, where the external demand is

$$E = \begin{bmatrix} 40{,}000 \\ 80{,}000 \end{bmatrix}.$$

72. An industrial system with three industries has the input-output matrix D and the external demand matrix E shown below.

$$D = \begin{bmatrix} 0.1 & 0.3 & 0.2 \\ 0.0 & 0.2 & 0.3 \\ 0.4 & 0.1 & 0.1 \end{bmatrix} \quad \text{and} \quad E = \begin{bmatrix} 3000 \\ 3500 \\ 8500 \end{bmatrix}$$

Solve for the output matrix X in the equation $X = DX + E$.

Least Squares Regression Analysis

In Exercises 73–76, find the least squares regression line for the points.

73. (1, 5), (2, 4), (3, 2)

74. (2, 1), (3, 3), (4, 2), (5, 4), (6, 4)

75. (−2, 4), (−1, 2), (0, 1), (1, −2), (2, −3)

76. (1, 1), (1, 3), (1, 2), (1, 4), (2, 5)

77. A farmer used four test plots to determine the relationship between wheat yield (in kilograms) per square kilometer and the amount of fertilizer (in hundreds of kilograms) per square kilometer. The results are shown in the table.

Fertilizer, x	1.0	1.5	2.0	2.5
Yield, y	32	41	48	53

(a) Find the least squares regression line for these data.
(b) Estimate the yield for a fertilizer application of 160 kilograms per square kilometer.

78. The Consumer Price Index (CPI) for all items for the years 2001 to 2005 is shown in the table. (Source: Bureau of Labor Statistics)

Year	2001	2002	2003	2004	2005
CPI, y	177.1	179.9	184.0	188.9	195.3

(a) Find the least squares regression line for these data. Let x represent the year, with $x = 0$ corresponding to 2000.
(b) Estimate the CPI for the years 2010 and 2015.

79. The table shows the average monthly cable television rates y in the United States (in dollars) for the years 2000 through 2005. (Source: *Broadband Cable Databook*)

Year	2000	2001	2002	2003	2004	2005
Rate, y	30.37	32.87	34.71	36.59	38.14	39.63

(a) Use the method of least squares to find the least squares regression line for the data. Let x represent the year, with $x = 0$ corresponding to 2000.
(b) Use the linear regression capabilities of a graphing utility to find a linear model for the data. How does this model compare with the model obtained in part (a)?
(c) Use the linear model to create a table of estimated values for y. Compare the estimated values with the actual data.
(d) Use the linear model to predict the average monthly rate in 2010.
(e) Use the linear model to predict when the average monthly rate will be $51.00.

80. The table shows the numbers of cellular phone subscribers y (in millions) in the United States for the years 2000 through 2005. (Source: Cellular Telecommunications and Internet Association)

Year	2000	2001	2002	2003	2004	2005
Subscribers, y	109.5	128.3	140.8	158.7	182.1	207.9

(a) Use the method of least squares to find the least squares regression line for the data. Let x represent the year, with $x = 0$ corresponding to 2000.
(b) Use the linear regression capabilities of a graphing utility to find a linear model for the data. How does this model compare with the model obtained in part (a)?
(c) Use the linear model to create a table of estimated values for y. Compare the estimated values with the actual data.
(d) Use the linear model to predict the number of subscribers in 2010.
(e) Use the linear model to predict when the number of subscribers will be 260 million.

81. The table shows the average salaries y (in millions of dollars) of Major League baseball players in the United States for the years 2000 through 2005. (Source: Major League Baseball)

Year	2000	2001	2002	2003	2004	2005
Salary, y	1.8	2.1	2.3	2.4	2.3	2.5

(a) Use the method of least squares to find the least squares regression line for the data. Let x represent the year, with $x = 0$ corresponding to 2000.
(b) Use the linear regression capabilities of a graphing utility or computer software program to find a linear model for the data. How does this model compare with the model obtained in part (a)?
(c) Use the linear model to create a table of estimated values for y. Compare the estimated values with the actual data.
(d) Use the linear model to predict the average salary in 2010.
(e) Use the linear model to predict when the average salary will be 3.7 million.

CHAPTER 2 Projects

1 Exploring Matrix Multiplication

The first two test scores for Anna, Bruce, Chris, and David are shown in the table. Use the table to create a matrix M to represent the data. Input matrix M into a graphing utility or computer software program and use it to answer the following questions.

	Test 1	*Test 2*
Anna	84	96
Bruce	56	72
Chris	78	83
David	82	91

1. Which test was more difficult? Which was easier? Explain.
2. How would you rank the performances of the four students?
3. Describe the meanings of the matrix products $M\begin{bmatrix}1\\0\end{bmatrix}$ and $M\begin{bmatrix}0\\1\end{bmatrix}$.
4. Describe the meanings of the matrix products $[1 \quad 0 \quad 0 \quad 0]M$ and $[0 \quad 0 \quad 1 \quad 0]M$.
5. Describe the meanings of the matrix products $M\begin{bmatrix}1\\1\end{bmatrix}$ and $\frac{1}{2}M\begin{bmatrix}1\\1\end{bmatrix}$.
6. Describe the meanings of the matrix products $[1 \quad 1 \quad 1 \quad 1]M$ and $\frac{1}{4}[1 \quad 1 \quad 1 \quad 1]M$.
7. Describe the meaning of the matrix product $[1 \quad 1 \quad 1 \quad 1]M\begin{bmatrix}1\\1\end{bmatrix}$.
8. Use matrix multiplication to express the combined overall average score on both tests.
9. How could you use matrix multiplication to scale the scores on test 1 by a factor of 1.1?

2 Nilpotent Matrices

Let A be a nonzero square $n \times n$ matrix. Is it possible that a positive integer k exists such that $A^k = 0$? For example, find A^3 for the matrix

$$A = \begin{bmatrix} 0 & 1 & 2 \\ 0 & 0 & 1 \\ 0 & 0 & 0 \end{bmatrix}.$$

A square matrix A is said to be **nilpotent of index k** if $A \neq 0$, $A^2 \neq 0, \ldots,$ $A^{k-1} \neq 0$, but $A^k = 0$. In this project you will explore the world of nilpotent matrices.

1. What is the index of the nilpotent matrix A?
2. Use a graphing utility or computer software program to determine which of the matrices below are nilpotent and to find their indices.

(a) $\begin{bmatrix} 0 & 1 \\ 0 & 0 \end{bmatrix}$ (b) $\begin{bmatrix} 0 & 1 \\ 1 & 0 \end{bmatrix}$ (c) $\begin{bmatrix} 0 & 0 \\ 1 & 0 \end{bmatrix}$

(d) $\begin{bmatrix} 1 & 0 \\ 1 & 0 \end{bmatrix}$ (e) $\begin{bmatrix} 0 & 0 & 1 \\ 0 & 0 & 0 \\ 0 & 0 & 0 \end{bmatrix}$ (f) $\begin{bmatrix} 0 & 0 & 0 \\ 1 & 0 & 0 \\ 1 & 1 & 0 \end{bmatrix}$

3. Find 3×3 nilpotent matrices of indices 2 and 3.
4. Find 4×4 nilpotent matrices of indices 2, 3, and 4.
5. Find a nilpotent matrix of index 5.
6. Are nilpotent matrices invertible? Prove your answer.
7. If A is nilpotent, what can you say about A^T? Prove your answer.
8. If A is nilpotent, show that $I - A$ is invertible.

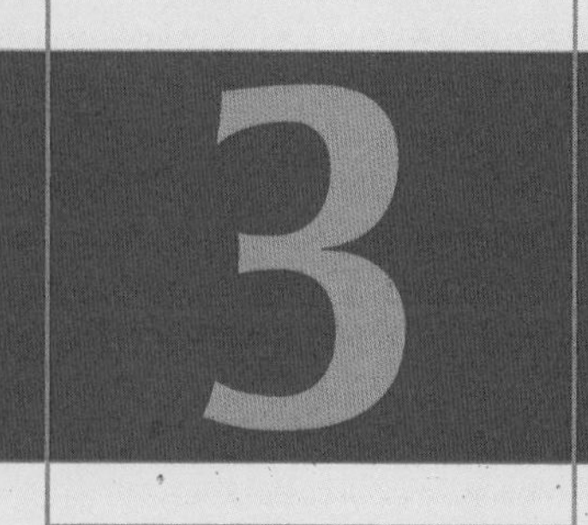

3 Determinants

CHAPTER OBJECTIVES

- Find the determinants of a 2×2 matrix and a triangular matrix.
- Find the minors and cofactors of a matrix and use expansion by cofactors to find the determinant of a matrix.
- Use elementary row or column operations to evaluate the determinant of a matrix.
- Recognize conditions that yield zero determinants.
- Find the determinant of an elementary matrix.
- Use the determinant and properties of the determinant to decide whether a matrix is singular or nonsingular, and recognize equivalent conditions for a nonsingular matrix.
- Verify and find an eigenvalue and an eigenvector of a matrix.
- Find and use the adjoint of a matrix to find its inverse.
- Use Cramer's Rule to solve a system of linear equations.
- Use determinants to find the area of a triangle defined by three distinct points, to find an equation of a line passing through two distinct points, to find the volume of a tetrahedron defined by four distinct points, and to find an equation of a plane passing through three distinct points.

3.1 The Determinant of a Matrix

Every *square* matrix can be associated with a real number called its *determinant.* Determinants have many uses, several of which will be explored in this chapter. The first two sections of this chapter concentrate on procedures for evaluating the determinant of a matrix.

Historically, the use of determinants arose from the recognition of special patterns that occur in the solutions of systems of linear equations. For instance, the general solution of the system

$$\begin{aligned} a_{11}x_1 + a_{12}x_2 &= b_1 \\ a_{21}x_1 + a_{22}x_2 &= b_2 \end{aligned}$$

can be shown to be

$$x_1 = \frac{b_1a_{22} - b_2a_{12}}{a_{11}a_{22} - a_{21}a_{12}} \quad \text{and} \quad x_2 = \frac{b_2a_{11} - b_1a_{21}}{a_{11}a_{22} - a_{21}a_{12}},$$

provided that $a_{11}a_{22} - a_{21}a_{12} \neq 0$. Note that both fractions have the same denominator, $a_{11}a_{22} - a_{21}a_{12}$. This quantity is called the determinant of the coefficient matrix A.

Definition of the Determinant of a 2×2 Matrix

The **determinant** of the matrix

$$A = \begin{bmatrix} a_{11} & a_{12} \\ a_{21} & a_{22} \end{bmatrix}$$

is given by

$$\det(A) = |A| = a_{11}a_{22} - a_{21}a_{12}.$$

REMARK: In this text, $\det(A)$ and $|A|$ are used interchangeably to represent the determinant of a matrix. Vertical bars are also used to denote the absolute value of a real number; the context will show which use is intended. Furthermore, it is common practice to delete the matrix brackets and write

$$\begin{vmatrix} a_{11} & a_{12} \\ a_{21} & a_{22} \end{vmatrix} \quad \text{instead of} \quad \left| \begin{bmatrix} a_{11} & a_{12} \\ a_{21} & a_{22} \end{bmatrix} \right|.$$

A convenient method for remembering the formula for the determinant of a 2×2 matrix is shown in the diagram below.

$$|A| = \begin{vmatrix} a_{11} & a_{12} \\ a_{21} & a_{22} \end{vmatrix} = a_{11}a_{22} - a_{21}a_{12}$$

The determinant is the difference of the products of the two diagonals of the matrix. Note that the order is important, as demonstrated above.

EXAMPLE 1 The Determinant of a Matrix of Order 2

Find the determinant of each matrix.

(a) $A = \begin{bmatrix} 2 & -3 \\ 1 & 2 \end{bmatrix}$ (b) $B = \begin{bmatrix} 2 & 1 \\ 4 & 2 \end{bmatrix}$ (c) $C = \begin{bmatrix} 0 & 3 \\ 2 & 4 \end{bmatrix}$

SOLUTION (a) $|A| = \begin{vmatrix} 2 & -3 \\ 1 & 2 \end{vmatrix} = 2(2) - 1(-3) = 4 + 3 = 7$

(b) $|B| = \begin{vmatrix} 2 & 1 \\ 4 & 2 \end{vmatrix} = 2(2) - 4(1) = 4 - 4 = 0$

(c) $|C| = \begin{vmatrix} 0 & 3 \\ 2 & 4 \end{vmatrix} = 0(4) - 2(3) = 0 - 6 = -6$

REMARK: The determinant of a matrix can be positive, zero, or negative.

The determinant of a matrix of order 1 is defined simply as the entry of the matrix. For instance, if $A = [-2]$, then

$$\det(A) = -2.$$

To define the determinant of a matrix of order higher than 2, it is convenient to use the notions of *minors* and *cofactors.*

Definitions of Minors and Cofactors of a Matrix

If A is a square matrix, then the **minor** M_{ij} of the element a_{ij} is the determinant of the matrix obtained by deleting the ith row and jth column of A. The **cofactor** C_{ij} is given by

$$C_{ij} = (-1)^{i+j}M_{ij}.$$

For example, if A is a 3×3 matrix, then the minors and cofactors of a_{21} and a_{22} are as shown in the diagram below.

Minor of a_{21}

$$\begin{bmatrix} a_{11} & a_{12} & a_{13} \\ a_{21} & a_{22} & a_{23} \\ a_{31} & a_{32} & a_{33} \end{bmatrix}, \quad M_{21} = \begin{vmatrix} a_{12} & a_{13} \\ a_{32} & a_{33} \end{vmatrix}$$

Delete row 2 and column 1.

Cofactor of a_{21}

$$C_{21} = (-1)^{2+1}M_{21} = -M_{21}$$

Minor of a_{22}

$$\begin{bmatrix} a_{11} & a_{12} & a_{13} \\ a_{21} & a_{22} & a_{23} \\ a_{31} & a_{32} & a_{33} \end{bmatrix}, \quad M_{22} = \begin{vmatrix} a_{11} & a_{13} \\ a_{31} & a_{33} \end{vmatrix}$$

Delete row 2 and column 2.

Cofactor of a_{22}

$$C_{22} = (-1)^{2+2}M_{22} = M_{22}$$

As you can see, the minors and cofactors of a matrix can differ only in sign. To obtain the cofactors of a matrix, first find the minors and then apply the checkerboard pattern of $+$'s and $-$'s shown below.

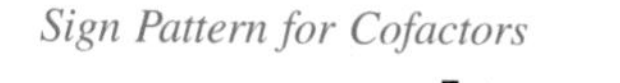

Sign Pattern for Cofactors

$$\begin{bmatrix} + & - & + \\ - & + & - \\ + & - & + \end{bmatrix}$$

3×3 matrix

$$\begin{bmatrix} + & - & + & - \\ - & + & - & + \\ + & - & + & - \\ - & + & - & + \end{bmatrix}$$

4×4 matrix

$$\begin{bmatrix} + & - & + & - & + & \cdots \\ - & + & - & + & - & \cdots \\ + & - & + & - & + & \cdots \\ - & + & - & + & - & \cdots \\ + & - & + & - & + & \cdots \\ \vdots & \vdots & \vdots & \vdots & \vdots & \end{bmatrix}$$

$n \times n$ matrix

Note that *odd* positions (where $i + j$ is odd) have negative signs, and even positions (where $i + j$ is even) have positive signs.

EXAMPLE 2 Find the Minors and Cofactors of a Matrix

Find all the minors and cofactors of

$$A = \begin{bmatrix} 0 & 2 & 1 \\ 3 & -1 & 2 \\ 4 & 0 & 1 \end{bmatrix}.$$

SOLUTION To find the minor M_{11}, delete the first row and first column of A and evaluate the determinant of the resulting matrix.

$$\begin{bmatrix} 0 & 2 & 1 \\ 3 & -1 & 2 \\ 4 & 0 & 1 \end{bmatrix}, \quad M_{11} = \begin{vmatrix} -1 & 2 \\ 0 & 1 \end{vmatrix} = -1(1) - 0(2) = -1$$

Similarly, to find M_{12}, delete the first row and second column.

$$\begin{bmatrix} 0 & 2 & 1 \\ 3 & -1 & 2 \\ 4 & 0 & 1 \end{bmatrix}, \quad M_{12} = \begin{vmatrix} 3 & 2 \\ 4 & 1 \end{vmatrix} = 3(1) - 4(2) = -5$$

Continuing this pattern, you obtain

$$\begin{array}{lll} M_{11} = -1 & M_{12} = -5 & M_{13} = 4 \\ M_{21} = 2 & M_{22} = -4 & M_{23} = -8 \\ M_{31} = 5 & M_{32} = -3 & M_{33} = -6. \end{array}$$

Now, to find the cofactors, combine the checkerboard pattern of signs with these minors to obtain

$$\begin{array}{lll} C_{11} = -1 & C_{12} = 5 & C_{13} = 4 \\ C_{21} = -2 & C_{22} = -4 & C_{23} = 8 \\ C_{31} = 5 & C_{32} = 3 & C_{33} = -6. \end{array}$$

Now that the minors and cofactors of a matrix have been defined, you are ready for a general definition of the determinant of a matrix. The next definition is called **inductive** because it uses determinants of matrices of order $n - 1$ to define the determinant of a matrix of order n.

Definition of the Determinant of a Matrix

If A is a square matrix (of order 2 or greater), then the determinant of A is the sum of the entries in the first row of A multiplied by their cofactors. That is,

$$\det(A) = |A| = \sum_{j=1}^{n} a_{1j}C_{1j} = a_{11}C_{11} + a_{12}C_{12} + \cdots + a_{1n}C_{1n}.$$

REMARK: Try checking that, for 2×2 matrices, this definition yields $|A| = a_{11}a_{22} - a_{21}a_{12}$, as previously defined.

When you use this definition to evaluate a determinant, you are **expanding by cofactors in the first row.** This procedure is demonstrated in Example 3.

EXAMPLE 3 The Determinant of a Matrix of Order 3

Find the determinant of

$$A = \begin{bmatrix} 0 & 2 & 1 \\ 3 & -1 & 2 \\ 4 & 0 & 1 \end{bmatrix}.$$

SOLUTION This matrix is the same as the one in Example 2. There you found the cofactors of the entries in the first row to be

$$C_{11} = -1, \qquad C_{12} = 5, \qquad C_{13} = 4.$$

By the definition of a determinant, you have

$$\begin{aligned} |A| &= a_{11}C_{11} + a_{12}C_{12} + a_{13}C_{13} \\ &= 0(-1) + 2(5) + 1(4) = 14. \end{aligned}$$ **First row expansion**

Although the determinant is defined as an expansion by the cofactors in the first row, it can be shown that the determinant can be evaluated by expanding by *any* row or column. For instance, you could expand the 3×3 matrix in Example 3 by the second row to obtain

$$\begin{aligned} |A| &= a_{21}C_{21} + a_{22}C_{22} + a_{23}C_{23} \\ &= 3(-2) + (-1)(-4) + 2(8) = 14 \end{aligned}$$ **Second row expansion**

or by the first column to obtain

$$\begin{aligned} |A| &= a_{11}C_{11} + a_{21}C_{21} + a_{31}C_{31} \\ &= 0(-1) + 3(-2) + 4(5) = 14. \end{aligned}$$ **First column expansion**

Try other possibilities to confirm that the determinant of A can be evaluated by expanding by *any* row or column. This is stated in the theorem below, Laplace's Expansion of a Determinant, named after the French mathematician Pierre Simon de Laplace (1749–1827).

THEOREM 3.1 Expansion by Cofactors

Let A be a square matrix of order n. Then the determinant of A is given by

$$\det(A) = |A| = \sum_{j=1}^{n} a_{ij}C_{ij} = a_{i1}C_{i1} + a_{i2}C_{i2} + \cdot\cdot\cdot + a_{in}C_{in}$$ ***i*th row expansion**

or

$$\det(A) = |A| = \sum_{i=1}^{n} a_{ij}C_{ij} = a_{1j}C_{1j} + a_{2j}C_{2j} + \cdot\cdot\cdot + a_{nj}C_{nj}.$$ ***j*th column expansion**

When expanding by cofactors, you do not need to evaluate the cofactors of zero entries, because a zero entry times its cofactor is zero.

$$a_{ij}C_{ij} = (0)C_{ij} = 0$$

The row (or column) containing the most zeros is usually the best choice for expansion by cofactors. This is demonstrated in the next example.

EXAMPLE 4 The Determinant of a Matrix of Order 4

Find the determinant of

$$A = \begin{bmatrix} 1 & -2 & 3 & 0 \\ -1 & 1 & 0 & 2 \\ 0 & 2 & 0 & 3 \\ 3 & 4 & 0 & -2 \end{bmatrix}.$$

SOLUTION By inspecting this matrix, you can see that three of the entries in the third column are zeros. You can eliminate some of the work in the expansion by using the third column.

$$|A| = 3(C_{13}) + 0(C_{23}) + 0(C_{33}) + 0(C_{43})$$

Because C_{23}, C_{33}, and C_{43} have zero coefficients, you need only find the cofactor C_{13}. To do this, delete the first row and third column of A and evaluate the determinant of the resulting matrix.

$$C_{13} = (-1)^{1+3}\begin{vmatrix} -1 & 1 & 2 \\ 0 & 2 & 3 \\ 3 & 4 & -2 \end{vmatrix} = \begin{vmatrix} -1 & 1 & 2 \\ 0 & 2 & 3 \\ 3 & 4 & -2 \end{vmatrix}$$

Expanding by cofactors in the second row yields

$$C_{13} = (0)(-1)^{2+1}\begin{vmatrix} 1 & 2 \\ 4 & -2 \end{vmatrix} + (2)(-1)^{2+2}\begin{vmatrix} -1 & 2 \\ 3 & -2 \end{vmatrix} + (3)(-1)^{2+3}\begin{vmatrix} -1 & 1 \\ 3 & 4 \end{vmatrix}$$

$$= 0 + 2(1)(-4) + 3(-1)(-7) = 13.$$

You obtain $|A| = 3(13) = 39$.

Technology Note

Most graphing utilities and computer software programs have the capability of calculating the determinant of a square matrix. If you use the determinant command of a graphing utility to verify that the determinant of the matrix A in Example 4 is 39, your screen may look like the one below. Keystrokes and programming syntax for these utilities/programs applicable to Example 4 are provided in the **Online Technology Guide,** available at *college.hmco.com/pic/larsonELA6e*.

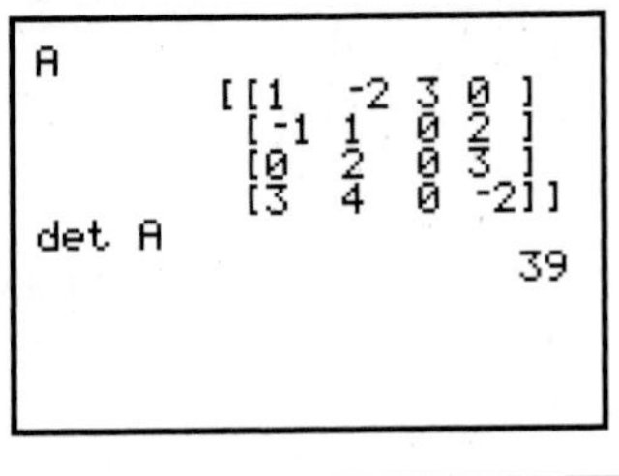

There is an alternative method commonly used for evaluating the determinant of a 3×3 matrix A. To apply this method, copy the first and second columns of A to form fourth and fifth columns. The determinant of A is then obtained by adding (or subtracting) the products of the six diagonals, as shown in the following diagram.

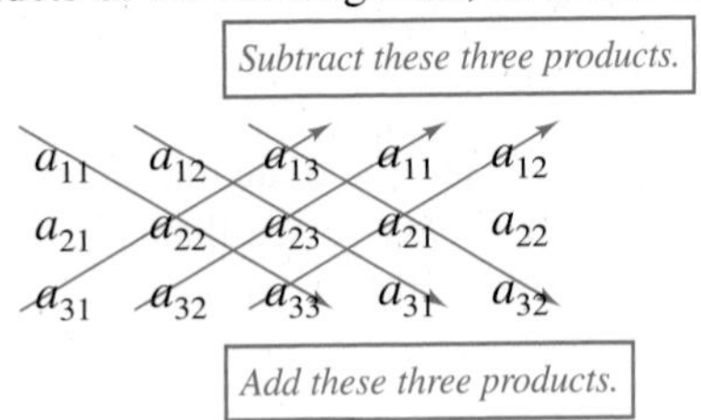

Try confirming that the determinant of A is

$$|A| = a_{11}a_{22}a_{33} + a_{12}a_{23}a_{31} + a_{13}a_{21}a_{32} - a_{31}a_{22}a_{13} - a_{32}a_{23}a_{11} - a_{33}a_{21}a_{12}.$$

EXAMPLE 5 The Determinant of a Matrix of Order 3

Find the determinant of

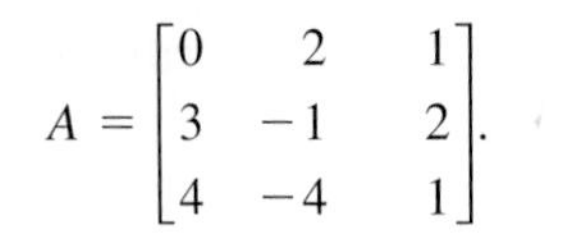

$$A = \begin{bmatrix} 0 & 2 & 1 \\ 3 & -1 & 2 \\ 4 & -4 & 1 \end{bmatrix}.$$

Simulation

To explore this concept further with an electronic simulation, and for keystrokes and programming syntax regarding specific graphing utilities and computer software programs involving Example 5, please visit *college.hmco.com/pic/larsonELA6e.* Similar exercises and projects are also available on this website.

SOLUTION Begin by recopying the first two columns and then computing the six diagonal products as follows.

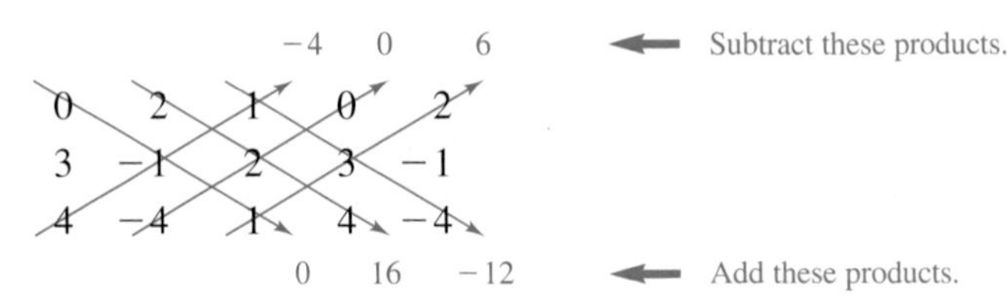

Now, by adding the lower three products and subtracting the upper three products, you can find the determinant of A to be $|A| = 0 + 16 + (-12) - (-4) - 0 - 6 = 2$.

REMARK: The diagonal process illustrated in Example 5 is valid *only* for matrices of order 3. For matrices of higher orders, another method must be used.

Triangular Matrices

Evaluating determinants of matrices of order 4 or higher can be tedious. There is, however, an important exception: the determinant of a *triangular* matrix. Recall from Section 2.4 that a square matrix is called *upper triangular* if it has all zero entries below its main diagonal, and *lower triangular* if it has all zero entries above its main diagonal. A matrix that is both upper and lower triangular is called **diagonal.** That is, a diagonal matrix is one in which all entries above and below the main diagonal are zero.

Upper Triangular Matrix

$$\begin{bmatrix} a_{11} & a_{12} & a_{13} & \cdots & a_{1n} \\ 0 & a_{22} & a_{23} & \cdots & a_{2n} \\ 0 & 0 & a_{33} & \cdots & a_{3n} \\ \vdots & \vdots & \vdots & & \vdots \\ 0 & 0 & 0 & \cdots & a_{nn} \end{bmatrix}$$

Lower Triangular Matrix

$$\begin{bmatrix} a_{11} & 0 & 0 & \cdots & 0 \\ a_{21} & a_{22} & 0 & \cdots & 0 \\ a_{31} & a_{32} & a_{33} & \cdots & 0 \\ \vdots & \vdots & \vdots & & \vdots \\ a_{n1} & a_{n2} & a_{n3} & \cdots & a_{nn} \end{bmatrix}$$

To find the determinant of a triangular matrix, simply form the product of the entries on the main diagonal. It is easy to see that this procedure is valid for triangular matrices of order 2 or 3. For instance, the determinant of

$$A = \begin{bmatrix} 2 & 3 & -1 \\ 0 & -1 & 2 \\ 0 & 0 & 3 \end{bmatrix}$$

can be found by expanding by the third row to obtain

$$|A| = 0(-1)^{3+1}\begin{vmatrix} 3 & -1 \\ -1 & 2 \end{vmatrix} + 0(-1)^{3+2}\begin{vmatrix} 2 & -1 \\ 0 & 2 \end{vmatrix} + 3(-1)^{3+3}\begin{vmatrix} 2 & 3 \\ 0 & -1 \end{vmatrix}$$

$$= 3(1)(-2) = -6,$$

which is the product of the entries on the main diagonal.

THEOREM 3.2
Determinant of a Triangular Matrix

If A is a triangular matrix of order n, then its determinant is the product of the entries on the main diagonal. That is,

$$\det(A) = |A| = a_{11}a_{22}a_{33} \cdot \cdot \cdot a_{nn}.$$

PROOF You can use *mathematical induction** to prove this theorem for the case in which A is an upper triangular matrix. The case in which A is lower triangular can be proven similarly. If A has order 1, then $A = [a_{11}]$ and the determinant is $|A| = a_{11}$. Assuming the theorem is true for any upper triangular matrix of order $k - 1$, consider an upper triangular matrix A of order k. Expanding by the kth row, you obtain

$$|A| = 0C_{k1} + 0C_{k2} + \cdot \cdot \cdot + 0C_{k(k-1)} + a_{kk}C_{kk} = a_{kk}C_{kk}.$$

Now, note that $C_{kk} = (-1)^{2k}M_{kk} = M_{kk}$, where M_{kk} is the determinant of the upper triangular matrix formed by deleting the kth row and kth column of A. Because this matrix is of order $k - 1$, you can apply the induction assumption to write

$$|A| = a_{kk}M_{kk} = a_{kk}(a_{11}a_{22}a_{33} \cdot \cdot \cdot a_{k-1\,k-1}) = a_{11}a_{22}a_{33} \cdot \cdot \cdot a_{kk}.$$

EXAMPLE 6 **The Determinant of a Triangular Matrix**

Find the determinant of each matrix.

(a) $A = \begin{bmatrix} 2 & 0 & 0 & 0 \\ 4 & -2 & 0 & 0 \\ -5 & 6 & 1 & 0 \\ 1 & 5 & 3 & 3 \end{bmatrix}$ (b) $B = \begin{bmatrix} -1 & 0 & 0 & 0 & 0 \\ 0 & 3 & 0 & 0 & 0 \\ 0 & 0 & 2 & 0 & 0 \\ 0 & 0 & 0 & 4 & 0 \\ 0 & 0 & 0 & 0 & -2 \end{bmatrix}$

*A discussion of mathematical induction can be found in Appendix A.

SOLUTION (a) The determinant of this lower triangular matrix is given by

$$|A| = (2)(-2)(1)(3) = -12.$$

(b) The determinant of this *diagonal* matrix is given by

$$|B| = (-1)(3)(2)(4)(-2) = 48.$$

SECTION 3.1 Exercises

In Exercises 1–12, find the determinant of the matrix.

1. $[1]$

2. $[-3]$

3. $\begin{bmatrix} 2 & 1 \\ 3 & 4 \end{bmatrix}$

4. $\begin{bmatrix} -3 & 1 \\ 5 & 2 \end{bmatrix}$

5. $\begin{bmatrix} 5 & 2 \\ -6 & 3 \end{bmatrix}$

6. $\begin{bmatrix} 2 & -2 \\ 4 & 3 \end{bmatrix}$

7. $\begin{bmatrix} -7 & 6 \\ \frac{1}{2} & 3 \end{bmatrix}$

8. $\begin{bmatrix} \frac{1}{3} & 5 \\ 4 & -9 \end{bmatrix}$

9. $\begin{bmatrix} 2 & 6 \\ 0 & 3 \end{bmatrix}$

10. $\begin{bmatrix} 2 & -3 \\ -6 & 9 \end{bmatrix}$

11. $\begin{bmatrix} \lambda - 3 & 2 \\ 4 & \lambda - 1 \end{bmatrix}$

12. $\begin{bmatrix} \lambda - 2 & 0 \\ 4 & \lambda - 4 \end{bmatrix}$

In Exercises 13–16, find (a) the minors and (b) the cofactors of the matrix.

13. $\begin{bmatrix} 1 & 2 \\ 3 & 4 \end{bmatrix}$

14. $\begin{bmatrix} -1 & 0 \\ 2 & 1 \end{bmatrix}$

15. $\begin{bmatrix} -3 & 2 & 1 \\ 4 & 5 & 6 \\ 2 & -3 & 1 \end{bmatrix}$

16. $\begin{bmatrix} -3 & 4 & 2 \\ 6 & 3 & 1 \\ 4 & -7 & -8 \end{bmatrix}$

17. Find the determinant of the matrix in Exercise 15 using the method of expansion by cofactors. Use (a) the second row and (b) the second column.

18. Find the determinant of the matrix in Exercise 16 using the method of expansion by cofactors. Use (a) the third row and (b) the first column.

In Exercises 19–34, use expansion by cofactors to find the determinant of the matrix.

19. $\begin{bmatrix} 1 & 4 & -2 \\ 3 & 2 & 0 \\ -1 & 4 & 3 \end{bmatrix}$

20. $\begin{bmatrix} 2 & -1 & 3 \\ 1 & 4 & 4 \\ 1 & 0 & 2 \end{bmatrix}$

21. $\begin{bmatrix} 2 & 4 & 6 \\ 0 & 3 & 1 \\ 0 & 0 & -5 \end{bmatrix}$

22. $\begin{bmatrix} -3 & 0 & 0 \\ 7 & 11 & 0 \\ 1 & 2 & 2 \end{bmatrix}$

23. $\begin{bmatrix} 0.1 & 0.2 & 0.3 \\ -0.3 & 0.2 & 0.2 \\ 0.5 & 0.4 & 0.4 \end{bmatrix}$

24. $\begin{bmatrix} -0.4 & 0.4 & 0.3 \\ 0.2 & 0.2 & 0.2 \\ 0.3 & 0.2 & 0.2 \end{bmatrix}$

25. $\begin{bmatrix} x & y & 1 \\ 2 & 3 & 1 \\ 0 & -1 & 1 \end{bmatrix}$

26. $\begin{bmatrix} x & y & 1 \\ -2 & -2 & 1 \\ 1 & 5 & 1 \end{bmatrix}$

27. $\begin{bmatrix} 2 & 6 & 6 & 2 \\ 2 & 7 & 3 & 6 \\ 1 & 5 & 0 & 1 \\ 3 & 7 & 0 & 7 \end{bmatrix}$

28. $\begin{bmatrix} 1 & 4 & 3 & 2 \\ -5 & 6 & 2 & 1 \\ 0 & 0 & 0 & 0 \\ 3 & -2 & 1 & 5 \end{bmatrix}$

29. $\begin{bmatrix} 5 & 3 & 0 & 6 \\ 4 & 6 & 4 & 12 \\ 0 & 2 & -3 & 4 \\ 0 & 1 & -2 & 2 \end{bmatrix}$

30. $\begin{bmatrix} 3 & 0 & 7 & 0 \\ 2 & 6 & 11 & 12 \\ 4 & 1 & -1 & 2 \\ 1 & 5 & 2 & 10 \end{bmatrix}$

31. $\begin{bmatrix} w & x & y & z \\ 21 & -15 & 24 & 30 \\ -10 & 24 & -32 & 18 \\ -40 & 22 & 32 & -35 \end{bmatrix}$

32. $\begin{bmatrix} w & x & y & z \\ 10 & 15 & -25 & 30 \\ -30 & 20 & -15 & -10 \\ 30 & 35 & -25 & -40 \end{bmatrix}$

33. $\begin{bmatrix} 5 & 2 & 0 & 0 & -2 \\ 0 & 1 & 4 & 3 & 2 \\ 0 & 0 & 2 & 6 & 3 \\ 0 & 0 & 3 & 4 & 1 \\ 0 & 0 & 0 & 0 & 2 \end{bmatrix}$ **34.** $\begin{bmatrix} 4 & 3 & -2 & 1 & 2 \\ 0 & 0 & 0 & 0 & 0 \\ 1 & 2 & -7 & 13 & 12 \\ 6 & -2 & 5 & 6 & 7 \\ 1 & 4 & 2 & 0 & 9 \end{bmatrix}$

In Exercises 35–40, use a graphing utility or computer software program with matrix capabilities to find the determinant of the matrix.

35. $\begin{bmatrix} \frac{1}{2} & 1 & -5 \\ 4 & -\frac{1}{4} & -4 \\ 3 & 2 & -2 \end{bmatrix}$ **36.** $\begin{bmatrix} 0.25 & -1 & 0.6 \\ 0.50 & 0.8 & -0.2 \\ 0.75 & 0.9 & -0.4 \end{bmatrix}$

37. $\begin{bmatrix} 4 & 3 & 2 & 5 \\ 1 & 6 & -1 & 2 \\ -3 & 2 & 4 & 5 \\ 6 & 1 & 3 & -2 \end{bmatrix}$ **38.** $\begin{bmatrix} 1 & 2 & 4 & -1 \\ 6 & 2 & 3 & -2 \\ 4 & -1 & 4 & 3 \\ 5 & -2 & 2 & 4 \end{bmatrix}$

39. $\begin{bmatrix} 1 & 2 & -1 & 4 & 2 & -1 \\ 0 & 1 & 2 & -2 & -3 & 1 \\ 0 & 3 & 2 & -1 & 3 & -2 \\ 1 & 2 & 0 & -2 & 3 & 2 \\ 1 & -2 & 3 & 1 & 2 & -1 \\ 2 & 0 & 2 & 3 & 1 & 1 \end{bmatrix}$

40. $\begin{bmatrix} 8 & 5 & 1 & -2 & 0 & 3 \\ -1 & 0 & 7 & 1 & 6 & -5 \\ 0 & 8 & 6 & 5 & -3 & 1 \\ 1 & 2 & 5 & -8 & 4 & 3 \\ 2 & 6 & -2 & 0 & 6 & 7 \\ 8 & -3 & 1 & 2 & -5 & 1 \end{bmatrix}$

In Exercises 41–46, find the determinant of the triangular matrix.

41. $\begin{bmatrix} -2 & 0 & 0 \\ 4 & 6 & 0 \\ -3 & 7 & 2 \end{bmatrix}$ **42.** $\begin{bmatrix} 5 & 0 & 0 \\ 0 & 6 & 0 \\ 0 & 0 & -3 \end{bmatrix}$

43. $\begin{bmatrix} 5 & 8 & -4 & 2 \\ 0 & 0 & 6 & 0 \\ 0 & 0 & 2 & 2 \\ 0 & 0 & 0 & -1 \end{bmatrix}$ **44.** $\begin{bmatrix} 4 & 0 & 0 & 0 \\ -1 & \frac{1}{2} & 0 & 0 \\ 3 & 5 & 3 & 0 \\ -8 & 7 & 0 & -2 \end{bmatrix}$

45. $\begin{bmatrix} -1 & 4 & 2 & 1 & -3 \\ 0 & 3 & -4 & 5 & 2 \\ 0 & 0 & 2 & 7 & 0 \\ 0 & 0 & 0 & 5 & -1 \\ 0 & 0 & 0 & 0 & 1 \end{bmatrix}$

46. $\begin{bmatrix} 7 & 0 & 0 & 0 & 0 \\ -8 & \frac{1}{4} & 0 & 0 & 0 \\ 4 & 5 & 2 & 0 & 0 \\ 3 & -3 & 5 & -1 & 0 \\ 1 & 13 & 4 & 1 & -2 \end{bmatrix}$

True or False? In Exercises 47 and 48, determine whether each statement is true or false. If a statement is true, give a reason or cite an appropriate statement from the text. If a statement is false, provide an example that shows the statement is not true in all cases or cite an appropriate statement from the text.

47. (a) The determinant of the 2×2 matrix A is $a_{21}a_{12} - a_{11}a_{22}$.

(b) The determinant of a matrix of order 1 is the entry of the matrix.

(c) The ij-cofactor of a square matrix A is the matrix defined by deleting the ith row and the jth column of A.

48. (a) To find the determinant of a triangular matrix, add the entries on the main diagonal.

(b) The determinant of a matrix can be evaluated using expansion by cofactors in any row or column.

(c) When expanding by cofactors, you need not evaluate the cofactors of zero entries.

In Exercises 49–54, solve for x.

49. $\begin{vmatrix} x+3 & 2 \\ 1 & x+2 \end{vmatrix} = 0$ **50.** $\begin{vmatrix} x-2 & -1 \\ -3 & x \end{vmatrix} = 0$

51. $\begin{vmatrix} x+1 & -2 \\ 1 & x-2 \end{vmatrix} = 0$ **52.** $\begin{vmatrix} x+3 & 1 \\ -4 & x-1 \end{vmatrix} = 0$

53. $\begin{vmatrix} x-1 & 2 \\ 3 & x-2 \end{vmatrix} = 0$ **54.** $\begin{vmatrix} x-2 & -1 \\ -3 & x \end{vmatrix} = 0$

In Exercises 55–58, find the values of λ for which the determinant is zero.

55. $\begin{vmatrix} \lambda+2 & 2 \\ 1 & \lambda \end{vmatrix}$ **56.** $\begin{vmatrix} \lambda-1 & 1 \\ 4 & \lambda-3 \end{vmatrix}$

57. $\begin{vmatrix} \lambda & 2 & 0 \\ 0 & \lambda+1 & 2 \\ 0 & 1 & \lambda \end{vmatrix}$ **58.** $\begin{vmatrix} \lambda & 0 & 1 \\ 0 & \lambda & 3 \\ 2 & 2 & \lambda-2 \end{vmatrix}$

In Exercises 59–64, evaluate the determinant, in which the entries are functions. Determinants of this type occur when changes of variables are made in calculus.

59. $\begin{vmatrix} 4u & -1 \\ -1 & 2v \end{vmatrix}$

60. $\begin{vmatrix} 3x^2 & -3y^2 \\ 1 & 1 \end{vmatrix}$

61. $\begin{vmatrix} e^{2x} & e^{3x} \\ 2e^{2x} & 3e^{3x} \end{vmatrix}$

62. $\begin{vmatrix} e^{-x} & xe^{-x} \\ -e^{-x} & (1-x)e^{-x} \end{vmatrix}$

63. $\begin{vmatrix} x & \ln x \\ 1 & 1/x \end{vmatrix}$

64. $\begin{vmatrix} x & x \ln x \\ 1 & 1 + \ln x \end{vmatrix}$

65. The determinant of a 2×2 matrix involves two products. The determinant of a 3×3 matrix involves six triple products. Show that the determinant of a 4×4 matrix involves 24 quadruple products. (In general, the determinant of an $n \times n$ matrix involves $n!$ n-fold products.)

66. Show that the system of linear equations

$$ax + by = e$$
$$cx + dy = f$$

has a unique solution if and only if the determinant of the coefficient matrix is nonzero.

In Exercises 67–74, evaluate the determinants to verify the equation.

67. $\begin{vmatrix} w & x \\ y & z \end{vmatrix} = -\begin{vmatrix} y & z \\ w & x \end{vmatrix}$

68. $\begin{vmatrix} w & cx \\ y & cz \end{vmatrix} = c\begin{vmatrix} w & x \\ y & z \end{vmatrix}$

69. $\begin{vmatrix} w & x \\ y & z \end{vmatrix} = \begin{vmatrix} w & x + cw \\ y & z + cy \end{vmatrix}$

70. $\begin{vmatrix} w & x \\ cw & cx \end{vmatrix} = 0$

71. $\begin{vmatrix} 1 & x & x^2 \\ 1 & y & y^2 \\ 1 & z & z^2 \end{vmatrix} = (y - x)(z - x)(z - y)$

72. $\begin{vmatrix} a + b & a & a \\ a & a + b & a \\ a & a & a + b \end{vmatrix} = b^2(3a + b)$

73. $\begin{vmatrix} 1 & 1 & 1 \\ a & b & c \\ a^2 & b^2 & c^2 \end{vmatrix} = (a - b)(b - c)(c - a)$

74. $\begin{vmatrix} 1 & 1 & 1 \\ a & b & c \\ a^3 & b^3 & c^3 \end{vmatrix} = (a - b)(b - c)(c - a)(a + b + c)$

75. You are given the equation

$$\begin{vmatrix} x & 0 & c \\ -1 & x & b \\ 0 & -1 & a \end{vmatrix} = ax^2 + bx + c.$$

(a) Verify the equation.

(b) Use the equation as a model to find a determinant that is equal to $ax^3 + bx^2 + cx + d$.

76. Writing Explain why it is easy to calculate the determinant of a matrix that has an entire row of zeros.

3.2 Evaluation of a Determinant Using Elementary Operations

Which of the two determinants shown below is easier to evaluate?

$$|A| = \begin{vmatrix} 1 & -2 & 3 & 1 \\ 4 & -6 & 3 & 2 \\ -2 & 4 & -9 & -3 \\ 3 & -6 & 9 & 2 \end{vmatrix} \quad \text{or} \quad |B| = \begin{vmatrix} 1 & -2 & 3 & 1 \\ 0 & 2 & -9 & -2 \\ 0 & 0 & -3 & -1 \\ 0 & 0 & 0 & -1 \end{vmatrix}$$

Given what you now know about the determinant of a triangular matrix, it is clear that the second determinant is *much* easier to evaluate. Its determinant is simply the product of the entries on the main diagonal. That is, $|B| = (1)(2)(-3)(-1) = 6$. On the other hand, using expansion by cofactors (the only technique discussed so far) to evaluate the first determinant is messy. For instance, if you expand by cofactors across the first row, you have

$$|A| = 1\begin{vmatrix} -6 & 3 & 2 \\ 4 & -9 & -3 \\ -6 & 9 & 2 \end{vmatrix} + 2\begin{vmatrix} 4 & 3 & 2 \\ -2 & -9 & -3 \\ 3 & 9 & 2 \end{vmatrix} + 3\begin{vmatrix} 4 & -6 & 2 \\ -2 & 4 & -3 \\ 3 & -6 & 2 \end{vmatrix} - 1\begin{vmatrix} 4 & -6 & 3 \\ -2 & 4 & -9 \\ 3 & -6 & 9 \end{vmatrix}.$$

Evaluating the determinants of these four 3×3 matrices produces

$$|A| = (1)(-60) + (2)(39) + (3)(-10) - (1)(-18) = 6.$$

It is not coincidental that these two determinants have the same value. In fact, you can obtain the matrix B by performing elementary row operations on matrix A. (Try verifying this.) In this section, you will see the effects of elementary row (and column) operations on the value of a determinant.

EXAMPLE 1 The Effects of Elementary Row Operations on a Determinant

(a) The matrix B was obtained from A by interchanging the rows of A.

$$|A| = \begin{vmatrix} 2 & -3 \\ 1 & 4 \end{vmatrix} = 11 \quad \text{and} \quad |B| = \begin{vmatrix} 1 & 4 \\ 2 & -3 \end{vmatrix} = -11$$

(b) The matrix B was obtained from A by adding -2 times the first row of A to the second row of A.

$$|A| = \begin{vmatrix} 1 & -3 \\ 2 & -4 \end{vmatrix} = 2 \quad \text{and} \quad |B| = \begin{vmatrix} 1 & -3 \\ 0 & 2 \end{vmatrix} = 2$$

(c) The matrix B was obtained from A by multiplying the first row of A by $\frac{1}{2}$.

$$|A| = \begin{vmatrix} 2 & -8 \\ -2 & 9 \end{vmatrix} = 2 \quad \text{and} \quad |B| = \begin{vmatrix} 1 & -4 \\ -2 & 9 \end{vmatrix} = 1$$

In Example 1, you can see that interchanging two rows of a matrix changed the sign of its determinant. Adding a multiple of one row to another did not change the determinant. Finally, multiplying a row by a nonzero constant multiplied the determinant by that same constant. The next theorem generalizes these observations. The proof of Property 1 follows the theorem, and the proofs of the other two properties are left as exercises. (See Exercises 54 and 55.)

THEOREM 3.3
Elementary Row Operations and Determinants

Let A and B be square matrices.

1. If B is obtained from A by interchanging two rows of A, then
$$\det(B) = -\det(A).$$
2. If B is obtained from A by adding a multiple of a row of A to another row of A, then
$$\det(B) = \det(A).$$
3. If B is obtained from A by multiplying a row of A by a nonzero constant c, then
$$\det(B) = c\det(A).$$

PROOF To prove Property 1, use mathematical induction, as follows. Assume that A and B are 2×2 matrices such that

$$A = \begin{bmatrix} a_{11} & a_{12} \\ a_{21} & a_{22} \end{bmatrix} \quad \text{and} \quad B = \begin{bmatrix} a_{21} & a_{22} \\ a_{11} & a_{12} \end{bmatrix}.$$

Then, you have $|A| = a_{11}a_{22} - a_{21}a_{12}$ and $|B| = a_{21}a_{12} - a_{11}a_{22}$. So $|B| = -|A|$. Now assume the property is true for matrices of order $(n - 1)$. Let A be an $n \times n$ matrix such that B is obtained from A by interchanging two rows of A. Then, to find $|A|$ and $|B|$, expand along a row other than the two interchanged rows. By the induction assumption, the cofactors of B will be the negatives of the cofactors of A because the corresponding $(n - 1) \times (n - 1)$ matrices have two rows interchanged. Finally, $|B| = -|A|$ and the proof is complete.

HISTORICAL NOTE

Augustin-Louis Cauchy (1789–1857)

was encouraged by Pierre Simon de Laplace, one of France's leading mathematicians, to study mathematics. Cauchy is often credited with bringing rigor to modern mathematics. To read about his work, visit *college.hmco.com/pic/larsonELA6e.*

REMARK: Note that the third property of Theorem 3.3 enables you to divide a row by the common factor. For instance,

$$\begin{vmatrix} 2 & 4 \\ 1 & 3 \end{vmatrix} = 2\begin{vmatrix} 1 & 2 \\ 1 & 3 \end{vmatrix}. \qquad \text{Factor 2 out of first row.}$$

Theorem 3.3 provides a practical way to evaluate determinants. (This method works particularly well with computers.) To find the determinant of a matrix A, use elementary row operations to obtain a triangular matrix B that is row-equivalent to A. For each step in the elimination process, use Theorem 3.3 to determine the effect of the elementary row operation on the determinant. Finally, find the determinant of B by multiplying the entries on its main diagonal. This process is demonstrated in the next example.

EXAMPLE 2 Evaluating a Determinant Using Elementary Row Operations

Find the determinant of

$$A = \begin{bmatrix} 2 & -3 & 10 \\ 1 & 2 & -2 \\ 0 & 1 & -3 \end{bmatrix}.$$

SOLUTION Using elementary row operations, rewrite A in triangular form as follows.

$$\begin{vmatrix} 2 & -3 & 10 \\ 1 & 2 & -2 \\ 0 & 1 & -3 \end{vmatrix} = -\begin{vmatrix} 1 & 2 & -2 \\ 2 & -3 & 10 \\ 0 & 1 & -3 \end{vmatrix}$$

Interchange the first two rows.

$$= -\begin{vmatrix} 1 & 2 & -2 \\ 0 & -7 & 14 \\ 0 & 1 & -3 \end{vmatrix}$$

Add -2 times the first row to the second row to produce a new second row.

$$= 7\begin{vmatrix} 1 & 2 & -2 \\ 0 & 1 & -2 \\ 0 & 1 & -3 \end{vmatrix}$$

Factor -7 out of the second row.

$$= 7\begin{vmatrix} 1 & 2 & -2 \\ 0 & 1 & -2 \\ 0 & 0 & -1 \end{vmatrix}$$

Add -1 times the second row to the third row to produce a new third row.

Now, because the final matrix is triangular, you can conclude that the determinant is

$$|A| = 7(1)(1)(-1) = -7.$$

Determinants and Elementary Column Operations

Although Theorem 3.3 is stated in terms of elementary *row* operations, the theorem remains valid if the word "column" replaces the word "row." Operations performed on the columns (rather than the rows) of a matrix are called **elementary column operations,** and two matrices are called **column-equivalent** if one can be obtained from the other by elementary column operations. The column version of Theorem 3.3 is illustrated as follows.

$$\begin{vmatrix} 2 & 1 & -3 \\ 4 & 0 & 1 \\ 0 & 0 & 2 \end{vmatrix} = -\begin{vmatrix} 1 & 2 & -3 \\ 0 & 4 & 1 \\ 0 & 0 & 2 \end{vmatrix} \qquad \begin{vmatrix} 2 & 3 & -5 \\ 4 & 1 & 0 \\ -2 & 4 & -3 \end{vmatrix} = 2\begin{vmatrix} 1 & 3 & -5 \\ 2 & 1 & 0 \\ -1 & 4 & -3 \end{vmatrix}$$

Interchange the first two columns. *Factor 2 out of the first column.*

In evaluating a determinant by hand, it is occasionally convenient to use elementary column operations, as shown in Example 3.

EXAMPLE 3 Evaluating a Determinant Using Elementary Column Operations

Find the determinant of

$$A = \begin{bmatrix} -1 & 2 & 2 \\ 3 & -6 & 4 \\ 5 & -10 & -3 \end{bmatrix}.$$

SOLUTION Because the first two columns of A are multiples of each other, you can obtain a column of zeros by adding 2 times the first column to the second column, as follows.

$$\begin{vmatrix} -1 & 2 & 2 \\ 3 & -6 & 4 \\ 5 & -10 & -3 \end{vmatrix} = \begin{vmatrix} -1 & 0 & 2 \\ 3 & 0 & 4 \\ 5 & 0 & -3 \end{vmatrix}$$

At this point you do not need to continue to rewrite the matrix in triangular form. Because there is an entire column of zeros, simply conclude that the determinant is zero. The validity of this conclusion follows from Theorem 3.1. Specifically, by expanding by cofactors along the second column, you have

$$|A| = (0)C_{12} + (0)C_{22} + (0)C_{32} = 0.$$

Example 3 shows that if one column of a matrix is a scalar multiple of another column, you can immediately conclude that the determinant of the matrix is zero. This is one of three conditions, listed next, that yield a determinant of zero.

THEOREM 3.4 Conditions That Yield a Zero Determinant

If A is a square matrix and any one of the following conditions is true, then $\det(A) = 0$.

1. An entire row (or an entire column) consists of zeros.
2. Two rows (or columns) are equal.
3. One row (or column) is a multiple of another row (or column).

PROOF Each part of this theorem is easily verified by using elementary row operations and expansion by cofactors. For example, if an entire row or column is zero, then each cofactor in the expansion is multiplied by zero. If condition 2 or 3 is true, you can use elementary row or column operations to create an entire row or column of zeros.

Recognizing the conditions listed in Theorem 3.4 can make evaluating a determinant much easier. For instance,

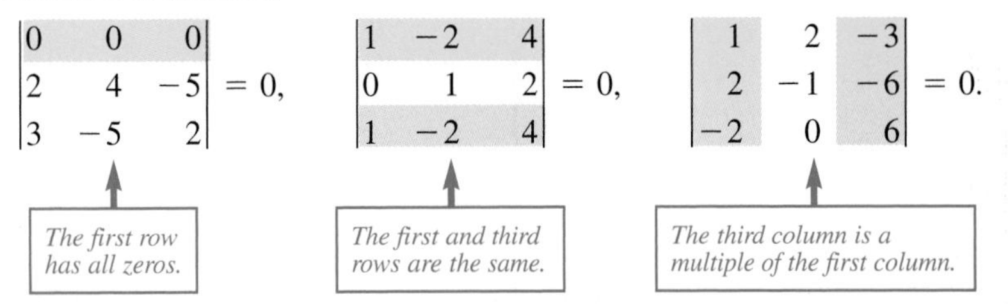

$$\begin{vmatrix} 0 & 0 & 0 \\ 2 & 4 & -5 \\ 3 & -5 & 2 \end{vmatrix} = 0, \qquad \begin{vmatrix} 1 & -2 & 4 \\ 0 & 1 & 2 \\ 1 & -2 & 4 \end{vmatrix} = 0, \qquad \begin{vmatrix} 1 & 2 & -3 \\ 2 & -1 & -6 \\ -2 & 0 & 6 \end{vmatrix} = 0.$$
The first row has all zeros.
The first and third rows are the same.
The third column is a multiple of the first column.

Do not conclude, however, that Theorem 3.4 gives the *only* conditions that produce a determinant of zero. This theorem is often used indirectly. That is, you can begin with a matrix that does not satisfy any of the conditions of Theorem 3.4 and, through elementary row or column operations, obtain a matrix that does satisfy one of the conditions. Then you can conclude that the original matrix has a determinant of zero. This process is demonstrated in Example 4.

EXAMPLE 4 A Matrix with a Zero Determinant

Find the determinant of

$$A = \begin{bmatrix} 1 & 4 & 1 \\ 2 & -1 & 0 \\ 0 & 18 & 4 \end{bmatrix}.$$

SOLUTION Adding -2 times the first row to the second row produces

$$|A| = \begin{vmatrix} 1 & 4 & 1 \\ 2 & -1 & 0 \\ 0 & 18 & 4 \end{vmatrix} = \begin{vmatrix} 1 & 4 & 1 \\ 0 & -9 & -2 \\ 0 & 18 & 4 \end{vmatrix}.$$

Now, because the second and third rows are multiples of each other, you can conclude that the determinant is zero.

In Example 4 you could have obtained a matrix with a row of all zeros by performing an additional elementary row operation (adding 2 times the second row to the third row). This is true in general. That is, a square matrix has a determinant of zero if and only if it is row- (or column-) equivalent to a matrix that has at least one row (or column) consisting entirely of zeros. This will be proved in the next section.

You have now surveyed two general methods for evaluating determinants. Of these, the method of using elementary row operations to reduce the matrix to triangular form is usually faster than cofactor expansion along a row or column. If the matrix is large, then the number of arithmetic operations needed for cofactor expansion can become extremely large. For this reason, most computer and calculator algorithms use the method involving elementary row operations. Table 3.1 (on page 138) shows the numbers of additions (plus subtractions) and multiplications (plus divisions) needed for each of these two methods for matrices of orders 3, 5, and 10.

TABLE 3.1

	Cofactor Expansion		Row Reduction	
Order n	*Additions*	*Multiplications*	*Additions*	*Multiplications*
3	5	9	5	10
5	119	205	30	45
10	3,628,799	6,235,300	285	339

In fact, the number of operations for the cofactor expansion of an $n \times n$ matrix grows like $n!$. Because $30! \approx 2.65 \times 10^{32}$, even a relatively small 30×30 matrix would require more than 10^{32} operations. If a computer could do one trillion operations per second, it would still take more than one trillion years to compute the determinant of this matrix using cofactor expansion. Yet, row reduction would take only a few seconds.

When evaluating a determinant *by hand,* you can sometimes save steps by using elementary row (or column) operations to create a row (or column) having zeros in all but one position and then using cofactor expansion to reduce the order of the matrix by 1. This approach is illustrated in the next two examples.

EXAMPLE 5 Evaluating a Determinant

Find the determinant of

$$A = \begin{bmatrix} -3 & 5 & 2 \\ 2 & -4 & -1 \\ -3 & 0 & 6 \end{bmatrix}.$$

SOLUTION Notice that the matrix A already has one zero in the third row. You can create another zero in the third row by adding 2 times the first column to the third column, as follows.

$$|A| = \begin{vmatrix} -3 & 5 & 2 \\ 2 & -4 & -1 \\ -3 & 0 & 6 \end{vmatrix} = \begin{vmatrix} -3 & 5 & -4 \\ 2 & -4 & 3 \\ -3 & 0 & 0 \end{vmatrix}$$

Expanding by cofactors along the third row produces

$$\begin{aligned} |A| &= \begin{vmatrix} -3 & 5 & -4 \\ 2 & -4 & 3 \\ -3 & 0 & 0 \end{vmatrix} \\ &= -3(-1)^4 \begin{vmatrix} 5 & -4 \\ -4 & 3 \end{vmatrix} \\ &= -3(1)(-1) = 3. \end{aligned}$$

EXAMPLE 6 **Evaluating a Determinant**

Evaluate the determinant of

$$A = \begin{bmatrix} 2 & 0 & 1 & 3 & -2 \\ -2 & 1 & 3 & 2 & -1 \\ 1 & 0 & -1 & 2 & 3 \\ 3 & -1 & 2 & 4 & -3 \\ 1 & 1 & 3 & 2 & 0 \end{bmatrix}.$$

SOLUTION Because the second column of this matrix already has two zeros, choose it for cofactor expansion. Two additional zeros can be created in the second column by adding the second row to the fourth row, and then adding -1 times the second row to the fifth row.

$$|A| = \begin{vmatrix} 2 & 0 & 1 & 3 & -2 \\ -2 & 1 & 3 & 2 & -1 \\ 1 & 0 & -1 & 2 & 3 \\ 3 & -1 & 2 & 4 & -3 \\ 1 & 1 & 3 & 2 & 0 \end{vmatrix}$$

$$= \begin{vmatrix} 2 & 0 & 1 & 3 & -2 \\ -2 & 1 & 3 & 2 & -1 \\ 1 & 0 & -1 & 2 & 3 \\ 1 & 0 & 5 & 6 & -4 \\ 3 & 0 & 0 & 0 & 1 \end{vmatrix}$$

$$= (1)(-1)^4 \begin{vmatrix} 2 & 1 & 3 & -2 \\ 1 & -1 & 2 & 3 \\ 1 & 5 & 6 & -4 \\ 3 & 0 & 0 & 1 \end{vmatrix}$$

(You have now reduced the problem of finding the determinant of a 5×5 matrix to finding the determinant of a 4×4 matrix.) Because you already have two zeros in the fourth row, it is chosen for the next cofactor expansion. Add -3 times the fourth column to the first column to produce the following.

$$|A| = \begin{vmatrix} 2 & 1 & 3 & -2 \\ 1 & -1 & 2 & 3 \\ 1 & 5 & 6 & -4 \\ 3 & 0 & 0 & 1 \end{vmatrix} = \begin{vmatrix} 8 & 1 & 3 & -2 \\ -8 & -1 & 2 & 3 \\ 13 & 5 & 6 & -4 \\ 0 & 0 & 0 & 1 \end{vmatrix}$$

$$= (1)(-1)^8 \begin{vmatrix} 8 & 1 & 3 \\ -8 & -1 & 2 \\ 13 & 5 & 6 \end{vmatrix}$$

Add the second row to the first row and then expand by cofactors along the first row.

$$|A| = \begin{vmatrix} 8 & 1 & 3 \\ -8 & -1 & 2 \\ 13 & 5 & 6 \end{vmatrix} = \begin{vmatrix} 0 & 0 & 5 \\ -8 & -1 & 2 \\ 13 & 5 & 6 \end{vmatrix}$$

$$= 5(-1)^4\begin{vmatrix} -8 & -1 \\ 13 & 5 \end{vmatrix} = 5(1)(-27) = -135$$

SECTION 3.2 Exercises

In Exercises 1–20, which property of determinants is illustrated by the equation?

1. $\begin{vmatrix} 2 & -6 \\ 1 & -3 \end{vmatrix} = 0$

2. $\begin{vmatrix} -4 & 5 \\ 12 & -15 \end{vmatrix} = 0$

3. $\begin{vmatrix} 1 & 4 & 2 \\ 0 & 0 & 0 \\ 5 & 6 & -7 \end{vmatrix} = 0$

4. $\begin{vmatrix} -4 & 3 & 2 \\ 8 & 0 & 0 \\ -4 & 3 & 2 \end{vmatrix} = 0$

5. $\begin{vmatrix} 1 & 3 & 4 \\ -7 & 2 & -5 \\ 6 & 1 & 2 \end{vmatrix} = -\begin{vmatrix} 1 & 4 & 3 \\ -7 & -5 & 2 \\ 6 & 2 & 1 \end{vmatrix}$

6. $\begin{vmatrix} 1 & 3 & 4 \\ -2 & 2 & 0 \\ 1 & 6 & 2 \end{vmatrix} = -\begin{vmatrix} 1 & 6 & 2 \\ -2 & 2 & 0 \\ 1 & 3 & 4 \end{vmatrix}$

7. $\begin{vmatrix} 5 & 10 \\ 2 & -7 \end{vmatrix} = 5\begin{vmatrix} 1 & 2 \\ 2 & -7 \end{vmatrix}$

8. $\begin{vmatrix} 4 & 1 \\ 2 & 8 \end{vmatrix} = 2\begin{vmatrix} 2 & 1 \\ 1 & 8 \end{vmatrix}$

9. $\begin{vmatrix} 1 & 8 & -3 \\ 3 & -12 & 6 \\ 7 & 4 & 9 \end{vmatrix} = 12\begin{vmatrix} 1 & 2 & -1 \\ 3 & -3 & 2 \\ 7 & 1 & 3 \end{vmatrix}$

10. $\begin{vmatrix} 1 & 2 & 3 \\ 4 & -8 & 6 \\ 5 & 4 & 12 \end{vmatrix} = 6\begin{vmatrix} 1 & 1 & 1 \\ 4 & -4 & 2 \\ 5 & 2 & 4 \end{vmatrix}$

11. $\begin{vmatrix} 5 & 0 & 10 \\ 25 & -30 & 40 \\ -15 & 5 & 20 \end{vmatrix} = 5^3\begin{vmatrix} 1 & 0 & 2 \\ 5 & -6 & 8 \\ -3 & 1 & 4 \end{vmatrix}$

12. $\begin{vmatrix} 6 & 0 & 0 & 0 \\ 0 & 6 & 0 & 0 \\ 0 & 0 & 6 & 0 \\ 0 & 0 & 0 & 6 \end{vmatrix} = 6^4\begin{vmatrix} 1 & 0 & 0 & 0 \\ 0 & 1 & 0 & 0 \\ 0 & 0 & 1 & 0 \\ 0 & 0 & 0 & 1 \end{vmatrix}$

13. $\begin{vmatrix} 2 & -3 \\ 8 & 7 \end{vmatrix} = \begin{vmatrix} 2 & -3 \\ 0 & 19 \end{vmatrix}$

14. $\begin{vmatrix} 2 & 1 \\ 0 & -1 \end{vmatrix} = \begin{vmatrix} 2 & 1 \\ 4 & 1 \end{vmatrix}$

15. $\begin{vmatrix} 1 & -3 & 2 \\ 5 & 2 & -1 \\ -1 & 0 & 6 \end{vmatrix} = \begin{vmatrix} 1 & -3 & 2 \\ 0 & 17 & -11 \\ -1 & 0 & 6 \end{vmatrix}$

16. $\begin{vmatrix} 3 & 2 & 4 \\ -2 & 1 & 5 \\ 5 & -7 & -20 \end{vmatrix} = \begin{vmatrix} 3 & 2 & -6 \\ -2 & 1 & 0 \\ 5 & -7 & 15 \end{vmatrix}$

17. $\begin{vmatrix} 5 & 4 & 2 \\ 4 & -3 & 4 \\ 7 & 6 & 3 \end{vmatrix} = -\begin{vmatrix} 5 & 4 & 2 \\ -4 & 3 & -4 \\ 7 & 6 & 3 \end{vmatrix}$

18. $\begin{vmatrix} 2 & 1 & -1 \\ 0 & 1 & 4 \\ 5 & 3 & 1 \end{vmatrix} = -\begin{vmatrix} 2 & 1 & -1 \\ 5 & 3 & 1 \\ 0 & 1 & 4 \end{vmatrix}$

19. $\begin{vmatrix} 2 & 1 & -1 & 0 & 3 & 4 \\ 1 & 0 & 1 & 3 & 5 & 2 \\ 3 & 6 & 1 & -3 & 2 & 6 \\ 0 & 4 & 0 & 2 & 4 & 0 \\ -1 & 8 & 5 & 3 & 4 & -2 \\ -2 & 0 & 2 & 6 & 4 & -4 \end{vmatrix} = 0$

20. $\begin{vmatrix} 4 & 3 & 1 & 9 & 6 & 9 \\ 9 & -1 & 2 & 3 & 0 & -3 \\ 3 & 4 & 6 & 9 & -6 & 12 \\ 5 & 2 & 0 & 6 & 12 & 6 \\ 6 & 0 & 3 & 0 & -3 & 0 \\ 1 & -2 & 1 & -3 & 9 & -6 \end{vmatrix} = 0$

In Exercises 21–24, use either elementary row or column operations, or cofactor expansion, to evaluate the determinant by hand. Then use a graphing utility or computer software program to verify the value of the determinant.

21. $\begin{vmatrix} 1 & 0 & 2 \\ -1 & 1 & 4 \\ 2 & 0 & 3 \end{vmatrix}$

22. $\begin{vmatrix} -1 & 3 & 2 \\ 0 & 2 & 0 \\ 1 & 1 & -1 \end{vmatrix}$

23. $\begin{vmatrix} 1 & 2 & 1 & -1 \\ 0 & 1 & 0 & 2 \\ 0 & 3 & -1 & 1 \\ 0 & 0 & 4 & 1 \end{vmatrix}$

24. $\begin{vmatrix} 3 & 2 & 1 & 1 \\ -1 & 0 & 2 & 0 \\ 4 & 1 & -1 & 0 \\ 3 & 1 & 1 & 0 \end{vmatrix}$

In Exercises 25–38, use elementary row or column operations to evaluate the determinant.

25. $\begin{vmatrix} 1 & 7 & -3 \\ 1 & 3 & 1 \\ 4 & 8 & 1 \end{vmatrix}$

26. $\begin{vmatrix} 1 & 1 & 1 \\ 2 & -1 & -2 \\ 1 & -2 & -1 \end{vmatrix}$

27. $\begin{vmatrix} 2 & -1 & -1 \\ 1 & 3 & 2 \\ 1 & 1 & 3 \end{vmatrix}$

28. $\begin{vmatrix} 3 & -1 & -3 \\ -1 & -4 & -2 \\ 3 & -1 & -1 \end{vmatrix}$

29. $\begin{vmatrix} 4 & 3 & -2 \\ 5 & 4 & 1 \\ -2 & 3 & 4 \end{vmatrix}$

30. $\begin{vmatrix} 3 & 8 & -7 \\ 0 & -5 & 4 \\ 6 & 1 & 6 \end{vmatrix}$

31. $\begin{vmatrix} 5 & -8 & 0 \\ 9 & 7 & 4 \\ -8 & 7 & 1 \end{vmatrix}$

32. $\begin{vmatrix} 4 & -8 & 5 \\ 8 & -5 & 3 \\ 8 & 5 & 2 \end{vmatrix}$

33. $\begin{vmatrix} 4 & -7 & 9 & 1 \\ 6 & 2 & 7 & 0 \\ 3 & 6 & -3 & 3 \\ 0 & 7 & 4 & -1 \end{vmatrix}$

34. $\begin{vmatrix} 9 & -4 & 2 & 5 \\ 2 & 7 & 6 & -5 \\ 4 & 1 & -2 & 0 \\ 7 & 3 & 4 & 10 \end{vmatrix}$

35. $\begin{vmatrix} 1 & -2 & 7 & 9 \\ 3 & -4 & 5 & 5 \\ 3 & 6 & 1 & -1 \\ 4 & 5 & 3 & 2 \end{vmatrix}$

36. $\begin{vmatrix} 0 & -3 & 8 & 2 \\ 8 & 1 & -1 & 6 \\ -4 & 6 & 0 & 9 \\ -7 & 0 & 0 & 14 \end{vmatrix}$

37. $\begin{vmatrix} 1 & -1 & 8 & 4 & 2 \\ 2 & 6 & 0 & -4 & 3 \\ 2 & 0 & 2 & 6 & 2 \\ 0 & 2 & 8 & 0 & 0 \\ 0 & 1 & 1 & 2 & 2 \end{vmatrix}$

38. $\begin{vmatrix} 3 & -2 & 4 & 3 & 1 \\ -1 & 0 & 2 & 1 & 0 \\ 5 & -1 & 0 & 3 & 2 \\ 4 & 7 & -8 & 0 & 0 \\ 1 & 2 & 3 & 0 & 2 \end{vmatrix}$

True or False? In Exercises 39 and 40, determine whether each statement is true or false. If a statement is true, give a reason or cite an appropriate statement from the text. If a statement is false, provide an example that shows the statement is not true in all cases or cite an appropriate statement from the text.

39. (a) Interchanging two rows of a given matrix changes the sign of its determinant.

(b) Multiplying a row of a matrix by a nonzero constant results in the determinant being multiplied by the same nonzero constant.

(c) If two rows of a square matrix are equal, then its determinant is 0.

40. (a) Adding a multiple of one row of a matrix to another row changes only the sign of the determinant.

(b) Two matrices are column-equivalent if one matrix can be obtained by performing elementary column operations on the other.

(c) If one row of a square matrix is a multiple of another row, then the determinant is 0.

In Exercises 41–46, find the determinant of the elementary matrix. (Assume $k \neq 0$.)

41. $\begin{bmatrix} 1 & 0 & 0 \\ 0 & k & 0 \\ 0 & 0 & 1 \end{bmatrix}$

42. $\begin{bmatrix} 1 & 0 & 0 \\ 0 & 1 & 0 \\ 0 & 0 & k \end{bmatrix}$

43. $\begin{bmatrix} 0 & 1 & 0 \\ 1 & 0 & 0 \\ 0 & 0 & 1 \end{bmatrix}$

44. $\begin{bmatrix} 0 & 0 & 1 \\ 0 & 1 & 0 \\ 1 & 0 & 0 \end{bmatrix}$

45. $\begin{bmatrix} 1 & 0 & 0 \\ k & 1 & 0 \\ 0 & 0 & 1 \end{bmatrix}$

46. $\begin{bmatrix} 1 & 0 & 0 \\ 0 & 1 & 0 \\ 0 & k & 1 \end{bmatrix}$

47. Prove the property.

$$\begin{vmatrix} a_{11} & a_{12} & a_{13} \\ a_{21} & a_{22} & a_{23} \\ a_{31} & a_{32} & a_{33} \end{vmatrix} + \begin{vmatrix} b_{11} & a_{12} & a_{13} \\ b_{21} & a_{22} & a_{23} \\ b_{31} & a_{32} & a_{33} \end{vmatrix} = \begin{vmatrix} (a_{11} + b_{11}) & a_{12} & a_{13} \\ (a_{21} + b_{21}) & a_{22} & a_{23} \\ (a_{31} + b_{31}) & a_{32} & a_{33} \end{vmatrix}$$

48. Prove the property.

$$\begin{vmatrix} 1 + a & 1 & 1 \\ 1 & 1 + b & 1 \\ 1 & 1 & 1 + c \end{vmatrix} = abc\left(1 + \frac{1}{a} + \frac{1}{b} + \frac{1}{c}\right),$$

$a \neq 0, \quad b \neq 0, \quad c \neq 0$

In Exercises 49–52, evaluate the determinant.

49. $\begin{vmatrix} \cos\theta & \sin\theta \\ -\sin\theta & \cos\theta \end{vmatrix}$

50. $\begin{vmatrix} \sec\theta & \tan\theta \\ \tan\theta & \sec\theta \end{vmatrix}$

51. $\begin{vmatrix} \sin\theta & 1 \\ 1 & \sin\theta \end{vmatrix}$

52. $\begin{vmatrix} \sec\theta & 1 \\ 1 & \sec\theta \end{vmatrix}$

53. Writing Solve the equation for x, if possible. Explain your result.

$$\begin{vmatrix} \cos x & 0 & \sin x \\ \sin x & 0 & -\cos x \\ \sin x - \cos x & 1 & \sin x + \cos x \end{vmatrix} = 0$$

54. Guided Proof Prove Property 2 of Theorem 3.3: If B is obtained from A by adding a multiple of a row of A to another row of A, then $\det(B) = \det(A)$.

Getting Started: To prove that the determinant of B is equal to the determinant of A, you need to show that their respective cofactor expansions are equal.

(i) Begin your proof by letting B be the matrix obtained by adding c times the jth row of A to the ith row of A.
(ii) Find the determinant of B by expanding along this ith row.
(iii) Distribute and then group the terms containing a coefficient of c and those not containing a coefficient of c.
(iv) Show that the sum of the terms not containing a coefficient of c is the determinant of A and the sum of the terms containing a coefficient of c is equal to 0.

55. Guided Proof Prove Property 3 of Theorem 3.3: If B is obtained from A by multiplying a row of A by a nonzero constant c, then $\det(B) = c\det(A)$.

Getting Started: To prove that the determinant of B is equal to c times the determinant of A, you need to show that the determinant of B is equal to c times the cofactor expansion of the determinant of A.

(i) Begin your proof by letting B be the matrix obtained by multiplying c times the ith row of A.
(ii) Find the determinant of B by expanding along this ith row.
(iii) Factor out the common factor c.
(iv) Show that the result is c times the determinant of A.

56. Writing A computer operator charges \$0.001 (one tenth of a cent) for each addition and subtraction, and \$0.003 for each multiplication and division. Compare and contrast the costs of calculating the determinant of a 10×10 matrix by cofactor expansion and then by row reduction. Which method would you prefer to use for calculating determinants?

3.3 Properties of Determinants

In this section you will learn several important properties of determinants. You will begin by considering the determinant of the product of two matrices.

EXAMPLE 1 The Determinant of a Matrix Product

Find $|A|$, $|B|$, and $|AB|$ for the matrices

$$A = \begin{bmatrix} 1 & -2 & 2 \\ 0 & 3 & 2 \\ 1 & 0 & 1 \end{bmatrix} \quad \text{and} \quad B = \begin{bmatrix} 2 & 0 & 1 \\ 0 & -1 & -2 \\ 3 & 1 & -2 \end{bmatrix}.$$

SOLUTION Using the techniques described in the preceding sections, you can show that $|A|$ and $|B|$ have the values

$$|A| = \begin{vmatrix} 1 & -2 & 2 \\ 0 & 3 & 2 \\ 1 & 0 & 1 \end{vmatrix} = -7 \quad \text{and} \quad |B| = \begin{vmatrix} 2 & 0 & 1 \\ 0 & -1 & -2 \\ 3 & 1 & -2 \end{vmatrix} = 11.$$

The matrix product AB is

$$AB = \begin{bmatrix} 1 & -2 & 2 \\ 0 & 3 & 2 \\ 1 & 0 & 1 \end{bmatrix} \begin{bmatrix} 2 & 0 & 1 \\ 0 & -1 & -2 \\ 3 & 1 & -2 \end{bmatrix} = \begin{bmatrix} 8 & 4 & 1 \\ 6 & -1 & -10 \\ 5 & 1 & -1 \end{bmatrix}.$$

Using the same techniques, you can show that $|AB|$ has the value

$$|AB| = \begin{vmatrix} 8 & 4 & 1 \\ 6 & -1 & -10 \\ 5 & 1 & -1 \end{vmatrix} = -77.$$

In Example 1, note that the determinant of the matrix product is equal to the product of the determinants. That is,

$$|AB| = |A||B|$$
$$-77 = (-7)(11).$$

This is true in general, as indicated in the next theorem.

THEOREM 3.5 Determinant of a Matrix Product

If A and B are square matrices of order n, then

$$\det(AB) = \det(A)\det(B).$$

PROOF To begin, observe that if E is an elementary matrix, then, by Theorem 3.3, the next few statements are true. If E is obtained from I by interchanging two rows, then $|E| = -1$. If E is obtained by multiplying a row of I by a nonzero constant c, then $|E| = c$. If E is obtained by adding a multiple of one row of I to another row of I, then $|E| = 1$. Additionally, by Theorem 2.12, if E results from performing an elementary row operation on I and the same elementary row operation is performed on B, then the matrix EB results. It follows that

$$|EB| = |E|\,|B|.$$

This can be generalized to conclude that $|E_k \cdots E_2E_1B| = |E_k| \cdots |E_2|\,|E_1|\,|B|$, where E_i is an elementary matrix. Now consider the matrix AB. If A is *nonsingular*, then, by Theorem 2.14, it can be written as the product of elementary matrices $A = E_k \cdots E_2E_1$ and you can write

$$\begin{aligned} |AB| &= |E_k \cdots E_2E_1B| \\ &= |E_k| \cdots |E_2|\,|E_1|\,|B| = |E_k \cdots E_2E_1|\,|B| = |A|\,|B|. \end{aligned}$$

If A is *singular,* then A is row-equivalent to a matrix with an entire row of zeros. From Theorem 3.4, you can conclude that $|A| = 0$. Moreover, because A is singular, it follows that AB is also singular. (If AB were nonsingular, then $A[B(AB)^{-1}] = I$ would imply that A is nonsingular.) So, $|AB| = 0$, and you can conclude that $|AB| = |A||B|$.

REMARK: Theorem 3.5 can be extended to include the product of any finite number of matrices. That is,

$$|A_1A_2A_3 \cdot \cdot \cdot A_k| = |A_1||A_2||A_3| \cdot \cdot \cdot |A_k|.$$

The relationship between $|A|$ and $|cA|$ is shown in the next theorem.

THEOREM 3.6 Determinant of a Scalar Multiple of a Matrix

If A is an $n \times n$ matrix and c is a scalar, then the determinant of cA is given by

$$\det(cA) = c^n \det(A).$$

PROOF This formula can be proven by repeated applications of Property 3 of Theorem 3.3. Factor the scalar c out of each of the n rows of $|cA|$ to obtain

$$|cA| = c^n|A|.$$

EXAMPLE 2 **The Determinant of a Scalar Multiple of a Matrix**

Find the determinant of the matrix.

$$A = \begin{bmatrix} 10 & -20 & 40 \\ 30 & 0 & 50 \\ -20 & -30 & 10 \end{bmatrix}$$

SOLUTION Because

$$A = 10\begin{bmatrix} 1 & -2 & 4 \\ 3 & 0 & 5 \\ -2 & -3 & 1 \end{bmatrix} \quad \text{and} \quad \begin{vmatrix} 1 & -2 & 4 \\ 3 & 0 & 5 \\ -2 & -3 & 1 \end{vmatrix} = 5,$$

you can apply Theorem 3.6 to conclude that

$$|A| = 10^3\begin{vmatrix} 1 & -2 & 4 \\ 3 & 0 & 5 \\ -2 & -3 & 1 \end{vmatrix} = 1000(5) = 5000.$$

Theorems 3.5 and 3.6 provide formulas for evaluating the determinants of the product of two matrices and a scalar multiple of a matrix. These theorems do not, however, list a formula for the determinant of the *sum* of two matrices. It is important to note that the sum of the determinants of two matrices usually does not equal the determinant of their sum. That is, in general, $|A| + |B| \neq |A + B|$. For instance, if

$$A = \begin{bmatrix} 6 & 2 \\ 2 & 1 \end{bmatrix} \quad \text{and} \quad B = \begin{bmatrix} 3 & 7 \\ 0 & -1 \end{bmatrix},$$

then $|A| = 2$ and $|B| = -3$, but $A + B = \begin{bmatrix} 9 & 9 \\ 2 & 0 \end{bmatrix}$ and $|A + B| = -18$.

Determinants and the Inverse of a Matrix

You saw in Chapter 2 that some square matrices are not invertible. It can also be difficult to tell simply by inspection whether or not a matrix has an inverse. Can you tell which of the two matrices shown below is invertible?

$$A = \begin{bmatrix} 0 & 2 & -1 \\ 3 & -2 & 1 \\ 3 & 2 & -1 \end{bmatrix} \quad \text{or} \quad B = \begin{bmatrix} 0 & 2 & -1 \\ 3 & -2 & 1 \\ 3 & 2 & 1 \end{bmatrix}$$

The next theorem shows that determinants are useful for classifying square matrices as invertible or noninvertible.

THEOREM 3.7 Determinant of an Invertible Matrix

A square matrix A is invertible (nonsingular) if and only if

$$\det(A) \neq 0.$$

PROOF To prove the theorem in one direction, assume A is invertible. Then $AA^{-1} = I$, and by Theorem 3.5 you can write $|A||A^{-1}| = |I|$. Now, because $|I| = 1$, you know that neither determinant on the left is zero. Specifically, $|A| \neq 0$.

To prove the theorem in the other direction, assume the determinant of A is nonzero. Then, using Gauss-Jordan elimination, find a matrix B, in reduced row-echelon form, that is row-equivalent to A. Because B is in reduced row-echelon form, it must be the identity matrix I or it must have at least one row that consists entirely of zeros. But if B has a row of all zeros, then by Theorem 3.4 you know that $|B| = 0$, which would imply that $|A| = 0$. Because you assumed that $|A|$ is nonzero, you can conclude that $B = I$. A is, therefore, row-equivalent to the identity matrix, and by Theorem 2.15 you know that A is invertible.

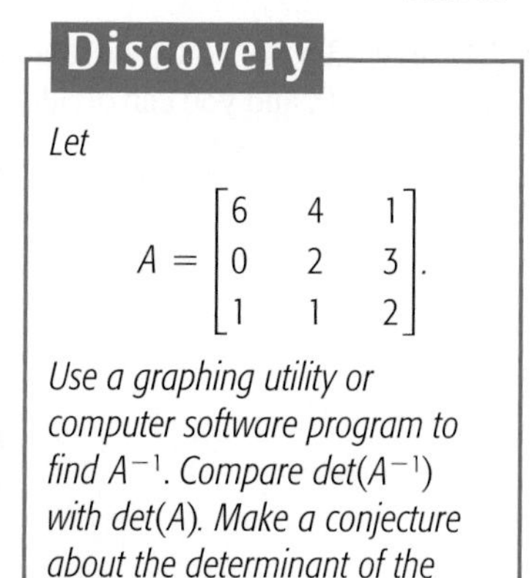

Discovery

Let

$$A = \begin{bmatrix} 6 & 4 & 1 \\ 0 & 2 & 3 \\ 1 & 1 & 2 \end{bmatrix}.$$

Use a graphing utility or computer software program to find A^{-1}. Compare $\det(A^{-1})$ with $\det(A)$. Make a conjecture about the determinant of the inverse of a matrix.

EXAMPLE 3 **Classifying Square Matrices as Singular or Nonsingular**

Which of the matrices has an inverse?

$$\text{(a)} \begin{bmatrix} 0 & 2 & -1 \\ 3 & -2 & 1 \\ 3 & 2 & -1 \end{bmatrix} \qquad \text{(b)} \begin{bmatrix} 0 & 2 & -1 \\ 3 & -2 & 1 \\ 3 & 2 & 1 \end{bmatrix}$$

SOLUTION (a) Because

$$\begin{vmatrix} 0 & 2 & -1 \\ 3 & -2 & 1 \\ 3 & 2 & -1 \end{vmatrix} = 0,$$

you can conclude that this matrix has no inverse (it is singular).

(b) Because

$$\begin{vmatrix} 0 & 2 & -1 \\ 3 & -2 & 1 \\ 3 & 2 & 1 \end{vmatrix} = -12 \neq 0,$$

you can conclude that this matrix has an inverse (it is nonsingular).

The next theorem provides a convenient way to find the determinant of the inverse of a matrix.

THEOREM 3.8 Determinant of an Inverse Matrix

If A is invertible, then

$$\det(A^{-1}) = \frac{1}{\det(A)}.$$

PROOF Because A is invertible, $AA^{-1} = I$, and you can apply Theorem 3.5 to conclude that $|A||A^{-1}| = |I| = 1$. Because A is invertible, you also know that $|A| \neq 0$, and you can divide each side by $|A|$ to obtain

$$|A^{-1}| = \frac{1}{|A|}.$$

EXAMPLE 4 The Determinant of the Inverse of a Matrix

Find $|A^{-1}|$ for the matrix

$$A = \begin{bmatrix} 1 & 0 & 3 \\ 0 & -1 & 2 \\ 2 & 1 & 0 \end{bmatrix}.$$

SOLUTION One way to solve this problem is to find A^{-1} and then evaluate its determinant. It is easier, however, to apply Theorem 3.8, as follows. Find the determinant of A,

$$|A| = \begin{vmatrix} 1 & 0 & 3 \\ 0 & -1 & 2 \\ 2 & 1 & 0 \end{vmatrix} = 4,$$

and then use the formula $|A^{-1}| = 1/|A|$ to conclude that $|A^{-1}| = \frac{1}{4}$.

REMARK: In Example 4, the inverse of A is

$$A^{-1} = \begin{bmatrix} -\frac{1}{2} & \frac{3}{4} & \frac{3}{4} \\ 1 & -\frac{3}{2} & -\frac{1}{2} \\ \frac{1}{2} & -\frac{1}{4} & -\frac{1}{4} \end{bmatrix}.$$

Try evaluating the determinant of this matrix directly. Then compare your answer with that obtained in Example 4.

Note that Theorem 3.7 (a matrix is invertible if and only if its determinant is nonzero) provides another equivalent condition that can be added to the list in Theorem 2.15. All six conditions are summarized below.

Equivalent Conditions for a Nonsingular Matrix

If A is an $n \times n$ matrix, then the following statements are equivalent.

1. A is invertible.
2. $A\mathbf{x} = \mathbf{b}$ has a unique solution for every $n \times 1$ column matrix $\mathbf{b}$.
3. $A\mathbf{x} = O$ has only the trivial solution.
4. A is row-equivalent to I_n.
5. A can be written as the product of elementary matrices.
6. $\det(A) \neq 0$

REMARK: In Section 3.2 you saw that a square matrix A can have a determinant of zero if A is row-equivalent to a matrix that has at least one row consisting entirely of zeros. The validity of this statement follows from the equivalence of Properties 4 and 6.

EXAMPLE 5 **Systems of Linear Equations**

Which of the systems has a unique solution?

(a)
$$\begin{aligned} 2x_2 - x_3 &= -1 \\ 3x_1 - 2x_2 + x_3 &= 4 \\ 3x_1 + 2x_2 - x_3 &= -4 \end{aligned}$$

(b)
$$\begin{aligned} 2x_2 - x_3 &= -1 \\ 3x_1 - 2x_2 + x_3 &= 4 \\ 3x_1 + 2x_2 + x_3 &= -4 \end{aligned}$$

SOLUTION From Example 3 you know that the coefficient matrices for these two systems have the determinants below.

(a) $$\begin{vmatrix} 0 & 2 & -1 \\ 3 & -2 & 1 \\ 3 & 2 & -1 \end{vmatrix} = 0$$

(b) $$\begin{vmatrix} 0 & 2 & -1 \\ 3 & -2 & 1 \\ 3 & 2 & 1 \end{vmatrix} = -12$$

Using the preceding list of equivalent conditions, you can conclude that only the second system has a unique solution.

Determinants and the Transpose of a Matrix

The next theorem tells you that the determinant of the transpose of a square matrix is equal to the determinant of the original matrix. This theorem can be proven using mathematical induction and Theorem 3.1, which states that a determinant can be evaluated using cofactor expansion along a row or a column. The details of the proof are left to you. (See Exercise 56.)

THEOREM 3.9 Determinant of a Transpose

If A is a square matrix, then

$$\det(A) = \det(A^T).$$

EXAMPLE 6 **The Determinant of a Transpose**

Show that $|A| = |A^T|$ for the matrix below.

$$A = \begin{bmatrix} 3 & 1 & -2 \\ 2 & 0 & 0 \\ -4 & -1 & 5 \end{bmatrix}$$

SOLUTION To find the determinant of A, expand by cofactors along the second *row* to obtain

$$|A| = 2(-1)^3 \begin{vmatrix} 1 & -2 \\ -1 & 5 \end{vmatrix} = (2)(-1)(3) = -6.$$

To find the determinant of

$$A^T = \begin{bmatrix} 3 & 2 & -4 \\ 1 & 0 & -1 \\ -2 & 0 & 5 \end{bmatrix},$$

expand by cofactors down the second *column* to obtain

$$|A^T| = 2(-1)^3\begin{vmatrix} 1 & -1 \\ -2 & 5 \end{vmatrix} = (2)(-1)(3) = -6. \text{ So, } |A| = |A^T|.$$

SECTION 3.3 Exercises

In Exercises 1–6, find (a) $|A|$, (b) $|B|$, (c) AB, and (d) $|AB|$. Then verify that $|A||B| = |AB|$.

1. $A = \begin{bmatrix} -2 & 1 \\ 4 & -2 \end{bmatrix}, \quad B = \begin{bmatrix} 1 & 1 \\ 0 & -1 \end{bmatrix}$

2. $A = \begin{bmatrix} 1 & 2 \\ 2 & 4 \end{bmatrix}, \quad B = \begin{bmatrix} -1 & 2 \\ 3 & 0 \end{bmatrix}$

3. $A = \begin{bmatrix} -1 & 2 & 1 \\ 1 & 0 & 1 \\ 0 & 1 & 0 \end{bmatrix}, \quad B = \begin{bmatrix} -1 & 0 & 0 \\ 0 & 2 & 0 \\ 0 & 0 & 3 \end{bmatrix}$

4. $A = \begin{bmatrix} 2 & 0 & 1 \\ 1 & -1 & 2 \\ 3 & 1 & 0 \end{bmatrix}, \quad B = \begin{bmatrix} 2 & -1 & 4 \\ 0 & 1 & 3 \\ 3 & -2 & 1 \end{bmatrix}$

5. $A = \begin{bmatrix} 2 & 0 & 1 & 1 \\ 1 & -1 & 0 & 1 \\ 2 & 3 & 1 & 0 \\ 1 & 2 & 3 & 0 \end{bmatrix}, \quad B = \begin{bmatrix} 1 & 0 & -1 & 1 \\ 2 & 1 & 0 & 2 \\ 1 & 1 & -1 & 0 \\ 3 & 2 & 1 & 0 \end{bmatrix}$

6. $A = \begin{bmatrix} 3 & 2 & 4 & 0 \\ 1 & -1 & 2 & 1 \\ 0 & 0 & 3 & 1 \\ -1 & 1 & 1 & 0 \end{bmatrix}, \quad B = \begin{bmatrix} 4 & 2 & -1 & 0 \\ 1 & 1 & 2 & -1 \\ 0 & 0 & 2 & 1 \\ -1 & 0 & 0 & 0 \end{bmatrix}$

In Exercises 7–10, use the fact that $|cA| = c^n|A|$ to evaluate the determinant of the $n \times n$ matrix.

7. $A = \begin{bmatrix} 4 & 2 \\ 6 & -8 \end{bmatrix}$

8. $A = \begin{bmatrix} 5 & 15 \\ 10 & -20 \end{bmatrix}$

9. $A = \begin{bmatrix} -3 & 6 & 9 \\ 6 & 9 & 12 \\ 9 & 12 & 15 \end{bmatrix}$

10. $A = \begin{bmatrix} 4 & 16 & 0 \\ 12 & -8 & 8 \\ 16 & 20 & -4 \end{bmatrix}$

In Exercises 11–14, find (a) $|A|$, (b) $|B|$, and (c) $|A + B|$. Then verify that $|A| + |B| \neq |A + B|$.

11. $A = \begin{bmatrix} -1 & 1 \\ 2 & 0 \end{bmatrix}, \quad B = \begin{bmatrix} 1 & -1 \\ -2 & 0 \end{bmatrix}$

12. $A = \begin{bmatrix} 1 & -2 \\ 1 & 0 \end{bmatrix}, \quad B = \begin{bmatrix} 3 & -2 \\ 0 & 0 \end{bmatrix}$

13. $A = \begin{bmatrix} 1 & 0 & 1 \\ -1 & 2 & 1 \\ 0 & 1 & 1 \end{bmatrix}, \quad B = \begin{bmatrix} -1 & 0 & 2 \\ 0 & 1 & 2 \\ 1 & 1 & 1 \end{bmatrix}$

14. $A = \begin{bmatrix} 0 & 1 & 2 \\ 1 & -1 & 0 \\ 2 & 1 & 1 \end{bmatrix}, \quad B = \begin{bmatrix} 0 & 1 & -1 \\ 2 & 1 & 1 \\ 0 & 1 & 1 \end{bmatrix}$

In Exercises 15–18, find (a) $|A^T|$, (b) $|A^2|$, (c) $|AA^T|$, (d) $|2A|$, and (e) $|A^{-1}|$.

15. $A = \begin{bmatrix} 6 & -11 \\ 4 & -5 \end{bmatrix}$

16. $A = \begin{bmatrix} -4 & 10 \\ 5 & 6 \end{bmatrix}$

17. $A = \begin{bmatrix} 2 & 0 & 5 \\ 4 & -1 & 6 \\ 3 & 2 & 1 \end{bmatrix}$

18. $A = \begin{bmatrix} 1 & 5 & 4 \\ 0 & -6 & 2 \\ 0 & 0 & -3 \end{bmatrix}$

In Exercises 19–22, use a graphing utility or computer software program with matrix capabilities to find (a) $|A|$, (b) $|A^T|$, (c) $|A^2|$, (d) $|2A|$, and (e) $|A^{-1}|$.

19. $A = \begin{bmatrix} 4 & 2 \\ -1 & 5 \end{bmatrix}$

20. $A = \begin{bmatrix} -2 & 4 \\ 6 & 8 \end{bmatrix}$

HM **21.** $A = \begin{bmatrix} 4 & -2 & 1 & 5 \\ 3 & 8 & 2 & -1 \\ 6 & 8 & 9 & 2 \\ 2 & 3 & -1 & 0 \end{bmatrix}$

HM **22.** $A = \begin{bmatrix} 6 & 5 & 1 & -1 \\ -2 & 4 & 3 & 5 \\ 6 & 1 & -4 & -2 \\ 2 & 2 & 1 & 3 \end{bmatrix}$

23. Let A and B be square matrices of order 4 such that $|A| = -5$ and $|B| = 3$. Find (a) $|AB|$, (b) $|A^3|$, (c) $|3B|$, (d) $|(AB)^T|$, and (e) $|A^{-1}|$.

24. Let A and B be square matrices of order 3 such that $|A| = 10$ and $|B| = 12$. Find (a) $|AB|$, (b) $|A^4|$, (c) $|2B|$, (d) $|(AB)^T|$, and (e) $|A^{-1}|$.

25. Let A and B be square matrices of order 4 such that $|A| = 4$ and $|B| = 2$. Find (a) $|BA|$, (b) $|B^2|$, (c) $|2A|$, (d) $|(AB)^T|$, and (e) $|B^{-1}|$.

26. Let A and B be square matrices of order 3 such that $|A| = -2$ and $|B| = 5$. Find (a) $|BA|$, (b) $|B^4|$, (c) $|2A|$, (d) $|(AB)^T|$, and (e) $|B^{-1}|$.

In Exercises 27–34, use a determinant to decide whether the matrix is singular or nonsingular.

27. $\begin{bmatrix} 5 & 4 \\ 10 & 8 \end{bmatrix}$

28. $\begin{bmatrix} 3 & -6 \\ 4 & 2 \end{bmatrix}$

29. $\begin{bmatrix} 14 & 5 & 7 \\ -2 & 0 & 3 \\ 1 & -5 & -10 \end{bmatrix}$

30. $\begin{bmatrix} 1 & 0 & 4 \\ 0 & 6 & 3 \\ 2 & -1 & 4 \end{bmatrix}$

31. $\begin{bmatrix} \frac{1}{2} & \frac{3}{2} & 2 \\ \frac{2}{3} & -\frac{1}{3} & 0 \\ 1 & 1 & 1 \end{bmatrix}$

32. $\begin{bmatrix} 2 & -\frac{1}{2} & 8 \\ 1 & -\frac{1}{4} & 4 \\ -\frac{5}{2} & \frac{3}{2} & 8 \end{bmatrix}$

33. $\begin{bmatrix} 1 & 0 & -8 & 2 \\ 0 & 8 & -1 & 10 \\ 0 & 0 & 0 & 1 \\ 0 & 0 & 0 & 2 \end{bmatrix}$

34. $\begin{bmatrix} 0.8 & 0.2 & -0.6 & 0.1 \\ -1.2 & 0.6 & 0.6 & 0 \\ 0.7 & -0.3 & 0.1 & 0 \\ 0.2 & -0.3 & 0.6 & 0 \end{bmatrix}$

In Exercises 35–40, find $|A^{-1}|$. Begin by finding A^{-1}, and then evaluate its determinant. Verify your result by finding $|A|$ and then applying the formula from Theorem 3.8, $|A^{-1}| = \dfrac{1}{|A|}$.

35. $A = \begin{bmatrix} 2 & 3 \\ 1 & 4 \end{bmatrix}$

36. $A = \begin{bmatrix} 1 & -2 \\ 2 & 2 \end{bmatrix}$

37. $A = \begin{bmatrix} 1 & 0 & 2 \\ 2 & -1 & 3 \\ 1 & -2 & 2 \end{bmatrix}$

38. $A = \begin{bmatrix} 1 & 0 & 1 \\ 2 & -1 & 2 \\ 1 & -2 & 3 \end{bmatrix}$

39. $A = \begin{bmatrix} 1 & 0 & -1 & 3 \\ 1 & 0 & 3 & -2 \\ 2 & 0 & 2 & -1 \\ 1 & -3 & 1 & 2 \end{bmatrix}$

40. $A = \begin{bmatrix} 0 & 1 & 0 & 3 \\ 1 & -2 & -3 & 1 \\ 0 & 0 & 2 & -2 \\ 1 & -2 & -4 & 1 \end{bmatrix}$

In Exercises 41–44, use the determinant of the coefficient matrix to determine whether the system of linear equations has a unique solution.

41. $\begin{aligned} x_1 - x_2 + x_3 &= 4 \\ 2x_1 - x_2 + x_3 &= 6 \\ 3x_1 - 2x_2 + 2x_3 &= 0 \end{aligned}$

42. $\begin{aligned} x_1 + x_2 - x_3 &= 4 \\ 2x_1 - x_2 + x_3 &= 6 \\ 3x_1 - 2x_2 + 2x_3 &= 0 \end{aligned}$

HM **43.** $\begin{aligned} 2x_1 + x_2 + 5x_3 + x_4 &= 5 \\ x_1 + x_2 - 3x_3 - 4x_4 &= -1 \\ 2x_1 + 2x_2 + 2x_3 - 3x_4 &= 2 \\ x_1 + 5x_2 - 6x_3 \quad &= 3 \end{aligned}$

HM **44.** $\begin{aligned} x_1 - x_2 - x_3 - x_4 &= 0 \\ x_1 + x_2 - x_3 - x_4 &= 0 \\ x_1 + x_2 + x_3 - x_4 &= 0 \\ x_1 + x_2 + x_3 + x_4 &= 6 \end{aligned}$

In Exercises 45–48, find the value(s) of k such that A is singular.

45. $A = \begin{bmatrix} k-1 & 3 \\ 2 & k-2 \end{bmatrix}$

46. $A = \begin{bmatrix} k-1 & 2 \\ 2 & k+2 \end{bmatrix}$

47. $A = \begin{bmatrix} 1 & 0 & 3 \\ 2 & -1 & 0 \\ 4 & 2 & k \end{bmatrix}$

48. $A = \begin{bmatrix} 1 & k & 2 \\ -2 & 0 & -k \\ 3 & 1 & -4 \end{bmatrix}$

49. Let A and B be $n \times n$ matrices such that $AB = I$. Prove that $|A| \neq 0$ and $|B| \neq 0$.

50. Let A and B be $n \times n$ matrices such that AB is singular. Prove that either A or B is singular.

51. Find two 2×2 matrices such that $|A| + |B| = |A + B|$.

52. Verify the equation.

$$\begin{vmatrix} a+b & a & a \\ a & a+b & a \\ a & a & a+b \end{vmatrix} = b^2(3a+b)$$

53. Let A be an $n \times n$ matrix in which the entries of each row add up to zero. Find $|A|$.

54. Illustrate the result of Exercise 53 with the matrix

$$A = \begin{bmatrix} 2 & -1 & -1 \\ -3 & 1 & 2 \\ 0 & -2 & 2 \end{bmatrix}.$$

55. Guided Proof Prove that the determinant of an invertible matrix A is equal to ± 1 if all of the entries of A and A^{-1} are integers.

Getting Started: Denote $\det(A)$ as x and $\det(A^{-1})$ as y. Note that x and y are real numbers. To prove that $\det(A)$ is equal to ± 1, you must show that both x and y are integers such that their product xy is equal to 1.

(i) Use the property for the determinant of a matrix product to show that $xy = 1$.

(ii) Use the definition of a determinant and the fact that the entries of A and A^{-1} are integers to show that both $x = \det(A)$ and $y = \det(A^{-1})$ are integers.

(iii) Conclude that $x = \det(A)$ must be either 1 or -1 because these are the only integer solutions to the equation $xy = 1$.

56. Guided Proof Prove Theorem 3.9: If A is a square matrix, then $\det(A) = \det(A^T)$.

Getting Started: To prove that the determinants of A and A^T are equal, you need to show that their cofactor expansions are equal. Because the cofactors are $\pm$ determinants of smaller matrices, you need to use mathematical induction.

(i) Initial step for induction: If A is of order 1, then $A = [a_{11}] = A^T$, so $\det(A) = \det(A^T) = a_{11}$.

(ii) Assume the inductive hypothesis holds for all matrices of order $n - 1$. Let A be a square matrix of order n. Write an expression for the determinant of A by expanding by the first row.

(iii) Write an expression for the determinant of A^T by expanding by the first column.

(iv) Compare the expansions in (i) and (ii). The entries of the first row of A are the same as the entries of the first column of A^T. Compare cofactors (these are the $\pm$ determinants of smaller matrices that are transposes of one another) and use the inductive hypothesis to conclude that they are equal as well.

57. Writing Let A and P be $n \times n$ matrices, where P is invertible. Does $P^{-1}AP = A$? Illustrate your conclusion with appropriate examples. What can you say about the two determinants $|P^{-1}AP|$ and $|A|$?

58. Writing Let A be an $n \times n$ nonzero matrix satisfying $A^{10} = O$. Explain why A must be singular. What properties of determinants are you using in your argument?

True or False? In Exercises 59 and 60, determine whether each statement is true or false. If a statement is true, give a reason or cite an appropriate statement from the text. If a statement is false, provide an example that shows that the statement is not true in all cases or cite an appropriate statement from the text.

59. (a) If A is an $n \times n$ matrix and c is a nonzero scalar, then the determinant of the matrix cA is given by $nc \cdot \det(A)$.

(b) If A is an invertible matrix, then the determinant of A^{-1} is equal to the reciprocal of the determinant of A.

(c) If A is an invertible $n \times n$ matrix, then $A\mathbf{x} = \mathbf{b}$ has a unique solution for every $\mathbf{b}$.

60. (a) In general, the determinant of the sum of two matrices equals the sum of the determinants of the matrices.

(b) If A is a square matrix, then the determinant of A is equal to the determinant of the transpose of A.

(c) If the determinant of an $n \times n$ matrix A is nonzero, then $A\mathbf{x} = 0$ has only the trivial solution.

61. A square matrix is called **skew-symmetric** if $A^T = -A$. Prove that if A is an $n \times n$ skew-symmetric matrix, then $|A| = (-1)^n|A|$.

62. Let A be a skew-symmetric matrix of odd order. Use the result of Exercise 61 to prove that $|A| = 0$.

In Exercises 63–68, determine whether the matrix is orthogonal. An invertible square matrix A is called **orthogonal** if $A^{-1} = A^T$.

63. $\begin{bmatrix} 0 & 1 \\ 1 & 0 \end{bmatrix}$

64. $\begin{bmatrix} 0 & 0 \\ 1 & 0 \end{bmatrix}$

65. $\begin{bmatrix} 1 & -1 \\ -1 & -1 \end{bmatrix}$

66. $\begin{bmatrix} 1/\sqrt{2} & -1/\sqrt{2} \\ -1/\sqrt{2} & -1/\sqrt{2} \end{bmatrix}$

67. $\begin{bmatrix} 1 & 0 & 0 \\ 0 & 0 & 1 \\ 0 & 1 & 0 \end{bmatrix}$

68. $\begin{bmatrix} 1/\sqrt{2} & 0 & -1/\sqrt{2} \\ 0 & 1 & 0 \\ 1/\sqrt{2} & 0 & 1/\sqrt{2} \end{bmatrix}$

69. Prove that if A is an orthogonal matrix, then $|A| = \pm 1$.

In Exercises 70 and 71, use a graphing utility with matrix capabilities to determine whether A is orthogonal. To test for orthogonality, find (a) A^{-1}, (b) A^T, and (c) $|A|$, and verify that $A^{-1} = A^T$ and $|A| = \pm 1$.

70. $A = \begin{bmatrix} \frac{3}{5} & 0 & -\frac{4}{5} \\ 0 & 1 & 0 \\ \frac{4}{5} & 0 & \frac{3}{5} \end{bmatrix}$ **71.** $A = \begin{bmatrix} \frac{2}{3} & -\frac{2}{3} & \frac{1}{3} \\ \frac{2}{3} & \frac{1}{3} & -\frac{2}{3} \\ \frac{1}{3} & \frac{2}{3} & \frac{2}{3} \end{bmatrix}$

72. If A is an idempotent matrix ($A^2 = A$), then prove that the determinant of A is either 0 or 1.

73. Let S be an $n \times n$ singular matrix. Prove that for any $n \times n$ matrix B, the matrix SB is also singular.

74. Let A_{11}, A_{12}, and A_{22} be $n \times n$ matrices. Find the determinant of the partitioned matrix

$$\begin{bmatrix} A_{11} & A_{12} \\ 0 & A_{22} \end{bmatrix}$$

in terms of the determinants of A_{11}, A_{12}, and A_{22}.

3.4 Introduction to Eigenvalues

This chapter continues with a look ahead to one of the most important topics of linear algebra—**eigenvalues.** One application of eigenvalues involves the study of population growth. For example, suppose that half of a population of rabbits raised in a laboratory survive their first year. Of those, half survive their second year. Their maximum life span is 3 years. Furthermore, during the first year the rabbits produce no offspring, whereas the average number of offspring is 6 during the second year and 8 during the third year. If there are 24 rabbits in each age class now, what will the distribution be in 1 year? In 20 years?

As you will find in Chapter 7, the solution of this application depends on the concept of eigenvalues. You will see later that eigenvalues are used in a wide variety of real-life applications of linear algebra. Aside from population growth, eigenvalues are used in solving systems of differential equations, in quadratic forms, and in engineering and science.

The central question of the **eigenvalue problem** can be stated as follows. If A is an $n \times n$ matrix, do there exist $n \times 1$ nonzero matrices $\mathbf{x}$ such that $A\mathbf{x}$ is a scalar multiple of $\mathbf{x}$? The scalar is usually denoted by λ (the Greek letter lambda) and is called an **eigenvalue** of A, and the nonzero column matrix $\mathbf{x}$ is called an **eigenvector** of A corresponding to λ. The fundamental equation for the eigenvalue problem is

$$A\mathbf{x} = \lambda\mathbf{x}.$$

EXAMPLE 1 Verifying Eigenvalues and Eigenvectors

Let $A = \begin{bmatrix} 1 & 4 \\ 2 & 3 \end{bmatrix}$, $\mathbf{x}_1 = \begin{bmatrix} 1 \\ 1 \end{bmatrix}$, and $\mathbf{x}_2 = \begin{bmatrix} 2 \\ -1 \end{bmatrix}$.

Verify that $\lambda_1 = 5$ is an eigenvalue of A corresponding to $\mathbf{x}_1$ and that $\lambda_2 = -1$ is an eigenvalue of A corresponding to $\mathbf{x}_2$.

SOLUTION To verify that $\lambda_1 = 5$ is an eigenvalue of A corresponding to $\mathbf{x}_1$, multiply the matrices A and $\mathbf{x}_1$, as follows.

$$A\mathbf{x}_1 = \begin{bmatrix} 1 & 4 \\ 2 & 3 \end{bmatrix}\begin{bmatrix} 1 \\ 1 \end{bmatrix} = \begin{bmatrix} 5 \\ 5 \end{bmatrix} = 5\begin{bmatrix} 1 \\ 1 \end{bmatrix} = \lambda_1\mathbf{x}_1$$

Similarly, to verify that $\lambda_2 = -1$ is an eigenvalue of A corresponding to $\mathbf{x}_2$, multiply A and $\mathbf{x}_2$.

$$A\mathbf{x}_2 = \begin{bmatrix} 1 & 4 \\ 2 & 3 \end{bmatrix}\begin{bmatrix} 2 \\ -1 \end{bmatrix} = \begin{bmatrix} -2 \\ 1 \end{bmatrix} = -1\begin{bmatrix} 2 \\ -1 \end{bmatrix} = \lambda_2\mathbf{x}_2.$$

From this example, you can see that it is easy to verify whether a scalar λ and an $n \times 1$ matrix $\mathbf{x}$ satisfy the equation $A\mathbf{x} = \lambda\mathbf{x}$. Notice also that if $\mathbf{x}$ is an eigenvector corresponding to λ, then so is any nonzero multiple of $\mathbf{x}$. For instance, the column matrices

$$\begin{bmatrix} 2 \\ 2 \end{bmatrix} \quad \text{and} \quad \begin{bmatrix} -7 \\ -7 \end{bmatrix}$$

are also eigenvectors of A corresponding to $\lambda_1 = 5$.

Provided with an $n \times n$ matrix A, how can you find the eigenvalues and corresponding eigenvectors? The key is to write the equation $A\mathbf{x} = \lambda\mathbf{x}$ in the equivalent form

$$(\lambda I - A)\mathbf{x} = \mathbf{0},$$

where I is the $n \times n$ identity matrix. This homogeneous system of equations has nonzero solutions if and only if the coefficient matrix $(\lambda I - A)$ is singular; that is, if and only if the determinant of $(\lambda I - A)$ is zero. The equation $\det(\lambda I - A) = 0$ is called the **characteristic equation** of A, and is a polynomial equation of degree n in the variable λ. Once you have found the eigenvalues of A, you can use Gaussian elimination to find the corresponding eigenvectors, as shown in the next two examples.

EXAMPLE 2 Finding Eigenvalues and Eigenvectors

Find the eigenvalues and corresponding eigenvectors of the matrix $A = \begin{bmatrix} 1 & 4 \\ 2 & 3 \end{bmatrix}$.

SOLUTION The characteristic equation of A is

$$\begin{aligned} |\lambda I - A| &= \left| \begin{bmatrix} \lambda & 0 \\ 0 & \lambda \end{bmatrix} - \begin{bmatrix} 1 & 4 \\ 2 & 3 \end{bmatrix} \right| \\ &= \begin{vmatrix} \lambda - 1 & -4 \\ -2 & \lambda - 3 \end{vmatrix} \\ &= \lambda^2 - 4\lambda + 3 - 8 \\ &= \lambda^2 - 4\lambda - 5 \\ &= (\lambda - 5)(\lambda + 1) = 0. \end{aligned}$$

This yields two eigenvalues, $\lambda_1 = 5$ and $\lambda_2 = -1$.

To find the corresponding eigenvectors, solve the homogeneous linear system $(\lambda I - A)\mathbf{x} = \mathbf{0}$. For $\lambda_1 = 5$, the coefficient matrix is

$$5I - A = \begin{bmatrix} 5 & 0 \\ 0 & 5 \end{bmatrix} - \begin{bmatrix} 1 & 4 \\ 2 & 3 \end{bmatrix} = \begin{bmatrix} 5 - 1 & -4 \\ -2 & 5 - 3 \end{bmatrix} = \begin{bmatrix} 4 & -4 \\ -2 & 2 \end{bmatrix},$$

which row reduces to

$$\begin{bmatrix} 1 & -1 \\ 0 & 0 \end{bmatrix}.$$

The solutions of the homogeneous system having this coefficient matrix are all of the form

$$\begin{bmatrix} t \\ t \end{bmatrix},$$

where t is a real number. So, the eigenvectors corresponding to the eigenvalue $\lambda_1 = 5$ are the nonzero scalar multiples of

$$\begin{bmatrix} 1 \\ 1 \end{bmatrix}.$$

Similarly, for $\lambda_2 = -1$, the corresponding coefficient matrix is

$$-I - A = \begin{bmatrix} -1 & 0 \\ 0 & -1 \end{bmatrix} - \begin{bmatrix} 1 & 4 \\ 2 & 3 \end{bmatrix} = \begin{bmatrix} -1-1 & -4 \\ -2 & -1-3 \end{bmatrix} = \begin{bmatrix} -2 & -4 \\ -2 & -4 \end{bmatrix},$$

which row reduces to

$$\begin{bmatrix} 1 & 2 \\ 0 & 0 \end{bmatrix}.$$

The solutions of the homogeneous system having this coefficient matrix are all of the form

$$\begin{bmatrix} 2t \\ -t \end{bmatrix},$$

where t is a real number. So, the eigenvectors corresponding to the eigenvalue $\lambda_2 = -1$ are the nonzero scalar multiples of

$$\begin{bmatrix} 2 \\ -1 \end{bmatrix}.$$

Technology Note

Most graphing utilities and computer software programs have the capability of calculating the eigenvalues of a square matrix. Using a graphing utility to verify Example 2, your screen may look like the one below. Keystrokes and programming syntax for these utilities/programs applicable to Example 2 are provided in the **Online Technology Guide,** available at *college.hmco.com/pic/larsonELA6e.*

```
A
                 [[1 4]
                  [2 3]]
eigVl A
                 {-1 5}
```

EXAMPLE 3 Finding Eigenvalues and Eigenvectors

Find the eigenvalues and corresponding eigenvectors of the matrix

$$A = \begin{bmatrix} 1 & 2 & -2 \\ 1 & 2 & 1 \\ -1 & -1 & 0 \end{bmatrix}.$$

SOLUTION The characteristic equation of A is

$$|\lambda I - A| = \left|\begin{bmatrix} \lambda & 0 & 0 \\ 0 & \lambda & 0 \\ 0 & 0 & \lambda \end{bmatrix} - \begin{bmatrix} 1 & 2 & -2 \\ 1 & 2 & 1 \\ -1 & -1 & 0 \end{bmatrix}\right| = \begin{vmatrix} \lambda - 1 & -2 & 2 \\ -1 & \lambda - 2 & -1 \\ 1 & 1 & \lambda \end{vmatrix}$$

$$= (\lambda - 1)\begin{vmatrix} \lambda - 2 & -1 \\ 1 & \lambda \end{vmatrix} - (-2)\begin{vmatrix} -1 & -1 \\ 1 & \lambda \end{vmatrix} + 2\begin{vmatrix} -1 & \lambda - 2 \\ 1 & 1 \end{vmatrix}$$

$$= (\lambda - 1)(\lambda^2 - 2\lambda + 1) + 2(-\lambda + 1) + 2(-1 - \lambda + 2)$$

$$= \lambda^3 - 3\lambda^2 - \lambda + 3 = (\lambda^2 - 1)(\lambda - 3) = 0.$$

This yields three eigenvalues, $\lambda_1 = 1$, $\lambda_2 = -1$, and $\lambda_3 = 3$.

To find the corresponding eigenvectors, solve the homogeneous linear system $(\lambda I - A)\mathbf{x} = \mathbf{0}$. For $\lambda_1 = 1$, the coefficient matrix is

$$I - A = \begin{bmatrix} 1 & 0 & 0 \\ 0 & 1 & 0 \\ 0 & 0 & 1 \end{bmatrix} - \begin{bmatrix} 1 & 2 & -2 \\ 1 & 2 & 1 \\ -1 & -1 & 0 \end{bmatrix}$$

$$= \begin{bmatrix} 1 - 1 & -2 & 2 \\ -1 & 1 - 2 & -1 \\ 1 & 1 & 1 \end{bmatrix} = \begin{bmatrix} 0 & -2 & 2 \\ -1 & -1 & -1 \\ 1 & 1 & 1 \end{bmatrix},$$

which row reduces to

$$\begin{bmatrix} 1 & 0 & 2 \\ 0 & 1 & -1 \\ 0 & 0 & 0 \end{bmatrix}.$$

The solutions of the homogenous system having this coefficient matrix are all of the form

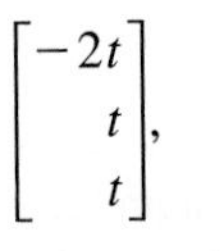

$$\begin{bmatrix} -2t \\ t \\ t \end{bmatrix},$$

where t is a real number. So, the eigenvectors corresponding to the eigenvalue $\lambda_1 = 1$ are the nonzero scalar multiples of

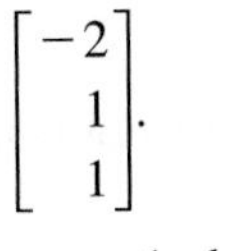

$$\begin{bmatrix} -2 \\ 1 \\ 1 \end{bmatrix}.$$

For $\lambda_2 = -1$, the coefficient matrix is

$$-I - A = \begin{bmatrix} -1 & 0 & 0 \\ 0 & -1 & 0 \\ 0 & 0 & -1 \end{bmatrix} - \begin{bmatrix} 1 & 2 & -2 \\ 1 & 2 & 1 \\ -1 & -1 & 0 \end{bmatrix}$$

$$= \begin{bmatrix} -1 - 1 & -2 & 2 \\ -1 & -1 - 2 & -1 \\ 1 & 1 & -1 \end{bmatrix} = \begin{bmatrix} -2 & -2 & 2 \\ -1 & -3 & -1 \\ 1 & 1 & -1 \end{bmatrix},$$

which row reduces to

$$\begin{bmatrix} 1 & 0 & -2 \\ 0 & 1 & 1 \\ 0 & 0 & 0 \end{bmatrix}.$$

The solutions of the homogeneous system having this coefficient matrix are all of the form

$$\begin{bmatrix} 2t \\ -t \\ t \end{bmatrix},$$

where t is a real number. So, the eigenvectors corresponding to the eigenvalue $\lambda_2 = -1$ are the nonzero scalar multiples of

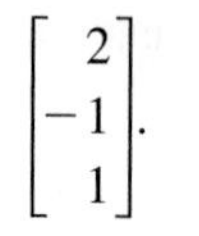

$$\begin{bmatrix} 2 \\ -1 \\ 1 \end{bmatrix}.$$

For $\lambda_3 = 3$, the coefficient matrix is

$$3I - A = \begin{bmatrix} 3 & 0 & 0 \\ 0 & 3 & 0 \\ 0 & 0 & 3 \end{bmatrix} - \begin{bmatrix} 1 & 2 & -2 \\ 1 & 2 & 1 \\ -1 & -1 & 0 \end{bmatrix}$$

$$= \begin{bmatrix} 3-1 & -2 & 2 \\ -1 & 3-2 & -1 \\ 1 & 1 & 3 \end{bmatrix} = \begin{bmatrix} 2 & -2 & 2 \\ -1 & 1 & -1 \\ 1 & 1 & 3 \end{bmatrix},$$

which row reduces to

$$\begin{bmatrix} 1 & 0 & 2 \\ 0 & 1 & 1 \\ 0 & 0 & 0 \end{bmatrix}.$$

The solutions of the homogeneous system having this coefficient matrix are all of the form

$$\begin{bmatrix} -2t \\ -t \\ t \end{bmatrix},$$

where t is a real number. So, the eigenvectors corresponding to the eigenvalue $\lambda_3 = 3$ are the nonzero scalar multiples of

$$\begin{bmatrix} -2 \\ -1 \\ 1 \end{bmatrix}.$$

Eigenvalues and eigenvectors have ample applications in mathematics, engineering, biology, and other sciences. You will see one such application to stochastic processes at the end of this chapter. In Chapter 7 you will study eigenvalues and their applications in more detail and learn how to solve the rabbit population problem presented earlier.

SECTION 3.4 Exercises

Eigenvalues and Eigenvectors

In Exercises 1–4, verify that λ_i is an eigenvalue of A and that $\mathbf{x}_i$ is a corresponding eigenvector.

1. $A = \begin{bmatrix} 1 & 2 \\ 0 & -3 \end{bmatrix}$; $\lambda_1 = 1$, $\mathbf{x}_1 = \begin{bmatrix} 1 \\ 0 \end{bmatrix}$;

$\lambda_2 = -3$, $\mathbf{x}_2 = \begin{bmatrix} -1 \\ 2 \end{bmatrix}$

2. $A = \begin{bmatrix} 4 & 3 \\ 1 & 2 \end{bmatrix}$; $\lambda_1 = 5$, $\mathbf{x}_1 = \begin{bmatrix} 3 \\ 1 \end{bmatrix}$;

$\lambda_2 = 1$, $\mathbf{x}_2 = \begin{bmatrix} -1 \\ 1 \end{bmatrix}$

3. $A = \begin{bmatrix} 1 & 1 & 1 \\ 0 & 1 & 0 \\ 1 & 1 & 1 \end{bmatrix}$; $\lambda_1 = 2$, $\mathbf{x}_1 = \begin{bmatrix} 1 \\ 0 \\ 1 \end{bmatrix}$;

$\lambda_2 = 0$, $\mathbf{x}_2 = \begin{bmatrix} -1 \\ 0 \\ 1 \end{bmatrix}$; $\lambda_3 = 1$, $\mathbf{x}_3 = \begin{bmatrix} -1 \\ 1 \\ -1 \end{bmatrix}$

4. $A = \begin{bmatrix} 1 & -2 & 1 \\ 0 & 1 & 4 \\ 0 & 0 & 2 \end{bmatrix}$; $\lambda_1 = 1$, $\mathbf{x}_1 = \begin{bmatrix} 1 \\ 0 \\ 0 \end{bmatrix}$;

$\lambda_2 = 2$, $\mathbf{x}_2 = \begin{bmatrix} -7 \\ 4 \\ 1 \end{bmatrix}$

In Exercises 5–14, find (a) the characteristic equation, (b) the eigenvalues, and (c) the corresponding eigenvectors of the matrix.

5. $\begin{bmatrix} 4 & -5 \\ 2 & -3 \end{bmatrix}$

6. $\begin{bmatrix} 2 & 2 \\ 2 & 2 \end{bmatrix}$

7. $\begin{bmatrix} 2 & 1 \\ 3 & 0 \end{bmatrix}$

8. $\begin{bmatrix} 2 & 5 \\ 4 & 3 \end{bmatrix}$

9. $\begin{bmatrix} -2 & 4 \\ 2 & 5 \end{bmatrix}$

10. $\begin{bmatrix} 3 & -1 \\ 5 & -3 \end{bmatrix}$

11. $\begin{bmatrix} 1 & -1 & -1 \\ 1 & 3 & 1 \\ -3 & 1 & -1 \end{bmatrix}$

12. $\begin{bmatrix} 2 & 0 & 1 \\ 0 & 3 & 4 \\ 0 & 0 & 1 \end{bmatrix}$

13. $\begin{bmatrix} 1 & 2 & 1 \\ 0 & 1 & 0 \\ 4 & 0 & 1 \end{bmatrix}$

14. $\begin{bmatrix} 1 & 0 & 1 \\ 0 & -1 & 0 \\ 2 & 1 & -1 \end{bmatrix}$

In Exercises 15–24, use a graphing utility or computer software program with matrix capabilities to find the eigenvalues of the matrix. Then find the corresponding eigenvectors.

15. $\begin{bmatrix} 2 & 5 \\ -1 & -4 \end{bmatrix}$

16. $\begin{bmatrix} 4 & 3 \\ -3 & -2 \end{bmatrix}$

17. $\begin{bmatrix} 4 & -2 & -2 \\ 0 & 1 & 0 \\ 1 & 0 & 1 \end{bmatrix}$

18. $\begin{bmatrix} 4 & 0 & 0 \\ 0 & 0 & -3 \\ 0 & -2 & 1 \end{bmatrix}$

19. $\begin{bmatrix} 1 & 0 & -1 \\ 0 & -2 & 0 \\ 0 & -2 & -2 \end{bmatrix}$

20. $\begin{bmatrix} 1 & 1 & 0 \\ 0 & -2 & 1 \\ 0 & -2 & 2 \end{bmatrix}$

HM **21.** $\begin{bmatrix} 3 & 0 & 0 & 0 \\ 0 & -1 & 0 & 0 \\ 0 & 0 & 2 & 5 \\ 0 & 0 & 3 & 0 \end{bmatrix}$

HM **22.** $\begin{bmatrix} 1 & 0 & 2 & 3 \\ 0 & 2 & 0 & 0 \\ 0 & 0 & 1 & 3 \\ 0 & -1 & 3 & 1 \end{bmatrix}$

23. $\begin{bmatrix} 1 & 0 & 1 & 0 \\ 0 & -2 & 0 & 0 \\ 0 & 0 & 2 & 1 \\ 0 & 0 & 3 & 0 \end{bmatrix}$

24. $\begin{bmatrix} 2 & 0 & -1 & -1 \\ 0 & 2 & 1 & 0 \\ 0 & -1 & 0 & 0 \\ 0 & 0 & 2 & 0 \end{bmatrix}$

True or False? In Exercises 25 and 26, determine whether each statement is true or false. If a statement is true, give a reason or cite an appropriate statement from the text. If a statement is false, provide an example that shows the statement is not true in all cases or cite an appropriate statement from the text.

25. (a) If $\mathbf{x}$ is an eigenvector corresponding to a given eigenvalue λ, then any multiple of $\mathbf{x}$ is also an eigenvector corresponding to that same λ.

(b) If $\lambda = a$ is an eigenvalue of the matrix A, then $\lambda = a$ is a solution of the characteristic equation $\lambda I - A = 0$.

26. (a) The characteristic equation of the matrix $A = \begin{bmatrix} 2 & -1 \\ 1 & 0 \end{bmatrix}$ yields eigenvalues $\lambda_1 = \lambda_2 = 1$.

(b) The matrix $A = \begin{bmatrix} 4 & -2 \\ -1 & 0 \end{bmatrix}$ has irrational eigenvalues $\lambda_1 = 2 + \sqrt{6}$ and $\lambda_2 = 2 - \sqrt{6}$.

3.5 Applications of Determinants

So far in this chapter, you have examined procedures for evaluating determinants, studied properties of determinants, and learned how determinants are used to find eigenvalues. In this section, you will study an explicit formula for the inverse of a nonsingular matrix and then use this formula to derive a theorem known as Cramer's Rule. You will then solve several applications of determinants using Cramer's Rule.

The Adjoint of a Matrix

Recall from Section 3.1 that the cofactor C_{ij} of a matrix A is defined as $(-1)^{i+j}$ times the determinant of the matrix obtained by deleting the ith row and the jth column of A. If A is a square matrix, then the **matrix of cofactors** of A has the form

$$\begin{bmatrix} C_{11} & C_{12} & \cdots & C_{1n} \\ C_{21} & C_{22} & \cdots & C_{2n} \\ \vdots & \vdots & & \vdots \\ C_{n1} & C_{n2} & \cdots & C_{nn} \end{bmatrix}.$$

The transpose of this matrix is called the **adjoint** of A and is denoted by adj(A). That is,

$$\text{adj}(A) = \begin{bmatrix} C_{11} & C_{21} & \cdots & C_{n1} \\ C_{12} & C_{22} & \cdots & C_{n2} \\ \vdots & \vdots & & \vdots \\ C_{1n} & C_{2n} & \cdots & C_{nn} \end{bmatrix}.$$

EXAMPLE 1 Finding the Adjoint of a Square Matrix

Find the adjoint of

$$A = \begin{bmatrix} -1 & 3 & 2 \\ 0 & -2 & 1 \\ 1 & 0 & -2 \end{bmatrix}.$$

SOLUTION The cofactor C_{11} is given by

$$\begin{bmatrix} -1 & 3 & 2 \\ 0 & -2 & 1 \\ 1 & 0 & -2 \end{bmatrix} \rightarrow C_{11} = (-1)^2 \begin{vmatrix} -2 & 1 \\ 0 & -2 \end{vmatrix} = 4.$$

Continuing this process produces the following matrix of cofactors of A.

$$\begin{bmatrix} \begin{vmatrix} -2 & 1 \\ 0 & -2 \end{vmatrix} & -\begin{vmatrix} 0 & 1 \\ 1 & -2 \end{vmatrix} & \begin{vmatrix} 0 & -2 \\ 1 & 0 \end{vmatrix} \\ -\begin{vmatrix} 3 & 2 \\ 0 & -2 \end{vmatrix} & \begin{vmatrix} -1 & 2 \\ 1 & -2 \end{vmatrix} & -\begin{vmatrix} -1 & 3 \\ 1 & 0 \end{vmatrix} \\ \begin{vmatrix} 3 & 2 \\ -2 & 1 \end{vmatrix} & -\begin{vmatrix} -1 & 2 \\ 0 & 1 \end{vmatrix} & \begin{vmatrix} -1 & 3 \\ 0 & -2 \end{vmatrix} \end{bmatrix} = \begin{bmatrix} 4 & 1 & 2 \\ 6 & 0 & 3 \\ 7 & 1 & 2 \end{bmatrix}$$

The transpose of this matrix is the adjoint of A. That is,

$$\text{adj}(A) = \begin{bmatrix} 4 & 6 & 7 \\ 1 & 0 & 1 \\ 2 & 3 & 2 \end{bmatrix}.$$

The adjoint of a matrix A can be used to find the inverse of A, as indicated in the next theorem.

THEOREM 3.10
The Inverse of a Matrix Given by Its Adjoint

If A is an $n \times n$ invertible matrix, then

$$A^{-1} = \frac{1}{\det(A)}\text{adj}(A).$$

PROOF Begin by proving that the product of A and its adjoint is equal to the product of the determinant of A and I_n.

Consider the product

$$A[\text{adj}(A)] = \begin{bmatrix} a_{11} & a_{12} & \cdots & a_{1n} \\ a_{21} & a_{22} & \cdots & a_{2n} \\ \vdots & \vdots & & \vdots \\ a_{i1} & a_{i2} & \cdots & a_{in} \\ \vdots & \vdots & & \vdots \\ a_{n1} & a_{n2} & \cdots & a_{nn} \end{bmatrix} \begin{bmatrix} C_{11} & C_{21} & \cdots & C_{j1} & \cdots & C_{n1} \\ C_{12} & C_{22} & \cdots & C_{j2} & \cdots & C_{n2} \\ \vdots & \vdots & & \vdots & & \vdots \\ C_{1n} & C_{2n} & \cdots & C_{jn} & \cdots & C_{nn} \end{bmatrix}.$$

The entry in the ith row and jth column of this product is

$$a_{i1}C_{j1} + a_{i2}C_{j2} + \cdot \cdot \cdot + a_{in}C_{jn}.$$

If $i = j$, then this sum is simply the cofactor expansion of A along its ith row, which means that the sum is the determinant of A. On the other hand, if $i \neq j$, then the sum is zero.

$$A[\text{adj}(A)] = \begin{bmatrix} \det(A) & 0 & \cdots & 0 \\ 0 & \det(A) & \cdots & 0 \\ \vdots & \vdots & & \vdots \\ 0 & 0 & \cdots & \det(A) \end{bmatrix} = \det(A)I$$

Because A is invertible, $\det(A) \neq 0$ and you can write

$$\frac{1}{\det(A)} A[\operatorname{adj}(A)] = I \qquad \text{or} \qquad A\left[\frac{1}{\det(A)} \operatorname{adj}(A)\right] = I.$$

Multiplying both sides of the equation by A^{-1} results in the equation

$$A^{-1}A\left[\frac{1}{\det(A)} \operatorname{adj}(A)\right] = A^{-1}I, \text{ which yields } \frac{1}{\det(A)} \operatorname{adj}(A) = A^{-1}.$$

If A is a 2×2 matrix $A = \begin{bmatrix} a & b \\ c & d \end{bmatrix}$, then the adjoint of A is simply

$$\operatorname{adj}(A) = \begin{bmatrix} d & -b \\ -c & a \end{bmatrix}.$$

Moreover, if A is invertible, then from Theorem 3.10 you have

$$A^{-1} = \frac{1}{|A|} \operatorname{adj}(A) = \frac{1}{ad - bc} \begin{bmatrix} d & -b \\ -c & a \end{bmatrix},$$

which agrees with the result in Section 2.3.

EXAMPLE 2 Using the Adjoint of a Matrix to Find Its Inverse

Use the adjoint of

$$A = \begin{bmatrix} -1 & 3 & 2 \\ 0 & -2 & 1 \\ 1 & 0 & -2 \end{bmatrix}$$

to find A^{-1}.

SOLUTION The determinant of this matrix is 3. Using the adjoint of A (found in Example 1), you can find the inverse of A to be

$$A^{-1} = \frac{1}{|A|} \operatorname{adj}(A) = \frac{1}{3} \begin{bmatrix} 4 & 6 & 7 \\ 1 & 0 & 1 \\ 2 & 3 & 2 \end{bmatrix} = \begin{bmatrix} \frac{4}{3} & 2 & \frac{7}{3} \\ \frac{1}{3} & 0 & \frac{1}{3} \\ \frac{2}{3} & 1 & \frac{2}{3} \end{bmatrix}.$$

You can check to see that this matrix is the inverse of A by multiplying to obtain

$$AA^{-1} = \begin{bmatrix} -1 & 3 & 2 \\ 0 & -2 & 1 \\ 1 & 0 & -2 \end{bmatrix} \begin{bmatrix} \frac{4}{3} & 2 & \frac{7}{3} \\ \frac{1}{3} & 0 & \frac{1}{3} \\ \frac{2}{3} & 1 & \frac{2}{3} \end{bmatrix} = \begin{bmatrix} 1 & 0 & 0 \\ 0 & 1 & 0 \\ 0 & 0 & 1 \end{bmatrix}.$$

REMARK: Theorem 3.10 is not particularly efficient for calculating inverses. The Gauss-Jordan elimination method discussed in Section 2.3 is much better. Theorem 3.10 is theoretically useful, however, because it provides a concise formula for the inverse of a matrix.

Cramer's Rule

Cramer's Rule, named after Gabriel Cramer (1704–1752), is a formula that uses determinants to solve a system of n linear equations in n variables. This rule can be applied only to systems of linear equations that have unique solutions.

To see how Cramer's Rule arises, look at the solution of a general system involving two linear equations in two variables.

$$\begin{aligned} a_{11}x_1 + a_{12}x_2 &= b_1 \\ a_{21}x_1 + a_{22}x_2 &= b_2 \end{aligned}$$

Multiplying the first equation by $-a_{21}$ and the second by a_{11} and adding the results produces

$$\begin{aligned} -a_{21}a_{11}x_1 - a_{21}a_{12}x_2 &= -a_{21}b_1 \\ a_{11}a_{21}x_1 + a_{11}a_{22}x_2 &= a_{11}b_2 \\ \hline (a_{11}a_{22} - a_{21}a_{12})x_2 &= a_{11}b_2 - a_{21}b_1. \end{aligned}$$

Solving for x_2 (provided that $a_{11}a_{22} - a_{21}a_{12} \neq 0$) produces

$$x_2 = \frac{a_{11}b_2 - a_{21}b_1}{a_{11}a_{22} - a_{21}a_{12}}.$$

In a similar way, you can solve for x_1 to obtain

$$x_1 = \frac{a_{22}b_1 - a_{12}b_2}{a_{11}a_{22} - a_{21}a_{12}}.$$

Finally, recognizing that the numerators and denominators of both x_1 and x_2 can be represented as determinants, you have

$$x_1 = \frac{\begin{vmatrix} b_1 & a_{12} \\ b_2 & a_{22} \end{vmatrix}}{\begin{vmatrix} a_{11} & a_{12} \\ a_{21} & a_{22} \end{vmatrix}}, \qquad x_2 = \frac{\begin{vmatrix} a_{11} & b_1 \\ a_{21} & b_2 \end{vmatrix}}{\begin{vmatrix} a_{11} & a_{12} \\ a_{21} & a_{22} \end{vmatrix}}, \qquad a_{11}a_{22} - a_{21}a_{12} \neq 0.$$

The denominator for both x_1 and x_2 is simply the determinant of the coefficient matrix A. The determinant forming the numerator of x_1 can be obtained from A by replacing its first column by the column representing the constants of the system. The determinant forming the numerator of x_2 can be obtained in a similar way. These two determinants are denoted by $|A_1|$ and $|A_2|$, as follows.

$$|A_1| = \begin{vmatrix} b_1 & a_{12} \\ b_2 & a_{22} \end{vmatrix} \quad \text{and} \quad |A_2| = \begin{vmatrix} a_{11} & b_1 \\ a_{21} & b_2 \end{vmatrix}$$

You have $x_1 = \dfrac{|A_1|}{|A|}$ and $x_2 = \dfrac{|A_2|}{|A|}$. This determinant form of the solution is called **Cramer's Rule.**

EXAMPLE 3 **Using Cramer's Rule**

Use Cramer's Rule to solve the system of linear equations.

$$\begin{aligned} 4x_1 - 2x_2 &= 10 \\ 3x_1 - 5x_2 &= 11 \end{aligned}$$

SOLUTION First find the determinant of the coefficient matrix.

$$|A| = \begin{vmatrix} 4 & -2 \\ 3 & -5 \end{vmatrix} = -14$$

Because $|A| \neq 0$, you know the system has a unique solution, and applying Cramer's Rule produces

$$x_1 = \frac{|A_1|}{|A|} = \frac{\begin{vmatrix} 10 & -2 \\ 11 & -5 \end{vmatrix}}{-14} = \frac{-28}{-14} = 2$$

$$x_2 = \frac{|A_2|}{|A|} = \frac{\begin{vmatrix} 4 & 10 \\ 3 & 11 \end{vmatrix}}{-14} = \frac{14}{-14} = -1.$$

The solution is $x_1 = 2$ and $x_2 = -1$.

Cramer's Rule generalizes easily to systems of n linear equations in n variables. The value of each variable is the quotient of two determinants. The denominator is the determinant of the coefficient matrix, and the numerator is the determinant of the matrix formed by replacing the column corresponding to the variable being solved for with the column representing the constants. For example, the solution for x_3 in the system

$$\begin{aligned} a_{11}x_1 + a_{12}x_2 + a_{13}x_3 &= b_1 \\ a_{21}x_1 + a_{22}x_2 + a_{23}x_3 &= b_2 \\ a_{31}x_1 + a_{32}x_2 + a_{33}x_3 &= b_3 \end{aligned}$$

is

$$x_3 = \frac{|A_3|}{|A|} = \frac{\begin{vmatrix} a_{11} & a_{12} & b_1 \\ a_{21} & a_{22} & b_2 \\ a_{31} & a_{32} & b_3 \end{vmatrix}}{\begin{vmatrix} a_{11} & a_{12} & a_{13} \\ a_{21} & a_{22} & a_{23} \\ a_{31} & a_{32} & a_{33} \end{vmatrix}}.$$

THEOREM 3.11
Cramer's Rule

If a system of n linear equations in n variables has a coefficient matrix with a nonzero determinant $|A|$, then the solution of the system is given by

$$x_1 = \frac{\det(A_1)}{\det(A)}, \quad x_2 = \frac{\det(A_2)}{\det(A)}, \quad \ldots, \quad x_n = \frac{\det(A_n)}{\det(A)},$$

where the ith column of A_i is the column of constants in the system of equations.

PROOF

Let the system be represented by $AX = B$. Because $|A|$ is nonzero, you can write

$$X = A^{-1}B = \frac{1}{|A|}\operatorname{adj}(A)B = \begin{bmatrix} x_1 \\ x_2 \\ \vdots \\ x_n \end{bmatrix}.$$

If the entries of B are $b_1, b_2, \ldots, b_n$, then x_i is

$$x_i = \frac{1}{|A|}(b_1C_{1i} + b_2C_{2i} + \cdots + b_nC_{ni}),$$

but the sum (in parentheses) is precisely the cofactor expansion of A_i, which means that $x_i = |A_i|/|A|$, and the proof is complete.

EXAMPLE 4 Using Cramer's Rule

Use Cramer's Rule to solve the system of linear equations for x.

$$\begin{aligned} -x + 2y - 3z &= 1 \\ 2x \qquad + z &= 0 \\ 3x - 4y + 4z &= 2 \end{aligned}$$

SOLUTION

The determinant of the coefficient matrix is

$$|A| = \begin{vmatrix} -1 & 2 & -3 \\ 2 & 0 & 1 \\ 3 & -4 & 4 \end{vmatrix} = 10.$$

Because $|A| \neq 0$, you know the solution is unique, and Cramer's Rule can be applied to solve for x, as follows.

$$x = \frac{\begin{vmatrix} 1 & 2 & -3 \\ 0 & 0 & 1 \\ 2 & -4 & 4 \end{vmatrix}}{10} = \frac{(1)(-1)^5\begin{vmatrix} 1 & 2 \\ 2 & -4 \end{vmatrix}}{10} = \frac{(1)(-1)(-8)}{10} = \frac{4}{5}$$

REMARK: Try applying Cramer's Rule in Example 4 to solve for y and z. You will see that the solution is $y = -\frac{3}{2}$ and $z = -\frac{8}{5}$.

Area, Volume, and Equations of Lines and Planes

Determinants have many applications in analytic geometry. Several are presented here. The first application is finding the area of a triangle in the xy-plane.

Area of a Triangle in the xy-Plane

The area of the triangle whose vertices are (x_1, y_1), (x_2, y_2), and (x_3, y_3) is given by

$$\text{Area} = \pm\frac{1}{2}\det\begin{bmatrix} x_1 & y_1 & 1 \\ x_2 & y_2 & 1 \\ x_3 & y_3 & 1 \end{bmatrix},$$

where the sign ($\pm$) is chosen to yield a positive area.

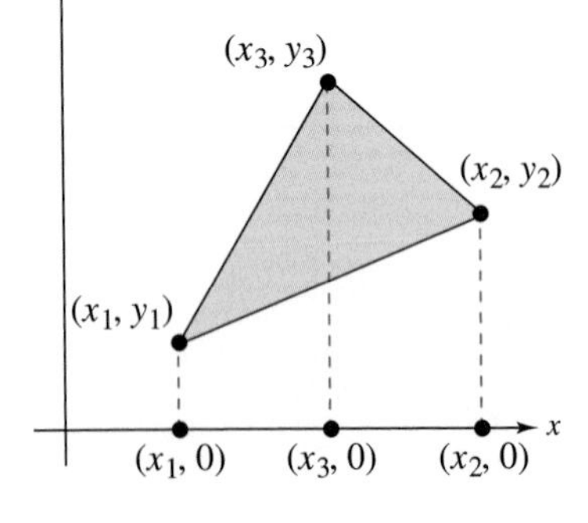

Figure 3.1

PROOF Prove the case for $y_i > 0$. Assume $x_1 \le x_3 \le x_2$ and that (x_3, y_3) lies above the line segment connecting (x_1, y_1) and (x_2, y_2), as shown in Figure 3.1. Consider the three trapezoids whose vertices are

Trapezoid 1: $(x_1, 0)$, (x_1, y_1), (x_3, y_3), $(x_3, 0)$
Trapezoid 2: $(x_3, 0)$, (x_3, y_3), (x_2, y_2), $(x_2, 0)$
Trapezoid 3: $(x_1, 0)$, (x_1, y_1), (x_2, y_2), $(x_2, 0)$.

The area of the triangle is equal to the sum of the areas of the first two trapezoids less the area of the third. So

$$\begin{aligned} \text{Area} &= \tfrac{1}{2}(y_1 + y_3)(x_3 - x_1) + \tfrac{1}{2}(y_3 + y_2)(x_2 - x_3) - \tfrac{1}{2}(y_1 + y_2)(x_2 - x_1) \\ &= \tfrac{1}{2}(x_1y_2 + x_2y_3 + x_3y_1 - x_1y_3 - x_2y_1 - x_3y_2) \\ &= \frac{1}{2}\begin{vmatrix} x_1 & y_1 & 1 \\ x_2 & y_2 & 1 \\ x_3 & y_3 & 1 \end{vmatrix}. \end{aligned}$$

If the vertices do not occur in the order $x_1 \le x_3 \le x_2$ or if the vertex (x_3, y_3) is not above the line segment connecting the other two vertices, then the formula may yield the negative value of the area.

EXAMPLE 5 Finding the Area of a Triangle

Find the area of the triangle whose vertices are $(1, 0)$, $(2, 2)$, and $(4, 3)$.

SOLUTION It is not necessary to know the relative positions of the three vertices. Simply evaluate the determinant

$$\frac{1}{2}\begin{vmatrix} 1 & 0 & 1 \\ 2 & 2 & 1 \\ 4 & 3 & 1 \end{vmatrix} = -\frac{3}{2}$$

and conclude that the area of the triangle is $\frac{3}{2}$.

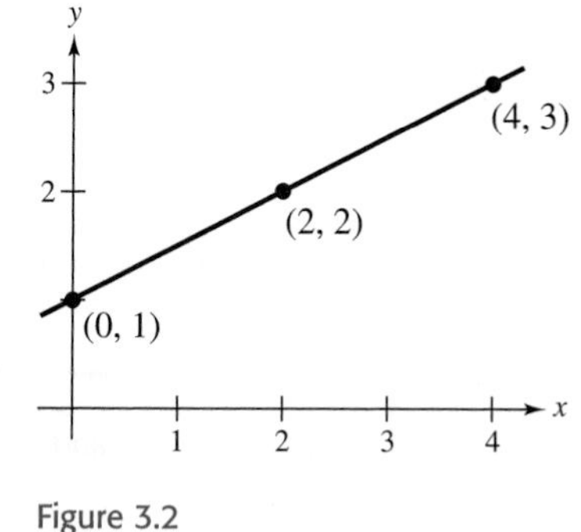

Figure 3.2

Suppose the three points in Example 5 had been on the same line. What would have happened had you applied the area formula to three such points? The answer is that the determinant would have been zero. Consider, for instance, the collinear points $(0, 1)$, $(2, 2)$, and $(4, 3)$, shown in Figure 3.2. The determinant that yields the area of the "triangle" having these three points as vertices is

$$\frac{1}{2}\begin{vmatrix} 0 & 1 & 1 \\ 2 & 2 & 1 \\ 4 & 3 & 1 \end{vmatrix} = 0.$$

If three points in the xy-plane lie on the same line, then the determinant in the formula for the area of a triangle turns out to be zero. This result is generalized in the test below.

Test for Collinear Points in the xy-Plane

Three points (x_1, y_1), (x_2, y_2), and (x_3, y_3) are collinear if and only if

$$\det\begin{bmatrix} x_1 & y_1 & 1 \\ x_2 & y_2 & 1 \\ x_3 & y_3 & 1 \end{bmatrix} = 0.$$

The next determinant form, for an equation of the line passing through two points in the xy-plane, is derived from the test for collinear points.

Two-Point Form of the Equation of a Line

An equation of the line passing through the distinct points (x_1, y_1) and (x_2, y_2) is given by

$$\det\begin{bmatrix} x & y & 1 \\ x_1 & y_1 & 1 \\ x_2 & y_2 & 1 \end{bmatrix} = 0.$$

EXAMPLE 6 **Finding an Equation of the Line Passing Through Two Points**

Find an equation of the line passing through the points (2, 4) and (−1, 3).

SOLUTION Applying the determinant formula for the equation of a line passing through two points produces

$$\begin{vmatrix} x & y & 1 \\ 2 & 4 & 1 \\ -1 & 3 & 1 \end{vmatrix} = 0.$$

To evaluate this determinant, expand by cofactors along the top row to obtain

$$x\begin{vmatrix} 4 & 1 \\ 3 & 1 \end{vmatrix} - y\begin{vmatrix} 2 & 1 \\ -1 & 1 \end{vmatrix} + 1\begin{vmatrix} 2 & 4 \\ -1 & 3 \end{vmatrix} = 0$$

$$x - 3y + 10 = 0.$$

An equation of the line is $x - 3y = -10$.

The formula for the area of a triangle in the plane has a straightforward generalization to three-dimensional space, which is presented without proof as follows.

Volume of a Tetrahedron

The volume of the tetrahedron whose vertices are (x_1, y_1, z_1), (x_2, y_2, z_2), (x_3, y_3, z_3), and (x_4, y_4, z_4) is given by

$$\text{Volume} = \pm\tfrac{1}{6}\det\begin{bmatrix} x_1 & y_1 & z_1 & 1 \\ x_2 & y_2 & z_2 & 1 \\ x_3 & y_3 & z_3 & 1 \\ x_4 & y_4 & z_4 & 1 \end{bmatrix},$$

where the sign (±) is chosen to yield a positive volume.

EXAMPLE 7 **Finding the Volume of a Tetrahedron**

Find the volume of the tetrahedron whose vertices are (0, 4, 1), (4, 0, 0), (3, 5, 2), and (2, 2, 5), as shown in Figure 3.3 (on page 167).

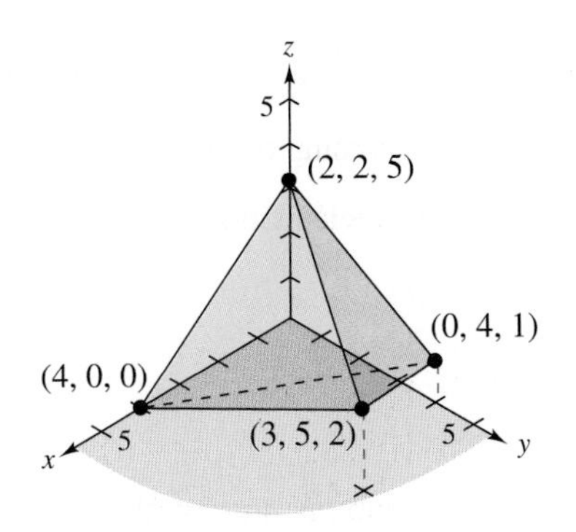

Figure 3.3

SOLUTION Using the determinant formula for volume produces

$$\frac{1}{6}\begin{vmatrix} 0 & 4 & 1 & 1 \\ 4 & 0 & 0 & 1 \\ 3 & 5 & 2 & 1 \\ 2 & 2 & 5 & 1 \end{vmatrix} = \frac{1}{6}(-72) = -12.$$

The volume of the tetrahedron is 12.

If four points in three-dimensional space happen to lie in the same plane, then the determinant in the formula for volume turns out to be zero. So, you have the test below.

Test for Coplanar Points in Space

Four points (x_1, y_1, z_1), (x_2, y_2, z_2), (x_3, y_3, z_3), and (x_4, y_4, z_4) are coplanar if and only if

$$\det\begin{bmatrix} x_1 & y_1 & z_1 & 1 \\ x_2 & y_2 & z_2 & 1 \\ x_3 & y_3 & z_3 & 1 \\ x_4 & y_4 & z_4 & 1 \end{bmatrix} = 0.$$

This test now provides the determinant form for an equation of a plane passing through three points in space, as shown below.

Three-Point Form of the Equation of a Plane

An equation of the plane passing through the distinct points (x_1, y_1, z_1), (x_2, y_2, z_2), and (x_3, y_3, z_3) is given by

$$\det\begin{bmatrix} x & y & z & 1 \\ x_1 & y_1 & z_1 & 1 \\ x_2 & y_2 & z_2 & 1 \\ x_3 & y_3 & z_3 & 1 \end{bmatrix} = 0.$$

EXAMPLE 8 **Finding an Equation of the Plane Passing Through Three Points**

Find an equation of the plane passing through the points $(0, 1, 0)$, $(-1, 3, 2)$, and $(-2, 0, 1)$.

SOLUTION Using the determinant form of the equation of a plane passing through three points produces

$$\begin{vmatrix} x & y & z & 1 \\ 0 & 1 & 0 & 1 \\ -1 & 3 & 2 & 1 \\ -2 & 0 & 1 & 1 \end{vmatrix} = 0.$$

To evaluate this determinant, subtract the fourth column from the second column to obtain

$$\begin{vmatrix} x & y-1 & z & 1 \\ 0 & 0 & 0 & 1 \\ -1 & 2 & 2 & 1 \\ -2 & -1 & 1 & 1 \end{vmatrix} = 0.$$

Now, expanding by cofactors along the second row yields

$$x\begin{vmatrix} 2 & 2 \\ -1 & 1 \end{vmatrix} - (y-1)\begin{vmatrix} -1 & 2 \\ -2 & 1 \end{vmatrix} + z\begin{vmatrix} -1 & 2 \\ -2 & -1 \end{vmatrix},$$

which produces the equation $4x - 3y + 5z = -3$.

SECTION 3.5 Exercises

The Adjoint of a Matrix

In Exercises 1–8, find the adjoint of the matrix A. Then use the adjoint to find the inverse of A, if possible.

1. $A = \begin{bmatrix} 1 & 2 \\ 3 & 4 \end{bmatrix}$ **2.** $A = \begin{bmatrix} -1 & 0 \\ 0 & 4 \end{bmatrix}$

3. $A = \begin{bmatrix} 1 & 0 & 0 \\ 0 & 2 & 6 \\ 0 & -4 & -12 \end{bmatrix}$ **4.** $A = \begin{bmatrix} 1 & 2 & 3 \\ 0 & 1 & -1 \\ 2 & 2 & 2 \end{bmatrix}$

5. $A = \begin{bmatrix} -3 & -5 & -7 \\ 2 & 4 & 3 \\ 0 & 1 & -1 \end{bmatrix}$ **6.** $A = \begin{bmatrix} 0 & 1 & 1 \\ 1 & 2 & 3 \\ -1 & -1 & -2 \end{bmatrix}$

7. $A = \begin{bmatrix} -1 & 2 & 0 & 1 \\ 3 & -1 & 4 & 1 \\ 0 & 0 & 1 & 2 \\ -1 & 1 & 1 & 2 \end{bmatrix}$ **8.** $A = \begin{bmatrix} 1 & 1 & 1 & 0 \\ 1 & 1 & 0 & 1 \\ 1 & 0 & 1 & 1 \\ 0 & 1 & 1 & 1 \end{bmatrix}$

9. Prove that if $|A| = 1$ and all entries of A are integers, then all entries of $|A^{-1}|$ must also be integers.

10. Prove that if an $n \times n$ matrix A is not invertible, then $A[\text{adj}(A)]$ is the zero matrix.

In Exercises 11 and 12, prove the formula for a nonsingular $n \times n$ matrix A. Assume $n \geq 3$.

11. $|\text{adj}(A)| = |A|^{n-1}$ **12.** $\text{adj}[\text{adj}(A)] = |A|^{n-2}A$

13. Illustrate the formula provided in Exercise 11 for the matrix

$$A = \begin{bmatrix} 1 & 0 \\ 1 & -2 \end{bmatrix}.$$

14. Illustrate the formula provided in Exercise 12 for the matrix

$$A = \begin{bmatrix} -1 & 3 \\ 1 & 2 \end{bmatrix}.$$

15. Prove that if A is an $n \times n$ invertible matrix, then $\text{adj}(A^{-1}) = [\text{adj}(A)]^{-1}$.

16. Illustrate the formula provided in Exercise 15 for the matrix
$$A = \begin{bmatrix} 1 & 3 \\ 1 & 2 \end{bmatrix}.$$

Cramer's Rule

In Exercises 17–32, use Cramer's Rule to solve the system of linear equations, if possible.

17. $\begin{aligned} x_1 + 2x_2 &= 5 \\ -x_1 + x_2 &= 1 \end{aligned}$

18. $\begin{aligned} 2x_1 - x_2 &= -10 \\ 3x_1 + 2x_2 &= -1 \end{aligned}$

19. $\begin{aligned} 3x_1 + 4x_2 &= -2 \\ 5x_1 + 3x_2 &= 4 \end{aligned}$

20. $\begin{aligned} 18x_1 + 12x_2 &= 13 \\ 30x_1 + 24x_2 &= 23 \end{aligned}$

21. $\begin{aligned} 20x_1 + 8x_2 &= 11 \\ 12x_1 - 24x_2 &= 21 \end{aligned}$

22. $\begin{aligned} 13x_1 - 6x_2 &= 17 \\ 26x_1 - 12x_2 &= 8 \end{aligned}$

23. $\begin{aligned} -0.4x_1 + 0.8x_2 &= 1.6 \\ 2x_1 - 4x_2 &= 5.0 \end{aligned}$

24. $\begin{aligned} -0.4x_1 + 0.8x_2 &= 1.6 \\ 0.2x_1 + 0.3x_2 &= 0.6 \end{aligned}$

25. $\begin{aligned} 3x_1 + 6x_2 &= 5 \\ 6x_1 + 12x_2 &= 10 \end{aligned}$

26. $\begin{aligned} 3x_1 + 2x_2 &= 1 \\ 2x_1 + 10x_2 &= 6 \end{aligned}$

27. $\begin{aligned} 4x_1 - x_2 - x_3 &= 1 \\ 2x_1 + 2x_2 + 3x_3 &= 10 \\ 5x_1 - 2x_2 - 2x_3 &= -1 \end{aligned}$

28. $\begin{aligned} 4x_1 - 2x_2 + 3x_3 &= -2 \\ 2x_1 + 2x_2 + 5x_3 &= 16 \\ 8x_1 - 5x_2 - 2x_3 &= 4 \end{aligned}$

29. $\begin{aligned} 3x_1 + 4x_2 + 4x_3 &= 11 \\ 4x_1 - 4x_2 + 6x_3 &= 11 \\ 6x_1 - 6x_2 &= 3 \end{aligned}$

30. $\begin{aligned} 14x_1 - 21x_2 - 7x_3 &= -21 \\ -4x_1 + 2x_2 - 2x_3 &= 2 \\ 56x_1 - 21x_2 + 7x_3 &= 7 \end{aligned}$

31. $\begin{aligned} 3x_1 + 3x_2 + 5x_3 &= 1 \\ 3x_1 + 5x_2 + 9x_3 &= 2 \\ 5x_1 + 9x_2 + 17x_3 &= 4 \end{aligned}$

32. $\begin{aligned} 2x_1 + 3x_2 + 5x_3 &= 4 \\ 3x_1 + 5x_2 + 9x_3 &= 7 \\ 5x_1 + 9x_2 + 17x_3 &= 13 \end{aligned}$

In Exercises 33–42, use a graphing utility or a computer software program with matrix capabilities and Cramer's Rule to solve for x_1, if possible.

33. $\begin{aligned} -0.4x_1 + 0.8x_2 &= 1.6 \\ 2x_1 - 4x_2 &= 5 \end{aligned}$

34. $\begin{aligned} 0.2x_1 - 0.6x_2 &= 2.4 \\ -x_1 + 1.4x_2 &= -8.8 \end{aligned}$

35. $\begin{aligned} -\tfrac{1}{4}x_1 + \tfrac{3}{8}x_2 &= -2 \\ \tfrac{3}{2}x_1 + \tfrac{3}{4}x_2 &= -12 \end{aligned}$

36. $\begin{aligned} \tfrac{5}{6}x_1 - x_2 &= -20 \\ \tfrac{4}{3}x_1 - \tfrac{7}{2}x_2 &= -51 \end{aligned}$

37. $\begin{aligned} 4x_1 - x_2 + x_3 &= -5 \\ 2x_1 + 2x_2 + 3x_3 &= 10 \\ 5x_1 - 2x_2 + 6x_3 &= 1 \end{aligned}$

38. $\begin{aligned} 5x_1 - 3x_2 + 2x_3 &= 2 \\ 2x_1 + 2x_2 - 3x_3 &= 3 \\ x_1 - 7x_2 + 8x_3 &= -4 \end{aligned}$

39. $\begin{aligned} 3x_1 - 2x_2 + x_3 &= -29 \\ -4x_1 + x_2 - 3x_3 &= 37 \\ x_1 - 5x_2 + x_3 &= -24 \end{aligned}$

40. $\begin{aligned} -8x_1 + 7x_2 - 10x_3 &= -151 \\ 12x_1 + 3x_2 - 5x_3 &= 86 \\ 15x_1 - 9x_2 + 2x_3 &= 187 \end{aligned}$

41. $\begin{aligned} 3x_1 - 2x_2 + 9x_3 + 4x_4 &= 35 \\ -x_1 - 9x_3 - 6x_4 &= -17 \\ 3x_3 + x_4 &= 5 \\ 2x_1 + 2x_2 + 8x_4 &= -4 \end{aligned}$

42. $\begin{aligned} -x_1 - x_2 + x_4 &= -8 \\ 3x_1 + 5x_2 + 5x_3 &= 24 \\ 2x_3 + x_4 &= -6 \\ -2x_1 - 3x_2 - 3x_3 &= -15 \end{aligned}$

43. Use Cramer's Rule to solve the system of linear equations for x and y.
$$\begin{aligned} kx + (1 - k)y &= 1 \\ (1 - k)x + ky &= 3 \end{aligned}$$
For what value(s) of k will the system be inconsistent?

44. Verify the following system of linear equations in $\cos A$, $\cos B$, and $\cos C$ for the triangle shown in Figure 3.4.
$$\begin{aligned} c\cos B + b\cos C &= a \\ c\cos A + a\cos C &= b \\ b\cos A + a\cos B &= c \end{aligned}$$
Then use Cramer's Rule to solve for $\cos C$, and use the result to verify the Law of Cosines, $c^2 = a^2 + b^2 - 2ab\cos C$.

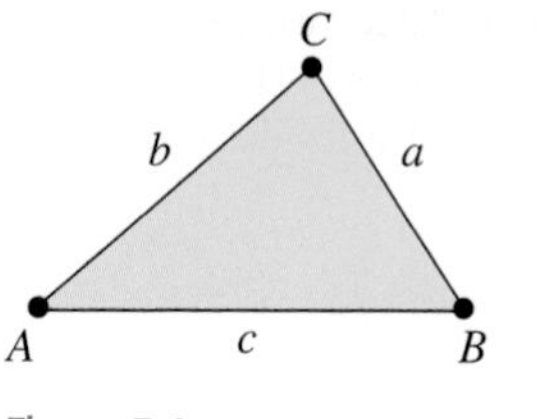

Figure 3.4

Area, Volume, and Equations of Lines and Planes

In Exercises 45–48, find the area of the triangle having the given vertices.

45. $(0, 0), (2, 0), (0, 3)$

46. $(1, 1), (2, 4), (4, 2)$

47. $(-1, 2), (2, 2), (-2, 4)$

48. $(1, 1), (-1, 1), (0, -2)$

In Exercises 49–52, determine whether the points are collinear.

49. $(1, 2), (3, 4), (5, 6)$

50. $(-1, 0), (1, 1), (3, 3)$

51. $(-2, 5), (0, -1), (3, -9)$

52. $(-1, -3), (-4, 7), (2, -13)$

In Exercises 53–56, find an equation of the line passing through the given points.

53. $(0, 0), (3, 4)$ **54.** $(-4, 7), (2, 4)$

55. $(-2, 3), (-2, -4)$ **56.** $(1, 4), (3, 4)$

In Exercises 57–60, find the volume of the tetrahedron having the given vertices.

57. $(1, 0, 0), (0, 1, 0), (0, 0, 1), (1, 1, 1)$

58. $(1, 1, 1), (0, 0, 0), (2, 1, -1), (-1, 1, 2)$

59. $(3, -1, 1), (4, -4, 4), (1, 1, 1), (0, 0, 1)$

60. $(0, 0, 0), (0, 2, 0), (3, 0, 0), (1, 1, 4)$

In Exercises 61–64, determine whether the points are coplanar.

61. $(-4, 1, 0), (0, 1, 2), (4, 3, -1), (0, 0, 1)$

62. $(1, 2, 3), (-1, 0, 1), (0, -2, -5), (2, 6, 11)$

63. $(0, 0, -1), (0, -1, 0), (1, 1, 0), (2, 1, 2)$

64. $(1, 2, 7), (-3, 6, 6), (4, 4, 2), (3, 3, 4)$

In Exercises 65–68, find an equation of the plane passing through the three points.

65. $(1, -2, 1), (-1, -1, 7), (2, -1, 3)$

66. $(0, -1, 0), (1, 1, 0), (2, 1, 2)$

67. $(0, 0, 0), (1, -1, 0), (0, 1, -1)$

68. $(1, 2, 7), (4, 4, 2), (3, 3, 4)$

In Exercises 69–71, Cramer's Rule has been used to solve for one of the variables in a system of equations. Determine whether Cramer's Rule was used correctly to solve for the variable. If not, identify the mistake.

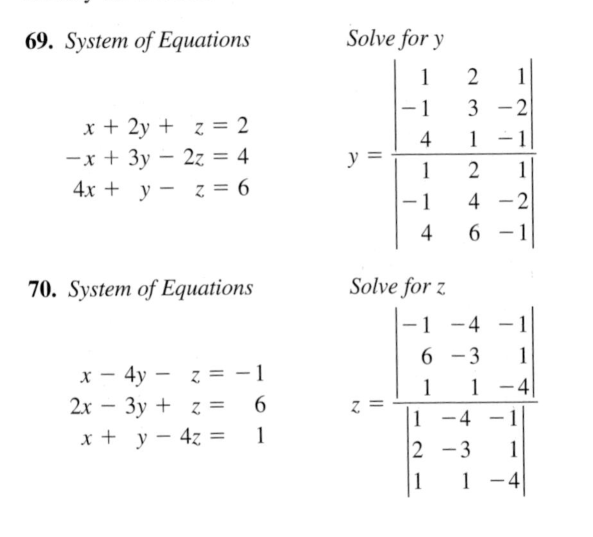

69. *System of Equations*

$$\begin{aligned} x + 2y + z &= 2 \\ -x + 3y - 2z &= 4 \\ 4x + y - z &= 6 \end{aligned}$$

Solve for y

$$y = \frac{\begin{vmatrix} 1 & 2 & 1 \\ -1 & 3 & -2 \\ 4 & 1 & -1 \end{vmatrix}}{\begin{vmatrix} 1 & 2 & 1 \\ -1 & 4 & -2 \\ 4 & 6 & -1 \end{vmatrix}}$$

70. *System of Equations*

$$\begin{aligned} x - 4y - z &= -1 \\ 2x - 3y + z &= 6 \\ x + y - 4z &= 1 \end{aligned}$$

Solve for z

$$z = \frac{\begin{vmatrix} -1 & -4 & -1 \\ 6 & -3 & 1 \\ 1 & 1 & -4 \end{vmatrix}}{\begin{vmatrix} 1 & -4 & -1 \\ 2 & -3 & 1 \\ 1 & 1 & -4 \end{vmatrix}}$$

71. *System of Equations*

$$\begin{aligned} 5x - 2y + z &= 15 \\ 3x - 3y - z &= -7 \\ 2x - y - 7z &= -3 \end{aligned}$$

Solve for x

$$x = \frac{\begin{vmatrix} 15 & -2 & 1 \\ -7 & -3 & -1 \\ -3 & -1 & -7 \end{vmatrix}}{\begin{vmatrix} 5 & -2 & 1 \\ 3 & -3 & -1 \\ 2 & -1 & -7 \end{vmatrix}}$$

72. The table below shows the numbers of subscribers y (in millions) of a cellular communications company in the United States for the years 2003 to 2005. (Source: U.S. Census Bureau)

Year	*Subscribers*
2003	158.7
2004	182.1
2005	207.9

(a) Create a system of linear equations for the data to fit the curve $y = at^2 + bt + c$, where t is the year and $t = 3$ corresponds to 2003, and y is the number of subscribers.
(b) Use Cramer's Rule to solve your system.
(c) Use a graphing utility to plot the data and graph your regression polynomial function.
(d) Briefly describe how well the polynomial function fits the data.

73. The table below shows the projected values (in millions of dollars) of hardback college textbooks sold in the United States for the years 2007 to 2009. (Source: U.S. Census Bureau)

Year	*Value*
2007	4380
2008	4439
2009	4524

(a) Create a system of linear equations for the data to fit the curve $y = at^2 + bt + c$, where t is the year and $t = 7$ corresponds to 2007, and y is the value of the textbooks.
(b) Use Cramer's Rule to solve your system.
(c) Use a graphing utility to plot the data and graph your regression polynomial function.
(d) Briefly describe how well the polynomial function fits the data.

CHAPTER 3 Review Exercises

In Exercises 1–18, find the determinant of the matrix.

1. $\begin{bmatrix} 4 & -1 \\ 2 & 2 \end{bmatrix}$

2. $\begin{bmatrix} 0 & -3 \\ 1 & 2 \end{bmatrix}$

3. $\begin{bmatrix} -3 & 1 \\ 6 & -2 \end{bmatrix}$

4. $\begin{bmatrix} -2 & 0 \\ 0 & 3 \end{bmatrix}$

5. $\begin{bmatrix} 1 & 4 & -2 \\ 0 & -3 & 1 \\ 1 & 1 & -1 \end{bmatrix}$

6. $\begin{bmatrix} 5 & 0 & 2 \\ 0 & -1 & 3 \\ 0 & 0 & 1 \end{bmatrix}$

7. $\begin{bmatrix} -2 & 0 & 0 \\ 0 & -3 & 0 \\ 0 & 0 & -1 \end{bmatrix}$

8. $\begin{bmatrix} -15 & 0 & 4 \\ 3 & 0 & -5 \\ 12 & 0 & 6 \end{bmatrix}$

9. $\begin{bmatrix} -3 & 6 & 9 \\ 9 & 12 & -3 \\ 0 & 15 & -6 \end{bmatrix}$

10. $\begin{bmatrix} -15 & 0 & 3 \\ 3 & 9 & -6 \\ 12 & -3 & 6 \end{bmatrix}$

11. $\begin{bmatrix} 2 & 0 & -1 & 4 \\ -1 & 2 & 0 & 3 \\ 3 & 0 & 1 & 2 \\ -2 & 0 & 3 & 1 \end{bmatrix}$

12. $\begin{bmatrix} 2 & 0 & 0 & 0 \\ -3 & 1 & 0 & 0 \\ 4 & -1 & 3 & 0 \\ 5 & 2 & 1 & -1 \end{bmatrix}$

13. $\begin{bmatrix} -4 & 1 & 2 & 3 \\ 1 & -2 & 1 & 2 \\ 2 & -1 & 3 & 4 \\ 1 & 2 & 2 & -1 \end{bmatrix}$

14. $\begin{bmatrix} 3 & -1 & 2 & 1 \\ -2 & 0 & 1 & -3 \\ -1 & 2 & -3 & 4 \\ -2 & 1 & -2 & 1 \end{bmatrix}$

15. $\begin{bmatrix} -1 & 1 & -1 & 0 & 0 \\ 0 & 1 & -1 & 0 & 1 \\ 1 & 0 & 1 & -1 & 0 \\ 0 & -1 & 0 & 1 & -1 \\ 0 & 1 & 1 & -1 & 1 \end{bmatrix}$

16. $\begin{bmatrix} 1 & 2 & -1 & 3 & 4 \\ 2 & 3 & -1 & 2 & -2 \\ 1 & 2 & 0 & 1 & -1 \\ 1 & 0 & 2 & -1 & 0 \\ 0 & -1 & 1 & 0 & 2 \end{bmatrix}$

17. $\begin{bmatrix} -1 & 0 & 0 & 0 & 0 \\ 0 & -1 & 0 & 0 & 0 \\ 0 & 0 & -1 & 0 & 0 \\ 0 & 0 & 0 & -1 & 0 \\ 0 & 0 & 0 & 0 & -1 \end{bmatrix}$

18. $\begin{bmatrix} 0 & 0 & 0 & 0 & 2 \\ 0 & 0 & 0 & 2 & 0 \\ 0 & 0 & 2 & 0 & 0 \\ 0 & 2 & 0 & 0 & 0 \\ 2 & 0 & 0 & 0 & 0 \end{bmatrix}$

In Exercises 19–22, determine which property of determinants is illustrated by the equation.

19. $\begin{vmatrix} 2 & -1 \\ 6 & -3 \end{vmatrix} = 0$

20. $\begin{vmatrix} 1 & 2 & -1 \\ 2 & 0 & 3 \\ 4 & -1 & 1 \end{vmatrix} = -\begin{vmatrix} 1 & -1 & 2 \\ 2 & 3 & 0 \\ 4 & 1 & -1 \end{vmatrix}$

21. $\begin{vmatrix} 2 & -4 & 3 & 2 \\ 0 & 4 & 6 & 1 \\ 1 & 8 & 9 & 0 \\ 6 & 12 & -6 & 1 \end{vmatrix} = -12\begin{vmatrix} 2 & 1 & 1 & 2 \\ 0 & -1 & 2 & 1 \\ 1 & -2 & 3 & 0 \\ 6 & -3 & -2 & 1 \end{vmatrix}$

22. $\begin{vmatrix} 1 & 3 & 1 \\ 0 & -1 & 2 \\ 1 & 2 & 1 \end{vmatrix} = \begin{vmatrix} 1 & 3 & 1 \\ 2 & 5 & 4 \\ 1 & 2 & 1 \end{vmatrix}$

In Exercises 23 and 24, find (a) $|A|$, (b) $|B|$, (c) AB, and (d) $|AB|$. Then verify that $|A||B| = |AB|$.

23. $A = \begin{bmatrix} -1 & 2 \\ 0 & 1 \end{bmatrix}$, $B = \begin{bmatrix} 3 & 4 \\ 2 & 1 \end{bmatrix}$

24. $A = \begin{bmatrix} 1 & 2 & 3 \\ 4 & 5 & 6 \\ 7 & 8 & 0 \end{bmatrix}$, $B = \begin{bmatrix} 1 & 2 & 1 \\ 0 & -1 & 1 \\ 0 & 2 & 3 \end{bmatrix}$

In Exercises 25 and 26, find (a) $|A^T|$, (b) $|A^3|$, (c) $|A^TA|$, and (d) $|5A|$.

25. $A = \begin{bmatrix} -2 & 6 \\ 1 & 3 \end{bmatrix}$

26. $A = \begin{bmatrix} 3 & 0 & 1 \\ -1 & 0 & 0 \\ 2 & 1 & 2 \end{bmatrix}$

In Exercises 27 and 28, find (a) $|A|$ and (b) $|A^{-1}|$.

27. $A = \begin{bmatrix} 1 & 0 & -4 \\ 0 & 3 & 2 \\ -2 & 7 & 6 \end{bmatrix}$

28. $A = \begin{bmatrix} 2 & -1 & 4 \\ 5 & 0 & 3 \\ 1 & -2 & 0 \end{bmatrix}$

In Exercises 29–32, find $|A^{-1}|$. Begin by finding A^{-1}, and then evaluate its determinant. Verify your result by finding $|A|$ and then applying the formula from Theorem 3.8, $|A^{-1}| = \dfrac{1}{|A|}$.

29. $\begin{bmatrix} 1 & -1 \\ 2 & 4 \end{bmatrix}$

30. $\begin{bmatrix} 10 & 2 \\ -2 & 7 \end{bmatrix}$

31. $\begin{bmatrix} 1 & 0 & 1 \\ 2 & -1 & 4 \\ 2 & 6 & 0 \end{bmatrix}$

32. $\begin{bmatrix} -1 & 1 & 2 \\ 2 & 4 & 8 \\ 1 & -1 & 0 \end{bmatrix}$

In Exercises 33–36, solve the system of linear equations by each of the methods shown.

(a) Gaussian elimination with back-substitution
(b) Gauss-Jordan elimination
(c) Cramer's Rule

33. $\begin{aligned} 3x_1 + 3x_2 + 5x_3 &= 1 \\ 3x_1 + 5x_2 + 9x_3 &= 2 \\ 5x_1 + 9x_2 + 17x_3 &= 4 \end{aligned}$

34. $\begin{aligned} 2x_1 + x_2 + 2x_3 &= 6 \\ -x_1 + 2x_2 - 3x_3 &= 0 \\ 3x_1 + 2x_2 - x_3 &= 6 \end{aligned}$

35. $\begin{aligned} x_1 + 2x_2 - x_3 &= -7 \\ 2x_1 - 2x_2 - 2x_3 &= -8 \\ -x_1 + 3x_2 + 4x_3 &= 8 \end{aligned}$

36. $\begin{aligned} 2x_1 + 3x_2 + 5x_3 &= 4 \\ 3x_1 + 5x_2 + 9x_3 &= 7 \\ 5x_1 + 9x_2 + 13x_3 &= 17 \end{aligned}$

In Exercises 37–42, use the determinant of the coefficient matrix to determine whether the system of linear equations has a unique solution.

37. $\begin{aligned} 5x + 4y &= 2 \\ -x + y &= -22 \end{aligned}$

38. $\begin{aligned} 2x - 5y &= 2 \\ 3x - 7y &= 1 \end{aligned}$

39. $\begin{aligned} -x + y + 2z &= 1 \\ 2x + 3y + z &= -2 \\ 5x + 4y + 2z &= 4 \end{aligned}$

40. $\begin{aligned} 2x + 3y + z &= 10 \\ 2x - 3y - 3z &= 22 \\ 8x + 6y &= -2 \end{aligned}$

41. $\begin{aligned} x_1 + 2x_2 + 6x_3 &= 1 \\ 2x_1 + 5x_2 + 15x_3 &= 4 \\ 3x_1 + x_2 + 3x_3 &= -6 \end{aligned}$

42. $\begin{aligned} x_1 + 5x_2 + 3x_3 &= 14 \\ 4x_1 + 2x_2 + 5x_3 &= 3 \\ 3x_3 + 8x_4 + 6x_5 &= 16 \\ 2x_1 + 4x_2 - 2x_5 &= 0 \\ 2x_1 - x_3 &= 0 \end{aligned}$

True or False? In Exercises 43 and 44, determine whether each statement is true or false. If a statement is true, give a reason or cite an appropriate statement from the text. If a statement is false, provide an example that shows the statement is not true in all cases or cite an appropriate statement from the text.

43. (a) The cofactor C_{22} of a given matrix is always a positive number.
(b) If a square matrix B is obtained from A by interchanging two rows, then $\det(B) = \det(A)$.
(c) If A is a square matrix of order n, then $\det(A) = -\det(A^T)$.

44. (a) If A and B are square matrices of order n such that $\det(AB) = -1$, then both A and B are nonsingular.
(b) If A is a 3×3 matrix with $\det(A) = 5$, then $\det(2A) = 10$.
(c) If A and B are square matrices of order n, then $\det(A + B) = \det(A) + \det(B)$.

45. If A is a 3×3 matrix such that $|A| = 2$, then what is the value of $|4A|$?

46. If A is a 4×4 matrix such that $|A| = -1$, then what is the value of $|2A|$?

47. Prove the property below.

$$\begin{vmatrix} a_{11} & a_{12} & a_{13} \\ a_{21} & a_{22} & a_{23} \\ a_{31} + c_{31} & a_{32} + c_{32} & a_{33} + c_{33} \end{vmatrix} = \begin{vmatrix} a_{11} & a_{12} & a_{13} \\ a_{21} & a_{22} & a_{23} \\ a_{31} & a_{32} & a_{33} \end{vmatrix} + \begin{vmatrix} a_{11} & a_{12} & a_{13} \\ a_{21} & a_{22} & a_{23} \\ c_{31} & c_{32} & c_{33} \end{vmatrix}$$

48. Illustrate the property shown in Exercise 47 for the following.

$$A = \begin{bmatrix} 1 & 0 & 2 \\ 1 & -1 & 2 \\ 2 & 1 & -1 \end{bmatrix}, \quad c_{31} = 3, \quad c_{32} = 0, \quad c_{33} = 1$$

49. Find the determinant of the $n \times n$ matrix.

$$\begin{bmatrix} 1-n & 1 & 1 & \cdots & 1 \\ 1 & 1-n & 1 & \cdots & 1 \\ \vdots & \vdots & \vdots & & \vdots \\ 1 & 1 & 1 & \cdots & 1-n \end{bmatrix}$$

50. Show that

$$\begin{vmatrix} a & 1 & 1 & 1 \\ 1 & a & 1 & 1 \\ 1 & 1 & a & 1 \\ 1 & 1 & 1 & a \end{vmatrix} = (a + 3)(a - 1)^3.$$

In Exercises 51–54, find the eigenvalues and corresponding eigenvectors of the matrix.

51. $\begin{bmatrix} -3 & 10 \\ 5 & 2 \end{bmatrix}$

52. $\begin{bmatrix} 5 & 2 \\ 4 & 1 \end{bmatrix}$

53. $\begin{bmatrix} 1 & 0 & 0 \\ -2 & 3 & 0 \\ 0 & 0 & 4 \end{bmatrix}$

54. $\begin{bmatrix} -3 & 0 & 4 \\ 2 & 1 & 1 \\ -1 & 0 & 1 \end{bmatrix}$

Calculus In Exercises 55–58, find the Jacobians of the functions. If x, y, and z are continuous functions of u, v, and w with continuous first partial derivatives, the **Jacobians** $J(u, v)$ and $J(u, v, w)$ are defined as

$$J(u, v) = \begin{vmatrix} \dfrac{\partial x}{\partial u} & \dfrac{\partial x}{\partial v} \\ \dfrac{\partial y}{\partial u} & \dfrac{\partial y}{\partial v} \end{vmatrix} \quad \text{and} \quad J(u, v, w) = \begin{vmatrix} \dfrac{\partial x}{\partial u} & \dfrac{\partial x}{\partial v} & \dfrac{\partial x}{\partial w} \\ \dfrac{\partial y}{\partial u} & \dfrac{\partial y}{\partial v} & \dfrac{\partial y}{\partial w} \\ \dfrac{\partial z}{\partial u} & \dfrac{\partial z}{\partial v} & \dfrac{\partial z}{\partial w} \end{vmatrix}.$$

55. $x = \frac{1}{2}(v - u), \quad y = \frac{1}{2}(v + u)$

56. $x = au + bv, \quad y = cu + dv$

57. $x = \frac{1}{2}(u + v), \quad y = \frac{1}{2}(u - v), \quad z = 2uvw$

58. $x = u - v + w, \quad y = 2uv, \quad z = u + v + w$

59. Writing Compare the various methods for calculating the determinant of a matrix. Which method requires the least amount of computation? Which method do you prefer if the matrix has very few zeros?

60. Prove that if $|A| = |B| \neq 0$ and A and B are of the same size, then there exists a matrix C such that $|C| = 1$ and $A = CB$.

The Adjoint of a Matrix

In Exercises 61 and 62, find the adjoint of the matrix.

61. $\begin{bmatrix} 0 & 1 \\ -2 & 1 \end{bmatrix}$

62. $\begin{bmatrix} 1 & -1 & 1 \\ 0 & 1 & 2 \\ 0 & 0 & -1 \end{bmatrix}$

Cramer's Rule

In Exercises 63–66, use the determinant of the coefficient matrix to determine whether the system of linear equations has a unique solution. If it does, use Cramer's Rule to find the solution.

63. $\begin{aligned} 0.2x - 0.1y &= 0.07 \\ 0.4x - 0.5y &= -0.01 \end{aligned}$

64. $\begin{aligned} 2x + y &= 0.3 \\ 3x - y &= -1.3 \end{aligned}$

65. $\begin{aligned} 2x_1 + 3x_2 + 3x_3 &= 3 \\ 6x_1 + 6x_2 + 12x_3 &= 13 \\ 12x_1 + 9x_2 - x_3 &= 2 \end{aligned}$

66. $\begin{aligned} 4x_1 + 4x_2 + 4x_3 &= 5 \\ 4x_1 - 2x_2 - 8x_3 &= 1 \\ 8x_1 + 2x_2 - 4x_3 &= 6 \end{aligned}$

67. The table shows the projected populations (in millions) of the United States for the years 2010, 2020, and 2030. (Source: U.S. Census Bureau)

Year	*Population*
2010	308.9
2020	335.8
2030	363.6

(a) Create a system of linear equations for the data to fit the curve $y = at^2 + bt + c$, where t is the year and $t = 10$ corresponds to 2010, and y is the population.

(b) Use Cramer's Rule to solve your system.

(c) Use a graphing utility to plot the data and graph your regression polynomial function.

(d) Briefly describe how well the polynomial function fits the data.

68. The table shows the projected amounts (in dollars) spent per person per year on basic cable and satellite television in the United States for the years 2007 through 2009. (Source: U.S. Census Bureau)

Year	*Amount*
2007	296
2008	308
2009	321

(a) Create a system of linear equations for the data to fit the curve $y = at^2 + bt + c$, where t is the year and $t = 7$ corresponds to 2007, and y is the number of subscribers.

(b) Use Cramer's Rule to solve your system.

(c) Use a graphing utility to plot the data and graph your regression polynomial function.

(d) Briefly describe how well the polynomial function fits the data.

Area, Volume, and Equations of Lines and Planes

In Exercises 69 and 70, use a determinant to find the area of the triangle with the given vertices.

69. $(1, 0), (5, 0), (5, 8)$

70. $(-4, 0), (4, 0), (0, 6)$

In Exercises 71 and 72, use the determinant to find an equation of the line passing through the given points.

71. $(-4, 0), (4, 4)$

72. $(2, 5), (6, -1)$

In Exercises 73 and 74, find an equation of the plane passing through the given points.

73. $(0, 0, 0), (1, 0, 3), (0, 3, 4)$

74. $(0, 0, 0), (2, -1, 1), (-3, 2, 5)$

True or False? In Exercises 75 and 76, determine whether each statement is true or false. If a statement is true, give a reason or cite an appropriate statement from the text. If a statement is false, provide an example that shows the statement is not true in all cases or cite an appropriate statement from the text.

75. (a) In Cramer's Rule, the value of x_i is the quotient of two determinants, where the numerator is the determinant of the coefficient matrix.

(b) Three points (x_1, y_1), (x_2, y_2), and (x_3, y_3) are collinear if the determinant of the matrix that has the coordinates as entries in the first two columns and 1's as entries in the third column is nonzero.

76. (a) If A is a square matrix, then the matrix of cofactors of A is called the adjoint of A.

(b) In Cramer's Rule, the denominator is the determinant of the matrix formed by replacing the column corresponding to the variable being solved for with the column representing the constants.

CHAPTER 3 Projects

1 Eigenvalues and Stochastic Matrices

In Section 2.5, you studied a consumer preference model for competing cable television companies. The matrix representing the transition probabilities was

$$P = \begin{bmatrix} 0.70 & 0.15 & 0.15 \\ 0.20 & 0.80 & 0.15 \\ 0.10 & 0.05 & 0.70 \end{bmatrix}.$$

When provided with the initial state matrix X, you observed that the number of subscribers after 1 year is the product PX.

$$X = \begin{bmatrix} 15{,}000 \\ 20{,}000 \\ 65{,}000 \end{bmatrix} \longrightarrow PX = \begin{bmatrix} 0.70 & 0.15 & 0.15 \\ 0.20 & 0.80 & 0.15 \\ 0.10 & 0.05 & 0.70 \end{bmatrix} \begin{bmatrix} 15{,}000 \\ 20{,}000 \\ 65{,}000 \end{bmatrix} = \begin{bmatrix} 23{,}250 \\ 28{,}750 \\ 48{,}000 \end{bmatrix}$$

After 10 years, the number of subscribers had nearly reached a **steady state.**

$$X = \begin{bmatrix} 15{,}000 \\ 20{,}000 \\ 65{,}000 \end{bmatrix} \longrightarrow P^{10}X = \begin{bmatrix} 33{,}287 \\ 47{,}147 \\ 19{,}566 \end{bmatrix}$$

That is, for large values of n, the product P^nX approaches a limit $\overline{X}$, $P\overline{X} = \overline{X}$.

From your knowledge of eigenvalues, this means that 1 is an eigenvalue of P with corresponding eigenvector $\overline{X}$.

1. Use a computer or calculator to show that the eigenvalues and eigenvectors of P are as follows.

Eigenvalues: $\lambda_1 = 1, \lambda_2 = 0.65, \lambda_3 = 0.55$

Eigenvectors: $\mathbf{x}_1 = \begin{bmatrix} 7 \\ 10 \\ 4 \end{bmatrix}, \mathbf{x}_2 = \begin{bmatrix} 0 \\ -1 \\ 1 \end{bmatrix}, \mathbf{x}_3 = \begin{bmatrix} -2 \\ 1 \\ 1 \end{bmatrix}$

2. Let S be the matrix whose columns are the eigenvectors of P. Show that $S^{-1}PS$ is a diagonal matrix D. What are the entries along the diagonal of D?
3. Show that $P^n = (SDS^{-1})^n = SD^nS^{-1}$. Use this result to calculate $P^{10}X$ and verify the result from Section 2.5.

2 The Cayley-Hamilton Theorem

The **characteristic polynomial** of a square matrix A is given by the determinant $|\lambda I - A|$. If the order of A is n, then the characteristic polynomial $p(\lambda)$ is an nth-degree polynomial in the variable λ.

$$p(\lambda) = \det(\lambda I - A) = \lambda^n + c_{n-1}\lambda^{n-1} + \cdots + c_2\lambda^2 + c_1\lambda + c_0$$

The Cayley-Hamilton Theorem asserts that every square matrix satisfies its characteristic polynomial. That is, for the $n \times n$ matrix A,

$$p(A) = A^n + c_{n-1}A^{n-1} + \cdots + c_2A^2 + c_1A + c_0I = O.$$

Note that this is a matrix equation. The zero on the right is the $n \times n$ zero matrix, and the coefficient c_0 has been multiplied by the $n \times n$ identity matrix I.

1. Verify the Cayley-Hamilton Theorem for the matrix

$$\begin{bmatrix} 2 & -2 \\ -2 & -1 \end{bmatrix}.$$

2. Verify the Cayley-Hamilton Theorem for the matrix

$$\begin{bmatrix} 6 & 0 & 4 \\ -2 & 1 & 3 \\ 2 & 0 & 4 \end{bmatrix}.$$

3. Prove the Cayley-Hamilton Theorem for an arbitrary 2×2 matrix A,

$$A = \begin{bmatrix} a & b \\ c & d \end{bmatrix}.$$

4. If A is nonsingular and $p(A) = A^n + c_{n-1}A^{n-1} + \cdots + c_1A + c_0I = O$, show that

$$A^{-1} = \frac{1}{c_0}(-A^{n-1} - c_{n-1}A^{n-2} - \cdots - c_2A - c_1I).$$

Use this result to find the inverse of the matrix

$$A = \begin{bmatrix} 1 & 2 \\ 3 & 5 \end{bmatrix}.$$

5. The Cayley-Hamilton Theorem can be used to calculate powers A^n of the square matrix A. For example, the characteristic polynomial of the matrix

$$A = \begin{bmatrix} 3 & -1 \\ 2 & -1 \end{bmatrix}$$

is $p(\lambda) = \lambda^2 - 2\lambda - 1$.

The Cayley-Hamilton Theorem implies that

$$A^2 - 2A - I = O \qquad \text{or} \qquad A^2 = 2A + I.$$

So, A^2 is shown in terms of lower powers of A.

$$A^2 = 2A + I = 2\begin{bmatrix} 3 & -1 \\ 2 & -1 \end{bmatrix} + \begin{bmatrix} 1 & 0 \\ 0 & 1 \end{bmatrix} = \begin{bmatrix} 7 & -2 \\ 4 & -1 \end{bmatrix}$$

Similarly, multiplying both sides of the equation $A^2 = 2A + I$ by A gives A^3 in terms of lower powers of A. Moreover, you can write A^3 in terms of just A and I after replacing A^2 with $2A + I$, as follows.

$$A^3 = 2A^2 + A = 2(2A + I) + A = 5A + 2I$$

(a) Use this method to find A^3 and A^4. (First write A^4 as a linear combination of A and I—that is, as a sum of scalar multiples of A and I.)

(b) Find A^5 for the matrix

$$\begin{bmatrix} 0 & 0 & 1 \\ 2 & 2 & -1 \\ 1 & 0 & 2 \end{bmatrix}.$$

(*Hint:* Find the characteristic polynomial of A, then use the Cayley-Hamilton Theorem to express A^3 as a linear combination of A^2, A, and I. Inductively express A^5 as a linear combination of A^2, A, and I.)

CHAPTERS 1–3 Cumulative Test

Take this test as you would take a test in class. When you are finished, check your work against the answers provided in the back of the book.

1. Solve the system of linear equations.
$$\begin{aligned} 4x_1 + x_2 - 3x_3 &= 11 \\ 2x_1 - 3x_2 + 2x_3 &= 9 \\ x_1 + x_2 + x_3 &= -3 \end{aligned}$$

2. Find the solution set of the system of linear equations represented by the augmented matrix.
$$\begin{bmatrix} 0 & 1 & -1 & 0 & 2 \\ 1 & 0 & 2 & -1 & 0 \\ 1 & 2 & 0 & -1 & 4 \end{bmatrix}$$

3. Solve the homogeneous linear system corresponding to the coefficient matrix below.
$$\begin{bmatrix} 1 & 2 & 1 & -2 \\ 0 & 0 & 2 & -4 \\ -2 & -4 & 1 & -2 \end{bmatrix}$$

4. Find conditions on k such that the system is consistent.
$$\begin{aligned} x + 2y - z &= 3 \\ -x - y + z &= 2 \\ -x + y + z &= k \end{aligned}$$

5. A manufacturer produces three different models of a product that are shipped to two different warehouses. The number of units of model i that are shipped to warehouse j is represented by a_{ij} in the matrix
$$A = \begin{bmatrix} 200 & 300 \\ 600 & 350 \\ 250 & 400 \end{bmatrix}.$$
The prices of the three models in dollars per unit are represented by the matrix $B = [12.50 \quad 9.00 \quad 21.50]$.

   Find the product BA and state what the entries of the product represent.

6. Solve for x and y in the matrix equation $2A - B = I$ if
$$A = \begin{bmatrix} -1 & 1 \\ 2 & 3 \end{bmatrix} \quad \text{and} \quad B = \begin{bmatrix} x & 2 \\ y & 5 \end{bmatrix}.$$

7. Find A^TA for the matrix
$$A = \begin{bmatrix} 1 & 2 & 3 \\ 4 & 5 & 6 \end{bmatrix}.$$

8. Find the inverses (if they exist) of the matrices.
$$\text{(a)} \begin{bmatrix} -2 & 3 \\ 4 & 6 \end{bmatrix} \qquad \text{(b)} \begin{bmatrix} -2 & 3 \\ 3 & 6 \end{bmatrix}$$

9. Find the inverse of the matrix
$$\begin{bmatrix} 1 & 1 & 0 \\ -3 & 6 & 5 \\ 0 & 1 & 0 \end{bmatrix}.$$

10. Factor the matrix

$$A = \begin{bmatrix} 2 & -4 \\ 1 & 0 \end{bmatrix}$$

into a product of elementary matrices.

11. Find the determinant of the matrix

$$\begin{bmatrix} 5 & 1 & 2 & 4 \\ 1 & 0 & -2 & -3 \\ 1 & 1 & 6 & 1 \\ 1 & 0 & 0 & -4 \end{bmatrix}.$$

12. Find each determinant if

$$A = \begin{bmatrix} 1 & -3 \\ 4 & 2 \end{bmatrix} \quad \text{and} \quad B = \begin{bmatrix} -2 & 1 \\ 0 & 5 \end{bmatrix}.$$

(a) $|A|$ (b) $|B|$ (c) $|AB|$ (d) $|A^{-1}|$

13. If $|A| = 7$ and A is of order 4, then find each determinant.

(a) $|3A|$ (b) $|A^T|$ (c) $|A^{-1}|$ (d) $|A^3|$

14. Use the adjoint of

$$A = \begin{bmatrix} 1 & -5 & -1 \\ 0 & -2 & 1 \\ 1 & 0 & 2 \end{bmatrix}$$

to find A^{-1}.

15. Let $\mathbf{x}_1$, $\mathbf{x}_2$, $\mathbf{x}_3$, and $\mathbf{b}$ be the column matrices below.

$$\mathbf{x}_1 = \begin{bmatrix} 1 \\ 0 \\ 1 \end{bmatrix} \quad \mathbf{x}_2 = \begin{bmatrix} 1 \\ 1 \\ 0 \end{bmatrix} \quad \mathbf{x}_3 = \begin{bmatrix} 0 \\ 1 \\ 1 \end{bmatrix} \quad \mathbf{b} = \begin{bmatrix} 1 \\ 2 \\ 3 \end{bmatrix}$$

Find constants a, b, and c such that $a\mathbf{x}_1 + b\mathbf{x}_2 + c\mathbf{x}_3 = \mathbf{b}$.

16. Use linear equations to find the parabola $y = ax^2 + bx + c$ that passes through the points $(-1, 2)$, $(0, 1)$, and $(2, 6)$. Sketch the points and the parabola.

17. Use a determinant to find an equation of the line passing through the points $(1, 4)$ and $(5, -2)$.

18. Use a determinant to find the area of the triangle with vertices $(3, 1)$, $(7, 1)$, and $(7, 9)$.

19. Find the eigenvalues and corresponding eigenvectors of the matrix below.

$$\begin{bmatrix} 1 & 4 & 6 \\ 1 & 2 & 2 \\ -1 & -2 & -4 \end{bmatrix}$$

20. Let A, B, and C be three nonzero $n \times n$ matrices such that $AC = BC$. Does it follow that $A = B$? Prove your answer.

21. For any matrix B, prove that the matrix B^TB is symmetric.

22. Prove that if the matrix A has an inverse, then the inverse is unique.

23. (a) Define row equivalence of matrices.

(b) Prove that if A is row-equivalent to B and B is row-equivalent to C, then A is row-equivalent to C.

Vector Spaces

CHAPTER OBJECTIVES

- Perform, recognize, and utilize vector operations on vectors in R^n.
- Determine whether a set of vectors with two operations is a vector space and recognize standard examples of vector spaces, such as: R^n, $M_{m,n}$, P_n, P, $C(-\infty, \infty)$, $C[a, b]$.
- Determine whether a subset W of a vector space V is a subspace.
- Write a linear combination of a finite set of vectors in V.
- Determine whether a set S of vectors in a vector space V is a spanning set of V.
- Determine whether a finite set of vectors in a vector space V is linearly independent.
- Recognize standard bases in the vector spaces R^n, $M_{m,n}$, and P_n.
- Determine if a vector space is finite dimensional or infinite dimensional.
- Find the dimension of a subspace of R^n, $M_{m,n}$, and P_n.
- Find a basis and dimension for the column or row space and a basis for the nullspace (nullity) of a matrix.
- Find a general solution of a consistent system $A\mathbf{x} = \mathbf{b}$ in the form $\mathbf{x}_p + \mathbf{x}_h$.
- Find $[\mathbf{x}]_B$ in R^n, $M_{m,n}$, and P_n.
- Find the transition matrix from the basis B to the basis B' in R^n.
- Find $[\mathbf{x}]_{B'}$ for a vector $\mathbf{x}$ in R^n.
- Determine whether a function is a solution of a differential equation and find the general solution of a given differential equation.
- Find the Wronskian for a set of functions and test a set of solutions for linear independence.
- Identify and sketch the graph of a conic or degenerate conic section and perform a rotation of axes.

4.1 Vectors in R^n

In physics and engineering, a vector is characterized by two quantities (length and direction) and is represented by a directed line segment. In this chapter you will see that these are only two special types of vectors. Their geometric representations can help you understand the more general definition of a vector.

This section begins with a short review of vectors in the plane, which is the way vectors were developed historically.

Vectors in the Plane

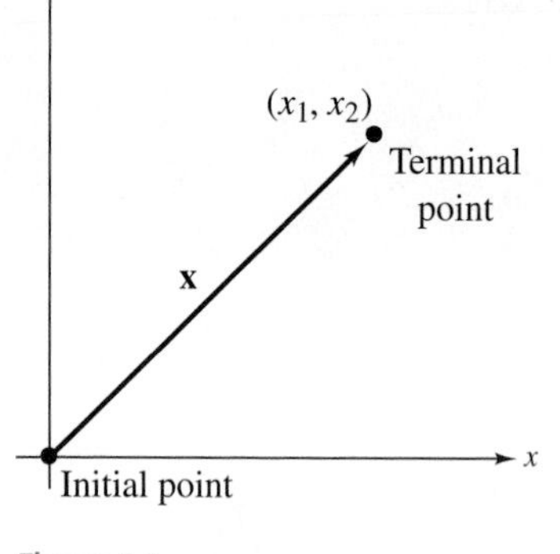

Figure 4.1

A **vector in the plane** is represented geometrically by a **directed line segment** whose **initial point** is the origin and whose **terminal point** is the point (x_1, x_2), as shown in Figure 4.1. This vector is represented by the same **ordered pair** used to represent its terminal point. That is,

$$\mathbf{x} = (x_1, x_2).$$

The coordinates x_1 and x_2 are called the **components** of the vector $\mathbf{x}$. Two vectors in the plane $\mathbf{u} = (u_1, u_2)$ and $\mathbf{v} = (v_1, v_2)$ are **equal** if and only if $u_1 = v_1$ and $u_2 = v_2$.

REMARK: The term *vector* derives from the Latin word *vectus,* meaning "to carry." The idea is that if you were to carry something from the origin to the point (x_1, x_2), the trip could be represented by the directed line segment from $(0, 0)$ to (x_1, x_2). Vectors are represented by lowercase letters set in boldface type (such as $\mathbf{u}$, $\mathbf{v}$, $\mathbf{w}$, and $\mathbf{x}$).

EXAMPLE 1 Vectors in the Plane

Use a directed line segment to represent each vector in the plane.
(a) $\mathbf{u} = (2, 3)$ (b) $\mathbf{v} = (-1, 2)$

SOLUTION To represent each vector, draw a directed line segment from the origin to the indicated terminal point, as shown in Figure 4.2.

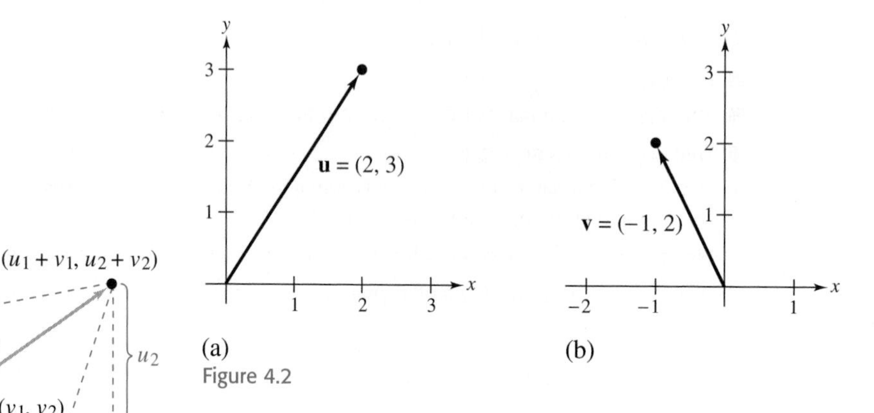

Figure 4.2

The first basic vector operation is **vector addition.** To add two vectors in the plane, add their corresponding components. That is, the **sum** of $\mathbf{u}$ and $\mathbf{v}$ is the vector

$$\mathbf{u} + \mathbf{v} = (u_1, u_2) + (v_1, v_2) = (u_1 + v_1, u_2 + v_2).$$

Geometrically, the sum of two vectors in the plane is represented as the diagonal of a parallelogram having $\mathbf{u}$ and $\mathbf{v}$ as its adjacent sides, as shown in Figure 4.3.

In the next example, one of the vectors you will add is the vector $(0, 0)$, called the **zero vector.** The zero vector is denoted by $\mathbf{0}$.

Vector Addition

Figure 4.3

EXAMPLE 2 Adding Two Vectors in the Plane

Simulation

Explore this concept further with an electronic simulation available on the website *college.hmco.com/pic/larsonELA6e.* Please visit this website for keystrokes and programming syntax for specific graphing utilities and computer software programs applicable to Example 2. Similar exercises and projects are also available on this website.

Find the sum of the vectors.

(a) $\mathbf{u} = (1, 4)$, $\mathbf{v} = (2, -2)$ (b) $\mathbf{u} = (3, -2)$, $\mathbf{v} = (-3, 2)$ (c) $\mathbf{u} = (2, 1)$, $\mathbf{v} = (0, 0)$

SOLUTION (a) $\mathbf{u} + \mathbf{v} = (1, 4) + (2, -2) = (3, 2)$

(b) $\mathbf{u} + \mathbf{v} = (3, -2) + (-3, 2) = (0, 0)$

(c) $\mathbf{u} + \mathbf{v} = (2, 1) + (0, 0) = (2, 1)$

Figure 4.4 gives the graphical representation of each sum.

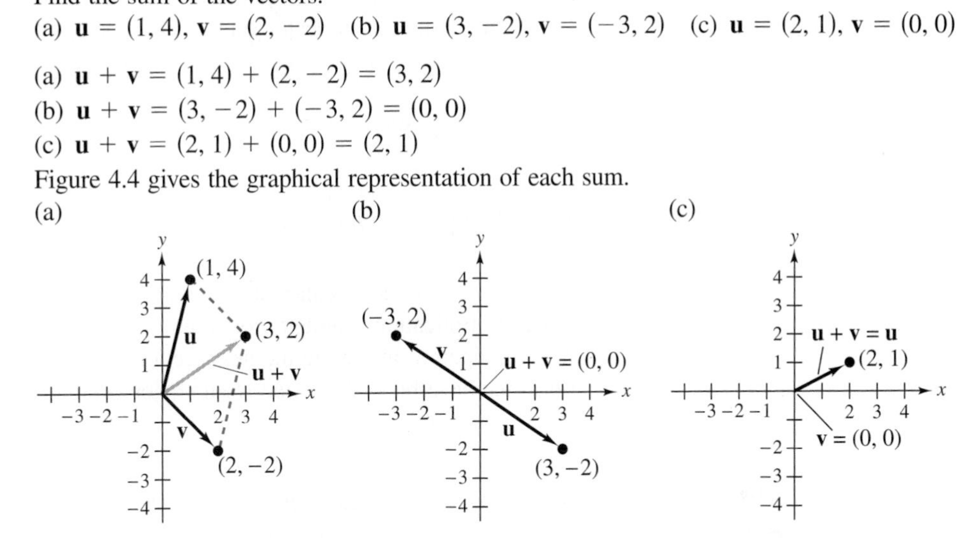

Figure 4.4

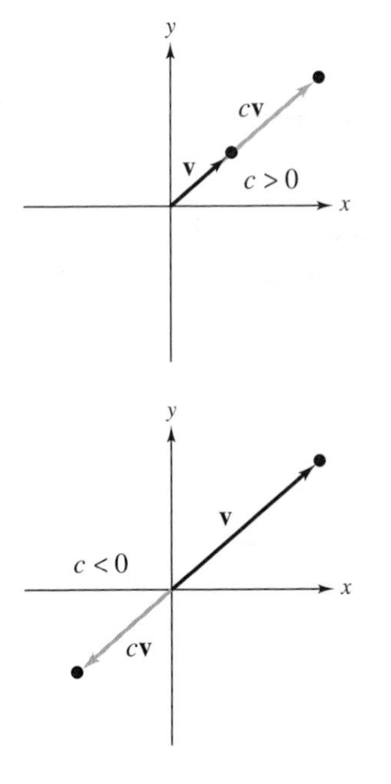

Figure 4.5

The second basic vector operation is called **scalar multiplication.** To multiply a vector $\mathbf{v}$ by a scalar c, multiply each of the components of $\mathbf{v}$ by c. That is,

$$c\mathbf{v} = c(v_1, v_2) = (cv_1, cv_2).$$

Recall from Chapter 2 that the word *scalar* is used to mean a real number. Historically, this usage arose from the fact that multiplying a vector by a real number changes the "scale" of the vector. For instance, if a vector $\mathbf{v}$ is multiplied by 2, the resulting vector $2\mathbf{v}$ is a vector having the same direction as $\mathbf{v}$ and twice the length. In general, for a scalar c, the vector $c\mathbf{v}$ will be $|c|$ times as long as $\mathbf{v}$. If c is positive, then $c\mathbf{v}$ and $\mathbf{v}$ have the same direction, and if c is negative, then $c\mathbf{v}$ and $\mathbf{v}$ have opposite directions. This is shown in Figure 4.5.

The product of a vector $\mathbf{v}$ and the scalar -1 is denoted by

$$-\mathbf{v} = (-1)\mathbf{v}.$$

The vector $-\mathbf{v}$ is called the **negative** of $\mathbf{v}$. The **difference** of $\mathbf{u}$ and $\mathbf{v}$ is defined as

$$\mathbf{u} - \mathbf{v} = \mathbf{u} + (-\mathbf{v}),$$

and you can say $\mathbf{v}$ is **subtracted** from $\mathbf{u}$.

EXAMPLE 3 Operations with Vectors in the Plane

Provided with $\mathbf{v} = (-2, 5)$ and $\mathbf{u} = (3, 4)$, find each vector.

(a) $\frac{1}{2}\mathbf{v}$ (b) $\mathbf{u} - \mathbf{v}$ (c) $\frac{1}{2}\mathbf{v} + \mathbf{u}$

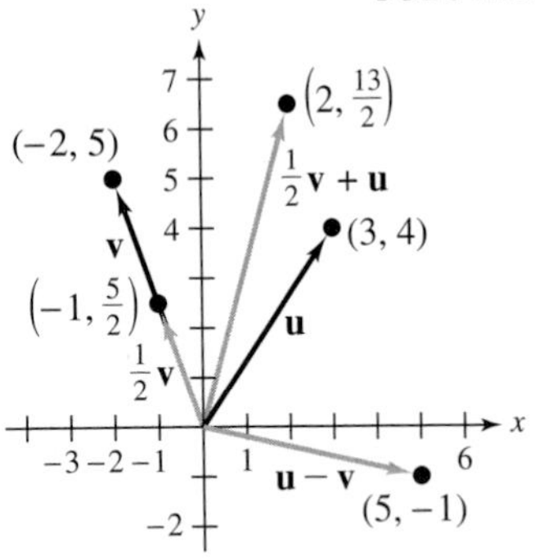

SOLUTION (a) Because $\mathbf{v} = (-2, 5)$, you have

$$\tfrac{1}{2}\mathbf{v} = \left(\tfrac{1}{2}(-2), \tfrac{1}{2}(5)\right) = \left(-1, \tfrac{5}{2}\right).$$

(b) By the definition of vector subtraction, you have

$$\mathbf{u} - \mathbf{v} = (3 - (-2), 4 - 5) = (5, -1).$$

(c) Using the result of part(a), you have

$$\tfrac{1}{2}\mathbf{v} + \mathbf{u} = \left(-1, \tfrac{5}{2}\right) + (3, 4) = \left(-1 + 3, \tfrac{5}{2} + 4\right) = \left(2, \tfrac{13}{2}\right).$$

Figure 4.6 gives a graphical representation of these vector operations.

Figure 4.6

Vector addition and scalar multiplication share many properties with matrix addition and scalar multiplication. The ten properties listed in the next theorem play a fundamental role in linear algebra. In fact, in the next section you will see that it is precisely these ten properties that have been abstracted from vectors in the plane to help define the general notion of a vector space.

THEOREM 4.1 Properties of Vector Addition and Scalar Multiplication in the Plane

Let $\mathbf{u}$, $\mathbf{v}$, and $\mathbf{w}$ be vectors in the plane, and let c and d be scalars.

1. $\mathbf{u} + \mathbf{v}$ is a vector in the plane. — **Closure under addition**
2. $\mathbf{u} + \mathbf{v} = \mathbf{v} + \mathbf{u}$ — **Commutative property of addition**
3. $(\mathbf{u} + \mathbf{v}) + \mathbf{w} = \mathbf{u} + (\mathbf{v} + \mathbf{w})$ — **Associative property of addition**
4. $\mathbf{u} + \mathbf{0} = \mathbf{u}$ — **Additive identity property**
5. $\mathbf{u} + (-\mathbf{u}) = \mathbf{0}$ — **Additive inverse property**
6. $c\mathbf{u}$ is a vector in the plane. — **Closure under scalar multiplication**
7. $c(\mathbf{u} + \mathbf{v}) = c\mathbf{u} + c\mathbf{v}$ — **Distributive property**
8. $(c + d)\mathbf{u} = c\mathbf{u} + d\mathbf{u}$ — **Distributive property**
9. $c(d\mathbf{u}) = (cd)\mathbf{u}$ — **Associative property of multiplication**
10. $1(\mathbf{u}) = \mathbf{u}$ — **Multiplicative identity property**

PROOF The proof of each property is a straightforward application of the definition of vector addition and scalar multiplication combined with the corresponding properties of addition and multiplication of real numbers. For instance, to prove the associative property of vector addition, you can write

$$\begin{aligned}
(\mathbf{u} + \mathbf{v}) + \mathbf{w} &= [(u_1, u_2) + (v_1, v_2)] + (w_1, w_2) \\
&= (u_1 + v_1, u_2 + v_2) + (w_1, w_2) \\
&= ((u_1 + v_1) + w_1, (u_2 + v_2) + w_2) \\
&= (u_1 + (v_1 + w_1), u_2 + (v_2 + w_2)) \\
&= (u_1, u_2) + (v_1 + w_1, v_2 + w_2) \\
&= (u_1, u_2) + [(v_1, v_2) + (w_1, w_2)] \\
&= \mathbf{u} + (\mathbf{v} + \mathbf{w}).
\end{aligned}$$

REMARK: Note that the associative property of vector addition allows you to write such expressions as $\mathbf{u} + \mathbf{v} + \mathbf{w}$ without ambiguity, because you obtain the same vector sum regardless of which addition is performed first.

Similarly, to prove the right distributive property of scalar multiplication over addition,

$$\begin{aligned}(c + d)\mathbf{u} &= (c + d)(u_1, u_2)\\ &= ((c + d)u_1, (c + d)u_2) = (cu_1 + du_1, cu_2 + du_2)\\ &= (cu_1, cu_2) + (du_1, du_2) = c(u_1, u_2) + d(u_1, u_2)\\ &= c\mathbf{u} + d\mathbf{u}.\end{aligned}$$

The proofs of the other eight properties are left as an exercise. (See Exercise 61.)

Vectors in R^n

HISTORICAL NOTE

William Rowan Hamilton (1805–1865)

is considered to be Ireland's most famous mathematician. His work led to the development of modern vector notation. We still use his **i**, **j**, and **k** notation today. To read about his work, visit *college.hmco.com/pic/larsonELA6e.*

The discussion of vectors in the plane can now be extended to a discussion of vectors in n-space. A vector in n-space is represented by an **ordered n-tuple.** For instance, an ordered triple has the form (x_1, x_2, x_3), an ordered quadruple has the form (x_1, x_2, x_3, x_4), and a general ordered n-tuple has the form $(x_1, x_2, x_3, \ldots, x_n)$. The set of all n-tuples is called **n-space** and is denoted by R^n.

R^1 = 1-space = set of all real numbers

R^2 = 2-space = set of all ordered pairs of real numbers

R^3 = 3-space = set of all ordered triples of real numbers

R^4 = 4-space = set of all ordered quadruples of real numbers

$\vdots$

R^n = n-space = set of all ordered n-tuples of real numbers

The practice of using an ordered pair to represent either a point or a vector in R^2 continues in R^n. That is, an n-tuple $(x_1, x_2, x_3, \ldots, x_n)$ can be viewed as a **point** in R^n with the x_i's as its coordinates or as a **vector**

$$\mathbf{x} = (x_1, x_2, x_3, \ldots, x_n) \qquad \text{Vector in } R^n$$

with the x_i's as its components. As with vectors in the plane, two vectors in R^n are **equal** if and only if corresponding components are equal. [In the case of $n = 2$ or $n = 3$, the familiar (x, y) or (x, y, z) notation is used occasionally.]

The sum of two vectors in R^n and the scalar multiple of a vector in R^n are called the **standard operations in R^n** and are defined as follows.

Definitions of Vector Addition and Scalar Multiplication in R^n

Let $\mathbf{u} = (u_1, u_2, u_3, \ldots, u_n)$ and $\mathbf{v} = (v_1, v_2, v_3, \ldots, v_n)$ be vectors in R^n and let c be a real number. Then the sum of $\mathbf{u}$ and $\mathbf{v}$ is defined as the vector

$$\mathbf{u} + \mathbf{v} = (u_1 + v_1, u_2 + v_2, u_3 + v_3, \ldots, u_n + v_n),$$

and the **scalar multiple** of $\mathbf{u}$ by c is defined as the vector

$$c\mathbf{u} = (cu_1, cu_2, cu_3, \ldots, cu_n).$$

As with 2-space, the **negative** of a vector in R^n is defined as

$$-\mathbf{u} = (-u_1, -u_2, -u_3, \ldots, -u_n)$$

and the **difference** of two vectors in R^n is defined as

$$\mathbf{u} - \mathbf{v} = (u_1 - v_1, u_2 - v_2, u_3 - v_3, \ldots, u_n - v_n).$$

The **zero vector** in R^n is denoted by $\mathbf{0} = (0, 0, \ldots, 0)$.

EXAMPLE 4 **Vector Operations in R^3**

Provided that $\mathbf{u} = (-1, 0, 1)$ and $\mathbf{v} = (2, -1, 5)$ in R^3, find each vector.

(a) $\mathbf{u} + \mathbf{v}$ (b) $2\mathbf{u}$ (c) $\mathbf{v} - 2\mathbf{u}$

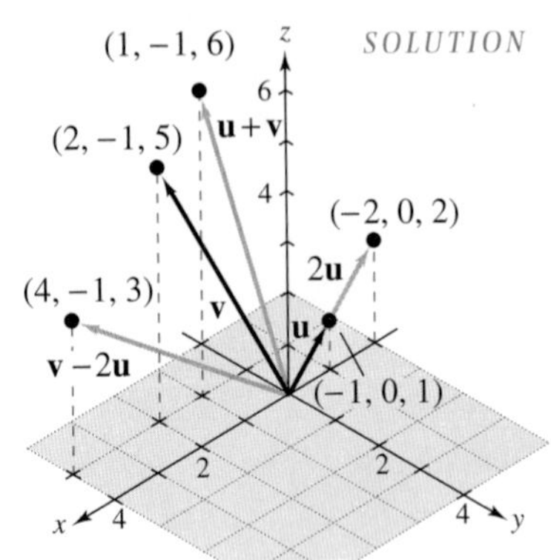

Figure 4.7

SOLUTION (a) To add two vectors, add their corresponding components, as follows.

$$\mathbf{u} + \mathbf{v} = (-1, 0, 1) + (2, -1, 5) = (1, -1, 6)$$

(b) To multiply a vector by a scalar, multiply each component by the scalar, as follows.

$$2\mathbf{u} = 2(-1, 0, 1) = (-2, 0, 2)$$

(c) Using the result of part (b), you have

$$\mathbf{v} - 2\mathbf{u} = (2, -1, 5) - (-2, 0, 2) = (4, -1, 3).$$

Figure 4.7 gives a graphical representation of these vector operations in R^3.

Technology Note

Some graphing utilities and computer software programs will perform vector addition and scalar multiplication. Using a graphing utility, you may verify Example 4 as follows. Keystrokes and programming syntax for these utilities/programs applicable to Example 4 are provided in the **Online Technology Guide,** available at *college.hmco.com/pic/larsonELA6e*.

```
VECTOR:U        3
 e1=-1
 e2=0
 e3=1
```

```
VECTOR:V        3
 e1=2
 e2=-1
 e3=5
```

```
U+V
             [1  -1  6]
2U
             [-2  0  2]
V-2U
             [4  -1  3]
```

The following properties of vector addition and scalar multiplication for vectors in R^n are the same as those listed in Theorem 4.1 for vectors in the plane. Their proofs, based on the definitions of vector addition and scalar multiplication in R^n, are left as an exercise. (See Exercise 62.)

THEOREM 4.2 Properties of Vector Addition and Scalar Multiplication in R^n

Let $\mathbf{u}$, $\mathbf{v}$, and $\mathbf{w}$ be vectors in R^n, and let c and d be scalars.

1. $\mathbf{u} + \mathbf{v}$ is a vector in R^n. — Closure under addition
2. $\mathbf{u} + \mathbf{v} = \mathbf{v} + \mathbf{u}$ — Commutative property of addition
3. $(\mathbf{u} + \mathbf{v}) + \mathbf{w} = \mathbf{u} + (\mathbf{v} + \mathbf{w})$ — Associative property addition
4. $\mathbf{u} + \mathbf{0} = \mathbf{u}$ — Additive identity property
5. $\mathbf{u} + (-\mathbf{u}) = \mathbf{0}$ — Additive inverse property
6. $c\mathbf{u}$ is a vector in R^n. — Closure under scalar multiplication
7. $c(\mathbf{u} + \mathbf{v}) = c\mathbf{u} + c\mathbf{v}$ — Distributive property
8. $(c + d)\mathbf{u} = c\mathbf{u} + d\mathbf{u}$ — Distributive property
9. $c(d\mathbf{u}) = (cd)\mathbf{u}$ — Associative property of multiplication
10. $1(\mathbf{u}) = \mathbf{u}$ — Multiplicative identity property

Using the ten properties from Theorem 4.2, you can perform algebraic manipulations with vectors in R^n in much the same way as you do with real numbers, as demonstrated in the next example.

EXAMPLE 5 Vector Operations in R^4

Let $\mathbf{u} = (2, -1, 5, 0)$, $\mathbf{v} = (4, 3, 1, -1)$, and $\mathbf{w} = (-6, 2, 0, 3)$ be vectors in R^4. Solve for $\mathbf{x}$.

(a) $\mathbf{x} = 2\mathbf{u} - (\mathbf{v} + 3\mathbf{w})$ (b) $3(\mathbf{x} + \mathbf{w}) = 2\mathbf{u} - \mathbf{v} + \mathbf{x}$

SOLUTION (a) Using the properties listed in Theorem 4.2, you have

$$\begin{aligned}
\mathbf{x} &= 2\mathbf{u} - (\mathbf{v} + 3\mathbf{w}) \\
&= 2\mathbf{u} - \mathbf{v} - 3\mathbf{w} \\
&= (4, -2, 10, 0) - (4, 3, 1, -1) - (-18, 6, 0, 9) \\
&= (4 - 4 + 18, -2 - 3 - 6, 10 - 1 - 0, 0 + 1 - 9) \\
&= (18, -11, 9, -8).
\end{aligned}$$

(b) Begin by solving for $\mathbf{x}$ as follows.

$$\begin{aligned}
3(\mathbf{x} + \mathbf{w}) &= 2\mathbf{u} - \mathbf{v} + \mathbf{x} \\
3\mathbf{x} + 3\mathbf{w} &= 2\mathbf{u} - \mathbf{v} + \mathbf{x} \\
3\mathbf{x} - \mathbf{x} &= 2\mathbf{u} - \mathbf{v} - 3\mathbf{w} \\
2\mathbf{x} &= 2\mathbf{u} - \mathbf{v} - 3\mathbf{w} \\
\mathbf{x} &= \tfrac{1}{2}(2\mathbf{u} - \mathbf{v} - 3\mathbf{w})
\end{aligned}$$

Using the result of part (a) produces

$$\mathbf{x} = \tfrac{1}{2}(18, -11, 9, -8)$$
$$= \left(9, -\tfrac{11}{2}, \tfrac{9}{2}, -4\right).$$

The zero vector $\mathbf{0}$ in R^n is called the **additive identity** in R^n. Similarly, the vector $-\mathbf{v}$ is called the **additive inverse** of $\mathbf{v}$. The theorem below summarizes several important properties of the additive identity and additive inverse in R^n.

THEOREM 4.3 Properties of Additive Identity and Additive Inverse

Let $\mathbf{v}$ be a vector in R^n, and let c be a scalar. Then the following properties are true.

1. The additive identity is unique. That is, if $\mathbf{v} + \mathbf{u} = \mathbf{v}$, then $\mathbf{u} = \mathbf{0}$.
2. The additive inverse of $\mathbf{v}$ is unique. That is, if $\mathbf{v} + \mathbf{u} = \mathbf{0}$, then $\mathbf{u} = -\mathbf{v}$.
3. $0\mathbf{v} = \mathbf{0}$
4. $c\mathbf{0} = \mathbf{0}$
5. If $c\mathbf{v} = \mathbf{0}$, then $c = 0$ or $\mathbf{v} = \mathbf{0}$.
6. $-(-\mathbf{v}) = \mathbf{v}$

PROOF To prove the first property, assume $\mathbf{v} + \mathbf{u} = \mathbf{v}$. Then the steps below are justified by Theorem 4.2.

$\mathbf{v} + \mathbf{u} = \mathbf{v}$	Given
$(\mathbf{v} + \mathbf{u}) + (-\mathbf{v}) = \mathbf{v} + (-\mathbf{v})$	Add $-\mathbf{v}$ to both sides.
$(\mathbf{v} + \mathbf{u}) + (-\mathbf{v}) = \mathbf{0}$	Additive inverse
$\mathbf{v} + [\mathbf{u} + (-\mathbf{v})] = \mathbf{0}$	Associative property
$\mathbf{v} + [(-\mathbf{v}) + \mathbf{u}] = \mathbf{0}$	Commutative property
$[\mathbf{v} + (-\mathbf{v})] + \mathbf{u} = \mathbf{0}$	Associative property
$\mathbf{0} + \mathbf{u} = \mathbf{0}$	Additive inverse
$\mathbf{u} + \mathbf{0} = \mathbf{0}$	Commutative property
$\mathbf{u} = \mathbf{0}$	Additive identity

As you gain experience in reading and writing proofs involving vector algebra, you will not need to list this many steps. For now, however, it's a good idea. The proofs of the other five properties are left as exercises. (See Exercises 63–67.)

REMARK: In Properties 3 and 5 of Theorem 4.3, note that two different zeros are used, the scalar 0 and the vector $\mathbf{0}$.

The next example illustrates an important type of problem in linear algebra—writing one vector $\mathbf{x}$ as the sum of scalar multiples of other vectors $\mathbf{v}_1, \mathbf{v}_2, \ldots,$ and $\mathbf{v}_n$. That is,

$$\mathbf{x} = c_1\mathbf{v}_1 + c_2\mathbf{v}_2 + \cdots + c_n\mathbf{v}_n.$$

The vector $\mathbf{x}$ is called a **linear combination** of the vectors $\mathbf{v}_1, \mathbf{v}_2, \ldots,$ and $\mathbf{v}_n$.

EXAMPLE 6 **Writing a Vector as a Linear Combination of Other Vectors**

Provided that $\mathbf{x} = (-1, -2, -2)$, $\mathbf{u} = (0, 1, 4)$, $\mathbf{v} = (-1, 1, 2)$, and $\mathbf{w} = (3, 1, 2)$ in R^3, find scalars a, b, and c such that

$$\mathbf{x} = a\mathbf{u} + b\mathbf{v} + c\mathbf{w}.$$

SOLUTION By writing

$$\overbrace{(-1, -2, -2)}^{\mathbf{x}} = a\overbrace{(0, 1, 4)}^{\mathbf{u}} + b\overbrace{(-1, 1, 2)}^{\mathbf{v}} + c\overbrace{(3, 1, 2)}^{\mathbf{w}}$$
$$= (-b + 3c, a + b + c, 4a + 2b + 2c),$$

you can equate corresponding components so that they form the system of three linear equations in a, b, and c shown below.

$$-b + 3c = -1 \quad \text{Equation from first component}$$
$$a + b + c = -2 \quad \text{Equation from second component}$$
$$4a + 2b + 2c = -2 \quad \text{Equation from third component}$$

Using the techniques of Chapter 1, solve for a, b, and c to get

$$a = 1, \quad b = -2, \quad \text{and} \quad c = -1.$$

$\mathbf{x}$ can be written as a linear combination of $\mathbf{u}$, $\mathbf{v}$, and $\mathbf{w}$.

$$\mathbf{x} = \mathbf{u} - 2\mathbf{v} - \mathbf{w}$$

Try using vector addition and scalar multiplication to check this result.

Discovery

Is the vector $(1, 1)$ a linear combination of the vectors $(1, 2)$ and $(-2, -4)$? Graph these vectors in the plane and explain your answer geometrically. Similarly, determine whether the vector $(1, 1)$ is a linear combination of the vectors $(1, 2)$ and $(2, 1)$. What is the geometric significance of these two questions? Is every vector in R^2 a linear combination of the vectors $(1, 2)$ and $(2, 1)$? Give a geometric explanation for your answer.

You will often find it useful to represent a vector $\mathbf{u} = (u_1, u_2, \ldots, u_n)$ in R^n as either a $1 \times n$ row matrix (row vector),

$$\mathbf{u} = [u_1 \quad u_2 \quad \cdots \quad u_n],$$

or an $n \times 1$ column matrix (column vector),

$$\mathbf{u} = \begin{bmatrix} u_1 \\ u_2 \\ \vdots \\ u_n \end{bmatrix}.$$

This approach is valid because the matrix operations of addition and scalar multiplication give the same results as the corresponding vector operations. That is, the matrix sums

$$\begin{aligned}\mathbf{u} + \mathbf{v} &= [u_1 \quad u_2 \quad \cdots \quad u_n] + [v_1 \quad v_2 \quad \cdots \quad v_n] \\ &= [u_1 + v_1 \quad u_2 + v_2 \quad \cdots \quad u_n + v_n]\end{aligned}$$

and

$$\mathbf{u} + \mathbf{v} = \begin{bmatrix} u_1 \\ u_2 \\ \vdots \\ u_n \end{bmatrix} + \begin{bmatrix} v_1 \\ v_2 \\ \vdots \\ v_n \end{bmatrix} = \begin{bmatrix} u_1 + v_1 \\ u_2 + v_2 \\ \vdots \\ u_n + v_n \end{bmatrix}$$

yield the same results as the vector operation of addition,

$$\mathbf{u} + \mathbf{v} = (u_1, u_2, \ldots, u_n) + (v_1, v_2, \ldots, v_n) = (u_1 + v_1, u_2 + v_2, \ldots, u_n + v_n).$$

The same argument applies to scalar multiplication. The only difference in the three notations for vectors is how the components are displayed; the underlying operations are the same.

SECTION 4.1 Exercises

In Exercises 1 and 2, find the component form of the vector shown.

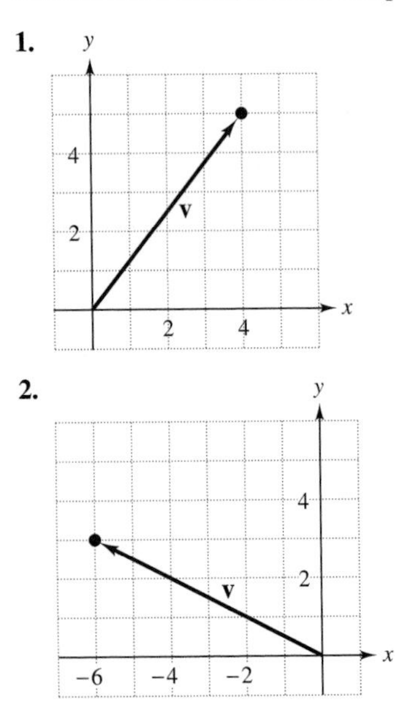

In Exercises 3–6, use a directed line segment to represent the vector.

3. $\mathbf{u} = (2, -4)$ **4.** $\mathbf{v} = (-2, 3)$

5. $\mathbf{u} = (-3, -4)$ **6.** $\mathbf{v} = (-2, -5)$

In Exercises 7–10, find the sum of the vectors and illustrate the indicated vector operations geometrically.

7. $\mathbf{u} = (1, 3), \mathbf{v} = (2, -2)$ **8.** $\mathbf{u} = (-1, 4), \mathbf{v} = (4, -3)$

9. $\mathbf{u} = (2, -3), \mathbf{v} = (-3, -1)$

10. $\mathbf{u} = (4, -2), \mathbf{v} = (-2, -3)$

In Exercises 11–16, find the vector $\mathbf{v}$ and illustrate the indicated vector operations geometrically, where $\mathbf{u} = (-2, 3)$ and $\mathbf{w} = (-3, -2)$.

11. $\mathbf{v} = \frac{3}{2}\mathbf{u}$ **12.** $\mathbf{v} = \mathbf{u} + \mathbf{w}$ **13.** $\mathbf{v} = \mathbf{u} + 2\mathbf{w}$

14. $\mathbf{v} = -\mathbf{u} + \mathbf{w}$ **15.** $\mathbf{v} = \frac{1}{2}(3\mathbf{u} + \mathbf{w})$ **16.** $\mathbf{v} = \mathbf{u} - 2\mathbf{w}$

17. Given the vector $\mathbf{v} = (2, 1)$, sketch (a) $2\mathbf{v}$, (b) $-3\mathbf{v}$, and (c) $\frac{1}{2}\mathbf{v}$.

18. Given the vector $\mathbf{v} = (3, -2)$, sketch (a) $4\mathbf{v}$, (b) $-\frac{1}{2}\mathbf{v}$, and (c) $0\mathbf{v}$.

In Exercises 19–24, let $\mathbf{u} = (1, 2, 3)$, $\mathbf{v} = (2, 2, -1)$, and $\mathbf{w} = (4, 0, -4)$.

19. Find $\mathbf{u} - \mathbf{v}$ and $\mathbf{v} - \mathbf{u}$. **20.** Find $\mathbf{u} - \mathbf{v} + 2\mathbf{w}$.

21. Find $2\mathbf{u} + 4\mathbf{v} - \mathbf{w}$. **22.** Find $5\mathbf{u} - 3\mathbf{v} - \frac{1}{2}\mathbf{w}$.

23. Find $\mathbf{z}$, where $2\mathbf{z} - 3\mathbf{u} = \mathbf{w}$.

24. Find $\mathbf{z}$, where $2\mathbf{u} + \mathbf{v} - \mathbf{w} + 3\mathbf{z} = \mathbf{0}$.

25. Given the vector $\mathbf{v} = (1, 2, 2)$, sketch (a) $2\mathbf{v}$, (b) $-\mathbf{v}$, and (c) $\frac{1}{2}\mathbf{v}$.

26. Given the vector $\mathbf{v} = (2, 0, 1)$, sketch (a) $-\mathbf{v}$, (b) $2\mathbf{v}$, and (c) $\frac{1}{2}\mathbf{v}$.

27. Which of the vectors below are scalar multiples of $\mathbf{z} = (3, 2, -5)$?

(a) $\mathbf{u} = (-6, -4, 10)$
(b) $\mathbf{v} = \left(2, \frac{4}{3}, -\frac{10}{3}\right)$
(c) $\mathbf{w} = (6, 4, 10)$

28. Which of the vectors below are scalar multiples of $\mathbf{z} = \left(\frac{1}{2}, -\frac{2}{3}, \frac{3}{4}\right)$?

(a) $\mathbf{u} = (6, -4, 9)$
(b) $\mathbf{v} = \left(-1, \frac{4}{3}, -\frac{3}{2}\right)$
(c) $\mathbf{w} = (12, 0, 9)$

In Exercises 29–32, find (a) $\mathbf{u} - \mathbf{v}$, (b) $2(\mathbf{u} + 3\mathbf{v})$, and (c) $2\mathbf{v} - \mathbf{u}$.

29. $\mathbf{u} = (4, 0, -3, 5)$, $\mathbf{v} = (0, 2, 5, 4)$

30. $\mathbf{u} = (0, 4, 3, 4, 4)$, $\mathbf{v} = (6, 8, -3, 3, -5)$

31. $\mathbf{u} = (-7, 0, 0, 0, 9)$, $\mathbf{v} = (2, -3, -2, 3, 3)$

32. $\mathbf{u} = (6, -5, 4, 3)$, $\mathbf{v} = \left(-2, \frac{5}{3}, -\frac{4}{3}, -1\right)$

In Exercises 33 and 34, use a graphing utility with matrix capabilities to find the following, where $\mathbf{u} = (1, 2, -3, 1)$, $\mathbf{v} = (0, 2, -1, -2)$, and $\mathbf{w} = (2, -2, 1, 3)$.

33. (a) $\mathbf{u} + 2\mathbf{v}$
(b) $\mathbf{w} - 3\mathbf{u}$
(c) $4\mathbf{v} + \frac{1}{2}\mathbf{u} - \mathbf{w}$
(d) $\frac{1}{4}(3\mathbf{u} + 2\mathbf{v} - \mathbf{w})$

34. (a) $\mathbf{v} + 3\mathbf{w}$
(b) $2\mathbf{w} - \frac{1}{2}\mathbf{u}$
(c) $2\mathbf{u} + \mathbf{w} - 3\mathbf{v}$
(d) $\frac{1}{2}(4\mathbf{v} - 3\mathbf{u} + \mathbf{w})$

In Exercises 35–38, solve for $\mathbf{w}$ provided that $\mathbf{u} = (1, -1, 0, 1)$ and $\mathbf{v} = (0, 2, 3, -1)$.

35. $2\mathbf{w} = \mathbf{u} - 3\mathbf{v}$

36. $\mathbf{w} + \mathbf{u} = -\mathbf{v}$

37. $\frac{1}{2}\mathbf{w} = 2\mathbf{u} + 3\mathbf{v}$

38. $\mathbf{w} + 3\mathbf{v} = -2\mathbf{u}$

In Exercises 39–44, write $\mathbf{v}$ as a linear combination of $\mathbf{u}$ and $\mathbf{w}$, if possible, where $\mathbf{u} = (1, 2)$ and $\mathbf{w} = (1, -1)$.

39. $\mathbf{v} = (2, 1)$

40. $\mathbf{v} = (0, 3)$

41. $\mathbf{v} = (3, 0)$

42. $\mathbf{v} = (1, -1)$

43. $\mathbf{v} = (-1, -2)$

44. $\mathbf{v} = (1, -4)$

In Exercises 45 and 46, find $\mathbf{w}$ such that $2\mathbf{u} + \mathbf{v} - 3\mathbf{w} = \mathbf{0}$.

45. $\mathbf{u} = (0, 2, 7, 5)$, $\mathbf{v} = (-3, 1, 4, -8)$

46. $\mathbf{u} = (0, 0, -8, 1)$, $\mathbf{v} = (1, -8, 0, 7)$

In Exercises 47–50, write $\mathbf{v}$ as a linear combination of $\mathbf{u}_1$, $\mathbf{u}_2$, and $\mathbf{u}_3$, if possible.

47. $\mathbf{u}_1 = (2, 3, 5)$, $\mathbf{u}_2 = (1, 2, 4)$, $\mathbf{u}_3 = (-2, 2, 3)$, $\mathbf{v} = (10, 1, 4)$

48. $\mathbf{u}_1 = (1, 3, 5)$, $\mathbf{u}_2 = (2, -1, 3)$, $\mathbf{u}_3 = (-3, 2, -4)$, $\mathbf{v} = (-1, 7, 2)$

49. $\mathbf{u}_1 = (1, 1, 2, 2)$, $\mathbf{u}_2 = (2, 3, 5, 6)$, $\mathbf{u}_3 = (-3, 1, -4, 2)$, $\mathbf{v} = (0, 5, 3, 0)$

50. $\mathbf{u}_1 = (1, 3, 2, 1)$, $\mathbf{u}_2 = (2, -2, -5, 4)$, $\mathbf{u}_3 = (2, -1, 3, 6)$, $\mathbf{v} = (2, 5, -4, 0)$

In Exercises 51–54, use a graphing utility or computer software program with matrix capabilities to write $\mathbf{v}$ as a linear combination of $\mathbf{u}_1, \mathbf{u}_2, \mathbf{u}_3, \mathbf{u}_4$, and $\mathbf{u}_5$, or of $\mathbf{u}_1, \mathbf{u}_2, \mathbf{u}_3, \mathbf{u}_4, \mathbf{u}_5$, and $\mathbf{u}_6$. Then verify your solution.

51. $\mathbf{u}_1 = (1, 2, -3, 4, -1)$
$\mathbf{u}_2 = (1, 2, 0, 2, 1)$
$\mathbf{u}_3 = (0, 1, 1, 1, -4)$
$\mathbf{u}_4 = (2, 1, -1, 2, 1)$
$\mathbf{u}_5 = (0, 2, 2, -1, -1)$
$\mathbf{v} = (5, 3, -11, 11, 9)$

52. $\mathbf{u}_1 = (1, 1, -1, 2, 1)$
$\mathbf{u}_2 = (2, 1, 2, -1, 1)$
$\mathbf{u}_3 = (1, 2, 0, 1, 2)$
$\mathbf{u}_4 = (0, 2, 0, 1, -4)$
$\mathbf{u}_5 = (1, 1, 2, -1, 2)$
$\mathbf{v} = (5, 8, 7, -2, 4)$

53. $\mathbf{u}_1 = (1, 2, -3, 4, -1, 2)$
$\mathbf{u}_2 = (1, -2, 1, -1, 2, 1)$
$\mathbf{u}_3 = (0, 2, -1, 2, -1, -1)$
$\mathbf{u}_4 = (1, 0, 3, -4, 1, 2)$
$\mathbf{u}_5 = (1, -2, 1, -1, 2, -3)$
$\mathbf{u}_6 = (3, 2, 1, -2, 3, 0)$
$\mathbf{v} = (10, 30, -13, 14, -7, 27)$

54. $\mathbf{u}_1 = (1, -3, 4, -5, 2, -1)$
$\mathbf{u}_2 = (3, -2, 4, -3, -2, 1)$
$\mathbf{u}_3 = (1, 1, 1, -1, 4, -1)$
$\mathbf{u}_4 = (3, -1, 3, -4, 2, 3)$
$\mathbf{u}_5 = (1, -2, 1, 5, -3, 4)$
$\mathbf{u}_6 = (4, 2, -1, 3, -1, 1)$
$\mathbf{v} = (8, 17, -16, 26, 0, -4)$

True or False? In Exercises 55 and 56, determine whether each statement is true or false. If a statement is true, give a reason or cite an appropriate statement from the text. If a statement is false, provide an example that shows the statement is not true in all cases or cite an appropriate statement from the text.

55. (a) Two vectors in R^n are equal if and only if their corresponding components are equal.
(b) For a nonzero scalar c, the vector $c\mathbf{v}$ is c times as long as $\mathbf{v}$ and has the same direction as $\mathbf{v}$ if $c > 0$ and the opposite direction if $c < 0$.

56. (a) To add two vectors in R^n, add their corresponding components.

(b) The zero vector $\mathbf{0}$ in R^n is defined as the additive inverse of a vector.

In Exercises 57 and 58, the zero vector $\mathbf{0} = (0, 0, 0)$ can be written as a linear combination of the vectors $\mathbf{v}_1$, $\mathbf{v}_2$, and $\mathbf{v}_3$ as $\mathbf{0} = 0\mathbf{v}_1 + 0\mathbf{v}_2 + 0\mathbf{v}_3$. This is called the *trivial* solution. Can you find a *nontrivial* way of writing $\mathbf{0}$ as a linear combination of the three vectors?

57. $\mathbf{v}_1 = (1, 0, 1)$, $\mathbf{v}_2 = (-1, 1, 2)$, $\mathbf{v}_3 = (0, 1, 4)$

58. $\mathbf{v}_1 = (1, 0, 1)$, $\mathbf{v}_2 = (-1, 1, 2)$, $\mathbf{v}_3 = (0, 1, 3)$

59. Illustrate properties 1–10 of Theorem 4.2 for $\mathbf{u} = (2, -1, 3, 6)$, $\mathbf{v} = (1, 4, 0, 1)$, $\mathbf{w} = (3, 0, 2, 0)$, $c = 5$, and $d = -2$.

60. Illustrate properties 1–10 of Theorem 4.2 for $\mathbf{u} = (2, -1, 3)$, $\mathbf{v} = (3, 4, 0)$, $\mathbf{w} = (7, 8, -4)$, $c = 2$, and $d = -1$.

61. Complete the proof of Theorem 4.1.

62. Prove each property of vector addition and scalar multiplication from Theorem 4.2.

(a) Property 1: $\mathbf{u} + \mathbf{v}$ is a vector in R^n.
(b) Property 2: $\mathbf{u} + \mathbf{v} = \mathbf{v} + \mathbf{u}$
(c) Property 3: $(\mathbf{u} + \mathbf{v}) + \mathbf{w} = \mathbf{u} + (\mathbf{v} + \mathbf{w})$
(d) Property 4: $\mathbf{u} + \mathbf{0} = \mathbf{u}$
(e) Property 5: $\mathbf{u} + (-\mathbf{u}) = \mathbf{0}$
(f) Property 6: $c\mathbf{u}$ is a vector in R^n.
(g) Property 7: $c(\mathbf{u} + \mathbf{v}) = c\mathbf{u} + c\mathbf{v}$
(h) Property 8: $(c + d)\mathbf{u} = c\mathbf{u} + d\mathbf{u}$
(i) Property 9: $c(d\mathbf{u}) = (cd)\mathbf{u}$
(j) Property 10: $1(\mathbf{u}) = \mathbf{u}$

In Exercises 63–67, complete the proofs of the remaining properties of Theorem 4.3 by supplying the justification for each step. Use the properties of vector addition and scalar multiplication from Theorem 4.2.

63. Property 2: The additive inverse of $\mathbf{v}$ is unique. That is, if $\mathbf{v} + \mathbf{u} = \mathbf{0}$, then $\mathbf{u} = -\mathbf{v}$.

$\mathbf{v} + \mathbf{u} = \mathbf{0}$ Given
$(-\mathbf{v}) + (\mathbf{v} + \mathbf{u}) = (-\mathbf{v}) + \mathbf{0}$ a. ________
$((-\mathbf{v}) + \mathbf{v}) + \mathbf{u} = -\mathbf{v}$ b. ________
$\mathbf{0} + \mathbf{u} = -\mathbf{v}$ c. ________
$\mathbf{u} + \mathbf{0} = -\mathbf{v}$ d. ________
$\mathbf{u} = -\mathbf{v}$ e. ________

64. Property 3: $0\mathbf{v} = \mathbf{0}$

$0\mathbf{v} = (0 + 0)\mathbf{v}$ a. ________
$0\mathbf{v} = 0\mathbf{v} + 0\mathbf{v}$ b. ________
$0\mathbf{v} + (-0\mathbf{v}) = (0\mathbf{v} + 0\mathbf{v}) + (-0\mathbf{v})$ c. ________
$\mathbf{0} = 0\mathbf{v} + (0\mathbf{v} + (-0\mathbf{v}))$ d. ________
$\mathbf{0} = 0\mathbf{v} + \mathbf{0}$ e. ________
$\mathbf{0} = 0\mathbf{v}$ f. ________

65. Property 4: $c\mathbf{0} = \mathbf{0}$

$c\mathbf{0} = c(\mathbf{0} + \mathbf{0})$ a. ________
$c\mathbf{0} = c\mathbf{0} + c\mathbf{0}$ b. ________
$c\mathbf{0} + (-c\mathbf{0}) = (c\mathbf{0} + c\mathbf{0}) + (-c\mathbf{0})$ c. ________
$\mathbf{0} = c\mathbf{0} + (c\mathbf{0} + (-c\mathbf{0}))$ d. ________
$\mathbf{0} = c\mathbf{0} + \mathbf{0}$ e. ________
$\mathbf{0} = c\mathbf{0}$ f. ________

66. Property 5: If $c\mathbf{v} = \mathbf{0}$, then $c = 0$ or $\mathbf{v} = \mathbf{0}$. If $c = 0$, you are done. If $c \neq 0$, then c^{-1} exists, and you have

$c^{-1}(c\mathbf{v}) = c^{-1}\mathbf{0}$ a. ________
$(c^{-1}c)\mathbf{v} = \mathbf{0}$ b. ________
$1\mathbf{v} = \mathbf{0}$ c. ________
$\mathbf{v} = \mathbf{0}.$ d. ________

67. Property 6: $-(-\mathbf{v}) = \mathbf{v}$

$-(-\mathbf{v}) + (-\mathbf{v}) = \mathbf{0}$ and $\mathbf{v} + (-\mathbf{v}) = \mathbf{0}$ a. ________
$-(-\mathbf{v}) + (-\mathbf{v}) = \mathbf{v} + (-\mathbf{v})$ b. ________
$-(-\mathbf{v}) + (-\mathbf{v}) + \mathbf{v} = \mathbf{v} + (-\mathbf{v}) + \mathbf{v}$ c. ________
$-(-\mathbf{v}) + ((-\mathbf{v}) + \mathbf{v}) = \mathbf{v} + ((-\mathbf{v}) + \mathbf{v})$ d. ________
$-(-\mathbf{v}) + \mathbf{0} = \mathbf{v} + \mathbf{0}$ e. ________
$-(-\mathbf{v}) = \mathbf{v}$ f. ________

In Exercises 68 and 69, determine if the third column can be written as a linear combination of the first two columns.

68. $\begin{bmatrix} 1 & 2 & 3 \\ 7 & 8 & 9 \\ 4 & 5 & 6 \end{bmatrix}$ **69.** $\begin{bmatrix} 1 & 2 & 3 \\ 7 & 8 & 9 \\ 4 & 5 & 7 \end{bmatrix}$

70. Writing Let $A\mathbf{x} = \mathbf{b}$ be a system of m linear equations in n variables. Designate the columns of A as $\mathbf{a}_1, \mathbf{a}_2, \ldots, \mathbf{a}_n$. If $\mathbf{b}$ is a linear combination of these n column vectors, explain why this implies that the linear system is consistent. Illustrate your answer with appropriate examples. What can you conclude about the linear system if $\mathbf{b}$ is not a linear combination of the columns of A?

71. Writing How could you describe vector subtraction geometrically? What is the relationship between vector subtraction and the basic vector operations of addition and scalar multiplication?

4.2 Vector Spaces

In Theorem 4.2, ten special properties of vector addition and scalar multiplication in R^n were listed. Suitable definitions of addition and scalar multiplication reveal that many other mathematical quantities (such as matrices, polynomials, and functions) also share these ten properties. *Any* set that satisfies these properties (or **axioms**) is called a **vector space,** and the objects in the set are called **vectors.**

It is important to realize that the next definition of vector space is precisely that—a *definition.* You do not need to prove anything because you are simply listing the axioms required of vector spaces. This type of definition is called an **abstraction** because you are abstracting a collection of properties from a particular setting R^n to form the axioms for a more general setting.

Definition of Vector Space

Let V be a set on which two operations (**vector addition** and **scalar multiplication**) are defined. If the listed axioms are satisfied for every $\mathbf{u}$, $\mathbf{v}$, and $\mathbf{w}$ in V and every scalar (real number) c and d, then V is called a **vector space.**

Addition:

1. $\mathbf{u} + \mathbf{v}$ is in V. — Closure under addition
2. $\mathbf{u} + \mathbf{v} = \mathbf{v} + \mathbf{u}$ — Commutative property
3. $\mathbf{u} + (\mathbf{v} + \mathbf{w}) = (\mathbf{u} + \mathbf{v}) + \mathbf{w}$ — Associative property
4. V has a **zero vector** $\mathbf{0}$ such that for every $\mathbf{u}$ in V, $\mathbf{u} + \mathbf{0} = \mathbf{u}$. — Additive identity
5. For every $\mathbf{u}$ in V, there is a vector in V denoted by $-\mathbf{u}$ such that $\mathbf{u} + (-\mathbf{u}) = \mathbf{0}$. — Additive inverse

Scalar Multiplication:

6. $c\mathbf{u}$ is in V. — Closure under scalar multiplication
7. $c(\mathbf{u} + \mathbf{v}) = c\mathbf{u} + c\mathbf{v}$ — Distributive property
8. $(c + d)\mathbf{u} = c\mathbf{u} + d\mathbf{u}$ — Distributive property
9. $c(d\mathbf{u}) = (cd)\mathbf{u}$ — Associative property
10. $1(\mathbf{u}) = \mathbf{u}$ — Scalar identity

It is important to realize that a vector space consists of four entities: a set of vectors, a set of scalars, and two operations. When you refer to a vector space V, be sure all four entities are clearly stated or understood. Unless stated otherwise, assume that the set of scalars is the set of real numbers.

The first two examples of vector spaces on the next page are not surprising. They are, in fact, the models used to form the ten vector space axioms.

EXAMPLE 1 **R^2 with the Standard Operations Is a Vector Space**

The set of all ordered pairs of real numbers R^2 with the standard operations is a vector space. To verify this, look back at Theorem 4.1. Vectors in this space have the form

$$\mathbf{v} = (v_1, v_2).$$

EXAMPLE 2 **R^n with the Standard Operations Is a Vector Space**

The set of all ordered n-tuples of real numbers R^n with the standard operations is a vector space. This is verified by Theorem 4.2. Vectors in this space are of the form

$$\mathbf{v} = (v_1, v_2, v_3, \ldots, v_n).$$

REMARK: From Example 2 you can conclude that R^1, the set of real numbers (with the usual operations of addition and multiplication), is a vector space.

The next three examples describe vector spaces in which the basic set V does not consist of ordered n-tuples. Each example describes the set V and defines the two vector operations. To show that the set is a vector space, you must verify all ten axioms.

EXAMPLE 3 **The Vector Space of All 2×3 Matrices**

Show that the set of all 2×3 matrices with the operations of matrix addition and scalar multiplication is a vector space.

SOLUTION If A and B are 2×3 matrices and c is a scalar, then $A + B$ and cA are also 2×3 matrices. The set is, therefore, closed under matrix addition and scalar multiplication. Moreover, the other eight vector space axioms follow directly from Theorems 2.1 and 2.2 (see Section 2.2). You can conclude that the set is a vector space. Vectors in this space have the form

$$\mathbf{a} = A = \begin{bmatrix} a_{11} & a_{12} & a_{13} \\ a_{21} & a_{22} & a_{23} \end{bmatrix}.$$

REMARK: In the same way you are able to show that the set of all 2×3 matrices is a vector space, you can show that the set of all $m \times n$ matrices, denoted by $M_{m,n}$, is a vector space.

EXAMPLE 4 **The Vector Space of All Polynomials of Degree 2 or Less**

Let P_2 be the set of all polynomials of the form

$$p(x) = a_2x^2 + a_1x + a_0,$$

where a_0, a_1, and a_2 are real numbers. The *sum* of two polynomials $p(x) = a_2x^2 + a_1x + a_0$ and $q(x) = b_2x^2 + b_1x + b_0$ is defined in the usual way by

$$p(x) + q(x) = (a_2 + b_2)x^2 + (a_1 + b_1)x + (a_0 + b_0),$$

and the *scalar multiple* of $p(x)$ by the scalar c is defined by

$$cp(x) = ca_2x^2 + ca_1x + ca_0.$$

Show that P_2 is a vector space.

SOLUTION Verification of each of the ten vector space axioms is a straightforward application of the properties of real numbers. For instance, because the set of real numbers is closed under addition, it follows that $a_2 + b_2, a_1 + b_1$, and $a_0 + b_0$ are real numbers, and

$$p(x) + q(x) = (a_2 + b_2)x^2 + (a_1 + b_1)x + (a_0 + b_0)$$

is in the set P_2 because it is a polynomial of degree 2 or less. P_2 is closed under addition. Similarly, you can use the fact that the set of real numbers is closed under multiplication to show that P_2 is closed under scalar multiplication. To verify the commutative axiom of addition, write

$$\begin{aligned}
p(x) + q(x) &= (a_2x^2 + a_1x + a_0) + (b_2x^2 + b_1x + b_0) \\
&= (a_2 + b_2)x^2 + (a_1 + b_1)x + (a_0 + b_0) \\
&= (b_2 + a_2)x^2 + (b_1 + a_1)x + (b_0 + a_0) \\
&= (b_2x^2 + b_1x + b_0) + (a_2x^2 + a_1x + a_0) \\
&= q(x) + p(x).
\end{aligned}$$

Can you see where the commutative property of addition of real numbers was used? The zero vector in this space is the zero polynomial given by $\mathbf{0}(x) = 0x^2 + 0x + 0$, for all x. Try verifying the other vector space axioms. You may then conclude that P_2 is a vector space.

REMARK: Even though the zero polynomial $\mathbf{0}(x) = 0$ has no degree, P_2 is often described as the set of all polynomials of degree 2 or *less*.

P_n is defined as the set of all polynomials of degree n or less (together with the zero polynomial). The procedure used to verify that P_2 is a vector space can be extended to show that P_n, with the usual operations of polynomial addition and scalar multiplication, is a vector space.

EXAMPLE 5 **The Vector Space of Continuous Functions (Calculus)**

Let $C(-\infty, \infty)$ be the set of all real-valued continuous functions defined on the entire real line. This set consists of all polynomial functions and all other continuous functions on the entire real line. For instance, $f(x) = \sin x$ and $g(x) = e^x$ are members of this set.

Addition is defined by

$$(f + g)(x) = f(x) + g(x),$$

as shown in Figure 4.8. Scalar multiplication is defined by

$$(cf)(x) = c[f(x)].$$

Show that $C(-\infty, \infty)$ is a vector space.

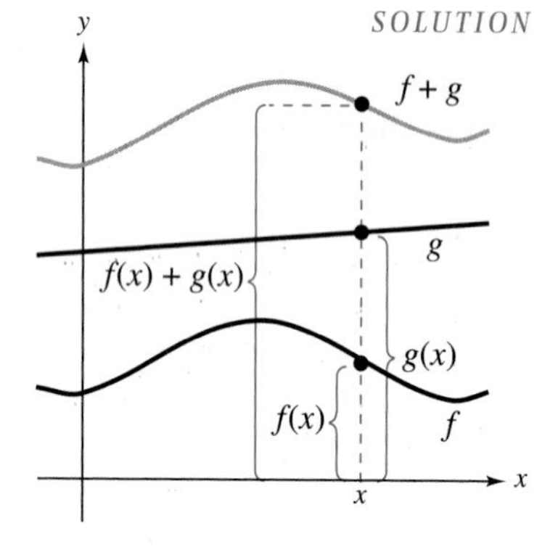

Figure 4.8

SOLUTION To verify that the set $C(-\infty, \infty)$ is closed under addition and scalar multiplication, you can use a result from calculus—the sum of two continuous functions is continuous and the product of a scalar and a continuous function is continuous. To verify that the set $C(-\infty, \infty)$ has an additive identity, consider the function f_0 that has a value of zero for all x, meaning that

$$f_0(x) = 0, \quad \text{where } x \text{ is any real number.}$$

This function is continuous on the entire real line (its graph is simply the line $y = 0$), which means that it is in the set $C(-\infty, \infty)$. Moreover, if f is any other function that is continuous on the entire real line, then

$$(f + f_0)(x) = f(x) + f_0(x) = f(x) + 0 = f(x).$$

This shows that f_0 is the additive identity in $C(-\infty, \infty)$. The verification of the other vector space axioms is left to you.

For convenience, the summary below lists some important vector spaces frequently referenced in the remainder of this text. The operations are the standard operations in each case.

Summary of Important Vector Spaces

R = set of all real numbers
R^2 = set of all ordered pairs
R^3 = set of all ordered triples
R^n = set of all n-tuples
$C(-\infty, \infty)$ = set of all continuous functions defined on the real number line
$C[a, b]$ = set of all continuous functions defined on a closed interval $[a, b]$
P = set of all polynomials
P_n = set of all polynomials of degree $\leq n$
$M_{m,n}$ = set of all $m \times n$ matrices
$M_{n,n}$ = set of all $n \times n$ square matrices

You have seen the versatility of the concept of a vector space. For instance, a vector can be a real number, an n-tuple, a matrix, a polynomial, a continuous function, and so on. But what is the purpose of this abstraction, and why bother to define it? There are several reasons, but the most important reason applies to efficiency. This abstraction turns out to be mathematically efficient because general results that apply to all vector spaces can now be derived. Once a theorem has been proved for an abstract vector space, you need not give separate proofs for n-tuples, matrices, and polynomials. You can simply point out that the theorem is true for any vector space, regardless of the particular form the vectors happen to take. This process is illustrated in Theorem 4.4.

THEOREM 4.4 Properties of Scalar Multiplication

Let $\mathbf{v}$ be any element of a vector space V, and let c be any scalar. Then the following properties are true.

1. $0\mathbf{v} = \mathbf{0}$
2. $c\mathbf{0} = \mathbf{0}$
3. If $c\mathbf{v} = \mathbf{0}$, then $c = 0$ or $\mathbf{v} = \mathbf{0}$.
4. $(-1)\mathbf{v} = -\mathbf{v}$

PROOF To prove these properties, you are restricted to using the ten vector space axioms. For instance, to prove the second property, note from axiom 4 that $\mathbf{0} = \mathbf{0} + \mathbf{0}$. This allows you to write the steps below.

$$c\mathbf{0} = c(\mathbf{0} + \mathbf{0}) \qquad \text{Additive identity}$$

$$c\mathbf{0} = c\mathbf{0} + c\mathbf{0} \qquad \text{Left distributive property}$$

$$c\mathbf{0} + (-c\mathbf{0}) = (c\mathbf{0} + c\mathbf{0}) + (-c\mathbf{0}) \qquad \text{Add } -c\mathbf{0} \text{ to both sides.}$$

$$c\mathbf{0} + (-c\mathbf{0}) = c\mathbf{0} + [c\mathbf{0} + (-c\mathbf{0})] \qquad \text{Associative property}$$

$$\mathbf{0} = c\mathbf{0} + \mathbf{0} \qquad \text{Additive inverse}$$

$$\mathbf{0} = c\mathbf{0} \qquad \text{Additive identity}$$

To prove the third property, suppose that $c\mathbf{v} = \mathbf{0}$. To show that this implies either $c = 0$ or $\mathbf{v} = \mathbf{0}$, assume that $c \neq 0$. (If $c = 0$, you have nothing more to prove.) Now, because $c \neq 0$, you can use the reciprocal $1/c$ to show that $\mathbf{v} = \mathbf{0}$, as follows.

$$\mathbf{v} = 1\mathbf{v} = \left(\frac{1}{c}\right)(c)\mathbf{v} = \frac{1}{c}(c\mathbf{v}) = \frac{1}{c}(\mathbf{0}) = \mathbf{0}$$

Note that the last step uses Property 2 (the one you just proved). The proofs of the first and fourth properties are left as exercises. (See Exercises 38 and 39.)

The remaining examples in this section describe some sets (with operations) that *do not* form vector spaces. To show that a set is not a vector space, you need only find one axiom that is not satisfied. For example, if you can find two members of V that do not commute ($\mathbf{u} + \mathbf{v} \neq \mathbf{v} + \mathbf{u}$), then regardless of how many other members of V do commute and how many of the other ten axioms are satisfied, you can still conclude that V is not a vector space.

EXAMPLE 6 The Set of Integers Is Not a Vector Space

The set of all integers (with the standard operations) does not form a vector space because it is not closed under scalar multiplication. For example,

$$\underset{\text{Scalar}}{\tfrac{1}{2}}\underset{\text{Integer}}{(1)} = \underset{\text{Noninteger}}{\tfrac{1}{2}}.$$

REMARK: In Example 6, notice that a single failure of one of the ten vector space axioms suffices to show that a set is not a vector space.

In Example 4 it was shown that the set of all polynomials of degree 2 or less forms a vector space. You will now see that the set of all polynomials whose degree is exactly 2 does not form a vector space.

EXAMPLE 7 The Set of Second-Degree Polynomials Is Not a Vector Space

The set of all second-degree polynomials is not a vector space because it is not closed under addition. To see this, consider the second-degree polynomials

$$p(x) = x^2 \quad \text{and} \quad q(x) = -x^2 + x + 1,$$

whose sum is the first-degree polynomial

$$p(x) + q(x) = x + 1.$$

The sets in Examples 6 and 7 are not vector spaces because they fail one or both closure axioms. In the next example you will look at a set that passes both tests for closure but still fails to be a vector space.

EXAMPLE 8 A Set That Is Not a Vector Space

Let $V = R^2$, the set of all ordered pairs of real numbers, with the standard operation of addition and the *nonstandard* definition of scalar multiplication listed below.

$$c(x_1, x_2) = (cx_1, 0)$$

Show that V is not a vector space.

SOLUTION In this example, the operation of scalar multiplication is not the standard one. For instance, the product of the scalar 2 and the ordered pair $(3, 4)$ does not equal $(6, 8)$. Instead, the second component of the product is 0,

$$2(3, 4) = (2 \cdot 3, 0) = (6, 0).$$

This example is interesting because it actually satisfies the first nine axioms of the definition of a vector space (try showing this). The tenth axiom is where you get into trouble. In attempting to verify that axiom, the nonstandard definition of scalar multiplication gives you

$$1(1, 1) = (1, 0) \neq (1, 1).$$

The tenth axiom is not verified and the set (together with the two operations) is not a vector space.

Do not be confused by the notation used for scalar multiplication in Example 8. In writing

$$c(x_1, x_2) = (cx_1, 0),$$

the scalar multiple of (x_1, x_2) by c is *defined* to be $(cx_1, 0)$. As it turns out, this nonstandard definition fails to satisfy the tenth vector space axiom.

SECTION 4.2 Exercises

In Exercises 1–6, describe the zero vector (the additive identity) of the vector space.

1. R^4 **2.** $C(-\infty, \infty)$ **3.** $M_{2,3}$

4. $M_{1,4}$ **5.** P_3 **6.** $M_{2,2}$

In Exercises 7–12, describe the additive inverse of a vector in the vector space.

7. R^4 **8.** $C(-\infty, \infty)$ **9.** $M_{2,3}$

10. $M_{1,4}$ **11.** P_3 **12.** $M_{2,2}$

In Exercises 13–28, determine whether the set, together with the indicated operations, is a vector space. If it is not, identify at least one of the ten vector space axioms that fails.

13. $M_{4,6}$ with the standard operations

14. $M_{1,1}$ with the standard operations

15. The set of all third-degree polynomials with the standard operations

16. The set of all fifth-degree polynomials with the standard operations

17. The set of all first-degree polynomial functions $ax + b$, $a \neq 0$, whose graphs pass through the origin with the standard operations

18. The set of all quadratic functions whose graphs pass through the origin with the standard operations

19. The set $\{(x, y): x \geq 0, y \text{ is a real number}\}$ with the standard operations in R^2

20. The set $\{(x, y): x \geq 0, y \geq 0\}$ with the standard operations in R^2

21. The set $\{(x, x): x \text{ is a real number}\}$ with the standard operations

22. The set $\{(x, \frac{1}{2}x): x \text{ is a real number}\}$ with the standard operations

23. The set of all 2×2 matrices of the form

$$\begin{bmatrix} a & b \\ c & 0 \end{bmatrix}$$

with the standard operations

24. The set of all 2×2 matrices of the form

$$\begin{bmatrix} a & b \\ c & 1 \end{bmatrix}$$

with the standard operations

25. The set of all 2×2 singular matrices with the standard operations

26. The set of all 2×2 nonsingular matrices with the standard operations

27. The set of all 2×2 diagonal matrices with the standard operations

28. $C[0, 1]$, the set of all continuous functions defined on the interval $[0, 1]$, with the standard operations

29. Rather than use the standard definitions of addition and scalar multiplication in R^2, suppose these two operations are defined as follows.

(a) $(x_1, y_1) + (x_2, y_2) = (x_1 + x_2, y_1 + y_2)$
$c(x, y) = (cx, y)$

(b) $(x_1, y_1) + (x_2, y_2) = (x_1, 0)$
$c(x, y) = (cx, cy)$

(c) $(x_1, y_1) + (x_2, y_2) = (x_1 + x_2, y_1 + y_2)$
$c(x, y) = (\sqrt{c}x, \sqrt{c}y)$

With these new definitions, is R^2 a vector space? Justify your answers.

30. Rather than use the standard definitions of addition and scalar multiplication in R^3, suppose these two operations are defined as follows.

(a) $(x_1, y_1, z_1) + (x_2, y_2, z_2) = (x_1 + x_2, y_1 + y_2, z_1 + z_2)$
$c(x, y, z) = (cx, cy, 0)$

(b) $(x_1, y_1, z_1) + (x_2, y_2, z_2) = (0, 0, 0)$
$c(x, y, z) = (cx, cy, cz)$

(c) $(x_1, y_1, z_1) + (x_2, y_2, z_2)$
$= (x_1 + x_2 + 1, y_1 + y_2 + 1, z_1 + z_2 + 1)$
$c(x, y, z) = (cx, cy, cz)$

(d) $(x_1, y_1, z_1) + (x_2, y_2, z_2)$

$$= (x_1 + x_2 + 1, y_1 + y_2 + 1, z_1 + z_2 + 1)$$

$$c(x, y, z) = (cx + c - 1, cy + c - 1, cz + c - 1)$$

With these new definitions, is R^3 a vector space? Justify your answers.

31. Prove in full detail that $M_{2,2}$, with the standard operations, is a vector space.

32. Prove in full detail, with the standard operations in R^2, that the set $\{(x, 2x): x \text{ is a real number}\}$ is a vector space.

33. Determine whether the set R^2, with the operations

$(x_1, y_1) + (x_2, y_2) = (x_1x_2, y_1y_2)$

and

$c(x_1, y_1) = (cx_1, cy_1)$,

is a vector space. If it is, verify each vector space axiom; if not, state all vector space axioms that fail.

34. Let V be the set of all positive real numbers. Determine whether V is a vector space with the operations below.

$x + y = xy$ Addition

$cx = x^c$ Scalar multiplication

If it is, verify each vector space axiom; if not, state all vector space axioms that fail.

True or False? In Exercises 35 and 36, determine whether each statement is true or false. If a statement is true, give a reason or cite an appropriate statement from the text. If a statement is false, provide an example that shows the statement is not true in all cases or cite an appropriate statement from the text.

35. (a) A vector space consists of four entities: a set of vectors, a set of scalars, and two operations.

(b) The set of all integers with the standard operations is a vector space.

(c) The set of all pairs of real numbers of the form (x, y), where $y \geq 0$, with the standard operations on R^2 is a vector space.

36. (a) To show that a set is not a vector space, it is sufficient to show that just one axiom is not satisfied.

(b) The set of all first-degree polynomials with the standard operations is a vector space.

(c) The set of all pairs of real numbers of the form $(0, y)$, with the standard operations on R^2, is a vector space.

37. Complete the proof of the cancellation property of vector addition by supplying the justification for each step.

Prove that if $\mathbf{u}$, $\mathbf{v}$, and $\mathbf{w}$ are vectors in a vector space V such that $\mathbf{u} + \mathbf{w} = \mathbf{v} + \mathbf{w}$, then $\mathbf{u} = \mathbf{v}$.

$\mathbf{u} + \mathbf{w} = \mathbf{v} + \mathbf{w}$	Given
$(\mathbf{u} + \mathbf{w}) + (-\mathbf{w}) = (\mathbf{v} + \mathbf{w}) + (-\mathbf{w})$	a. ________
$\mathbf{u} + (\mathbf{w} + (-\mathbf{w})) = \mathbf{v} + (\mathbf{w} + (-\mathbf{w}))$	b. ________
$\mathbf{u} + \mathbf{0} = \mathbf{v} + \mathbf{0}$	c. ________
$\mathbf{u} = \mathbf{v}$	d. ________

38. Prove Property 1 of Theorem 4.4.

39. Prove Property 4 of Theorem 4.4.

40. Prove that in a given vector space V, the zero vector is unique.

41. Prove that in a given vector space V, the additive inverse of a vector is unique.

4.3 Subspaces of Vector Spaces

In most important applications in linear algebra, vector spaces occur as **subspaces** of larger spaces. For instance, you will see that the solution set of a homogeneous system of linear equations in n variables is a subspace of R^n. (See Theorem 4.16.)

A subset of a vector space is a subspace if it is a vector space (with the *same* operations), as stated in the next definition.

Definition of Subspace of a Vector Space

A nonempty subset W of a vector space V is called a **subspace** of V if W is a vector space under the operations of addition and scalar multiplication defined in V.

REMARK: Note that if W is a subspace of V, it must be closed under the operations inherited from V.

EXAMPLE 1 A Subspace of R^3

Show that the set $W = \{(x_1, 0, x_3): x_1 \text{ and } x_3 \text{ are real numbers}\}$ is a subspace of R^3 with the standard operations.

SOLUTION The set W is nonempty because it contains the zero vector $(0, 0, 0)$.

Graphically, the set W can be interpreted as simply the xz-plane, as shown in Figure 4.9. The set W is closed under addition because the sum of any two vectors in the xz-plane must also lie in the xz-plane. That is, if $(x_1, 0, x_3)$ and $(y_1, 0, y_3)$ are in W, then their sum $(x_1 + y_1, 0, x_3 + y_3)$ is also in W (because the second component is zero). Similarly, to see that W is closed under scalar multiplication, let $(x_1, 0, x_3)$ be in W and let c be a scalar. Then $c(x_1, 0, x_3) = (cx_1, 0, cx_3)$ has zero as its second component and must be in W. The other eight vector space axioms can be verified as well, and these verifications are left to you.

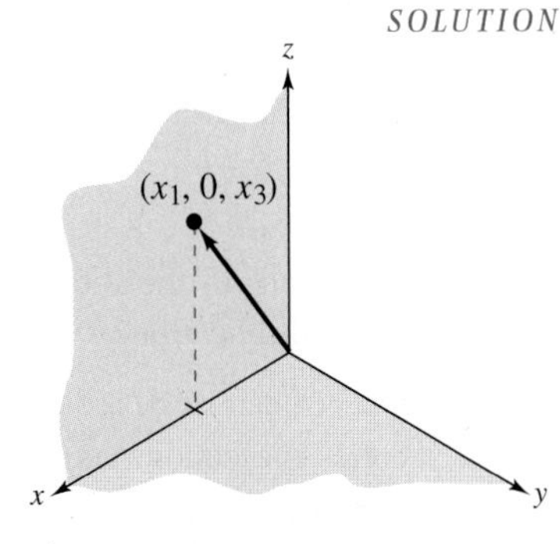

Figure 4.9

To establish that a set W is a vector space, you must verify all ten vector space properties. If W is a subset of a larger vector space V (and the operations defined on W are the *same* as those defined on V), however, then most of the ten properties are *inherited* from the larger space and need no verification. The next theorem tells us it is sufficient to test for closure in order to establish that a nonempty subset of a vector space is a subspace.

THEOREM 4.5 Test for a Subspace

If W is a nonempty subset of a vector space V, then W is a subspace of V if and only if the following closure conditions hold.

1. If $\mathbf{u}$ and $\mathbf{v}$ are in W, then $\mathbf{u} + \mathbf{v}$ is in W.
2. If $\mathbf{u}$ is in W and c is any scalar, then $c\mathbf{u}$ is in W.

PROOF The proof of this theorem in one direction is straightforward. That is, if W is a subspace of V, then W is a vector space and must be closed under addition and scalar multiplication.

To prove the theorem in the other direction, assume that W is closed under addition and scalar multiplication. Note that if $\mathbf{u}, \mathbf{v}$, and $\mathbf{w}$ are in W, then they are also in V. Consequently, vector space axioms 2, 3, 7, 8, 9, and 10 are satisfied automatically. Because W is closed under addition and scalar multiplication, it follows that for any $\mathbf{v}$ in W and scalar $c = 0$,

$$c\mathbf{v} = \mathbf{0}$$

and

$$(-1)\mathbf{v} = -\mathbf{v}$$

both lie in W, which satisfies axioms 4 and 5.

REMARK: Note that if W is a subspace of a vector space V, then both W and V must have the same zero vector $\mathbf{0}$. (See Exercise 43.)

Because a subspace of a vector space is a vector space, it must contain the zero vector. In fact, the simplest subspace of a vector space is the one consisting of only the zero vector,

$$W = \{\mathbf{0}\}.$$

This subspace is called the **zero subspace.** Another obvious subspace of V is V itself. Every vector space contains these two trivial subspaces, and subspaces other than these two are called **proper** (or nontrivial) subspaces.

EXAMPLE 2 The Subspace of $M_{2,2}$

Let W be the set of all 2×2 symmetric matrices. Show that W is a subspace of the vector space $M_{2,2}$, with the standard operations of matrix addition and scalar multiplication.

SOLUTION Recall that a matrix is called *symmetric* if it is equal to its own transpose. Because $M_{2,2}$ is a vector space, you only need to show that W (a subset of $M_{2,2}$) satisfies the conditions of Theorem 4.5. Begin by observing that W is *nonempty*. W is closed under addition because $A_1 = A_1^T$ and $A_2 = A_2^T$, which implies that

$$(A_1 + A_2)^T = A_1^T + A_2^T = A_1 + A_2.$$

So, if A_1 and A_2 are symmetric matrices of order 2, then so is $A_1 + A_2$. Similarly, W is closed under scalar multiplication because $A = A^T$ implies that $(cA)^T = cA^T = cA$. If A is a symmetric matrix of order 2, then so is cA.

The result of Example 2 can be generalized. That is, for any positive integer n, the set of symmetric matrices of order n is a subspace of the vector space $M_{n,n}$ with the standard operations. The next example describes a subset of $M_{n,n}$ that is *not* a subspace.

EXAMPLE 3 The Set of Singular Matrices Is Not a Subspace of $M_{n,n}$

Let W be the set of singular matrices of order 2. Show that W is not a subspace of $M_{2,2}$ with the standard operations.

SOLUTION By Theorem 4.5, you can show that a subset W is not a subspace by showing that W is empty, W is not closed under addition, or W is not closed under scalar multiplication. For this particular set, W is nonempty and closed under scalar multiplication, but it is not closed under addition. To see this, let A and B be

$$A = \begin{bmatrix} 1 & 0 \\ 0 & 0 \end{bmatrix} \quad \text{and} \quad B = \begin{bmatrix} 0 & 0 \\ 0 & 1 \end{bmatrix}.$$

Then A and B are both singular (noninvertible), but their sum

$$A + B = \begin{bmatrix} 1 & 0 \\ 0 & 1 \end{bmatrix}$$

is nonsingular (invertible). So W is not closed under addition, and by Theorem 4.5 you can conclude that it is not a subspace of $M_{2,2}$.

EXAMPLE 4 **The Set of First Quadrant Vectors Is Not a Subspace of R^2**

Show that $W = \{(x_1, x_2): x_1 \geq 0 \text{ and } x_2 \geq 0\}$, with the standard operations, is not a subspace of R^2.

SOLUTION This set is nonempty and closed under addition. It is not, however, closed under scalar multiplication. To see this, note that $(1, 1)$ is in W, but the scalar multiple

$$(-1)(1, 1) = (-1, -1)$$

is not in W. So W is not a subspace of R^2.

You will often encounter sequences of subspaces nested within each other. For instance, consider the vector spaces

$$P_0, P_1, P_2, P_3, \ldots, \quad \text{and} \quad P_n,$$

where P_k is the set of all polynomials of degree less than or equal to k, with the standard operations. It is easy to show that if $j \leq k$, then P_j is a subspace of P_k. You can write $P_0 \subset P_1 \subset P_2 \subset P_3 \subset \cdots \subset P_n$. Another nesting of subspaces is described in Example 5.

EXAMPLE 5 **Subspaces of Functions (Calculus)**

Let W_5 be the *vector space* of all functions defined on $[0, 1]$, and let W_1, W_2, W_3, and W_4 be defined as follows.

W_1 = set of all polynomial functions defined on the interval $[0, 1]$
W_2 = set of all functions that are differentiable on $[0, 1]$
W_3 = set of all functions that are continuous on $[0, 1]$
W_4 = set of all functions that are integrable on $[0, 1]$

Show that $W_1 \subset W_2 \subset W_3 \subset W_4 \subset W_5$ and that W_i is a subspace of W_j for $i \leq j$.

SOLUTION From calculus you know that every polynomial function is differentiable on $[0, 1]$. So, $W_1 \subset W_2$. Moreover, $W_2 \subset W_3$ because every differentiable function is continuous, $W_3 \subset W_4$ because every continuous function is integrable, and $W_4 \subset W_5$ because every integrable function is a function. Resulting from the previous remarks, you have $W_1 \subset W_2 \subset W_3 \subset W_4 \subset W_5$, as shown in Figure 4.10. The verification that W_i is a subspace of W_j for $i \leq j$ is left to you.

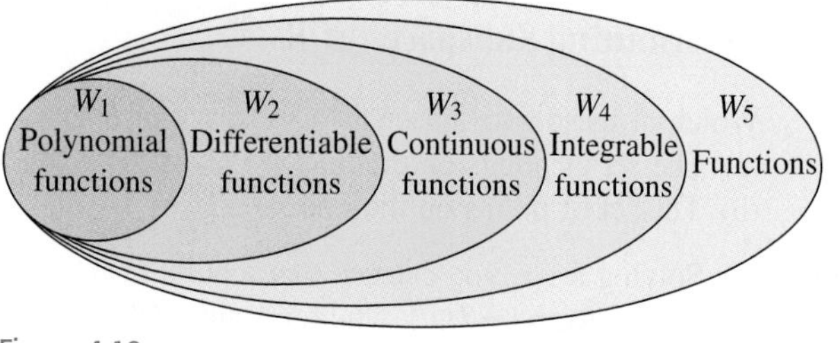

Figure 4.10

Note in Example 5 that if U, V, and W are vector spaces such that W is a subspace of V and V is a subspace of U, then W is also a subspace of U. This special case of the next theorem tells us that the intersection of two subspaces is a subspace, as shown in Figure 4.11.

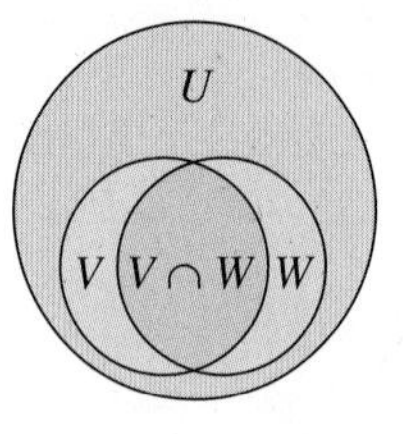

Figure 4.11 The intersection of two subspaces is a subspace.

THEOREM 4.6
The Intersection of Two Subspaces Is a Subspace

If V and W are both subspaces of a vector space U, then the intersection of V and W (denoted by $V \cap W$) is also a subspace of U.

PROOF Because V and W are both subspaces of U, you know that both contain the zero vector, which means that $V \cap W$ is nonempty. To show that $V \cap W$ is closed under addition, let $\mathbf{v}_1$ and $\mathbf{v}_2$ be any two vectors in $V \cap W$. Then, because V and W are both subspaces of U, you know that both are closed under addition. Because $\mathbf{v}_1$ and $\mathbf{v}_2$ are both in V, their sum $\mathbf{v}_1 + \mathbf{v}_2$ must be in V. Similarly, $\mathbf{v}_1 + \mathbf{v}_2$ is in W because $\mathbf{v}_1$ and $\mathbf{v}_2$ are both in W. But this implies that $\mathbf{v}_1 + \mathbf{v}_2$ is in $V \cap W$, and it follows that $V \cap W$ is closed under addition. It is left to you to show (by a similar argument) that $V \cap W$ is closed under scalar multiplication. (See Exercise 48.)

REMARK: Theorem 4.6 states that the *intersection* of two subspaces is a subspace. In Exercise 44 you are asked to show that the *union* of two subspaces is not (in general) a subspace.

Subspace of R^n

R^n is a convenient source for examples of vector spaces, and the remainder of this section is devoted to looking at subspaces of R^n.

EXAMPLE 6 **Determining Subspaces of R^2**

Which of these two subsets is a subspace of R^2?
(a) The set of points on the line $x + 2y = 0$
(b) The set of points on the line $x + 2y = 1$

SOLUTION (a) Solving for x, you can see that a point in R^2 is on the line $x + 2y = 0$ if and only if it has the form $(-2t, t)$, where t is any real number. (See Figure 4.12.)

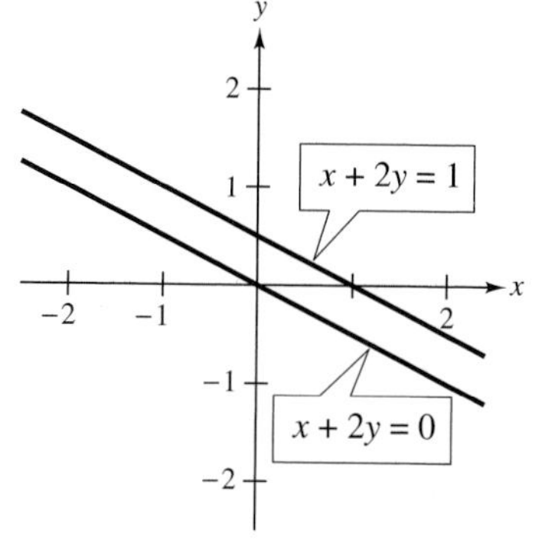

Figure 4.12

To show that this set is closed under addition, let

$$\mathbf{v}_1 = (-2t_1, t_1) \quad \text{and} \quad \mathbf{v}_2 = (-2t_2, t_2)$$

be any two points on the line. Then you have

$$\begin{aligned}\mathbf{v}_1 + \mathbf{v}_2 &= (-2t_1, t_1) + (-2t_2, t_2)\\ &= (-2(t_1 + t_2), t_1 + t_2)\\ &= (-2t_3, t_3),\end{aligned}$$

where $t_3 = t_1 + t_2$. $\mathbf{v}_1 + \mathbf{v}_2$ lies on the line, and the set is closed under addition. In a similar way, you can show that the set is closed under scalar multiplication. So, this set is a subspace of R^2.

(b) This subset of R^2 is *not* a subspace of R^2 because every subspace must contain the zero vector, and the zero vector $(0, 0)$ is not on the line. (See Figure 4.12.)

Of the two lines in Example 6, the one that is a subspace of R^2 is the one that passes through the origin. This is characteristic of subspaces of R^2. That is, if W is a subset of R^2, then it is a subspace if and only if one of the three possibilities listed below is true.

1. W consists of the *single point* $(0, 0)$.
2. W consists of all points on a *line* that pass through the origin.
3. W consists of all of R^2.

These three possibilities are shown graphically in Figure 4.13.

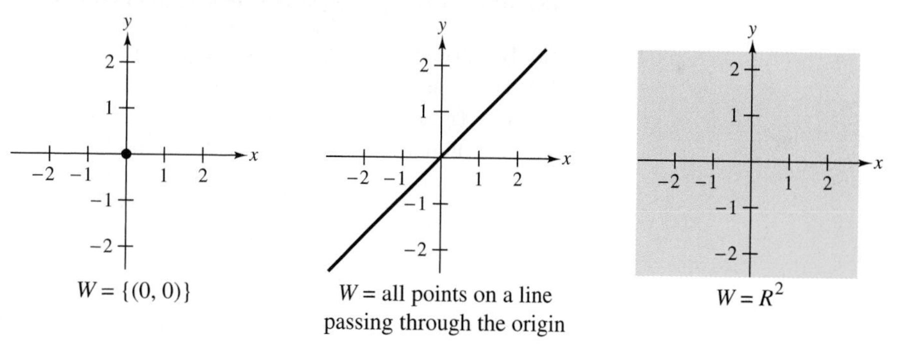

Figure 4.13

EXAMPLE 7 **A Subset of R^2 That Is Not a Subspace**

Show that the subset of R^2 consisting of all points on the unit circle $x^2 + y^2 = 1$ is not a subspace.

SOLUTION This subset of R^2 is *not* a subspace because the points $(1, 0)$ and $(0, 1)$ are in the subset, but their sum $(1, 1)$ is not. (See Figure 4.14.) So, this subset is not closed under addition.

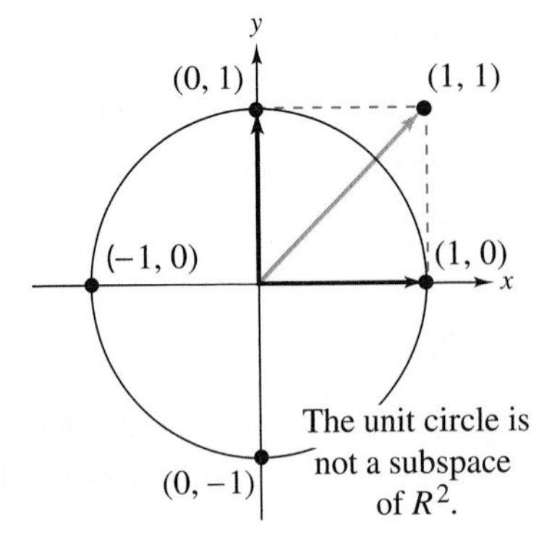

Figure 4.14

REMARK: Another way you can tell that the subset shown in Figure 4.14 is not a subspace of R^2 is by noting that it does not contain the zero vector (the origin).

EXAMPLE 8 Determining Subspaces of R^3

Which of the subsets below is a subspace of R^3?

(a) $W = \{(x_1, x_2, 1): x_1 \text{ and } x_2 \text{ are real numbers}\}$

(b) $W = \{(x_1, x_1 + x_3, x_3): x_1 \text{ and } x_3 \text{ are real numbers}\}$

SOLUTION

(a) Because $\mathbf{0} = (0, 0, 0)$ is not in W, you know that W is *not* a subspace of R^3.

(b) This set is nonempty because it contains the zero vector $(0, 0, 0)$. Let $\mathbf{v} = (v_1, v_1 + v_3, v_3)$ and $\mathbf{u} = (u_1, u_1 + u_3, u_3)$ be two vectors in W, and let c be any real number. Show that W is closed under addition, as follows.

$$\begin{aligned}\mathbf{v} + \mathbf{u} &= (v_1 + u_1, v_1 + v_3 + u_1 + u_3, v_3 + u_3)\\ &= (v_1 + u_1, (v_1 + u_1) + (v_3 + u_3), v_3 + u_3)\\ &= (x_1, x_1 + x_3, x_3)\end{aligned}$$

where $x_1 = v_1 + u_1$ and $x_3 = v_3 + u_3$. $\mathbf{v} + \mathbf{u}$ is in W because it is of the proper form. Similarly, W is closed under scalar multiplication because

$$\begin{aligned}c\mathbf{v} &= (cv_1, c(v_1 + v_3), cv_3)\\ &= (cv_1, cv_1 + cv_3, cv_3)\\ &= (x_1, x_1 + x_3, x_3),\end{aligned}$$

where $x_1 = cv_1$ and $x_3 = cv_3$, which means that $c\mathbf{v}$ is in W. Finally, because W is closed under addition and scalar multiplication, you can conclude that it is a subspace of R^3.

In Example 8, note that the graph of each subset is a plane in R^3. But the only subset that is a *subspace* is the one represented by a plane that passes through the origin. (See Figure 4.15.)

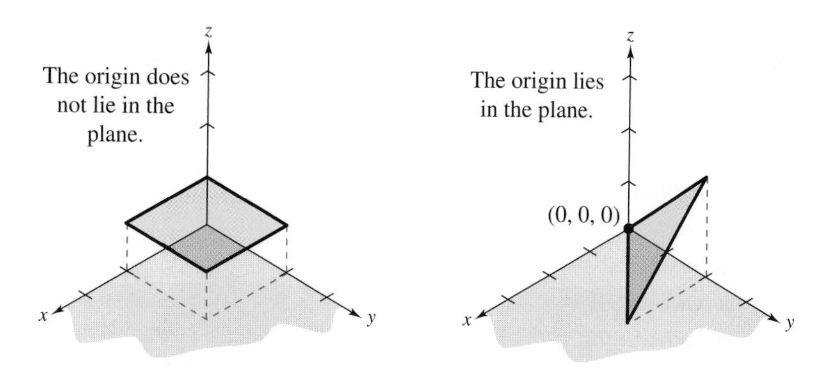

Figure 4.15

In general, you can show that a subset W of R^3 is a subspace of R^3 (with the standard operations) if and only if it has one of the forms listed below.

1. W consists of the *single point* $(0, 0, 0)$.
2. W consists of all points on a *line* that pass through the origin.
3. W consists of all points in a *plane* that pass through the origin.
4. W consists of all of R^3.

SECTION 4.3 Exercises

In Exercises 1–6, verify that W is a subspace of V. In each case, assume that V has the standard operations.

1. $W = \{(x_1, x_2, x_3, 0): x_1, x_2, \text{ and } x_3 \text{ are real numbers}\}$
$V = R^4$

2. $W = \{(x, y, 2x - 3y): x \text{ and } y \text{ are real numbers}\}$
$V = R^3$

3. W is the set of all 2×2 matrices of the form

$$\begin{bmatrix} 0 & a \\ b & 0 \end{bmatrix}.$$

$V = M_{2,2}$

4. W is the set of all 3×2 matrices of the form

$$\begin{bmatrix} a & b \\ a + b & 0 \\ 0 & c \end{bmatrix}.$$

$V = M_{3,2}$

5. Calculus W is the set of all functions that are continuous on $[0, 1]$. V is the set of all functions that are integrable on $[0, 1]$.

6. Calculus W is the set of all functions that are differentiable on $[0, 1]$. V is the set of all functions that are continuous on $[0, 1]$.

In Exercises 7–18, W is not a subspace of the vector space. Verify this by giving a specific example that violates the test for a vector subspace (Theorem 4.5).

7. W is the set of all vectors in R^3 whose third component is -1.

8. W is the set of all vectors in R^2 whose second component is 1.

9. W is the set of all vectors in R^2 whose components are rational numbers.

10. W is the set of all vectors in R^2 whose components are integers.

11. W is the set of all nonnegative functions in $C(-\infty, \infty)$.

12. W is the set of all linear functions $ax + b$, $a \neq 0$, in $C(-\infty, \infty)$.

13. W is the set of all vectors in R^3 whose components are nonnegative.

14. W is the set of all vectors in R^3 whose components are Pythagorean triples.

15. W is the set of all matrices in $M_{n,n}$ with zero determinants.

16. W is the set of all matrices in $M_{n,n}$ such that $A^2 = A$.

17. W is the set of all vectors in R^2 whose second component is the cube of the first.

18. W is the set of all vectors in R^2 whose second component is the square of the first.

In Exercises 19–24, determine if the subset of $C(-\infty, \infty)$ is a subspace of $C(-\infty, \infty)$.

19. The set of all nonnegative functions: $f(x) \geq 0$

20. The set of all even functions: $f(-x) = f(x)$

21. The set of all odd functions: $f(-x) = -f(x)$

22. The set of all constant functions: $f(x) = c$

23. The set of all functions such that $f(0) = 0$

24. The set of all functions such that $f(0) = 1$

In Exercises 25–30, determine if the subset of $M_{n,n}$ is a subspace of $M_{n,n}$ with the standard operations.

25. The set of all $n \times n$ upper triangular matrices

26. The set of all $n \times n$ matrices with integer entries

27. The set of all $n \times n$ matrices A that commute with a given matrix B

28. The set of all $n \times n$ singular matrices

29. The set of all $n \times n$ invertible matrices

30. The set of all $n \times n$ matrices whose entries add up to zero

In Exercises 31–36, determine whether the set W is a subspace of R^3 with the standard operations. Justify your answer.

31. $W = \{(x_1, 0, x_3): x_1 \text{ and } x_3 \text{ are real numbers}\}$

32. $W = \{(x_1, x_2, 4): x_1 \text{ and } x_2 \text{ are real numbers}\}$

33. $W = \{(a, b, a + 2b): a \text{ and } b \text{ are real numbers}\}$

34. $W = \{(s, s - t, t): s \text{ and } t \text{ are real numbers}\}$

35. $W = \{(x_1, x_2, x_1x_2): x_1 \text{ and } x_2 \text{ are real numbers}\}$

36. $W = \{(x_1, 1/x_1, x_3): x_1 \text{ and } x_3 \text{ are real numbers}, x_1 \neq 0\}$

True or False? In Exercises 37 and 38, determine whether each statement is true or false. If a statement is true, give a reason or cite an appropriate statement from the text. If a statement is false, provide an example that shows the statement is not true in all cases or cite an appropriate statement from the text.

37. (a) Every vector space V contains at least one subspace that is the zero subspace.

(b) If V and W are both subspaces of a vector space U, then the intersection of V and W is also a subspace.

(c) If U, V, and W are vector spaces such that W is a subspace of V and U is a subspace of V, then $W = U$.

38. (a) Every vector space V contains two proper subspaces that are the zero subspace and itself.

(b) If W is a subspace of R^2, then W must contain the vector $(0, 0)$.

(c) If W and U are subspaces of a vector space V, then the union of W and U is a subspace of V.

39. Guided Proof Prove that a nonempty set W is a subspace of a vector space V if and only if $a\mathbf{x} + b\mathbf{y}$ is an element of W for all scalars a and b and all vectors $\mathbf{x}$ and $\mathbf{y}$ in W.

Getting Started: In one direction, assume W is a subspace, and show by using closure axioms that $a\mathbf{x} + b\mathbf{y}$ lies in W. In the other direction, assume $a\mathbf{x} + b\mathbf{y}$ is an element of W for all real numbers a and b and elements $\mathbf{x}$ and $\mathbf{y}$ in W, and verify that W is closed under addition and scalar multiplication.

(i) If W is a subspace of V, then use scalar multiplication closure to show that $a\mathbf{x}$ and $b\mathbf{y}$ are in W. Now use additive closure to get the desired result.

(ii) Conversely, assume $a\mathbf{x} + b\mathbf{y}$ is in W. By cleverly assigning specific values to a and b, show that W is closed under addition and scalar multiplication.

40. Let $\mathbf{x}$, $\mathbf{y}$, and $\mathbf{z}$ be vectors in a vector space V. Show that the set of all linear combinations of $\mathbf{x}$, $\mathbf{y}$, and $\mathbf{z}$

$$W = \{a\mathbf{x} + b\mathbf{y} + c\mathbf{z}: a, b, \text{ and } c \text{ are scalars}\}$$

is a subspace of V. This subspace is called the **span** of $\{\mathbf{x}, \mathbf{y}, \mathbf{z}\}$.

41. Let A be a fixed 2×3 matrix. Prove that the set

$$W = \left\{\mathbf{x} \in R^3: A\mathbf{x} = \begin{bmatrix} 1 \\ 2 \end{bmatrix}\right\}$$

is not a subspace of R^3.

42. Let A be a fixed $m \times n$ matrix. Prove that the set

$$W = \{\mathbf{x} \in R^n : A\mathbf{x} = \mathbf{0}\}$$

is a subspace of R^n.

43. Let W be a subspace of the vector space V. Prove that the zero vector in V is also the zero vector in W.

44. Give an example showing that the union of two subspaces of a vector space V is not necessarily a subspace of V.

45. Let A be a fixed 2×2 matrix. Prove that the set

$$W = \{X : XA = AX\}$$

is a subspace of $M_{2,2}$.

46. Calculus Determine whether the set

$$S = \left\{f \in C[0, 1]: \int_0^1 f(x)\, dx = 0\right\}$$

is a subspace of $C[0, 1]$. Prove your answer.

47. Let V and W be two subspaces of a vector space U. Prove that the set

$$V + W = \{\mathbf{u} : \mathbf{u} = \mathbf{v} + \mathbf{w}, \text{ where } \mathbf{v} \in V \text{ and } \mathbf{w} \in W\}$$

is a subspace of U. Describe $V + W$ if V and W are the subspaces of $U = R^2$:

$V = \{(x, 0) : x \text{ is a real number}\}$ and $W = \{(0, y) : y \text{ is a real number}\}$.

48. Complete the proof of Theorem 4.6 by showing that the intersection of two subspaces of a vector space is closed under scalar multiplication.

4.4 Spanning Sets and Linear Independence

This section begins to develop procedures for representing each vector in a vector space as a **linear combination** of a select number of vectors in the space.

Definition of Linear Combination of Vectors

A vector $\mathbf{v}$ in a vector space V is called a **linear combination** of the vectors $\mathbf{u}_1, \mathbf{u}_2, \ldots, \mathbf{u}_k$ in V if $\mathbf{v}$ can be written in the form

$$\mathbf{v} = c_1\mathbf{u}_1 + c_2\mathbf{u}_2 + \cdots + c_k\mathbf{u}_k,$$

where $c_1, c_2, \ldots, c_k$ are scalars.

Often, one or more of the vectors in a set can be written as linear combinations of other vectors in the set. Examples 1, 2, and 3 illustrate this possibility.

EXAMPLE 1 Examples of Linear Combinations

(a) For the set of vectors in R^3,

$$S = \{\overset{\mathbf{v}_1}{(1, 3, 1)}, \overset{\mathbf{v}_2}{(0, 1, 2)}, \overset{\mathbf{v}_3}{(1, 0, -5)}\},$$

$\mathbf{v}_1$ is a linear combination of $\mathbf{v}_2$ and $\mathbf{v}_3$ because

$$\begin{aligned}\mathbf{v}_1 = 3\mathbf{v}_2 + \mathbf{v}_3 &= 3(0, 1, 2) + (1, 0, -5)\\ &= (1, 3, 1).\end{aligned}$$

(b) For the set of vectors in $M_{2,2}$,

$$S = \left\{\overset{\mathbf{v}_1}{\begin{bmatrix} 0 & 8 \\ 2 & 1 \end{bmatrix}}, \overset{\mathbf{v}_2}{\begin{bmatrix} 0 & 2 \\ 1 & 0 \end{bmatrix}}, \overset{\mathbf{v}_3}{\begin{bmatrix} -1 & 3 \\ 1 & 2 \end{bmatrix}}, \overset{\mathbf{v}_4}{\begin{bmatrix} -2 & 0 \\ 1 & 3 \end{bmatrix}}\right\},$$

$\mathbf{v}_1$ is a linear combination of $\mathbf{v}_2$, $\mathbf{v}_3$, and $\mathbf{v}_4$ because

$$\begin{aligned}\mathbf{v}_1 &= \mathbf{v}_2 + 2\mathbf{v}_3 - \mathbf{v}_4\\ &= \begin{bmatrix}0 & 2\\ 1 & 0\end{bmatrix} + 2\begin{bmatrix}-1 & 3\\ 1 & 2\end{bmatrix} - \begin{bmatrix}-2 & 0\\ 1 & 3\end{bmatrix}\\ &= \begin{bmatrix}0 & 8\\ 2 & 1\end{bmatrix}.\end{aligned}$$

In Example 1 it was easy to verify that one of the vectors in the set S was a linear combination of the other vectors because you were provided with the appropriate coefficients to form the linear combination. In the next example, a procedure for finding the coefficients is demonstrated.

EXAMPLE 2 Finding a Linear Combination

Write the vector $\mathbf{w} = (1, 1, 1)$ as a linear combination of vectors in the set S.

$$S = \{\overset{\mathbf{v}_1}{(1, 2, 3)}, \overset{\mathbf{v}_2}{(0, 1, 2)}, \overset{\mathbf{v}_3}{(-1, 0, 1)}\}$$

SOLUTION You need to find scalars c_1, c_2, and c_3 such that

$$\begin{aligned}(1, 1, 1) &= c_1(1, 2, 3) + c_2(0, 1, 2) + c_3(-1, 0, 1)\\ &= (c_1 - c_3, 2c_1 + c_2, 3c_1 + 2c_2 + c_3).\end{aligned}$$

Equating corresponding components yields the system of linear equations below.

$$\begin{aligned}c_1 \qquad\quad - c_3 &= 1\\ 2c_1 + c_2 \qquad\quad &= 1\\ 3c_1 + 2c_2 + c_3 &= 1\end{aligned}$$

Using Gauss-Jordan elimination, you can show that this system has an infinite number of solutions, each of the form

$$c_1 = 1 + t, \qquad c_2 = -1 - 2t, \qquad c_3 = t.$$

To obtain one solution, you could let $t = 1$. Then $c_3 = 1$, $c_2 = -3$, and $c_1 = 2$, and you have

$$\mathbf{w} = 2\mathbf{v}_1 - 3\mathbf{v}_2 + \mathbf{v}_3.$$

Other choices for t would yield other ways to write $\mathbf{w}$ as a linear combination of $\mathbf{v}_1$, $\mathbf{v}_2$, and $\mathbf{v}_3$.

EXAMPLE 3 **Finding a Linear Combination**

If possible, write the vector $\mathbf{w} = (1, -2, 2)$ as a linear combination of vectors in the set S from Example 2.

SOLUTION Following the procedure from Example 2 results in the system

$$\begin{aligned} c_1 \qquad - c_3 &= 1 \\ 2c_1 + c_2 \qquad &= -2 \\ 3c_1 + 2c_2 + c_3 &= 2. \end{aligned}$$

The augmented matrix of this system row reduces to

$$\left[\begin{array}{ccc|c} 1 & 0 & -1 & 0 \\ 0 & 1 & 2 & 0 \\ 0 & 0 & 0 & 1 \end{array}\right].$$

From the third row you can conclude that the system of equations is inconsistent, and that means that there is no solution. Consequently, $\mathbf{w}$ *cannot* be written as a linear combination of $\mathbf{v}_1$, $\mathbf{v}_2$, and $\mathbf{v}_3$.

Spanning Sets

If every vector in a vector space can be written as a linear combination of vectors in a set S, then S is called a **spanning set** of the vector space.

Definition of Spanning Set of a Vector Space

Let $S = \{\mathbf{v}_1, \mathbf{v}_2, \ldots, \mathbf{v}_k\}$ be a subset of a vector space V. The set S is called a **spanning set** of V if *every* vector in V can be written as a linear combination of vectors in S. In such cases it is said that S **spans** V.

EXAMPLE 4 **Examples of Spanning Sets**

(a) The set $S = \{(1, 0, 0), (0, 1, 0), (0, 0, 1)\}$ spans R^3 because any vector $\mathbf{u} = (u_1, u_2, u_3)$ in R^3 can be written as

$$\mathbf{u} = u_1(1, 0, 0) + u_2(0, 1, 0) + u_3(0, 0, 1) = (u_1, u_2, u_3).$$

(b) The set $S = \{1, x, x^2\}$ spans P_2 because any polynomial function $p(x) = a + bx + cx^2$ in P_2 can be written as

$$\begin{aligned} p(x) &= a(1) + b(x) + c(x^2) \\ &= a + bx + cx^2. \end{aligned}$$

The spanning sets in Example 4 are called the **standard spanning sets** of R^3 and P_2, respectively. (You will learn more about standard spanning sets in the next section.) In the next example you will look at a nonstandard spanning set of R^3.

EXAMPLE 5 A Spanning Set of R^3

Show that the set $S = \{(1, 2, 3), (0, 1, 2), (-2, 0, 1)\}$ spans R^3.

SOLUTION Let $\mathbf{u} = (u_1, u_2, u_3)$ be *any* vector in R^3. You need to find scalars c_1, c_2, and c_3 such that

$$\begin{aligned}(u_1, u_2, u_3) &= c_1(1, 2, 3) + c_2(0, 1, 2) + c_3(-2, 0, 1)\\ &= (c_1 - 2c_3, 2c_1 + c_2, 3c_1 + 2c_2 + c_3).\end{aligned}$$

This vector equation produces the system

$$\begin{aligned} c_1 \qquad\quad - 2c_3 &= u_1\\ 2c_1 + c_2 \qquad\quad &= u_2\\ 3c_1 + 2c_2 + c_3 &= u_3.\end{aligned}$$

The coefficient matrix for this system has a nonzero determinant, and it follows from the list of equivalent conditions given in Section 3.3 that the system has a unique solution. So, any vector in R^3 can be written as a linear combination of the vectors in S, and you can conclude that the set S spans R^3.

EXAMPLE 6 A Set That Does Not Span R^3

From Example 3 you know that the set

$$S = \{(1, 2, 3), (0, 1, 2), (-1, 0, 1)\}$$

does not span R^3 because $\mathbf{w} = (1, -2, 2)$ is in R^3 and cannot be expressed as a linear combination of the vectors in S.

Comparing the sets of vectors in Examples 5 and 6, note that the sets are the same except for a seemingly insignificant difference in the third vector.

$$S_1 = \{(1, 2, 3), (0, 1, 2), (-2, 0, 1)\} \qquad \textbf{Example 5}$$
$$S_2 = \{(1, 2, 3), (0, 1, 2), (-1, 0, 1)\} \qquad \textbf{Example 6}$$

The difference, however, is significant, because the set S_1 spans R^3 whereas the set S_2 does not. The reason for this difference can be seen in Figure 4.16. The vectors in S_2 lie in a common plane; the vectors in S_1 do not.

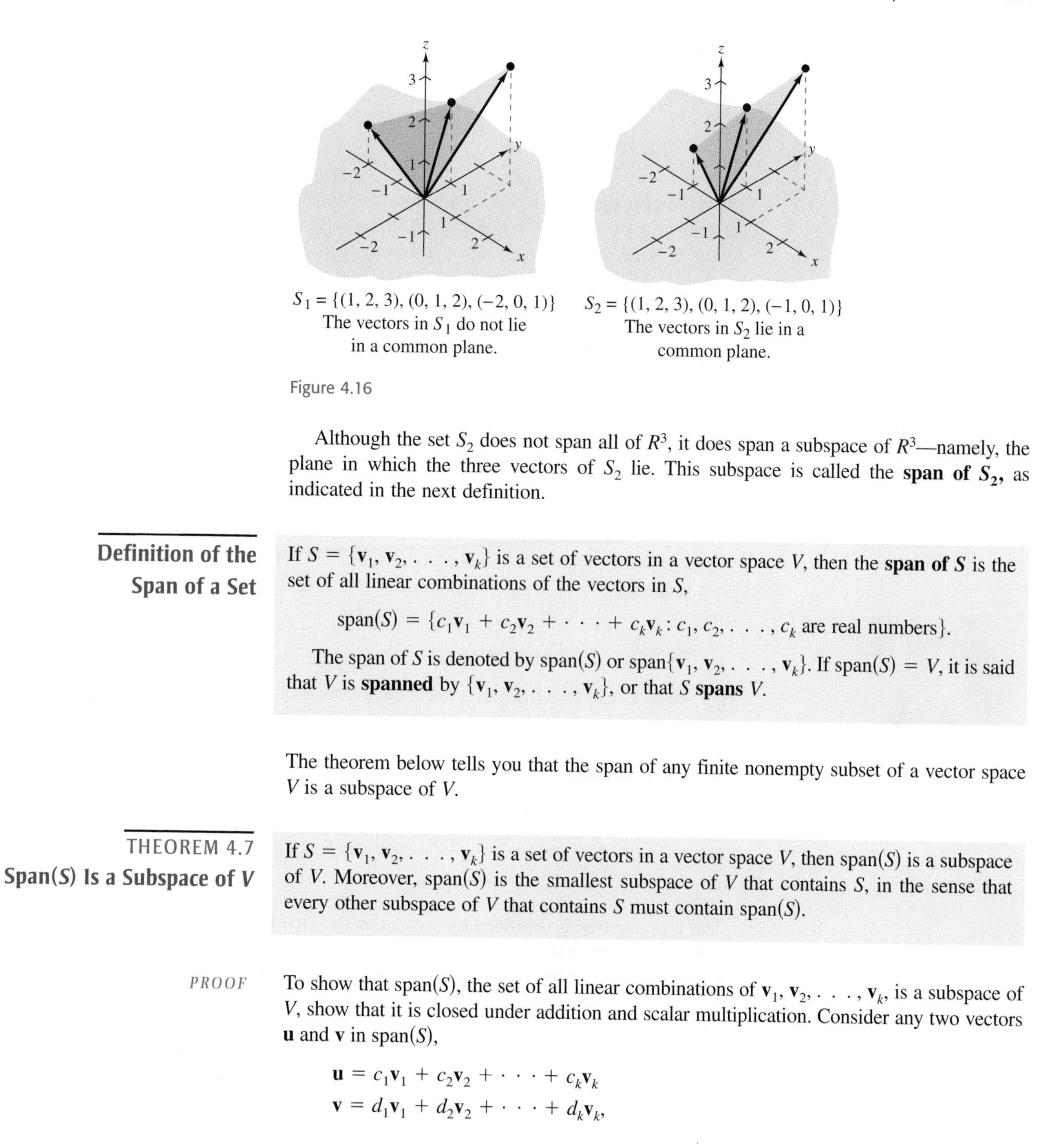

$S_1 = \{(1, 2, 3), (0, 1, 2), (-2, 0, 1)\}$
The vectors in S_1 do not lie in a common plane.

$S_2 = \{(1, 2, 3), (0, 1, 2), (-1, 0, 1)\}$
The vectors in S_2 lie in a common plane.

Figure 4.16

Although the set S_2 does not span all of R^3, it does span a subspace of R^3—namely, the plane in which the three vectors of S_2 lie. This subspace is called the **span of S_2,** as indicated in the next definition.

Definition of the Span of a Set

If $S = \{\mathbf{v}_1, \mathbf{v}_2, \ldots, \mathbf{v}_k\}$ is a set of vectors in a vector space V, then the **span of S** is the set of all linear combinations of the vectors in S,

$$\text{span}(S) = \{c_1\mathbf{v}_1 + c_2\mathbf{v}_2 + \cdots + c_k\mathbf{v}_k : c_1, c_2, \ldots, c_k \text{ are real numbers}\}.$$

The span of S is denoted by span(S) or span$\{\mathbf{v}_1, \mathbf{v}_2, \ldots, \mathbf{v}_k\}$. If span($S$) $= V$, it is said that V is **spanned** by $\{\mathbf{v}_1, \mathbf{v}_2, \ldots, \mathbf{v}_k\}$, or that S **spans** V.

The theorem below tells you that the span of any finite nonempty subset of a vector space V is a subspace of V.

THEOREM 4.7 Span(S) Is a Subspace of V

If $S = \{\mathbf{v}_1, \mathbf{v}_2, \ldots, \mathbf{v}_k\}$ is a set of vectors in a vector space V, then span(S) is a subspace of V. Moreover, span(S) is the smallest subspace of V that contains S, in the sense that every other subspace of V that contains S must contain span(S).

PROOF To show that span(S), the set of all linear combinations of $\mathbf{v}_1, \mathbf{v}_2, \ldots, \mathbf{v}_k$, is a subspace of V, show that it is closed under addition and scalar multiplication. Consider any two vectors $\mathbf{u}$ and $\mathbf{v}$ in span(S),

$$\mathbf{u} = c_1\mathbf{v}_1 + c_2\mathbf{v}_2 + \cdots + c_k\mathbf{v}_k$$
$$\mathbf{v} = d_1\mathbf{v}_1 + d_2\mathbf{v}_2 + \cdots + d_k\mathbf{v}_k,$$

where $c_1, c_2, \ldots, c_k$ and $d_1, d_2, \ldots, d_k$ are scalars. Then

$$\mathbf{u} + \mathbf{v} = (c_1 + d_1)\mathbf{v}_1 + (c_2 + d_2)\mathbf{v}_2 + \cdots + (c_k + d_k)\mathbf{v}_k$$

and

$$c\mathbf{u} = (cc_1)\mathbf{v}_1 + (cc_2)\mathbf{v}_2 + \cdots + (cc_k)\mathbf{v}_k,$$

which means that $\mathbf{u} + \mathbf{v}$ and $c\mathbf{u}$ are also in span(S) because they can be written as linear combinations of vectors in S. So, span(S) is a subspace of V. It is left to you to prove that span(S) is the smallest subspace of V that contains S. (See Exercise 50.)

Linear Dependence and Linear Independence

For a given set of vectors $S = \{\mathbf{v}_1, \mathbf{v}_2, \ldots, \mathbf{v}_k\}$ in a vector space V, the vector equation

$$c_1\mathbf{v}_1 + c_2\mathbf{v}_2 + \cdots + c_k\mathbf{v}_k = \mathbf{0}$$

always has the **trivial solution**

$$c_1 = 0, c_2 = 0, \ldots, c_k = 0.$$

Often, however, there are also **nontrivial** solutions. For instance, in Example 1(a) you saw that in the set

$$S = \{\underset{\mathbf{v}_1}{(1, 3, 1)}, \underset{\mathbf{v}_2}{(0, 1, 2)}, \underset{\mathbf{v}_3}{(1, 0, -5)}\},$$

the vector $\mathbf{v}_1$ can be written as a linear combination of the other two as follows.

$$\mathbf{v}_1 = 3\mathbf{v}_2 + \mathbf{v}_3$$

The vector equation

$$c_1\mathbf{v}_1 + c_2\mathbf{v}_2 + c_3\mathbf{v}_3 = \mathbf{0}$$

has a nontrivial solution in which the coefficients are *not all zero:*

$$c_1 = 1, \quad c_2 = -3, \quad c_3 = -1.$$

This characteristic is described by saying that the set S is **linearly dependent.** Had the *only* solution been the trivial one ($c_1 = c_2 = c_3 = 0$), then the set S would have been **linearly independent.** This notion is essential to the study of linear algebra, and is formally stated in the next definition.

Definition of Linear Dependence and Linear Independence

A set of vectors $S = \{\mathbf{v}_1, \mathbf{v}_2, \ldots, \mathbf{v}_k\}$ in a vector space V is called **linearly independent** if the vector equation

$$c_1\mathbf{v}_1 + c_2\mathbf{v}_2 + \cdots + c_k\mathbf{v}_k = \mathbf{0}$$

has only the trivial solution, $c_1 = 0, c_2 = 0, \ldots, c_k = 0$. If there are also nontrivial solutions, then S is called **linearly dependent.**

EXAMPLE 7 Examples of Linearly Dependent Sets

(a) The set $S = \{(1, 2), (2, 4)\}$ in R^2 is linearly dependent because

$$-2(1, 2) + (2, 4) = (0, 0).$$

(b) The set $S = \{(1, 0), (0, 1), (-2, 5)\}$ in R^2 is linearly dependent because

$$2(1, 0) - 5(0, 1) + (-2, 5) = (0, 0).$$

(c) The set $S = \{(0, 0), (1, 2)\}$ in R^2 is linearly dependent because

$$1(0, 0) + 0(1, 2) = (0, 0).$$

The next example demonstrates a testing procedure for determining whether a set of vectors is linearly independent or linearly dependent.

EXAMPLE 8 Testing for Linear Independence

Determine whether the set of vectors in R^3 is linearly independent or linearly dependent.

$$S = \{\overset{\mathbf{v}_1}{(1, 2, 3)}, \overset{\mathbf{v}_2}{(0, 1, 2)}, \overset{\mathbf{v}_3}{(-2, 0, 1)}\}$$

SOLUTION To test for linear independence or linear dependence, form the vector equation

$$c_1\mathbf{v}_1 + c_2\mathbf{v}_2 + c_3\mathbf{v}_3 = \mathbf{0}.$$

If the only solution of this equation is

$$c_1 = c_2 = c_3 = 0,$$

then the set S is linearly independent. Otherwise, S is linearly dependent. Expanding this equation, you have

$$c_1(1, 2, 3) + c_2(0, 1, 2) + c_3(-2, 0, 1) = (0, 0, 0)$$
$$(c_1 - 2c_3, 2c_1 + c_2, 3c_1 + 2c_2 + c_3) = (0, 0, 0),$$

which yields the homogeneous system of linear equations in c_1, c_2, and c_3 shown below.

$$\begin{aligned} c_1 \qquad\quad - 2c_3 &= 0 \\ 2c_1 + c_2 \qquad\quad &= 0 \\ 3c_1 + 2c_2 + c_3 &= 0 \end{aligned}$$

The augmented matrix of this system reduces by Gauss-Jordan elimination as follows.

$$\begin{bmatrix} 1 & 0 & -2 & 0 \\ 2 & 1 & 0 & 0 \\ 3 & 2 & 1 & 0 \end{bmatrix} \longrightarrow \begin{bmatrix} 1 & 0 & 0 & 0 \\ 0 & 1 & 0 & 0 \\ 0 & 0 & 1 & 0 \end{bmatrix}$$

This implies that the only solution is the trivial solution

$$c_1 = c_2 = c_3 = 0.$$

So, S is linearly independent.

The steps shown in Example 8 are summarized as follows.

Testing for Linear Independence and Linear Dependence

Let $S = \{\mathbf{v}_1, \mathbf{v}_2, \ldots, \mathbf{v}_k\}$ be a set of vectors in a vector space V. To determine whether S is linearly independent or linearly dependent, perform the following steps.

1. From the vector equation $c_1\mathbf{v}_1 + c_2\mathbf{v}_2 + \cdots + c_k\mathbf{v}_k = \mathbf{0}$, write a homogeneous system of linear equations in the variables $c_1, c_2, \ldots,$ and c_k.
2. Use Gaussian elimination to determine whether the system has a unique solution.
3. If the system has only the trivial solution, $c_1 = 0, c_2 = 0, \ldots, c_k = 0$, then the set S is linearly independent. If the system also has nontrivial solutions, then S is linearly dependent.

EXAMPLE 9 Testing for Linear Independence

Determine whether the set of vectors in P_2 is linearly independent or linearly dependent.

$$S = \{\overset{\mathbf{v}_1}{1 + x - 2x^2}, \overset{\mathbf{v}_2}{2 + 5x - x^2}, \overset{\mathbf{v}_3}{x + x^2}\}$$

SOLUTION Expanding the equation $c_1\mathbf{v}_1 + c_2\mathbf{v}_2 + c_3\mathbf{v}_3 = \mathbf{0}$ produces

$$c_1(1 + x - 2x^2) + c_2(2 + 5x - x^2) + c_3(x + x^2) = 0 + 0x + 0x^2$$
$$(c_1 + 2c_2) + (c_1 + 5c_2 + c_3)x + (-2c_1 - c_2 + c_3)x^2 = 0 + 0x + 0x^2.$$

Equating corresponding coefficients of equal powers of x produces the homogeneous system of linear equations in c_1, c_2, and c_3 shown below.

$$\begin{aligned} c_1 + 2c_2 &= 0 \\ c_1 + 5c_2 + c_3 &= 0 \\ -2c_1 - c_2 + c_3 &= 0 \end{aligned}$$

The augmented matrix of this system reduces by Gaussian elimination as follows.

$$\begin{bmatrix} 1 & 2 & 0 & 0 \\ 1 & 5 & 1 & 0 \\ -2 & -1 & 1 & 0 \end{bmatrix} \longrightarrow \begin{bmatrix} 1 & 2 & 0 & 0 \\ 0 & 1 & \frac{1}{3} & 0 \\ 0 & 0 & 0 & 0 \end{bmatrix}$$

This implies that the system has an infinite number of solutions. So, the system must have nontrivial solutions, and you can conclude that the set S is linearly dependent.

One nontrivial solution is

$$c_1 = 2, \quad c_2 = -1, \quad \text{and} \quad c_3 = 3,$$

which yields the nontrivial linear combination

$$(2)(1 + x - 2x^2) + (-1)(2 + 5x - x^2) + (3)(x + x^2) = 0.$$

EXAMPLE 10 **Testing for Linear Independence**

Determine whether the set of vectors in $M_{2,2}$ is linearly independent or linearly dependent.

$$S = \left\{ \overset{\mathbf{v}_1}{\begin{bmatrix} 2 & 1 \\ 0 & 1 \end{bmatrix}}, \overset{\mathbf{v}_2}{\begin{bmatrix} 3 & 0 \\ 2 & 1 \end{bmatrix}}, \overset{\mathbf{v}_3}{\begin{bmatrix} 1 & 0 \\ 2 & 0 \end{bmatrix}} \right\}$$

SOLUTION From the equation

$$c_1\mathbf{v}_1 + c_2\mathbf{v}_2 + c_3\mathbf{v}_3 = \mathbf{0},$$

you have

$$c_1\begin{bmatrix} 2 & 1 \\ 0 & 1 \end{bmatrix} + c_2\begin{bmatrix} 3 & 0 \\ 2 & 1 \end{bmatrix} + c_3\begin{bmatrix} 1 & 0 \\ 2 & 0 \end{bmatrix} = \begin{bmatrix} 0 & 0 \\ 0 & 0 \end{bmatrix},$$

which produces the system of linear equations in c_1, c_2, and c_3 shown below.

$$\begin{aligned} 2c_1 + 3c_2 + c_3 &= 0 \\ c_1 &= 0 \\ 2c_2 + 2c_3 &= 0 \\ c_1 + c_2 &= 0 \end{aligned}$$

Using Gaussian elimination, the augmented matrix of this system reduces as follows.

$$\begin{bmatrix} 2 & 3 & 1 & 0 \\ 1 & 0 & 0 & 0 \\ 0 & 2 & 2 & 0 \\ 1 & 1 & 0 & 0 \end{bmatrix} \longrightarrow \begin{bmatrix} 1 & 0 & 0 & 0 \\ 0 & 1 & 0 & 0 \\ 0 & 0 & 1 & 0 \\ 0 & 0 & 0 & 0 \end{bmatrix}$$

The system has only the trivial solution and you can conclude that the set S is linearly independent.

EXAMPLE 11 **Testing for Linear Independence**

Determine whether the set of vectors in $M_{4,1}$ is linearly independent or linearly dependent.

$$S = \left\{ \overset{\mathbf{v}_1}{\begin{bmatrix} 1 \\ 0 \\ -1 \\ 0 \end{bmatrix}}, \overset{\mathbf{v}_2}{\begin{bmatrix} 1 \\ 1 \\ 0 \\ 2 \end{bmatrix}}, \overset{\mathbf{v}_3}{\begin{bmatrix} 0 \\ 3 \\ 1 \\ -2 \end{bmatrix}}, \overset{\mathbf{v}_4}{\begin{bmatrix} 0 \\ 1 \\ -1 \\ 2 \end{bmatrix}} \right\}$$

SOLUTION From the equation

$$c_1\mathbf{v}_1 + c_2\mathbf{v}_2 + c_3\mathbf{v}_3 + c_4\mathbf{v}_4 = \mathbf{0},$$

you obtain

$$c_1\begin{bmatrix} 1 \\ 0 \\ -1 \\ 0 \end{bmatrix} + c_2\begin{bmatrix} 1 \\ 1 \\ 0 \\ 2 \end{bmatrix} + c_3\begin{bmatrix} 0 \\ 3 \\ 1 \\ -2 \end{bmatrix} + c_4\begin{bmatrix} 0 \\ 1 \\ -1 \\ 2 \end{bmatrix} = \begin{bmatrix} 0 \\ 0 \\ 0 \\ 0 \end{bmatrix}.$$

This equation produces the system of linear equations in c_1, c_2, c_3, and c_4 shown below.

$$\begin{aligned} c_1 + c_2 \qquad\qquad\quad &= 0 \\ c_2 + 3c_3 + c_4 &= 0 \\ -c_1 \qquad + c_3 - c_4 &= 0 \\ 2c_2 - 2c_3 + 2c_4 &= 0 \end{aligned}$$

Using Gaussian elimination, you can row reduce the augmented matrix of this system as follows.

$$\begin{bmatrix} 1 & 1 & 0 & 0 & 0 \\ 0 & 1 & 3 & 1 & 0 \\ -1 & 0 & 1 & -1 & 0 \\ 0 & 2 & -2 & 2 & 0 \end{bmatrix} \longrightarrow \begin{bmatrix} 1 & 0 & 0 & 0 & 0 \\ 0 & 1 & 0 & 0 & 0 \\ 0 & 0 & 1 & 0 & 0 \\ 0 & 0 & 0 & 1 & 0 \end{bmatrix}$$

The system has only the trivial solution, and you can conclude that the set S is linearly independent.

If a set of vectors is linearly dependent, then by definition the equation

$$c_1\mathbf{v}_1 + c_2\mathbf{v}_2 + \cdots + c_k\mathbf{v}_k = \mathbf{0}$$

has a nontrivial solution (a solution for which not all the c_i's are zero). For instance, if $c_1 \neq 0$, then you can solve this equation for $\mathbf{v}_1$ and write $\mathbf{v}_1$ as a linear combination of the other vectors $\mathbf{v}_2, \mathbf{v}_3, \ldots,$ and $\mathbf{v}_k$. In other words, the vector $\mathbf{v}_1$ *depends* on the other vectors in the set. This property is characteristic of a linearly dependent set.

THEOREM 4.8
A Property of Linearly Dependent Sets

A set $S = \{\mathbf{v}_1, \mathbf{v}_2, . . . , \mathbf{v}_k\}$, $k \geq 2$, is linearly dependent if and only if at least one of the vectors $\mathbf{v}_j$ can be written as a linear combination of the other vectors in S.

PROOF To prove the theorem in one direction, assume S is a linearly dependent set. Then there exist scalars $c_1, c_2, c_3, . . . , c_k$ (not all zero) such that

$$c_1\mathbf{v}_1 + c_2\mathbf{v}_2 + c_3\mathbf{v}_3 + \cdots + c_k\mathbf{v}_k = \mathbf{0}.$$

Because one of the coefficients must be nonzero, no generality is lost by assuming $c_1 \neq 0$. Then solving for $\mathbf{v}_1$ as a linear combination of the other vectors produces

$$c_1\mathbf{v}_1 = -c_2\mathbf{v}_2 - c_3\mathbf{v}_3 - \cdots - c_k\mathbf{v}_k$$

$$\mathbf{v}_1 = -\frac{c_2}{c_1}\mathbf{v}_2 - \frac{c_3}{c_1}\mathbf{v}_3 - \cdots - \frac{c_k}{c_1}\mathbf{v}_k.$$

Conversely, suppose the vector $\mathbf{v}_1$ in S is a linear combination of the other vectors. That is,

$$\mathbf{v}_1 = c_2\mathbf{v}_2 + c_3\mathbf{v}_3 + \cdots + c_k\mathbf{v}_k.$$

Then the equation $-\mathbf{v}_1 + c_2\mathbf{v}_2 + c_3\mathbf{v}_3 + \cdots + c_k\mathbf{v}_k = \mathbf{0}$ has at least one coefficient, -1, that is nonzero, and you can conclude that S is linearly dependent.

EXAMPLE 12 Writing a Vector as a Linear Combination of Other Vectors

In Example 9, you determined that the set

$$S = \{\overset{\mathbf{v}_1}{1 + x - 2x^2}, \overset{\mathbf{v}_2}{2 + 5x - x^2}, \overset{\mathbf{v}_3}{x + x^2}\}$$

is linearly dependent. Show that one of the vectors in this set can be written as a linear combination of the other two.

SOLUTION In Example 9, the equation $c_1\mathbf{v}_1 + c_2\mathbf{v}_2 + c_3\mathbf{v}_3 = \mathbf{0}$ produced the system

$$\begin{aligned} c_1 + 2c_2 &= 0 \\ c_1 + 5c_2 + c_3 &= 0 \\ -2c_1 - c_2 + c_3 &= 0. \end{aligned}$$

This system has an infinite number of solutions represented by $c_3 = 3t$, $c_2 = -t$, and $c_1 = 2t$. Letting $t = 1$ results in the equation $2\mathbf{v}_1 - \mathbf{v}_2 + 3\mathbf{v}_3 = \mathbf{0}$. So, $\mathbf{v}_2$ can be written as a linear combination of $\mathbf{v}_1$ and $\mathbf{v}_3$ as follows.

$$\mathbf{v}_2 = 2\mathbf{v}_1 + 3\mathbf{v}_3$$

A check yields

$$\begin{aligned} 2 + 5x - x^2 &= 2(1 + x - 2x^2) + 3(x + x^2) \\ &= 2 + 2x - 4x^2 + 3x + 3x^2 \\ &= 2 + 5x - x^2. \end{aligned}$$

Theorem 4.8 has a practical corollary that provides a simple test for determining whether *two* vectors are linearly dependent. In Exercise 69 you are asked to prove this corollary.

THEOREM 4.8 Corollary

Two vectors $\mathbf{u}$ and $\mathbf{v}$ in a vector space V are linearly dependent if and only if one is a scalar multiple of the other.

REMARK: The zero vector is always a scalar multiple of another vector in a vector space.

EXAMPLE 13 Testing for Linear Dependence of Two Vectors

(a) The set

$$S = \{\overset{\mathbf{v}_1}{(1, 2, 0)}, \overset{\mathbf{v}_2}{(-2, 2, 1)}\}$$

is linearly independent because $\mathbf{v}_1$ and $\mathbf{v}_2$ are not scalar multiples of each other, as shown in Figure 4.17(a).

(b) The set

$$S = \{\overset{\mathbf{v}_1}{(4, -4, -2)}, \overset{\mathbf{v}_2}{(-2, 2, 1)}\}$$

is linearly dependent because $\mathbf{v}_1 = -2\mathbf{v}_2$, as shown in Figure 4.17(b).

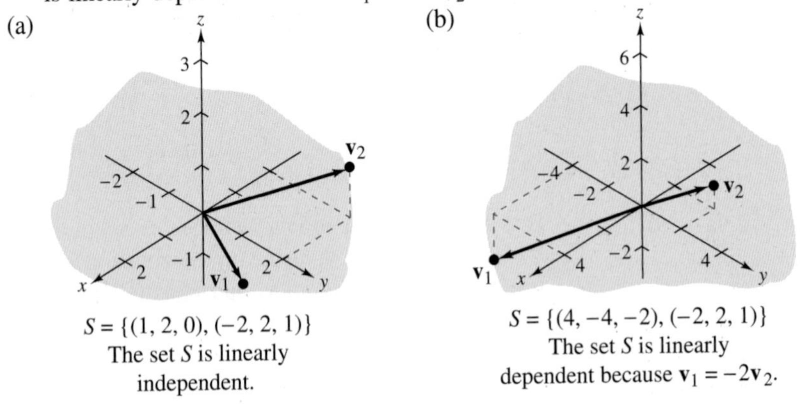

(a) $S = \{(1, 2, 0), (-2, 2, 1)\}$ The set S is linearly independent.

(b) $S = \{(4, -4, -2), (-2, 2, 1)\}$ The set S is linearly dependent because $\mathbf{v}_1 = -2\mathbf{v}_2$.

Figure 4.17

SECTION 4.4 Exercises

In Exercises 1–4, determine whether each vector can be written as a linear combination of the vectors in S.

1. $S = \{(2, -1, 3), (5, 0, 4)\}$

(a) $\mathbf{u} = (1, 1, -1)$ (b) $\mathbf{v} = \left(8, -\frac{1}{4}, \frac{27}{4}\right)$

(c) $\mathbf{w} = (1, -8, 12)$ (d) $\mathbf{z} = (-1, -2, 2)$

2. $S = \{(1, 2, -2), (2, -1, 1)\}$

(a) $\mathbf{u} = (1, -5, -5)$ (b) $\mathbf{v} = (-2, -6, 6)$

(c) $\mathbf{w} = (-1, -22, 22)$ (d) $\mathbf{z} = (-4, -3, 3)$

3. $S = \{(2, 0, 7), (2, 4, 5), (2, -12, 13)\}$

(a) $\mathbf{u} = (-1, 5, -6)$ (b) $\mathbf{v} = (-3, 15, 18)$

(c) $\mathbf{w} = \left(\frac{1}{3}, \frac{4}{3}, \frac{1}{2}\right)$ (d) $\mathbf{z} = (2, 20, -3)$

4. $S = \{(6, -7, 8, 6), (4, 6, -4, 1)\}$

(a) $\mathbf{u} = (-42, 113, -112, -60)$

(b) $\mathbf{v} = \left(\frac{49}{2}, \frac{99}{4}, -14, \frac{19}{2}\right)$

(c) $\mathbf{w} = \left(-4, -14, \frac{27}{2}, \frac{53}{8}\right)$

(d) $\mathbf{z} = \left(8, 4, -1, \frac{17}{4}\right)$

In Exercises 5–16, determine whether the set S spans R^2. If the set does not span R^2, give a geometric description of the subspace that it does span.

5. $S = \{(2, 1), (-1, 2)\}$ **6.** $S = \{(1, -1), (2, 1)\}$

7. $S = \{(5, 0), (5, -4)\}$ **8.** $S = \{(2, 0), (0, 1)\}$

9. $S = \{(-3, 5)\}$ **10.** $S = \{(1, 1)\}$

11. $S = \{(1, 3), (-2, -6), (4, 12)\}$

12. $S = \left\{(1, 2), (-2, -4), \left(\frac{1}{2}, 1\right)\right\}$

13. $S = \{(-1, 2), (2, -4)\}$

14. $S = \{(0, 2), (1, 4)\}$

15. $S = \{(-1, 4), (4, -1), (1, 1)\}$

16. $S = \{(-1, 2), (2, -1), (1, 1)\}$

In Exercises 17–22, determine whether the set S spans R^3. If the set does not span R^3, give a geometric description of the subspace that it does span.

17. $S = \{(4, 7, 3), (-1, 2, 6), (2, -3, 5)\}$

18. $S = \{(6, 7, 6), (3, 2, -4), (1, -3, 2)\}$

19. $S = \{(-2, 5, 0), (4, 6, 3)\}$

20. $S = \{(1, 0, 1), (1, 1, 0), (0, 1, 1)\}$

21. $S = \{(1, -2, 0), (0, 0, 1), (-1, 2, 0)\}$

22. $S = \{(1, 0, 3), (2, 0, -1), (4, 0, 5), (2, 0, 6)\}$

In Exercises 23–34, determine whether the set S is linearly independent or linearly dependent.

23. $S = \{(-2, 2), (3, 5)\}$ **24.** $S = \{(-2, 4), (1, -2)\}$

25. $S = \{(0, 0), (1, -1)\}$ **26.** $S = \{(1, 0), (1, 1), (2, -1)\}$

27. $S = \{(1, -4, 1), (6, 3, 2)\}$

28. $S = \{(6, 2, 1), (-1, 3, 2)\}$

29. $S = \{(1, 1, 1), (2, 2, 2), (3, 3, 3)\}$

30. $S = \left\{\left(\frac{3}{4}, \frac{5}{2}, \frac{3}{2}\right), \left(3, 4, \frac{7}{2}\right), \left(-\frac{3}{2}, 6, 2\right)\right\}$

31. $S = \{(-4, -3, 4), (1, -2, 3), (6, 0, 0)\}$

32. $S = \{(1, 0, 0), (0, 4, 0), (0, 0, -6), (1, 5, -3)\}$

33. $S = \{(4, -3, 6, 2), (1, 8, 3, 1), (3, -2, -1, 0)\}$

34. $S = \{(0, 0, 0, 1), (0, 0, 1, 1), (0, 1, 1, 1), (1, 1, 1, 1)\}$

In Exercises 35–38, show that the set is linearly dependent by finding a nontrivial linear combination (of vectors in the set) whose sum is the zero vector. Then express one of the vectors in the set as a linear combination of the other vectors in the set.

35. $S = \{(3, 4), (-1, 1), (2, 0)\}$

36. $S = \{(2, 4), (-1, -2), (0, 6)\}$

37. $S = \{(1, 1, 1), (1, 1, 0), (0, 1, 1), (0, 0, 1)\}$

38. $S = \{(1, 2, 3, 4), (1, 0, 1, 2), (1, 4, 5, 6)\}$

39. For which values of t is each set linearly independent?

(a) $S = \{(t, 1, 1), (1, t, 1), (1, 1, t)\}$

(b) $S = \{(t, 1, 1), (1, 0, 1), (1, 1, 3t)\}$

40. For which values of t is each set linearly independent?

(a) $S = \{(t, 0, 0), (0, 1, 0), (0, 0, 1)\}$

(b) $S = \{(t, t, t), (t, 1, 0), (t, 0, 1)\}$

41. Given the matrices

$$A = \begin{bmatrix} 2 & -3 \\ 4 & 1 \end{bmatrix} \quad \text{and} \quad B = \begin{bmatrix} 0 & 5 \\ 1 & -2 \end{bmatrix}$$

in $M_{2,2}$, determine which of the matrices listed below are linear combinations of A and B.

(a) $\begin{bmatrix} 6 & -19 \\ 10 & 7 \end{bmatrix}$ (b) $\begin{bmatrix} 6 & 2 \\ 9 & 11 \end{bmatrix}$

(c) $\begin{bmatrix} -2 & 28 \\ 1 & -11 \end{bmatrix}$ (d) $\begin{bmatrix} 0 & 0 \\ 0 & 0 \end{bmatrix}$

42. Determine whether the following matrices from $M_{2,2}$ form a linearly independent set.

$$A = \begin{bmatrix} 1 & -1 \\ 4 & 5 \end{bmatrix}, \quad B = \begin{bmatrix} 4 & 3 \\ -2 & 3 \end{bmatrix}, \quad C = \begin{bmatrix} 1 & -8 \\ 22 & 23 \end{bmatrix}$$

In Exercises 43–46, determine whether each set in P_2 is linearly independent.

43. $S = \{2 - x, 2x - x^2, 6 - 5x + x^2\}$

44. $S = \{x^2 - 1, 2x + 5\}$

45. $S = \{x^2 + 3x + 1, 2x^2 + x - 1, 4x\}$

46. $S = \{x^2, x^2 + 1\}$

47. Determine whether the set $S = \{1, x^2, x^2 + 2\}$ spans P_2.

48. Determine whether the set $S = \{x^2 - 2x, x^3 + 8, x^3 - x^2, x^2 - 4\}$ spans P_3.

49. By inspection, determine why each of the sets is linearly dependent.

(a) $S = \{(1, -2), (2, 3), (-2, 4)\}$

(b) $S = \{(1, -6, 2), (2, -12, 4)\}$

(c) $S = \{(0, 0), (1, 0)\}$

50. Complete the proof of Theorem 4.7.

In Exercises 51 and 52, determine whether the sets S_1 and S_2 span the same subspace of R^3.

51. $S_1 = \{(1, 2, -1), (0, 1, 1), (2, 5, -1)\}$
$S_2 = \{(-2, -6, 0), (1, 1, -2)\}$

52. $S_1 = \{(0, 0, 1), (0, 1, 1), (2, 1, 1)\}$
$S_2 = \{(1, 1, 1), (1, 1, 2), (2, 1, 1)\}$

True or False? In Exercises 53 and 54, determine whether each statement is true or false. If a statement is true, give a reason or cite an appropriate statement from the text. If a statement is false, provide an example that shows the statement is not true in all cases or cite an appropriate statement from the text.

53. (a) A set of vectors $S = \{\mathbf{v}_1, \mathbf{v}_2, . . . , \mathbf{v}_k\}$ in a vector space is called linearly dependent if the vector equation $c_1\mathbf{v}_1 + c_2\mathbf{v}_2 + \cdot \cdot \cdot + c_k\mathbf{v}_k = \mathbf{0}$ has only the trivial solution.

(b) Two vectors $\mathbf{u}$ and $\mathbf{v}$ in a vector space V are linearly dependent if and only if one is a scalar multiple of the other.

54. (a) A set $S = \{\mathbf{v}_1, \mathbf{v}_2, . . . , \mathbf{v}_k\}$, $k \geq 2$, is linearly independent if and only if at least one of the vectors $\mathbf{v}_j$ can be written as a linear combination of the other vectors.

(b) If a subset S spans a vector space V, then every vector in V can be written as a linear combination of the vectors in S.

In Exercises 55 and 56, prove that the set of vectors is linearly independent and spans R^3.

55. $B = \{(1, 1, 1), (1, 1, 0), (1, 0, 0)\}$

56. $B = \{(1, 2, 3), (3, 2, 1), (0, 0, 1)\}$

57. Guided Proof Prove that a nonempty subset of a finite set of linearly independent vectors is linearly independent.

Getting Started: You need to show that a subset of a linearly independent set of vectors cannot be linearly dependent.

(i) Suppose S is a set of linearly independent vectors. Let T be a subset of S.

(ii) If T is linearly dependent, then there exist constants not all zero satisfying the vector equation $c_1\mathbf{v}_1 + c_2\mathbf{v}_2 + \cdot \cdot \cdot + c_k\mathbf{v}_k = \mathbf{0}$.

(iii) Use this fact to derive a contradiction and conclude that T is linearly independent.

58. Prove that if S_1 is a nonempty subset of the finite set S_2, and S_1 is linearly dependent, then so is S_2.

59. Prove that any set of vectors containing the zero vector is linearly dependent.

60. Provided that $\{\mathbf{u}_1, \mathbf{u}_2, . . . , \mathbf{u}_n\}$ is a linearly independent set of vectors and that the set $\{\mathbf{u}_1, \mathbf{u}_2, . . . , \mathbf{u}_n, \mathbf{v}\}$ is linearly dependent, prove that $\mathbf{v}$ is a linear combination of the $\mathbf{u}_i$'s.

61. Let $\{\mathbf{v}_1, \mathbf{v}_2, . . . , \mathbf{v}_k\}$ be a linearly independent set of vectors in a vector space V. Delete the vector $\mathbf{v}_k$ from this set and prove that the set $\{\mathbf{v}_1, \mathbf{v}_2, . . . , \mathbf{v}_{k-1}\}$ cannot span V.

62. If V is spanned by $\{\mathbf{v}_1, \mathbf{v}_2, . . . , \mathbf{v}_k\}$ and one of these vectors can be written as a linear combination of the other $k - 1$ vectors, prove that the span of these $k - 1$ vectors is also V.

63. Writing The set $\{(1, 2, 3), (1, 0, -2), (-1, 0, 2)\}$ is linearly dependent, but $(1, 2, 3)$ cannot be written as a linear combination of $(1, 0, -2)$ and $(-1, 0, 2)$. Why does this statement not contradict Theorem 4.8?

64. Writing Under what conditions will a set consisting of a single vector be linearly independent?

65. Let $S = \{\mathbf{u}, \mathbf{v}\}$ be a linearly independent set. Prove that the set $\{\mathbf{u} + \mathbf{v}, \mathbf{u} - \mathbf{v}\}$ is linearly independent.

66. Let $\mathbf{u}$, $\mathbf{v}$, and $\mathbf{w}$ be any three vectors from a vector space V. Determine whether the set of vectors $\{\mathbf{v} - \mathbf{u}, \mathbf{w} - \mathbf{v}, \mathbf{u} - \mathbf{w}\}$ is linearly independent or linearly dependent.

67. Let $f_1(x) = 3x$ and $f_2(x) = |x|$. Graph both functions on the interval $-2 \leq x \leq 2$. Show that these functions are linearly dependent in the vector space $C[0, 1]$, but linearly independent in $C[-1, 1]$.

68. Writing Let A be a nonsingular matrix of order 3. Prove that if $\{\mathbf{v}_1, \mathbf{v}_2, \mathbf{v}_3\}$ is a linearly independent set in $M_{3,1}$, then the set $\{A\mathbf{v}_1, A\mathbf{v}_2, A\mathbf{v}_3\}$ is also linearly independent. Explain, by means of an example, why this is not true if A is singular.

69. Prove the corollary to Theorem 4.8: Two vectors $\mathbf{u}$ and $\mathbf{v}$ are linearly dependent if and only if one is a scalar multiple of the other.

4.5 Basis and Dimension

In this section you will continue your study of spanning sets. In particular, you will look at spanning sets (in a vector space) that both are linearly independent *and* span the entire space. Such a set forms a **basis** for the vector space. (The plural of *basis* is *bases.*)

Definition of Basis

A set of vectors $S = \{\mathbf{v}_1, \mathbf{v}_2, \ldots, \mathbf{v}_n\}$ in a vector space V is called a **basis** for V if the following conditions are true.

1. S spans V. 2. S is linearly independent.

REMARK: This definition tells you that a basis has two features. A basis S must have *enough vectors* to span V, but *not so many vectors* that one of them could be written as a linear combination of the other vectors in S.

This definition does not imply that every vector space has a basis consisting of a finite number of vectors. In this text, however, the discussion of bases is restricted to those consisting of a finite number of vectors. Moreover, if a vector space V has a basis consisting of a finite number of vectors, then V is **finite dimensional.** Otherwise, V is called **infinite dimensional.** [The vector space P of *all* polynomials is infinite dimensional, as is the vector space $C(-\infty, \infty)$ of all continuous functions defined on the real line.] The vector space $V = \{\mathbf{0}\}$, consisting of the zero vector alone, is finite dimensional.

EXAMPLE 1 **The Standard Basis for R^3**

Show that the following set is a basis for R^3.

$$S = \{(1, 0, 0), (0, 1, 0), (0, 0, 1)\}$$

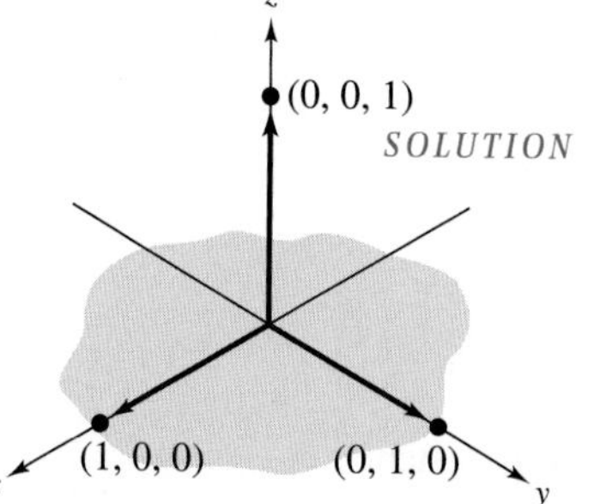

Figure 4.18

SOLUTION Example 4(a) in Section 4.4 showed that S spans R^3. Furthermore, S is linearly independent because the vector equation

$$c_1(1, 0, 0) + c_2(0, 1, 0) + c_3(0, 0, 1) = (0, 0, 0)$$

has only the trivial solution $c_1 = c_2 = c_3 = 0$. (Try verifying this.) So, S is a basis for R^3. (See Figure 4.18.)

The basis $S = \{(1, 0, 0), (0, 1, 0), (0, 0, 1)\}$ is called the **standard basis** for R^3. This result can be generalized to n-space. That is, the vectors

$$\begin{aligned}
\mathbf{e}_1 &= (1, 0, \ldots, 0) \\
\mathbf{e}_2 &= (0, 1, \ldots, 0) \\
&\vdots \\
\mathbf{e}_n &= (0, 0, \ldots, 1)
\end{aligned}$$

form a basis for R^n called the **standard basis** for R^n.

The next two examples describe nonstandard bases for R^2 and R^3.

EXAMPLE 2 The Nonstandard Basis for R^2

Show that the set

$$S = \{\overset{\mathbf{v}_1}{(1, 1)}, \overset{\mathbf{v}_2}{(1, -1)}\}$$

is a basis for R^2.

SOLUTION According to the definition of a basis for a vector space, you must show that S spans R^2 and S is linearly independent.

To verify that S spans R^2, let

$$\mathbf{x} = (x_1, x_2)$$

represent an arbitrary vector in R^2. To show that $\mathbf{x}$ can be written as a linear combination of $\mathbf{v}_1$ and $\mathbf{v}_2$, consider the equation

$$\begin{aligned}
c_1\mathbf{v}_1 + c_2\mathbf{v}_2 &= \mathbf{x} \\
c_1(1, 1) + c_2(1, -1) &= (x_1, x_2) \\
(c_1 + c_2, c_1 - c_2) &= (x_1, x_2).
\end{aligned}$$

Equating corresponding components yields the system of linear equations shown below.

$$\begin{aligned}
c_1 + c_2 &= x_1 \\
c_1 - c_2 &= x_2
\end{aligned}$$

Because the coefficient matrix of this system has a nonzero determinant, you know that the system has a unique solution. You can now conclude that S spans R^2.

To show that S is linearly independent, consider the linear combination

$$\begin{aligned}
c_1\mathbf{v}_1 + c_2\mathbf{v}_2 &= \mathbf{0} \\
c_1(1, 1) + c_2(1, -1) &= (0, 0) \\
(c_1 + c_2, c_1 - c_2) &= (0, 0).
\end{aligned}$$

Equating corresponding components yields the homogeneous system

$$\begin{aligned}
c_1 + c_2 &= 0 \\
c_1 - c_2 &= 0.
\end{aligned}$$

Because the coefficient matrix of this system has a nonzero determinant, you know that the system has only the trivial solution

$$c_1 = c_2 = 0.$$

So, you can conclude that S is linearly independent.

You can conclude that S is a basis for R^2 because it is a linearly independent spanning set for R^2.

EXAMPLE 3 A Nonstandard Basis for R^3

From Examples 5 and 8 in the preceding section, you know that

$$S = \{(1, 2, 3), (0, 1, 2), (-2, 0, 1)\}$$

spans R^3 and is linearly independent. So, S is a basis for R^3.

EXAMPLE 4 A Basis for Polynomials

Show that the vector space P_3 has the basis

$$S = \{1, x, x^2, x^3\}.$$

SOLUTION It is clear that S spans P_3 because the span of S consists of all polynomials of the form

$$a_0 + a_1x + a_2x^2 + a_3x^3, \quad a_0, a_1, a_2, \text{ and } a_3 \text{ are real,}$$

which is precisely the form of all polynomials in P_3.

To verify the linear independence of S, recall that the zero vector $\mathbf{0}$ in P_3 is the polynomial $\mathbf{0}(x) = 0$ for all x. The test for linear independence yields the equation

$$a_0 + a_1x + a_2x^2 + a_3x^3 = \mathbf{0}(x) = 0, \quad \text{for all } x.$$

This third-degree polynomial is said to be *identically equal to zero.* From algebra you know that for a polynomial to be identically equal to zero, all of its coefficients must be zero; that is,

$$a_0 = a_1 = a_2 = a_3 = 0.$$

So, S is linearly independent and is a basis for P_3.

REMARK: The basis $S = \{1, x, x^2, x^3\}$ is called the **standard basis** for P_3. Similarly, the **standard basis** for P_n is

$$S = \{1, x, x^2, \ldots, x^n\}.$$

EXAMPLE 5 **A Basis for $M_{2,2}$**

The set

$$S = \left\{ \begin{bmatrix} 1 & 0 \\ 0 & 0 \end{bmatrix}, \begin{bmatrix} 0 & 1 \\ 0 & 0 \end{bmatrix}, \begin{bmatrix} 0 & 0 \\ 1 & 0 \end{bmatrix}, \begin{bmatrix} 0 & 0 \\ 0 & 1 \end{bmatrix} \right\}$$

is a basis for $M_{2,2}$. This set is called the **standard basis** for $M_{2,2}$. In a similar manner, the standard basis for the vector space $M_{m,n}$ consists of the mn distinct $m \times n$ matrices having a single 1 and all the other entries equal to zero.

THEOREM 4.9 Uniqueness of Basis Representation

If $S = \{\mathbf{v}_1, \mathbf{v}_2, \ldots, \mathbf{v}_n\}$ is a basis for a vector space V, then every vector in V can be written in one and only one way as a linear combination of vectors in S.

PROOF The existence portion of the proof is straightforward. That is, because S spans V, you know that an arbitrary vector $\mathbf{u}$ in V can be expressed as $\mathbf{u} = c_1\mathbf{v}_1 + c_2\mathbf{v}_2 + \cdots + c_n\mathbf{v}_n$.

To prove uniqueness (that a vector can be represented in only one way), suppose $\mathbf{u}$ has another representation $\mathbf{u} = b_1\mathbf{v}_1 + b_2\mathbf{v}_2 + \cdots + b_n\mathbf{v}_n$. Subtracting the second representation from the first produces

$$\mathbf{u} - \mathbf{u} = (c_1 - b_1)\mathbf{v}_1 + (c_2 - b_2)\mathbf{v}_2 + \cdots + (c_n - b_n)\mathbf{v}_n = \mathbf{0}.$$

Because S is linearly independent, however, the only solution to this equation is the trivial solution

$$c_1 - b_1 = 0, \quad c_2 - b_2 = 0, \quad \ldots, \quad c_n - b_n = 0,$$

which means that $c_i = b_i$ for all $i = 1, 2, \ldots, n$. So, $\mathbf{u}$ has only one representation for the basis S.

EXAMPLE 6 **Uniqueness of Basis Representation**

Let $\mathbf{u} = (u_1, u_2, u_3)$ be any vector in R^3. Show that the equation $\mathbf{u} = c_1\mathbf{v}_1 + c_2\mathbf{v}_2 + c_3\mathbf{v}_3$ has a unique solution for the basis $S = \{\mathbf{v}_1, \mathbf{v}_2, \mathbf{v}_3\} = \{(1, 2, 3), (0, 1, 2), (-2, 0, 1)\}$.

SOLUTION From the equation

$$\begin{aligned}(u_1, u_2, u_3) &= c_1(1, 2, 3) + c_2(0, 1, 2) + c_3(-2, 0, 1) \\ &= (c_1 - 2c_3, 2c_1 + c_2, 3c_1 + 2c_2 + c_3),\end{aligned}$$

the following system of linear equations is obtained.

$$\begin{aligned} c_1 \qquad - 2c_3 &= u_1 \\ 2c_1 + c_2 \qquad &= u_2 \\ 3c_1 + 2c_2 + c_3 &= u_3 \end{aligned} \qquad \underbrace{\begin{bmatrix} 1 & 0 & -2 \\ 2 & 1 & 0 \\ 3 & 2 & 1 \end{bmatrix}}_{A} \underbrace{\begin{bmatrix} c_1 \\ c_2 \\ c_3 \end{bmatrix}}_{c} = \underbrace{\begin{bmatrix} u_1 \\ u_2 \\ u_3 \end{bmatrix}}_{u}$$

Because the matrix A is invertible, you know this system has a unique solution $\mathbf{c} = A^{-1}\mathbf{u}$. Solving for A^{-1} yields

$$A^{-1} = \begin{bmatrix} -1 & 4 & -2 \\ 2 & -7 & 4 \\ -1 & 2 & -1 \end{bmatrix},$$

which implies

$$\begin{aligned} c_1 &= -u_1 + 4u_2 - 2u_3 \\ c_2 &= 2u_1 - 7u_2 + 4u_3 \\ c_3 &= -u_1 + 2u_2 - u_3. \end{aligned}$$

For instance, the vector $\mathbf{u} = (1, 0, 0)$ can be represented uniquely as a linear combination of $\mathbf{v}_1$, $\mathbf{v}_2$, and $\mathbf{v}_3$ as follows.

$$(1, 0, 0) = -\mathbf{v}_1 + 2\mathbf{v}_2 - \mathbf{v}_3$$

You will now study two important theorems concerning bases.

THEOREM 4.10 Bases and Linear Dependence

If $S = \{\mathbf{v}_1, \mathbf{v}_2, . . . , \mathbf{v}_n\}$ is a basis for a vector space V, then every set containing more than n vectors in V is linearly dependent.

PROOF Let $S_1 = \{\mathbf{u}_1, \mathbf{u}_2, . . . , \mathbf{u}_m\}$ be any set of m vectors in V, where $m > n$. To show that S_1 is linearly *dependent,* you need to find scalars $k_1, k_2, . . . , k_m$ (not all zero) such that

$$k_1\mathbf{u}_1 + k_2\mathbf{u}_2 + \cdot \cdot \cdot + k_m\mathbf{u}_m = \mathbf{0}. \qquad \textbf{Equation 1}$$

Because S is a basis for V, it follows that each $\mathbf{u}_i$ is a linear combination of vectors in S, and you can write

$$\begin{aligned} \mathbf{u}_1 &= c_{11}\mathbf{v}_1 + c_{21}\mathbf{v}_2 + \cdot \cdot \cdot + c_{n1}\mathbf{v}_n \\ \mathbf{u}_2 &= c_{12}\mathbf{v}_1 + c_{22}\mathbf{v}_2 + \cdot \cdot \cdot + c_{n2}\mathbf{v}_n \\ &\vdots \\ \mathbf{u}_m &= c_{1m}\mathbf{v}_1 + c_{2m}\mathbf{v}_2 + \cdot \cdot \cdot + c_{nm}\mathbf{v}_n. \end{aligned}$$

Substituting each of these representations of $\mathbf{u}_i$ into Equation 1 and regrouping terms produces

$$d_1\mathbf{v}_1 + d_2\mathbf{v}_2 + \cdot \cdot \cdot + d_n\mathbf{v}_n = \mathbf{0},$$

where $d_i = c_{i1}k_1 + c_{i2}k_2 + \cdot \cdot \cdot + c_{im}k_m$. Because the $\mathbf{v}_i$'s form a linearly independent set, you can conclude that each $d_i = 0$. So, the system of equations shown below is obtained.

$$\begin{aligned}
c_{11}k_1 + c_{12}k_2 + \cdot \cdot \cdot + c_{1m}k_m &= 0\\
c_{21}k_1 + c_{22}k_2 + \cdot \cdot \cdot + c_{2m}k_m &= 0\\
\vdots \qquad \vdots \qquad\qquad\quad \vdots \quad &\quad \vdots\\
c_{n1}k_1 + c_{n2}k_2 + \cdot \cdot \cdot + c_{nm}k_m &= 0
\end{aligned}$$

But this homogeneous system has fewer equations than variables $k_1, k_2, \ldots, k_m$, and from Theorem 1.1 you know it must have *nontrivial* solutions. Consequently, S_1 is linearly dependent.

EXAMPLE 7 Linearly Dependent Sets in R^3 and P_3

(a) Because R^3 has a basis consisting of three vectors, the set

$$S = \{(1, 2, -1), (1, 1, 0), (2, 3, 0), (5, 9, -1)\}$$

must be linearly dependent.

(b) Because P_3 has a basis consisting of four vectors, the set

$$S = \{1, 1 + x, 1 - x, 1 + x + x^2, 1 - x + x^2\}$$

must be linearly dependent.

Because R^n has the standard basis consisting of n vectors, it follows from Theorem 4.10 that every set of vectors in R^n containing more than n vectors must be linearly dependent. Another significant consequence of Theorem 4.10 is shown in the next theorem.

THEOREM 4.11 Number of Vectors in a Basis

If a vector space V has one basis with n vectors, then every basis for V has n vectors.

PROOF Let

$$S_1 = \{\mathbf{v}_1, \mathbf{v}_2, \ldots, \mathbf{v}_n\}$$

be the basis for V, and let

$$S_2 = \{\mathbf{u}_1, \mathbf{u}_2, \ldots, \mathbf{u}_m\}$$

be any other basis for V. Because S_1 is a basis and S_2 is linearly independent, Theorem 4.10 implies that $m \leq n$. Similarly, $n \leq m$ because S_1 is linearly independent and S_2 is a basis. Consequently, $n = m$.

EXAMPLE 8 **Spanning Sets and Bases**

Use Theorem 4.11 to explain why each of the statements below is true.

(a) The set $S_1 = \{(3, 2, 1), (7, -1, 4)\}$ is not a basis for R^3.

(b) The set

$$S_2 = \{x + 2, x^2, x^3 - 1, 3x + 1, x^2 - 2x + 3\}$$

is not a basis for P_3.

SOLUTION (a) The standard basis for R^3 has three vectors, and S_1 has only two. By Theorem 4.11, S_1 cannot be a basis for R^3.

(b) The standard basis for P_3, $S = \{1, x, x^2, x^3\}$, has four elements. By Theorem 4.11, the set S_2 has too many elements to be a basis for P_3.

The Dimension of a Vector Space

The discussion of spanning sets, linear independence, and bases leads to an important notion in the study of vector spaces. By Theorem 4.11, you know that if a vector space V has a basis consisting of n vectors, then every other basis for the space also has n vectors. The number n is called the **dimension** of V.

Definition of Dimension of a Vector Space

If a vector space V has a basis consisting of n vectors, then the number n is called the **dimension** of V, denoted by $\dim(V) = n$. If V consists of the zero vector alone, the dimension of V is defined as zero.

This definition allows you to observe the characteristics of the dimensions of the familiar vector spaces listed below. In each case, the dimension is determined by simply counting the number of vectors in the standard basis.

1. The dimension of R^n with the standard operations is n.
2. The dimension of P_n with the standard operations is $n + 1$.
3. The dimension of $M_{m,n}$ with the standard operations is mn.

If W is a subspace of an n-dimensional vector space, then it can be shown that W is finite dimensional and the dimension of W is less than or equal to n. (See Exercise 81.) In the next three examples, you will look at a technique for determining the dimension of a subspace. Basically, you determine the dimension by finding a set of linearly independent vectors that spans the subspace. This set is a basis for the subspace, and the dimension of the subspace is the number of vectors in the basis.

EXAMPLE 9 Finding the Dimension of a Subspace

Determine the dimension of each subspace of R^3.

(a) $W = \{(d, c - d, c)\text{: } c \text{ and } d \text{ are real numbers}\}$
(b) $W = \{(2b, b, 0)\text{: } b \text{ is a real number}\}$

SOLUTION The goal in each example is to find a set of linearly independent vectors that spans the subspace.

(a) By writing the representative vector $(d, c - d, c)$ as

$$\begin{aligned}(d, c - d, c) &= (0, c, c) + (d, -d, 0) \\ &= c(0, 1, 1) + d(1, -1, 0),\end{aligned}$$

you can see that W is spanned by the set

$$S = \{(0, 1, 1), (1, -1, 0)\}.$$

Using the techniques described in the preceding section, you can show that this set is linearly independent. So, it is a basis for W, and you can conclude that W is a two-dimensional subspace of R^3.

(b) By writing the representative vector $(2b, b, 0)$ as

$$(2b, b, 0) = b(2, 1, 0),$$

you can see that W is spanned by the set $S = \{(2, 1, 0)\}$. So, W is a one-dimensional subspace of R^3.

REMARK: In Example 9(a), the subspace W is a two-dimensional plane in R^3 determined by the vectors $(0, 1, 1)$ and $(1, -1, 0)$. In Example 9(b), the subspace is a one-dimensional line.

EXAMPLE 10 Finding the Dimension of a Subspace

Find the dimension of the subspace W of R^4 spanned by

$$S = \{\underset{\mathbf{v}_1}{(-1, 2, 5, 0)}, \underset{\mathbf{v}_2}{(3, 0, 1, -2)}, \underset{\mathbf{v}_3}{(-5, 4, 9, 2)}\}.$$

SOLUTION Although W is spanned by the set S, S is not a basis for W because S is a linearly dependent set. Specifically, $\mathbf{v}_3$ can be written as a linear combination of $\mathbf{v}_1$ and $\mathbf{v}_2$ as follows.

$$\mathbf{v}_3 = 2\mathbf{v}_1 - \mathbf{v}_2$$

This means that W is spanned by the set $S_1 = \{\mathbf{v}_1, \mathbf{v}_2\}$. Moreover, S_1 is linearly independent because neither vector is a scalar multiple of the other, and you can conclude that the dimension of W is 2.

EXAMPLE 11 Finding the Dimension of a Subspace

Let W be the subspace of all symmetric matrices in $M_{2,2}$. What is the dimension of W?

SOLUTION Every 2×2 symmetric matrix has the form listed below.

$$A = \begin{bmatrix} a & b \\ b & c \end{bmatrix} = \begin{bmatrix} a & 0 \\ 0 & 0 \end{bmatrix} + \begin{bmatrix} 0 & b \\ b & 0 \end{bmatrix} + \begin{bmatrix} 0 & 0 \\ 0 & c \end{bmatrix}$$

$$= a\begin{bmatrix} 1 & 0 \\ 0 & 0 \end{bmatrix} + b\begin{bmatrix} 0 & 1 \\ 1 & 0 \end{bmatrix} + c\begin{bmatrix} 0 & 0 \\ 0 & 1 \end{bmatrix}$$

So, the set

$$S = \left\{\begin{bmatrix} 1 & 0 \\ 0 & 0 \end{bmatrix}, \begin{bmatrix} 0 & 1 \\ 1 & 0 \end{bmatrix}, \begin{bmatrix} 0 & 0 \\ 0 & 1 \end{bmatrix}\right\}$$

spans W. Moreover, S can be shown to be linearly independent, and you can conclude that the dimension of W is 3.

Usually, to conclude that a set

$$S = \{\mathbf{v}_1, \mathbf{v}_2, . . . , \mathbf{v}_n\}$$

is a basis for a vector space V, you must show that S satisfies two conditions: S spans V and is linearly independent. If V is known to have a dimension of n, however, then the next theorem tells you that you do not need to check both conditions: either one will suffice. The proof is left as an exercise. (See Exercise 82.)

THEOREM 4.12 Basis Tests in an n-Dimensional Space

Let V be a vector space of dimension n.

1. If $S = \{\mathbf{v}_1, \mathbf{v}_2, . . . , \mathbf{v}_n\}$ is a linearly independent set of vectors in V, then S is a basis for V.
2. If $S = \{\mathbf{v}_1, \mathbf{v}_2, . . . , \mathbf{v}_n\}$ spans V, then S is a basis for V.

EXAMPLE 12 Testing for a Basis in an n-Dimensional Space

Show that the set of vectors is a basis for $M_{5,1}$.

$$S = \left\{\overset{\mathbf{v}_1}{\begin{bmatrix} 1 \\ 2 \\ -1 \\ 3 \\ 4 \end{bmatrix}}, \overset{\mathbf{v}_2}{\begin{bmatrix} 0 \\ 1 \\ 3 \\ -2 \\ 3 \end{bmatrix}}, \overset{\mathbf{v}_3}{\begin{bmatrix} 0 \\ 0 \\ 2 \\ -1 \\ 5 \end{bmatrix}}, \overset{\mathbf{v}_4}{\begin{bmatrix} 0 \\ 0 \\ 0 \\ 2 \\ -3 \end{bmatrix}}, \overset{\mathbf{v}_5}{\begin{bmatrix} 0 \\ 0 \\ 0 \\ 0 \\ -2 \end{bmatrix}}\right\}$$

SOLUTION Because S has five vectors and the dimension of $M_{5,1}$ is five, you can apply Theorem 4.12 to verify that S is a basis by showing either that S is linearly independent or that S spans $M_{5,1}$. To show the first of these, form the vector equation

$$c_1\mathbf{v}_1 + c_2\mathbf{v}_2 + c_3\mathbf{v}_3 + c_4\mathbf{v}_4 + c_5\mathbf{v}_5 = \mathbf{0},$$

which yields the homogeneous system of linear equations shown below.

$$\begin{aligned} c_1 &= 0 \\ 2c_1 + c_2 &= 0 \\ -c_1 + 3c_2 + 2c_3 &= 0 \\ 3c_1 - 2c_2 - c_3 + 2c_4 &= 0 \\ 4c_1 + 3c_2 + 5c_3 - 3c_4 - 2c_5 &= 0 \end{aligned}$$

Because this system has only the trivial solution, S must be linearly independent. So, by Theorem 4.12, S is a basis for $M_{5,1}$.

SECTION 4.5 Exercises

In Exercises 1–6, write the standard basis for the vector space.

1. R^6 **2.** R^4 **3.** $M_{2,4}$

4. $M_{4,1}$ **5.** P_4 **6.** P_2

Writing In Exercises 7–14, explain why S is not a basis for R^2.

7. $S = \{(1, 2), (1, 0), (0, 1)\}$

8. $S = \{(-1, 2), (1, -2), (2, 4)\}$

9. $S = \{(-4, 5), (0, 0)\}$

10. $S = \{(2, 3), (6, 9)\}$

11. $S = \{(6, -5), (12, -10)\}$

12. $S = \{(4, -3), (8, -6)\}$

13. $S = \{(-3, 2)\}$

14. $S = \{(-1, 2)\}$

Writing In Exercises 15–20, explain why S is not a basis for R^3.

15. $S = \{(1, 3, 0), (4, 1, 2), (-2, 5, -2)\}$

16. $S = \{(2, 1, -2), (-2, -1, 2), (4, 2, -4)\}$

17. $S = \{(7, 0, 3), (8, -4, 1)\}$

18. $S = \{(1, 1, 2), (0, 2, 1)\}$

19. $S = \{(0, 0, 0), (1, 0, 0), (0, 1, 0)\}$

20. $S = \{(6, 4, 1), (3, -5, 1), (8, 13, 6), (0, 6, 9)\}$

Writing In Exercises 21–24, explain why S is not a basis for P_2.

21. $S = \{1, 2x, x^2 - 4, 5x\}$

22. $S = \{2, x, x + 3, 3x^2\}$

23. $S = \{1 - x, 1 - x^2, 3x^2 - 2x - 1\}$

24. $S = \{6x - 3, 3x^2, 1 - 2x - x^2\}$

Writing In Exercises 25–28, explain why S is not a basis for $M_{2,2}$.

25. $S = \left\{\begin{bmatrix} 1 & 0 \\ 0 & 1 \end{bmatrix}, \begin{bmatrix} 0 & 1 \\ 1 & 0 \end{bmatrix}\right\}$

26. $S = \left\{\begin{bmatrix} 1 & 1 \\ 0 & 0 \end{bmatrix}, \begin{bmatrix} 0 & 1 \\ 1 & 0 \end{bmatrix}\right\}$

27. $S = \left\{\begin{bmatrix} 1 & 0 \\ 0 & 0 \end{bmatrix}, \begin{bmatrix} 0 & 1 \\ 1 & 0 \end{bmatrix}, \begin{bmatrix} 1 & 0 \\ 0 & 1 \end{bmatrix}, \begin{bmatrix} 8 & -4 \\ -4 & 3 \end{bmatrix}\right\}$

28. $S = \left\{\begin{bmatrix} 1 & 0 \\ 0 & 1 \end{bmatrix}, \begin{bmatrix} 0 & 1 \\ 1 & 0 \end{bmatrix}, \begin{bmatrix} 1 & 1 \\ 0 & 0 \end{bmatrix}\right\}$

In Exercises 29–34, determine whether the set $\{\mathbf{v}_1, \mathbf{v}_2\}$ is a basis for R^2.

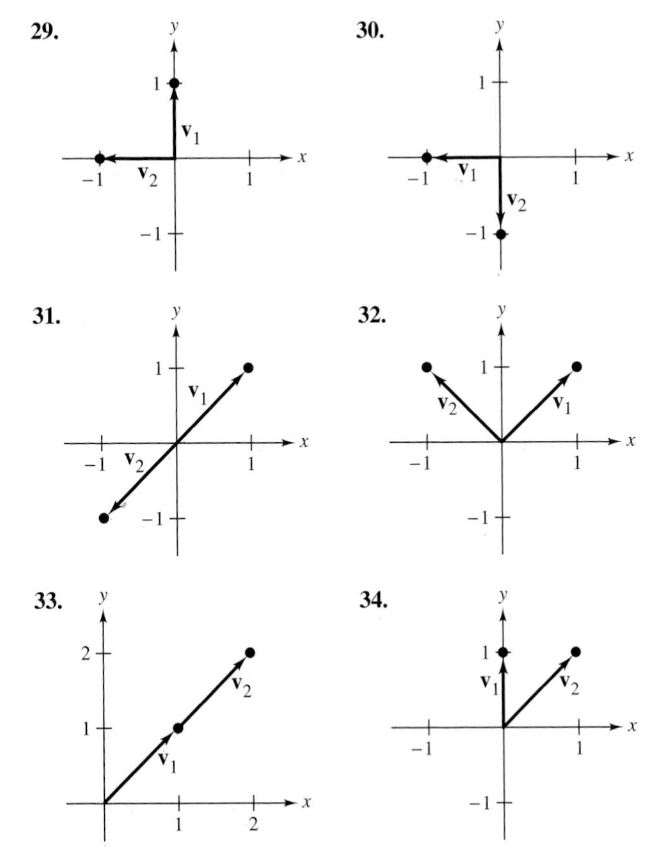

In Exercises 35–42, determine whether S is a basis for the indicated vector space.

35. $S = \{(3, -2), (4, 5)\}$ for R^2

36. $S = \{(1, 2), (1, -1)\}$ for R^2

37. $S = \{(1, 5, 3), (0, 1, 2), (0, 0, 6)\}$ for R^3

38. $S = \{(2, 1, 0), (0, -1, 1)\}$ for R^3

39. $S = \{(0, 3, -2), (4, 0, 3), (-8, 15, -16)\}$ for R^3

40. $S = \{(0, 0, 0), (1, 5, 6), (6, 2, 1)\}$ for R^3

41. $S = \{(-1, 2, 0, 0), (2, 0, -1, 0), (3, 0, 0, 4), (0, 0, 5, 0)\}$ for R^4

42. $S = \{(1, 0, 0, 1), (0, 2, 0, 2), (1, 0, 1, 0), (0, 2, 2, 0)\}$ for R^4

In Exercises 43 and 44, determine whether S is a basis for $M_{2,2}$.

43. $S = \left\{\begin{bmatrix} 2 & 0 \\ 0 & 3 \end{bmatrix}, \begin{bmatrix} 1 & 4 \\ 0 & 1 \end{bmatrix}, \begin{bmatrix} 0 & 1 \\ 3 & 2 \end{bmatrix}, \begin{bmatrix} 0 & 1 \\ 2 & 0 \end{bmatrix}\right\}$

44. $S = \left\{\begin{bmatrix} 1 & 2 \\ -5 & 4 \end{bmatrix}, \begin{bmatrix} 2 & -7 \\ 6 & 2 \end{bmatrix}, \begin{bmatrix} 4 & -9 \\ 11 & 12 \end{bmatrix}, \begin{bmatrix} 12 & -16 \\ 17 & 42 \end{bmatrix}\right\}$

In Exercises 45–48, determine whether S is a basis for P_3.

45. $S = \{t^3 - 2t^2 + 1, t^2 - 4, t^3 + 2t, 5t\}$

46. $S = \{4t - t^2, 5 + t^3, 3t + 5, 2t^3 - 3t^2\}$

47. $S = \{4 - t, t^3, 6t^2, t^3 + 3t, 4t - 1\}$

48. $S = \{t^3 - 1, 2t^2, t + 3, 5 + 2t + 2t^2 + t^3\}$

In Exercises 49–54, determine whether S is a basis for R^3. If it is, write $\mathbf{u} = (8, 3, 8)$ as a linear combination of the vectors in S.

49. $S = \{(4, 3, 2), (0, 3, 2), (0, 0, 2)\}$

50. $S = \{(1, 0, 0), (1, 1, 0), (1, 1, 1)\}$

51. $S = \{(0, 0, 0), (1, 3, 4), (6, 1, -2)\}$

52. $S = \{(1, 0, 1), (0, 0, 0), (0, 1, 0)\}$

53. $S = \left\{\left(\frac{2}{3}, \frac{5}{2}, 1\right), \left(1, \frac{3}{2}, 0\right), (2, 12, 6)\right\}$

54. $S = \{(1, 4, 7), (3, 0, 1), (2, 1, 2)\}$

In Exercises 55–62, determine the dimension of the vector space.

55. R^6 **56.** R^4 **57.** R **58.** R^3

59. P_7 **60.** P_4 **61.** $M_{2,3}$ **62.** $M_{3,2}$

63. Find a basis for $D_{3,3}$ (the vector space of all 3×3 diagonal matrices). What is the dimension of this vector space?

64. Find a basis for the vector space of all 3×3 symmetric matrices. What is the dimension of this vector space?

65. Find all subsets of the set that forms a basis for R^2.
$S = \{(1, 0), (0, 1), (1, 1)\}$

66. Find all subsets of the set that forms a basis for R^3.
$S = \{(1, 3, -2), (-4, 1, 1), (-2, 7, -3), (2, 1, 1)\}$

67. Find a basis for R^2 that includes the vector $(1, 1)$.

68. Find a basis for R^3 that includes the set $S = \{(1, 0, 2), (0, 1, 1)\}$.

In Exercises 69 and 70, (a) give a geometric description of, (b) find a basis for, and (c) determine the dimension of the subspace W of R^2.

69. $W = \{(2t, t): t \text{ is a real number}\}$

70. $W = \{(0, t): t \text{ is a real number}\}$

In Exercises 71 and 72, (a) give a geometric description of, (b) find a basis for, and (c) determine the dimension of the subspace W of R^3.

71. $W = \{(2t, t, -t): t \text{ is a real number}\}$

72. $W = \{(2t - t, s, t): s \text{ and } t \text{ are real numbers}\}$

In Exercises 73–76, find (a) a basis for and (b) the dimension of the subspace W of R^4.

73. $W = \{(2s - t, s, t, s)$: s and t are real numbers$\}$

74. $W = \{(5t, -3t, t, t)$: t is a real number$\}$

75. $W = \{(0, 6t, t, -t)$: t is a real number$\}$

76. $W = \{(s + 4t, t, s, 2s - t)$: s and t are real numbers$\}$

True or False? In Exercises 77 and 78, determine whether each statement is true or false. If a statement is true, give a reason or cite an appropriate statement from the text. If a statement is false, provide an example that shows the statement is not true in all cases or cite an appropriate statement from the text.

77. (a) If $\dim(V) = n$, then there exists a set of $n - 1$ vectors in V that will span V.

(b) If $\dim(V) = n$, then there exists a set of $n + 1$ vectors in V that will span V.

78. (a) If $\dim(V) = n$, then any set of $n + 1$ vectors in V must be linearly dependent.

(b) If $\dim(V) = n$, then any set of $n - 1$ vectors in V must be linearly independent.

79. Prove that if $S = \{\mathbf{v}_1, \mathbf{v}_2, \ldots, \mathbf{v}_n\}$ is a basis for a vector space V and c is a nonzero scalar, then the set $S_1 = \{c\mathbf{v}_1, c\mathbf{v}_2, \ldots, c\mathbf{v}_n\}$ is also a basis for V.

80. Prove that the vector space P of all polynomials is infinite dimensional.

81. Prove that if W is a subspace of a finite-dimensional vector space V, then (dimension of W) $\leq$ (dimension of V).

82. Prove Theorem 4.12.

83. Writing

(a) Let $S_1 = \text{span}((1, 0, 0), (1, 1, 0))$ and $S_2 = \text{span}((0, 0, 1), (0, 1, 0))$ be subspaces of R^3. Find a basis for and the dimension of each of the subspaces S_1, S_2, $S_1 \cap S_2$, and $S_1 + S_2$. (See Exercise 47 in Section 4.3)

(b) Let S_1 and S_2 be two-dimensional subspaces of R^3. Is it possible that $S_1 \cap S_2 = \{(0, 0, 0)\}$? Explain.

84. Guided Proof Let S be a spanning set for the finite dimensional vector space V. Prove that there exists a subset S' of S that forms a basis for V.

Getting Started: S is a spanning set, but it may not be a basis because it may be linearly dependent. You need to remove extra vectors so that a subset S' is a spanning set and is also linearly independent.

(i) If S is a linearly independent set, you are done. If not, remove some vector $\mathbf{v}$ from S that is a linear combination of the other vectors in S.

(ii) Call this set S_1. If S_1 is a linearly independent set, you are done. If not, continue to remove dependent vectors until you produce a linearly independent subset S'.

(iii) Conclude that this subset is the minimal spanning set S'.

85. Let S be a linearly independent set of vectors from the finite dimensional vector space V. Prove that there exists a basis for V containing S.

86. Let V be a vector space of dimension n. Prove that any set of less than n vectors cannot span V.

4.6 Rank of a Matrix and Systems of Linear Equations

In this section you will investigate the vector space spanned by the **row vectors** (or **column vectors**) of a matrix. Then you will see how such spaces relate to solutions of systems of linear equations.

To begin, you need to know some terminology. For an $m \times n$ matrix A, the n-tuples corresponding to the rows of A are called the **row vectors** of A.

$$A = \begin{bmatrix} a_{11} & a_{12} & \cdots & a_{1n} \\ a_{21} & a_{22} & \cdots & a_{2n} \\ \vdots & \vdots & & \vdots \\ a_{m1} & a_{m2} & \cdots & a_{mn} \end{bmatrix}$$

Row Vectors of A

$(a_{11}, a_{12}, \ldots, a_{1n})$
$(a_{21}, a_{22}, \ldots, a_{2n})$
$\vdots$
$(a_{m1}, a_{m2}, \ldots, a_{mn})$

Similarly, the columns of A are called the **column vectors** of A. You will find it useful to preserve the column notation for these column vectors.

Column Vectors of A

$$A = \begin{bmatrix} a_{11} & a_{12} & \cdots & a_{1n} \\ a_{21} & a_{22} & \cdots & a_{2n} \\ \vdots & \vdots & & \vdots \\ a_{m1} & a_{m2} & \cdots & a_{mn} \end{bmatrix} \qquad \begin{bmatrix} a_{11} \\ a_{21} \\ \vdots \\ a_{m1} \end{bmatrix} \begin{bmatrix} a_{12} \\ a_{22} \\ \vdots \\ a_{m2} \end{bmatrix} \cdots \begin{bmatrix} a_{1n} \\ a_{2n} \\ \vdots \\ a_{mn} \end{bmatrix}$$

EXAMPLE 1 Row Vectors and Column Vectors

For the matrix

$$A = \begin{bmatrix} 0 & 1 & -1 \\ -2 & 3 & 4 \end{bmatrix},$$

the row vectors are $(0, 1, -1)$ and $(-2, 3, 4)$ and the column vectors are

$$\begin{bmatrix} 0 \\ -2 \end{bmatrix}, \quad \begin{bmatrix} 1 \\ 3 \end{bmatrix}, \quad \text{and} \quad \begin{bmatrix} -1 \\ 4 \end{bmatrix}.$$

In Example 1, note that for an $m \times n$ matrix A, the row vectors are vectors in R^n and the column vectors are vectors in R^m. This leads to the two definitions of the **row space** and **column space** of a matrix listed below.

Definitions of Row Space and Column Space of a Matrix

Let A be an $m \times n$ matrix.

1. The **row space** of A is the subspace of R^n spanned by the row vectors of A.
2. The **column space** of A is the subspace of R^m spanned by the column vectors of A.

As it turns out, the row and column spaces of A share many properties. Because of your familiarity with elementary row operations, however, you will begin by looking at the row space of a matrix. Recall that two matrices are row-equivalent if one can be obtained from the other by elementary row operations. The next theorem tells you that row-equivalent matrices have the same row space.

THEOREM 4.13 Row-Equivalent Matrices Have the Same Row Space

If an $m \times n$ matrix A is row-equivalent to an $m \times n$ matrix B, then the row space of A is equal to the row space of B.

PROOF Because the rows of B can be obtained from the rows of A by elementary row operations (scalar multiplication and addition), it follows that the row vectors of B can be written as linear combinations of the row vectors of A. The row vectors of B lie in the row space of A, and the subspace spanned by the row vectors of B is contained in the row space of A. But it is also true that the rows of A can be obtained from the rows of B by elementary row operations. So, you can conclude that the two row spaces are subspaces of each other, making them equal.

REMARK: Note that Theorem 4.13 says that the row space of a matrix is not changed by elementary row operations. Elementary row operations can, however, change the *column* space.

If a matrix B is in row-echelon form, then its nonzero row vectors form a linearly independent set. (Try verifying this.) Consequently, they form a basis for the row space of B, and by Theorem 4.13 they also form a basis for the row space of A. This important result is stated in the next theorem.

THEOREM 4.14
Basis for the Row Space of a Matrix

If a matrix A is row-equivalent to a matrix B in row-echelon form, then the nonzero row vectors of B form a basis for the row space of A.

EXAMPLE 2 Finding a Basis for a Row Space

Find a basis for the row space of

$$A = \begin{bmatrix} 1 & 3 & 1 & 3 \\ 0 & 1 & 1 & 0 \\ -3 & 0 & 6 & -1 \\ 3 & 4 & -2 & 1 \\ 2 & 0 & -4 & -2 \end{bmatrix}.$$

SOLUTION Using elementary *row* operations, rewrite A in row-echelon form as follows.

$$B = \begin{bmatrix} 1 & 3 & 1 & 3 \\ 0 & 1 & 1 & 0 \\ 0 & 0 & 0 & 1 \\ 0 & 0 & 0 & 0 \\ 0 & 0 & 0 & 0 \end{bmatrix} \begin{matrix} \mathbf{w}_1 \\ \mathbf{w}_2 \\ \mathbf{w}_3 \\ \\ \\ \end{matrix}$$

By Theorem 4.14, you can conclude that the nonzero row vectors of B,

$$\mathbf{w}_1 = (1, 3, 1, 3), \quad \mathbf{w}_2 = (0, 1, 1, 0), \quad \text{and} \quad \mathbf{w}_3 = (0, 0, 0, 1),$$

form a basis for the row space of A.

The technique used in Example 2 to find the row space of a matrix can be used to solve the next type of problem. Suppose you are asked to find a basis for the subspace spanned by the set $S = \{\mathbf{v}_1, \mathbf{v}_2, \ldots, \mathbf{v}_k\}$ in R^n. By using the vectors in S to form the rows of a matrix A, you can use elementary row operations to rewrite A in row-echelon form. The nonzero rows of this matrix will then form a basis for the subspace spanned by S. This is demonstrated in Example 3.

EXAMPLE 3 Finding a Basis for a Subspace

Find a basis for the subspace of R^3 spanned by

$$S = \{\underset{\mathbf{v}_1}{(-1, 2, 5)}, \underset{\mathbf{v}_2}{(3, 0, 3)}, \underset{\mathbf{v}_3}{(5, 1, 8)}\}.$$

SOLUTION Use $\mathbf{v}_1$, $\mathbf{v}_2$, and $\mathbf{v}_3$ to form the rows of a matrix A. Then write A in row-echelon form.

$$A = \begin{bmatrix} -1 & 2 & 5 \\ 3 & 0 & 3 \\ 5 & 1 & 8 \end{bmatrix} \begin{matrix} \mathbf{v}_1 \\ \mathbf{v}_2 \\ \mathbf{v}_3 \end{matrix} \longrightarrow B = \begin{bmatrix} 1 & -2 & -5 \\ 0 & 1 & 3 \\ 0 & 0 & 0 \end{bmatrix} \begin{matrix} \mathbf{w}_1 \\ \mathbf{w}_2 \\ \end{matrix}$$

So, the nonzero row vectors of B,

$$\mathbf{w}_1 = (1, -2, -5) \quad \text{and} \quad \mathbf{w}_2 = (0, 1, 3),$$

form a basis for the row space of A. That is, they form a basis for the subspace spanned by $S = \{\mathbf{v}_1, \mathbf{v}_2, \mathbf{v}_3\}$.

To find a basis for the column space of a matrix A, you have two options. On the one hand, you could use the fact that the column space of A is equal to the row space of A^T and apply the technique of Example 2 to the matrix A^T. On the other hand, observe that although row operations can change the column space of a matrix, they do not change the dependency relationships between columns. You are asked to prove this fact in Exercise 71.

For example, consider the row-equivalent matrices A and B from Example 2.

$$A = \begin{bmatrix} 1 & 3 & 1 & 3 \\ 0 & 1 & 1 & 0 \\ -3 & 0 & 6 & -1 \\ 3 & 4 & -2 & 1 \\ 2 & 0 & -4 & -2 \end{bmatrix} \qquad B = \begin{bmatrix} 1 & 3 & 1 & 3 \\ 0 & 1 & 1 & 0 \\ 0 & 0 & 0 & 1 \\ 0 & 0 & 0 & 0 \\ 0 & 0 & 0 & 0 \end{bmatrix}$$
$$\qquad \mathbf{a}_1 \quad \mathbf{a}_2 \quad \mathbf{a}_3 \quad \mathbf{a}_4 \qquad\qquad \mathbf{b}_1 \quad \mathbf{b}_2 \quad \mathbf{b}_3 \quad \mathbf{b}_4$$

Notice that columns 1, 2, and 3 of matrix B satisfy the equation $\mathbf{b}_3 = -2\mathbf{b}_1 + \mathbf{b}_2$, and so do the corresponding columns of matrix A; that is,

$$\mathbf{a}_3 = -2\mathbf{a}_1 + \mathbf{a}_2.$$

Similarly, the column vectors $\mathbf{b}_1$, $\mathbf{b}_2$, and $\mathbf{b}_4$ of matrix B are linearly independent, and so are the corresponding columns of matrix A.

The next example shows how to find a basis for the column space of a matrix using both of these methods.

EXAMPLE 4 **Finding a Basis for the Column Space of a Matrix**

Find a basis for the column space of matrix A from Example 2.

$$A = \begin{bmatrix} 1 & 3 & 1 & 3 \\ 0 & 1 & 1 & 0 \\ -3 & 0 & 6 & -1 \\ 3 & 4 & -2 & 1 \\ 2 & 0 & -4 & -2 \end{bmatrix}$$

SOLUTION 1 Take the transpose of A and use elementary row operations to write A^T in row-echelon form.

$$A^T = \begin{bmatrix} 1 & 0 & -3 & 3 & 2 \\ 3 & 1 & 0 & 4 & 0 \\ 1 & 1 & 6 & -2 & -4 \\ 3 & 0 & -1 & 1 & -2 \end{bmatrix} \to \begin{bmatrix} 1 & 0 & -3 & 3 & 2 \\ 0 & 1 & 9 & -5 & -6 \\ 0 & 0 & 1 & -1 & -1 \\ 0 & 0 & 0 & 0 & 0 \end{bmatrix} \begin{matrix} \mathbf{w}_1 \\ \mathbf{w}_2 \\ \mathbf{w}_3 \\ \\ \end{matrix}$$

So, $\mathbf{w}_1 = (1, 0, -3, 3, 2)$, $\mathbf{w}_2 = (0, 1, 9, -5, -6)$, and $\mathbf{w}_3 = (0, 0, 1, -1, -1)$ form a basis for the row space of A^T. This is equivalent to saying that the column vectors

$$\begin{bmatrix} 1 \\ 0 \\ -3 \\ 3 \\ 2 \end{bmatrix}, \quad \begin{bmatrix} 0 \\ 1 \\ 9 \\ -5 \\ -6 \end{bmatrix}, \quad \text{and} \quad \begin{bmatrix} 0 \\ 0 \\ 1 \\ -1 \\ -1 \end{bmatrix}$$

form a basis for the column space of A.

SOLUTION 2 In Example 2, row operations were used on the original matrix A to obtain its row-echelon form B. It is easy to see that in matrix B, the first, second, and fourth column vectors are linearly independent (these columns have the leading 1's). The corresponding columns of matrix A are linearly independent, and a basis for the column space consists of the vectors

$$\begin{bmatrix} 1 \\ 0 \\ -3 \\ 3 \\ 2 \end{bmatrix}, \quad \begin{bmatrix} 3 \\ 1 \\ 0 \\ 4 \\ 0 \end{bmatrix}, \quad \text{and} \quad \begin{bmatrix} 3 \\ 0 \\ -1 \\ 1 \\ -2 \end{bmatrix}.$$

Notice that this is a different basis for the column space than that obtained in the first solution. Verify that these bases span the same subspace of R^5.

REMARK: Notice that in the second solution, the row-echelon form B indicates which columns of A form the basis of the column space. You do not use the column vectors of B to form the basis.

Notice in Examples 2 and 4 that both the row space and the column space of A have a dimension of 3 (because there are *three* vectors in both bases). This is generalized in the next theorem.

THEOREM 4.15 Row and Column Spaces Have Equal Dimensions

If A is an $m \times n$ matrix, then the row space and column space of A have the same dimension.

PROOF Let $\mathbf{v}_1, \mathbf{v}_2, \ldots,$ and $\mathbf{v}_m$ be the row vectors and $\mathbf{u}_1, \mathbf{u}_2, \ldots,$ and $\mathbf{u}_n$ be the column vectors of the matrix

$$A = \begin{bmatrix} a_{11} & a_{12} & \cdots & a_{1n} \\ a_{21} & a_{22} & \cdots & a_{2n} \\ \vdots & \vdots & & \vdots \\ a_{m1} & a_{m2} & \cdots & a_{mn} \end{bmatrix}.$$

Suppose the row space of A has dimension r and basis $S = \{\mathbf{b}_1, \mathbf{b}_2, \ldots, \mathbf{b}_r\}$, where $\mathbf{b}_i = (b_{i1}, b_{i2}, \ldots, b_{in})$. Using this basis, you can write the row vectors of A as

$$\begin{aligned} \mathbf{v}_1 &= c_{11}\mathbf{b}_1 + c_{12}\mathbf{b}_2 + \cdots + c_{1r}\mathbf{b}_r \\ \mathbf{v}_2 &= c_{21}\mathbf{b}_1 + c_{22}\mathbf{b}_2 + \cdots + c_{2r}\mathbf{b}_r \\ &\vdots \\ \mathbf{v}_m &= c_{m1}\mathbf{b}_1 + c_{m2}\mathbf{b}_2 + \cdots + c_{mr}\mathbf{b}_r. \end{aligned}$$

Rewrite this system of vector equations as follows.

$$\begin{aligned} [a_{11}a_{12}\cdots a_{1n}] &= c_{11}[b_{11}b_{12}\cdots b_{1n}] + c_{12}[b_{21}b_{22}\cdots b_{2n}] + \cdots + c_{1r}[b_{r1}b_{r2}\cdots b_{rn}] \\ [a_{21}a_{22}\cdots a_{2n}] &= c_{21}[b_{11}b_{12}\cdots b_{1n}] + c_{22}[b_{21}b_{22}\cdots b_{2n}] + \cdots + c_{2r}[b_{r1}b_{r2}\cdots b_{rn}] \\ &\vdots \\ [a_{m1}a_{m2}\cdots a_{mn}] &= c_{m1}[b_{11}b_{12}\cdots b_{1n}] + c_{m2}[b_{21}b_{22}\cdots b_{2n}] + \cdots + c_{mr}[b_{r1}b_{r2}\cdots b_{rn}] \end{aligned}$$

Now, take only entries corresponding to the first column of matrix A to obtain the system of scalar equations shown below.

$$\begin{aligned} a_{11} &= c_{11}b_{11} + c_{12}b_{21} + c_{13}b_{31} + \cdots + c_{1r}b_{r1} \\ a_{21} &= c_{21}b_{11} + c_{22}b_{21} + c_{23}b_{31} + \cdots + c_{2r}b_{r1} \\ a_{31} &= c_{31}b_{11} + c_{32}b_{21} + c_{33}b_{31} + \cdots + c_{3r}b_{r1} \\ &\vdots \\ a_{m1} &= c_{m1}b_{11} + c_{m2}b_{21} + c_{m3}b_{31} + \cdots + c_{mr}b_{r1} \end{aligned}$$

Similarly, for the entries of the jth column you can obtain the system below.

$$\begin{aligned}
a_{1j} &= c_{11}b_{1j} + c_{12}b_{2j} + c_{13}b_{3j} + \cdots + c_{1r}b_{rj} \\
a_{2j} &= c_{21}b_{1j} + c_{22}b_{2j} + c_{23}b_{3j} + \cdots + c_{2r}b_{rj} \\
a_{3j} &= c_{31}b_{1j} + c_{32}b_{2j} + c_{33}b_{3j} + \cdots + c_{3r}b_{rj} \\
&\vdots \\
a_{mj} &= c_{m1}b_{1j} + c_{m2}b_{2j} + c_{m3}b_{3j} + \cdots + c_{mr}b_{rj}
\end{aligned}$$

Now, let the vectors $\mathbf{c}_i = [c_{1i}\, c_{2i} \cdots c_{mi}]^T$. Then the system for the jth column can be rewritten in a vector form as

$$\mathbf{u}_j = b_{1j}\mathbf{c}_1 + b_{2j}\mathbf{c}_2 + \cdots + b_{rj}\mathbf{c}_r.$$

Put all column vectors together to obtain

$$\begin{aligned}
\mathbf{u}_1 &= [a_{11} \;\; a_{21} \cdots a_{m1}]^T = b_{11}\mathbf{c}_1 + b_{21}\mathbf{c}_2 + \cdots + b_{r1}\mathbf{c}_r \\
\mathbf{u}_2 &= [a_{12} \;\; a_{22} \cdots a_{m2}]^T = b_{12}\mathbf{c}_1 + b_{22}\mathbf{c}_2 + \cdots + b_{r2}\mathbf{c}_r \\
&\vdots \\
\mathbf{u}_n &= [a_{1n} \;\; a_{2n} \cdots a_{mn}]^T = b_{1n}\mathbf{c}_1 + b_{2n}\mathbf{c}_2 + \cdots + b_{rn}\mathbf{c}_r.
\end{aligned}$$

Because each column vector of A is a linear combination of r vectors, you know that the dimension of the column space of A is less than or equal to r (the dimension of the row space of A). That is,

$$\dim(\text{column space of } A) \le \dim(\text{row space of } A).$$

Repeating this procedure for A^T, you can conclude that the dimension of the column space of A^T is less than or equal to the dimension of the row space of A^T. But this implies that the dimension of the row space of A is less than or equal to the dimension of the column space of A. That is,

$$\dim(\text{row space of } A) \le \dim(\text{column space of } A).$$

So, the two dimensions must be equal.

The dimension of the row (or column) space of a matrix has the special name provided in the next definition.

Definition of the Rank of a Matrix

The dimension of the row (or column) space of a matrix A is called the **rank** of A and is denoted by rank(A).

REMARK: Some texts distinguish between the *row rank* and the *column rank* of a matrix. But because these ranks are equal (Theorem 4.15), this text will not distinguish between them.

EXAMPLE 5 **Finding the Rank of a Matrix**

Find the rank of the matrix

$$A = \begin{bmatrix} 1 & -2 & 0 & 1 \\ 2 & 1 & 5 & -3 \\ 0 & 1 & 3 & 5 \end{bmatrix}.$$

SOLUTION Convert to row-echelon form as follows.

$$A = \begin{bmatrix} 1 & -2 & 0 & 1 \\ 2 & 1 & 5 & -3 \\ 0 & 1 & 3 & 5 \end{bmatrix} \longrightarrow B = \begin{bmatrix} 1 & -2 & 0 & 1 \\ 0 & 1 & 1 & -1 \\ 0 & 0 & 1 & 3 \end{bmatrix}$$

Because B has three nonzero rows, the rank of A is 3.

The Nullspace of a Matrix

The notions of row and column spaces and rank have some important applications to systems of linear equations. Consider first the homogeneous linear system

$$A\mathbf{x} = \mathbf{0}$$

where A is an $m \times n$ matrix, $\mathbf{x} = [x_1 \ \ x_2 \ \ . \ . \ . \ \ x_n]^T$ is the column vector of unknowns, and $\mathbf{0} = [0 \ \ 0 \ \ . \ . \ . \ \ 0]^T$ is the zero vector in R^m.

$$\begin{bmatrix} a_{11} \, a_{12} \, . \, . \, . \, a_{1n} \\ a_{21} \, a_{22} \, . \, . \, . \, a_{2n} \\ \vdots \quad \vdots \qquad \vdots \\ a_{m1} \, a_{m2} \, . \, . \, . \, a_{mn} \end{bmatrix} \begin{bmatrix} x_1 \\ x_2 \\ \vdots \\ x_n \end{bmatrix} = \begin{bmatrix} 0 \\ 0 \\ \vdots \\ 0 \end{bmatrix}$$

The next theorem tells you that the set of all solutions of this homogeneous system is a subspace of R^n.

THEOREM 4.16 Solutions of a Homogeneous System

If A is an $m \times n$ matrix, then the set of all solutions of the homogeneous system of linear equations

$$A\mathbf{x} = \mathbf{0}$$

is a subspace of R^n called the **nullspace** of A and is denoted by $N(A)$. So,

$$N(A) = \{\mathbf{x} \in R^n : A\mathbf{x} = \mathbf{0}\}.$$

The dimension of the nullspace of A is called the **nullity** of A.

PROOF Because A is an $m \times n$ matrix, you know that $\mathbf{x}$ has size $n \times 1$. So, the set of all solutions of the system is a *subset* of R^n. This set is clearly nonempty, because $A\mathbf{0} = \mathbf{0}$. You can verify that it is a subspace by showing that it is closed under the operations of addition and scalar multiplication. Let $\mathbf{x}_1$ and $\mathbf{x}_2$ be two solution vectors of the system $A\mathbf{x} = \mathbf{0}$, and let c be a scalar. Because $A\mathbf{x}_1 = \mathbf{0}$ and $A\mathbf{x}_2 = \mathbf{0}$, you know that

$$A(\mathbf{x}_1 + \mathbf{x}_2) = A\mathbf{x}_1 + A\mathbf{x}_2 = \mathbf{0} + \mathbf{0} = \mathbf{0} \qquad \text{Addition}$$

and

$$A(c\mathbf{x}_1) = c(A\mathbf{x}_1) = c(\mathbf{0}) = \mathbf{0}. \qquad \text{Scalar multiplication}$$

So, both $(\mathbf{x}_1 + \mathbf{x}_2)$ and $c\mathbf{x}_1$ are solutions of $A\mathbf{x} = \mathbf{0}$, and you can conclude that the set of all solutions forms a subspace of R^n.

REMARK: The nullspace of A is also called the **solution space** of the system $A\mathbf{x} = \mathbf{0}$.

EXAMPLE 6 Finding the Solution Space of a Homogeneous System

Find the nullspace of the matrix.

$$A = \begin{bmatrix} 1 & 2 & -2 & 1 \\ 3 & 6 & -5 & 4 \\ 1 & 2 & 0 & 3 \end{bmatrix}$$

SOLUTION The nullspace of A is the solution space of the homogeneous system $A\mathbf{x} = \mathbf{0}$.

To solve this system, you need to write the augmented matrix $[A \,\vdots\, \mathbf{0}]$ in reduced row-echelon form. Because the system of equations is homogeneous, the right-hand column of the augmented matrix consists entirely of zeros and will not change as you do row operations. It is sufficient to find the reduced row-echelon form of A.

$$A = \begin{bmatrix} 1 & 2 & -2 & 1 \\ 3 & 6 & -5 & 4 \\ 1 & 2 & 0 & 3 \end{bmatrix} \longrightarrow \begin{bmatrix} 1 & 2 & 0 & 3 \\ 0 & 0 & 1 & 1 \\ 0 & 0 & 0 & 0 \end{bmatrix}$$

The system of equations corresponding to the reduced row-echelon form is

$$\begin{aligned} x_1 + 2x_2 \qquad + 3x_4 &= 0 \\ x_3 + x_4 &= 0. \end{aligned}$$

Choose x_2 and x_4 as free variables to represent the solutions in this parametric form.

$$x_1 = -2s - 3t, \quad x_2 = s, \quad x_3 = -t, \quad x_4 = t$$

This means that the solution space of $A\mathbf{x} = \mathbf{0}$ consists of all solution vectors $\mathbf{x}$ of the form shown below.

$$\mathbf{x} = \begin{bmatrix} x_1 \\ x_2 \\ x_3 \\ x_4 \end{bmatrix} = \begin{bmatrix} -2s - 3t \\ s \\ -t \\ t \end{bmatrix} = s\begin{bmatrix} -2 \\ 1 \\ 0 \\ 0 \end{bmatrix} + t\begin{bmatrix} -3 \\ 0 \\ -1 \\ 1 \end{bmatrix}$$

A basis for the nullspace of A consists of the vectors

$$\begin{bmatrix} -2 \\ 1 \\ 0 \\ 0 \end{bmatrix} \quad \text{and} \quad \begin{bmatrix} -3 \\ 0 \\ -1 \\ 1 \end{bmatrix}.$$

In other words, these two vectors are solutions of $A\mathbf{x} = \mathbf{0}$, and all solutions of this homogeneous system are linear combinations of these two vectors.

REMARK: Although the basis in Example 6 proved that the vectors spanned the solution set, it did not prove that they were linearly independent. When homogeneous systems are solved from the reduced row-echelon form, the spanning set is always independent.

In Example 6, matrix A has four columns. Furthermore, the rank of the matrix is 2, and the dimension of the nullspace is 2. So, you can see that

$$\text{Number of columns} = \text{rank} + \text{nullity}.$$

One way to see this is to look at the reduced row-echelon form of A.

$$\begin{bmatrix} 1 & 2 & 0 & 3 \\ 0 & 0 & 1 & 1 \\ 0 & 0 & 0 & 0 \end{bmatrix}$$

The columns with the leading 1's (columns 1 and 3) determine the rank of the matrix. The other columns (2 and 4) determine the nullity of the matrix because they correspond to the free variables. This relationship is generalized in the next theorem.

THEOREM 4.17 Dimension of the Solution Space

If A is an $m \times n$ matrix of rank r, then the dimension of the solution space of $A\mathbf{x} = \mathbf{0}$ is $n - r$. That is,

$$n = \text{rank}(A) + \text{nullity}(A).$$

PROOF Because A has rank r, you know it is row-equivalent to a reduced row-echelon matrix B with r nonzero rows. No generality is lost by assuming that the upper left corner of B has the form of the $r \times r$ identity matrix I_r. Moreover, because the zero rows of B contribute nothing to the solution, you can discard them to form the $r \times n$ matrix B', where $B' = [I_r \vdots C]$. The matrix C has $n - r$ columns corresponding to the variables $x_{r+1}, x_{r+2}, \ldots, x_n$. So, the solution space of $A\mathbf{x} = \mathbf{0}$ can be represented by the system

$$\begin{array}{lllllll} x_1 + & & c_{11}x_{r+1} & + c_{12}x_{r+2} & + \cdots & + c_{1,n-r}x_n & = 0 \\ & x_2 + & c_{21}x_{r+1} & + c_{22}x_{r+2} & + \cdots & + c_{2,n-r}x_n & = 0 \\ & & \vdots & \vdots & & \vdots & \vdots \\ & & x_r + c_{r1}x_{r+1} & + c_{r2}x_{r+2} & + \cdots & + c_{r,n-r}x_n & = 0. \end{array}$$

Solving for the first r variables in terms of the last $n - r$ variables produces $n - r$ vectors in the basis of the solution space. Consequently, the solution space has dimension $n - r$.

Example 7 illustrates this theorem and further explores the column space of a matrix.

EXAMPLE 7 Rank and Nullity of a Matrix

Let the column vectors of the matrix A be denoted by $\mathbf{a}_1$, $\mathbf{a}_2$, $\mathbf{a}_3$, $\mathbf{a}_4$, and $\mathbf{a}_5$.

$$A = \begin{bmatrix} 1 & 0 & -2 & 1 & 0 \\ 0 & -1 & -3 & 1 & 3 \\ -2 & -1 & 1 & -1 & 3 \\ 0 & 3 & 9 & 0 & -12 \end{bmatrix}$$
$$\quad \mathbf{a}_1 \quad \mathbf{a}_2 \quad \mathbf{a}_3 \quad \mathbf{a}_4 \quad \mathbf{a}_5$$

(a) Find the rank and nullity of A.
(b) Find a subset of the column vectors of A that forms a basis for the column space of A.
(c) If possible, write the third column of A as a linear combination of the first two columns.

SOLUTION Let B be the reduced row-echelon form of A.

$$A = \begin{bmatrix} 1 & 0 & -2 & 1 & 0 \\ 0 & -1 & -3 & 1 & 3 \\ -2 & -1 & 1 & -1 & 3 \\ 0 & 3 & 9 & 0 & -12 \end{bmatrix} \longrightarrow B = \begin{bmatrix} 1 & 0 & -2 & 0 & 1 \\ 0 & 1 & 3 & 0 & -4 \\ 0 & 0 & 0 & 1 & -1 \\ 0 & 0 & 0 & 0 & 0 \end{bmatrix}$$
$$\mathbf{a}_1 \quad \mathbf{a}_2 \quad \mathbf{a}_3 \quad \mathbf{a}_4 \quad \mathbf{a}_5 \qquad\qquad \mathbf{b}_1 \quad \mathbf{b}_2 \quad \mathbf{b}_3 \quad \mathbf{b}_4 \quad \mathbf{b}_5$$

(a) Because B has three nonzero rows, the rank of A is 3. Also, the number of columns of A is $n = 5$, which implies that the nullity of A is $n - \text{rank} = 5 - 3 = 2$.

(b) Because the first, second, and fourth column vectors of B are linearly independent, the corresponding column vectors of A,

$$\mathbf{a}_1 = \begin{bmatrix} 1 \\ 0 \\ -2 \\ 0 \end{bmatrix}, \quad \mathbf{a}_2 = \begin{bmatrix} 0 \\ -1 \\ -1 \\ 3 \end{bmatrix}, \quad \text{and} \quad \mathbf{a}_4 = \begin{bmatrix} 1 \\ 1 \\ -1 \\ 0 \end{bmatrix},$$

form a basis for the column space of A.

(c) The third column of B is a linear combination of the first two columns: $\mathbf{b}_3 = -2\mathbf{b}_1 + 3\mathbf{b}_2$. The same dependency relationship holds for the corresponding columns of matrix A.

$$\mathbf{a}_3 = \begin{bmatrix} -2 \\ -3 \\ 1 \\ 9 \end{bmatrix} = -2\begin{bmatrix} 1 \\ 0 \\ -2 \\ 0 \end{bmatrix} + 3\begin{bmatrix} 0 \\ -1 \\ -1 \\ 3 \end{bmatrix} = -2\mathbf{a}_1 + 3\mathbf{a}_2$$

Solutions of Systems of Linear Equations

You now know that the set of all solution vectors of the *homogeneous* linear system $A\mathbf{x} = \mathbf{0}$ is a subspace. Is this true also of the set of all solution vectors of the *nonhomogeneous* system $A\mathbf{x} = \mathbf{b}$, where $\mathbf{b} \neq \mathbf{0}$? The answer is "no," because the zero vector is never a solution of a nonhomogeneous system. There is a relationship, however, between the sets of solutions of the two systems $A\mathbf{x} = \mathbf{0}$ and $A\mathbf{x} = \mathbf{b}$. Specifically, if $\mathbf{x}_p$ is a *particular* solution of the nonhomogeneous system $A\mathbf{x} = \mathbf{b}$, then *every* solution of this system can be written in the form

$$\mathbf{x} = \mathbf{x}_p + \mathbf{x}_h,$$

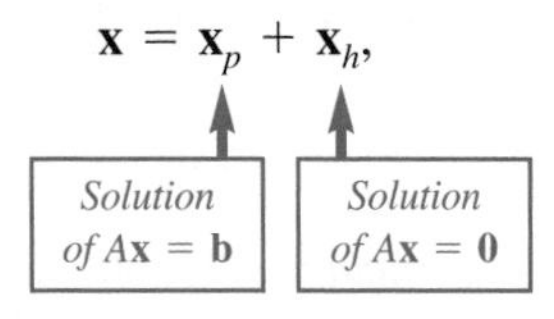

where $\mathbf{x}_h$ is a solution of the corresponding homogeneous system $A\mathbf{x} = \mathbf{0}$. The next theorem states this important concept.

THEOREM 4.18 Solutions of a Nonhomogeneous Linear System

If $\mathbf{x}_p$ is a particular solution of the nonhomogeneous system $A\mathbf{x} = \mathbf{b}$, then every solution of this system can be written in the form $\mathbf{x} = \mathbf{x}_p + \mathbf{x}_h$, where $\mathbf{x}_h$ is a solution of the corresponding homogeneous system $A\mathbf{x} = \mathbf{0}$.

PROOF Let $\mathbf{x}$ be any solution of $A\mathbf{x} = \mathbf{b}$. Then $(\mathbf{x} - \mathbf{x}_p)$ is a solution of the homogeneous system $A\mathbf{x} = \mathbf{0}$, because

$$A(\mathbf{x} - \mathbf{x}_p) = A\mathbf{x} - A\mathbf{x}_p = \mathbf{b} - \mathbf{b} = \mathbf{0}.$$

Letting $\mathbf{x}_h = \mathbf{x} - \mathbf{x}_p$, you have $\mathbf{x} = \mathbf{x}_p + \mathbf{x}_h$.

EXAMPLE 8 Finding the Solution Set of a Nonhomogeneous System

Find the set of all solution vectors of the system of linear equations.

$$\begin{aligned} x_1 \qquad - 2x_3 + x_4 &= 5 \\ 3x_1 + x_2 - 5x_3 \qquad &= 8 \\ x_1 + 2x_2 \qquad - 5x_4 &= -9 \end{aligned}$$

SOLUTION The augmented matrix for the system $A\mathbf{x} = \mathbf{b}$ reduces as follows.

$$\begin{bmatrix} 1 & 0 & -2 & 1 & 5 \\ 3 & 1 & -5 & 0 & 8 \\ 1 & 2 & 0 & -5 & -9 \end{bmatrix} \rightarrow \begin{bmatrix} 1 & 0 & -2 & 1 & 5 \\ 0 & 1 & 1 & -3 & -7 \\ 0 & 0 & 0 & 0 & 0 \end{bmatrix}$$

The system of linear equations corresponding to the reduced row-echelon matrix is

$$\begin{aligned} x_1 \qquad - 2x_3 + x_4 &= 5 \\ x_2 + x_3 - 3x_4 &= -7. \end{aligned}$$

Letting $x_3 = s$ and $x_4 = t$, you can write a representative solution vector of $A\mathbf{x} = \mathbf{b}$ as follows.

$$\mathbf{x} = \begin{bmatrix} x_1 \\ x_2 \\ x_3 \\ x_4 \end{bmatrix} = \begin{bmatrix} 2s - t + 5 \\ -s + 3t - 7 \\ s + 0t + 0 \\ 0s + t + 0 \end{bmatrix} = s\begin{bmatrix} 2 \\ -1 \\ 1 \\ 0 \end{bmatrix} + t\begin{bmatrix} -1 \\ 3 \\ 0 \\ 1 \end{bmatrix} + \begin{bmatrix} 5 \\ -7 \\ 0 \\ 0 \end{bmatrix}$$

$$= s\mathbf{u}_1 + t\mathbf{u}_2 + \mathbf{x}_p$$

You can see that $\mathbf{x}_p$ is a *particular* solution vector of $A\mathbf{x} = \mathbf{b}$, and $\mathbf{x}_h = s\mathbf{u}_1 + t\mathbf{u}_2$ represents an arbitrary vector in the solution space of $A\mathbf{x} = \mathbf{0}$.

The final theorem in this section describes how the column space of a matrix can be used to determine whether a system of linear equations is consistent.

THEOREM 4.19 Solutions of a System of Linear Equations

The system of linear equations $A\mathbf{x} = \mathbf{b}$ is consistent if and only if $\mathbf{b}$ is in the column space of A.

PROOF Let

$$A = \begin{bmatrix} a_{11} & a_{12} & \cdots & a_{1n} \\ a_{21} & a_{22} & \cdots & a_{2n} \\ \vdots & \vdots & & \vdots \\ a_{m1} & a_{m2} & \cdots & a_{mn} \end{bmatrix}, \quad \mathbf{x} = \begin{bmatrix} x_1 \\ x_2 \\ \vdots \\ x_n \end{bmatrix}, \quad \text{and} \quad \mathbf{b} = \begin{bmatrix} b_1 \\ b_2 \\ \vdots \\ b_m \end{bmatrix}$$

be the coefficient matrix, the column matrix of unknowns, and the right-hand side, respectively, of the system $A\mathbf{x} = \mathbf{b}$. Then

$$A\mathbf{x} = \begin{bmatrix} a_{11} & a_{12} & \cdots & a_{1n} \\ a_{21} & a_{22} & \cdots & a_{2n} \\ \vdots & \vdots & & \vdots \\ a_{m1} & a_{m2} & \cdots & a_{mn} \end{bmatrix} \begin{bmatrix} x_1 \\ x_2 \\ \vdots \\ x_n \end{bmatrix} = \begin{bmatrix} a_{11}x_1 + a_{12}x_2 + \cdots + a_{1n}x_n \\ a_{21}x_1 + a_{22}x_2 + \cdots + a_{2n}x_n \\ \vdots \\ a_{m1}x_1 + a_{m2}x_2 + \cdots + a_{mn}x_n \end{bmatrix}$$

$$= x_1 \begin{bmatrix} a_{11} \\ a_{21} \\ \vdots \\ a_{m1} \end{bmatrix} + x_2 \begin{bmatrix} a_{12} \\ a_{22} \\ \vdots \\ a_{m2} \end{bmatrix} + \cdots + x_n \begin{bmatrix} a_{1n} \\ a_{2n} \\ \vdots \\ a_{mn} \end{bmatrix}.$$

So, $A\mathbf{x} = \mathbf{b}$ if and only if $\mathbf{b}$ is a linear combination of the columns of A. That is, the system is consistent if and only if $\mathbf{b}$ is in the subspace of R^m spanned by the columns of A.

EXAMPLE 9 Consistency of a System of Linear Equations

Consider the system of linear equations

$$\begin{aligned} x_1 + x_2 - x_3 &= -1 \\ x_1 \qquad + x_3 &= 3 \\ 3x_1 + 2x_2 - x_3 &= 1. \end{aligned}$$

The rank of the coefficient matrix is equal to the rank of the augmented matrix.

$$A = \begin{bmatrix} 1 & 1 & -1 \\ 1 & 0 & 1 \\ 3 & 2 & -1 \end{bmatrix} \quad \rightarrow \quad \begin{bmatrix} 1 & 0 & 1 \\ 0 & 1 & -2 \\ 0 & 0 & 0 \end{bmatrix}$$

$$[A \ \vdots \ \mathbf{b}] = \begin{bmatrix} 1 & 1 & -1 & -1 \\ 1 & 0 & 1 & 3 \\ 3 & 2 & -1 & 1 \end{bmatrix} \quad \rightarrow \quad \begin{bmatrix} 1 & 0 & 1 & 3 \\ 0 & 1 & -2 & -4 \\ 0 & 0 & 0 & 0 \end{bmatrix}$$

As shown above, $\mathbf{b}$ is in the column space of A, and the system of linear equations is consistent.

Systems of Linear Equations with Square Coefficient Matrices

The final summary in this section presents several major results involving systems of linear equations, matrices, determinants, and vector spaces.

Summary of Equivalent Conditions for Square Matrices

If A is an $n \times n$ matrix, then the following conditions are equivalent.

1. A is invertible.
2. $A\mathbf{x} = \mathbf{b}$ has a unique solution for any $n \times 1$ matrix $\mathbf{b}$.
3. $A\mathbf{x} = \mathbf{0}$ has only the trivial solution.
4. A is row-equivalent to I_n.
5. $|A| \neq 0$
6. $\text{Rank}(A) = n$
7. The n row vectors of A are linearly independent.
8. The n column vectors of A are linearly independent.

SECTION 4.6 Exercises

In Exercises 1–12, find (a) the rank of the matrix, (b) a basis for the row space, and (c) a basis for the column space.

1. $\begin{bmatrix} 1 & 0 \\ 0 & 2 \end{bmatrix}$

2. $\begin{bmatrix} 2 & 4 \\ 1 & 6 \end{bmatrix}$

3. $\begin{bmatrix} 1 & 2 & 3 \end{bmatrix}$

4. $\begin{bmatrix} 0 & 1 & -2 \end{bmatrix}$

5. $\begin{bmatrix} 1 & -3 & 2 \\ 4 & 2 & 1 \end{bmatrix}$

6. $\begin{bmatrix} 1 & 2 & 4 \\ -1 & 2 & 1 \end{bmatrix}$

7. $\begin{bmatrix} 4 & 20 & 31 \\ 6 & -5 & -6 \\ 2 & -11 & -16 \end{bmatrix}$

8. $\begin{bmatrix} 2 & -3 & 1 \\ 5 & 10 & 6 \\ 8 & -7 & 5 \end{bmatrix}$

9. $\begin{bmatrix} -2 & -4 & 4 & 5 \\ 3 & 6 & -6 & -4 \\ -2 & -4 & 4 & 9 \end{bmatrix}$

10. $\begin{bmatrix} 2 & 4 & -3 & -6 \\ 7 & 14 & -6 & -3 \\ -2 & -4 & 1 & -2 \\ 2 & 4 & -2 & -2 \end{bmatrix}$

11. $\begin{bmatrix} 2 & 4 & -2 & 1 & 1 \\ 2 & 5 & 4 & -2 & 2 \\ 4 & 3 & 1 & 1 & 2 \\ 2 & -4 & 2 & -1 & 1 \\ 0 & 1 & 4 & 2 & -1 \end{bmatrix}$

12. $\begin{bmatrix} 4 & 0 & 2 & 3 & 1 \\ 2 & -1 & 2 & 0 & 1 \\ 5 & 2 & 2 & 1 & -1 \\ 4 & 0 & 2 & 2 & 1 \\ 2 & -2 & 0 & 0 & 1 \end{bmatrix}$

In Exercises 13–16, find a basis for the subspace of R^3 spanned by S.

13. $S = \{(1, 2, 4), (-1, 3, 4), (2, 3, 1)\}$

14. $S = \{(4, 2, -1), (1, 2, -8), (0, 1, 2)\}$

15. $S = \{(4, 4, 8), (1, 1, 2), (1, 1, 1)\}$

16. $S = \{(1, 2, 2), (-1, 0, 0), (1, 1, 1)\}$

In Exercises 17–20, find a basis for the subspace of R^4 spanned by S.

17. $S = \{(2, 9, -2, 53), (-3, 2, 3, -2), (8, -3, -8, 17), (0, -3, 0, 15)\}$

18. $S = \{(6, -3, 6, 34), (3, -2, 3, 19), (8, 3, -9, 6), (-2, 0, 6, -5)\}$

19. $S = \{(-3, 2, 5, 28), (-6, 1, -8, -1), (14, -10, 12, -10), (0, 5, 12, 50)\}$

20. $S = \{(2, 5, -3, -2), (-2, -3, 2, -5), (1, 3, -2, 2), (-1, -5, 3, 5)\}$

In Exercises 21–32, find a basis for, and the dimension of, the solution space of $A\mathbf{x} = \mathbf{0}$.

21. $A = \begin{bmatrix} 2 & -1 \\ 1 & 3 \end{bmatrix}$ **22.** $A = \begin{bmatrix} 2 & -1 \\ -6 & 3 \end{bmatrix}$

23. $A = \begin{bmatrix} 1 & 2 & 3 \end{bmatrix}$ **24.** $A = \begin{bmatrix} 1 & 4 & 2 \end{bmatrix}$

25. $A = \begin{bmatrix} 1 & 2 & 3 \\ 0 & 1 & 0 \end{bmatrix}$ **26.** $A = \begin{bmatrix} 1 & 4 & 2 \\ 0 & 0 & 1 \end{bmatrix}$

27. $A = \begin{bmatrix} 1 & 2 & -3 \\ 2 & -1 & 4 \\ 4 & 3 & -2 \end{bmatrix}$ **28.** $A = \begin{bmatrix} 3 & -6 & 21 \\ -2 & 4 & -14 \\ 1 & -2 & 7 \end{bmatrix}$

29. $A = \begin{bmatrix} 1 & 3 & -2 & 4 \\ 0 & 1 & -1 & 2 \\ -2 & -6 & 4 & -8 \end{bmatrix}$

30. $A = \begin{bmatrix} 1 & 4 & 2 & 1 \\ 0 & 1 & 1 & -1 \\ -2 & -8 & -4 & -2 \end{bmatrix}$

31. $A = \begin{bmatrix} 2 & 6 & 3 & 1 \\ 2 & 1 & 0 & -2 \\ 3 & -2 & 1 & 1 \\ 0 & 6 & 2 & 0 \end{bmatrix}$ **32.** $A = \begin{bmatrix} 1 & 4 & 2 & 1 \\ 2 & -1 & 1 & 1 \\ 4 & 2 & 1 & 1 \\ 0 & 4 & 2 & 0 \end{bmatrix}$

In Exercises 33–40, find (a) a basis for and (b) the dimension of the solution space of the homogeneous system of linear equations.

33. $\begin{aligned} -x + y + z &= 0 \\ 3x - y &= 0 \\ 2x - 4y - 5z &= 0 \end{aligned}$ **34.** $\begin{aligned} 4x - y + 2z &= 0 \\ 2x + 3y - z &= 0 \\ 3x + y + z &= 0 \end{aligned}$

35. $\begin{aligned} x - 2y + 3z &= 0 \\ -3x + 6y - 9z &= 0 \end{aligned}$ **36.** $\begin{aligned} x + 2y - 4z &= 0 \\ -3x - 6y + 12z &= 0 \end{aligned}$

37. $\begin{aligned} 3x_1 + 3x_2 + 15x_3 + 11x_4 &= 0 \\ x_1 - 3x_2 + x_3 + x_4 &= 0 \\ 2x_1 + 3x_2 + 11x_3 + 8x_4 &= 0 \end{aligned}$

38. $\begin{aligned} 2x_1 + 2x_2 + 4x_3 - 2x_4 &= 0 \\ x_1 + 2x_2 + x_3 + 2x_4 &= 0 \\ -x_1 + x_2 + 4x_3 - 2x_4 &= 0 \end{aligned}$

39. $\begin{aligned} 9x_1 - 4x_2 - 2x_3 - 20x_4 &= 0 \\ 12x_1 - 6x_2 - 4x_3 - 29x_4 &= 0 \\ 3x_1 - 2x_2 - 7x_4 &= 0 \\ 3x_1 - 2x_2 - x_3 - 8x_4 &= 0 \end{aligned}$

40. $\begin{aligned} x_1 + 3x_2 + 2x_3 + 22x_4 + 13x_5 &= 0 \\ x_1 + x_3 - 2x_4 + x_5 &= 0 \\ 3x_1 + 6x_2 + 5x_3 + 42x_4 + 27x_5 &= 0 \end{aligned}$

In Exercises 41–46, (a) determine whether the nonhomogeneous system $A\mathbf{x} = \mathbf{b}$ is consistent, and (b) if the system is consistent, write the solution in the form $\mathbf{x} = \mathbf{x}_h + \mathbf{x}_p$, where $\mathbf{x}_h$ is a solution of $A\mathbf{x} = \mathbf{0}$ and $\mathbf{x}_p$ is a particular solution of $A\mathbf{x} = \mathbf{b}$.

41. $\begin{aligned} x + 3y + 10z &= 18 \\ -2x + 7y + 32z &= 29 \\ -x + 3y + 14z &= 12 \\ x + y + 2z &= 8 \end{aligned}$ **42.** $\begin{aligned} 3x - 8y + 4z &= 19 \\ -6y + 2z + 4w &= 5 \\ 5x + 22z + w &= 29 \\ x - 2y + 2z &= 8 \end{aligned}$

43. $\begin{aligned} 3w - 2x + 16y - 2z &= -7 \\ -w + 5x - 14y + 18z &= 29 \\ 3w - x + 14y + 2z &= 1 \end{aligned}$

44. $\begin{aligned} 2x - 4y + 5z &= 8 \\ -7x + 14y + 4z &= -28 \\ 3x - 6y + z &= 12 \end{aligned}$

45. $\begin{aligned} x_1 + 2x_2 + x_3 + x_4 + 5x_5 &= 0 \\ -5x_1 - 10x_2 + 3x_3 + 3x_4 + 55x_5 &= -8 \\ x_1 + 2x_2 + 2x_3 - 3x_4 - 5x_5 &= 14 \\ -x_1 - 2x_2 + x_3 + x_4 + 15x_5 &= -2 \end{aligned}$

46. $\begin{aligned} 5x_1 - 4x_2 + 12x_3 - 33x_4 + 14x_5 &= -4 \\ -2x_1 + x_2 - 6x_3 + 12x_4 - 8x_5 &= 1 \\ 2x_1 - x_2 + 6x_3 - 12x_4 + 8x_5 &= -1 \end{aligned}$

In Exercises 47–50, determine whether $\mathbf{b}$ is in the column space of A. If it is, write $\mathbf{b}$ as a linear combination of the column vectors of A.

47. $A = \begin{bmatrix} -1 & 2 \\ 4 & 0 \end{bmatrix}, \quad \mathbf{b} = \begin{bmatrix} 3 \\ 4 \end{bmatrix}$

48. $A = \begin{bmatrix} -1 & 2 \\ 2 & -4 \end{bmatrix}, \quad \mathbf{b} = \begin{bmatrix} 2 \\ 4 \end{bmatrix}$

49. $A = \begin{bmatrix} 1 & 3 & 0 \\ -1 & 1 & 0 \\ 2 & 0 & 1 \end{bmatrix}, \quad \mathbf{b} = \begin{bmatrix} 1 \\ 2 \\ -3 \end{bmatrix}$

50. $A = \begin{bmatrix} 1 & 3 & 2 \\ -1 & 1 & 2 \\ 0 & 1 & 1 \end{bmatrix}, \quad \mathbf{b} = \begin{bmatrix} 1 \\ 1 \\ 0 \end{bmatrix}$

51. Writing Explain why the row vectors of a 4×3 matrix form a linearly dependent set. (Assume all matrix entries are distinct.)

52. Writing Explain why the column vectors of a 3×4 matrix form a linearly dependent set. (Assume all matrix entries are distinct.)

53. Prove that if A is not square, then either the row vectors of A or the column vectors of A form a linearly dependent set.

54. Give an example showing that the rank of the product of two matrices can be less than the rank of either matrix.

55. Give examples of matrices A and B of the same size such that

(a) $\text{rank}(A + B) < \text{rank}(A)$ and $\text{rank}(A + B) < \text{rank}(B)$
(b) $\text{rank}(A + B) = \text{rank}(A)$ and $\text{rank}(A + B) = \text{rank}(B)$
(c) $\text{rank}(A + B) > \text{rank}(A)$ and $\text{rank}(A + B) > \text{rank}(B)$.

56. Prove that the nonzero row vectors of a matrix in row-echelon form are linearly independent.

57. Let A be an $m \times n$ matrix (where $m < n$) whose rank is r.

(a) What is the largest value r can be?
(b) How many vectors are in a basis for the row space of A?
(c) How many vectors are in a basis for the column space of A?
(d) Which vector space R^k has the row space as a subspace?
(e) Which vector space R^k has the column space as a subspace?

58. Show that the three points (x_1, y_1), (x_2, y_2), and (x_3, y_3) in a plane are collinear if and only if the matrix

$$\begin{bmatrix} x_1 & y_1 & 1 \\ x_2 & y_2 & 1 \\ x_3 & y_3 & 1 \end{bmatrix}$$

has rank less than 3.

59. Given matrices A and B, show that the row vectors of AB are in the row space of B and the column vectors of AB are in the column space of A.

60. Find the ranks of the matrix

$$\begin{bmatrix} 1 & 2 & 3 & \cdots & n \\ n+1 & n+2 & n+3 & \cdots & 2n \\ 2n+1 & 2n+2 & 2n+3 & \cdots & 3n \\ \vdots & \vdots & \vdots & & \vdots \\ n^2-n+1 & n^2-n+2 & n^2-n+3 & \cdots & n^2 \end{bmatrix}$$

for $n = 2$, 3, and 4. Can you find a pattern in these ranks?

61. Prove each property of the system of linear equations in n variables $A\mathbf{x} = \mathbf{b}$.

(a) If $\text{rank}(A) = \text{rank}([A \; \vdots \; \mathbf{b}]) = n$, then the system has a unique solution.
(b) If $\text{rank}(A) = \text{rank}([A \; \vdots \; \mathbf{b}]) < n$, then the system has an infinite number of solutions.
(c) If $\text{rank}(A) < \text{rank}([A \; \vdots \; \mathbf{b}])$, then the system is inconsistent.

True or False? In Exercises 62–65, determine whether each statement is true or false. If a statement is true, give a reason or cite an appropriate statement from the text. If a statement is false, provide an example that shows the statement is not true in all cases or cite an appropriate statement from the text.

62. (a) The nullspace of A is also called the solution space of A.
(b) The nullspace of A is the solution space of the homogeneous system $A\mathbf{x} = \mathbf{0}$.

63. (a) If an $m \times n$ matrix A is row-equivalent to an $m \times n$ matrix B, then the row space of A is equivalent to the row space of B.
(b) If A is an $m \times n$ matrix of rank r, then the dimension of the solution space of $A\mathbf{x} = \mathbf{0}$ is $m - r$.

64. (a) If an $m \times n$ matrix B can be obtained from elementary row operations on an $m \times n$ matrix A, then the column space of B is equal to the column space of A.
(b) The system of linear equations $A\mathbf{x} = \mathbf{b}$ is inconsistent if and only if $\mathbf{b}$ is in the column space of A.

65. (a) The column space of a matrix A is equal to the row space of A^T.
(b) Row operations on a matrix A may change the dependency relationships among the columns of A.

In Exercises 66 and 67, use the fact that matrices A and B are row-equivalent.

(a) Find the rank and nullity of A.
(b) Find a basis for the nullspace of A.
(c) Find a basis for the row space of A.
(d) Find a basis for the column space of A.
(e) Determine whether or not the rows of A are linearly independent.
(f) Let the columns of A be denoted by $\mathbf{a}_1, \mathbf{a}_2, \mathbf{a}_3, \mathbf{a}_4$, and $\mathbf{a}_5$. Which of the following sets is (are) linearly independent?

(i) $\{\mathbf{a}_1, \mathbf{a}_2, \mathbf{a}_4\}$ (ii) $\{\mathbf{a}_1, \mathbf{a}_2, \mathbf{a}_3\}$ (iii) $\{\mathbf{a}_1, \mathbf{a}_3, \mathbf{a}_5\}$

66. $A = \begin{bmatrix} 1 & 2 & 1 & 0 & 0 \\ 2 & 5 & 1 & 1 & 0 \\ 3 & 7 & 2 & 2 & -2 \\ 4 & 9 & 3 & -1 & 4 \end{bmatrix}$

$B = \begin{bmatrix} 1 & 0 & 3 & 0 & -4 \\ 0 & 1 & -1 & 0 & 2 \\ 0 & 0 & 0 & 1 & -2 \\ 0 & 0 & 0 & 0 & 0 \end{bmatrix}$

67. $A = \begin{bmatrix} -2 & -5 & 8 & 0 & -17 \\ 1 & 3 & -5 & 1 & 5 \\ 3 & 11 & -19 & 7 & 1 \\ 1 & 7 & -13 & 5 & -3 \end{bmatrix}$

$B = \begin{bmatrix} 1 & 0 & 1 & 0 & 1 \\ 0 & 1 & -2 & 0 & 3 \\ 0 & 0 & 0 & 1 & -5 \\ 0 & 0 & 0 & 0 & 0 \end{bmatrix}$

68. Let A be an $m \times n$ matrix.

(a) Prove that the system of linear equations $A\mathbf{x} = \mathbf{b}$ is consistent for all column vectors $\mathbf{b}$ if and only if the rank of A is m.

(b) Prove that the homogeneous system of linear equations $A\mathbf{x} = \mathbf{0}$ has only the trivial solution if and only if the columns of A are linearly independent.

69. Let A and B be square matrices of order n satisfying $A\mathbf{x} = B\mathbf{x}$ for all $\mathbf{x} \in R^n$.

(a) Find the rank and nullity of $A - B$.

(b) Show that A and B must be identical.

70. Let A be an $m \times n$ matrix. Prove that $N(A) \subset N(A^TA)$.

71. Prove that row operations do not change the dependency relationships among the columns of an $m \times n$ matrix.

4.7 Coordinates and Change of Basis

In Theorem 4.9, you saw that if B is a basis for a vector space V, then every vector $\mathbf{x}$ in V can be expressed in one and only one way as a linear combination of vectors in B. The coefficients in the linear combination are the **coordinates of x relative to *B*.** In the context of coordinates, the order of the vectors in this basis is important, and this will sometimes be emphasized by referring to the basis B as an *ordered* basis.

Coordinate Representation Relative to a Basis

Let $B = \{\mathbf{v}_1, \mathbf{v}_2, \ldots, \mathbf{v}_n\}$ be an ordered basis for a vector space V and let $\mathbf{x}$ be a vector in V such that

$$\mathbf{x} = c_1\mathbf{v}_1 + c_2\mathbf{v}_2 + \cdots + c_n\mathbf{v}_n.$$

The scalars $c_1, c_2, \ldots, c_n$ are called the **coordinates of x relative to the basis *B*.** The **coordinate matrix** (or **coordinate vector**) **of x relative to *B*** is the column matrix in R^n whose components are the coordinates of $\mathbf{x}$.

$$[\mathbf{x}]_B = \begin{bmatrix} c_1 \\ c_2 \\ \vdots \\ c_n \end{bmatrix}$$

Coordinate Representation in R^n

In R^n, the notation for coordinate matrices conforms to the usual component notation, except that column notation is used for the coordinate matrix. In other words, writing a vector in R^n as $\mathbf{x} = (x_1, x_2, \ldots, x_n)$ means that the x_i's are the coordinates of $\mathbf{x}$ *relative to the standard basis S* in R^n. So, you have

$$[\mathbf{x}]_S = \begin{bmatrix} x_1 \\ x_2 \\ \vdots \\ x_n \end{bmatrix}.$$

EXAMPLE 1 **Coordinates and Components in R^n**

Find the coordinate matrix of $\mathbf{x} = (-2, 1, 3)$ in R^3 relative to the standard basis

$$S = \{(1, 0, 0), (0, 1, 0), (0, 0, 1)\}.$$

SOLUTION Because $\mathbf{x}$ can be written as

$$\mathbf{x} = (-2, 1, 3) = -2(1, 0, 0) + 1(0, 1, 0) + 3(0, 0, 1),$$

you can see that the coordinate matrix of $\mathbf{x}$ relative to the standard basis is simply

$$[\mathbf{x}]_S = \begin{bmatrix} -2 \\ 1 \\ 3 \end{bmatrix}.$$

So, the components of $\mathbf{x}$ are the same as its coordinates relative to the standard basis.

EXAMPLE 2 **Finding a Coordinate Matrix Relative to a Standard Basis**

The coordinate matrix of $\mathbf{x}$ in R^2 relative to the (nonstandard) ordered basis $B = \{\mathbf{v}_1, \mathbf{v}_2\} = \{(1, 0), (1, 2)\}$ is

$$[\mathbf{x}]_B = \begin{bmatrix} 3 \\ 2 \end{bmatrix}.$$

Find the coordinates of $\mathbf{x}$ relative to the standard basis $B' = \{\mathbf{u}_1, \mathbf{u}_2\} = \{(1, 0), (0, 1)\}$.

SOLUTION Because $[\mathbf{x}]_B = \begin{bmatrix} 3 \\ 2 \end{bmatrix}$, you can write

$$\mathbf{x} = 3\mathbf{v}_1 + 2\mathbf{v}_2 = 3(1, 0) + 2(1, 2) = (5, 4).$$

Moreover, because $(5, 4) = 5(1, 0) + 4(0, 1)$, it follows that the coordinates of $\mathbf{x}$ relative to B' are

$$[\mathbf{x}]_{B'} = \begin{bmatrix} 5 \\ 4 \end{bmatrix}.$$

Figure 4.19 compares these two coordinate representations.

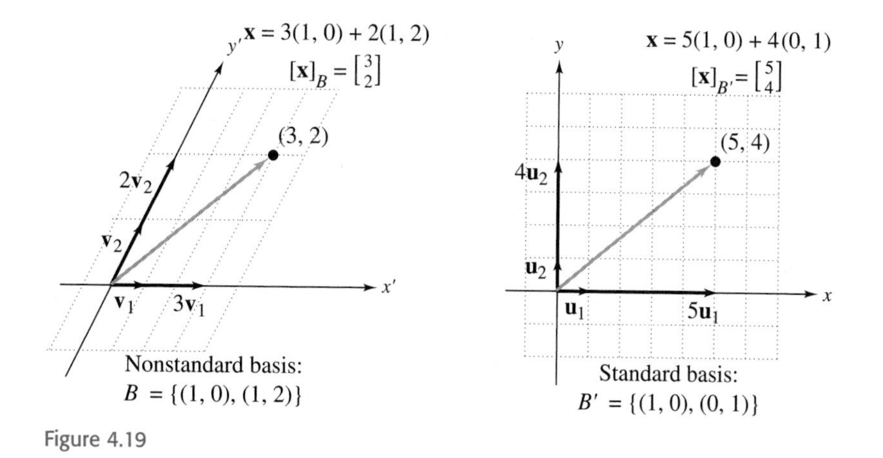

Figure 4.19

Example 2 shows that the procedure for finding the coordinate matrix relative to a *standard* basis is straightforward. The problem becomes a bit tougher, however, when you must find the coordinate matrix relative to a *nonstandard* basis. Here is an example.

EXAMPLE 3 Finding a Coordinate Matrix Relative to a Nonstandard Basis

Find the coordinate matrix of $\mathbf{x} = (1, 2, -1)$ in R^3 relative to the (nonstandard) basis

$$B' = \{\mathbf{u}_1, \mathbf{u}_2, \mathbf{u}_3\} = \{(1, 0, 1), (0, -1, 2), (2, 3, -5)\}.$$

SOLUTION Begin by writing $\mathbf{x}$ as a linear combination of $\mathbf{u}_1$, $\mathbf{u}_2$, and $\mathbf{u}_3$.

$$\begin{aligned} \mathbf{x} &= c_1\mathbf{u}_1 + c_2\mathbf{u}_2 + c_3\mathbf{u}_3 \\ (1, 2, -1) &= c_1(1, 0, 1) + c_2(0, -1, 2) + c_3(2, 3, -5) \end{aligned}$$

Equating corresponding components produces the following system of linear equations.

$$\begin{aligned} c_1 \qquad\quad + 2c_3 &= 1 \\ -c_2 + 3c_3 &= 2 \\ c_1 + 2c_2 - 5c_3 &= -1 \end{aligned}$$

$$\begin{bmatrix} 1 & 0 & 2 \\ 0 & -1 & 3 \\ 1 & 2 & -5 \end{bmatrix} \begin{bmatrix} c_1 \\ c_2 \\ c_3 \end{bmatrix} = \begin{bmatrix} 1 \\ 2 \\ -1 \end{bmatrix}$$

The solution of this system is $c_1 = 5$, $c_2 = -8$, and $c_3 = -2$. So,

$$\mathbf{x} = 5(1, 0, 1) + (-8)(0, -1, 2) + (-2)(2, 3, -5),$$

and the coordinate matrix of $\mathbf{x}$ relative to B' is

$$[\mathbf{x}]_{B'} = \begin{bmatrix} 5 \\ -8 \\ -2 \end{bmatrix}.$$

REMARK: Note that the solution in Example 3 is written as

$$[\mathbf{x}]_{B'} = \begin{bmatrix} 5 \\ -8 \\ -2 \end{bmatrix}.$$

It would be incorrect to write the solution as

$$\mathbf{x} = \begin{bmatrix} 5 \\ -8 \\ -2 \end{bmatrix}.$$

Do you see why?

Change of Basis in R^n

The procedure demonstrated in Examples 2 and 3 is called a **change of basis.** That is, you were provided with the coordinates of a vector relative to one basis B and were asked to find the coordinates relative to another basis B'.

For instance, if in Example 3 you let B be the standard basis, then the problem of finding the coordinate matrix of $\mathbf{x} = (1, 2, -1)$ relative to the basis B' becomes one of solving for c_1, c_2, and c_3 in the matrix equation

$$\underset{P}{\begin{bmatrix} 1 & 0 & 2 \\ 0 & -1 & 3 \\ 1 & 2 & -5 \end{bmatrix}} \underset{[\mathbf{x}]_{B'}}{\begin{bmatrix} c_1 \\ c_2 \\ c_3 \end{bmatrix}} = \underset{[\mathbf{x}]_B}{\begin{bmatrix} 1 \\ 2 \\ -1 \end{bmatrix}}.$$

The matrix P is called the **transition matrix from B' to B,** where $[\mathbf{x}]_{B'}$ is the coordinate matrix of $\mathbf{x}$ relative to B', and $[\mathbf{x}]_B$ is the coordinate matrix of $\mathbf{x}$ relative to B. Multiplication by the transition matrix P changes a coordinate matrix relative to B' into a coordinate matrix relative to B. That is,

$$P[\mathbf{x}]_{B'} = [\mathbf{x}]_B. \qquad \text{Change of basis from } B' \text{ to } B$$

To perform a change of basis from B to B', use the matrix P^{-1} (the **transition matrix from B to B'**) and write

$$[\mathbf{x}]_{B'} = P^{-1}[\mathbf{x}]_B. \qquad \text{Change of basis from } B \text{ to } B'$$

This means that the change of basis problem in Example 3 can be represented by the matrix equation

$$\begin{bmatrix} c_1 \\ c_2 \\ c_3 \end{bmatrix} = \underset{P^{-1}}{\begin{bmatrix} -1 & 4 & 2 \\ 3 & -7 & -3 \\ 1 & -2 & -1 \end{bmatrix}} \underset{[\mathbf{x}]_B}{\begin{bmatrix} 1 \\ 2 \\ -1 \end{bmatrix}} = \underset{[\mathbf{x}]_{B'}}{\begin{bmatrix} 5 \\ -8 \\ -2 \end{bmatrix}}.$$

This discussion generalizes as follows. Suppose that

$$B = \{\mathbf{v}_1, \mathbf{v}_2, \ldots, \mathbf{v}_n\} \quad \text{and} \quad B' = \{\mathbf{u}_1, \mathbf{u}_2, \ldots, \mathbf{u}_n\}$$

are two ordered bases for R^n. If $\mathbf{x}$ is a vector in R^n and

$$[\mathbf{x}]_B = \begin{bmatrix} c_1 \\ c_2 \\ \vdots \\ c_n \end{bmatrix} \quad \text{and} \quad [\mathbf{x}]_{B'} = \begin{bmatrix} d_1 \\ d_2 \\ \vdots \\ d_n \end{bmatrix}$$

are the coordinate matrices of $\mathbf{x}$ relative to B and B', then the **transition matrix P from B' to B** is the matrix P such that

$$[\mathbf{x}]_B = P[\mathbf{x}]_{B'}.$$

The next theorem tells you that the transition matrix P is invertible and its inverse is the **transition matrix from B to B'.** That is,

$$[\mathbf{x}]_{B'} = P^{-1}[\mathbf{x}]_B.$$

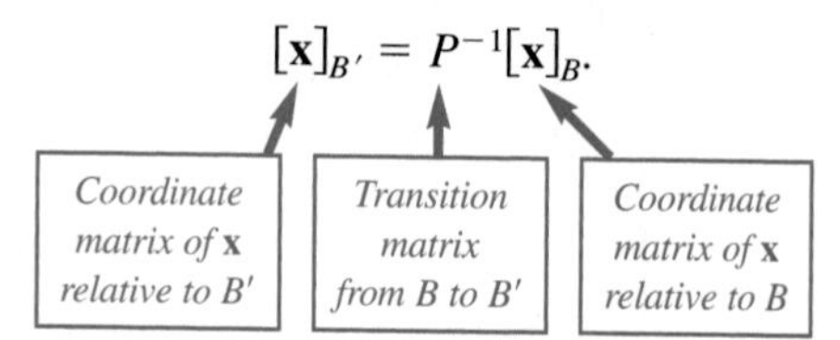

THEOREM 4.20
The Inverse of a Transition Matrix

If P is the transition matrix from a basis B' to a basis B in R^n, then P is invertible and the transition matrix from B to B' is given by P^{-1}.

Before proving Theorem 4.20, you will prove a preliminary lemma.

LEMMA Let $B = \{\mathbf{v}_1, \mathbf{v}_2, \ldots, \mathbf{v}_n\}$ and $B' = \{\mathbf{u}_1, \mathbf{u}_2, \ldots, \mathbf{u}_n\}$ be two bases for a vector space V. If

$$\begin{aligned}
\mathbf{v}_1 &= c_{11}\mathbf{u}_1 + c_{21}\mathbf{u}_2 + \cdots + c_{n1}\mathbf{u}_n \\
\mathbf{v}_2 &= c_{12}\mathbf{u}_1 + c_{22}\mathbf{u}_2 + \cdots + c_{n2}\mathbf{u}_n \\
&\vdots \\
\mathbf{v}_n &= c_{1n}\mathbf{u}_1 + c_{2n}\mathbf{u}_2 + \cdots + c_{nn}\mathbf{u}_n,
\end{aligned}$$

then the transition matrix from B to B' is

$$Q = \begin{bmatrix} c_{11} & c_{12} & \cdots & c_{1n} \\ c_{21} & c_{22} & \cdots & c_{2n} \\ \vdots & \vdots & & \vdots \\ c_{n1} & c_{n2} & \cdots & c_{nn} \end{bmatrix}.$$

PROOF (OF LEMMA) Let

$$\mathbf{v} = d_1\mathbf{v}_1 + d_2\mathbf{v}_2 + \cdots + d_n\mathbf{v}_n$$

be an arbitrary vector in V. The coordinate matrix of $\mathbf{v}$ with respect to the basis B is

$$[\mathbf{v}]_B = \begin{bmatrix} d_1 \\ d_2 \\ \vdots \\ d_n \end{bmatrix}.$$

Then you have

$$Q[\mathbf{v}]_B = \begin{bmatrix} c_{11} & c_{12} & \cdots & c_{1n} \\ c_{21} & c_{22} & \cdots & c_{2n} \\ \vdots & \vdots & & \vdots \\ c_{n1} & c_{n2} & \cdots & c_{nn} \end{bmatrix} \begin{bmatrix} d_1 \\ d_2 \\ \vdots \\ d_n \end{bmatrix} = \begin{bmatrix} c_{11}d_1 + c_{12}d_2 + \cdots + c_{1n}d_n \\ c_{21}d_1 + c_{22}d_2 + \cdots + c_{2n}d_n \\ \vdots \\ c_{n1}d_1 + c_{n2}d_2 + \cdots + c_{nn}d_n \end{bmatrix}.$$

On the other hand, you can write

$$\begin{aligned}
\mathbf{v} &= d_1\mathbf{v}_1 + d_2\mathbf{v}_2 + \cdots + d_n\mathbf{v}_n \\
&= d_1(c_{11}\mathbf{u}_1 + \cdots + c_{n1}\mathbf{u}_n) + \cdots + d_n(c_{1n}\mathbf{u}_1 + \cdots + c_{nn}\mathbf{u}_n) \\
&= (d_1c_{11} + \cdots + d_nc_{1n})\mathbf{u}_1 + \cdots + (d_1c_{n1} + \cdots + d_nc_{nn})\mathbf{u}_n,
\end{aligned}$$

which implies

$$[\mathbf{v}]_{B'} = \begin{bmatrix} c_{11}d_1 + c_{12}d_2 + \cdots + c_{1n}d_n \\ c_{21}d_1 + c_{22}d_2 + \cdots + c_{2n}d_n \\ \vdots \\ c_{n1}d_1 + c_{n2}d_2 + \cdots + c_{nn}d_n \end{bmatrix}.$$

So, $Q[\mathbf{v}]_B = [\mathbf{v}]_{B'}$ and you can conclude that Q is the transition matrix from B to B'.

PROOF (OF THEOREM 4.20) From the preceding lemma, let Q be the transition matrix from B to B'. Then

$$[\mathbf{v}]_B = P[\mathbf{v}]_{B'} \quad \text{and} \quad [\mathbf{v}]_{B'} = Q[\mathbf{v}]_B,$$

which implies that $[\mathbf{v}]_B = PQ[\mathbf{v}]_B$ for every vector $\mathbf{v}$ in R^n. From this it follows that $PQ = I$. So, P is invertible and P^{-1} is equal to Q, the transition matrix from B to B'.

There is a nice way to use Gauss-Jordan elimination to find the transition matrix P^{-1}. First define two matrices B and B' whose columns correspond to the vectors in B and B'. That is,

$$B = \begin{bmatrix} v_{11} & v_{12} & \cdots & v_{1n} \\ v_{21} & v_{22} & \cdots & v_{2n} \\ \vdots & \vdots & & \vdots \\ v_{n1} & v_{n2} & \cdots & v_{nn} \end{bmatrix} \quad \text{and} \quad B' = \begin{bmatrix} u_{11} & u_{12} & \cdots & u_{1n} \\ u_{21} & u_{22} & \cdots & u_{2n} \\ \vdots & \vdots & & \vdots \\ u_{n1} & u_{n2} & \cdots & u_{nn} \end{bmatrix}.$$

$$\mathbf{v}_1 \quad \mathbf{v}_2 \quad \cdots \quad \mathbf{v}_n \qquad\qquad \mathbf{u}_1 \quad \mathbf{u}_2 \quad \cdots \quad \mathbf{u}_n$$

Then, by reducing the $n \times 2n$ matrix $[B' \ \vdots \ B]$ so that the identity matrix I_n occurs in place of B', you obtain the matrix $[I_n \ \vdots \ P^{-1}]$. This procedure is stated formally in the next theorem.

THEOREM 4.21 Transition Matrix from B to B'

Let $B = \{\mathbf{v}_1, \mathbf{v}_2, \ldots, \mathbf{v}_n\}$ and $B' = \{\mathbf{u}_1, \mathbf{u}_2, \ldots, \mathbf{u}_n\}$ be two bases for R^n. Then the transition matrix P^{-1} from B to B' can be found by using Gauss-Jordan elimination on the $n \times 2n$ matrix $[B' \ \vdots \ B]$, as follows.

$$[B' \ \vdots \ B] \longrightarrow [I_n \ \vdots \ P^{-1}]$$

PROOF To begin, let

$$\begin{aligned} \mathbf{v}_1 &= c_{11}\mathbf{u}_1 + c_{21}\mathbf{u}_2 + \cdots + c_{n1}\mathbf{u}_n \\ \mathbf{v}_2 &= c_{12}\mathbf{u}_1 + c_{22}\mathbf{u}_2 + \cdots + c_{n2}\mathbf{u}_n \\ &\vdots \\ \mathbf{v}_n &= c_{1n}\mathbf{u}_1 + c_{2n}\mathbf{u}_2 + \cdots + c_{nn}\mathbf{u}_n, \end{aligned}$$

which implies that

$$c_{1i}\begin{bmatrix} u_{11} \\ u_{12} \\ \vdots \\ u_{1n} \end{bmatrix} + c_{2i}\begin{bmatrix} u_{21} \\ u_{22} \\ \vdots \\ u_{2n} \end{bmatrix} + \cdots + c_{ni}\begin{bmatrix} u_{n1} \\ u_{n2} \\ \vdots \\ u_{nn} \end{bmatrix} = \begin{bmatrix} v_{i1} \\ v_{i2} \\ \vdots \\ v_{in} \end{bmatrix}$$

for $i = 1, 2, \ldots, n$. From these vector equations you can write the n systems of linear equations

$$\begin{aligned}
u_{11}c_{1i} + u_{21}c_{2i} + \cdot \cdot \cdot + u_{n1}c_{ni} &= v_{i1}\\
u_{12}c_{1i} + u_{22}c_{2i} + \cdot \cdot \cdot + u_{n2}c_{ni} &= v_{i2}\\
&\vdots\\
u_{1n}c_{1i} + u_{2n}c_{2i} + \cdot \cdot \cdot + u_{nn}c_{ni} &= v_{in}
\end{aligned}$$

for $i = 1, 2, . . . , n$. Because each of the n systems has the same coefficient matrix, you can reduce all n systems simultaneously using the augmented matrix below.

$$\left[\begin{array}{cccc} u_{11} & u_{21} & \cdots & u_{n1} \\ u_{12} & u_{22} & \cdots & u_{n2} \\ \vdots & \vdots & & \vdots \\ u_{1n} & u_{2n} & \cdots & u_{nn} \end{array}\right.\!\!\!\!\!\!\underbrace{\phantom{u_{11}\;u_{21}\;\cdots\;u_{n1}}}_{B'}
\begin{array}{c} \vdots \\ \vdots \\ \vdots \\ \vdots \end{array}
\left.\begin{array}{cccc} v_{11} & v_{21} & \cdots & v_{n1} \\ v_{12} & v_{22} & \cdots & v_{n2} \\ \vdots & \vdots & & \vdots \\ v_{1n} & v_{2n} & \cdots & v_{nn} \end{array}\right]\!\!\!\!\!\!\underbrace{\phantom{v_{11}\;v_{21}\;\cdots\;v_{n1}}}_{B}$$

Applying Gauss-Jordan elimination to this matrix produces

$$\left[\begin{array}{cccc:cccc} 1 & 0 & \cdots & 0 & c_{11} & c_{12} & \cdots & c_{1n} \\ 0 & 1 & \cdots & 0 & c_{21} & c_{22} & \cdots & c_{2n} \\ \vdots & \vdots & & \vdots & \vdots & \vdots & & \vdots \\ 0 & 0 & \cdots & 1 & c_{n1} & c_{n2} & \cdots & c_{nn} \end{array}\right].$$

By the lemma following Theorem 4.20, however, the right-hand side of this matrix is $Q = P^{-1}$, which implies that the matrix has the form

$[I \ \vdots \ P^{-1}]$, which proves the theorem.

In the next example, you will apply this procedure to the change of basis problem from Example 3.

EXAMPLE 4 Finding a Transition Matrix

Find the transition matrix from B to B' for the following bases in R^3.

$$B = \{(1, 0, 0), (0, 1, 0), (0, 0, 1)\} \quad \text{and} \quad B' = \{(1, 0, 1), (0, -1, 2), (2, 3, -5)\}$$

SOLUTION First use the vectors in the two bases to form the matrices B and B'.

$$B = \begin{bmatrix} 1 & 0 & 0 \\ 0 & 1 & 0 \\ 0 & 0 & 1 \end{bmatrix} \quad \text{and} \quad B' = \begin{bmatrix} 1 & 0 & 2 \\ 0 & -1 & 3 \\ 1 & 2 & -5 \end{bmatrix}$$

Then form the matrix $[B' \ \vdots \ B]$ and use Gauss-Jordan elimination to rewrite $[B' \ \vdots \ B]$ as $[I_3 \ \vdots \ P^{-1}]$.

$$\left[\begin{array}{ccc:ccc} 1 & 0 & 2 & 1 & 0 & 0 \\ 0 & -1 & 3 & 0 & 1 & 0 \\ 1 & 2 & -5 & 0 & 0 & 1 \end{array}\right] \longrightarrow \left[\begin{array}{ccc:ccc} 1 & 0 & 0 & -1 & 4 & 2 \\ 0 & 1 & 0 & 3 & -7 & -3 \\ 0 & 0 & 1 & 1 & -2 & -1 \end{array}\right]$$

From this you can conclude that the transition matrix from B to B' is

$$P^{-1} = \begin{bmatrix} -1 & 4 & 2 \\ 3 & -7 & -3 \\ 1 & -2 & -1 \end{bmatrix}.$$

Try multiplying P^{-1} by the coordinate matrix of

$$\mathbf{x} = \begin{bmatrix} 1 \\ 2 \\ -1 \end{bmatrix}$$

to see that the result is the same as the one obtained in Example 3.

Discovery

Let $B = \{(1, 0), (1, 2)\}$ and $B' = \{(1, 0), (0, 1)\}$. Form the matrix $[B' \vdots B]$. Make a conjecture about the necessity of using Gauss-Jordan elimination to obtain the transition matrix P^{-1} if the change of basis is from a nonstandard basis to a standard basis.

Note that when B is the standard basis, as in Example 4, the process of changing $[B' \vdots B]$ to $[I_n \vdots P^{-1}]$ becomes

$$[B' \vdots I_n] \longrightarrow [I_n \vdots P^{-1}].$$

But this is the same process that was used to find inverse matrices in Section 2.3. In other words, if B is the standard basis in R^n, then the transition matrix from B to B' is

$$P^{-1} = (B')^{-1}. \qquad \text{Standard basis to nonstandard basis}$$

The process is even simpler if B' is the standard basis, because the matrix $[B' \vdots B]$ is already in the form

$$[I_n \vdots B] = [I_n \vdots P^{-1}].$$

In this case, the transition matrix is simply

$$P^{-1} = B. \qquad \text{Nonstandard basis to standard basis}$$

For instance, the transition matrix in Example 2 from $B = \{(1, 0), (1, 2)\}$ to $B' = \{(1, 0), (0, 1)\}$ is

$$P^{-1} = B = \begin{bmatrix} 1 & 1 \\ 0 & 2 \end{bmatrix}.$$

EXAMPLE 5 Finding a Transition Matrix

Find the transition matrix from B to B' for the following bases for R^2.

$$B = \{(-3, 2), (4, -2)\} \qquad \text{and} \qquad B' = \{(-1, 2), (2, -2)\}$$

SOLUTION Begin by forming the matrix

$$[B' \;\vdots\; B] = \begin{bmatrix} -1 & 2 & \vdots & -3 & 4 \\ 2 & -2 & \vdots & 2 & -2 \end{bmatrix}$$

and use Gauss-Jordan elimination to obtain the transition matrix P^{-1} from B to B':

$$[I_2 \;\vdots\; P^{-1}] = \begin{bmatrix} 1 & 0 & \vdots & -1 & 2 \\ 0 & 1 & \vdots & -2 & 3 \end{bmatrix}.$$

So, you have

$$P^{-1} = \begin{bmatrix} -1 & 2 \\ -2 & 3 \end{bmatrix}.$$

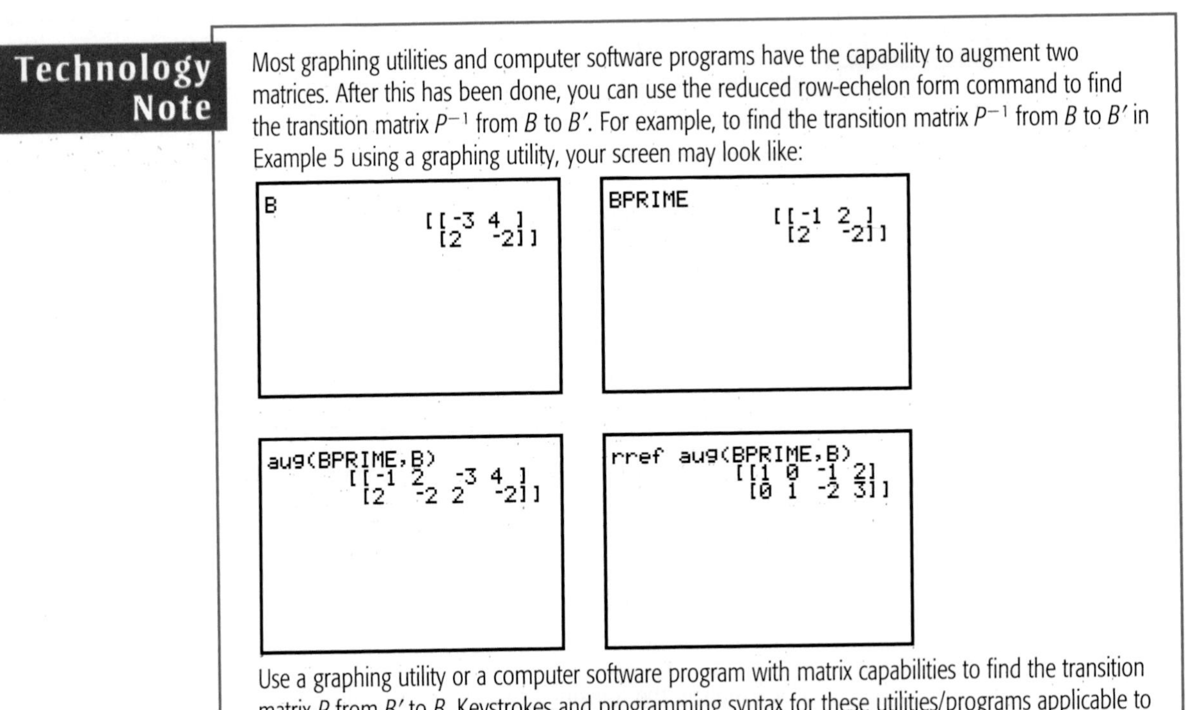

Technology Note

Most graphing utilities and computer software programs have the capability to augment two matrices. After this has been done, you can use the reduced row-echelon form command to find the transition matrix P^{-1} from B to B'. For example, to find the transition matrix P^{-1} from B to B' in Example 5 using a graphing utility, your screen may look like:

Use a graphing utility or a computer software program with matrix capabilities to find the transition matrix P from B' to B. Keystrokes and programming syntax for these utilities/programs applicable to Example 5 are provided in the **Online Technology Guide,** available at *college.hmco.com/pic/larsonELA6e*.

In Example 5, if you had found the transition matrix from B' to B (rather than from B to B'), you would have obtained

$$[B \;\vdots\; B'] = \begin{bmatrix} -3 & 4 & \vdots & -1 & 2 \\ 2 & -2 & \vdots & 2 & -2 \end{bmatrix},$$

which reduces to

$$[I_2 \ \vdots \ P] = \begin{bmatrix} 1 & 0 & \vdots & 3 & -2 \\ 0 & 1 & \vdots & 2 & -1 \end{bmatrix}.$$

The transition matrix from B' to B is

$$P = \begin{bmatrix} 3 & -2 \\ 2 & -1 \end{bmatrix}.$$

You can verify that this is the inverse of the transition matrix found in Example 5 by multiplication:

$$PP^{-1} = \begin{bmatrix} 3 & -2 \\ 2 & -1 \end{bmatrix}\begin{bmatrix} -1 & 2 \\ -2 & 3 \end{bmatrix} = \begin{bmatrix} 1 & 0 \\ 0 & 1 \end{bmatrix} = I_2.$$

Coordinate Representation in General n-Dimensional Spaces

One benefit of coordinate representation is that it enables you to represent vectors in an arbitrary n-dimensional space using the same notation used in R^n. For instance, in Example 6, note that the coordinate matrix of a vector in P_3 is a vector in R^4.

EXAMPLE 6 Coordinate Representation in P_3

Find the coordinate matrix of $p = 3x^3 - 2x^2 + 4$ relative to the standard basis in P_3,

$$S = \{1, x, x^2, x^3\}.$$

SOLUTION First write p as a linear combination of the basis vectors (in the order provided).

$$p = 4(1) + 0(x) + (-2)(x^2) + 3(x^3)$$

This tells you that the coordinate matrix of p relative to S is

$$[p]_S = \begin{bmatrix} 4 \\ 0 \\ -2 \\ 3 \end{bmatrix}.$$

In the preceding section, you saw that it is sometimes convenient to represent $n \times 1$ matrices as n-tuples. The next example presents some justification for this practice.

EXAMPLE 7 Coordinate Representation in $M_{3,1}$

Find the coordinate matrix of

$$X = \begin{bmatrix} -1 \\ 4 \\ 3 \end{bmatrix}$$

relative to the standard basis in $M_{3,1}$,

$$S = \left\{ \begin{bmatrix} 1 \\ 0 \\ 0 \end{bmatrix}, \begin{bmatrix} 0 \\ 1 \\ 0 \end{bmatrix}, \begin{bmatrix} 0 \\ 0 \\ 1 \end{bmatrix} \right\}.$$

SOLUTION Because X can be written as

$$X = \begin{bmatrix} -1 \\ 4 \\ 3 \end{bmatrix} = (-1)\begin{bmatrix} 1 \\ 0 \\ 0 \end{bmatrix} + 4\begin{bmatrix} 0 \\ 1 \\ 0 \end{bmatrix} + 3\begin{bmatrix} 0 \\ 0 \\ 1 \end{bmatrix},$$

the coordinate matrix of X relative to S is

$$[X]_S = \begin{bmatrix} -1 \\ 4 \\ 3 \end{bmatrix}.$$

REMARK: In Section 6.2 you will learn more about the use of R^n to represent an arbitrary n-dimensional vector space.

Theorems 4.20 and 4.21 can be generalized to cover arbitrary n-dimensional spaces. This text, however, does not cover this.

SECTION 4.7 Exercises

In Exercises 1–6, you are provided with the coordinate matrix of $\mathbf{x}$ relative to a (nonstandard) basis B. Find the coordinate vector of $\mathbf{x}$ relative to the standard basis in R^n.

1. $B = \{(2, -1), (0, 1)\}, [\mathbf{x}]_B = [4, 1]^T$
2. $B = \{(-1, 4), (4, -1)\}, [\mathbf{x}]_B = [-2, 3]^T$
3. $B = \{(1, 0, 1), (1, 1, 0), (0, 1, 1)\}, [\mathbf{x}]_B = [2, 3, 1]^T$
4. $B = \left\{\left(\frac{3}{4}, \frac{5}{2}, \frac{3}{2}\right), \left(3, 4, \frac{7}{2}\right), \left(-\frac{3}{2}, 6, 2\right)\right\}, [\mathbf{x}]_B = [2, 0, 4]^T$
5. $B = \{(0, 0, 0, 1), (0, 0, 1, 1), (0, 1, 1, 1), (1, 1, 1, 1)\},$ $[\mathbf{x}]_B = [1, -2, 3, -1]^T$
6. $B = \{(4, 0, 7, 3), (0, 5, -1, -1), (-3, 4, 2, 1), (0, 1, 5, 0)\},$ $[\mathbf{x}]_B = [-2, 3, 4, 1]^T$

In Exercises 7–12, find the coordinate matrix of $\mathbf{x}$ in R^n relative to the basis B.

7. $B = \{(4, 0), (0, 3)\}, \mathbf{x} = (12, 6)$
8. $B = \{(-6, 7), (4, -3)\}, \mathbf{x} = (-26, 32)$
9. $B = \{(8, 11, 0), (7, 0, 10), (1, 4, 6)\}, \mathbf{x} = (3, 19, 2)$
10. $B = \left\{\left(\frac{3}{2}, 4, 1\right), \left(\frac{3}{4}, \frac{5}{2}, 0\right), \left(1, \frac{1}{2}, 2\right)\right\}, \mathbf{x} = \left(3, -\frac{1}{2}, 8\right)$
11. $B = \{(4, 3, 3), (-11, 0, 11), (0, 9, 2)\}, \mathbf{x} = (11, 18, -7)$
12. $B = \{(9, -3, 15, 4), (3, 0, 0, 1), (0, -5, 6, 8), (3, -4, 2, -3)\},$ $\mathbf{x} = (0, -20, 7, 15)$

In Exercises 13–18, find the transition matrix from B to B' by hand.

13. $B = \{(1, 0), (0, 1)\}, B' = \{(2, 4), (1, 3)\}$
14. $B = \{(1, 0), (0, 1)\}, B' = \{(1, 1), (5, 6)\}$
15. $B = \{(2, 4), (-1, 3)\}, B' = \{(1, 0), (0, 1)\}$
16. $B = \{(1, 1), (1, 0)\}, B' = \{(1, 0), (0, 1)\}$
17. $B = \{(1, 0, 0), (0, 1, 0), (0, 0, 1)\},$ $B' = \{(1, 0, 0), (0, 2, 8), (6, 0, 12)\}$
18. $B = \{(1, 0, 0), (0, 1, 0), (0, 0, 1)\},$ $B' = \{(1, 3, -1), (2, 7, -4), (2, 9, -7)\}$

In Exercises 19–28, use a graphing utility or computer software program with matrix capabilities to find the transition matrix from B to B'.

19. $B = \{(2, 5), (1, 2)\}, B' = \{(2, 1), (-1, 2)\}$
20. $B = \{(-2, 1), (3, 2)\}, B' = \{(1, 2), (-1, 0)\}$

21. $B = \{(1, 3, 3), (1, 5, 6), (1, 4, 5)\},$
$B' = \{(1, 0, 0), (0, 1, 0), (0, 0, 1)\}$

22. $B = \{(2, -1, 4), (0, 2, 1), (-3, 2, 1)\},$
$B' = \{(1, 0, 0), (0, 1, 0), (0, 0, 1)\}$

23. $B = \{(1, 2, 4), (-1, 2, 0), (2, 4, 0)\},$
$B' = \{(0, 2, 1), (-2, 1, 0), (1, 1, 1)\}$

24. $B = \{(3, 2, 1), (1, 1, 2), (1, 2, 0)\},$
$B' = \{(1, 1, -1), (0, 1, 2), (-1, 4, 0)\}$

25. $B = \{(1, 0, 0, 0), (0, 1, 0, 0), (0, 0, 1, 0), (0, 0, 0, 1)\},$
$B' = \{(1, 3, 2, -1), (-2, -5, -5, 4), (-1, -2, -2, 4), (-2, -3, -5, 11)\}$

26. $B = \{(1, 1, 1, 1), (0, 1, 1, 1), (0, 0, 1, 1), (0, 0, 0, 1)\},$
$B' = \{(1, 0, 0, 0), (0, 1, 0, 0), (0, 0, 1, 0), (0, 0, 0, 1)\}$

HM **27.** $B = \{(1, 0, 0, 0, 0), (0, 1, 0, 0, 0), (0, 0, 1, 0, 0), (0, 0, 0, 1, 0), (0, 0, 0, 0, 1)\},$
$B' = \{(1, 2, 4, -1, 2), (-2, -3, 4, 2, 1), (0, 1, 2, -2, 1), (0, 1, 2, 2, 1), (1, -1, 0, 1, 2)\}$

HM **28.** $B = \{(1, 0, 0, 0, 0), (0, 1, 0, 0, 0), (0, 0, 1, 0, 0), (0, 0, 0, 1, 0), (0, 0, 0, 0, 1)\},$
$B' = \{(2, 4, -2, 1, 0), (3, -1, 0, 1, 2), (0, 0, -2, 4, 5), (2, -1, 2, 1, 1), (0, 1, 2, -3, 1)\}$

In Exercises 29–32, use Theorem 4.21 to (a) find the transition matrix from B to B', (b) find the transition matrix from B' to B, (c) verify that the two transition matrices are inverses of each other, and (d) find $[\mathbf{x}]_B$ when provided with $[\mathbf{x}]_{B'}$.

29. $B = \{(1, 3), (-2, -2)\}, \quad B' = \{(-12, 0), (-4, 4)\},$
$[\mathbf{x}]_{B'} = \begin{bmatrix} -1 \\ 3 \end{bmatrix}$

30. $B = \{(2, -2), (6, 3)\}, \quad B' = \{(1, 1), (32, 31)\},$
$[\mathbf{x}]_{B'} = \begin{bmatrix} 2 \\ -1 \end{bmatrix}$

31. $B = \{(1, 0, 2), (0, 1, 3), (1, 1, 1)\},$
$B' = \{(2, 1, 1), (1, 0, 0), (0, 2, 1)\},$
$[\mathbf{x}]_{B'} = \begin{bmatrix} 1 \\ 2 \\ -1 \end{bmatrix}$

32. $B = \{(1, 1, 1), (1, -1, 1), (0, 0, 1)\},$
$B' = \{(2, 2, 0), (0, 1, 1), (1, 0, 1)\},$
$[\mathbf{x}]_{B'} = \begin{bmatrix} 2 \\ 3 \\ 1 \end{bmatrix}$

HM In Exercises 33 and 34, use a graphing utility with matrix capabilities to (a) find the transition matrix from B to B', (b) find the transition matrix from B' to B, (c) verify that the two transition matrices are inverses of one another, and (d) find $[\mathbf{x}]_B$ when provided with $[\mathbf{x}]_{B'}$.

33. $B = \{(4, 2, -4), (6, -5, -6), (2, -1, 8)\},$
$B' = \{(1, 0, 4), (4, 2, 8), (2, 5, -2)\},$
$[\mathbf{x}]_{B'} = \begin{bmatrix} 1 \\ -1 \\ 2 \end{bmatrix}$

34. $B = \{(1, 3, 4), (2, -5, 2), (-4, 2, -6)\},$
$B' = \{(1, 2, -2), (4, 1, -4), (-2, 5, 8)\},$
$[\mathbf{x}]_{B'} = \begin{bmatrix} -1 \\ 0 \\ 2 \end{bmatrix}$

In Exercises 35–38, find the coordinate matrix of p relative to the standard basis in P_2.

35. $p = x^2 + 11x + 4$

36. $p = 3x^2 + 114x + 13$

37. $p = -2x^2 + 5x + 1$

38. $p = 4x^2 - 3x - 2$

In Exercises 39–42, find the coordinate matrix of X relative to the standard basis in $M_{3,1}$.

39. $X = \begin{bmatrix} 0 \\ 3 \\ 2 \end{bmatrix}$

40. $X = \begin{bmatrix} 2 \\ -1 \\ 4 \end{bmatrix}$

41. $X = \begin{bmatrix} 1 \\ 2 \\ -1 \end{bmatrix}$

42. $X = \begin{bmatrix} 1 \\ 0 \\ -4 \end{bmatrix}$

True or False? In Exercises 43 and 44, determine whether each statement is true or false. If a statement is true, give a reason or cite an appropriate statement from the text. If a statement is false, provide an example that shows the statement is not true in all cases or cite an appropriate statement from the text.

43. (a) If P is the transition matrix from a basis B to B', then the equation $P[\mathbf{x}]_{B'} = [\mathbf{x}]_B$ represents the change of basis from B to B'.

(b) If B is the standard basis in R^n, then the transition matrix from B to B' is $P^{-1} = (B')^{-1}$.

44. (a) If P is the transition matrix used to perform a change of basis from B' to B, then P^{-1} is the transition matrix from B to B'.

(b) To perform the change of basis from a nonstandard basis B' to the standard basis B, the transition matrix P^{-1} is simply B.

45. Let P be the transition matrix from B'' to B', and let Q be the transition matrix from B' to B. What is the transition matrix from B'' to B?

46. Let P be the transition matrix from B'' to B', and let Q be the transition matrix from B' to B. What is the transition matrix from B to B''?

47. Writing Let B and B' be two bases for the vector space R^n. Discuss the nature of the transition matrix from B to B' if one of the bases is the standard basis.

48. Writing Is it possible for a transition matrix to equal the identity matrix? Illustrate your answer with appropriate examples.

4.8 Applications of Vector Spaces

Linear Differential Equations (Calculus)

A **linear differential equation of order n** is of the form

$$y^{(n)} + g_{n-1}(x)y^{(n-1)} + \cdot \cdot \cdot + g_1(x)y' + g_0(x)y = f(x),$$

where $g_0, g_1, \ldots, g_{n-1}$, and f are functions of x with a common domain. If $f(x) = 0$, the equation is **homogeneous.** Otherwise it is **nonhomogeneous.** A function y is called a **solution** of the linear differential equation if the equation is satisfied when y and its first n derivatives are substituted into the equation.

EXAMPLE 1 A Second-Order Linear Differential Equation

Show that both $y_1 = e^x$ and $y_2 = e^{-x}$ are solutions of the second-order linear differential equation $y'' - y = 0$.

SOLUTION For the function $y_1 = e^x$, you have $y_1' = e^x$ and $y_1'' = e^x$. So,

$$y_1'' - y_1 = e^x - e^x = 0,$$

and $y_1 = e^x$ is a solution of the differential equation. Similarly, for $y_2 = e^{-x}$, you have

$$y_2' = -e^{-x} \quad \text{and} \quad y_2'' = e^{-x}.$$

This implies that

$$y_2'' - y_2 = e^{-x} - e^{-x} = 0.$$

So, $y_2 = e^{-x}$ is also a solution of the linear differential equation.

There are two important observations you can make about Example 1. The first is that in the vector space $C''(-\infty, \infty)$ of all twice differentiable functions defined on the entire real line, the two solutions $y_1 = e^x$ and $y_2 = e^{-x}$ are *linearly independent.* This means that the only solution of

$$C_1y_1 + C_2y_2 = 0$$

that is valid for all x is $C_1 = C_2 = 0$. The second observation is that every *linear combination* of y_1 and y_2 is also a solution of the linear differential equation. To see this, let $y = C_1y_1 + C_2y_2$. Then

$$\begin{aligned} y &= C_1e^x + C_2e^{-x} \\ y' &= C_1e^x - C_2e^{-x} \\ y'' &= C_1e^x + C_2e^{-x}. \end{aligned}$$

Substituting into the differential equation $y'' - y = 0$ produces

$$y'' - y = (C_1e^x + C_2e^{-x}) - (C_1e^x + C_2e^{-x}) = 0.$$

So, $y = C_1e^x + C_2e^{-x}$ is a solution.

These two observations are generalized in the next theorem, which is stated without proof.

Solutions of a Linear Homogeneous Differential Equation

Every nth-order linear homogeneous differential equation

$$y^{(n)} + g_{n-1}(x)y^{(n-1)} + \cdots + g_1(x)y' + g_0(x)y = 0$$

has n linearly independent solutions. Moreover, if $\{y_1, y_2, \ldots, y_n\}$ is a set of linearly independent solutions, then every solution is of the form

$$y = C_1y_1 + C_2y_2 + \cdots + C_ny_n,$$

where $C_1, C_2, \ldots,$ and C_n are real numbers.

REMARK: The solution $y = C_1y_1 + C_2y_2 + \cdots + C_ny_n$ is called the **general solution** of the given differential equation.

In light of the preceding theorem, you can see the importance of being able to determine whether a set of solutions is linearly independent. Before describing a way of testing for linear independence, you are provided with a preliminary definition.

Definition of the Wronskian of a Set of Functions

Let $\{y_1, y_2, \ldots, y_n\}$ be a set of functions, each of which has $n - 1$ derivatives on an interval I. The determinant

$$W(y_1, y_2, \ldots, y_n) = \begin{vmatrix} y_1 & y_2 & \cdots & y_n \\ y_1' & y_2' & \cdots & y_n' \\ \vdots & \vdots & & \vdots \\ y_1^{(n-1)} & y_2^{(n-1)} & \cdots & y_n^{(n-1)} \end{vmatrix}$$

is called the **Wronskian** of the given set of functions.

REMARK: The Wronskian of a set of functions is named after the Polish mathematician Josef Maria Wronski (1778–1853).

EXAMPLE 2 Finding the Wronskian of a Set of Functions

(a) The Wronskian of the set $\{1 - x, 1 + x, 2 - x\}$ is

$$W = \begin{vmatrix} 1-x & 1+x & 2-x \\ -1 & 1 & -1 \\ 0 & 0 & 0 \end{vmatrix} = 0.$$

(b) The Wronskian of the set $\{x, x^2, x^3\}$ is

$$W = \begin{vmatrix} x & x^2 & x^3 \\ 1 & 2x & 3x^2 \\ 0 & 2 & 6x \end{vmatrix} = 2x^3.$$

The Wronskian in part (a) of Example 2 is said to be **identically equal to zero,** because it is zero for any value of x. The Wronskian in part (b) is not identically equal to zero because values of x exist for which this Wronskian is nonzero.

The next theorem shows how the Wronskian of a set of functions can be used to test for linear independence.

Wronskian Test for Linear Independence

Let $\{y_1, y_2, . . . , y_n\}$ be a set of n solutions of an nth-order linear homogeneous differential equation. This set is linearly independent if and only if the Wronskian is not identically equal to zero.

REMARK: This test does *not* apply to an arbitrary set of functions. Each of the functions $y_1, y_2, . . . ,$ and y_n must be a solution of the same linear homogeneous differential equation of order n.

EXAMPLE 3 Testing a Set of Solutions for Linear Independence

Determine whether $\{1, \cos x, \sin x\}$ is a set of linearly independent solutions of the linear homogeneous differential equation

$$y''' + y' = 0.$$

SOLUTION Begin by observing that each of the functions is a solution of $y''' + y' = 0$. (Try checking this.) Next, testing for linear independence produces the Wronskian of the three functions, as follows.

$$W = \begin{vmatrix} 1 & \cos x & \sin x \\ 0 & -\sin x & \cos x \\ 0 & -\cos x & -\sin x \end{vmatrix}$$
$$= \sin^2 x + \cos^2 x = 1$$

Because W is not identically equal to zero, you can conclude that the set $\{1, \cos x, \sin x\}$ is linearly independent. Moreover, because this set consists of three linearly independent solutions of a third-order linear homogeneous differential equation, you can conclude that the general solution is $y = C_1 + C_2 \cos x + C_3 \sin x$.

EXAMPLE 4 **Testing a Set of Solutions for Linear Independence**

Determine whether $\{e^x, xe^x, (x + 1)e^x\}$ is a set of linearly independent solutions of the linear homogeneous differential equation

$$y''' - 3y'' + 3y' - y = 0.$$

SOLUTION As in Example 3, begin by verifying that each of the functions is actually a solution of $y''' - 3y'' + 3y' - y = 0$. (This verification is left to you.) Testing for linear independence produces the Wronskian of the three functions as follows.

$$W = \begin{vmatrix} e^x & xe^x & (x+1)e^x \\ e^x & (x+1)e^x & (x+2)e^x \\ e^x & (x+2)e^x & (x+3)e^x \end{vmatrix} = 0$$

So, the set $\{e^x, xe^x, (x + 1)e^x\}$ is linearly dependent.

In Example 4, the Wronskian was used to determine that the set

$$\{e^x, xe^x, (x + 1)e^x\}$$

is linearly dependent. Another way to determine the linear dependence of this set is to observe that the third function is a linear combination of the first two. That is,

$$(x + 1)e^x = e^x + xe^x.$$

Try showing that the different set $\{e^x, xe^x, x^2e^x\}$ forms a linearly independent set of solutions of the differential equation

$$y''' - 3y'' + 3y' - y = 0.$$

Conic Sections and Rotation

Every conic section in the xy-plane has an equation of the form

$$ax^2 + bxy + cy^2 + dx + ey + f = 0.$$

Identifying the graph of this equation is fairly simple as long as b, the coefficient of the xy-term, is zero. In such cases the conic axes are parallel to the coordinate axes, and the identification is accomplished by writing the equation in standard (completed square) form. The standard forms of the equations of the four basic conics are provided in the next summary. For circles, ellipses, and hyperbolas, the point (h, k) is the center. For parabolas, the point (h, k) is the vertex.

Standard Forms of Equations of Conics

Circle (r = radius): $(x - h)^2 + (y - k)^2 = r^2$

Ellipse (2α = major axis length, 2β = minor axis length):

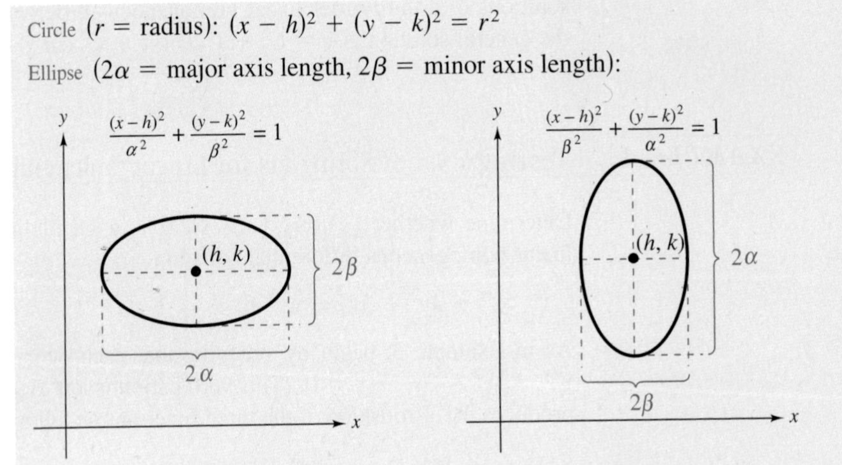

Hyperbola (2α = transverse axis length, 2β = minor axis length):

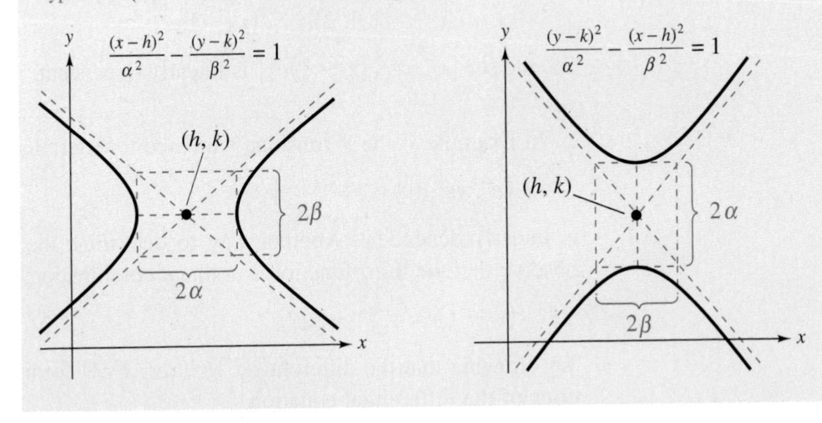

Standard Forms of Equations of Conics (cont.)

Parabola (p = directed distance from vertex to focus):

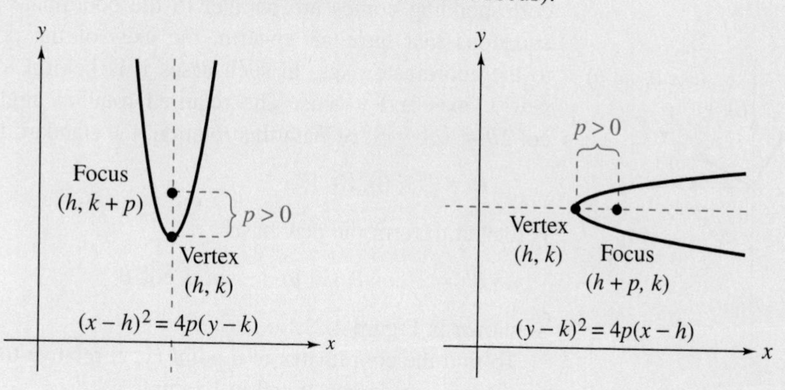

EXAMPLE 5 **Identifying Conic Sections**

(a) The standard form of $x^2 - 2x + 4y - 3 = 0$ is $(x - 1)^2 = 4(-1)(y - 1)$. The graph of this equation is a parabola with the vertex at $(h, k) = (1, 1)$. The axis of the parabola is vertical. Because $p = -1$, the focus is the point $(1, 0)$. Finally, because the focus lies below the vertex, the parabola opens downward, as shown in Figure 4.20(a).

(b) The standard form of $x^2 + 4y^2 + 6x - 8y + 9 = 0$ is

$$\frac{(x + 3)^2}{4} + \frac{(y - 1)^2}{1} = 1.$$

The graph of this equation is an ellipse with its center at $(h, k) = (-3, 1)$. The major axis is horizontal, and its length is $2\alpha = 4$. The length of the minor axis is $2\beta = 2$. The vertices of this ellipse occur at $(-5, 1)$ and $(-1, 1)$, and the endpoints of the minor axis occur at $(-3, 2)$ and $(-3, 0)$, as shown in Figure 4.20(b).

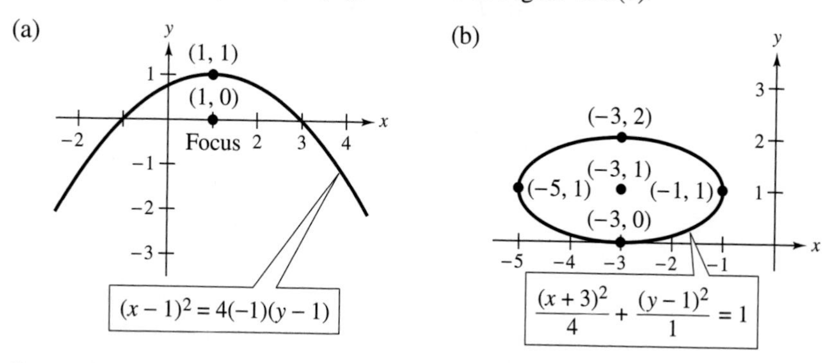

Figure 4.20

The equations of the conics in Example 5 have no xy-term. Consequently, the axes of the corresponding conics are parallel to the coordinate axes. For second-degree polynomial equations that have an xy-term, the axes of the corresponding conics are not parallel to the coordinate axes. In such cases it is helpful to *rotate* the standard axes to form a new x'-axis and y'-axis. The required rotation angle θ (measured counterclockwise) is $\cot 2\theta = (a - c)/b$. With this rotation, the standard basis in the plane

$$B = \{(1, 0), (0, 1)\}$$

is rotated to form the new basis

$$B' = \{(\cos \theta, \sin \theta), (-\sin \theta, \cos \theta)\},$$

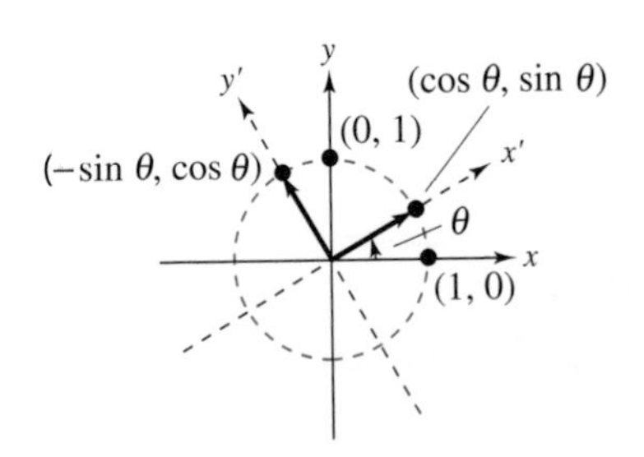

Figure 4.21

as shown in Figure 4.21.

To find the coordinates of a point (x, y) relative to this new basis, you can use a transition matrix, as demonstrated in Example 6.

EXAMPLE 6 A Transition Matrix for Rotation in the Plane

Find the coordinates of a point (x, y) in R^2 relative to the basis

$$B' = \{(\cos \theta, \sin \theta), (-\sin \theta, \cos \theta)\}.$$

SOLUTION By Theorem 4.21 you have

$$[B' \vdots B] = \left[\begin{array}{cc|cc} \cos \theta & -\sin \theta & 1 & 0 \\ \sin \theta & \cos \theta & 0 & 1 \end{array}\right].$$

Because B is the standard basis in R^2, P^{-1} is represented by $(B')^{-1}$. You can use Theorem 3.10 to find $(B')^{-1}$. This results in

$$[I \vdots P^{-1}] = \left[\begin{array}{cc|cc} 1 & 0 & \cos \theta & \sin \theta \\ 0 & 1 & -\sin \theta & \cos \theta \end{array}\right].$$

By letting $[x', y']^T$ be the coordinates of (x, y) relative to B', you can use the transition matrix P^{-1} as follows.

$$\begin{bmatrix} \cos \theta & \sin \theta \\ -\sin \theta & \cos \theta \end{bmatrix} \begin{bmatrix} x \\ y \end{bmatrix} = \begin{bmatrix} x' \\ y' \end{bmatrix}$$

The x'- and y'-coordinates are

$$x' = x \cos \theta + y \sin \theta$$
$$y' = -x \sin \theta + y \cos \theta.$$

The last two equations in Example 6 give the $x'y'$-coordinates in terms of the xy-coordinates. To perform a rotation of axes for a second-degree polynomial equation, it is helpful to express the xy-coordinates in terms of the $x'y'$-coordinates. To do this, solve the last two equations in Example 6 for x and y to obtain

$$x = x' \cos \theta - y' \sin \theta \qquad \text{and} \qquad y = x' \sin \theta + y' \cos \theta.$$

Substituting these expressions for x and y into the given second-degree equation produces a second-degree polynomial equation in x' and y' that has no $x'y'$-term.

Rotation of Axes

The second-degree equation $ax^2 + bxy + cy^2 + dx + ey + f = 0$ can be written in the form

$$a'(x')^2 + c'(y')^2 + d'x' + e'y' + f' = 0$$

by rotating the coordinate axes counterclockwise through the angle θ, where θ is defined by $\cot 2\theta = \dfrac{a-c}{b}$. The coefficients of the new equation are obtained from the substitutions

$$x = x' \cos \theta - y' \sin \theta$$
$$y = x' \sin \theta + y' \cos \theta.$$

REMARK: When you solve for $\sin \theta$ and $\cos \theta$, the trigonometric identity $\cot 2\theta = \dfrac{\cot^2 \theta - 1}{2 \cot \theta}$ is often useful.

Example 7 demonstrates how to identify the graph of a second-degree polynomial by rotating the coordinate axes.

EXAMPLE 7 Rotation of a Conic Section

Perform a rotation of axes to eliminate the xy-term in

$$5x^2 - 6xy + 5y^2 + 14\sqrt{2}x - 2\sqrt{2}y + 18 = 0,$$

and sketch the graph of the resulting equation in the $x'y'$-plane.

SOLUTION The angle of rotation is represented by

$$\cot 2\theta = \frac{a-c}{b} = \frac{5-5}{-6} = 0.$$

This implies that $\theta = \pi/4$. So,

$$\sin \theta = \frac{1}{\sqrt{2}} \quad \text{and} \quad \cos \theta = \frac{1}{\sqrt{2}}.$$

By substituting

$$x = x' \cos \theta - y' \sin \theta = \frac{1}{\sqrt{2}}(x' - y')$$

and

$$y = x' \sin \theta + y' \cos \theta = \frac{1}{\sqrt{2}}(x' + y')$$

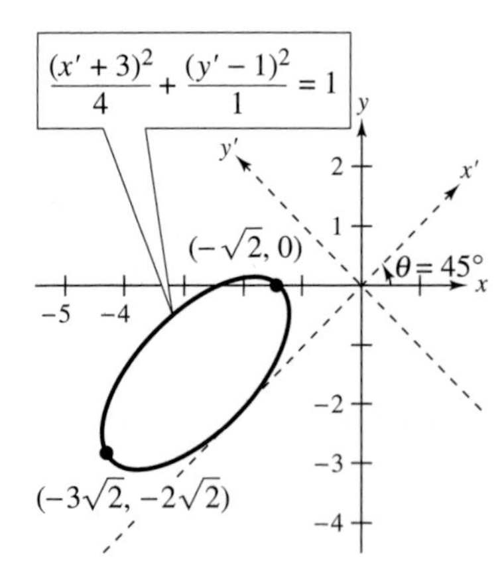

Figure 4.22

into the original equation and simplifying, you obtain

$$(x')^2 + 4(y')^2 + 6x' - 8y' + 9 = 0.$$

Finally, by completing the square, you find the standard form of this equation to be

$$\frac{(x'+3)^2}{2^2} + \frac{(y'-1)^2}{1^2} = \frac{(x'+3)^2}{4} + \frac{(y'-1)^2}{1} = 1,$$

which is the equation of an ellipse, as shown in Figure 4.22.

In Example 7 the new (rotated) basis for R^2 is

$$B' = \left\{\left(\frac{1}{\sqrt{2}}, \frac{1}{\sqrt{2}}\right), \left(-\frac{1}{\sqrt{2}}, \frac{1}{\sqrt{2}}\right)\right\},$$

and the coordinates of the vertices of the ellipse relative to B' are $[-5, 1]^T$ and $[-1, 1]^T$.

To find the coordinates of the vertices relative to the standard basis $B = \{(1, 0), (0, 1)\}$, use the equations

$$x = \frac{1}{\sqrt{2}}(x' - y')$$

and

$$y = \frac{1}{\sqrt{2}}(x' + y')$$

to obtain $\left(-3\sqrt{2}, -2\sqrt{2}\right)$ and $\left(-\sqrt{2}, 0\right)$, as shown in Figure 4.22.

SECTION 4.8 Exercises

Linear Differential Equations (Calculus)

In Exercises 1–8, determine which functions are solutions of the linear differential equation.

1. $y'' + y = 0$

(a) e^x (b) $\sin x$ (c) $\cos x$ (d) $\sin x - \cos x$

2. $y''' + 3y'' + 3y' + y = 0$

(a) x (b) e^x (c) e^{-x} (d) xe^{-x}

3. $y'' + 4y' + 4y = 0$

(a) e^{-2x} (b) xe^{-2x} (c) x^2e^{-2x} (d) $(x + 2)e^{-2x}$

4. $y'''' - 2y''' + y'' = 0$

(a) 1 (b) x (c) x^2 (d) e^x

5. $x^2y'' - 2y = 0$

(a) $y = \frac{1}{x^2}$ (b) $y = x^2$ (c) $y = e^{x^2}$ (d) $y = e^{-x^2}$

6. $xy'' + 2y' = 0$

(a) $y = x$ (b) $y = \frac{1}{x}$ (c) $y = xe^x$ (d) $y = xe^{-x}$

7. $y'' - y' - 2y = 0$

(a) $y = xe^{2x}$ (b) $y = 2e^{2x}$ (c) $y = 2e^{-2x}$ (d) $y = xe^{-x}$

8. $y' - 2xy = 0$

(a) $y = 3e^{x^2}$ (b) $y = xe^{x^2}$ (c) $y = x^2e^x$ (d) $y = xe^{-x}$

In Exercises 9–16, find the Wronskian for the set of functions.

9. $\{e^x, e^{-x}\}$ **10.** $\{e^{x^2}, e^{-x^2}\}$

11. $\{x, \sin x, \cos x\}$ **12.** $\{x, -\sin x, \cos x\}$

13. $\{e^{-x}, xe^{-x}, (x+3)e^{-x}\}$ **14.** $\{x, e^{-x}, e^x\}$

15. $\{1, e^x, e^{2x}\}$ **16.** $\{x^2, e^{x^2}, x^2e^x\}$

In Exercises 17–24, test the given set of solutions for linear independence.

	Differential Equation	*Solutions*
17.	$y'' + y = 0$	$\{\sin x, \cos x\}$
18.	$y'' + 4y' + 4y = 0$	$\{e^{-2x}, xe^{-2x}\}$
19.	$y''' + 4y'' + 4y' = 0$	$\{e^{-2x}, xe^{-2x}, (2x+1)e^{-2x}\}$
20.	$y''' + y' = 0$	$\{1, \sin x, \cos x\}$
21.	$y''' + y' = 0$	$\{2, -1 + 2\sin x, 1 + \sin x\}$
22.	$y''' + 3y'' + 3y' + y = 0$	$\{e^{-x}, xe^{-x}, x^2e^{-x}\}$
23.	$y''' + 3y'' + 3y' + y = 0$	$\{e^{-x}, xe^{-x}, e^{-x} + xe^{-x}\}$
24.	$y'''' - 2y''' + y'' = 0$	$\{1, x, e^x, xe^x\}$

25. Find the general solution of the differential equation from Exercise 17.

26. Find the general solution of the differential equation from Exercise 18.

27. Find the general solution of the differential equation from Exercise 20.

28. Find the general solution of the differential equation from Exercise 24.

29. Prove that $y = C_1 \cos ax + C_2 \sin ax$ is the general solution of $y'' + a^2y = 0$, $a \neq 0$.

30. Prove that the set $\{e^{ax}, e^{bx}\}$ is linearly independent if and only if $a \neq b$.

31. Prove that the set $\{e^{ax}, xe^{ax}\}$ is linearly independent.

32. Prove that the set $\{e^{ax}\cos bx, e^{ax}\sin bx\}$, where $b \neq 0$, is linearly independent.

33. Writing Is the sum of two solutions of a nonhomogeneous linear differential equation also a solution? Explain your answer.

34. Writing Is the scalar multiple of a solution of a nonhomogeneous linear differential equation also a solution? Explain your answer.

Conic Sections and Rotation

In Exercises 35–52, identify and sketch the graph.

35. $y^2 + x = 0$ **36.** $y^2 + 8x = 0$

37. $x^2 + 4y^2 - 16 = 0$ **38.** $5x^2 + 3y^2 - 15 = 0$

39. $\frac{x^2}{9} - \frac{y^2}{16} - 1 = 0$ **40.** $\frac{x^2}{16} - \frac{y^2}{25} = 1$

41. $x^2 - 2x + 8y + 17 = 0$

42. $y^2 - 6y - 4x + 21 = 0$

43. $9x^2 + 25y^2 - 36x - 50y + 61 = 0$

44. $4x^2 + y^2 - 8x + 3 = 0$

45. $9x^2 - y^2 + 54x + 10y + 55 = 0$

46. $4y^2 - 2x^2 - 4y - 8x - 15 = 0$

47. $x^2 + 4y^2 + 4x + 32y + 64 = 0$

48. $4y^2 + 4x^2 - 24x + 35 = 0$

49. $2x^2 - y^2 + 4x + 10y - 22 = 0$

50. $4x^2 - y^2 + 4x + 2y - 1 = 0$

51. $x^2 + 4x + 6y - 2 = 0$

52. $y^2 + 8x + 6y + 25 = 0$

In Exercises 53–62, perform a rotation of axes to eliminate the xy-term, and sketch the graph of the conic.

53. $xy + 1 = 0$ **54.** $xy - 2 = 0$

55. $4x^2 + 2xy + 4y^2 - 15 = 0$

56. $x^2 + 2xy + y^2 - 8x + 8y = 0$

57. $5x^2 - 2xy + 5y^2 - 24 = 0$

58. $5x^2 - 6xy + 5y^2 - 12 = 0$

59. $13x^2 + 6\sqrt{3}xy + 7y^2 - 16 = 0$

60. $3x^2 - 2\sqrt{3}xy + y^2 + 2x + 2\sqrt{3}y = 0$

61. $x^2 + 2\sqrt{3}xy + 3y^2 - 2\sqrt{3}x + 2y + 16 = 0$

62. $7x^2 - 2\sqrt{3}xy + 5y^2 = 16$

In Exercises 63–66, perform a rotation of axes to eliminate the xy-term, and sketch the graph of the "degenerate" conic.

63. $x^2 - 2xy + y^2 = 0$ **64.** $5x^2 - 2xy + 5y^2 = 0$

65. $x^2 + 2xy + y^2 - 1 = 0$ **66.** $x^2 - 10xy + y^2 = 0$

67. Prove that a rotation of $\theta = \pi/4$ will eliminate the xy-term from the equation

$ax^2 + bxy + ay^2 + dx + ey + f = 0.$

68. Prove that a rotation of θ, where $\cot 2\theta = (a - c)/b$, will eliminate the xy-term from the equation

$ax^2 + bxy + cy^2 + dx + ey + f = 0.$

69. For the equation $ax^2 + bxy + cy^2 = 0$, define the matrix A as

$$A = \begin{bmatrix} a & b/2 \\ b/2 & c \end{bmatrix}.$$

Prove that if $|A| \neq 0$, then the graph of $ax^2 + bxy + cy^2 = 0$ is two intersecting lines.

70. For the equation in Exercise 69, define the matrix A as $|A| = 0$, and describe the graph of $ax^2 + bxy + cy^2 = 0$.

CHAPTER 4 Review Exercises

In Exercises 1–4, find (a) $\mathbf{u} + \mathbf{v}$, (b) $2\mathbf{v}$, (c) $\mathbf{u} - \mathbf{v}$, and (d) $3\mathbf{u} - 2\mathbf{v}$.

1. $\mathbf{u} = (-1, 2, 3)$, $\mathbf{v} = (1, 0, 2)$

2. $\mathbf{u} = (-1, 2, 1)$, $\mathbf{v} = (0, 1, 1)$

3. $\mathbf{u} = (3, -1, 2, 3)$, $\mathbf{v} = (0, 2, 2, 1)$

4. $\mathbf{u} = (0, 1, -1, 2)$, $\mathbf{v} = (1, 0, 0, 2)$

In Exercises 5–8, solve for $\mathbf{x}$ provided that $\mathbf{u} = (1, -1, 2)$, $\mathbf{v} = (0, 2, 3)$, and $\mathbf{w} = (0, 1, 1)$.

5. $2\mathbf{x} - \mathbf{u} + 3\mathbf{v} + \mathbf{w} = \mathbf{0}$

6. $3\mathbf{x} + 2\mathbf{u} - \mathbf{v} + 2\mathbf{w} = \mathbf{0}$

7. $5\mathbf{u} - 2\mathbf{x} = 3\mathbf{v} + \mathbf{w}$

8. $2\mathbf{u} + 3\mathbf{x} = 2\mathbf{v} - \mathbf{w}$

In Exercises 9–12, write $\mathbf{v}$ as a linear combination of $\mathbf{u}_1$, $\mathbf{u}_2$, and $\mathbf{u}_3$, if possible.

9. $\mathbf{v} = (3, 0, -6)$, $\mathbf{u}_1 = (1, -1, 2)$, $\mathbf{u}_2 = (2, 4, -2)$, $\mathbf{u}_3 = (1, 2, -4)$

10. $\mathbf{v} = (4, 4, 5)$, $\mathbf{u}_1 = (1, 2, 3)$, $\mathbf{u}_2 = (-2, 0, 1)$, $\mathbf{u}_3 = (1, 0, 0)$

11. $\mathbf{v} = (1, 2, 3, 5)$, $\mathbf{u}_1 = (1, 2, 3, 4)$, $\mathbf{u}_2 = (-1, -2, -3, 4)$, $\mathbf{u}_3 = (0, 0, 1, 1)$

12. $\mathbf{v} = (4, -13, -5, -4)$, $\mathbf{u}_1 = (1, -2, 1, 1)$, $\mathbf{u}_2 = (-1, 2, 3, 2)$, $\mathbf{u}_3 = (0, -1, -1, -1)$

In Exercises 13–16, describe the zero vector and the additive inverse of a vector in the vector space.

13. $M_{3,4}$

14. P_8

15. R^3

16. $M_{2,3}$

In Exercises 17–24, determine whether W is a subspace of the vector space.

17. $W = \{(x, y): x = 2y\}$, $V = R^2$

18. $W = \{(x, y): x - y = 1\}$, $V = R^2$

19. $W = \{(x, y): y = ax, a \text{ is an integer}\}$, $V = R^2$

20. $W = \{(x, y): y = ax^2\}$, $V = R^2$

21. $W = \{(x, 2x, 3x): x \text{ is a real number}\}$, $V = R^3$

22. $W = \{(x, y, z): x \geq 0\}$, $V = R^3$

23. $W = \{f: f(0) = -1\}$, $V = C[-1, 1]$

24. $W = \{f: f(-1) = 0\}$, $V = C[-1, 1]$

25. Which of the subsets of R^3 is a subspace of R^3?

(a) $W = \{(x_1, x_2, x_3): x_1^2 + x_2^2 + x_3^2 = 0\}$

(b) $W = \{(x_1, x_2, x_3): x_1^2 + x_2^2 + x_3^2 = 1\}$

26. Which of the subsets of R^3 is a subspace of R^3?

(a) $W = \{(x_1, x_2, x_3): x_1 + x_2 + x_3 = 0\}$

(b) $W = \{(x_1, x_2, x_3): x_1 + x_2 + x_3 = 1\}$

In Exercises 27–32, determine whether S (a) spans R^3, (b) is linearly independent, and (c) is a basis for R^3.

27. $S = \{(1, -5, 4), (11, 6, -1), (2, 3, 5)\}$

28. $S = \{(4, 0, 1), (0, -3, 2), (5, 10, 0)\}$

29. $S = \left\{\left(-\frac{1}{2}, \frac{3}{4}, -1\right), (5, 2, 3), (-4, 6, -8)\right\}$

30. $S = \{(2, 0, 1), (2, -1, 1), (4, 2, 0)\}$

31. $S = \{(1, 0, 0), (0, 1, 0), (0, 0, 1), (-1, 2, -3)\}$

32. $S = \{(1, 0, 0), (0, 1, 0), (0, 0, 1), (2, -1, 0)\}$

33. Determine whether $S = \{1 - t, 2t + 3t^2, t^2 - 2t^3, 2 + t^3\}$ is a basis for P_3.

34. Determine whether $S = \{1, t, 1 + t^2\}$ is a basis for P_2.

In Exercises 35 and 36, determine whether the set is a basis for $M_{2,2}$.

35. $S = \left\{\begin{bmatrix} 1 & 0 \\ 2 & 3 \end{bmatrix}, \begin{bmatrix} -2 & 1 \\ -1 & 0 \end{bmatrix}, \begin{bmatrix} 3 & 4 \\ 2 & 3 \end{bmatrix}, \begin{bmatrix} -3 & -3 \\ 1 & 3 \end{bmatrix}\right\}$

36. $S = \left\{\begin{bmatrix} 1 & 0 \\ 0 & 1 \end{bmatrix}, \begin{bmatrix} -1 & 0 \\ 1 & 1 \end{bmatrix}, \begin{bmatrix} 2 & 1 \\ 1 & 0 \end{bmatrix}, \begin{bmatrix} 1 & 1 \\ 0 & 1 \end{bmatrix}\right\}$

In Exercises 37–40, find (a) a basis for and (b) the dimension of the solution space of the homogeneous system of equations.

37. $\begin{aligned} 2x_1 + 4x_2 + 3x_3 - 6x_4 &= 0 \\ x_1 + 2x_2 + 2x_3 - 5x_4 &= 0 \\ 3x_1 + 6x_2 + 5x_3 - 11x_4 &= 0 \end{aligned}$

38. $\begin{aligned} 3x_1 + 8x_2 + 2x_3 + 3x_4 &= 0 \\ 4x_1 + 6x_2 + 2x_3 - x_4 &= 0 \\ 3x_1 + 4x_2 + x_3 - 3x_4 &= 0 \end{aligned}$

39. $\begin{aligned} x_1 - 3x_2 + x_3 + x_4 &= 0 \\ 2x_1 + x_2 - x_3 + 2x_4 &= 0 \\ x_1 + 4x_2 - 2x_3 + x_4 &= 0 \\ 5x_1 - 8x_2 + 2x_3 + 5x_4 &= 0 \end{aligned}$

40. $\begin{aligned} -x_1 + 2x_2 - x_3 + 2x_4 &= 0 \\ -2x_1 + 2x_2 + x_3 + 4x_4 &= 0 \\ 3x_1 + 2x_2 + 2x_3 + 5x_4 &= 0 \\ -3x_1 + 8x_2 + 5x_3 + 17x_4 &= 0 \end{aligned}$

In Exercises 41–46, find a basis for the solution space of $A\mathbf{x} = \mathbf{0}$. Then verify that $\text{rank}(A) + \text{nullity}(A) = n$.

41. $A = \begin{bmatrix} 5 & -8 \\ -10 & 16 \end{bmatrix}$

42. $A = \begin{bmatrix} 1 & 4 \\ 3 & 2 \end{bmatrix}$

43. $A = \begin{bmatrix} 2 & -3 & -6 & -4 \\ 1 & 5 & -3 & 11 \\ 2 & 7 & -6 & 16 \end{bmatrix}$

44. $A = \begin{bmatrix} 1 & 0 & -2 & 0 \\ 4 & -2 & 4 & -2 \\ -2 & 0 & 1 & 3 \end{bmatrix}$

45. $A = \begin{bmatrix} 1 & 3 & 2 \\ 4 & -1 & -18 \\ -1 & 3 & 10 \\ 1 & 2 & 0 \end{bmatrix}$

46. $A = \begin{bmatrix} 1 & 2 & 1 & 2 \\ 1 & 4 & 0 & 3 \\ -2 & 3 & 0 & 2 \\ 1 & 2 & 6 & 1 \end{bmatrix}$

In Exercises 47–52, find (a) the rank and (b) a basis for the row space of the matrix.

47. $\begin{bmatrix} 1 & 2 \\ -4 & 3 \\ 6 & 1 \end{bmatrix}$

48. $\begin{bmatrix} 2 & -1 & 4 \\ 1 & 5 & 6 \\ 1 & 16 & 14 \end{bmatrix}$

49. $\begin{bmatrix} 1 & -4 & 0 & 4 \end{bmatrix}$

50. $\begin{bmatrix} 1 & 2 & -1 \end{bmatrix}$

51. $\begin{bmatrix} 7 & 0 & 2 \\ 4 & 1 & 6 \\ -1 & 16 & 14 \end{bmatrix}$

52. $\begin{bmatrix} 1 & 2 & 0 \\ -1 & 4 & 1 \\ 0 & 1 & 3 \end{bmatrix}$

In Exercises 53–58, find the coordinate matrix of $\mathbf{x}$ relative to the standard basis.

53. $B = \{(1, 1), (-1, 1)\}, \quad [\mathbf{x}]_B = [3, 5]^T$

54. $B = \{(2, 0), (3, 3)\}, \quad [\mathbf{x}]_B = [1, 1]^T$

55. $B = \left\{\left(\frac{1}{2}, \frac{1}{2}\right), (1, 0)\right\}, \quad [\mathbf{x}]_B = \left[\frac{1}{2}, \frac{1}{2}\right]^T$

56. $B = \{(4, 2), (1, -1)\}, \quad [\mathbf{x}]_B = [2, 1]^T$

57. $B = \{(1, 0, 0), (1, 1, 0), (0, 1, 1)\}, \quad [\mathbf{x}]_B = [2, 0, -1]^T$

58. $B = \{(1, 0, 1), (0, 1, 0), (0, 1, 1)\}, \quad [\mathbf{x}]_B = [4, 0, 2]^T$

In Exercises 59–64, find the coordinate matrix of $\mathbf{x}$ relative to the (nonstandard) basis for R^n.

59. $B' = \{(5, 0), (0, -8)\}, \quad \mathbf{x} = (2, 2)$

60. $B' = \{(1, 1), (0, -2)\}, \quad \mathbf{x} = (2, -1)$

61. $B' = \{(1, 2, 3), (1, 2, 0), (0, -6, 2)\}, \quad \mathbf{x} = (3, -3, 0)$

62. $B' = \{(1, 0, 0), (0, 1, 0), (1, 1, 1)\}, \quad \mathbf{x} = (4, -2, 9)$

63. $B' = \{(9, -3, 15, 4), (-3, 0, 0, -1), (0, -5, 6, 8), (-3, 4, -2, 3)\}, \quad \mathbf{x} = (21, -5, 43, 14)$

64. $B' = \{(1, -1, 2, 1), (1, 1, -4, 3), (1, 2, 0, 3), (1, 2, -2, 0)\}, \quad \mathbf{x} = (5, 3, -6, 2)$

In Exercises 65–68, find the coordinate matrix of $\mathbf{x}$ relative to the basis B'.

65. $B = \{(1, 1), (-1, 1)\}, \quad B' = \{(0, 1), (1, 2)\}, \quad [\mathbf{x}]_B = [3, -3]^T$

66. $B = \{(1, 0), (1, -1)\}, \quad B' = \{(1, 1), (1, -1)\}, \quad [\mathbf{x}]_B = [2, -2]^T$

67. $B = \{(1, 0, 0), (1, 1, 0), (1, 1, 1)\}, \quad B' = \{(0, 0, 1), (0, 1, 1), (1, 1, 1)\}, \quad [\mathbf{x}]_B = [-1, 2, -3]^T$

68. $B = \{(1, 1, -1), (1, 1, 0), (1, -1, 0)\}, \quad B' = \{(1, -1, 2), (2, 2, -1), (2, 2, 2)\}, \quad [\mathbf{x}]_B = [2, 2, -1]^T$

In Exercises 69–72, find the transition matrix from B to B'.

69. $B = \{(1, -1), (3, 1)\}, \quad B' = \{(1, 0), (0, 1)\}$

70. $B = \{(1, -1), (3, 1)\}, \quad B' = \{(1, 2), (-1, 0)\}$

71. $B = \{(1, 0, 0), (0, 1, 0), (0, 0, 1)\},$
$B' = \{(0, 0, 1), (0, 1, 0), (1, 0, 0)\}$

72. $B = \{(1, 1, 1), (1, 1, 0), (1, 0, 0)\},$
$B' = \{(1, 2, 3), (0, 1, 0), (1, 0, 1)\}$

73. Let W be the subspace of P_3 (all third-degree polynomials) such that $p(0) = 0$, and let U be the subspace of all polynomials such that $p(1) = 0$. Find a basis for W, a basis for U, and a basis for their intersection $W \cap U$.

74. **Calculus** Let $V = C'(-\infty, \infty)$, the vector space of all continuously differentiable functions on the real line.

(a) Prove that $W = \{f: f' = 3f\}$ is a subspace of V.

(b) Prove that $U = \{f: f' = f + 1\}$ is not a subspace of V.

75. **Writing** Let $B = \{p_1(x), p_2(x), . . . , p_n(x), p_{n+1}(x)\}$ be a basis for P_n. Must B contain a polynomial of each degree $0, 1, 2, . . . , n$? Explain your reasoning.

76. Let A and B be $n \times n$ matrices with $A \neq O$ and $B \neq O$. Prove that if A is symmetric and B is skew-symmetric $(B^T = -B)$, then $\{A, B\}$ is a linearly independent set.

77. Let $V = P_5$ and consider the set W of all polynomials of the form $(x^3 + x)p(x)$, where $p(x)$ is in P_2. Is W a subspace of V? Prove your answer.

78. Let $\mathbf{v}_1$, $\mathbf{v}_2$, and $\mathbf{v}_3$ be three linearly independent vectors in a vector space V. Is the set $\{\mathbf{v}_1 - \mathbf{v}_2, \mathbf{v}_2 - \mathbf{v}_3, \mathbf{v}_3 - \mathbf{v}_1\}$ linearly dependent or linearly independent?

79. Let A be an $n \times n$ square matrix. Prove that the row vectors of A are linearly dependent if and only if the column vectors of A are linearly dependent.

80. Let A be an $n \times n$ square matrix, and let λ be a scalar. Prove that the set

$$S = \{\mathbf{x}: A\mathbf{x} = \lambda\mathbf{x}\}$$

is a subspace of R^n. Determine the dimension of S if $\lambda = 3$ and

$$A = \begin{bmatrix} 3 & 1 & 0 \\ 0 & 3 & 0 \\ 0 & 0 & 1 \end{bmatrix}.$$

81. Let $f(x) = x$ and $g(x) = |x|$.

(a) Show that f and g are linearly independent in $C[-1, 1]$.

(b) Show that f and g are linearly dependent in $C[0, 1]$.

82. Given a set of functions, describe how its domain can influence whether the set is linearly independent or dependent.

True or False? In Exercises 83–86, determine whether each statement is true or false. If a statement is true, give a reason or cite an appropriate statement from the text. If a statement is false, provide an example that shows the statement is not true in all cases or cite an appropriate statement from the text.

83. (a) The standard operations in R^n are vector addition and scalar multiplication.

(b) The additive inverse of a vector is not unique.

(c) A vector space consists of four entities: a set of vectors, a set of scalars, and two operations.

84. (a) The set $W = \{(0, x^2, x^3): x^2 \text{ and } x^3 \text{ are real numbers}\}$ is a subspace of R^3.

(b) A linearly independent spanning set S is called a basis of a vector space V.

(c) If A is an invertible $n \times n$ matrix, then the n row vectors of A are linearly dependent.

85. (a) The set of all n-tuples is called n-space and is denoted by R^n.

(b) The additive identity of a vector is not unique.

(c) Once a theorem has been proved for an abstract vector space, you need not give separate proofs for n-tuples, matrices, and polynomials.

86. (a) The set of points on the line represented by $x + y = 0$ is a subspace of R^2.

(b) A set of vectors $S = \{\mathbf{v}_1, \mathbf{v}_2, . . . , \mathbf{v}_n\}$ in a vector space V is linearly independent if the vector equation $c_1\mathbf{v}_1 + c_2\mathbf{v}_2 + \cdot \cdot \cdot + c_n\mathbf{v}_n = \mathbf{0}$ has only the trivial solution.

(c) Elementary row operations preserve the column space of the matrix A.

Linear Differential Equations (Calculus)

In Exercises 87–90, determine whether each function is a solution of the linear differential equation.

87. $y'' - y' - 6y = 0$

(a) e^{3x} (b) e^{2x} (c) e^{-3x} (d) e^{-2x}

88. $y'''' - y = 0$

(a) e^x (b) e^{-x} (c) $\cos x$ (d) $\sin x$

89. $y' + 2y = 0$

(a) e^{-2x} (b) xe^{-2x} (c) x^2e^{-x} (d) $2xe^{-2x}$

90. $y'' + 9y = 0$

(a) $\sin 3x + \cos 3x$ (b) $3 \sin x + 3 \cos x$

(c) $\sin 3x$ (d) $\cos 3x$

In Exercises 91–94, find the Wronskian for the set of functions.

91. $\{1, x, e^x\}$ **92.** $\{1, x, 2 + x\}$

93. $\{1, \sin 2x, \cos 2x\}$ **94.** $\{x, \sin^2 x, \cos^2 x\}$

In Exercises 95–98, test the set of solutions for linear independence.

	Differential Equation	*Solutions*
95.	$y'' + 6y' + 9y = 0$	$\{e^{-3x}, xe^{-3x}\}$
96.	$y'' + 6y' + 9y = 0$	$\{e^{-3x}, 3e^{-3x}\}$
97.	$y''' - 6y'' + 11y' - 6y = 0$	$\{e^x, e^{2x}, e^x - e^{2x}\}$
98.	$y'' + 4y = 0$	$\{\sin 2x, \cos 2x\}$

Conic Sections and Rotation

In Exercises 99–106, identify and sketch the graph of the equation.

99. $x^2 + y^2 - 4x + 2y - 4 = 0$

100. $9x^2 + 9y^2 + 18x - 18y + 14 = 0$

101. $x^2 - y^2 + 2x - 3 = 0$

102. $4x^2 - y^2 + 8x - 6y + 4 = 0$

103. $2x^2 - 20x - y + 46 = 0$

104. $y^2 - 4x - 4 = 0$

105. $4x^2 + y^2 + 32x + 4y + 63 = 0$

106. $16x^2 + 25y^2 - 32x - 50y + 16 = 0$

In Exercises 107–110, perform a rotation of axes to eliminate the xy-term, and sketch the graph of the conic.

107. $xy = 3$ **108.** $9x^2 + 4xy + 9y^2 - 20 = 0$

109. $16x^2 - 24xy + 9y^2 - 60x - 80y + 100 = 0$

110. $7x^2 + 6\sqrt{3}xy + 13y^2 - 16 = 0$

CHAPTER 4 Projects

1 Solutions of Linear Systems

Write a short paragraph to answer each of the following questions about solutions of systems of linear equations. You should not perform any calculations, but instead base your explanations on the appropriate properties from the text.

1. One solution of the homogeneous linear system

$$\begin{aligned} x + 2y + z + 3w &= 0 \\ x - y \phantom{{}+z} + w &= 0 \\ y - z + 2w &= 0 \end{aligned}$$

is $x = -2$, $y = -1$, $z = 1$, and $w = 1$. Explain why $x = 4$, $y = 2$, $z = -2$, and $w = -2$ must also be a solution. Do not perform any row operations.

2. The vectors $\mathbf{x}_1$ and $\mathbf{x}_2$ are solutions of the homogeneous linear system $A\mathbf{x} = \mathbf{0}$. Explain why the vector $2\mathbf{x}_1 - 3\mathbf{x}_2$ must also be a solution.

3. Consider the two systems represented by the augmented matrices.

$$\begin{bmatrix} 1 & 1 & -5 & 3 \\ 1 & 0 & -2 & 1 \\ 2 & -1 & -1 & 0 \end{bmatrix} \qquad \begin{bmatrix} 1 & 1 & -5 & -9 \\ 1 & 0 & -2 & -3 \\ 2 & -1 & -1 & 0 \end{bmatrix}$$

If the first system is known to be consistent, explain why the second system is also consistent. Do not perform any row operations.

4. The vectors $\mathbf{x}_1$ and $\mathbf{x}_2$ are solutions of the linear system $A\mathbf{x} = \mathbf{b}$. Is the vector $2\mathbf{x}_1 - 3\mathbf{x}_2$ also a solution? Why or why not?
5. The linear systems $A\mathbf{x} = \mathbf{b}_1$ and $A\mathbf{x} = \mathbf{b}_2$ are consistent. Is the system $A\mathbf{x} = \mathbf{b}_1 + \mathbf{b}_2$ necessarily consistent? Why or why not?
6. Consider the linear system $A\mathbf{x} = \mathbf{b}$. If the rank of A equals the rank of the augmented matrix for the system, explain why the system must be consistent. Contrast this to the case in which the rank of A is less than the rank of the augmented matrix.

2 Direct Sum

Let U and W be subspaces of the vector space V. You learned in Section 4.3 that the intersection $U \cap W$ is also a subspace of V, whereas the union $U \cup W$ is, in general, not a subspace. In this project you will explore the **sum** and **direct sum** of subspaces, focusing especially on their geometric interpretation in R^n.

1. Define the sum of the subspaces U and W as follows.

   $U + W = \{\mathbf{u} + \mathbf{w} : \mathbf{u} \in U, \mathbf{w} \in W\}$

   Prove that $U + W$ is a subspace of V.
2. Consider the subspaces of $V = R^3$ listed below.

   $U = \{(x, y, x - y) : x, y \in R\}$
   $W = \{(x, 0, x) : x \in R\}$
   $Z = \{(x, x, x) : x \in R\}$

   Find $U + W$, $U + Z$, and $W + Z$.
3. If U and W are subspaces of V such that $V = U + W$ and $U \cap W = \{\mathbf{0}\}$, prove that every vector in V has a *unique* representation of the form $\mathbf{u} + \mathbf{w}$, where $\mathbf{u}$ is in U and $\mathbf{w}$ is in W. In this case, we say that V is the **direct sum** of U and W, and write

   $V = U \oplus W$. **Direct sum**

   Which of the sums in part (2) of this project are direct sums?
4. Let $V = U \oplus W$ and suppose that $\{\mathbf{u}_1, \mathbf{u}_2, \ldots, \mathbf{u}_k\}$ is a basis for the subspace U and $\{\mathbf{w}_1, \mathbf{w}_2, \ldots, \mathbf{w}_m\}$ is a basis for the subspace W. Prove that the set $\{\mathbf{u}_1, \ldots, \mathbf{u}_k, \mathbf{w}_1, \ldots, \mathbf{w}_m\}$ is a basis for V.
5. Consider the subspaces of $V = R^3$ listed below.

   $U = \{(x, 0, y) : x, y \in R\}$
   $W = \{(0, x, y) : x, y \in R\}$

   Show that $R^3 = U + W$. Is R^3 the *direct* sum of U and W? What are the dimensions of U, W, $U \cap W$, and $U + W$? In general, formulate a conjecture that relates the dimensions of U, W, $U \cap W$, and $U + W$.
6. Can you find two 2-dimensional subspaces of R^3 whose intersection is just the zero vector? Why or why not?

Inner Product Spaces

CHAPTER OBJECTIVES

- Find the length of $\mathbf{v}$, a vector $\mathbf{u}$ with the same length in the same direction as $\mathbf{v}$, and a unit vector in the same or opposite direction as $\mathbf{v}$.
- Find the distance between two vectors, the dot product, and the angle θ between $\mathbf{u}$ and $\mathbf{v}$.
- Verify the Cauchy-Schwarz Inequality, the Triangle Inequality, and the Pythagorean Theorem.
- Determine whether two vectors are orthogonal, parallel, or neither.
- Determine whether a function defines an inner product on R^n, $M_{m,n}$, or P_n, and find the inner product as defined for two vectors $\langle \mathbf{u}, \mathbf{v} \rangle$ in R^n, $M_{m,n}$, and P_n.
- Find the projection of a vector onto a vector or subspace.
- Determine whether a set of vectors in R^n is orthogonal, orthonormal, or neither.
- Find the coordinates of $\mathbf{x}$ relative to the orthonormal basis R^n.
- Use the Gram-Schmidt orthonormalization process.
- Find an orthonormal basis for the solution space of a homogeneous system.
- Determine whether subspaces are orthogonal and, if so, find the orthogonal complement of a subspace.
- Find the least squares solution of a system $A\mathbf{x} = \mathbf{b}$.
- Find the cross product of two vectors $\mathbf{u}$ and $\mathbf{v}$.
- Find the linear or quadratic least squares approximating function for a known function.
- Find the nth-order Fourier approximation for a known function.

5.1 Length and Dot Product in R^n

Section 4.1 mentioned that vectors in the plane can be characterized as directed line segments having a certain *length* and *direction.* In this section, R^2 will be used as a model for defining these and other geometric properties (such as distance and angle) of vectors in R^n. In the next section, these ideas will be extended to general vector spaces.

You will begin by reviewing the definition of the length of a vector in R^2. If $\mathbf{v} = (v_1, v_2)$ is a vector in the plane, then the **length,** or **magnitude,** of $\mathbf{v}$, denoted by $\|\mathbf{v}\|$, is defined as

$$\|\mathbf{v}\| = \sqrt{v_1^2 + v_2^2}.$$

This definition corresponds to the usual notion of length in Euclidean geometry. That is, the vector $\mathbf{v}$ is thought of as the hypotenuse of a right triangle whose sides have lengths of $|v_1|$ and $|v_2|$, as shown in Figure 5.1. Applying the Pythagorean Theorem produces

$$\|\mathbf{v}\|^2 = |v_1|^2 + |v_2|^2 = v_1^2 + v_2^2.$$

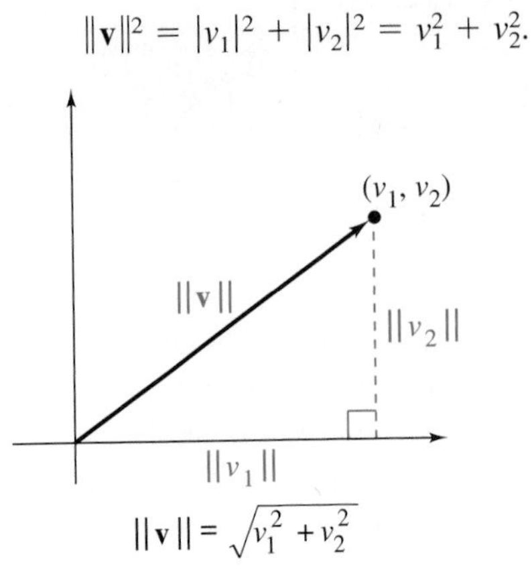

Figure 5.1

Using R^2 as a model, the length of a vector in R^n is defined as follows.

Definition of Length of a Vector in R^n

The **length,** or **magnitude,** of a vector $\mathbf{v} = (v_1, v_2, \ldots, v_n)$ in R^n is given by

$$\|\mathbf{v}\| = \sqrt{v_1^2 + v_2^2 + \cdots + v_n^2}.$$

REMARK: The length of a vector is also called its **norm.** If $\|\mathbf{v}\| = 1$, then the vector $\mathbf{v}$ is called a **unit vector.**

This definition shows that the length of a vector cannot be negative. That is, $\|\mathbf{v}\| \geq 0$. Moreover, $\|\mathbf{v}\| = 0$ if and only if $\mathbf{v}$ is the zero vector $\mathbf{0}$.

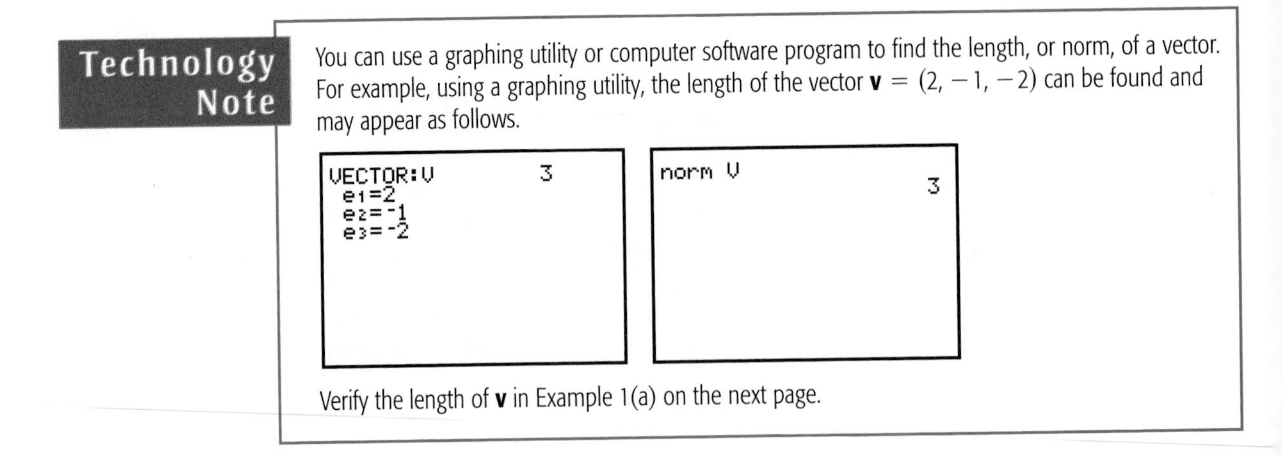

Technology Note
You can use a graphing utility or computer software program to find the length, or norm, of a vector. For example, using a graphing utility, the length of the vector $\mathbf{v} = (2, -1, -2)$ can be found and may appear as follows.
VECTOR:V 3
e1=2
e2=-1
e3=-2
norm V
3
Verify the length of $\mathbf{v}$ in Example 1(a) on the next page.

EXAMPLE 1 The Length of a Vector in R^n

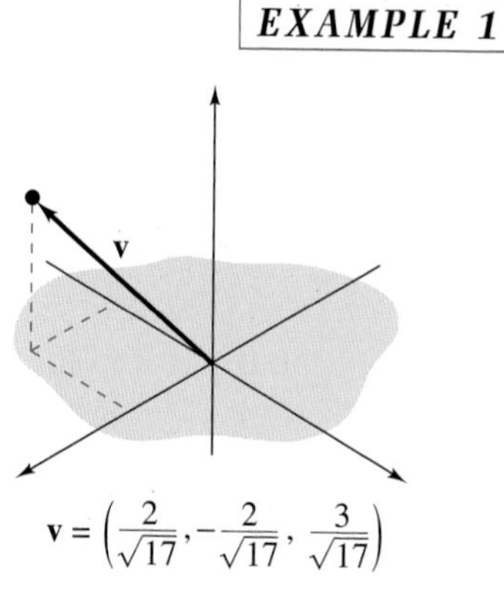

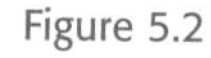
Figure 5.2

(a) In R^5, the length of $\mathbf{v} = (0, -2, 1, 4, -2)$ is

$$\|\mathbf{v}\| = \sqrt{0^2 + (-2)^2 + 1^2 + 4^2 + (-2)^2} = \sqrt{25} = 5.$$

(b) In R^3, the length of $\mathbf{v} = \left(2/\sqrt{17}, -2/\sqrt{17}, 3/\sqrt{17}\right)$ is

$$\|\mathbf{v}\| = \sqrt{\left(\frac{2}{\sqrt{17}}\right)^2 + \left(-\frac{2}{\sqrt{17}}\right)^2 + \left(\frac{3}{\sqrt{17}}\right)^2} = \sqrt{\frac{17}{17}} = 1.$$

Because its length is 1, $\mathbf{v}$ is a unit vector, as shown in Figure 5.2.

Each vector in the standard basis for R^n has length 1 and is called a **standard unit vector** in R^n. In physics and engineering it is common to denote the standard unit vectors in R^2 and R^3 as follows.

$$\{\mathbf{i}, \mathbf{j}\} = \{(1, 0), (0, 1)\}$$

and

$$\{\mathbf{i}, \mathbf{j}, \mathbf{k}\} = \{(1, 0, 0), (0, 1, 0), (0, 0, 1)\}$$

Two nonzero vectors $\mathbf{u}$ and $\mathbf{v}$ in R^n are **parallel** if one is a scalar multiple of the other—that is, $\mathbf{u} = c\mathbf{v}$. Moreover, if $c > 0$, then $\mathbf{u}$ and $\mathbf{v}$ have the **same direction,** and if $c < 0$, $\mathbf{u}$ and $\mathbf{v}$ have **opposite directions.** The next theorem gives a formula for finding the length of a scalar multiple of a vector.

THEOREM 5.1 Length of a Scalar Multiple

Let $\mathbf{v}$ be a vector in R^n and let c be a scalar. Then

$$\|c\mathbf{v}\| = |c|\,\|\mathbf{v}\|,$$

where $|c|$ is the absolute value of c.

PROOF Because $c\mathbf{v} = (cv_1, cv_2, \ldots, cv_n)$, it follows that

$$\begin{aligned}\|c\mathbf{v}\| &= \|(cv_1, cv_2, \ldots, cv_n)\| \\ &= \sqrt{(cv_1)^2 + (cv_2)^2 + \cdots + (cv_n)^2} \\ &= \sqrt{c^2(v_1^2 + v_2^2 + \cdots + v_n^2)} \\ &= |c|\sqrt{v_1^2 + v_2^2 + \cdots + v_n^2} \\ &= |c|\,\|\mathbf{v}\|.\end{aligned}$$

One important use of Theorem 5.1 is in finding a unit vector having the same direction as a given vector. The theorem below provides a procedure for doing this.

THEOREM 5.2 Unit Vector in the Direction of v

If $\mathbf{v}$ is a nonzero vector in R^n, then the vector

$$\mathbf{u} = \frac{\mathbf{v}}{\|\mathbf{v}\|}$$

has length 1 and has the same direction as $\mathbf{v}$. This vector $\mathbf{u}$ is called the **unit vector in the direction of v**.

PROOF Because $\mathbf{v}$ is nonzero, you know $\|\mathbf{v}\| \neq 0$. $1/\|\mathbf{v}\|$ is positive, and you can write $\mathbf{u}$ as a positive scalar multiple of $\mathbf{v}$.

$$\mathbf{u} = \left(\frac{1}{\|\mathbf{v}\|}\right)\mathbf{v}$$

So, it follows that $\mathbf{u}$ has the same direction as $\mathbf{v}$. Finally, $\mathbf{u}$ has length 1 because

$$\|\mathbf{u}\| = \left\|\frac{\mathbf{v}}{\|\mathbf{v}\|}\right\| = \frac{1}{\|\mathbf{v}\|}\|\mathbf{v}\| = 1.$$

The process of finding the unit vector in the direction of $\mathbf{v}$ is called **normalizing** the vector $\mathbf{v}$. This procedure is demonstrated in the next example.

EXAMPLE 2 Finding a Unit Vector

Find the unit vector in the direction of $\mathbf{v} = (3, -1, 2)$, and verify that this vector has length 1.

SOLUTION The unit vector in the direction of $\mathbf{v}$ is

$$\frac{\mathbf{v}}{\|\mathbf{v}\|} = \frac{(3, -1, 2)}{\sqrt{3^2 + (-1)^2 + 2^2}}$$

$$= \frac{1}{\sqrt{14}}(3, -1, 2)$$

$$= \left(\frac{3}{\sqrt{14}}, -\frac{1}{\sqrt{14}}, \frac{2}{\sqrt{14}}\right),$$

which is a unit vector because

$$\sqrt{\left(\frac{3}{\sqrt{14}}\right)^2 + \left(-\frac{1}{\sqrt{14}}\right)^2 + \left(\frac{2}{\sqrt{14}}\right)^2} = \sqrt{\frac{14}{14}} = 1.$$

(See Figure 5.3.)

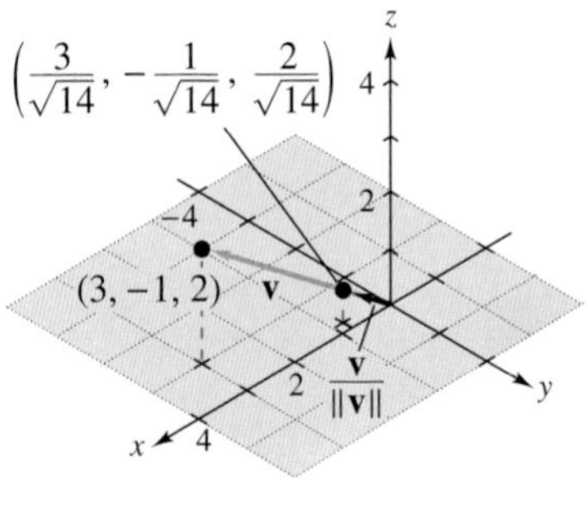

Figure 5.3

Technology Note

You can use a graphing utility or computer software program to find the unit vector for a given vector. For example, you can use a graphing utility to find the unit vector for $\mathbf{v} = (-3, 4)$, which may appear as:

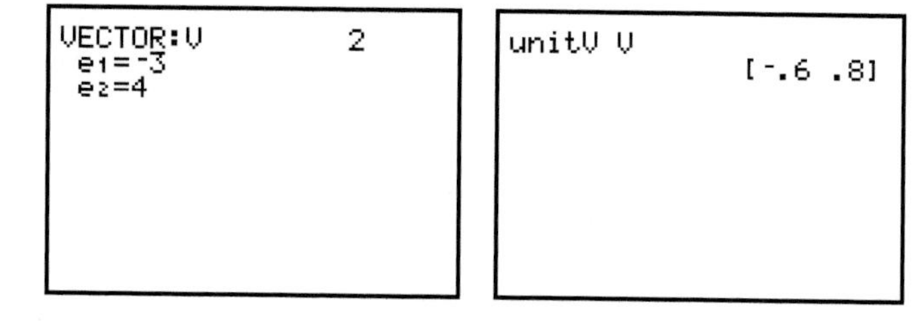

Distance Between Two Vectors in R^n

To define the **distance between two vectors** in R^n, R^2 will be used as the model. The Distance Formula from analytic geometry tells you that the distance d between two points in the plane, (u_1, u_2) and (v_1, v_2), is

$$d = \sqrt{(u_1 - v_1)^2 + (u_2 - v_2)^2}.$$

In vector terminology, this distance can be viewed as the length of $\mathbf{u} - \mathbf{v}$, where $\mathbf{u} = (u_1, u_2)$ and $\mathbf{v} = (v_1, v_2)$, as shown in Figure 5.4. That is,

$$\|\mathbf{u} - \mathbf{v}\| = \sqrt{(u_1 - v_1)^2 + (u_2 - v_2)^2},$$

which leads to the next definition.

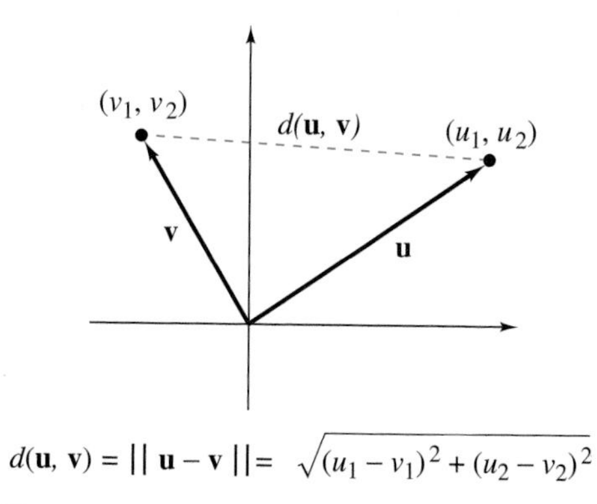

Figure 5.4

HISTORICAL NOTE

Olga Taussky-Todd (1906–1995)

became interested in mathematics at an early age. She heavily studied algebraic number theory and wrote a paper on the sum of squares, which earned her the Ford Prize from the Mathematical Association of America. To read about her work, visit *college.hmco.com/pic/larsonELA6e*.

Definition of Distance Between Two Vectors

The **distance between two vectors u** and **v** in R^n is

$$d(\mathbf{u}, \mathbf{v}) = \|\mathbf{u} - \mathbf{v}\|.$$

You can easily verify the three properties of distance listed below.

1. $d(\mathbf{u}, \mathbf{v}) \geq 0$
2. $d(\mathbf{u}, \mathbf{v}) = 0$ if and only if $\mathbf{u} = \mathbf{v}$.
3. $d(\mathbf{u}, \mathbf{v}) = d(\mathbf{v}, \mathbf{u})$

EXAMPLE 3 Finding the Distance Between Two Vectors

The distance between $\mathbf{u} = (0, 2, 2)$ and $\mathbf{v} = (2, 0, 1)$ is

$$\begin{aligned} d(\mathbf{u}, \mathbf{v}) &= \|\mathbf{u} - \mathbf{v}\| = \|(0 - 2, 2 - 0, 2 - 1)\| \\ &= \sqrt{(-2)^2 + 2^2 + 1^2} = 3. \end{aligned}$$

Dot Product and the Angle Between Two Vectors

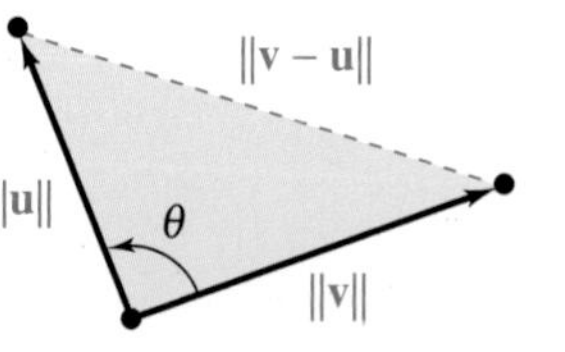

Angle Between Two Vectors

Figure 5.5

To find the angle θ $(0 \leq \theta \leq \pi)$ between two nonzero vectors $\mathbf{u} = (u_1, u_2)$ and $\mathbf{v} = (v_1, v_2)$ in R^2, the Law of Cosines can be applied to the triangle shown in Figure 5.5 to obtain

$$\|\mathbf{v} - \mathbf{u}\|^2 = \|\mathbf{u}\|^2 + \|\mathbf{v}\|^2 - 2\|\mathbf{u}\|\,\|\mathbf{v}\| \cos \theta.$$

Expanding and solving for $\cos \theta$ yields

$$\cos \theta = \frac{u_1 v_1 + u_2 v_2}{\|\mathbf{u}\|\,\|\mathbf{v}\|}.$$

The numerator of the quotient above is defined as the **dot product** of **u** and **v** and is denoted by

$$\mathbf{u} \cdot \mathbf{v} = u_1 v_1 + u_2 v_2.$$

This definition is generalized to R^n as follows.

Definition of Dot Product in R^n

The **dot product** of $\mathbf{u} = (u_1, u_2, \ldots, u_n)$ and $\mathbf{v} = (v_1, v_2, \ldots, v_n)$ is the *scalar* quantity

$$\mathbf{u} \cdot \mathbf{v} = u_1 v_1 + u_2 v_2 + \cdots + u_n v_n.$$

REMARK: Notice that the dot product of two vectors is a scalar, not another vector.

EXAMPLE 4 Finding the Dot Product of Two Vectors

The dot product of $\mathbf{u} = (1, 2, 0, -3)$ and $\mathbf{v} = (3, -2, 4, 2)$ is

$$\mathbf{u} \cdot \mathbf{v} = (1)(3) + (2)(-2) + (0)(4) + (-3)(2) = -7.$$

Technology Note

You can use a graphing utility or computer software program to find the dot product of two vectors. Using a graphing utility, you can verify Example 4, and it may appear as follows.

```
VECTOR:U        4
 e1=1
 e2=2
 e3=0
 e4=-3
```

```
VECTOR:V        4
 e1=3
 e2=-2
 e3=4
 e4=2
```

```
dot(U,V)
              -7
```

Keystrokes and programming syntax for these utilities/programs applicable to Example 4 are provided in the **Online Technology Guide,** available at *college.hmco.com/pic/larsonELA6e*.

THEOREM 5.3 Properties of the Dot Product

If $\mathbf{u}$, $\mathbf{v}$, and $\mathbf{w}$ are vectors in R^n and c is a scalar, then the following properties are true.

1. $\mathbf{u} \cdot \mathbf{v} = \mathbf{v} \cdot \mathbf{u}$
2. $\mathbf{u} \cdot (\mathbf{v} + \mathbf{w}) = \mathbf{u} \cdot \mathbf{v} + \mathbf{u} \cdot \mathbf{w}$
3. $c(\mathbf{u} \cdot \mathbf{v}) = (c\mathbf{u}) \cdot \mathbf{v} = \mathbf{u} \cdot (c\mathbf{v})$
4. $\mathbf{v} \cdot \mathbf{v} = \|\mathbf{v}\|^2$
5. $\mathbf{v} \cdot \mathbf{v} \geq 0$, and $\mathbf{v} \cdot \mathbf{v} = 0$ if and only if $\mathbf{v} = \mathbf{0}$.

PROOF

The proofs of these properties follow easily from the definition of dot product. For example, to prove the first property, you can write

$$\begin{aligned}\mathbf{u} \cdot \mathbf{v} &= u_1v_1 + u_2v_2 + \cdot \cdot \cdot + u_nv_n \\ &= v_1u_1 + v_2u_2 + \cdot \cdot \cdot + v_nu_n \\ &= \mathbf{v} \cdot \mathbf{u}.\end{aligned}$$

In Section 4.1, R^n was defined as the *set* of all ordered n-tuples of real numbers. When R^n is combined with the standard operations of vector addition, scalar multiplication, vector length, and the dot product, the resulting vector space is called **Euclidean n-space.** In the remainder of this text, unless stated otherwise, you may assume R^n to have the standard Euclidean operations.

EXAMPLE 5 **Finding Dot Products**

Given $\mathbf{u} = (2, -2)$, $\mathbf{v} = (5, 8)$, and $\mathbf{w} = (-4, 3)$, find

(a) $\mathbf{u} \cdot \mathbf{v}$. (b) $(\mathbf{u} \cdot \mathbf{v})\mathbf{w}$. (c) $\mathbf{u} \cdot (2\mathbf{v})$. (d) $\|\mathbf{w}\|^2$. (e) $\mathbf{u} \cdot (\mathbf{v} - 2\mathbf{w})$.

SOLUTION (a) By definition, you have

$$\mathbf{u} \cdot \mathbf{v} = 2(5) + (-2)(8) = -6.$$

(b) Using the result in part (a), you have

$$(\mathbf{u} \cdot \mathbf{v})\mathbf{w} = -6\mathbf{w} = -6(-4, 3) = (24, -18).$$

(c) By Property 3 of Theorem 5.3, you have

$$\mathbf{u} \cdot (2\mathbf{v}) = 2(\mathbf{u} \cdot \mathbf{v}) = 2(-6) = -12.$$

(d) By Property 4 of Theorem 5.3, you have

$$\|\mathbf{w}\|^2 = \mathbf{w} \cdot \mathbf{w} = (-4)(-4) + (3)(3) = 25.$$

(e) Because $2\mathbf{w} = (-8, 6)$, you have

$$\mathbf{v} - 2\mathbf{w} = (5 - (-8), 8 - 6) = (13, 2).$$

Consequently,

$$\mathbf{u} \cdot (\mathbf{v} - 2\mathbf{w}) = 2(13) + (-2)(2) = 26 - 4 = 22.$$

EXAMPLE 6 **Using Properties of the Dot Product**

Provided with two vectors $\mathbf{u}$ and $\mathbf{v}$ in R^n such that $\mathbf{u} \cdot \mathbf{u} = 39$, $\mathbf{u} \cdot \mathbf{v} = -3$, and $\mathbf{v} \cdot \mathbf{v} = 79$, evaluate $(\mathbf{u} + 2\mathbf{v}) \cdot (3\mathbf{u} + \mathbf{v})$.

SOLUTION Using Theorem 5.3, rewrite the dot product as

$$\begin{aligned}
(\mathbf{u} + 2\mathbf{v}) \cdot (3\mathbf{u} + \mathbf{v}) &= \mathbf{u} \cdot (3\mathbf{u} + \mathbf{v}) + (2\mathbf{v}) \cdot (3\mathbf{u} + \mathbf{v}) \\
&= \mathbf{u} \cdot (3\mathbf{u}) + \mathbf{u} \cdot \mathbf{v} + (2\mathbf{v}) \cdot (3\mathbf{u}) + (2\mathbf{v}) \cdot \mathbf{v} \\
&= 3(\mathbf{u} \cdot \mathbf{u}) + \mathbf{u} \cdot \mathbf{v} + 6(\mathbf{v} \cdot \mathbf{u}) + 2(\mathbf{v} \cdot \mathbf{v}) \\
&= 3(\mathbf{u} \cdot \mathbf{u}) + 7(\mathbf{u} \cdot \mathbf{v}) + 2(\mathbf{v} \cdot \mathbf{v}) \\
&= 3(39) + 7(-3) + 2(79) = 254.
\end{aligned}$$

Discovery

How does the dot product of two vectors compare with the product of their lengths? For instance, let $\mathbf{u} = (1, 1)$ *and* $\mathbf{v} = (-4, -3)$. *Calculate* $\mathbf{u} \cdot \mathbf{v}$ *and* $\|\mathbf{u}\|\,\|\mathbf{v}\|$. *Repeat this experiment with other choices for* $\mathbf{u}$ *and* $\mathbf{v}$. *Formulate a conjecture about the relationship between* $\mathbf{u} \cdot \mathbf{v}$ *and* $\|\mathbf{u}\|\,\|\mathbf{v}\|$.

To define the angle θ between two vectors $\mathbf{u}$ and $\mathbf{v}$ in R^n, you can use the formula in R^2

$$\cos \theta = \frac{\mathbf{u} \cdot \mathbf{v}}{\|\mathbf{u}\| \|\mathbf{v}\|}.$$

For such a definition to make sense, however, the value of the right-hand side of this formula cannot exceed 1 in absolute value. This fact comes from a famous theorem named after the French mathematician Augustin-Louis Cauchy (1789–1857) and the German mathematician Hermann Schwarz (1843–1921).

THEOREM 5.4 The Cauchy-Schwarz Inequality

If $\mathbf{u}$ and $\mathbf{v}$ are vectors in R^n, then

$$|\mathbf{u} \cdot \mathbf{v}| \leq \|\mathbf{u}\| \|\mathbf{v}\|,$$

where $|\mathbf{u} \cdot \mathbf{v}|$ denotes the *absolute value* of $\mathbf{u} \cdot \mathbf{v}$.

PROOF *Case 1.* If $\mathbf{u} = \mathbf{0}$, then it follows that

$$|\mathbf{u} \cdot \mathbf{v}| = |\mathbf{0} \cdot \mathbf{v}| = 0 \quad \text{and} \quad \|\mathbf{u}\| \|\mathbf{v}\| = 0\|\mathbf{v}\| = 0.$$

So, the theorem is true if $\mathbf{u} = \mathbf{0}$.
Case 2. If $\mathbf{u} \neq \mathbf{0}$, let t be any real number and consider the vector $t\mathbf{u} + \mathbf{v}$. Because $(t\mathbf{u} + \mathbf{v}) \cdot (t\mathbf{u} + \mathbf{v}) \geq 0$, it follows that

$$(t\mathbf{u} + \mathbf{v}) \cdot (t\mathbf{u} + \mathbf{v}) = t^2(\mathbf{u} \cdot \mathbf{u}) + 2t(\mathbf{u} \cdot \mathbf{v}) + \mathbf{v} \cdot \mathbf{v} \geq 0.$$

Now, let $a = \mathbf{u} \cdot \mathbf{u}$, $b = 2(\mathbf{u} \cdot \mathbf{v})$, and $c = \mathbf{v} \cdot \mathbf{v}$ to obtain the quadratic inequality $at^2 + bt + c \geq 0$. Because this quadratic is never negative, it has either no real roots or a single repeated real root. But by the Quadratic Formula, this implies that the discriminant, $b^2 - 4ac$, is less than or equal to zero.

$$\begin{aligned} b^2 - 4ac &\leq 0 \\ b^2 &\leq 4ac \\ 4(\mathbf{u} \cdot \mathbf{v})^2 &\leq 4(\mathbf{u} \cdot \mathbf{u})(\mathbf{v} \cdot \mathbf{v}) \\ (\mathbf{u} \cdot \mathbf{v})^2 &\leq (\mathbf{u} \cdot \mathbf{u})(\mathbf{v} \cdot \mathbf{v}) \end{aligned}$$

Taking the square root of both sides produces

$$|\mathbf{u} \cdot \mathbf{v}| \leq \sqrt{\mathbf{u} \cdot \mathbf{u}}\,\sqrt{\mathbf{v} \cdot \mathbf{v}} = \|\mathbf{u}\| \|\mathbf{v}\|.$$

EXAMPLE 7 An Example of the Cauchy-Schwarz Inequality

Verify the Cauchy-Schwarz Inequality for $\mathbf{u} = (1, -1, 3)$ and $\mathbf{v} = (2, 0, -1)$.

SOLUTION Because $\mathbf{u} \cdot \mathbf{v} = -1$, $\mathbf{u} \cdot \mathbf{u} = 11$, and $\mathbf{v} \cdot \mathbf{v} = 5$, you have

$$|\mathbf{u} \cdot \mathbf{v}| = |-1| = 1$$

and

$$\|\mathbf{u}\|\,\|\mathbf{v}\| = \sqrt{\mathbf{u} \cdot \mathbf{u}}\,\sqrt{\mathbf{v} \cdot \mathbf{v}} = \sqrt{11}\sqrt{5} = \sqrt{55}.$$

The inequality holds, and you have $|\mathbf{u} \cdot \mathbf{v}| \leq \|\mathbf{u}\|\,\|\mathbf{v}\|$.

The Cauchy-Schwarz Inequality leads to the definition of the angle between two nonzero vectors in R^n.

Definition of the Angle Between Two Vectors in R^n

The **angle** θ between two nonzero vectors in R^n is given by

$$\cos \theta = \frac{\mathbf{u} \cdot \mathbf{v}}{\|\mathbf{u}\|\,\|\mathbf{v}\|}, \quad 0 \leq \theta \leq \pi.$$

REMARK: The angle between the zero vector and another vector is not defined.

EXAMPLE 8 Finding the Angle Between Two Vectors

The angle between $\mathbf{u} = (-4, 0, 2, -2)$ and $\mathbf{v} = (2, 0, -1, 1)$ is

$$\cos \theta = \frac{\mathbf{u} \cdot \mathbf{v}}{\|\mathbf{u}\|\,\|\mathbf{v}\|} = \frac{-12}{\sqrt{24}\sqrt{6}} = -\frac{12}{\sqrt{144}} = -1.$$

Consequently, $\theta = \pi$. It makes sense that $\mathbf{u}$ and $\mathbf{v}$ should have opposite directions, because $\mathbf{u} = -2\mathbf{v}$.

Note that because $\|\mathbf{u}\|$ and $\|\mathbf{v}\|$ are always positive, $\mathbf{u} \cdot \mathbf{v}$ and $\cos \theta$ will always have the same sign. Moreover, because the cosine is positive in the first quadrant and negative in the second quadrant, the sign of the dot product of two vectors can be used to determine whether the angle between them is acute or obtuse, as shown in Figure 5.6.

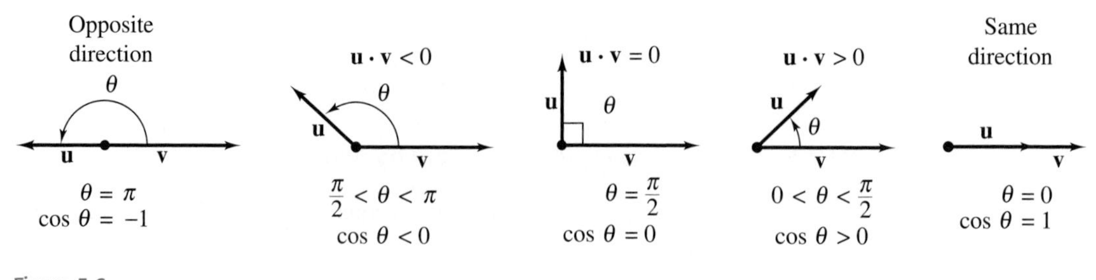

Figure 5.6

Figure 5.6 shows that two nonzero vectors meet at a right angle if and only if their dot product is zero. Two such vectors are said to be **orthogonal** (or perpendicular).

Definition of Orthogonal Vectors

Two vectors $\mathbf{u}$ and $\mathbf{v}$ in R^n are **orthogonal** if

$$\mathbf{u} \cdot \mathbf{v} = 0.$$

REMARK: Even though the angle between the zero vector and another vector is not defined, it is convenient to extend the definition of orthogonality to include the zero vector. In other words, the vector $\mathbf{0}$ is said to be orthogonal to every vector.

EXAMPLE 9 Orthogonal Vectors in R^n

(a) The vectors $\mathbf{u} = (1, 0, 0)$ and $\mathbf{v} = (0, 1, 0)$ are orthogonal because

$$\mathbf{u} \cdot \mathbf{v} = (1)(0) + (0)(1) + (0)(0) = 0.$$

(b) The vectors $\mathbf{u} = (3, 2, -1, 4)$ and $\mathbf{v} = (1, -1, 1, 0)$ are orthogonal because

$$\mathbf{u} \cdot \mathbf{v} = (3)(1) + (2)(-1) + (-1)(1) + (4)(0) = 0.$$

EXAMPLE 10 Finding Orthogonal Vectors

Determine all vectors in R^2 that are orthogonal to $\mathbf{u} = (4, 2)$.

SOLUTION Let $\mathbf{v} = (v_1, v_2)$ be orthogonal to $\mathbf{u}$. Then

$$\mathbf{u} \cdot \mathbf{v} = (4, 2) \cdot (v_1, v_2) = 4v_1 + 2v_2 = 0,$$

which implies that $2v_2 = -4v_1$ and $v_2 = -2v_1$. So, every vector that is orthogonal to (4, 2) is of the form

$$\mathbf{v} = (t, -2t) = t(1, -2),$$

where t is a real number. (See Figure 5.7.)

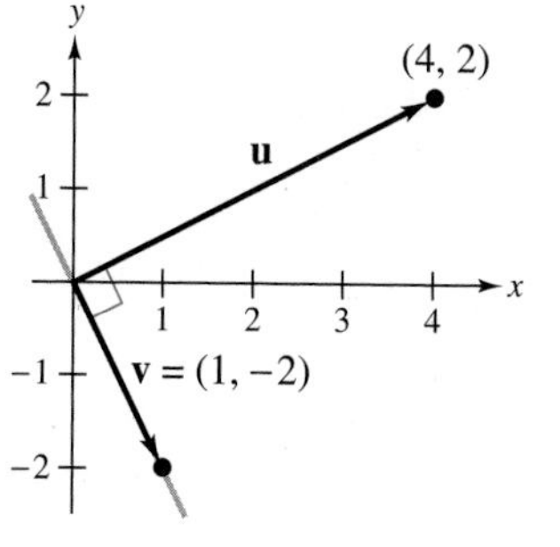

Figure 5.7

The Cauchy-Schwarz Inequality can be used to prove another well-known inequality called the **Triangle Inequality** (Theorem 5.5, page 288). The name "Triangle Inequality" is derived from the interpretation of the theorem in R^2, illustrated for the vectors $\mathbf{u}$ and $\mathbf{v}$ in Figure 5.8(a). If you consider $\|\mathbf{u}\|$ and $\|\mathbf{v}\|$ to be the lengths of two sides of a triangle, you can see that the length of the third side is $\|\mathbf{u} + \mathbf{v}\|$. Moreover, because the length of any side of a triangle cannot be greater than the sum of the lengths of the other two sides, you have

$$\|\mathbf{u} + \mathbf{v}\| \leq \|\mathbf{u}\| + \|\mathbf{v}\|.$$

Figure 5.8(b) illustrates the Triangle Inequality for the vectors $\mathbf{u}$ and $\mathbf{v}$ in R^3.

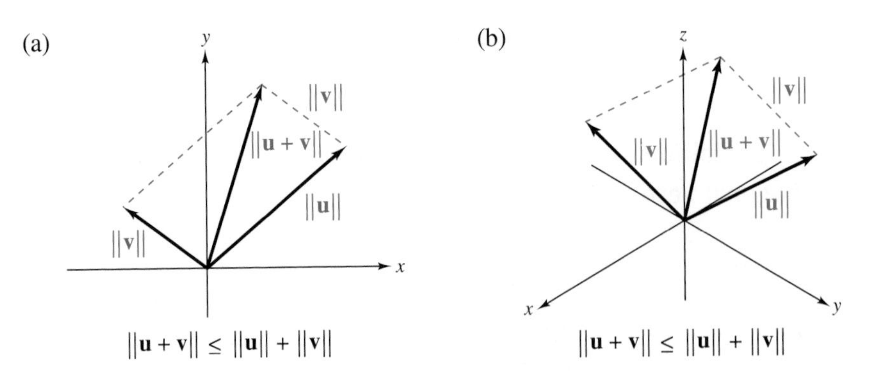

Figure 5.8

These results can be generalized to R^n in the following theorem.

THEOREM 5.5
The Triangle Inequality

If $\mathbf{u}$ and $\mathbf{v}$ are vectors in R^n, then

$$\|\mathbf{u} + \mathbf{v}\| \leq \|\mathbf{u}\| + \|\mathbf{v}\|.$$

PROOF Using the properties of the dot product, you have

$$\begin{aligned}\|\mathbf{u} + \mathbf{v}\|^2 &= (\mathbf{u} + \mathbf{v}) \cdot (\mathbf{u} + \mathbf{v}) \\ &= \mathbf{u} \cdot (\mathbf{u} + \mathbf{v}) + \mathbf{v} \cdot (\mathbf{u} + \mathbf{v}) \\ &= \mathbf{u} \cdot \mathbf{u} + 2(\mathbf{u} \cdot \mathbf{v}) + \mathbf{v} \cdot \mathbf{v} \\ &= \|\mathbf{u}\|^2 + 2(\mathbf{u} \cdot \mathbf{v}) + \|\mathbf{v}\|^2 \\ &\leq \|\mathbf{u}\|^2 + 2|\mathbf{u} \cdot \mathbf{v}| + \|\mathbf{v}\|^2.\end{aligned}$$

Now, by the Cauchy-Schwarz Inequality, $|\mathbf{u} \cdot \mathbf{v}| \leq \|\mathbf{u}\|\,\|\mathbf{v}\|$, and you can write

$$\begin{aligned}\|\mathbf{u} + \mathbf{v}\|^2 &\leq \|\mathbf{u}\|^2 + 2|\mathbf{u} \cdot \mathbf{v}| + \|\mathbf{v}\|^2 \\ &\leq \|\mathbf{u}\|^2 + 2\|\mathbf{u}\|\,\|\mathbf{v}\| + \|\mathbf{v}\|^2 \\ &= \left(\|\mathbf{u}\| + \|\mathbf{v}\|\right)^2.\end{aligned}$$

Because both $\|\mathbf{u} + \mathbf{v}\|$ and $\left(\|\mathbf{u}\| + \|\mathbf{v}\|\right)$ are nonnegative, taking the square root of both sides yields

$$\|\mathbf{u} + \mathbf{v}\| \leq \|\mathbf{u}\| + \|\mathbf{v}\|.$$

REMARK: Equality occurs in the Triangle Inequality if and only if the vectors $\mathbf{u}$ and $\mathbf{v}$ have the same direction. (See Exercise 117.)

From the proof of the Triangle Inequality, you have

$$\|\mathbf{u} + \mathbf{v}\|^2 = \|\mathbf{u}\|^2 + 2(\mathbf{u} \cdot \mathbf{v}) + \|\mathbf{v}\|^2.$$

If **u** and **v** are orthogonal, then $\mathbf{u} \cdot \mathbf{v} = 0$, and you have the extension of the **Pythagorean Theorem** to R^n shown below.

THEOREM 5.6
The Pythagorean Theorem

If **u** and **v** are vectors in R^n, then **u** and **v** are orthogonal if and only if

$$\|\mathbf{u} + \mathbf{v}\|^2 = \|\mathbf{u}\|^2 + \|\mathbf{v}\|^2.$$

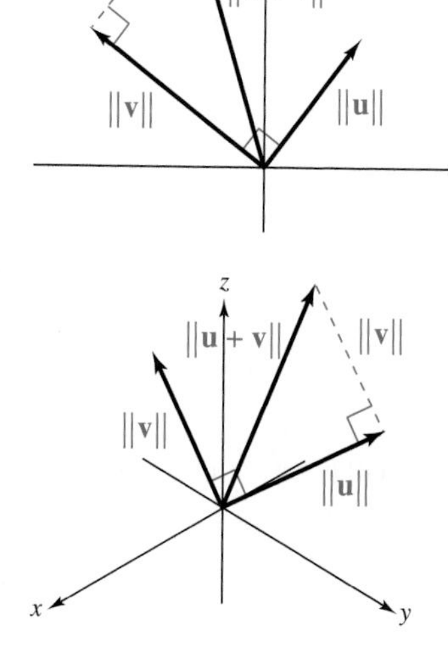

Figure 5.9

This relationship is illustrated graphically for R^2 and R^3 in Figure 5.9.

The Dot Product and Matrix Multiplication

It is often useful to represent a vector $\mathbf{u} = (u_1, u_2, \ldots, u_n)$ in R^n as an $n \times 1$ column matrix. In this notation, the dot product of two vectors

$$\mathbf{u} = \begin{bmatrix} u_1 \\ u_2 \\ \vdots \\ u_n \end{bmatrix} \quad \text{and} \quad \mathbf{v} = \begin{bmatrix} v_1 \\ v_2 \\ \vdots \\ v_n \end{bmatrix}$$

can be represented as the matrix product of the transpose of **u** multiplied by **v**.

$$\mathbf{u} \cdot \mathbf{v} = \mathbf{u}^T\mathbf{v} = [u_1 \ u_2 \ \ldots \ u_n] \begin{bmatrix} v_1 \\ v_2 \\ \vdots \\ v_n \end{bmatrix} = [u_1v_1 + u_2v_2 + \cdots + u_nv_n]$$

For example, the dot product of the vectors

$$\mathbf{u} = \begin{bmatrix} 1 \\ 2 \\ -1 \end{bmatrix} \quad \text{and} \quad \mathbf{v} = \begin{bmatrix} 3 \\ -2 \\ 4 \end{bmatrix}$$

is

$$\mathbf{u} \cdot \mathbf{v} = \mathbf{u}^T\mathbf{v} = [1 \quad 2 \quad -1] \begin{bmatrix} 3 \\ -2 \\ 4 \end{bmatrix} = [(1)(3) + (2)(-2) + (-1)(4)] = -5.$$

In this light, many of the properties of the dot product are direct consequences of the corresponding properties of matrix multiplication.

SECTION 5.1 Exercises

In Exercises 1–6, find the length of the vector.

1. $\mathbf{v} = (4, 3)$
2. $\mathbf{v} = (0, 1)$
3. $\mathbf{v} = (1, 2, 2)$
4. $\mathbf{v} = (2, 0, 6)$
5. $\mathbf{v} = (2, 0, -5, 5)$
6. $\mathbf{v} = (2, -4, 5, -1, 1)$

In Exercises 7–12, find (a) $\|\mathbf{u}\|$, (b) $\|\mathbf{v}\|$, and (c) $\|\mathbf{u} + \mathbf{v}\|$.

7. $\mathbf{u} = \left(-1, \frac{1}{4}\right), \quad \mathbf{v} = \left(4, -\frac{1}{8}\right)$
8. $\mathbf{u} = \left(1, \frac{1}{2}\right), \quad \mathbf{v} = \left(2, -\frac{1}{2}\right)$
9. $\mathbf{u} = (0, 4, 3), \quad \mathbf{v} = (1, -2, 1)$
10. $\mathbf{u} = (1, 2, 1), \quad \mathbf{v} = (0, 2, -2)$
11. $\mathbf{u} = (0, 1, -1, 2), \quad \mathbf{v} = (1, 1, 3, 0)$
12. $\mathbf{u} = (1, 0, 0, 0), \quad \mathbf{v} = (0, 1, 0, 0)$

In Exercises 13–18, find a unit vector (a) in the direction of $\mathbf{u}$ and (b) in the direction opposite that of $\mathbf{u}$.

13. $\mathbf{u} = (-5, 12)$
14. $\mathbf{u} = (1, -1)$
15. $\mathbf{u} = (3, 2, -5)$
16. $\mathbf{u} = (-1, 3, 4)$
17. $\mathbf{u} = (1, 0, 2, 2)$
18. $\mathbf{u} = (-1, 1, 2, 0)$

19. For what values of c is $\|c(1, 2, 3)\| = 1$?
20. For what values of c is $\|c(2, 2, -1)\| = 3$?

In Exercises 21–26, find the vector $\mathbf{v}$ with the given length that has the same direction as the vector $\mathbf{u}$.

21. $\|\mathbf{v}\| = 4, \quad \mathbf{u} = (1, 1)$
22. $\|\mathbf{v}\| = 4, \quad \mathbf{u} = (-1, 1)$
23. $\|\mathbf{v}\| = 2, \quad \mathbf{u} = \left(\sqrt{3}, 3, 0\right)$
24. $\|\mathbf{v}\| = 4, \quad \mathbf{u} = (-1, 2, 1)$
25. $\|\mathbf{v}\| = 3, \quad \mathbf{u} = (0, 2, 1, -1)$
26. $\|\mathbf{v}\| = 2, \quad \mathbf{u} = (1, -1, 4, 0)$

27. Given the vector $\mathbf{v} = (8, 8, 6)$, find $\mathbf{u}$ such that
    (a) $\mathbf{u}$ has the same direction as $\mathbf{v}$ and one-half its length.
    (b) $\mathbf{u}$ has the direction opposite that of $\mathbf{v}$ and one-fourth its length.
    (c) $\mathbf{u}$ has the direction opposite that of $\mathbf{v}$ and twice its length.
28. Given the vector $\mathbf{v} = (-1, 3, 0, 4)$, find $\mathbf{u}$ such that
    (a) $\mathbf{u}$ has the same direction as $\mathbf{v}$ and one-half its length.
    (b) $\mathbf{u}$ has the direction opposite that of $\mathbf{v}$ and one-fourth its length.
    (c) $\mathbf{u}$ has the direction opposite that of $\mathbf{v}$ and twice its length.

In Exercises 29–34, find the distance between $\mathbf{u}$ and $\mathbf{v}$.

29. $\mathbf{u} = (1, -1), \quad \mathbf{v} = (-1, 1)$
30. $\mathbf{u} = (3, 4), \quad \mathbf{v} = (7, 1)$
31. $\mathbf{u} = (1, 1, 2), \quad \mathbf{v} = (-1, 3, 0)$
32. $\mathbf{u} = (1, 2, 0), \quad \mathbf{v} = (-1, 4, 1)$
33. $\mathbf{u} = (0, 1, 2, 3), \quad \mathbf{v} = (1, 0, 4, -1)$
34. $\mathbf{u} = (0, 1, -1, 2), \quad \mathbf{v} = (1, 1, 2, 2)$

In Exercises 35–40, find (a) $\mathbf{u} \cdot \mathbf{v}$, (b) $\mathbf{u} \cdot \mathbf{u}$, (c) $\|\mathbf{u}\|^2$, (d) $(\mathbf{u} \cdot \mathbf{v})\mathbf{v}$, and (e) $\mathbf{u} \cdot (5\mathbf{v})$.

35. $\mathbf{u} = (3, 4), \quad \mathbf{v} = (2, -3)$
36. $\mathbf{u} = (-1, 2), \quad \mathbf{v} = (2, -2)$
37. $\mathbf{u} = (-1, 1, -2), \quad \mathbf{v} = (1, -3, -2)$
38. $\mathbf{u} = (2, -1, 1), \quad \mathbf{v} = (0, 2, -1)$
39. $\mathbf{u} = (4, 0, -3, 5), \quad \mathbf{v} = (0, 2, 5, 4)$
40. $\mathbf{u} = (0, 4, 3, 4, 4), \quad \mathbf{v} = (6, 8, -3, 3, -5)$
41. Find $(\mathbf{u} + \mathbf{v}) \cdot (2\mathbf{u} - \mathbf{v})$, given that $\mathbf{u} \cdot \mathbf{u} = 4$, $\mathbf{u} \cdot \mathbf{v} = -5$, and $\mathbf{v} \cdot \mathbf{v} = 10$.
42. Find $(3\mathbf{u} - \mathbf{v}) \cdot (\mathbf{u} - 3\mathbf{v})$, given that $\mathbf{u} \cdot \mathbf{u} = 8$, $\mathbf{u} \cdot \mathbf{v} = 7$, and $\mathbf{v} \cdot \mathbf{v} = 6$.

In Exercises 43–58, use a graphing utility or computer software program with vector capabilities to find (a)–(f).

(a) Norm of $\mathbf{u}$ and $\mathbf{v}$
(b) A unit vector in the direction of $\mathbf{v}$
(c) A unit vector in the direction opposite that of $\mathbf{u}$
(d) $\mathbf{u} \cdot \mathbf{v}$
(e) $\mathbf{u} \cdot \mathbf{u}$
(f) $\mathbf{v} \cdot \mathbf{v}$

43. $\mathbf{u} = (5, -12), \quad \mathbf{v} = (-8, -15)$
44. $\mathbf{u} = (3, -4), \quad \mathbf{v} = (5, 12)$
45. $\mathbf{u} = (5, 12), \quad \mathbf{v} = (-12, 5)$
46. $\mathbf{u} = (3, -4), \quad \mathbf{v} = (4, 3)$
47. $\mathbf{u} = (10, -24), \quad \mathbf{v} = (-5, -12)$
48. $\mathbf{u} = (9, 12), \quad \mathbf{v} = (-7, 24)$
49. $\mathbf{u} = (0, 5, 12), \quad \mathbf{v} = (0, -5, -12)$
50. $\mathbf{u} = (3, 0, -4), \quad \mathbf{v} = (-3, -4, 0)$
51. $\mathbf{u} = \left(1, \frac{1}{8}, \frac{2}{5}\right), \quad \mathbf{v} = \left(0, \frac{1}{4}, \frac{1}{5}\right)$
52. $\mathbf{u} = \left(-1, \frac{1}{2}, \frac{1}{4}\right), \quad \mathbf{v} = \left(0, \frac{1}{4}, -\frac{1}{2}\right)$

53. $\mathbf{u} = (0, 1, \sqrt{2}), \quad \mathbf{v} = (-1, \sqrt{2}, -1)$

54. $\mathbf{u} = (-1, \sqrt{3}, 2), \quad \mathbf{v} = (\sqrt{2}, -1, -\sqrt{2})$

55. $\mathbf{u} = (0, 2, 2, -1, 1, -2), \quad \mathbf{v} = (2, 0, 1, 1, 2, -2)$

56. $\mathbf{u} = (1, 2, 3, -2, -1, -3), \quad \mathbf{v} = (-1, 0, 2, 1, 2, -3)$

57. $\mathbf{u} = (-1, 1, 2, -1, 1, 1, -2, 1)$,
$\mathbf{v} = (-1, 0, 1, 2, -2, 1, 1, -2)$

58. $\mathbf{u} = (3, -1, 2, 1, 0, 1, 2, -1)$,
$\mathbf{v} = (1, 2, 0, -1, 2, -2, 1, 0)$

In Exercises 59–62, verify the Cauchy-Schwarz Inequality for the given vectors.

59. $\mathbf{u} = (3, 4), \quad \mathbf{v} = (2, -3)$

60. $\mathbf{u} = (-1, 0), \quad \mathbf{v} = (1, 1)$

61. $\mathbf{u} = (1, 1, -2), \quad \mathbf{v} = (1, -3, -2)$

62. $\mathbf{u} = (1, -1, 0), \quad \mathbf{v} = (0, 1, -1)$

In Exercises 63–72, find the angle θ between the vectors.

63. $\mathbf{u} = (3, 1), \quad \mathbf{v} = (-2, 4)$

64. $\mathbf{u} = (2, -1), \quad \mathbf{v} = (2, 0)$

65. $\mathbf{u} = \left(\cos\frac{\pi}{6}, \sin\frac{\pi}{6}\right), \quad \mathbf{v} = \left(\cos\frac{3\pi}{4}, \sin\frac{3\pi}{4}\right)$

66. $\mathbf{u} = \left(\cos\frac{\pi}{3}, \sin\frac{\pi}{3}\right), \quad \mathbf{v} = \left(\cos\frac{\pi}{4}, \sin\frac{\pi}{4}\right)$

67. $\mathbf{u} = (1, 1, 1), \quad \mathbf{v} = (2, 1, -1)$

68. $\mathbf{u} = (2, 3, 1), \quad \mathbf{v} = (-3, 2, 0)$

69. $\mathbf{u} = (0, 1, 0, 1), \quad \mathbf{v} = (3, 3, 3, 3)$

70. $\mathbf{u} = (1, -1, 0, 1), \quad \mathbf{v} = (-1, 2, -1, 0)$

71. $\mathbf{u} = (1, 3, -1, 2, 0), \quad \mathbf{v} = (-1, 4, 5, -3, 2)$

72. $\mathbf{u} = (1, -1, 1, 0, 1), \quad \mathbf{v} = (1, 0, -1, 0, 1)$

In Exercises 73–80, determine all vectors $\mathbf{v}$ that are orthogonal to $\mathbf{u}$.

73. $\mathbf{u} = (0, 5)$

74. $\mathbf{u} = (2, 7)$

75. $\mathbf{u} = (-3, 2)$

76. $\mathbf{u} = (0, 0)$

77. $\mathbf{u} = (4, -1, 0)$

78. $\mathbf{u} = (2, -1, 1)$

79. $\mathbf{u} = (0, 1, 0, 0, 0)$

80. $\mathbf{u} = (0, 0, -1, 0)$

In Exercises 81–88, determine whether $\mathbf{u}$ and $\mathbf{v}$ are orthogonal, parallel, or neither.

81. $\mathbf{u} = (2, 18), \quad \mathbf{v} = \left(\frac{3}{2}, -\frac{1}{6}\right)$

82. $\mathbf{u} = (4, 3), \quad \mathbf{v} = \left(\frac{1}{2}, -\frac{2}{3}\right)$

83. $\mathbf{u} = \left(-\frac{1}{3}, \frac{2}{3}\right), \quad \mathbf{v} = (2, -4)$

84. $\mathbf{u} = (1, -1), \quad \mathbf{v} = (0, -1)$

85. $\mathbf{u} = (0, 1, 0), \quad \mathbf{v} = (1, -2, 0)$

86. $\mathbf{u} = (0, 1, 6), \quad \mathbf{v} = (1, -2, -1)$

87. $\mathbf{u} = (-2, 5, 1, 0), \quad \mathbf{v} = \left(\frac{1}{4}, -\frac{5}{4}, 0, 1\right)$

88. $\mathbf{u} = \left(4, \frac{3}{2}, -1, \frac{1}{2}\right), \quad \mathbf{v} = \left(-2, -\frac{3}{4}, \frac{1}{2}, -\frac{1}{4}\right)$

In Exercises 89–92, use a graphing utility or computer software program with vector capabilities to determine whether $\mathbf{u}$ and $\mathbf{v}$ are orthogonal, parallel, or neither.

89. $\mathbf{u} = \left(-2, \frac{1}{2}, -1, 3\right), \quad \mathbf{v} = \left(\frac{3}{2}, 1, -\frac{5}{2}, 0\right)$

90. $\mathbf{u} = \left(-\frac{21}{2}, \frac{43}{2}, -12, \frac{3}{2}\right), \quad \mathbf{v} = \left(0, 6, \frac{21}{2}, -\frac{9}{2}\right)$

91. $\mathbf{u} = \left(-\frac{3}{4}, \frac{3}{2}, -\frac{9}{2}, -6\right), \quad \mathbf{v} = \left(\frac{3}{8}, -\frac{3}{4}, \frac{9}{8}, 3\right)$

92. $\mathbf{u} = \left(-\frac{4}{3}, \frac{8}{3}, -4, -\frac{32}{3}\right), \quad \mathbf{v} = \left(-\frac{16}{3}, -2, \frac{4}{3}, -\frac{2}{3}\right)$

Writing In Exercises 93 and 94, determine if the vectors are orthogonal, parallel, or neither. Then explain your reasoning.

93. $\mathbf{u} = (\cos\theta, \sin\theta, -1), \quad \mathbf{v} = (\sin\theta, -\cos\theta, 0)$

94. $\mathbf{u} = (-\sin\theta, \cos\theta, 1), \quad \mathbf{v} = (\sin\theta, -\cos\theta, 0)$

True or False? In Exercises 95 and 96, determine whether each statement is true or false. If a statement is true, give a reason or cite an appropriate statement from the text. If a statement is false, provide an example that shows the statement is not true in all cases or cite an appropriate statement from the text.

95. (a) The length or norm of a vector is
$\|\mathbf{v}\| = |v_1 + v_2 + v_3 + \cdot\cdot\cdot + v_n|$.

(b) The dot product of two vectors $\mathbf{u}$ and $\mathbf{v}$ is another vector represented by $\mathbf{u}\cdot\mathbf{v} = (u_1v_1, u_2v_2, u_3v_3, \ldots, u_nv_n)$.

96. (a) If $\mathbf{v}$ is a nonzero vector in R^n, the unit vector is $\mathbf{u} = \mathbf{v}/|\mathbf{v}|$.

(b) If $\mathbf{u}\cdot\mathbf{v} < 0$, then the angle θ between $\mathbf{u}$ and $\mathbf{v}$ is acute.

Writing In Exercises 97 and 98, write a short paragraph explaining why each expression is meaningless. Assume that $\mathbf{u}$ and $\mathbf{v}$ are vectors in R^n, and that c is a scalar.

97. (a) $\|\mathbf{u}\cdot\mathbf{v}\|$ (b) $\mathbf{u} + (\mathbf{u}\cdot\mathbf{v})$

98. (a) $(\mathbf{u}\cdot\mathbf{v})\cdot\mathbf{u}$ (b) $c\cdot(\mathbf{u}\cdot\mathbf{v})$

In Exercises 99–102, verify the Triangle Inequality for the vectors **u** and **v**.

99. $\mathbf{u} = (4, 0),\ \mathbf{v} = (1, 1)$ **100.** $\mathbf{u} = (1, 1, 1),\ \mathbf{v} = (0, 1, -2)$

101. $\mathbf{u} = (-1, 1),\ \mathbf{v} = (2, 0)$

102. $\mathbf{u} = (1, -1, 0),\ \mathbf{v} = (0, 1, 2)$

In Exercises 103–106, verify the Pythagorean Theorem for the vectors **u** and **v**.

103. $\mathbf{u} = (1, -1),\quad \mathbf{v} = (1, 1)$

104. $\mathbf{u} = (3, -2),\quad \mathbf{v} = (4, 6)$

105. $\mathbf{u} = (3, 4, -2),\quad \mathbf{v} = (4, -3, 0)$

106. $\mathbf{u} = (4, 1, -5),\quad \mathbf{v} = (2, -3, 1)$

107. Writing Explain what is known about θ, the angle between **u** and **v**, if

(a) $\mathbf{u} \cdot \mathbf{v} = 0.$ (b) $\mathbf{u} \cdot \mathbf{v} > 0.$ (c) $\mathbf{u} \cdot \mathbf{v} < 0.$

In Exercises 108 and 109, let $\mathbf{v} = (v_1, v_2)$ be a vector in R^2. Show that $(v_2, -v_1)$ is orthogonal to **v**, and use this fact to find two unit vectors orthogonal to the given vector.

108. $\mathbf{v} = (12, 5)$ **109.** $\mathbf{v} = (8, 15)$

110. Find the angle between the diagonal of a cube and one of its edges.

111. Find the angle between the diagonal of a cube and the diagonal of one of its sides.

112. Prove that if **u**, **v**, and **w** are vectors in R^n, then $(\mathbf{u} + \mathbf{v}) \cdot \mathbf{w} = \mathbf{u} \cdot \mathbf{w} + \mathbf{v} \cdot \mathbf{w}$.

113. Guided Proof Prove that if **u** is orthogonal to **v** and **w**, then **u** is orthogonal to $c\mathbf{v} + d\mathbf{w}$ for any scalars c and d.

Getting Started: To prove that **u** is orthogonal to $c\mathbf{v} + d\mathbf{w}$, you need to show that the dot product of **u** and $c\mathbf{v} + d\mathbf{w}$ is 0.

(i) Rewrite the dot product of **u** and $c\mathbf{v} + d\mathbf{w}$ as a linear combination of $(\mathbf{u} \cdot \mathbf{v})$ and $(\mathbf{u} \cdot \mathbf{w})$ using Properties 2 and 3 of Theorem 5.3.

(ii) Use the fact that **u** is orthogonal to **v** and **w**, and the result of part (i) to lead to the conclusion that **u** is orthogonal to $c\mathbf{v} + d\mathbf{w}$.

114. Prove that if **u** and **v** are vectors in R^n, then

$$\mathbf{u} \cdot \mathbf{v} = \tfrac{1}{4}\|\mathbf{u} + \mathbf{v}\|^2 - \tfrac{1}{4}\|\mathbf{u} - \mathbf{v}\|^2.$$

115. Prove that if **u** and **v** are vectors in R^n, then

$$\|\mathbf{u} + \mathbf{v}\|^2 + \|\mathbf{u} - \mathbf{v}\|^2 = 2\|\mathbf{u}\|^2 + 2\|\mathbf{v}\|^2.$$

116. Prove that the vectors $\mathbf{u} = (\cos\theta, -\sin\theta)$ and $\mathbf{v} = (\sin\theta, \cos\theta)$ are orthogonal unit vectors for any value of θ. Graph **u** and **v** for $\theta = \pi/3$.

117. Prove that $\|\mathbf{u} + \mathbf{v}\| = \|\mathbf{u}\| + \|\mathbf{v}\|$ if and only if **u** and **v** have the same direction.

118. Writing The vector $\mathbf{u} = (3240, 1450, 2235)$ gives the numbers of bushels of corn, oats, and wheat raised by a farmer in a certain year. The vector $\mathbf{v} = (2.22, 1.85, 3.25)$ gives the prices in dollars per bushel of the three crops. Find the dot product $\mathbf{u} \cdot \mathbf{v}$, and explain what information it gives.

119. Writing Let **x** be a solution to the $m \times n$ homogeneous linear system of equations $A\mathbf{x} = \mathbf{0}$. Explain why **x** is orthogonal to the row vectors of A.

120. The vector $\mathbf{u} = (1245, 2600)$ gives the numbers of units of two models of mountain bikes produced by a company. The vector $\mathbf{v} = (225, 275)$ gives the prices in dollars of the two models, respectively. Find the dot product $\mathbf{u} \cdot \mathbf{v}$ and explain what information it gives.

121. Use the matrix multiplication interpretation of the dot product, $\mathbf{u} \cdot \mathbf{v} = \mathbf{u}^T\mathbf{v}$, to prove the first three properties of Theorem 5.3.

5.2 Inner Product Spaces

In Section 5.1, the concepts of length, distance, and angle were extended from R^2 to R^n. This section extends these concepts one step further—to general vector spaces. This is accomplished by using the notion of an **inner product** of two vectors.

You already have one example of an inner product: the *dot product* in R^n. The dot product, called the **Euclidean inner product,** is only one of several inner products that can be defined on R^n. To distinguish between the standard inner product and other possible inner products, use the following notation.

$\mathbf{u} \cdot \mathbf{v} =$ dot product (Euclidean inner product for R^n)

$\langle \mathbf{u}, \mathbf{v} \rangle =$ general inner product for vector space V

A general inner product is defined in much the same way that a general vector space is defined—that is, in order for a function to qualify as an inner product, it must satisfy a set of axioms. The axioms parallel Properties 1, 2, 3, and 5 of the dot product given in Theorem 5.3.

Definition of Inner Product

Let $\mathbf{u}$, $\mathbf{v}$, and $\mathbf{w}$ be vectors in a vector space V, and let c be any scalar. An **inner product** on V is a function that associates a real number $\langle \mathbf{u}, \mathbf{v} \rangle$ with each pair of vectors $\mathbf{u}$ and $\mathbf{v}$ and satisfies the following axioms.

1. $\langle \mathbf{u}, \mathbf{v} \rangle = \langle \mathbf{v}, \mathbf{u} \rangle$
2. $\langle \mathbf{u}, \mathbf{v} + \mathbf{w} \rangle = \langle \mathbf{u}, \mathbf{v} \rangle + \langle \mathbf{u}, \mathbf{w} \rangle$
3. $c\langle \mathbf{u}, \mathbf{v} \rangle = \langle c\mathbf{u}, \mathbf{v} \rangle$
4. $\langle \mathbf{v}, \mathbf{v} \rangle \geq 0$, and $\langle \mathbf{v}, \mathbf{v} \rangle = 0$ if and only if $\mathbf{v} = \mathbf{0}$.

REMARK: A vector space V with an inner product is called an **inner product space.** Whenever an inner product space is referred to, assume that the set of scalars is the set of real numbers.

EXAMPLE 1 The Euclidean Inner Product for R^n

Show that the dot product in R^n satisfies the four axioms of an inner product.

SOLUTION In R^n, the dot product of two vectors $\mathbf{u} = (u_1, u_2, \ldots, u_n)$ and $\mathbf{v} = (v_1, v_2, \ldots, v_n)$ is

$$\mathbf{u} \cdot \mathbf{v} = u_1v_1 + u_2v_2 + \cdots + u_nv_n.$$

By Theorem 5.3, you know that this dot product satisfies the required four axioms, which verifies that it is an inner product on R^n.

The Euclidean inner product is not the only inner product that can be defined on R^n. A different inner product is illustrated in Example 2. To show that a function is an inner product, you must show that the four inner product axioms are satisfied.

EXAMPLE 2 A Different Inner Product for R^2

Show that the following function defines an inner product on R^2, where $\mathbf{u} = (u_1, u_2)$ and $\mathbf{v} = (v_1, v_2)$.

$$\langle \mathbf{u}, \mathbf{v} \rangle = u_1v_1 + 2u_2v_2$$

SOLUTION 1. Because the product of real numbers is commutative,

$$\langle \mathbf{u}, \mathbf{v} \rangle = u_1v_1 + 2u_2v_2 = v_1u_1 + 2v_2u_2 = \langle \mathbf{v}, \mathbf{u} \rangle.$$

2. Let $\mathbf{w} = (w_1, w_2)$. Then

$$\begin{aligned}\langle \mathbf{u}, \mathbf{v} + \mathbf{w} \rangle &= u_1(v_1 + w_1) + 2u_2(v_2 + w_2)\\ &= u_1v_1 + u_1w_1 + 2u_2v_2 + 2u_2w_2\\ &= (u_1v_1 + 2u_2v_2) + (u_1w_1 + 2u_2w_2)\\ &= \langle \mathbf{u}, \mathbf{v} \rangle + \langle \mathbf{u}, \mathbf{w} \rangle.\end{aligned}$$

3. If c is any scalar, then

$$c\langle \mathbf{u}, \mathbf{v} \rangle = c(u_1v_1 + 2u_2v_2) = (cu_1)v_1 + 2(cu_2)v_2 = \langle c\mathbf{u}, \mathbf{v} \rangle.$$

4. Because the square of a real number is nonnegative,

$$\langle \mathbf{v}, \mathbf{v} \rangle = v_1^2 + 2v_2^2 \geq 0.$$

Moreover, this expression is equal to zero if and only if $\mathbf{v} = \mathbf{0}$ (that is, if and only if $v_1 = v_2 = 0$).

Example 2 can be generalized to show that

$$\langle \mathbf{u}, \mathbf{v} \rangle = c_1u_1v_1 + c_2u_2v_2 + \cdot \cdot \cdot + c_nu_nv_n, \quad c_i > 0$$

is an inner product on R^n. The positive constants $c_1, \ldots, c_n$ are called **weights.** If any c_i is negative or 0, then this function does not define an inner product.

EXAMPLE 3 A Function That Is Not an Inner Product

Show that the following function is not an inner product on R^3, where $\mathbf{u} = (u_1, u_2, u_3)$ and $\mathbf{v} = (v_1, v_2, v_3)$.

$$\langle \mathbf{u}, \mathbf{v} \rangle = u_1v_1 - 2u_2v_2 + u_3v_3$$

SOLUTION Observe that Axiom 4 is not satisfied. For example, let $\mathbf{v} = (1, 2, 1)$. Then $\langle \mathbf{v}, \mathbf{v} \rangle = (1)(1) - 2(2)(2) + (1)(1) = -6$, which is less than zero.

EXAMPLE 4 An Inner Product on $M_{2,2}$

Let

$$A = \begin{bmatrix} a_{11} & a_{12} \\ a_{21} & a_{22} \end{bmatrix} \quad \text{and} \quad B = \begin{bmatrix} b_{11} & b_{12} \\ b_{21} & b_{22} \end{bmatrix}$$

be matrices in the vector space $M_{2,2}$. The function

$$\langle A, B \rangle = a_{11}b_{11} + a_{21}b_{21} + a_{12}b_{12} + a_{22}b_{22}$$

is an inner product on $M_{2,2}$. The verification of the four inner product axioms is left to you.

You obtain the inner product described in the next example from calculus. The verification of the inner product properties depends on the properties of the definite integral.

EXAMPLE 5 **An Inner Product Defined by a Definite Integral (Calculus)**

Let f and g be real-valued continuous functions in the vector space $C[a, b]$. Show that

$$\langle f, g \rangle = \int_a^b f(x)g(x)\, dx$$

defines an inner product on $C[a, b]$.

SOLUTION You can use familiar properties from calculus to verify the four parts of the definition.

1. $\langle f, g \rangle = \int_a^b f(x)g(x)\, dx = \int_a^b g(x)f(x)\, dx = \langle g, f \rangle$

2. $$\langle f, g + h \rangle = \int_a^b f(x)[g(x) + h(x)]\, dx = \int_a^b [f(x)g(x) + f(x)h(x)]\, dx$$
$$= \int_a^b f(x)g(x)\, dx + \int_a^b f(x)h(x)\, dx = \langle f, g \rangle + \langle f, h \rangle$$

3. $c\langle f, g \rangle = c\int_a^b f(x)g(x)\, dx = \int_a^b cf(x)g(x)\, dx = \langle cf, g \rangle$

4. Because $[f(x)]^2 \geq 0$ for all x, you know from calculus that

$$\langle f, f \rangle = \int_a^b [f(x)]^2\, dx \geq 0$$

with

$$\langle f, f \rangle = \int_a^b [f(x)]^2\, dx = 0$$

if and only if f is the zero function in $C[a, b]$, or if $a = b$.

The next theorem lists some easily verified properties of inner products.

THEOREM 5.7 Properties of Inner Products

Let $\mathbf{u}$, $\mathbf{v}$, and $\mathbf{w}$ be vectors in an inner product space V, and let c be any real number.

1. $\langle \mathbf{0}, \mathbf{v} \rangle = \langle \mathbf{v}, \mathbf{0} \rangle = 0$
2. $\langle \mathbf{u} + \mathbf{v}, \mathbf{w} \rangle = \langle \mathbf{u}, \mathbf{w} \rangle + \langle \mathbf{v}, \mathbf{w} \rangle$
3. $\langle \mathbf{u}, c\mathbf{v} \rangle = c\langle \mathbf{u}, \mathbf{v} \rangle$

PROOF The proof of the first property follows. The proofs of the other two properties are left as exercises. (See Exercises 85 and 86.) From the definition of an inner product, you know $\langle \mathbf{0}, \mathbf{v} \rangle = \langle \mathbf{v}, \mathbf{0} \rangle$, so you only need to show one of these to be zero. Using the fact that $0(\mathbf{v}) = \mathbf{0}$,

$$\begin{aligned} \langle \mathbf{0}, \mathbf{v} \rangle &= \langle 0(\mathbf{v}), \mathbf{v} \rangle \\ &= 0\langle \mathbf{v}, \mathbf{v} \rangle \\ &= 0. \end{aligned}$$

The definitions of norm (or length), distance, and angle for general inner product spaces closely parallel those for Euclidean n-space. Note that the definition of the angle θ between $\mathbf{u}$ and $\mathbf{v}$ presumes that

$$-1 \leq \frac{\langle \mathbf{u}, \mathbf{v} \rangle}{\|\mathbf{u}\| \, \|\mathbf{v}\|} \leq 1$$

for a general inner product, which follows from the Cauchy-Schwarz Inequality shown later in Theorem 5.8.

Definitions of Norm, Distance, and Angle

Let $\mathbf{u}$ and $\mathbf{v}$ be vectors in an inner product space V.

1. The **norm** (or **length**) of $\mathbf{u}$ is $\|\mathbf{u}\| = \sqrt{\langle \mathbf{u}, \mathbf{u} \rangle}$.
2. The **distance** between $\mathbf{u}$ and $\mathbf{v}$ is $d(\mathbf{u}, \mathbf{v}) = \|\mathbf{u} - \mathbf{v}\|$.
3. The **angle** between two nonzero vectors $\mathbf{u}$ and $\mathbf{v}$ is given by

$$\cos \theta = \frac{\langle \mathbf{u}, \mathbf{v} \rangle}{\|\mathbf{u}\| \, \|\mathbf{v}\|}, \quad 0 \leq \theta \leq \pi.$$

4. $\mathbf{u}$ and $\mathbf{v}$ are **orthogonal** if $\langle \mathbf{u}, \mathbf{v} \rangle = 0$.

REMARK: If $\|\mathbf{v}\| = 1$, then $\mathbf{v}$ is called a **unit vector.** Moreover, if $\mathbf{v}$ is any nonzero vector in an inner product space V, then the vector $\mathbf{u} = \mathbf{v}/\|\mathbf{v}\|$ is a unit vector and is called the **unit vector in the direction of v.**

EXAMPLE 6 Finding Inner Products

For polynomials $p = a_0 + a_1x + \cdots + a_nx^n$ and $q = b_0 + b_1x + \cdots + b_nx^n$ in the vector space P_n, the function

$$\langle p, q \rangle = a_0b_0 + a_1b_1 + \cdots + a_nb_n$$

is an inner product. Let $p(x) = 1 - 2x^2$, $q(x) = 4 - 2x + x^2$, and $r(x) = x + 2x^2$ be polynomials in P_2, and determine

(a) $\langle p, q \rangle$. (b) $\langle q, r \rangle$. (c) $\|q\|$. (d) $d(p, q)$.

SOLUTION (a) The inner product of p and q is

$$\begin{aligned}\langle p, q\rangle &= a_0b_0 + a_1b_1 + a_2b_2\\ &= (1)(4) + (0)(-2) + (-2)(1) = 2.\end{aligned}$$

(b) The inner product of q and r is

$$\langle q, r\rangle = (4)(0) + (-2)(1) + (1)(2) = 0.$$

(c) The norm of q is

$$\|q\| = \sqrt{\langle q, q\rangle} = \sqrt{4^2 + (-2)^2 + 1^2} = \sqrt{21}.$$

(d) The distance between p and q is

$$\begin{aligned}d(p, q) = \|p - q\| &= \|(1 - 2x^2) - (4 - 2x + x^2)\|\\ &= \|-3 + 2x - 3x^2\|\\ &= \sqrt{(-3)^2 + 2^2 + (-3)^2} = \sqrt{22}.\end{aligned}$$

Notice that the vectors q and r are orthogonal.

Orthogonality depends on the particular inner product used. That is, two vectors may be orthogonal with respect to one inner product but not to another. Try reworking Example 6 using the inner product $\langle p, q\rangle = a_0b_0 + a_1b_1 + 2a_2b_2$. With this inner product the only orthogonal pair is p and q.

EXAMPLE 7 Using the Inner Product on $C[0, 1]$ (Calculus)

Use the inner product defined in Example 5 and the functions $f(x) = x$ and $g(x) = x^2$ in $C[0, 1]$ to find

(a) $\|f\|$. (b) $d(f, g)$.

SOLUTION (a) Because $f(x) = x$, you have

$$\|f\|^2 = \langle f, f\rangle = \int_0^1 (x)(x)\,dx = \int_0^1 x^2\,dx = \left[\frac{x^3}{3}\right]_0^1 = \frac{1}{3}.\text{ So, } \|f\| = \frac{1}{\sqrt{3}}.$$

(b) To find $d(f, g)$, write

$$\begin{aligned}[d(f, g)]^2 &= \langle f - g, f - g\rangle\\ &= \int_0^1 [f(x) - g(x)]^2\,dx = \int_0^1 [x - x^2]^2\,dx\\ &= \int_0^1 [x^2 - 2x^3 + x^4]\,dx = \left[\frac{x^3}{3} - \frac{x^4}{2} + \frac{x^5}{5}\right]_0^1 = \frac{1}{30}.\end{aligned}$$

So, $d(f, g) = \dfrac{1}{\sqrt{30}}$.

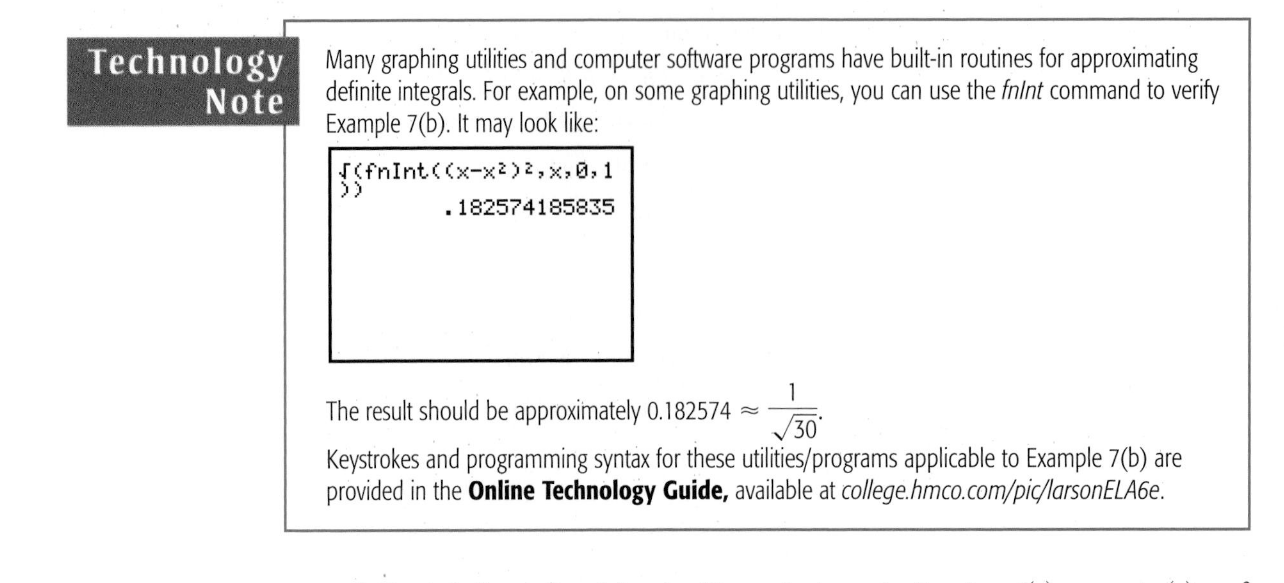

Technology Note

Many graphing utilities and computer software programs have built-in routines for approximating definite integrals. For example, on some graphing utilities, you can use the *fnInt* command to verify Example 7(b). It may look like:

```
√(fnInt((x-x²)²,x,0,1
))
            .182574185835
```

The result should be approximately $0.182574 \approx \dfrac{1}{\sqrt{30}}$.

Keystrokes and programming syntax for these utilities/programs applicable to Example 7(b) are provided in the **Online Technology Guide,** available at *college.hmco.com/pic/larsonELA6e*.

In Example 7, you found that the distance between the functions $f(x) = x$ and $g(x) = x^2$ in $C[0, 1]$ is $1/\sqrt{30} \approx 0.183$. In practice, the *actual* distance between a pair of vectors is not as useful as the *relative* distance between several pairs. For instance, the distance between $g(x) = x^2$ and $h(x) = x^2 + 1$ in $C[0, 1]$ is 1. From Figure 5.10, this seems reasonable. That is, whatever norm is defined on $C[0, 1]$, it seems reasonable that you would want to say that f and g are closer than g and h.

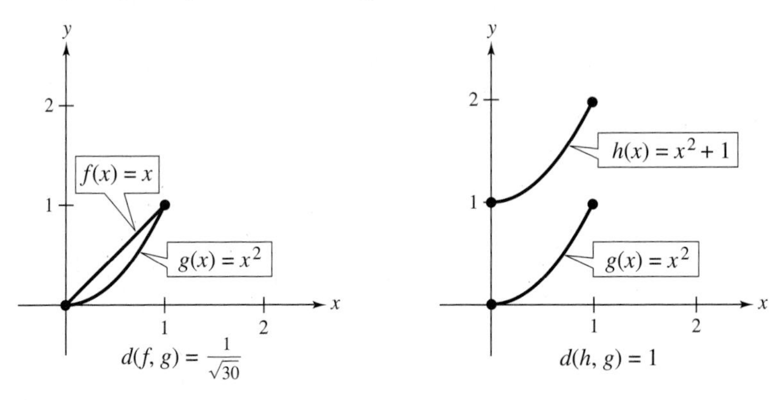

Figure 5.10

The properties of length and distance listed for R^n in the preceding section also hold for general inner product spaces. For instance, if $\mathbf{u}$ and $\mathbf{v}$ are vectors in an inner product space, then the following three properties are true.

Properties of Norm

1. $\|\mathbf{u}\| \geq 0$
2. $\|\mathbf{u}\| = 0$ if and only if $\mathbf{u} = \mathbf{0}$.
3. $\|c\mathbf{u}\| = |c| \, \|\mathbf{u}\|$

Properties of Distance

1. $d(\mathbf{u}, \mathbf{v}) \geq 0$
2. $d(\mathbf{u}, \mathbf{v}) = 0$ if and only if $\mathbf{u} = \mathbf{v}$.
3. $d(\mathbf{u}, \mathbf{v}) = d(\mathbf{v}, \mathbf{u})$

Theorem 5.8 lists the general inner product space versions of the Cauchy-Schwarz Inequality, the Triangle Inequality, and the general Pythagorean Theorem.

THEOREM 5.8

Let $\mathbf{u}$ and $\mathbf{v}$ be vectors in an inner product space V.

1. Cauchy-Schwarz Inequality: $|\langle \mathbf{u}, \mathbf{v} \rangle| \leq \|\mathbf{u}\| \, \|\mathbf{v}\|$
2. Triangle Inequality: $\|\mathbf{u} + \mathbf{v}\| \leq \|\mathbf{u}\| + \|\mathbf{v}\|$
3. Pythagorean Theorem: $\mathbf{u}$ and $\mathbf{v}$ are orthogonal if and only if

$$\|\mathbf{u} + \mathbf{v}\|^2 = \|\mathbf{u}\|^2 + \|\mathbf{v}\|^2.$$

The proofs of these three axioms parallel those for Theorems 5.4, 5.5, and 5.6. Simply substitute $\langle \mathbf{u}, \mathbf{v} \rangle$ for the Euclidean inner product $\mathbf{u} \cdot \mathbf{v}$.

EXAMPLE 8 An Example of the Cauchy-Schwarz Inequality (Calculus)

Let $f(x) = 1$ and $g(x) = x$ be functions in the vector space $C[0, 1]$, with the inner product defined in Example 5,

$$\langle f, g \rangle = \int_a^b f(x)g(x) \, dx.$$

Verify that $|\langle f, g \rangle| \leq \|f\| \, \|g\|$.

SOLUTION For the left side of this inequality you have

$$\langle f, g \rangle = \int_0^1 f(x)g(x) \, dx = \int_0^1 x \, dx = \frac{x^2}{2}\bigg]_0^1 = \frac{1}{2}.$$

For the right side of the inequality you have

$$\|f\|^2 = \int_0^1 f(x)f(x) \, dx = \int_0^1 dx = x\bigg]_0^1 = 1$$

and

$$\|g\|^2 = \int_0^1 g(x)g(x) \, dx = \int_0^1 x^2 \, dx = \frac{x^3}{3}\bigg]_0^1 = \frac{1}{3}.$$

So,

$$\|f\| \, \|g\| = \sqrt{(1)\left(\frac{1}{3}\right)} = \frac{1}{\sqrt{3}} \approx 0.577, \text{ and } |\langle f, g \rangle| \leq \|f\| \, \|g\|.$$

Orthogonal Projections in Inner Product Spaces

Let **u** and **v** be vectors in the plane. If **v** is nonzero, then **u** can be orthogonally projected onto **v**, as shown in Figure 5.11. This projection is denoted by $\text{proj}_{\mathbf{v}}\mathbf{u}$. Because $\text{proj}_{\mathbf{v}}\mathbf{u}$ is a scalar multiple of **v**, you can write

$$\text{proj}_{\mathbf{v}}\mathbf{u} = a\mathbf{v}.$$

If $a > 0$, as shown in Figure 5.11(a), then $\cos\theta > 0$ and the length of $\text{proj}_{\mathbf{v}}\mathbf{u}$ is

$$\|a\mathbf{v}\| = |a|\,\|\mathbf{v}\| = a\|\mathbf{v}\| = \|\mathbf{u}\|\cos\theta = \frac{\|\mathbf{u}\|\,\|\mathbf{v}\|\cos\theta}{\|\mathbf{v}\|} = \frac{\mathbf{u}\cdot\mathbf{v}}{\|\mathbf{v}\|},$$

which implies that $a = (\mathbf{u}\cdot\mathbf{v})/\|\mathbf{v}\|^2 = (\mathbf{u}\cdot\mathbf{v})/(\mathbf{v}\cdot\mathbf{v})$. So,

$$\text{proj}_{\mathbf{v}}\mathbf{u} = \frac{\mathbf{u}\cdot\mathbf{v}}{\mathbf{v}\cdot\mathbf{v}}\mathbf{v}.$$

If $a < 0$, as shown in Figure 5.11(b), then it can be shown that the orthogonal projection of **u** onto **v** is the same formula.

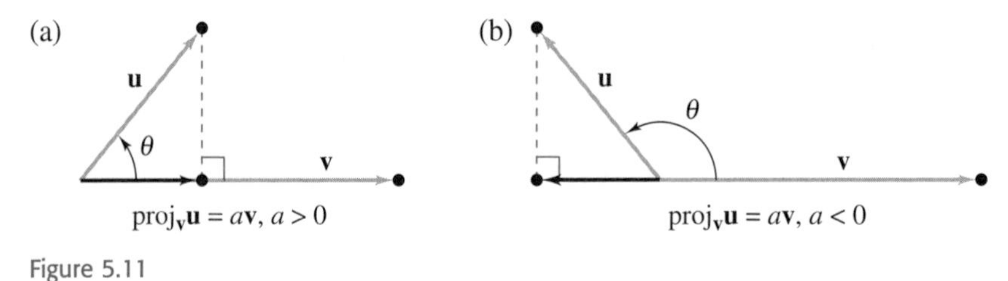

Figure 5.11

EXAMPLE 9 **Finding the Orthogonal Projection of u onto v**

In R^2, the orthogonal projection of $\mathbf{u} = (4, 2)$ onto $\mathbf{v} = (3, 4)$ is

$$\begin{aligned}\text{proj}_{\mathbf{v}}\mathbf{u} = \frac{\mathbf{u}\cdot\mathbf{v}}{\mathbf{v}\cdot\mathbf{v}}\mathbf{v} &= \frac{(4, 2)\cdot(3, 4)}{(3, 4)\cdot(3, 4)}(3, 4)\\ &= \frac{20}{25}(3, 4)\\ &= \left(\frac{12}{5}, \frac{16}{5}\right),\end{aligned}$$

as shown in Figure 5.12.

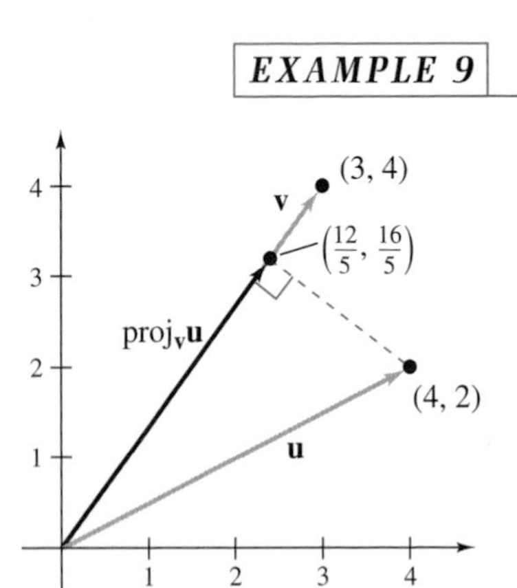

Figure 5.12

The notion of orthogonal projection extends naturally to a general inner product space.

Definition of Orthogonal Projection

Let **u** and **v** be vectors in an inner product space V, such that $\mathbf{v} \neq \mathbf{0}$. Then the **orthogonal projection** of **u** unto **v** is given by

$$\text{proj}_{\mathbf{v}}\mathbf{u} = \frac{\langle \mathbf{u}, \mathbf{v} \rangle}{\langle \mathbf{v}, \mathbf{v} \rangle}\mathbf{v}.$$

REMARK: If **v** is a unit vector, then $\langle \mathbf{v}, \mathbf{v} \rangle = \|\mathbf{v}\|^2 = 1$, and the formula for the orthogonal projection of **u** onto **v** takes the simpler form

$$\text{proj}_{\mathbf{v}}\mathbf{u} = \langle \mathbf{u}, \mathbf{v} \rangle \mathbf{v}.$$

EXAMPLE 10 Finding an Orthogonal Projection in R^3

Use the Euclidean inner product in R^3 to find the orthogonal projection of $\mathbf{u} = (6, 2, 4)$ onto $\mathbf{v} = (1, 2, 0)$.

SOLUTION Because $\mathbf{u} \cdot \mathbf{v} = 10$ and $\|\mathbf{v}\|^2 = \mathbf{v} \cdot \mathbf{v} = 5$, the orthogonal projection of **u** onto **v** is

$$\begin{aligned}\text{proj}_{\mathbf{v}}\mathbf{u} &= \frac{\mathbf{u} \cdot \mathbf{v}}{\mathbf{v} \cdot \mathbf{v}}\mathbf{v} \\ &= \frac{10}{5}(1, 2, 0) \\ &= 2(1, 2, 0) \\ &= (2, 4, 0),\end{aligned}$$

as shown in Figure 5.13.

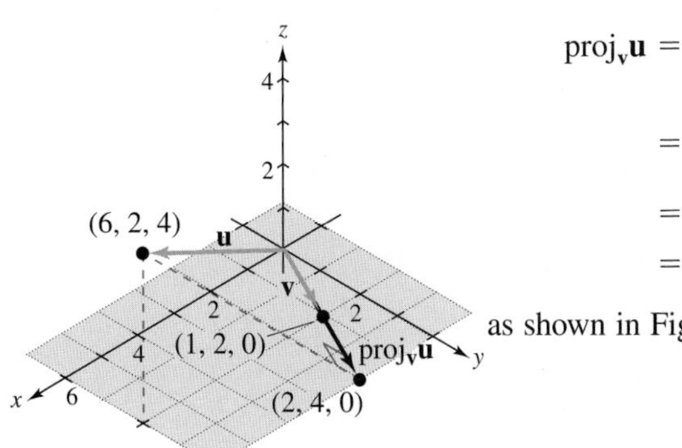

Figure 5.13

REMARK: Notice in Example 10 that $\mathbf{u} - \text{proj}_{\mathbf{v}}\mathbf{u} = (6, 2, 4) - (2, 4, 0) = (4, -2, 4)$ is orthogonal to $\mathbf{v} = (1, 2, 0)$. This is true in general: if **u** and **v** are nonzero vectors in an inner product space, then $\mathbf{u} - \text{proj}_{\mathbf{v}}\mathbf{u}$ is orthogonal to **v**. (See Exercise 84.)

An important property of orthogonal projections used in approximation problems (see Section 5.4) is shown in the next theorem. It states that, of all possible scalar multiples of a vector **v**, the orthogonal projection of **u** onto **v** is the one closest to **u**, as shown in Figure 5.14. For instance, in Example 10, this theorem implies that, of all the scalar multiples of the vector $\mathbf{v} = (1, 2, 0)$, the vector $\text{proj}_{\mathbf{v}}\mathbf{u} = (2, 4, 0)$ is closest to $\mathbf{u} = (6, 2, 4)$. You are asked to prove this explicitly in Exercise 90.

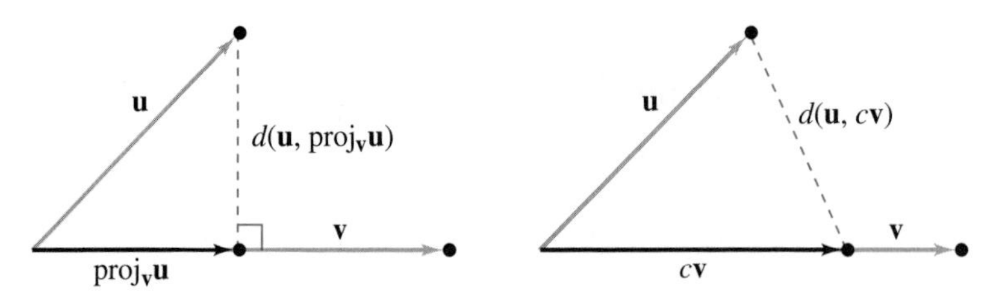

Figure 5.14

THEOREM 5.9
Orthogonal Projection and Distance

Let $\mathbf{u}$ and $\mathbf{v}$ be two vectors in an inner product space V, such that $\mathbf{v} \neq \mathbf{0}$. Then

$$d(\mathbf{u}, \text{proj}_{\mathbf{v}}\mathbf{u}) < d(\mathbf{u}, c\mathbf{v}), \quad c \neq \frac{\langle \mathbf{u}, \mathbf{v} \rangle}{\langle \mathbf{v}, \mathbf{v} \rangle}.$$

PROOF Let $b = \langle \mathbf{u}, \mathbf{v} \rangle / \langle \mathbf{v}, \mathbf{v} \rangle$. Then you can write

$$\|\mathbf{u} - c\mathbf{v}\|^2 = \|(\mathbf{u} - b\mathbf{v}) + (b - c)\mathbf{v}\|^2,$$

where $(\mathbf{u} - b\mathbf{v})$ and $(b - c)\mathbf{v}$ are orthogonal. You can verify this by using the inner product axioms to show that

$$\langle (\mathbf{u} - b\mathbf{v}), (b - c)\mathbf{v} \rangle = 0.$$

Now, by the Pythagorean Theorem you can write

$$\|(\mathbf{u} - b\mathbf{v}) + (b - c)\mathbf{v}\|^2 = \|\mathbf{u} - b\mathbf{v}\|^2 + \|(b - c)\mathbf{v}\|^2,$$

which implies that

$$\|\mathbf{u} - c\mathbf{v}\|^2 = \|\mathbf{u} - b\mathbf{v}\|^2 + (b - c)^2\|\mathbf{v}\|^2.$$

Because $b \neq c$ and $\mathbf{v} \neq \mathbf{0}$, you know that $(b - c)^2\|\mathbf{v}\|^2 > 0$. So,

$$\|\mathbf{u} - b\mathbf{v}\|^2 < \|\mathbf{u} - c\mathbf{v}\|^2,$$

and it follows that

$$d(\mathbf{u}, b\mathbf{v}) < d(\mathbf{u}, c\mathbf{v}).$$

The next example discusses a type of orthogonal projection in the inner product space $C[a, b]$.

EXAMPLE 11 **Finding an Orthogonal Projection in $C[a, b]$ (Calculus)**

Let $f(x) = 1$ and $g(x) = x$ be functions in $C[0, 1]$. Use the inner product defined in Example 5,

$$\langle f, g\rangle = \int_b^a f(x)g(x)\,dx,$$

to find the orthogonal projection of f onto g.

SOLUTION From Example 8 you know that

$$\langle f, g\rangle = \frac{1}{2} \quad \text{and} \quad \|g\|^2 = \langle g, g\rangle = \frac{1}{3}.$$

So, the orthogonal projection of f onto g is

$$\begin{aligned}\text{proj}_g f &= \frac{\langle f, g\rangle}{\langle g, g\rangle}g \\ &= \frac{1/2}{1/3}x \\ &= \frac{3}{2}x.\end{aligned}$$

SECTION 5.2 Exercises

In Exercises 1–10, find (a) $\langle \mathbf{u}, \mathbf{v}\rangle$, (b) $\|\mathbf{u}\|$, (c) $\|\mathbf{v}\|$, and (d) $d(\mathbf{u}, \mathbf{v})$ for the given inner product defined in R^n.

1. $\mathbf{u} = (3, 4),\ \ \mathbf{v} = (5, -12),\ \ \langle \mathbf{u}, \mathbf{v}\rangle = \mathbf{u} \cdot \mathbf{v}$

2. $\mathbf{u} = (1, 1),\ \ \mathbf{v} = (7, 9),\ \ \langle \mathbf{u}, \mathbf{v}\rangle = \mathbf{u} \cdot \mathbf{v}$

3. $\mathbf{u} = (-4, 3),\ \ \mathbf{v} = (0, 5),\ \ \langle \mathbf{u}, \mathbf{v}\rangle = 3u_1v_1 + u_2v_2$

4. $\mathbf{u} = (0, -6),\ \ \mathbf{v} = (-1, 1),\ \ \langle \mathbf{u}, \mathbf{v}\rangle = u_1v_1 + 2u_2v_2$

5. $\mathbf{u} = (0, 9, 4),\ \ \mathbf{v} = (9, -2, -4),\ \ \langle \mathbf{u}, \mathbf{v}\rangle = \mathbf{u} \cdot \mathbf{v}$

6. $\mathbf{u} = (0, 1, 2),\ \ \mathbf{v} = (1, 2, 0),\ \ \langle \mathbf{u}, \mathbf{v}\rangle = \mathbf{u} \cdot \mathbf{v}$

7. $\mathbf{u} = (8, 0, -8),\ \ \mathbf{v} = (8, 3, 16),$
$\langle \mathbf{u}, \mathbf{v}\rangle = 2u_1v_1 + 3u_2v_2 + u_3v_3$

8. $\mathbf{u} = (1, 1, 1),\ \ \mathbf{v} = (2, 5, 2),$
$\langle \mathbf{u}, \mathbf{v}\rangle = u_1v_1 + 2u_2v_2 + u_3v_3$

9. $\mathbf{u} = (2, 0, 1, -1),\ \ \mathbf{v} = (2, 2, 0, 1),$
$\langle \mathbf{u}, \mathbf{v}\rangle = \mathbf{u} \cdot \mathbf{v}$

10. $\mathbf{u} = (1, -1, 2, 0),\ \ \mathbf{v} = (2, 1, 0, -1),$
$\langle \mathbf{u}, \mathbf{v}\rangle = \mathbf{u} \cdot \mathbf{v}$

Calculus In Exercises 11–16, use the functions f and g in $C[-1, 1]$ to find (a) $\langle f, g\rangle$, (b) $\|f\|$, (c) $\|g\|$, and (d) $d(f, g)$ for the inner product

$$\langle f, g\rangle = \int_{-1}^{1} f(x)g(x)\,dx.$$

11. $f(x) = x^2,\ \ g(x) = x^2 + 1$

12. $f(x) = -x,\ \ g(x) = x^2 - x + 2$

13. $f(x) = x,\ \ g(x) = e^x$

14. $f(x) = x,\ \ g(x) = e^{-x}$

15. $f(x) = 1,\ \ g(x) = 3x^2 - 1$

16. $f(x) = -1,\ \ g(x) = 1 - 2x^2$

In Exercises 17–20, use the inner product $\langle A, B\rangle = 2a_{11}b_{11} + a_{12}b_{12} + a_{21}b_{21} + 2a_{22}b_{22}$ to find (a) $\langle A, B\rangle$, (b) $\|A\|$, (c) $\|B\|$, and (d) $d(A, B)$ for the matrices in $M_{2,2}$.

17. $A = \begin{bmatrix} -1 & 3 \\ 4 & -2 \end{bmatrix}$, $B = \begin{bmatrix} 0 & -2 \\ 1 & 1 \end{bmatrix}$

18. $A = \begin{bmatrix} 1 & 0 \\ 0 & 1 \end{bmatrix}$, $B = \begin{bmatrix} 0 & 1 \\ 1 & 0 \end{bmatrix}$

19. $A = \begin{bmatrix} 1 & -1 \\ 2 & 4 \end{bmatrix}$, $B = \begin{bmatrix} 0 & 1 \\ -2 & 0 \end{bmatrix}$

20. $A = \begin{bmatrix} 1 & 0 \\ 0 & -1 \end{bmatrix}$, $B = \begin{bmatrix} 1 & 1 \\ 0 & -1 \end{bmatrix}$

In Exercises 21–24, use the inner product $\langle p, q\rangle = a_0b_0 + a_1b_1 + a_2b_2$ to find (a) $\langle p, q\rangle$, (b) $\|p\|$, (c) $\|q\|$, and (d) $d(p, q)$ for the polynomials in P_2.

21. $p(x) = 1 - x + 3x^2$, $q(x) = x - x^2$

22. $p(x) = 1 + x + \frac{1}{2}x^2$, $q(x) = 1 + 2x^2$

23. $p(x) = 1 + x^2$, $q(x) = 1 - x^2$

24. $p(x) = 1 - 2x - x^2$, $q(x) = x - x^2$

In Exercises 25–28, prove that the indicated function is an inner product.

25. $\langle \mathbf{u}, \mathbf{v}\rangle$ as shown in Exercise 3

26. $\langle \mathbf{u}, \mathbf{v}\rangle$ as shown in Exercise 7

27. $\langle A, B\rangle$ as shown in Exercises 17 and 18

28. $\langle p, q\rangle$ as shown in Exercises 21 and 23

Writing In Exercises 29–36, state why $\langle \mathbf{u}, \mathbf{v}\rangle$ is not an inner product for $\mathbf{u} = (u_1, u_2)$ and $\mathbf{v} = (v_1, v_2)$ in R^2.

29. $\langle \mathbf{u}, \mathbf{v}\rangle = u_1v_1$

30. $\langle \mathbf{u}, \mathbf{v}\rangle = -u_2v_2$

31. $\langle \mathbf{u}, \mathbf{v}\rangle = u_1v_1 - u_2v_2$

32. $\langle \mathbf{u}, \mathbf{v}\rangle = u_1v_1 - 2u_2v_2$

33. $\langle \mathbf{u}, \mathbf{v}\rangle = u_1^2v_1^2 + u_2^2v_2^2$

34. $\langle \mathbf{u}, \mathbf{v}\rangle = u_1^2v_1^2 - u_2^2v_2^2$

35. $\langle \mathbf{u}, \mathbf{v}\rangle = 3u_1v_2 - u_2v_1$

36. $\langle \mathbf{u}, \mathbf{v}\rangle = u_1u_2 + v_1v_2$

In Exercises 37–46, find the angle between the vectors.

37. $\mathbf{u} = (3, 4)$, $\mathbf{v} = (5, -12)$, $\langle \mathbf{u}, \mathbf{v}\rangle = \mathbf{u}\cdot\mathbf{v}$

38. $\mathbf{u} = (2, -1)$, $\mathbf{v} = \left(\frac{1}{2}, 1\right)$, $\langle \mathbf{u}, \mathbf{v}\rangle = \mathbf{u}\cdot\mathbf{v}$

39. $\mathbf{u} = (-4, 3)$, $\mathbf{v} = (0, 5)$, $\langle \mathbf{u}, \mathbf{v}\rangle = 3u_1v_1 + u_2v_2$

40. $\mathbf{u} = \left(\frac{1}{4}, -1\right)$, $\mathbf{v} = (2, 1)$, $\langle \mathbf{u}, \mathbf{v}\rangle = 2u_1v_1 + u_2v_2$

41. $\mathbf{u} = (1, 1, 1)$, $\mathbf{v} = (2, -2, 2)$,
$\langle \mathbf{u}, \mathbf{v}\rangle = u_1v_1 + 2u_2v_2 + u_3v_3$

42. $\mathbf{u} = (0, 1, -1)$, $\mathbf{v} = (1, 2, 3)$, $\langle \mathbf{u}, \mathbf{v}\rangle = \mathbf{u}\cdot\mathbf{v}$

43. $p(x) = 1 - x + x^2$, $q(x) = 1 + x + x^2$,
$\langle p, q\rangle = a_0b_0 + a_1b_1 + a_2b_2$

44. $p(x) = 1 + x^2$, $q(x) = x - x^2$,
$\langle p, q\rangle = a_0b_0 + 2a_1b_1 + a_2b_2$

45. Calculus $f(x) = x$, $g(x) = x^2$,

$$\langle f, g\rangle = \int_{-1}^{1} f(x)g(x)\,dx$$

46. Calculus $f(x) = 1$, $g(x) = x^2$,

$$\langle f, g\rangle = \int_{-1}^{1} f(x)g(x)\,dx$$

In Exercises 47–58, verify (a) the Cauchy-Schwarz Inequality and (b) the Triangle Inequality.

47. $\mathbf{u} = (5, 12)$, $\mathbf{v} = (3, 4)$, $\langle \mathbf{u}, \mathbf{v}\rangle = \mathbf{u}\cdot\mathbf{v}$

48. $\mathbf{u} = (-1, 1)$, $\mathbf{v} = (1, -1)$, $\langle \mathbf{u}, \mathbf{v}\rangle = \mathbf{u}\cdot\mathbf{v}$

49. $\mathbf{u} = (1, 0, 4)$, $\mathbf{v} = (-5, 4, 1)$, $\langle \mathbf{u}, \mathbf{v}\rangle = \mathbf{u}\cdot\mathbf{v}$

50. $\mathbf{u} = (1, 0, 2)$, $\mathbf{v} = (1, 2, 0)$, $\langle \mathbf{u}, \mathbf{v}\rangle = \mathbf{u}\cdot\mathbf{v}$

51. $p(x) = 2x$, $q(x) = 3x^2 + 1$, $\langle p, q\rangle = a_0b_0 + a_1b_1 + a_2b_2$

52. $p(x) = x$, $q(x) = 1 - x^2$, $\langle p, q\rangle = a_0b_0 + 2a_1b_1 + a_2b_2$

53. $A = \begin{bmatrix} 0 & 3 \\ 2 & 1 \end{bmatrix}$, $B = \begin{bmatrix} -3 & 1 \\ 4 & 3 \end{bmatrix}$,
$\langle A, B\rangle = a_{11}b_{11} + a_{12}b_{12} + a_{21}b_{21} + a_{22}b_{22}$

54. $A = \begin{bmatrix} 0 & 1 \\ 2 & -1 \end{bmatrix}$, $B = \begin{bmatrix} 1 & 1 \\ 2 & -2 \end{bmatrix}$,
$\langle A, B\rangle = a_{11}b_{11} + a_{12}b_{12} + a_{21}b_{21} + a_{22}b_{22}$

55. Calculus $f(x) = \sin x$, $g(x) = \cos x$,

$$\langle f, g\rangle = \int_{-\pi}^{\pi} f(x)g(x)\,dx$$

56. Calculus $f(x) = x$, $g(x) = \cos \pi x$,

$$\langle f, g\rangle = \int_{0}^{2} f(x)g(x)\,dx$$

57. Calculus $f(x) = x$, $g(x) = e^x$,

$$\langle f, g\rangle = \int_{0}^{1} f(x)g(x)\,dx$$

58. Calculus $f(x) = x, \quad g(x) = e^{-x},$

$$\langle f, g\rangle = \int_0^1 f(x)g(x)\,dx$$

Calculus In Exercises 59–62, show that f and g are orthogonal in the inner product space $C[a, b]$ with the inner product

$$\langle f, g\rangle = \int_a^b f(x)g(x)\,dx.$$

59. $C[-\pi, \pi], \quad f(x) = \cos x, \quad g(x) = \sin x$

60. $C[-1, 1], \quad f(x) = x, \quad g(x) = \frac{1}{2}(3x^2 - 1)$

61. $C[-1, 1], \quad f(x) = x, \quad g(x) = \frac{1}{2}(5x^3 - 3x)$

62. $C[0, \pi], \quad f(x) = 1, \quad g(x) = \cos(2nx), \quad n = 1, 2, 3, \ldots$

In Exercises 63–66, (a) find $\text{proj}_{\mathbf{v}}\mathbf{u}$, (b) find $\text{proj}_{\mathbf{u}}\mathbf{v}$, and (c) sketch a graph of both $\text{proj}_{\mathbf{v}}\mathbf{u}$ and $\text{proj}_{\mathbf{u}}\mathbf{v}$.

63. $\mathbf{u} = (1, 2), \quad \mathbf{v} = (2, 1)$

64. $\mathbf{u} = (-1, -2), \quad \mathbf{v} = (4, 2)$

65. $\mathbf{u} = (-1, 3), \quad \mathbf{v} = (4, 4)$

66. $\mathbf{u} = (2, -2), \quad \mathbf{v} = (3, 1)$

In Exercises 67–70, find (a) $\text{proj}_{\mathbf{v}}\mathbf{u}$ and (b) $\text{proj}_{\mathbf{u}}\mathbf{v}$.

67. $\mathbf{u} = (1, 3, -2), \quad \mathbf{v} = (0, -1, 1)$

68. $\mathbf{u} = (1, 2, -1), \quad \mathbf{v} = (-1, 2, -1)$

69. $\mathbf{u} = (0, 1, 3, -6), \quad \mathbf{v} = (-1, 1, 2, 2)$

70. $\mathbf{u} = (-1, 4, -2, 3), \quad \mathbf{v} = (2, -1, 2, -1)$

Calculus In Exercises 71–78, find the orthogonal projection of f onto g. Use the inner product in $C[a, b]$

$$\langle f, g\rangle = \int_a^b f(x)g(x)\,dx.$$

71. $C[-1, 1], \quad f(x) = x, \quad g(x) = 1$

72. $C[-1, 1], \quad f(x) = x^3 - x, \quad g(x) = 2x - 1$

73. $C[0, 1], \quad f(x) = x, \quad g(x) = e^x$

74. $C[0, 1], \quad f(x) = x, \quad g(x) = e^{-x}$

75. $C[-\pi, \pi], \quad f(x) = \sin x, \quad g(x) = \cos x$

76. $C[-\pi, \pi], \quad f(x) = \sin 2x, \quad g(x) = \cos 2x$

77. $C[-\pi, \pi], \quad f(x) = x, \quad g(x) = \sin 2x$

78. $C[-\pi, \pi], \quad f(x) = x, \quad g(x) = \cos 2x$

True or False? In Exercises 79 and 80, determine whether each statement is true or false. If a statement is true, give a reason or cite an appropriate statement from the text. If a statement is false, provide an example that shows the statement is not true in all cases or cite an appropriate statement from the text.

79. (a) The dot product is the only inner product that can be defined in R^n.

(b) Of all the possible scalar multiples of a vector $\mathbf{v}$, the orthogonal projection of $\mathbf{u}$ onto $\mathbf{v}$ is the vector closest to $\mathbf{u}$.

80. (a) The norm of the vector $\mathbf{u}$ is defined as the angle between the vector $\mathbf{u}$ and the positive x-axis.

(b) The angle θ between a vector $\mathbf{v}$ and the projection of $\mathbf{u}$ onto $\mathbf{v}$ is obtuse if the scalar $a < 0$ and acute if $a > 0$, where $a\mathbf{v} = \text{proj}_{\mathbf{v}}\mathbf{u}$.

81. Let $\mathbf{u} = (4, 2)$ and $\mathbf{v} = (2, -2)$ be vectors in R^2 with the inner product $\langle \mathbf{u}, \mathbf{v}\rangle = u_1v_1 + 2u_2v_2$.

(a) Show that $\mathbf{u}$ and $\mathbf{v}$ are orthogonal.

(b) Sketch the vectors $\mathbf{u}$ and $\mathbf{v}$. Are they orthogonal in the Euclidean sense?

82. Prove that $\|\mathbf{u} + \mathbf{v}\|^2 + \|\mathbf{u} - \mathbf{v}\|^2 = 2\|\mathbf{u}\|^2 + 2\|\mathbf{v}\|^2$ for any vectors $\mathbf{u}$ and $\mathbf{v}$ in an inner product space V.

83. Prove that the function is an inner product for R^n.

$\langle \mathbf{u}, \mathbf{v}\rangle = c_1u_1v_1 + c_2u_2v_2 + \cdots + c_nu_nv_n, \quad c_i > 0$

84. Let $\mathbf{u}$ and $\mathbf{v}$ be nonzero vectors in an inner product space V. Prove that $\mathbf{u} - \text{proj}_{\mathbf{v}}\mathbf{u}$ is orthogonal to $\mathbf{v}$.

85. Prove Property 2 of Theorem 5.7: If $\mathbf{u}$, $\mathbf{v}$, and $\mathbf{w}$ are vectors in an inner product space, then $\langle \mathbf{u} + \mathbf{v}, \mathbf{w}\rangle = \langle \mathbf{u}, \mathbf{w}\rangle + \langle \mathbf{v}, \mathbf{w}\rangle$.

86. Prove Property 3 of Theorem 5.7: If $\mathbf{u}$ and $\mathbf{v}$ are vectors in an inner product space and c is a scalar, then $\langle \mathbf{u}, c\mathbf{v}\rangle = c\langle \mathbf{u}, \mathbf{v}\rangle$.

87. Guided Proof Let W be a subspace of the inner product space V. Prove that the set $W^\perp$ is a subspace of V.

$W^\perp = \{\mathbf{v} \in V : \langle \mathbf{v}, \mathbf{w}\rangle = 0 \text{ for all } \mathbf{w} \in W\}$

Getting Started: To prove that $W^\perp$ is a subspace of V, you must show that $W^\perp$ is nonempty and that the closure conditions for a subspace hold (Theorem 4.5).

(i) Find an obvious vector in $W^\perp$ to conclude that it is nonempty.

(ii) To show the closure of $W^\perp$ under addition, you need to show that $\langle \mathbf{v}_1 + \mathbf{v}_2, \mathbf{w}\rangle = 0$ for all $\mathbf{w} \in W$ and for any $\mathbf{v}_1, \mathbf{v}_2 \in W^\perp$. Use the properties of inner products and the fact that $\langle \mathbf{v}_1, \mathbf{w}\rangle$ and $\langle \mathbf{v}_2, \mathbf{w}\rangle$ are both zero to show this.

(iii) To show closure under multiplication by a scalar, proceed as in part (ii). You need to use the properties of inner products and the condition of belonging to $W^{\perp}$.

88. Use the result of Exercise 87 to find $W^{\perp}$ if W is a span of $(1, 2, 3)$ in $V = R^3$.

89. Guided Proof Let $\langle \mathbf{u}, \mathbf{v} \rangle$ be the Euclidean inner product on R^n. Use the fact that $\langle \mathbf{u}, \mathbf{v} \rangle = \mathbf{u}^T\mathbf{v}$ to prove that for any $n \times n$ matrix A

(a) $\langle A^T\mathbf{u}, \mathbf{v} \rangle = \langle \mathbf{u}, A\mathbf{v} \rangle$ and (b) $\langle A^TA\mathbf{u}, \mathbf{u} \rangle = \|A\mathbf{u}\|^2$.

Getting Started: To prove (a) and (b), you can make use of both the properties of transposes (Theorem 2.6) and the properties of dot products (Theorem 5.3).

(i) To prove part (a), you can make repeated use of the property $\langle \mathbf{u}, \mathbf{v} \rangle = \mathbf{u}^T\mathbf{v}$ and Property 4 of Theorem 2.6.

(ii) To prove part (b), you can make use of the property $\langle \mathbf{u}, \mathbf{v} \rangle = \mathbf{u}^T\mathbf{v}$, Property 4 of Theorem 2.6, and Property 4 of Theorem 5.3.

90. The two vectors from Example 10 are $\mathbf{u} = (6, 2, 4)$ and $\mathbf{v} = (1, 2, 0)$. Without using Theorem 5.9, show that among all the scalar multiples $c\mathbf{v}$ of the vector $\mathbf{v}$, the projection of $\mathbf{u}$ onto $\mathbf{v}$ is the vector closest to $\mathbf{u}$—that is, show that $d(\mathbf{u}, \text{proj}_{\mathbf{v}}\mathbf{u})$ is a minimum.

5.3 Orthonormal Bases: Gram-Schmidt Process

You saw in Section 4.7 that a vector space can have many different bases. While studying that section, you should have noticed that certain bases are more convenient than others. For example, R^3 has the convenient standard basis $B = \{(1, 0, 0), (0, 1, 0), (0, 0, 1)\}$. This set is the *standard* basis for R^3 because it has special characteristics that are particularly useful. One important characteristic is that the three vectors in the basis are *mutually orthogonal.* That is,

$$(1, 0, 0) \cdot (0, 1, 0) = 0$$
$$(1, 0, 0) \cdot (0, 0, 1) = 0$$
$$(0, 1, 0) \cdot (0, 0, 1) = 0.$$

A second important characteristic is that each vector in the basis is a *unit* vector.

This section identifies some advantages of bases consisting of mutually orthogonal unit vectors and develops a procedure for constructing such bases, known as the *Gram-Schmidt orthonormalization process.*

Definitions of Orthogonal and Orthonormal Sets

A set S of vectors in an inner product space V is called **orthogonal** if every pair of vectors in S is orthogonal. If, in addition, each vector in the set is a unit vector, then S is called **orthonormal.**

For $S = \{\mathbf{v}_1, \mathbf{v}_2, . . . , \mathbf{v}_n\}$, this definition has the form shown below.

Orthogonal	*Orthonormal*
1. $\langle \mathbf{v}_i, \mathbf{v}_j \rangle = 0, \ i \neq j$	1. $\langle \mathbf{v}_i, \mathbf{v}_j \rangle = 0, \ i \neq j$
	2. $\|\mathbf{v}_i\| = 1, \ i = 1, 2, . . . , n$

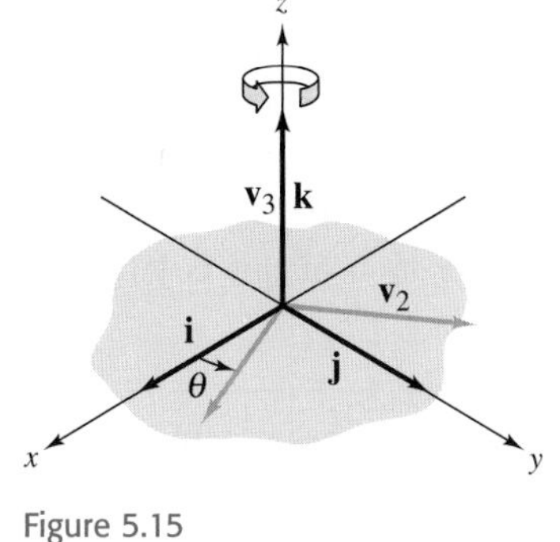

Figure 5.15

If S is a *basis,* then it is called an **orthogonal basis** or an **orthonormal basis,** respectively.

The standard basis for R^n is orthonormal, but it is not the only orthonormal basis for R^n. For instance, a nonstandard orthonormal basis for R^3 can be formed by rotating the standard basis about the z-axis to form

$$B = \{(\cos\theta, \sin\theta, 0), (-\sin\theta, \cos\theta, 0), (0, 0, 1)\},$$

as shown in Figure 5.15. Try verifying that the dot product of any two distinct vectors in B is zero, and that each vector in B is a unit vector. Example 1 describes another nonstandard orthonormal basis for R^3.

EXAMPLE 1 A Nonstandard Orthonormal Basis for R^3

Show that the set is an orthonormal basis for R^3.

$$S = \Bigg\{\underbrace{\left(\frac{1}{\sqrt{2}}, \frac{1}{\sqrt{2}}, 0\right)}_{\mathbf{v}_1}, \underbrace{\left(-\frac{\sqrt{2}}{6}, \frac{\sqrt{2}}{6}, \frac{2\sqrt{2}}{3}\right)}_{\mathbf{v}_2}, \underbrace{\left(\frac{2}{3}, -\frac{2}{3}, \frac{1}{3}\right)}_{\mathbf{v}_3}\Bigg\}$$

SOLUTION First show that the three vectors are mutually orthogonal.

$$\mathbf{v}_1 \cdot \mathbf{v}_2 = -\frac{1}{6} + \frac{1}{6} + 0 = 0$$

$$\mathbf{v}_1 \cdot \mathbf{v}_3 = \frac{2}{3\sqrt{2}} - \frac{2}{3\sqrt{2}} + 0 = 0$$

$$\mathbf{v}_2 \cdot \mathbf{v}_3 = -\frac{\sqrt{2}}{9} - \frac{\sqrt{2}}{9} + \frac{2\sqrt{2}}{9} = 0$$

Now, each vector is of length 1 because

$$\|\mathbf{v}_1\| = \sqrt{\mathbf{v}_1 \cdot \mathbf{v}_1} = \sqrt{\tfrac{1}{2} + \tfrac{1}{2} + 0} = 1$$

$$\|\mathbf{v}_2\| = \sqrt{\mathbf{v}_2 \cdot \mathbf{v}_2} = \sqrt{\tfrac{2}{36} + \tfrac{2}{36} + \tfrac{8}{9}} = 1$$

$$\|\mathbf{v}_3\| = \sqrt{\mathbf{v}_3 \cdot \mathbf{v}_3} = \sqrt{\tfrac{4}{9} + \tfrac{4}{9} + \tfrac{1}{9}} = 1.$$

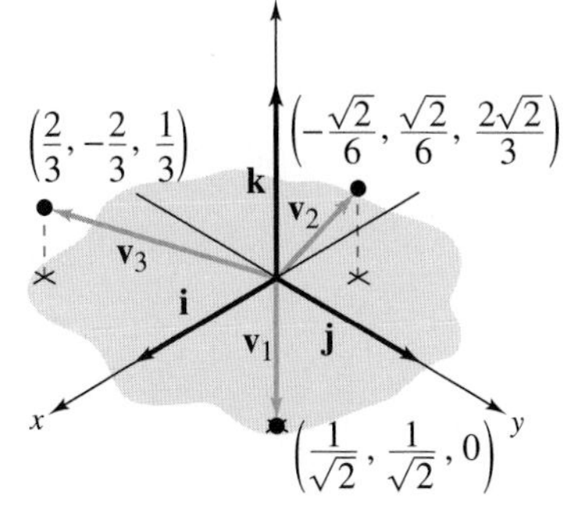

Figure 5.16

So, S is an orthonormal set. Because the three vectors do not lie in the same plane (see Figure 5.16), you know that they span R^3. By Theorem 4.12, they form a (nonstandard) orthonormal basis for R^3.

EXAMPLE 2 **An Orthonormal Basis for P_3**

In P_3, with the inner product

$$\langle p, q \rangle = a_0b_0 + a_1b_1 + a_2b_2 + a_3b_3,$$

the standard basis $B = \{1, x, x^2, x^3\}$ is orthonormal. The verification of this is left as an exercise. (See Exercise 19.)

The orthogonal set in the next example is used to construct Fourier approximations of continuous functions. (See Section 5.5.)

EXAMPLE 3 **An Orthogonal Set in $C[0, 2\pi]$ (Calculus)**

In $C[0, 2\pi]$, with the inner product

$$\langle f, g \rangle = \int_0^{2\pi} f(x)g(x)\,dx,$$

show that the set $S = \{1, \sin x, \cos x, \sin 2x, \cos 2x, \ldots, \sin nx, \cos nx\}$ is orthogonal.

SOLUTION To show that this set is orthogonal, you need to verify the inner products shown below, where m and n are positive integers.

$$\langle 1, \sin nx \rangle = \int_0^{2\pi} \sin nx\,dx = 0$$

$$\langle 1, \cos nx \rangle = \int_0^{2\pi} \cos nx\,dx = 0$$

$$\langle \sin mx, \cos nx \rangle = \int_0^{2\pi} \sin mx \cos nx\,dx = 0$$

$$\langle \sin mx, \sin nx \rangle = \int_0^{2\pi} \sin mx \sin nx\,dx = 0, \quad m \neq n$$

$$\langle \cos mx, \cos nx \rangle = \int_0^{2\pi} \cos mx \cos nx\,dx = 0, \quad m \neq n$$

One of these products is verified, and the others are left to you. If $m \neq n$, then the formula for rewriting a product of trigonometric functions as a sum can be used to obtain

$$\int_0^{2\pi} \sin mx \cos nx\,dx = \frac{1}{2}\int_0^{2\pi} [\sin(m+n)x + \sin(m-n)x]\,dx$$

$$= -\frac{1}{2}\left[\frac{\cos(m+n)x}{m+n} + \frac{\cos(m-n)x}{m-n}\right]_0^{2\pi}$$

$$= 0.$$

HISTORICAL NOTE

Jean-Baptiste Joseph Fourier (1768–1830)

is credited as a significant contributor to the field of education for scientists, mathematicians, and engineers. His research led to important results pertaining to eigenvalues, differential equations, and Fourier series (functions by trigonometric series). His work forced mathematicians of that day to accept the definition of a function, which at that time was very narrow. To read about his work, visit *college.hmco.com/pic/larsonELA6e*.

If $m = n$, then

$$\int_0^{2\pi} \sin mx \cos mx\, dx = \frac{1}{2m}\left[\sin^2 mx\right]_0^{2\pi} = 0.$$

Note that Example 3 shows only that the set S is orthogonal. This particular set is not orthonormal. An orthonormal set can be formed, however, by normalizing each vector in S. That is, because

$$\|1\|^2 = \int_0^{2\pi} dx = 2\pi$$

$$\|\sin nx\|^2 = \int_0^{2\pi} \sin^2 nx\, dx = \pi$$

$$\|\cos nx\|^2 = \int_0^{2\pi} \cos^2 nx\, dx = \pi,$$

it follows that the set

$$\left\{\frac{1}{\sqrt{2\pi}}, \frac{1}{\sqrt{\pi}}\sin x, \frac{1}{\sqrt{\pi}}\cos x, \ldots, \frac{1}{\sqrt{\pi}}\sin nx, \frac{1}{\sqrt{\pi}}\cos nx\right\}$$

is orthonormal.

Each set in Examples 1, 2, and 3 is linearly independent. Linear independence is a characteristic of any orthogonal set of nonzero vectors, as stated in the next theorem.

THEOREM 5.10 **Orthogonal Sets Are Linearly Independent**

If $S = \{\mathbf{v}_1, \mathbf{v}_2, \ldots, \mathbf{v}_n\}$ is an orthogonal set of *nonzero* vectors in an inner product space V, then S is linearly independent.

PROOF You need to show that the vector equation

$$c_1\mathbf{v}_1 + c_2\mathbf{v}_2 + \cdots + c_n\mathbf{v}_n = \mathbf{0}$$

implies $c_1 = c_2 = \cdots = c_n = 0$. To do this, form the inner product of the left side of the equation with each vector in S. That is, for each i,

$$\langle(c_1\mathbf{v}_1 + c_2\mathbf{v}_2 + \cdots + c_i\mathbf{v}_i + \cdots + c_n\mathbf{v}_n), \mathbf{v}_i\rangle = \langle\mathbf{0}, \mathbf{v}_i\rangle$$

$$c_1\langle\mathbf{v}_1, \mathbf{v}_i\rangle + c_2\langle\mathbf{v}_2, \mathbf{v}_i\rangle + \cdots + c_i\langle\mathbf{v}_i, \mathbf{v}_i\rangle + \cdots + c_n\langle\mathbf{v}_n, \mathbf{v}_i\rangle = 0.$$

Now, because S is orthogonal, $\langle\mathbf{v}_i, \mathbf{v}_j\rangle = 0$ for $j \neq i$, and now the equation reduces to

$$c_i\langle\mathbf{v}_i, \mathbf{v}_i\rangle = 0.$$

But because each vector in S is nonzero, you know that

$$\langle\mathbf{v}_i, \mathbf{v}_i\rangle = \|\mathbf{v}_i\|^2 \neq 0.$$

So every c_i must be zero and the set must be linearly independent.

As a consequence of Theorems 4.12 and 5.10, you have the result shown next.

COROLLARY TO THEOREM 5.10

If V is an inner product space of dimension n, then any orthogonal set of n nonzero vectors is a basis for V.

EXAMPLE 4 **Using Orthogonality to Test for a Basis**

Show that the following set is a basis for R^4.

$$S = \{\overset{\mathbf{v}_1}{(2, 3, 2, -2)}, \overset{\mathbf{v}_2}{(1, 0, 0, 1)}, \overset{\mathbf{v}_3}{(-1, 0, 2, 1)}, \overset{\mathbf{v}_4}{(-1, 2, -1, 1)}\}$$

SOLUTION The set S has four nonzero vectors. By the corollary to Theorem 5.10, you can show that S is a basis for R^4 by showing that it is an orthogonal set, as follows.

$$\mathbf{v}_1 \cdot \mathbf{v}_2 = 2 + 0 + 0 - 2 = 0$$
$$\mathbf{v}_1 \cdot \mathbf{v}_3 = -2 + 0 + 4 - 2 = 0$$
$$\mathbf{v}_1 \cdot \mathbf{v}_4 = -2 + 6 - 2 - 2 = 0$$
$$\mathbf{v}_2 \cdot \mathbf{v}_3 = -1 + 0 + 0 + 1 = 0$$
$$\mathbf{v}_2 \cdot \mathbf{v}_4 = -1 + 0 + 0 + 1 = 0$$
$$\mathbf{v}_3 \cdot \mathbf{v}_4 = 1 + 0 - 2 + 1 = 0$$

S is orthogonal, and by the corollary to Theorem 5.10, it is a basis for R^4.

Section 4.7 discussed a technique for finding a coordinate representation relative to a nonstandard basis. If the basis is *orthonormal,* this procedure can be streamlined.

Before presenting this procedure, you will look at an example in R^2. Figure 5.17 shows that $\mathbf{i} = (1, 0)$ and $\mathbf{j} = (0, 1)$ form an orthonormal basis for R^2. Any vector $\mathbf{w}$ in R^2 can be represented as $\mathbf{w} = \mathbf{w}_1 + \mathbf{w}_2$, where $\mathbf{w}_1 = \text{proj}_\mathbf{i}\mathbf{w}$ and $\mathbf{w}_2 = \text{proj}_\mathbf{j}\mathbf{w}$. Because $\mathbf{i}$ and $\mathbf{j}$ are unit vectors, it follows that $\mathbf{w}_1 = (\mathbf{w} \cdot \mathbf{i})\mathbf{i}$ and $\mathbf{w}_2 = (\mathbf{w} \cdot \mathbf{j})\mathbf{j}$. Consequently,

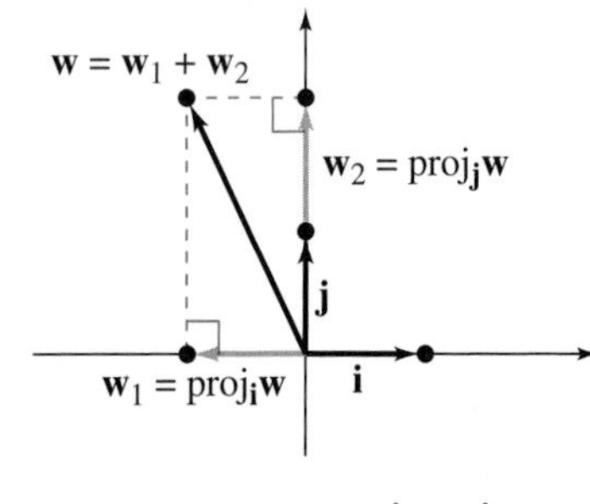

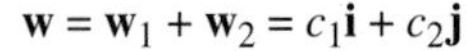

Figure 5.17

$$\begin{aligned}\mathbf{w} &= \mathbf{w}_1 + \mathbf{w}_2 \\ &= (\mathbf{w} \cdot \mathbf{i})\mathbf{i} + (\mathbf{w} \cdot \mathbf{j})\mathbf{j} \\ &= c_1\mathbf{i} + c_2\mathbf{j},\end{aligned}$$

which shows that the coefficients c_1 and c_2 are simply the dot products of $\mathbf{w}$ with the respective basis vectors. This is generalized in the next theorem.

THEOREM 5.11 Coordinates Relative to an Orthonormal Basis

If $B = \{\mathbf{v}_1, \mathbf{v}_2, . . . , \mathbf{v}_n\}$ is an orthonormal basis for an inner product space V, then the coordinate representation of a vector $\mathbf{w}$ with respect to B is

$$\mathbf{w} = \langle\mathbf{w}, \mathbf{v}_1\rangle\mathbf{v}_1 + \langle\mathbf{w}, \mathbf{v}_2\rangle\mathbf{v}_2 + \cdot \cdot \cdot + \langle\mathbf{w}, \mathbf{v}_n\rangle\mathbf{v}_n.$$

PROOF Because B is a basis for V, there must exist unique scalars $c_1, c_2, \ldots, c_n$ such that

$$\mathbf{w} = c_1\mathbf{v}_1 + c_2\mathbf{v}_2 + \cdots + c_n\mathbf{v}_n.$$

Taking the inner product (with $\mathbf{v}_i$) of both sides of this equation, you have

$$\begin{aligned}\langle \mathbf{w}, \mathbf{v}_i\rangle &= \langle (c_1\mathbf{v}_1 + c_2\mathbf{v}_2 + \cdots + c_n\mathbf{v}_n), \mathbf{v}_i\rangle \\ &= c_1\langle \mathbf{v}_1, \mathbf{v}_i\rangle + c_2\langle \mathbf{v}_2, \mathbf{v}_i\rangle + \cdots + c_n\langle \mathbf{v}_n, \mathbf{v}_i\rangle\end{aligned}$$

and by the orthogonality of B this equation reduces to

$$\langle \mathbf{w}, \mathbf{v}_i\rangle = c_i\langle \mathbf{v}_i, \mathbf{v}_i\rangle.$$

Because B is orthonormal, you have $\langle \mathbf{v}_i, \mathbf{v}_i\rangle = \|\mathbf{v}_i\|^2 = 1$, and it follows that $\langle \mathbf{w}, \mathbf{v}_i\rangle = c_i$.

In Theorem 5.11 the coordinates of $\mathbf{w}$ relative to the *orthonormal* basis B are called the **Fourier coefficients** of $\mathbf{w}$ relative to B, after the French mathematician Jean-Baptiste Joseph Fourier (1768–1830). The corresponding coordinate matrix of $\mathbf{w}$ relative to B is

$$\begin{aligned}[\mathbf{w}]_B &= [c_1\ c_2\ \cdots\ c_n]^T \\ &= [\langle \mathbf{w}, \mathbf{v}_1\rangle\ \langle \mathbf{w}, \mathbf{v}_2\rangle\ \cdots\ \langle \mathbf{w}, \mathbf{v}_n\rangle]^T.\end{aligned}$$

EXAMPLE 5 Representing Vectors Relative to an Orthonormal Basis

Find the coordinates of $\mathbf{w} = (5, -5, 2)$ relative to the orthonormal basis for R^3 shown below.

$$B = \{\underbrace{\left(\tfrac{3}{5}, \tfrac{4}{5}, 0\right)}_{\mathbf{v}_1}, \underbrace{\left(-\tfrac{4}{5}, \tfrac{3}{5}, 0\right)}_{\mathbf{v}_2}, \underbrace{(0, 0, 1)}_{\mathbf{v}_3}\}$$

SOLUTION Because B is orthonormal, you can use Theorem 5.11 to find the required coordinates.

$$\begin{aligned}\mathbf{w} \cdot \mathbf{v}_1 &= (5, -5, 2) \cdot \left(\tfrac{3}{5}, \tfrac{4}{5}, 0\right) = -1 \\ \mathbf{w} \cdot \mathbf{v}_2 &= (5, -5, 2) \cdot \left(-\tfrac{4}{5}, \tfrac{3}{5}, 0\right) = -7 \\ \mathbf{w} \cdot \mathbf{v}_3 &= (5, -5, 2) \cdot (0, 0, 1) = 2\end{aligned}$$

So, the coordinate matrix relative to B is

$$[\mathbf{w}]_B = \begin{bmatrix} -1 \\ -7 \\ 2 \end{bmatrix}.$$

Gram-Schmidt Orthonormalization Process

Having seen one of the advantages of orthonormal bases (the straightforwardness of coordinate representation), you will now look at a procedure for finding such a basis. This procedure is called the **Gram-Schmidt orthonormalization process,** after the Danish mathematician Jorgen Pederson Gram (1850–1916) and the German mathematician Erhardt Schmidt (1876–1959). It has three steps.

1. Begin with a basis for the inner product space. It need not be orthogonal nor consist of unit vectors.
2. Convert the given basis to an orthogonal basis.
3. Normalize each vector in the orthogonal basis to form an orthonormal basis.

REMARK: The Gram-Schmidt orthonormalization process leads to a matrix factorization similar to the *LU*-factorization you studied in Chapter 2. You are asked to investigate this ***QR*-factorization** in Project 1 at the end of this chapter.

THEOREM 5.12 Gram-Schmidt Orthonormalization Process

1. Let $B = \{\mathbf{v}_1, \mathbf{v}_2, \ldots, \mathbf{v}_n\}$ be a basis for an inner product space V.
2. Let $B' = \{\mathbf{w}_1, \mathbf{w}_2, \ldots, \mathbf{w}_n\}$, where $\mathbf{w}_i$ is given by

$$\mathbf{w}_1 = \mathbf{v}_1$$

$$\mathbf{w}_2 = \mathbf{v}_2 - \frac{\langle \mathbf{v}_2, \mathbf{w}_1 \rangle}{\langle \mathbf{w}_1, \mathbf{w}_1 \rangle}\mathbf{w}_1$$

$$\mathbf{w}_3 = \mathbf{v}_3 - \frac{\langle \mathbf{v}_3, \mathbf{w}_1 \rangle}{\langle \mathbf{w}_1, \mathbf{w}_1 \rangle}\mathbf{w}_1 - \frac{\langle \mathbf{v}_3, \mathbf{w}_2 \rangle}{\langle \mathbf{w}_2, \mathbf{w}_2 \rangle}\mathbf{w}_2$$

$$\vdots$$

$$\mathbf{w}_n = \mathbf{v}_n - \frac{\langle \mathbf{v}_n, \mathbf{w}_1 \rangle}{\langle \mathbf{w}_1, \mathbf{w}_1 \rangle}\mathbf{w}_1 - \frac{\langle \mathbf{v}_n, \mathbf{w}_2 \rangle}{\langle \mathbf{w}_2, \mathbf{w}_2 \rangle}\mathbf{w}_2 - \cdots - \frac{\langle \mathbf{v}_n, \mathbf{w}_{n-1} \rangle}{\langle \mathbf{w}_{n-1}, \mathbf{w}_{n-1} \rangle}\mathbf{w}_{n-1}.$$

Then B' is an *orthogonal* basis for V.

3. Let $\mathbf{u}_i = \dfrac{\mathbf{w}_i}{\|\mathbf{w}_i\|}$. Then the set $B'' = \{\mathbf{u}_1, \mathbf{u}_2, \ldots, \mathbf{u}_n\}$ is an *orthonormal* basis for V. Moreover, span $\{\mathbf{v}_1, \mathbf{v}_2, \ldots, \mathbf{v}_k\} = \text{span}\{\mathbf{u}_1, \mathbf{u}_2, \ldots, \mathbf{u}_k\}$ for $k = 1, 2, \ldots, n$.

Rather than give a general proof of this theorem, it seems more instructive to discuss a special case for which you can use a geometric model. Let $\{\mathbf{v}_1, \mathbf{v}_2\}$ be a basis for R^2, as shown in Figure 5.18. To determine an orthogonal basis for R^2, first choose one of the original vectors, say $\mathbf{v}_1$. Now you want to find a second vector orthogonal to $\mathbf{v}_1$. Figure 5.19 shows that $\mathbf{v}_2 - \text{proj}_{\mathbf{v}_1}\mathbf{v}_2$ has this property.

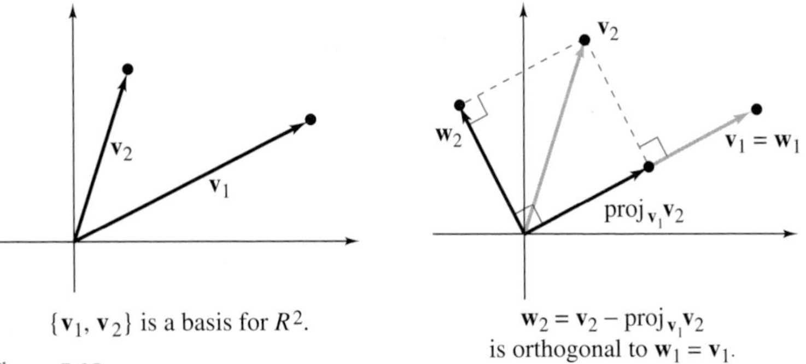

$\{\mathbf{v}_1, \mathbf{v}_2\}$ is a basis for R^2.

Figure 5.18

$\mathbf{w}_2 = \mathbf{v}_2 - \text{proj}_{\mathbf{v}_1}\mathbf{v}_2$
is orthogonal to $\mathbf{w}_1 = \mathbf{v}_1$.

Figure 5.19

By letting

$$\mathbf{w}_1 = \mathbf{v}_1 \qquad \text{and} \qquad \mathbf{w}_2 = \mathbf{v}_2 - \text{proj}_{\mathbf{v}_1}\mathbf{v}_2 = \mathbf{v}_2 - \frac{\mathbf{v}_2 \cdot \mathbf{w}_1}{\mathbf{w}_1 \cdot \mathbf{w}_1}\mathbf{w}_1,$$

you can conclude that the set $\{\mathbf{w}_1, \mathbf{w}_2\}$ is orthogonal. By the corollary to Theorem 5.10, it is a basis for R^2. Finally, by normalizing $\mathbf{w}_1$ and $\mathbf{w}_2$, you obtain the orthonormal basis for R^2 shown below.

$$\{\mathbf{u}_1, \mathbf{u}_2\} = \left\{\frac{\mathbf{w}_1}{\|\mathbf{w}_1\|}, \frac{\mathbf{w}_2}{\|\mathbf{w}_2\|}\right\}$$

EXAMPLE 6 Applying the Gram-Schmidt Orthonormalization Process

Apply the Gram-Schmidt orthonormalization process to the basis for R^2 shown below.

$$B = \{\overset{\mathbf{v}_1}{(1, 1)}, \overset{\mathbf{v}_2}{(0, 1)}\}$$

SOLUTION The Gram-Schmidt orthonormalization process produces

$$\mathbf{w}_1 = \mathbf{v}_1 = (1, 1)$$

$$\mathbf{w}_2 = \mathbf{v}_2 - \frac{\mathbf{v}_2 \cdot \mathbf{w}_1}{\mathbf{w}_1 \cdot \mathbf{w}_1}\mathbf{w}_1$$

$$= (0, 1) - \tfrac{1}{2}(1, 1) = \left(-\tfrac{1}{2}, \tfrac{1}{2}\right).$$

The set $B' = \{\mathbf{w}_1, \mathbf{w}_2\}$ is an orthogonal basis for R^2. By normalizing each vector in B', you obtain

$$\mathbf{u}_1 = \frac{\mathbf{w}_1}{\|\mathbf{w}_1\|} = \frac{1}{\sqrt{2}}(1, 1) = \frac{\sqrt{2}}{2}(1, 1) = \left(\frac{\sqrt{2}}{2}, \frac{\sqrt{2}}{2}\right)$$

$$\mathbf{u}_2 = \frac{\mathbf{w}_2}{\|\mathbf{w}_2\|} = \frac{1}{1/\sqrt{2}}\left(-\frac{1}{2}, \frac{1}{2}\right) = \sqrt{2}\left(-\frac{1}{2}, \frac{1}{2}\right) = \left(-\frac{\sqrt{2}}{2}, \frac{\sqrt{2}}{2}\right).$$

So, $B'' = \{\mathbf{u}_1, \mathbf{u}_2\}$ is an orthonormal basis for R^2. See Figure 5.20.

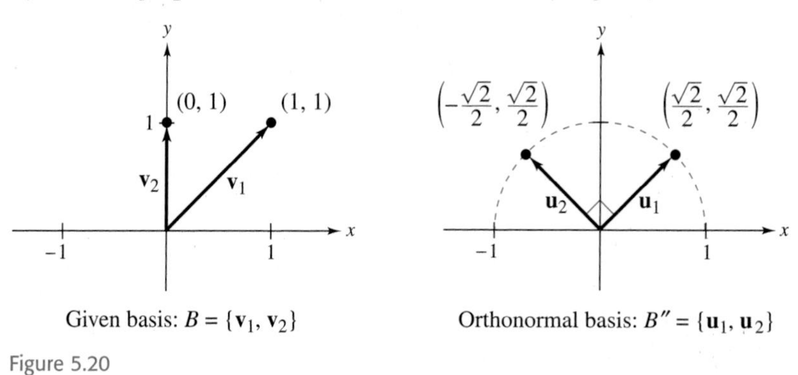

Figure 5.20

REMARK: An orthonormal set derived by the Gram-Schmidt orthonormalization process depends on the order of the vectors in the basis. For instance, try reworking Example 6 with the original basis ordered as $\{\mathbf{v}_2, \mathbf{v}_1\}$ rather than $\{\mathbf{v}_1, \mathbf{v}_2\}$.

EXAMPLE 7 Applying the Gram-Schmidt Orthonormalization Process

Apply the Gram-Schmidt orthonormalization process to the basis for R^3 shown below.

$$B = \{\overset{\mathbf{v}_1}{(1, 1, 0)}, \overset{\mathbf{v}_2}{(1, 2, 0)}, \overset{\mathbf{v}_3}{(0, 1, 2)}\}$$

SOLUTION Applying the Gram-Schmidt orthonormalization process produces

$$\mathbf{w}_1 = \mathbf{v}_1 = (1, 1, 0)$$

$$\mathbf{w}_2 = \mathbf{v}_2 - \frac{\mathbf{v}_2 \cdot \mathbf{w}_1}{\mathbf{w}_1 \cdot \mathbf{w}_1}\mathbf{w}_1 = (1, 2, 0) - \frac{3}{2}(1, 1, 0) = \left(-\frac{1}{2}, \frac{1}{2}, 0\right)$$

$$\mathbf{w}_3 = \mathbf{v}_3 - \frac{\mathbf{v}_3 \cdot \mathbf{w}_1}{\mathbf{w}_1 \cdot \mathbf{w}_1}\mathbf{w}_1 - \frac{\mathbf{v}_3 \cdot \mathbf{w}_2}{\mathbf{w}_2 \cdot \mathbf{w}_2}\mathbf{w}_2$$

$$= (0, 1, 2) - \frac{1}{2}(1, 1, 0) - \frac{1/2}{1/2}\left(-\frac{1}{2}, \frac{1}{2}, 0\right) = (0, 0, 2).$$

The set $B' = \{\mathbf{w}_1, \mathbf{w}_2, \mathbf{w}_3\}$ is an orthogonal basis for R^3. Normalizing each vector in B' produces

$$\mathbf{u}_1 = \frac{\mathbf{w}_1}{\|\mathbf{w}_1\|} = \frac{1}{\sqrt{2}}(1, 1, 0) = \left(\frac{\sqrt{2}}{2}, \frac{\sqrt{2}}{2}, 0\right)$$

$$\mathbf{u}_2 = \frac{\mathbf{w}_2}{\|\mathbf{w}_2\|} = \frac{1}{1/\sqrt{2}}\left(-\frac{1}{2}, \frac{1}{2}, 0\right) = \left(-\frac{\sqrt{2}}{2}, \frac{\sqrt{2}}{2}, 0\right)$$

$$\mathbf{u}_3 = \frac{\mathbf{w}_3}{\|\mathbf{w}_3\|} = \frac{1}{2}(0, 0, 2) = (0, 0, 1).$$

So, $B'' = \{\mathbf{u}_1, \mathbf{u}_2, \mathbf{u}_3\}$ is an orthonormal basis for R^3.

Examples 6 and 7 applied the Gram-Schmidt orthonormalization process to bases for R^2 and R^3. The process works equally well for a subspace of an inner product space. This procedure is demonstrated in the next example.

EXAMPLE 8 Applying the Gram-Schmidt Orthonormalization Process

The vectors $\mathbf{v}_1 = (0, 1, 0)$ and $\mathbf{v}_2 = (1, 1, 1)$ span a plane in R^3. Find an orthonormal basis for this subspace.

SOLUTION Applying the Gram-Schmidt orthonormalization process produces

$$\mathbf{w}_1 = \mathbf{v}_1 = (0, 1, 0)$$

$$\mathbf{w}_2 = \mathbf{v}_2 - \frac{\mathbf{v}_2 \cdot \mathbf{w}_1}{\mathbf{w}_1 \cdot \mathbf{w}_1}\mathbf{w}_1$$

$$= (1, 1, 1) - \frac{1}{1}(0, 1, 0) = (1, 0, 1).$$

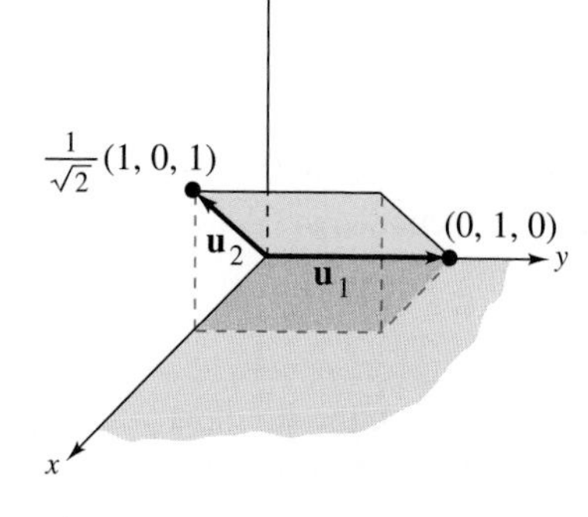

Figure 5.21

Normalizing $\mathbf{w}_1$ and $\mathbf{w}_2$ produces the orthonormal set

$$\mathbf{u}_1 = \frac{\mathbf{w}_1}{\|\mathbf{w}_1\|} = (0, 1, 0)$$

$$\mathbf{u}_2 = \frac{\mathbf{w}_2}{\|\mathbf{w}_2\|}$$

$$= \frac{1}{\sqrt{2}}(1, 0, 1)$$

$$= \left(\frac{\sqrt{2}}{2}, 0, \frac{\sqrt{2}}{2}\right).$$

See Figure 5.21.

EXAMPLE 9 Applying the Gram-Schmidt Orthonormalization Process (Calculus)

Apply the Gram-Schmidt orthonormalization process to the basis $B = \{1, x, x^2\}$ in P_2, using the inner product

$$\langle p, q \rangle = \int_{-1}^{1} p(x)q(x)\, dx.$$

SOLUTION Let $B = \{1, x, x^2\} = \{\mathbf{v}_1, \mathbf{v}_2, \mathbf{v}_3\}$. Then you have

$$\mathbf{w}_1 = \mathbf{v}_1 = 1$$

$$\mathbf{w}_2 = \mathbf{v}_2 - \frac{\langle \mathbf{v}_2, \mathbf{w}_1 \rangle}{\langle \mathbf{w}_1, \mathbf{w}_1 \rangle}\mathbf{w}_1 = x - \frac{0}{2}(1) = x$$

$$\begin{aligned}\mathbf{w}_3 &= \mathbf{v}_3 - \frac{\langle \mathbf{v}_3, \mathbf{w}_1 \rangle}{\langle \mathbf{w}_1, \mathbf{w}_1 \rangle}\mathbf{w}_1 - \frac{\langle \mathbf{v}_3, \mathbf{w}_2 \rangle}{\langle \mathbf{w}_2, \mathbf{w}_2 \rangle}\mathbf{w}_2 \\ &= x^2 - \frac{2/3}{2}(1) - \frac{0}{2/3}(x) \\ &= x^2 - \frac{1}{3}.\end{aligned}$$

(In Exercises 43–46 you are asked to verify these calculations.) Now, by normalizing $B' = \{\mathbf{w}_1, \mathbf{w}_2, \mathbf{w}_3\}$, you have

$$\mathbf{u}_1 = \frac{\mathbf{w}_1}{\|\mathbf{w}_1\|} = \frac{1}{\sqrt{2}}(1) = \frac{1}{\sqrt{2}}$$

$$\mathbf{u}_2 = \frac{\mathbf{w}_2}{\|\mathbf{w}_2\|} = \frac{1}{\sqrt{2/3}}(x) = \frac{\sqrt{3}}{\sqrt{2}}x$$

$$\mathbf{u}_3 = \frac{\mathbf{w}_3}{\|\mathbf{w}_3\|} = \frac{1}{\sqrt{8/45}}\left(x^2 - \frac{1}{3}\right) = \frac{\sqrt{5}}{2\sqrt{2}}(3x^2 - 1).$$

REMARK: The polynomials $\mathbf{u}_1$, $\mathbf{u}_2$, and $\mathbf{u}_3$ in Example 9 are called the first three **normalized Legendre polynomials,** after the French mathematician Adrien-Marie Legendre (1752–1833).

The computations in the Gram-Schmidt orthonormalization process are sometimes simpler when each vector $\mathbf{w}_i$ is normalized *before* it is used to determine the next vector. This **alternative form of the Gram-Schmidt orthonormalization process** has the steps shown below.

$$\mathbf{u}_1 = \frac{\mathbf{w}_1}{\|\mathbf{w}_1\|} = \frac{\mathbf{v}_1}{\|\mathbf{v}_1\|}$$

$$\mathbf{u}_2 = \frac{\mathbf{w}_2}{\|\mathbf{w}_2\|}, \text{ where } \mathbf{w}_2 = \mathbf{v}_2 - \langle \mathbf{v}_2, \mathbf{u}_1 \rangle \mathbf{u}_1$$

$$\mathbf{u}_3 = \frac{\mathbf{w}_3}{\|\mathbf{w}_3\|}, \text{ where } \mathbf{w}_3 = \mathbf{v}_3 - \langle \mathbf{v}_3, \mathbf{u}_1 \rangle \mathbf{u}_1 - \langle \mathbf{v}_3, \mathbf{u}_2 \rangle \mathbf{u}_2$$

$$\vdots$$

$$\mathbf{u}_n = \frac{\mathbf{w}_n}{\|\mathbf{w}_n\|}, \text{ where } \mathbf{w}_n = \mathbf{v}_n - \langle \mathbf{v}_n, \mathbf{u}_1 \rangle \mathbf{u}_1 - \cdots - \langle \mathbf{v}_n, \mathbf{u}_{n-1} \rangle \mathbf{u}_{n-1}$$

EXAMPLE 10 Alternative Form of Gram-Schmidt Orthonormalization Process

Find an orthonormal basis for the solution space of the homogeneous system of linear equations.

$$\begin{aligned} x_1 + x_2 \quad\quad\quad + 7x_4 &= 0 \\ 2x_1 + x_2 + 2x_3 + 6x_4 &= 0 \end{aligned}$$

SOLUTION The augmented matrix for this system reduces as follows.

$$\begin{bmatrix} 1 & 1 & 0 & 7 & 0 \\ 2 & 1 & 2 & 6 & 0 \end{bmatrix} \rightarrow \begin{bmatrix} 1 & 0 & 2 & -1 & 0 \\ 0 & 1 & -2 & 8 & 0 \end{bmatrix}$$

If you let $x_3 = s$ and $x_4 = t$, each solution of the system has the form

$$\begin{bmatrix} x_1 \\ x_2 \\ x_3 \\ x_4 \end{bmatrix} = \begin{bmatrix} -2s + t \\ 2s - 8t \\ s \\ t \end{bmatrix} = s\begin{bmatrix} -2 \\ 2 \\ 1 \\ 0 \end{bmatrix} + t\begin{bmatrix} 1 \\ -8 \\ 0 \\ 1 \end{bmatrix}.$$

So, one basis for the solution space is

$$B = \{\mathbf{v}_1, \mathbf{v}_2\} = \{(-2, 2, 1, 0), (1, -8, 0, 1)\}.$$

To find an orthonormal basis $B' = \{\mathbf{u}_1, \mathbf{u}_2\}$, use the alternative form of the Gram-Schmidt orthonormalization process, as follows.

$$\begin{aligned} \mathbf{u}_1 &= \frac{\mathbf{v}_1}{\|\mathbf{v}_1\|} \\ &= \frac{1}{3}(-2, 2, 1, 0) \\ &= \left(-\frac{2}{3}, \frac{2}{3}, \frac{1}{3}, 0\right) \\ \mathbf{w}_2 &= \mathbf{v}_2 - \langle \mathbf{v}_2, \mathbf{u}_1 \rangle \mathbf{u}_1 \\ &= (1, -8, 0, 1) - \left[(1, -8, 0, 1) \cdot \left(-\frac{2}{3}, \frac{2}{3}, \frac{1}{3}, 0\right)\right]\left(-\frac{2}{3}, \frac{2}{3}, \frac{1}{3}, 0\right) \\ &= (-3, -4, 2, 1) \\ \mathbf{u}_2 &= \frac{\mathbf{w}_2}{\|\mathbf{w}_2\|} \\ &= \frac{1}{\sqrt{30}}(-3, -4, 2, 1) \\ &= \left(-\frac{3}{\sqrt{30}}, -\frac{4}{\sqrt{30}}, \frac{2}{\sqrt{30}}, \frac{1}{\sqrt{30}}\right) \end{aligned}$$

SECTION 5.3 Exercises

In Exercises 1–14, determine whether the set of vectors in R^n is orthogonal, orthonormal, or neither.

1. $\{(2, -4), (2, 1)\}$
2. $\{(3, -2), (-4, -6)\}$
3. $\{(-4, 6), (5, 0)\}$
4. $\{(11, 4), (8, -3)\}$
5. $\{(\frac{3}{5}, \frac{4}{5}), (-\frac{4}{5}, \frac{3}{5})\}$
6. $\{(1, 2), (-\frac{2}{5}, \frac{1}{5})\}$
7. $\{(4, -1, 1), (-1, 0, 4), (-4, -17, -1)\}$
8. $\{(2, -4, 2), (0, 2, 4), (-10, -4, 2)\}$
9. $\left\{\left(\frac{\sqrt{2}}{2}, 0, \frac{\sqrt{2}}{2}\right), \left(-\frac{\sqrt{6}}{6}, \frac{\sqrt{6}}{3}, \frac{\sqrt{6}}{6}\right), \left(\frac{\sqrt{3}}{3}, \frac{\sqrt{3}}{3}, -\frac{\sqrt{3}}{3}\right)\right\}$
10. $\left\{\left(\frac{\sqrt{2}}{3}, 0, -\frac{\sqrt{2}}{6}\right), \left(0, \frac{2\sqrt{5}}{5}, -\frac{\sqrt{5}}{5}\right), \left(\frac{\sqrt{5}}{5}, 0, \frac{1}{2}\right)\right\}$
11. $\{(2, -5, -3), (4, -2, 6)\}$
12. $\{(-6, 3, 2, 1), (2, 0, 6, 0)\}$
13. $\left\{\left(\frac{\sqrt{2}}{2}, 0, 0, \frac{\sqrt{2}}{2}\right), \left(0, \frac{\sqrt{2}}{2}, \frac{\sqrt{2}}{2}, 0\right), \left(-\frac{1}{2}, \frac{1}{2}, -\frac{1}{2}, \frac{1}{2}\right)\right\}$
14. $\left\{\left(\frac{\sqrt{10}}{10}, 0, 0, \frac{3\sqrt{10}}{10}\right), (0, 0, 1, 0), (0, 1, 0, 0), \left(-\frac{3\sqrt{10}}{10}, 0, 0, \frac{\sqrt{10}}{10}\right)\right\}$

In Exercises 15–18, determine if the set of vectors in R^n is orthogonal and orthonormal. If the set is only orthogonal, normalize the set to produce an orthonormal set.

15. $\{(-1, 4), (8, 2)\}$
16. $\{(2, -5), (10, 4)\}$
17. $\{(\sqrt{3}, \sqrt{3}, \sqrt{3}), (-\sqrt{2}, 0, \sqrt{2})\}$
18. $\{(-\frac{2}{15}, \frac{1}{15}, \frac{2}{15}), (\frac{1}{15}, \frac{2}{15}, 0)\}$
19. Complete Example 2 by verifying that $\{1, x, x^2, x^3\}$ is an orthonormal basis for P_3 with the inner product $\langle p, q\rangle = a_0b_0 + a_1b_1 + a_2b_2 + a_3b_3$.
20. Verify that $\{(\sin\theta, \cos\theta), (\cos\theta, -\sin\theta)\}$ is an orthonormal basis for R^2.

In Exercises 21–26, find the coordinates of **x** relative to the orthonormal basis B in R^n.

21. $B = \left\{\left(-\frac{2\sqrt{13}}{13}, \frac{3\sqrt{13}}{13}\right), \left(\frac{3\sqrt{13}}{13}, \frac{2\sqrt{13}}{13}\right)\right\}$, $\mathbf{x} = (1, 2)$
22. $B = \left\{\left(\frac{\sqrt{5}}{5}, \frac{2\sqrt{5}}{5}\right), \left(-\frac{2\sqrt{5}}{5}, \frac{\sqrt{5}}{5}\right)\right\}$, $\mathbf{x} = (-3, 4)$
23. $B = \left\{\left(\frac{\sqrt{10}}{10}, 0, \frac{3\sqrt{10}}{10}\right), (0, 1, 0), \left(-\frac{3\sqrt{10}}{10}, 0, \frac{\sqrt{10}}{10}\right)\right\}$, $\mathbf{x} = (2, -2, 1)$
24. $B = \{(1, 0, 0), (0, 1, 0), (0, 0, 1)\}$, $\mathbf{x} = (3, -5, 11)$
25. $B = \{(\frac{3}{5}, \frac{4}{5}, 0), (-\frac{4}{5}, \frac{3}{5}, 0), (0, 0, 1)\}$, $\mathbf{x} = (5, 10, 15)$
26. $B = \{(\frac{5}{13}, 0, \frac{12}{13}, 0), (0, 1, 0, 0), (-\frac{12}{13}, 0, \frac{5}{13}, 0), (0, 0, 0, 1)\}$, $\mathbf{x} = (2, -1, 4, 3)$

In Exercises 27–36, use the Gram-Schmidt orthonormalization process to transform the given basis for R^n into an orthonormal basis. Use the Euclidean inner product for R^n and use the vectors in the order in which they are shown.

27. $B = \{(3, 4), (1, 0)\}$
28. $B = \{(1, 2), (-1, 0)\}$
29. $B = \{(0, 1), (2, 5)\}$
30. $B = \{(4, -3), (3, 2)\}$
31. $B = \{(1, -2, 2), (2, 2, 1), (2, -1, -2)\}$
32. $B = \{(1, 0, 0), (1, 1, 1), (1, 1, -1)\}$
33. $B = \{(4, -3, 0), (1, 2, 0), (0, 0, 4)\}$
34. $B = \{(0, 1, 2), (2, 0, 0), (1, 1, 1)\}$
35. $B = \{(0, 1, 1), (1, 1, 0), (1, 0, 1)\}$
36. $B = \{(3, 4, 0, 0), (-1, 1, 0, 0), (2, 1, 0, -1), (0, 1, 1, 0)\}$

In Exercises 37–42, use the Gram-Schmidt orthonormalization process to transform the given basis for a subspace of R^n into an orthonormal basis for the subspace. Use the Euclidean inner product for R^n and use the vectors in the order in which they are shown.

37. $B = \{(-8, 3, 5)\}$
38. $B = \{(4, -7, 6)\}$
39. $B = \{(3, 4, 0), (1, 0, 0)\}$
40. $B = \{(1, 2, 0), (2, 0, -2)\}$
41. $B = \{(1, 2, -1, 0), (2, 2, 0, 1), (1, 1, -1, 0)\}$
42. $B = \{(7, 24, 0, 0), (0, 0, 1, 1), (0, 0, 1, -2)\}$

Calculus In Exercises 43–46, let $B = \{1, x, x^2\}$ be a basis for P_2 with the inner product

$$\langle p, q\rangle = \int_{-1}^{1} p(x)q(x)\,dx.$$

Complete Example 9 by verifying the indicated inner products.

43. $\langle x, 1\rangle = 0$
44. $\langle 1, 1\rangle = 2$
45. $\langle x^2, 1\rangle = \frac{2}{3}$
46. $\langle x^2, x\rangle = 0$

True or False? In Exercises 47 and 48, determine whether each statement is true or false. If a statement is true, give a reason or cite an appropriate statement from the text. If a statement is false, provide an example that shows the statement is not true in all cases or cite an appropriate statement from the text.

47. (a) A set S of vectors in an inner product space V is orthogonal if every pair of vectors in S is orthogonal.

(b) To show that a set of nonzero vectors is a basis for R^n, it is sufficient to show that the set is an orthogonal set.

(c) An orthonormal basis derived by the Gram-Schmidt orthonormalization process does not depend on the order of the vectors in the basis.

48. (a) A set S of vectors in an inner product space V is orthonormal if every vector is a unit vector and each pair of vectors is orthogonal.

(b) If a set of nonzero vectors S in an inner product space V is orthogonal, then S is linearly independent.

(c) The Gram-Schmidt orthonormalization process is a procedure for finding an orthonormal basis for an inner product space V.

In Exercises 49–54, find an orthonormal basis for the solution space of the homogeneous system of linear equations.

49. $$\begin{aligned} 2x_1 + x_2 - 6x_3 + 2x_4 &= 0 \\ x_1 + 2x_2 - 3x_3 + 4x_4 &= 0 \\ x_1 + x_2 - 3x_3 + 2x_4 &= 0 \end{aligned}$$

50. $$\begin{aligned} x_1 + x_2 - 3x_3 - 2x_4 &= 0 \\ 2x_1 - x_2 \quad\quad - x_4 &= 0 \\ 3x_1 + x_2 - 5x_3 - 4x_4 &= 0 \end{aligned}$$

51. $$\begin{aligned} x_1 + x_2 - x_3 - x_4 &= 0 \\ 2x_1 + x_2 - 2x_3 - 2x_4 &= 0 \end{aligned}$$

52. $$\begin{aligned} x_1 - x_2 + x_3 + x_4 &= 0 \\ x_1 - 2x_2 + x_3 + x_4 &= 0 \end{aligned}$$

53. $x_1 + 3x_2 - 3x_3 = 0$

54. $x_1 - 2x_2 + x_3 = 0$

In Exercises 55–60, let $p(x) = a_0 + a_1x + a_2x^2$ and $q(x) = b_0 + b_1x + b_2x^2$ be vectors in P_2 with

$\langle p, q\rangle = a_0b_0 + a_1b_1 + a_2b_2$.

Determine whether the given second-degree polynomials form an orthonormal set, and if not, use the Gram-Schmidt orthonormalization process to form an orthonormal set.

55. $\left\{\dfrac{x^2+1}{\sqrt{2}}, \dfrac{x^2+x-1}{\sqrt{3}}\right\}$

56. $\{\sqrt{2}(x^2-1), \sqrt{2}(x^2+x+2)\}$

57. $\{x^2, x^2+2x, x^2+2x+1\}$

58. $\{1, x, x^2\}$

59. $\{x^2-1, x-1\}$

60. $\left\{\dfrac{3x^2+4x}{5}, \dfrac{-4x^2+3x}{5}, 1\right\}$

61. Use the inner product $\langle \mathbf{u}, \mathbf{v}\rangle = 2u_1v_1 + u_2v_2$ in R^2 and the Gram-Schmidt orthonormalization process to transform $\{(2, -1), (-2, 10)\}$ into an orthonormal basis.

62. **Writing** Explain why the result of Exercise 61 is not an orthonormal basis when the Euclidean inner product on R^2 is used.

63. Let $\{\mathbf{u}_1, \mathbf{u}_2, \ldots, \mathbf{u}_n\}$ be an orthonormal basis for R^n. Prove that $\|\mathbf{v}\|^2 = |\mathbf{v}\cdot\mathbf{u}_1|^2 + |\mathbf{v}\cdot\mathbf{u}_2|^2 + \cdots + |\mathbf{v}\cdot\mathbf{u}_n|^2$ for any vector $\mathbf{v}$ in R^n. This equation is called **Parseval's equality.**

64. **Guided Proof** Prove that if $\mathbf{w}$ is orthogonal to each vector in $S = \{\mathbf{v}_1, \mathbf{v}_2, \ldots, \mathbf{v}_n\}$, then $\mathbf{w}$ is orthogonal to every linear combination of vectors in S.

Getting Started: To prove that $\mathbf{w}$ is orthogonal to every linear combination of vectors in S, you need to show that their dot product is 0.

(i) Write $\mathbf{v}$ as a linear combination of vectors, with arbitrary scalars $c_1, \ldots, c_n$, in S.

(ii) Form the inner product of $\mathbf{w}$ and $\mathbf{v}$.

(iii) Use the properties of inner products to rewrite the inner product $\langle \mathbf{w}, \mathbf{v}\rangle$ as a linear combination of the inner products $\langle \mathbf{w}, \mathbf{v}_i\rangle$, $i = 1, \ldots, n$.

(iv) Use the fact that $\mathbf{w}$ is orthogonal to each vector in S to lead to the conclusion that $\mathbf{w}$ is orthogonal to $\mathbf{v}$.

65. Let P be an $n \times n$ matrix. Prove that the following conditions are equivalent.

(a) $P^{-1} = P^T$. (Such a matrix is called *orthogonal.*)

(b) The row vectors of P form an orthonormal basis for R^n.

(c) The column vectors of P form an orthonormal basis for R^n.

66. Use each matrix to illustrate the result of Exercise 65.

(a) $P = \begin{bmatrix} -1 & 0 & 0 \\ 0 & 0 & 1 \\ 0 & -1 & 0 \end{bmatrix}$

(b) $P = \begin{bmatrix} 1/\sqrt{2} & 1/\sqrt{2} & 0 \\ 1/\sqrt{2} & -1/\sqrt{2} & 0 \\ 0 & 0 & 1 \end{bmatrix}$

67. Find an orthonormal basis for R^4 that includes the vectors

$$\mathbf{v}_1 = \left(\frac{1}{\sqrt{2}}, 0, \frac{1}{\sqrt{2}}, 0\right) \quad \text{and} \quad \mathbf{v}_2 = \left(0, -\frac{1}{\sqrt{2}}, 0, \frac{1}{\sqrt{2}}\right).$$

68. Let W be a subspace of R^n. Prove that the set shown below is a subspace of R^n in W.

Then prove that the intersection of W and $W^{\perp}$ is $\{0\}$.

In Exercises 69–72, find bases for the four **fundamental subspaces** of the matrix A shown below.

$N(A)$ = nullspace of A $\quad$ $N(A^T)$ = nullspace of A^T

$R(A)$ = column space of A $\quad$ $R(A^T)$ = column space of A^T

Then show that $N(A) = R(A^T)^{\perp}$ and $N(A^T) = R(A)^{\perp}$.

69. $\begin{bmatrix} 1 & 1 & -1 \\ 0 & 2 & 1 \\ 1 & 3 & 0 \end{bmatrix}$

70. $\begin{bmatrix} 0 & 1 & -1 \\ 0 & -2 & 2 \\ 0 & -1 & 1 \end{bmatrix}$

71. $\begin{bmatrix} 1 & 0 & 1 \\ 1 & 1 & 1 \end{bmatrix}$

72. $\begin{bmatrix} 0 & 0 & 1 & 2 & 0 \\ 1 & -2 & 0 & 2 & 0 \\ -1 & 2 & 1 & 0 & 0 \\ 0 & 0 & 1 & 2 & 1 \end{bmatrix}$

73. Let A be an $m \times n$ matrix.

(a) Explain why $R(A^T)$ is the same as the row space of A.

(b) Prove that $N(A) \subset R(A^T)^{\perp}$.

(c) Prove that $N(A) = R(A^T)^{\perp}$.

(d) Prove that $N(A^T) = R(A)^{\perp}$.

5.4 Mathematical Models and Least Squares Analysis

In this section, you will study *inconsistent* systems of linear equations and learn how to find the "best possible solution" of such a system. The necessity of "solving" inconsistent systems arises in the computation of least squares regression lines, as illustrated in Example 1.

EXAMPLE 1 Least Squares Regression Line

Let $(1, 0)$, $(2, 1)$, and $(3, 3)$ be three points in the plane, as shown in Figure 5.22. How can you find the line $y = c_0 + c_1 x$ that "best fits" these points? One way is to note that if the three points were collinear, then the following system of equations would be consistent.

$$\begin{aligned} c_0 + c_1 &= 0 \\ c_0 + 2c_1 &= 1 \\ c_0 + 3c_1 &= 3 \end{aligned}$$

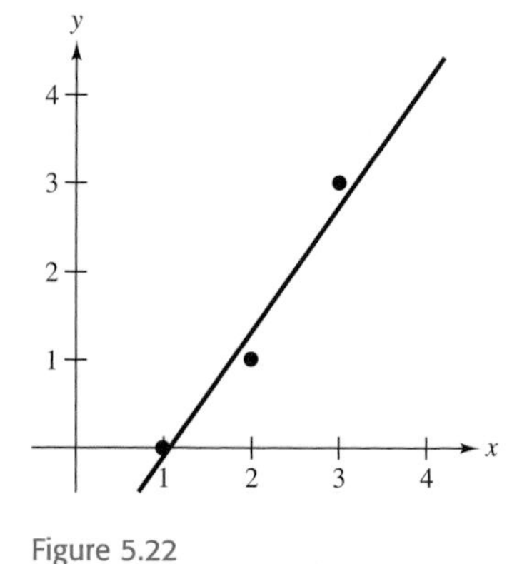

Figure 5.22

This system can be written in the matrix form $A\mathbf{x} = \mathbf{b}$, where

$$A = \begin{bmatrix} 1 & 1 \\ 1 & 2 \\ 1 & 3 \end{bmatrix}, \quad \mathbf{b} = \begin{bmatrix} 0 \\ 1 \\ 3 \end{bmatrix}, \quad \text{and} \quad \mathbf{x} = \begin{bmatrix} c_0 \\ c_1 \end{bmatrix}.$$

Because the points are not collinear, however, the system is inconsistent. Although it is impossible to find $\mathbf{x}$ such that $A\mathbf{x} = \mathbf{b}$, you can look for an $\mathbf{x}$ that *minimizes* the norm of the error $\|A\mathbf{x} - \mathbf{b}\|$. The solution

$$\mathbf{x} = \begin{bmatrix} c_0 \\ c_1 \end{bmatrix}$$

of this minimization problem is called the **least squares regression line** $y = c_0 + c_1x$.

In Section 2.5, you briefly studied the least squares regression line and how to calculate it using matrices. Now you will combine the ideas of orthogonality and projection to develop this concept in more generality. To begin, consider the linear system $A\mathbf{x} = \mathbf{b}$, where A is an $m \times n$ matrix and $\mathbf{b}$ is a column vector in R^m. You already know how to use Gaussian elimination with back-substitution to solve for $\mathbf{x}$ if the system is consistent. If the system is inconsistent, however, it is still useful to find the "best possible" solution; that is, the value of $\mathbf{x}$ for which the difference between $A\mathbf{x}$ and $\mathbf{b}$ is smallest. One way to define "best possible" is to require that the norm of $A\mathbf{x} - \mathbf{b}$ be minimized. This definition is the heart of the **least squares problem.**

Least Squares Problem

Given an $m \times n$ matrix A and a vector $\mathbf{b}$ in R^m, the **least squares problem** is to find $\mathbf{x}$ in R^n such that $\|A\mathbf{x} - \mathbf{b}\|^2$ is minimized.

REMARK: The term **least squares** comes from the fact that minimizing $\|A\mathbf{x} - \mathbf{b}\|$ is equivalent to minimizing $\|A\mathbf{x} - \mathbf{b}\|^2$, which is a sum of squares.

Orthogonal Subspaces

To solve the least squares problem, you first need to develop the concept of orthogonal subspaces. Two subspaces of R^n are said to be **orthogonal** if the vectors in each subspace are orthogonal to the vectors in the other subspace.

Definition of Orthogonal Subspaces

The subspaces S_1 and S_2 of R^n are **orthogonal** if $\mathbf{v}_1 \cdot \mathbf{v}_2 = 0$ for all $\mathbf{v}_1$ in S_1 and all $\mathbf{v}_2$ in S_2.

EXAMPLE 2 **Orthogonal Subspaces**

The subspaces

$$S_1 = \text{span}\left(\begin{bmatrix} 1 \\ 0 \\ 1 \end{bmatrix}, \begin{bmatrix} 1 \\ 1 \\ 0 \end{bmatrix}\right) \quad \text{and} \quad S_2 = \text{span}\left(\begin{bmatrix} -1 \\ 1 \\ 1 \end{bmatrix}\right)$$

are orthogonal because the dot product of any vector in S_1 and any vector in S_2 is zero.

Notice in Example 2 that the zero vector is the only vector common to both S_1 and S_2. This is true in general. If S_1 and S_2 are orthogonal subspaces of R^n, then their intersection consists only of the zero vector. You are asked to prove this fact in Exercise 45.

Provided with a subspace S of R^n, the set of all vectors orthogonal to every vector in S is called the **orthogonal complement** of S, as shown in the next definition.

Definition of Orthogonal Complement

If S is a subspace of R^n, then the **orthogonal complement of** $\mathbf{S}$ is the set

$$S^{\perp} = \{\mathbf{u} \in R^n : \mathbf{v} \cdot \mathbf{u} = 0 \text{ for all vectors } \mathbf{v} \in S\}.$$

The orthogonal complement of the trivial subspace $\{\mathbf{0}\}$ is all of R^n, and, conversely, the orthogonal complement of R^n is the trivial subspace $\{\mathbf{0}\}$. In Example 2, the subspace S_1 is the orthogonal complement of S_2, and the subspace S_2 is the orthogonal complement of S_1. In general, the orthogonal complement of a subspace of R^n is itself a subspace of R^n (see Exercise 46). You can find the orthogonal complement of a subspace of R^n by finding the nullspace of a matrix, as illustrated in the next example.

EXAMPLE 3 **Finding the Orthogonal Complement**

Find the orthogonal complement of the subspace S of R^4 spanned by the two column vectors $\mathbf{v}_1$ and $\mathbf{v}_2$ of the matrix A.

$$A = \begin{bmatrix} 1 & 0 \\ 2 & 0 \\ 1 & 0 \\ 0 & 1 \end{bmatrix}$$
$$\quad\;\; \mathbf{v}_1 \;\; \mathbf{v}_2$$

SOLUTION A vector $\mathbf{u} \in R^4$ will be in the orthogonal complement of S if its dot product with the two columns of A, $\mathbf{v}_1$ and $\mathbf{v}_2$, is zero. If you take the transpose of A, then you will see that the orthogonal complement of S consists of all the vectors $\mathbf{u}$ such that $A^T\mathbf{u} = \mathbf{0}$.

$$A^T\mathbf{u} = \mathbf{0}$$

$$\begin{bmatrix} 1 & 2 & 1 & 0 \\ 0 & 0 & 0 & 1 \end{bmatrix} \begin{bmatrix} x_1 \\ x_2 \\ x_3 \\ x_4 \end{bmatrix} = \begin{bmatrix} 0 \\ 0 \end{bmatrix}$$

That is, the orthogonal complement of S is the nullspace of the matrix A^T:

$$S^{\perp} = N(A^T).$$

Using the techniques for solving homogeneous linear systems, you can find that a possible basis for the orthogonal complement can consist of the two vectors

$$\mathbf{u}_1 = \begin{bmatrix} -2 \\ 1 \\ 0 \\ 0 \end{bmatrix} \quad \text{and} \quad \mathbf{u}_2 = \begin{bmatrix} -1 \\ 0 \\ 1 \\ 0 \end{bmatrix}.$$

Notice that R^4 in Example 3 is split into two subspaces, $S = \text{span}(\mathbf{v}_1, \mathbf{v}_2)$ and $S^{\perp} = \text{span}(\mathbf{u}_1, \mathbf{u}_2)$. In fact, the four vectors $\mathbf{v}_1$, $\mathbf{v}_2$, $\mathbf{u}_1$, and $\mathbf{u}_2$ form a basis for R^4. Each vector in R^4 can be *uniquely* written as a sum of a vector from S and a vector from $S^{\perp}$. This concept is generalized in the next definition.

Definition of Direct Sum

Let S_1 and S_2 be two subspaces of R^n. If each vector $\mathbf{x} \in R^n$ can be uniquely written as a sum of a vector $\mathbf{s}_1$ from S_1 and a vector $\mathbf{s}_2$ from S_2, $\mathbf{x} = \mathbf{s}_1 + \mathbf{s}_2$, then R^n is the **direct sum** of S_1 and S_2, and you can write $R^n = S_1 \oplus S_2$.

EXAMPLE 4 Direct Sum

(a) From Example 2, you can see that R^3 is the direct sum of the subspaces

$$S_1 = \text{span}\left(\begin{bmatrix} 1 \\ 0 \\ 1 \end{bmatrix}, \begin{bmatrix} 1 \\ 1 \\ 0 \end{bmatrix}\right) \quad \text{and} \quad S_2 = \text{span}\left(\begin{bmatrix} -1 \\ 1 \\ 1 \end{bmatrix}\right).$$

(b) From Example 3, you can see that $R^4 = S \oplus S^{\perp}$, where

$$S = \text{span}\left(\begin{bmatrix} 1 \\ 2 \\ 1 \\ 0 \end{bmatrix}, \begin{bmatrix} 0 \\ 0 \\ 0 \\ 1 \end{bmatrix}\right) \quad \text{and} \quad S^{\perp} = \text{span}\left(\begin{bmatrix} -2 \\ 1 \\ 0 \\ 0 \end{bmatrix}, \begin{bmatrix} -1 \\ 0 \\ 1 \\ 0 \end{bmatrix}\right).$$

The next theorem collects some important facts about orthogonal complements and direct sums.

THEOREM 5.13 Properties of Orthogonal Subspaces

Let S be a subspace of R^n. Then the following properties are true.

1. $\dim(S) + \dim(S^{\perp}) = n$
2. $R^n = S \oplus S^{\perp}$
3. $(S^{\perp})^{\perp} = S$

PROOF

1. If $S = R^n$ or $S = \{\mathbf{0}\}$, then Property 1 is trivial. So let $\{\mathbf{v}_1, \mathbf{v}_2, . . . , \mathbf{v}_t\}$ be a basis for S, $0 < t < n$. Let A be the $n \times t$ matrix whose columns are the basis vectors $\mathbf{v}_i$. Then $S = R(A)$, which implies that $S^\perp = N(A^T)$, where A^T is a $t \times n$ matrix of rank t (see Section 5.3, Exercise 73). Because the dimension of $N(A^T)$ is $n - t$, you have shown that

$$\dim(S) + \dim(S^\perp) = t + (n - t) = n.$$

2. If $S = R^n$ or $S = \{\mathbf{0}\}$, then Property 2 is trivial. So let $\{\mathbf{v}_1, \mathbf{v}_2, . . . , \mathbf{v}_t\}$ be a basis for S and let $\{\mathbf{v}_{t+1}, \mathbf{v}_{t+2}, . . . , \mathbf{v}_n\}$ be a basis for $S^\perp$. It can be shown that the set $\{\mathbf{v}_1, \mathbf{v}_2, . . . , \mathbf{v}_t, \mathbf{v}_{t+1}, . . . , \mathbf{v}_n\}$ is linearly independent and forms a basis for R^n. Let $\mathbf{x} \in R^n$, $\mathbf{x} = c_1\mathbf{v}_1 + \cdot \cdot \cdot + c_t\mathbf{v}_t + c_{t+1}\mathbf{v}_{t+1} + \cdot \cdot \cdot + c_n\mathbf{v}_n$. If you write $\mathbf{v} = c_1\mathbf{v}_1 + \cdot \cdot \cdot + c_t\mathbf{v}_t$ and $\mathbf{w} = c_{t+1}\mathbf{v}_{t+1} + \cdot \cdot \cdot + c_n\mathbf{v}_n$, then you have expressed an arbitrary vector $\mathbf{x}$ as the sum of a vector from S and a vector from $S^\perp$, $\mathbf{x} = \mathbf{v} + \mathbf{w}$.

   To show the uniqueness of this representation, assume $\mathbf{x} = \mathbf{v} + \mathbf{w} = \hat{\mathbf{v}} + \hat{\mathbf{w}}$ (where $\hat{\mathbf{r}}$ denotes a vector that has all zero entries except for one, which is 1). This implies that $\hat{\mathbf{v}} - \mathbf{v} = \mathbf{w} - \hat{\mathbf{w}}$. So, the two vectors $\hat{\mathbf{v}} - \mathbf{v}$ and $\mathbf{w} - \hat{\mathbf{w}}$ are in both S and $S^\perp$. Because $S \cap S^\perp = \{\mathbf{0}\}$, you must have $\hat{\mathbf{v}} = \mathbf{v}$ and $\mathbf{w} = \hat{\mathbf{w}}$.

3. Let $\mathbf{v} \in S$. Then $\mathbf{v} \cdot \mathbf{u} = 0$ for all $\mathbf{u} \in S^\perp$, which implies that $\mathbf{v} \in (S^\perp)^\perp$. On the other hand, if $\mathbf{v} \in (S^\perp)^\perp$, then, because $R^n = S \oplus S^\perp$, you can write $\mathbf{v}$ as the unique sum of the vector from S and a vector from $S^\perp$, $\mathbf{v} = \mathbf{s} + \mathbf{w}$, $\mathbf{s} \in S$, $\mathbf{w} \in S^\perp$. Because $\mathbf{w}$ is in $S^\perp$, it is orthogonal to every vector in S, and in particular to $\mathbf{v}$. So,

$$\begin{aligned} 0 = \mathbf{w} \cdot \mathbf{v} &= \mathbf{w} \cdot (\mathbf{s} + \mathbf{w}) \\ &= \mathbf{w} \cdot \mathbf{s} + \mathbf{w} \cdot \mathbf{w} \\ &= \mathbf{w} \cdot \mathbf{w}. \end{aligned}$$

   This implies that $\mathbf{w} = \mathbf{0}$ and $\mathbf{v} = \mathbf{s} + \mathbf{w} = \mathbf{s} \in S$.

You studied the projection of one vector onto another in Section 5.3. This is now generalized to projections of a vector $\mathbf{v}$ onto a subspace S. Because $R^n = S \oplus S^\perp$, every vector $\mathbf{v}$ in R^n can be uniquely written as a sum of a vector from S and a vector from $S^\perp$:

$$\mathbf{v} = \mathbf{v}_1 + \mathbf{v}_2, \qquad \mathbf{v}_1 \in S, \qquad \mathbf{v}_2 \in S^\perp.$$

The vector $\mathbf{v}_1$ is called the **projection** of $\mathbf{v}$ onto the subspace S, and is denoted by $\mathbf{v}_1 = \text{proj}_S\mathbf{v}$. So, $\mathbf{v}_2 = \mathbf{v} - \mathbf{v}_1 = \mathbf{v} - \text{proj}_S\mathbf{v}$, which implies that the vector $\mathbf{v} - \text{proj}_S\mathbf{v}$ is orthogonal to the subspace S.

Provided with a subspace S of R^n, you can use the Gram-Schmidt orthonormalization process to calculate an orthonormal basis for S. It is then an easy matter to compute the projection of a vector $\mathbf{v}$ onto S using the next theorem. (You are asked to prove this theorem in Exercise 47.)

THEOREM 5.14 Projection onto a Subspace

If $\{\mathbf{u}_1, \mathbf{u}_2, . . . , \mathbf{u}_t\}$ is an orthonormal basis for the subspace S of R^n, and $\mathbf{v} \in R^n$, then

$$\text{proj}_S\mathbf{v} = (\mathbf{v} \cdot \mathbf{u}_1)\mathbf{u}_1 + (\mathbf{v} \cdot \mathbf{u}_2)\mathbf{u}_2 + \cdot \cdot \cdot + (\mathbf{v} \cdot \mathbf{u}_t)\mathbf{u}_t.$$

EXAMPLE 5 **Projection onto a Subspace**

Find the projection of the vector $\mathbf{v} = \begin{bmatrix} 1 \\ 1 \\ 3 \end{bmatrix}$ onto the subspace S of R^3 spanned by the vectors

$$\mathbf{w}_1 = \begin{bmatrix} 0 \\ 3 \\ 1 \end{bmatrix} \quad \text{and} \quad \mathbf{w}_2 = \begin{bmatrix} 2 \\ 0 \\ 0 \end{bmatrix}.$$

SOLUTION By normalizing $\mathbf{w}_1$ and $\mathbf{w}_2$, you obtain an orthonormal basis for S.

$$\{\mathbf{u}_1, \mathbf{u}_2\} = \left\{\frac{1}{\sqrt{10}}\mathbf{w}_1, \frac{1}{2}\mathbf{w}_2\right\} = \left\{\begin{bmatrix} 0 \\ \frac{3}{\sqrt{10}} \\ \frac{1}{\sqrt{10}} \end{bmatrix}, \begin{bmatrix} 1 \\ 0 \\ 0 \end{bmatrix}\right\}$$

Use Theorem 5.14 to find the projection of $\mathbf{v}$ onto S.

$$\begin{aligned} \text{proj}_S\mathbf{v} &= (\mathbf{v} \cdot \mathbf{u}_1)\mathbf{u}_1 + (\mathbf{v} \cdot \mathbf{u}_2)\mathbf{u}_2 \\ &= \frac{6}{\sqrt{10}}\begin{bmatrix} 0 \\ \frac{3}{\sqrt{10}} \\ \frac{1}{\sqrt{10}} \end{bmatrix} + 1\begin{bmatrix} 1 \\ 0 \\ 0 \end{bmatrix} = \begin{bmatrix} 1 \\ \frac{9}{5} \\ \frac{3}{5} \end{bmatrix} \end{aligned}$$

The projection of $\mathbf{v}$ onto the plane S is illustrated in Figure 5.23.

Theorem 5.9 said that among all the scalar multiples of a vector $\mathbf{u}$, the orthogonal projection of $\mathbf{v}$ onto $\mathbf{u}$ is the one closest to $\mathbf{v}$. Example 5 suggests that this property is also true for projections onto subspaces. That is, among all the vectors in the subspace S, the vector $\text{proj}_S\mathbf{v}$ is the closest vector to $\mathbf{v}$. These two results are illustrated in Figure 5.24.

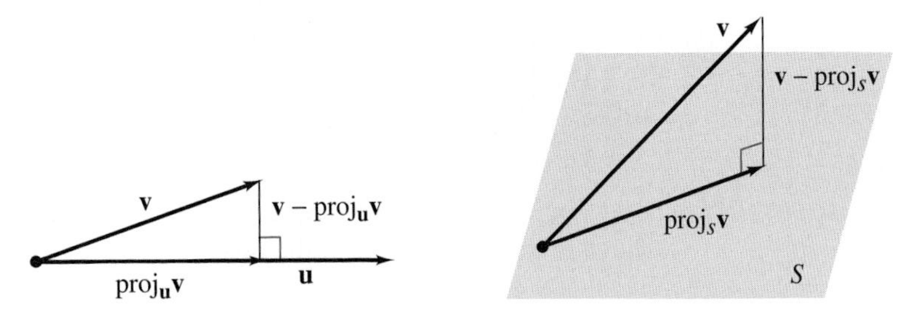

Figure 5.24

THEOREM 5.15 Orthogonal Projection and Distance

Let S be a subspace of R^n and let $\mathbf{v} \in R^n$. Then, for all $\mathbf{u} \in S$, $\mathbf{u} \neq \text{proj}_S\mathbf{v}$,

$$\|\mathbf{v} - \text{proj}_S\mathbf{v}\| < \|\mathbf{v} - \mathbf{u}\|.$$

PROOF Let $\mathbf{u} \in S$, $\mathbf{u} \neq \text{proj}_S\mathbf{v}$. By adding and subtracting the same quantity $\text{proj}_S\mathbf{v}$ to and from the vector $\mathbf{v} - \mathbf{u}$, you obtain

$$\mathbf{v} - \mathbf{u} = (\mathbf{v} - \text{proj}_S\mathbf{v}) + (\text{proj}_S\mathbf{v} - \mathbf{u}).$$

Observe that $(\text{proj}_S\mathbf{v} - \mathbf{u})$ is in S and $(\mathbf{v} - \text{proj}_S\mathbf{v})$ is orthogonal to S. So, $(\mathbf{v} - \text{proj}_S\mathbf{v})$ and $(\text{proj}_S\mathbf{v} - \mathbf{u})$ are orthogonal vectors, and you can use the Pythagorean Theorem (Theorem 5.6) to obtain

$$\|\mathbf{v} - \mathbf{u}\|^2 = \|\mathbf{v} - \text{proj}_S\mathbf{v}\|^2 + \|\text{proj}_S\mathbf{v} - \mathbf{u}\|^2.$$

Because $\mathbf{u} \neq \text{proj}_S\mathbf{v}$, the second term on the right is positive, and you have

$$\|\mathbf{v} - \text{proj}_S\mathbf{v}\| < \|\mathbf{v} - \mathbf{u}\|.$$

Fundamental Subspaces of a Matrix

You need to develop one more concept before solving the least squares problem. Recall that if A is an $m \times n$ matrix, the column space of A is a subspace of R^m consisting of all vectors of the form $A\mathbf{x}$, $\mathbf{x} \in R^n$. The four **fundamental subspaces** of the matrix A are defined as follows (see Exercises 69–73 in Section 5.3).

$$N(A) = \text{nullspace of } A \qquad N(A^T) = \text{nullspace of } A^T$$
$$R(A) = \text{column space of } A \qquad R(A^T) = \text{column space of } A^T$$

These subspaces play a crucial role in the solution of the least squares problem.

EXAMPLE 6 Fundamental Subspaces

Find the four fundamental subspaces of the matrix

$$A = \begin{bmatrix} 1 & 2 & 0 \\ 0 & 0 & 1 \\ 0 & 0 & 0 \\ 0 & 0 & 0 \end{bmatrix}.$$

SOLUTION The column space of A is simply the span of the first and third columns, because the second column is a scalar multiple of the first.

$$R(A) = \text{span}\left(\begin{bmatrix}1\\0\\0\\0\end{bmatrix}, \begin{bmatrix}0\\1\\0\\0\end{bmatrix}\right)$$

The column space of A^T is equivalent to the row space of A, which is spanned by the first two rows.

$$R(A^T) = \text{span}\left(\begin{bmatrix}1\\2\\0\end{bmatrix}, \begin{bmatrix}0\\0\\1\end{bmatrix}\right)$$

The nullspace of A is a solution space of the homogeneous system $A\mathbf{x} = \mathbf{0}$.

$$N(A) = \text{span}\left(\begin{bmatrix}-2\\1\\0\end{bmatrix}\right)$$

Finally, the nullspace of A^T is a solution space of the homogeneous system whose coefficient matrix is A^T.

$$N(A^T) = \text{span}\left(\begin{bmatrix}0\\0\\1\\0\end{bmatrix}, \begin{bmatrix}0\\0\\0\\1\end{bmatrix}\right)$$

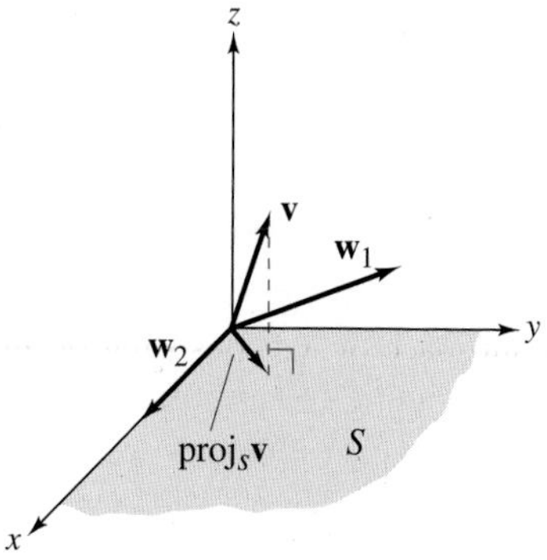

In Example 6, observe that $R(A)$ and $N(A^T)$ are orthogonal subspaces of R^4, and $R(A^T)$ and $N(A)$ are orthogonal subspaces of R^3. These and other properties of these subspaces are stated in the next theorem.

Figure 5.23

THEOREM 5.16 Fundamental Subspaces of a Matrix

If A is an $m \times n$ matrix, then

1. $R(A)$ and $N(A^T)$ are orthogonal subspaces of R^m.
2. $R(A^T)$ and $N(A)$ are orthogonal subspaces of R^n.
3. $R(A) \oplus N(A^T) = R^m$.
4. $R(A^T) \oplus N(A) = R^n$.

PROOF To prove Property 1, let $\mathbf{v} \in R(A)$ and $\mathbf{u} \in N(A^T)$. Because the column space of A is equal to the row space of A^T, you can see that $A^T\mathbf{u} = \mathbf{0}$ implies $\mathbf{u} \cdot \mathbf{v} = 0$. Property 2 follows from applying Property 1 to A^T.

To prove Property 3, observe that $R(A)^{\perp} = N(A^T)$. So,

$$R^m = R(A) \oplus R(A)^{\perp} = R(A) \oplus N(A^T).$$

A similar argument applied to $R(A^T)$ proves Property 4.

Least Squares

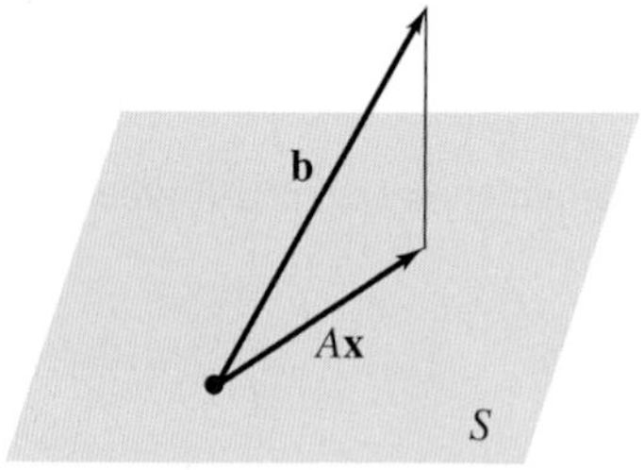

You have now developed all the tools needed to solve the least squares problem. Recall that you are attempting to find a vector $\mathbf{x}$ that minimizes $\|A\mathbf{x} - \mathbf{b}\|$, where A is an $m \times n$ matrix and $\mathbf{b}$ is a vector in R^m. Let S be the column space of A: $S = R(A)$. You can assume that $\mathbf{b}$ is not in S, because otherwise the system $A\mathbf{x} = \mathbf{b}$ would be consistent. You are looking for a vector $A\mathbf{x}$ in S that is as close as possible to $\mathbf{b}$, as indicated in Figure 5.25.

Figure 5.25

From Theorem 5.15 you know that the desired vector is the projection of $\mathbf{b}$ onto S. Letting $A\hat{\mathbf{x}} = \text{proj}_S\mathbf{b}$ be that projection, you can see that $A\hat{\mathbf{x}} - \mathbf{b} = \text{proj}_S\mathbf{b} - \mathbf{b}$ is orthogonal to $S = R(A)$. But this implies that $A\hat{\mathbf{x}} - \mathbf{b}$ is in $R(A)^\perp$, which equals $N(A^T)$ according to Theorem 5.16. This is the crucial observation: $A\hat{\mathbf{x}} - \mathbf{b}$ is in the nullspace of A^T. So, you have

$$\begin{aligned} A^T(A\hat{\mathbf{x}} - \mathbf{b}) &= \mathbf{0} \\ A^TA\hat{\mathbf{x}} - A^T\mathbf{b} &= \mathbf{0} \\ A^TA\hat{\mathbf{x}} &= A^T\mathbf{b}. \end{aligned}$$

The solution of the least squares problem comes down to solving the $n \times n$ linear system of equations $A^TA\mathbf{x} = A^T\mathbf{b}$. These equations are called the **normal equations** of the least squares problem $A\mathbf{x} = \mathbf{b}$.

EXAMPLE 7 Solving the Normal Equations

Find the solution of the least squares problem

$$A\mathbf{x} = \mathbf{b}$$

$$\begin{bmatrix} 1 & 1 \\ 1 & 2 \\ 1 & 3 \end{bmatrix} \begin{bmatrix} c_0 \\ c_1 \end{bmatrix} = \begin{bmatrix} 0 \\ 1 \\ 3 \end{bmatrix}$$

presented in Example 1.

SOLUTION Begin by calculating the matrix products shown below.

$$A^TA = \begin{bmatrix} 1 & 1 & 1 \\ 1 & 2 & 3 \end{bmatrix} \begin{bmatrix} 1 & 1 \\ 1 & 2 \\ 1 & 3 \end{bmatrix} = \begin{bmatrix} 3 & 6 \\ 6 & 14 \end{bmatrix}$$

$$A^T\mathbf{b} = \begin{bmatrix} 1 & 1 & 1 \\ 1 & 2 & 3 \end{bmatrix} \begin{bmatrix} 0 \\ 1 \\ 3 \end{bmatrix} = \begin{bmatrix} 4 \\ 11 \end{bmatrix}$$

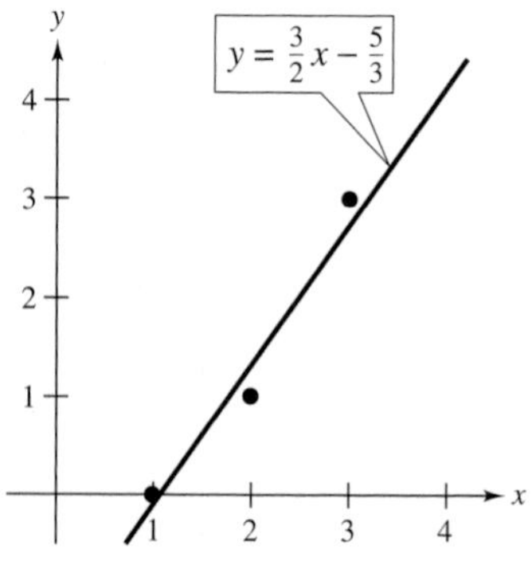

The normal equations are

$$A^TA\mathbf{x} = A^T\mathbf{b}$$

$$\begin{bmatrix} 3 & 6 \\ 6 & 14 \end{bmatrix} \begin{bmatrix} c_0 \\ c_1 \end{bmatrix} = \begin{bmatrix} 4 \\ 11 \end{bmatrix}.$$

Figure 5.26

The solution of this system of equations is $\mathbf{x} = \begin{bmatrix} -\frac{5}{3} \\ \frac{3}{2} \end{bmatrix}$, which implies that the least squares regression line for the data is $y = \frac{3}{2}x - \frac{5}{3}$, as indicated in Figure 5.26.

Technology Note

Many graphing utilities and computer software programs have built-in programs for finding the least squares regression line for a set of data points. If you have access to such tools, try verifying the result of Example 7. Keystrokes and programming syntax for these utilities/programs applicable to Example 7 are provided in the **Online Technology Guide,** available at *college.hmco.com/pic/larsonELA6e*.

REMARK: For an $m \times n$ matrix A, the normal equations form an $n \times n$ system of linear equations. This system is always consistent, but it may have an infinite number of solutions. It can be shown, however, that there is a unique solution if the rank of A is n.

The next example illustrates how to solve the projection problem from Example 5 using normal equations.

EXAMPLE 8 Orthogonal Projection onto a Subspace

Find the orthogonal projection of the vector $\mathbf{b} = \begin{bmatrix} 1 \\ 1 \\ 3 \end{bmatrix}$ onto the column space S of the matrix

$$A = \begin{bmatrix} 0 & 2 \\ 3 & 0 \\ 1 & 0 \end{bmatrix}.$$

SOLUTION To find the orthogonal projection of $\mathbf{b}$ onto S, first solve the least squares problem $A\mathbf{x} = \mathbf{b}$. As in Example 7, calculate the matrix products A^TA and $A^T\mathbf{b}$.

$$A^TA = \begin{bmatrix} 0 & 3 & 1 \\ 2 & 0 & 0 \end{bmatrix} \begin{bmatrix} 0 & 2 \\ 3 & 0 \\ 1 & 0 \end{bmatrix} = \begin{bmatrix} 10 & 0 \\ 0 & 4 \end{bmatrix}$$

$$A^T\mathbf{b} = \begin{bmatrix} 0 & 3 & 1 \\ 2 & 0 & 0 \end{bmatrix} \begin{bmatrix} 1 \\ 1 \\ 3 \end{bmatrix} = \begin{bmatrix} 6 \\ 2 \end{bmatrix}$$

The normal equations are

$$A^T A\mathbf{x} = A^T\mathbf{b}$$

$$\begin{bmatrix} 10 & 0 \\ 0 & 4 \end{bmatrix} \begin{bmatrix} x_1 \\ x_2 \end{bmatrix} = \begin{bmatrix} 6 \\ 2 \end{bmatrix}.$$

The solution of these equations is easily seen to be

$$\mathbf{x} = \begin{bmatrix} x_1 \\ x_2 \end{bmatrix} = \begin{bmatrix} \frac{3}{5} \\ \frac{1}{2} \end{bmatrix}.$$

Finally, the projection of $\mathbf{b}$ onto S is

$$A\mathbf{x} = \begin{bmatrix} 0 & 2 \\ 3 & 0 \\ 1 & 0 \end{bmatrix} \begin{bmatrix} \frac{3}{5} \\ \frac{1}{2} \end{bmatrix} = \begin{bmatrix} 1 \\ \frac{9}{5} \\ \frac{3}{5} \end{bmatrix},$$

which agrees with the solution obtained in Example 5.

Mathematical Modeling

Least squares problems play a fundamental role in mathematical modeling of real-life phenomena. The next example shows how to model the world population using a least squares quadratic polynomial.

EXAMPLE 9 **World Population**

Table 5.1 shows the world population (in billions) for six different years. (Source: U.S. Census Bureau)

TABLE 5.1

Year	1980	1985	1990	1995	2000	2005
Population (y)	4.5	4.8	5.3	5.7	6.1	6.5

Let $x = 0$ represent the year 1980. Find the least squares regression quadratic polynomial $y = c_0 + c_1x + c_2x^2$ for these data and use the model to estimate the population for the year 2010.

SOLUTION By substituting the data points (0, 4.5), (5, 4.8), (10, 5.3), (15, 5.7), (20, 6.1), and (25, 6.5) into the quadratic polynomial $y = c_0 + c_1x + c_2x^2$, you obtain the following system of linear equations.

$$\begin{aligned} c_0 \qquad\qquad\qquad\quad &= 4.5 \\ c_0 + 5c_1 + 25c_2 &= 4.8 \\ c_0 + 10c_1 + 100c_2 &= 5.3 \\ c_0 + 15c_1 + 225c_2 &= 5.7 \\ c_0 + 20c_1 + 400c_2 &= 6.1 \\ c_0 + 25c_1 + 625c_2 &= 6.5 \end{aligned}$$

This produces the least squares problem

$$A\mathbf{x} = \mathbf{b}$$

$$\begin{bmatrix} 1 & 0 & 0 \\ 1 & 5 & 25 \\ 1 & 10 & 100 \\ 1 & 15 & 225 \\ 1 & 20 & 400 \\ 1 & 25 & 625 \end{bmatrix} \begin{bmatrix} c_0 \\ c_1 \\ c_2 \end{bmatrix} = \begin{bmatrix} 4.5 \\ 4.8 \\ 5.3 \\ 5.7 \\ 6.1 \\ 6.5 \end{bmatrix}.$$

The normal equations are

$$A^TA\mathbf{x} = A^T\mathbf{b}$$

$$\begin{bmatrix} 6 & 75 & 1375 \\ 75 & 1375 & 28{,}125 \\ 1375 & 28{,}125 & 611{,}875 \end{bmatrix} \begin{bmatrix} c_0 \\ c_1 \\ c_2 \end{bmatrix} = \begin{bmatrix} 32.9 \\ 447 \\ 8435 \end{bmatrix}$$

and their solution is

$$\mathbf{x} = \begin{bmatrix} c_0 \\ c_1 \\ c_2 \end{bmatrix} \approx \begin{bmatrix} 4.5 \\ 0.08 \\ 0 \end{bmatrix}.$$

Note that $c_2 \approx 0$. So, the least squares polynomial for these data is the linear polynomial:

$$y = 4.5 + 0.08x.$$

Evaluating this polynomial at $x = 30$ gives the estimate of the world population for the year 2010:

$$y = 4.5 + 0.08(30) \approx 6.9 \text{ billion.}$$

Least squares models can arise in many other contexts. Section 5.5 explores some applications of least squares models to approximation of functions. In the final example of this section, a nonlinear model is used to find a relationship between the period of a planet and its mean distance from the sun.

EXAMPLE 10 Application to Astronomy

Table 5.2 shows the mean distances x and the periods y of the six planets that are closest to the sun. The mean distance is given in terms of astronomical units (where the Earth's mean distance is defined as 1.0), and the period is in years. Find a model for these data. (Source: *CRC Handbook of Chemistry and Physics*)

TABLE 5.2

Planet	*Mercury*	*Venus*	*Earth*	*Mars*	*Jupiter*	*Saturn*
Distance, x	0.387	0.723	1.0	1.523	5.203	9.541
Period, y	0.241	0.615	1.0	1.881	11.861	29.457

If you plot the data as shown, they do not seem to lie in a straight line. By taking the logarithm of each coordinate, however, you obtain points of the form $(\ln x, \ln y)$, as shown in Table 5.3.

TABLE 5.3

Planet	*Mercury*	*Venus*	*Earth*	*Mars*	*Jupiter*	*Saturn*
$\ln x$	−0.949	−0.324	0.0	0.421	1.649	2.256
$\ln y$	−1.423	−0.486	0.0	0.632	2.473	3.383

Figure 5.27 shows a plot of the transformed points and suggests that the least squares regression line would be a good fit. Using the techniques of this section, you can find that the equation of the line is

$$\ln y = \frac{3}{2} \ln x \qquad \text{or} \qquad y = x^{3/2}.$$

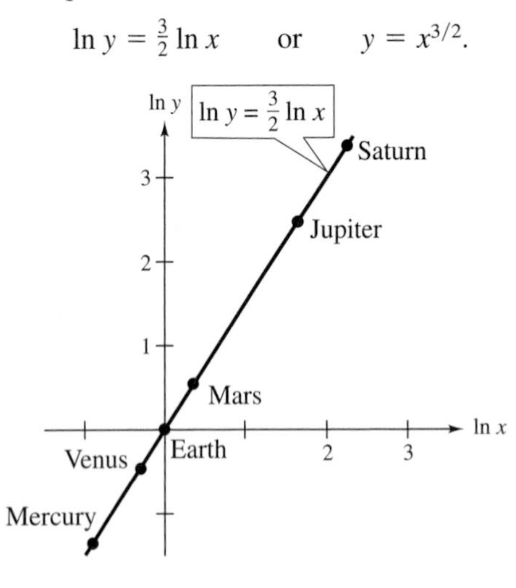

Figure 5.27

Technology Note

You can use a computer software program or graphing utility with a built-in power regression program to verify the result of Example 10. For example, using the data in Table 5.2 and a graphing utility, a power fit program would result in an answer of (or very similar to) $y \approx 1.00042x^{1.49954}$. Keystrokes and programming syntax for these utilities/programs applicable to Example 10 are provided in the **Online Technology Guide,** available at *college.hmco.com/pic/larsonELA6e*.

SECTION 5.4 Exercises

In Exercises 1–4, determine whether the sets are orthogonal.

1. $S_1 = \text{span}\left\{\begin{bmatrix} 2 \\ 1 \\ -1 \end{bmatrix}, \begin{bmatrix} 0 \\ 1 \\ 1 \end{bmatrix}\right\}$ $\quad S_2 = \text{span}\left\{\begin{bmatrix} -1 \\ 2 \\ 0 \end{bmatrix}\right\}$

2. $S_1 = \text{span}\left\{\begin{bmatrix} -3 \\ 0 \\ 1 \end{bmatrix}\right\}$ $\quad S_2 = \text{span}\left\{\begin{bmatrix} 2 \\ 1 \\ 6 \end{bmatrix}, \begin{bmatrix} 0 \\ 1 \\ 0 \end{bmatrix}\right\}$

3. $S_1 = \text{span}\left\{\begin{bmatrix} 1 \\ 1 \\ 1 \\ 1 \end{bmatrix}\right\}$ $\quad S_2 = \text{span}\left\{\begin{bmatrix} -1 \\ 1 \\ -1 \\ 1 \end{bmatrix}, \begin{bmatrix} 0 \\ 2 \\ -2 \\ 0 \end{bmatrix}\right\}$

4. $S_1 = \text{span}\left\{\begin{bmatrix} 0 \\ 0 \\ 2 \\ 1 \end{bmatrix}, \begin{bmatrix} 0 \\ 0 \\ 1 \\ -2 \end{bmatrix}\right\}$ $\quad S_2 = \text{span}\left\{\begin{bmatrix} 3 \\ 2 \\ 0 \\ 0 \end{bmatrix}, \begin{bmatrix} 0 \\ 1 \\ -2 \\ 2 \end{bmatrix}\right\}$

In Exercises 5–8, find the orthogonal complement $S^{\perp}$.

5. S is the subspace of R^3 consisting of the xz-plane.

6. S is the subspace of R^5 consisting of all vectors whose third and fourth components are zero.

7. $S = \text{span}\left\{\begin{bmatrix} 1 \\ 2 \\ 0 \\ 0 \end{bmatrix}, \begin{bmatrix} 0 \\ 1 \\ 0 \\ 1 \end{bmatrix}\right\}$ $\quad$ **8.** $S = \text{span}\left\{\begin{bmatrix} 0 \\ 1 \\ -1 \\ 1 \end{bmatrix}\right\}$

9. Find the orthogonal complement of the solution of Exercise 7.

10. Find the orthogonal complement of the solution of Exercise 8.

In Exercises 11–14, find the projection of the vector $\mathbf{v}$ onto the subspace S.

11. $S = \text{span}\left\{\begin{bmatrix} 0 \\ 0 \\ -1 \\ 1 \end{bmatrix}, \begin{bmatrix} 0 \\ 1 \\ 1 \\ 1 \end{bmatrix}\right\}, \quad \mathbf{v} = \begin{bmatrix} 1 \\ 0 \\ 1 \\ 1 \end{bmatrix}$

12. $S = \text{span}\left\{\begin{bmatrix} -1 \\ 2 \\ 0 \\ 0 \end{bmatrix}, \begin{bmatrix} 0 \\ 0 \\ 1 \\ 0 \end{bmatrix}, \begin{bmatrix} 0 \\ 0 \\ 0 \\ 1 \end{bmatrix}\right\}, \quad \mathbf{v} = \begin{bmatrix} 1 \\ 1 \\ 1 \\ 1 \end{bmatrix}$

13. $S = \text{span}\left\{\begin{bmatrix} 1 \\ 0 \\ 1 \end{bmatrix}, \begin{bmatrix} 0 \\ 1 \\ 1 \end{bmatrix}\right\}, \quad \mathbf{v} = \begin{bmatrix} 2 \\ 3 \\ 4 \end{bmatrix}$

14. $S = \text{span}\left\{\begin{bmatrix} 1 \\ 1 \\ 1 \\ 1 \end{bmatrix}, \begin{bmatrix} 0 \\ 1 \\ -1 \\ 0 \end{bmatrix}, \begin{bmatrix} 0 \\ 1 \\ 1 \\ 0 \end{bmatrix}\right\}, \quad \mathbf{v} = \begin{bmatrix} 1 \\ 2 \\ 3 \\ 4 \end{bmatrix}$

In Exercises 15 and 16, find the orthogonal projection of $\mathbf{b} = [2 \quad -2 \quad 1]^T$ onto the column space of the matrix A.

15. $A = \begin{bmatrix} 1 & 2 \\ 0 & 1 \\ 1 & 1 \end{bmatrix}$ $\quad$ **16.** $A = \begin{bmatrix} 0 & 2 \\ 1 & 1 \\ 1 & 3 \end{bmatrix}$

In Exercises 17–20, find bases for the four fundamental subspaces of the matrix A.

17. $A = \begin{bmatrix} 1 & 2 & 3 \\ 0 & 1 & 0 \end{bmatrix}$ $\quad$ **18.** $A = \begin{bmatrix} 0 & -1 & 1 \\ 1 & 2 & 0 \\ 1 & 1 & 1 \end{bmatrix}$

19. $A = \begin{bmatrix} 1 & 0 & 0 & 1 \\ 0 & 1 & 1 & 1 \\ 1 & 1 & 1 & 2 \\ 1 & 2 & 2 & 3 \end{bmatrix}$

20. $A = \begin{bmatrix} 1 & 0 & -1 \\ 0 & -1 & 1 \\ 1 & 1 & 0 \\ 1 & 0 & 1 \end{bmatrix}$

In Exercises 21–26, find the least squares solution of the system $A\mathbf{x} = \mathbf{b}$.

21. $A = \begin{bmatrix} 2 & 1 \\ 1 & 2 \\ 1 & 1 \end{bmatrix} \quad \mathbf{b} = \begin{bmatrix} 2 \\ 0 \\ -3 \end{bmatrix}$

22. $A = \begin{bmatrix} 0 & 1 \\ 1 & 0 \\ 1 & 2 \end{bmatrix} \quad \mathbf{b} = \begin{bmatrix} -1 \\ -1 \\ 3 \end{bmatrix}$

23. $A = \begin{bmatrix} 1 & 0 & 1 \\ 1 & 1 & 1 \\ 0 & 1 & 1 \\ 1 & 1 & 0 \end{bmatrix} \quad \mathbf{b} = \begin{bmatrix} 4 \\ -1 \\ 0 \\ 1 \end{bmatrix}$

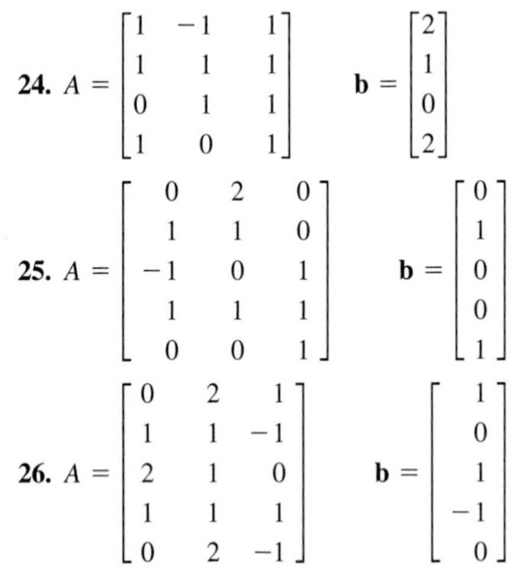

24. $A = \begin{bmatrix} 1 & -1 & 1 \\ 1 & 1 & 1 \\ 0 & 1 & 1 \\ 1 & 0 & 1 \end{bmatrix} \quad \mathbf{b} = \begin{bmatrix} 2 \\ 1 \\ 0 \\ 2 \end{bmatrix}$

25. $A = \begin{bmatrix} 0 & 2 & 0 \\ 1 & 1 & 0 \\ -1 & 0 & 1 \\ 1 & 1 & 1 \\ 0 & 0 & 1 \end{bmatrix} \quad \mathbf{b} = \begin{bmatrix} 0 \\ 1 \\ 0 \\ 0 \\ 1 \end{bmatrix}$

26. $A = \begin{bmatrix} 0 & 2 & 1 \\ 1 & 1 & -1 \\ 2 & 1 & 0 \\ 1 & 1 & 1 \\ 0 & 2 & -1 \end{bmatrix} \quad \mathbf{b} = \begin{bmatrix} 1 \\ 0 \\ 1 \\ -1 \\ 0 \end{bmatrix}$

In Exercises 27–32, find the least squares regression line for the data points. Graph the points and the line on the same set of axes.

27. $(-1, 1), (1, 0), (3, -3)$

28. $(1, 1), (2, 3), (4, 5)$

29. $(-2, -1), (-1, 0), (1, 0), (2, 2)$

30. $(-3, -3), (-2, -2), (0, 0), (1, 2)$

31. $(-2, 1), (-1, 2), (0, 1), (1, 2), (2, 1)$

32. $(-2, 0), (-1, 2), (0, 3), (1, 5), (2, 6)$

In Exercises 33–36, find the least squares quadratic polynomial for the data points.

33. $(0, 0), (2, 2), (3, 6), (4, 12)$

34. $(0, 2), \left(1, \frac{3}{2}\right), \left(2, \frac{5}{2}\right), (3, 4)$

35. $(-2, 0), (-1, 0), (0, 1), (1, 2), (2, 5)$

36. $(-2, 6), (-1, 5), \left(0, \frac{7}{2}\right), (1, 2), (2, -1)$

37. The table shows the annual sales (in millions of dollars) for Advanced Auto Parts and Auto Zone for 2000 through 2007. Find an appropriate regression line, quadratic regression polynomial, or cubic regression polynomial for each company. Then use the model to predict sales for the year 2010. Let t represent the year, with $t = 0$ corresponding to 2000. (Source: Advanced Auto Parts and Auto Zone)

Year	*2000*	*2001*	*2002*	*2003*	*2004*	*2005*	*2006*	*2007*
Advanced Auto Parts Sales, y	2288	2518	3288	3494	3770	4265	4625	5050
Auto Zone Sales, y	4483	4818	5326	5457	5637	5711	5948	6230

38. The table shows the numbers of doctorate degrees y awarded in the education fields in the United States during the years 2001 to 2004. Find the least squares regression line for the data. Let t represent the year, with $t = 1$ corresponding to 2001. (Source: U.S. National Science Foundation)

Year	2001	2002	2003	2004
Doctorate degrees, y	6337	6487	6627	6635

39. The table shows the world carbon dioxide emissions y (in millions of metric tons) during the years 1999 to 2004. Find the least squares regression quadratic polynomial for the data. Let t represent the year, with $t = -1$ corresponding to 1999. (Source: U.S. Energy Information Administration)

Year	1999	2000	2001	2002	2003	2004
CO_2, y	6325	6505	6578	6668	6999	7376

40. The table shows the sales y (in millions of dollars) for Gateway, Incorporated during the years 2000 to 2007. Find a least squares regression quadratic polynomial that best fits the data. Let t represent the year, with $t = 0$ corresponding to 2000. (Source: Gateway Inc.)

Year	2000	2001	2002	2003	2004	2005	2006	2007
Sales, y	9600.6	6079.2	4171.3	3402.4	3649.7	3854.1	4075.0	4310.0

41. The table shows the sales y (in millions of dollars) for Dell Incorporated during the years 1996 to 2007. Find the least squares regression line and the least squares cubic regression polynomial for the data. Let t represent the year, with $t = -4$ corresponding to 1996. Which model is the better fit for the data? Why? (Source: Dell Inc.)

Year	1996	1997	1998	1999
Sales, y	7759	12,327	18,243	25,265

Year	2000	2001	2002	2003
Sales, y	31,888	31,168	35,404	41,444

Year	2004	2005	2006	2007
Sales, y	49,205	55,908	58,200	61,000

42. The table shows the net profits y (in millions of dollars) for Polo Ralph Lauren during the years 1996 to 2007. Find the least squares regression line and the least squares cubic regression polynomial for the data. Let t represent the year, with $t = -4$ corresponding to 1996. Which model is the better fit for the data? Why? (Source: Polo Ralph Lauren)

Year	1996	1997	1998	1999
Net Profit, y	81.3	120.1	125.3	147.5

Year	2000	2001	2002	2003
Net Profit, y	166.3	168.6	183.7	184.4

Year	2004	2005	2006	2007
Net Profit, y	257.2	308.0	385.0	415.0

True or False? In Exercises 43 and 44, determine whether each statement is true or false. If a statement is true, give a reason or cite an appropriate statement from the text. If a statement is false, provide an example that shows the statement is not true in all cases or cite an appropriate statement from the text.

43. (a) If S_1 and S_2 are orthogonal subspaces of R^n, then their intersection is an empty set.

(b) If each vector $\mathbf{v} \in R^n$ can be uniquely written as a sum of a vector $\mathbf{s}_1$ from S_1 and a vector $\mathbf{s}_2$ from S_2, then R^n is called the direct sum of S_1 and S_2.

(c) The solution of the least squares problem consists essentially of solving the normal equations—that is, solving the $n \times n$ linear system of equations $A^TA\mathbf{x} = A^T\mathbf{b}$.

44. (a) If A is an $m \times n$ matrix, then $R(A)$ and $N(A^T)$ are orthogonal subspaces of R^n.

(b) The set of all vectors orthogonal to every vector in a subspace S is called the orthogonal complement of S.

(c) Given an $m \times n$ matrix A and a vector $\mathbf{b}$ in R^m, the least squares problem is to find $\mathbf{x}$ in R^n such that $\|A\mathbf{x} - \mathbf{b}\|^2$ is minimized.

45. Prove that if S_1 and S_2 are orthogonal subspaces of R^n, then their intersection consists only of the zero vector.

46. Prove that the orthogonal complement of a subspace of R^n is itself a subspace of R^n.

47. Prove Theorem 5.14.

48. Prove that if S_1 and S_2 are subspaces of R^n and if $R^n = S_1 \oplus S_2$, then $S_1 \cap S_2 = \{\mathbf{0}\}$.

49. Writing Describe the normal equations for the least squares problem if the $m \times n$ matrix A has orthonormal columns.

5.5 Applications of Inner Product Spaces

The Cross Product of Two Vectors in Space

Many problems in linear algebra involve finding a vector orthogonal to each vector in a set. Here you will look at a vector product that yields a vector in R^3 orthogonal to two vectors. This vector product is called the **cross product,** and it is most conveniently defined and calculated with vectors written in standard unit vector form.

$$\mathbf{v} = (v_1, v_2, v_3) = v_1\mathbf{i} + v_2\mathbf{j} + v_3\mathbf{k}$$

Definition of Cross Product of Two Vectors

Let $\mathbf{u} = u_1\mathbf{i} + u_2\mathbf{j} + u_3\mathbf{k}$ and $\mathbf{v} = v_1\mathbf{i} + v_2\mathbf{j} + v_3\mathbf{k}$ be vectors in R^3. The **cross product** of $\mathbf{u}$ and $\mathbf{v}$ is the vector

$$\mathbf{u} \times \mathbf{v} = (u_2v_3 - u_3v_2)\mathbf{i} - (u_1v_3 - u_3v_1)\mathbf{j} + (u_1v_2 - u_2v_1)\mathbf{k}.$$

REMARK: The cross product is defined only for vectors in R^3. The cross product of two vectors in R^2 or of vectors in R^n, $n > 3$, is not defined here.

A convenient way to remember the formula for the cross product $\mathbf{u} \times \mathbf{v}$ is to use the determinant form shown below.

$$\mathbf{u} \times \mathbf{v} = \begin{vmatrix} \mathbf{i} & \mathbf{j} & \mathbf{k} \\ u_1 & u_2 & u_3 \\ v_1 & v_2 & v_3 \end{vmatrix} \quad \begin{matrix} \\ \longleftarrow \text{Components of } \mathbf{u} \\ \longleftarrow \text{Components of } \mathbf{v} \end{matrix}$$

Technically this is not a determinant because the entries are not all real numbers. Nevertheless, it is useful because it provides an easy way to remember the cross product formula. Using cofactor expansion along the first row, you obtain

$$\begin{aligned} \mathbf{u} \times \mathbf{v} &= \begin{vmatrix} u_2 & u_3 \\ v_2 & v_3 \end{vmatrix}\mathbf{i} - \begin{vmatrix} u_1 & u_3 \\ v_1 & v_3 \end{vmatrix}\mathbf{j} + \begin{vmatrix} u_1 & u_2 \\ v_1 & v_2 \end{vmatrix}\mathbf{k} \\ &= (u_2v_3 - u_3v_2)\mathbf{i} - (u_1v_3 - u_3v_1)\mathbf{j} + (u_1v_2 - u_2v_1)\mathbf{k}, \end{aligned}$$

which yields the formula in the definition. Be sure to note that the **j**-component is preceded by a minus sign.

EXAMPLE 1 Finding the Cross Product of Two Vectors

Provided that $\mathbf{u} = \mathbf{i} - 2\mathbf{j} + \mathbf{k}$ and $\mathbf{v} = 3\mathbf{i} + \mathbf{j} - 2\mathbf{k}$, find

(a) $\mathbf{u} \times \mathbf{v}$. (b) $\mathbf{v} \times \mathbf{u}$. (c) $\mathbf{v} \times \mathbf{v}$.

SOLUTION

$$\text{(a)}\ \mathbf{u} \times \mathbf{v} = \begin{vmatrix} \mathbf{i} & \mathbf{j} & \mathbf{k} \\ 1 & -2 & 1 \\ 3 & 1 & -2 \end{vmatrix}$$

$$= \begin{vmatrix} -2 & 1 \\ 1 & -2 \end{vmatrix}\mathbf{i} - \begin{vmatrix} 1 & 1 \\ 3 & -2 \end{vmatrix}\mathbf{j} + \begin{vmatrix} 1 & -2 \\ 3 & 1 \end{vmatrix}\mathbf{k}$$

$$= 3\mathbf{i} + 5\mathbf{j} + 7\mathbf{k}$$

Simulation

To explore this concept further with an electronic simulation and for keystrokes and programming syntax for specific graphing utilities and computer software programs applicable to Example 1, please visit *college.hmco.com/pic/larsonELA6e*. Similar exercises and projects are also available on this website.

$$\text{(b)}\ \mathbf{v} \times \mathbf{u} = \begin{vmatrix} \mathbf{i} & \mathbf{j} & \mathbf{k} \\ 3 & 1 & -2 \\ 1 & -2 & 1 \end{vmatrix}$$

$$= \begin{vmatrix} 1 & -2 \\ -2 & 1 \end{vmatrix}\mathbf{i} - \begin{vmatrix} 3 & -2 \\ 1 & 1 \end{vmatrix}\mathbf{j} + \begin{vmatrix} 3 & 1 \\ 1 & -2 \end{vmatrix}\mathbf{k}$$

$$= -3\mathbf{i} - 5\mathbf{j} - 7\mathbf{k}$$

Note that this result is the negative of that in part (a).

$$\text{(c)}\ \mathbf{v} \times \mathbf{v} = \begin{vmatrix} \mathbf{i} & \mathbf{j} & \mathbf{k} \\ 3 & 1 & -2 \\ 3 & 1 & -2 \end{vmatrix}$$

$$= \begin{vmatrix} 1 & -2 \\ 1 & -2 \end{vmatrix}\mathbf{i} - \begin{vmatrix} 3 & -2 \\ 3 & -2 \end{vmatrix}\mathbf{j} + \begin{vmatrix} 3 & 1 \\ 3 & 1 \end{vmatrix}\mathbf{k}$$

$$= 0\mathbf{i} + 0\mathbf{j} + 0\mathbf{k} = \mathbf{0}$$

Technology Note

Some graphing utilities and computer software programs have vector capabilities that include finding a cross product. For example, on a graphing utility, you can verify $\mathbf{u} \times \mathbf{v}$ in Example 1(a) and it could appear as shown below. Keystrokes and programming syntax for these utilities/programs applicable to Example 1(a) are provided in the **Online Technology Guide,** available at *college.hmco.com/pic/larsonELA6e*.

```
VECTOR:U        3
 e1=1
 e2=-2
 e3=1
```

```
VECTOR:V        3
 e1=3
 e2=1
 e3=-2
```

```
cross(U,V)
            [3 5 7]
```

The results obtained in Example 1 suggest some interesting *algebraic* properties of the cross product. For instance,

$$\mathbf{u} \times \mathbf{v} = -(\mathbf{v} \times \mathbf{u}) \qquad \text{and} \qquad \mathbf{v} \times \mathbf{v} = \mathbf{0}.$$

These properties, along with several others, are shown in Theorem 5.17.

THEOREM 5.17 Algebraic Properties of the Cross Product

If $\mathbf{u}$, $\mathbf{v}$, and $\mathbf{w}$ are vectors in R^3 and c is a scalar, then the following properties are true.

1. $\mathbf{u} \times \mathbf{v} = -(\mathbf{v} \times \mathbf{u})$
2. $\mathbf{u} \times (\mathbf{v} + \mathbf{w}) = (\mathbf{u} \times \mathbf{v}) + (\mathbf{u} \times \mathbf{w})$
3. $c(\mathbf{u} \times \mathbf{v}) = c\mathbf{u} \times \mathbf{v} = \mathbf{u} \times c\mathbf{v}$
4. $\mathbf{u} \times \mathbf{0} = \mathbf{0} \times \mathbf{u} = \mathbf{0}$
5. $\mathbf{u} \times \mathbf{u} = \mathbf{0}$
6. $\mathbf{u} \cdot (\mathbf{v} \times \mathbf{w}) = (\mathbf{u} \times \mathbf{v}) \cdot \mathbf{w}$

PROOF The proof of the first property is shown here. The proofs of the other properties are left to you. (See Exercises 40–44.) Let $\mathbf{u}$ and $\mathbf{v}$ be

$$\mathbf{u} = u_1\mathbf{i} + u_2\mathbf{j} + u_3\mathbf{k}$$

and

$$\mathbf{v} = v_1\mathbf{i} + v_2\mathbf{j} + v_3\mathbf{k}.$$

Then $\mathbf{u} \times \mathbf{v}$ is

$$\begin{aligned} \mathbf{u} \times \mathbf{v} &= \begin{vmatrix} \mathbf{i} & \mathbf{j} & \mathbf{k} \\ u_1 & u_2 & u_3 \\ v_1 & v_2 & v_3 \end{vmatrix} \\ &= (u_2v_3 - u_3v_2)\mathbf{i} - (u_1v_3 - u_3v_1)\mathbf{j} + (u_1v_2 - u_2v_1)\mathbf{k}, \end{aligned}$$

and $\mathbf{v} \times \mathbf{u}$ is

$$\begin{aligned} \mathbf{v} \times \mathbf{u} &= \begin{vmatrix} \mathbf{i} & \mathbf{j} & \mathbf{k} \\ v_1 & v_2 & v_3 \\ u_1 & u_2 & u_3 \end{vmatrix} \\ &= (v_2u_3 - v_3u_2)\mathbf{i} - (v_1u_3 - v_3u_1)\mathbf{j} + (v_1u_2 - v_2u_1)\mathbf{k} \\ &= -(u_2v_3 - u_3v_2)\mathbf{i} + (u_1v_3 - u_3v_1)\mathbf{j} - (u_1v_2 - u_2v_1)\mathbf{k} \\ &= -(\mathbf{v} \times \mathbf{u}). \end{aligned}$$

Property 1 of Theorem 5.17 tells you that the vectors $\mathbf{u} \times \mathbf{v}$ and $\mathbf{v} \times \mathbf{u}$ have equal lengths but opposite directions. The geometric implication of this will be discussed after some geometric properties of the cross product of two vectors have been established.

THEOREM 5.18
Geometric Properties of the Cross Product

If $\mathbf{u}$ and $\mathbf{v}$ are nonzero vectors in R^3, then the following properties are true.

1. $\mathbf{u} \times \mathbf{v}$ is orthogonal to both $\mathbf{u}$ and $\mathbf{v}$.
2. The angle θ between $\mathbf{u}$ and $\mathbf{v}$ is given by

$$\|\mathbf{u} \times \mathbf{v}\| = \|\mathbf{u}\|\,\|\mathbf{v}\| \sin \theta.$$

3. $\mathbf{u}$ and $\mathbf{v}$ are parallel if and only if $\mathbf{u} \times \mathbf{v} = \mathbf{0}$.
4. The parallelogram having $\mathbf{u}$ and $\mathbf{v}$ as adjacent sides has an area of $\|\mathbf{u} \times \mathbf{v}\|$.

PROOF The proof of Property 4 follows. The proofs of the other properties are left to you. (See Exercises 45–47.) To prove Property 4, let $\mathbf{u}$ and $\mathbf{v}$ represent adjacent sides of a parallelogram, as shown in Figure 5.28. By Property 2, the area of the parallelogram is

$$\text{Area} = \overbrace{\|\mathbf{u}\|}^{\text{Base}}\,\overbrace{\|\mathbf{v}\| \sin \theta}^{\text{Height}} = \|\mathbf{u} \times \mathbf{v}\|.$$

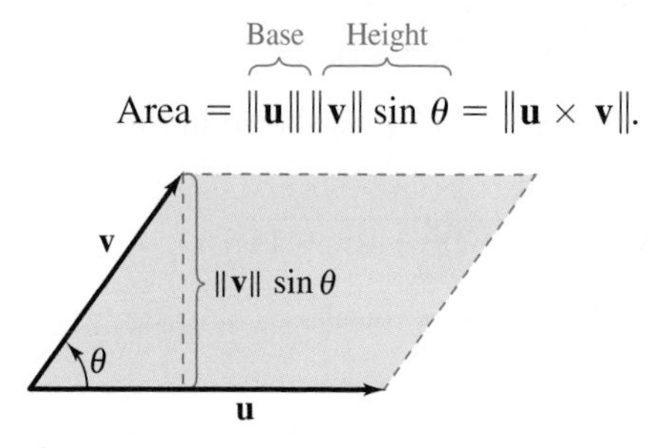

Figure 5.28

Property 1 states that the vector $\mathbf{u} \times \mathbf{v}$ is orthogonal to both $\mathbf{u}$ and $\mathbf{v}$. This implies that $\mathbf{u} \times \mathbf{v}$ (and $\mathbf{v} \times \mathbf{u}$) is orthogonal to the plane determined by $\mathbf{u}$ and $\mathbf{v}$. One way to remember the orientation of the vectors $\mathbf{u}$, $\mathbf{v}$, and $\mathbf{u} \times \mathbf{v}$ is to compare them with the unit vectors $\mathbf{i}$, $\mathbf{j}$, and $\mathbf{k}$, as shown in Figure 5.29. The three vectors $\mathbf{u}$, $\mathbf{v}$, and $\mathbf{u} \times \mathbf{v}$ form a *right-handed system,* whereas the three vectors $\mathbf{u}$, $\mathbf{v}$, and $\mathbf{v} \times \mathbf{u}$ form a *left-handed system.*

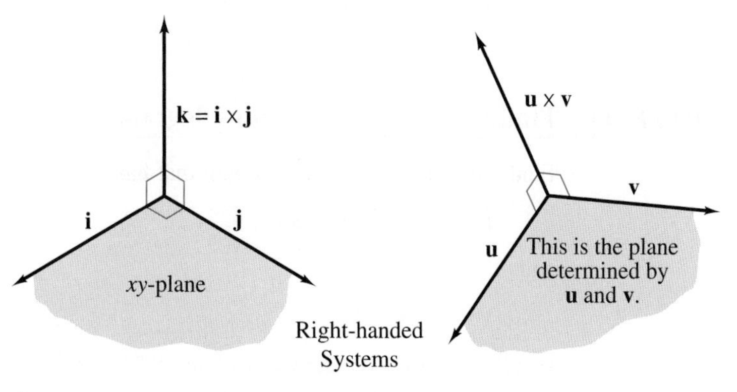

Figure 5.29

EXAMPLE 2 Finding a Vector Orthogonal to Two Given Vectors

Find a unit vector orthogonal to both

$$\mathbf{u} = \mathbf{i} - 4\mathbf{j} + \mathbf{k}$$

and

$$\mathbf{v} = 2\mathbf{i} + 3\mathbf{j}.$$

SOLUTION From Property 1 of Theorem 5.18, you know that the cross product

$$\mathbf{u} \times \mathbf{v} = \begin{vmatrix} \mathbf{i} & \mathbf{j} & \mathbf{k} \\ 1 & -4 & 1 \\ 2 & 3 & 0 \end{vmatrix}$$
$$= -3\mathbf{i} + 2\mathbf{j} + 11\mathbf{k}$$

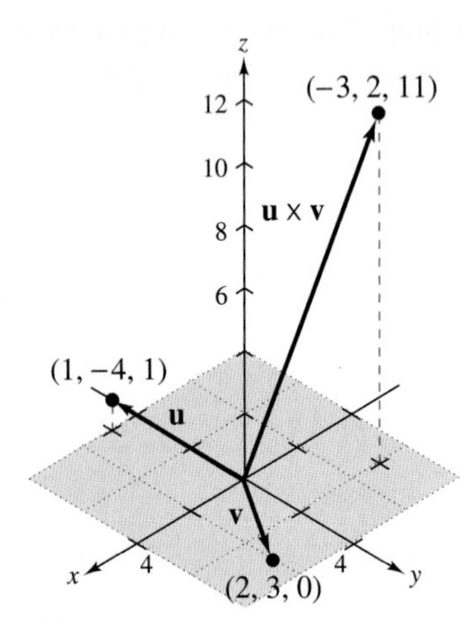

Figure 5.30

is orthogonal to both $\mathbf{u}$ and $\mathbf{v}$, as shown in Figure 5.30. Then, by dividing by the length of $\mathbf{u} \times \mathbf{v}$,

$$\|\mathbf{u} \times \mathbf{v}\| = \sqrt{(-3)^2 + 2^2 + 11^2}$$
$$= \sqrt{134},$$

you obtain the unit vector

$$\frac{\mathbf{u} \times \mathbf{v}}{\|\mathbf{u} \times \mathbf{v}\|} = -\frac{3}{\sqrt{134}}\mathbf{i} + \frac{2}{\sqrt{134}}\mathbf{j} + \frac{11}{\sqrt{134}}\mathbf{k},$$

which is orthogonal to both $\mathbf{u}$ and $\mathbf{v}$, as follows.

$$\left(-\frac{3}{\sqrt{134}}, \frac{2}{\sqrt{134}}, \frac{11}{\sqrt{134}}\right) \cdot (1, -4, 1) = 0$$
$$\left(-\frac{3}{\sqrt{134}}, \frac{2}{\sqrt{134}}, \frac{11}{\sqrt{134}}\right) \cdot (2, 3, 0) = 0$$

EXAMPLE 3 Finding the Area of a Parallelogram

Find the area of the parallelogram that has

$$\mathbf{u} = -3\mathbf{i} + 4\mathbf{j} + \mathbf{k}$$

and

$$\mathbf{v} = -2\mathbf{j} + 6\mathbf{k}$$

as adjacent sides, as shown in Figure 5.31.

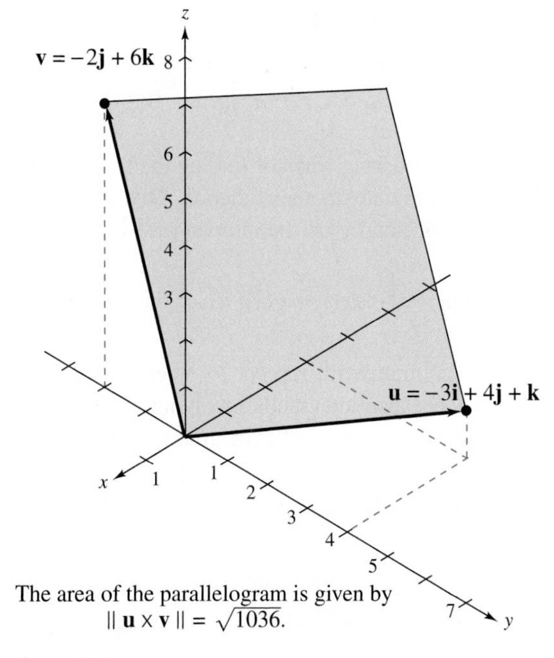

Figure 5.31

SOLUTION From Property 4 of Theorem 5.18, you know that the area of this parallelogram is $\|\mathbf{u} \times \mathbf{v}\|$. Because

$$\mathbf{u} \times \mathbf{v} = \begin{vmatrix} \mathbf{i} & \mathbf{j} & \mathbf{k} \\ -3 & 4 & 1 \\ 0 & -2 & 6 \end{vmatrix} = 26\mathbf{i} + 18\mathbf{j} + 6\mathbf{k},$$

the area of the parallelogram is

$$\|\mathbf{u} \times \mathbf{v}\| = \sqrt{26^2 + 18^2 + 6^2} = \sqrt{1036} \approx 32.19.$$

Least Squares Approximations (Calculus)

Many problems in the physical sciences and engineering involve an approximation of a function f by another function g. If f is in $C[a, b]$ (the inner product space of all continuous functions on $[a, b]$), then g usually is chosen from a subspace W of $C[a, b]$. For instance, to approximate the function

$$f(x) = e^x, \quad 0 \leq x \leq 1,$$

you could choose one of the following forms of g.

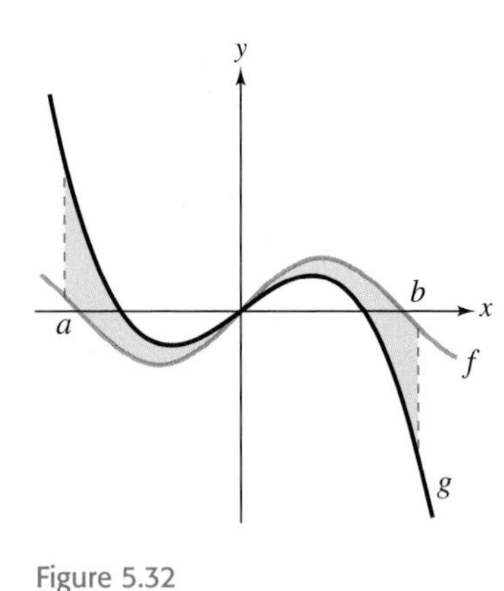

Figure 5.32

1. $g(x) = a_0 + a_1x, \quad 0 \le x \le 1$ **Linear**
2. $g(x) = a_0 + a_1x + a_2x^2, \quad 0 \le x \le 1$ **Quadratic**
3. $g(x) = a_0 + a_1 \cos x + a_2 \sin x, \quad 0 \le x \le 1$ **Trigonometric**

Before discussing ways of finding the function g, you must define how one function can "best" approximate another function. One natural way would require the area bounded by the graphs of f and g on the interval $[a, b]$,

$$\text{Area} = \int_a^b |f(x) - g(x)|\, dx,$$

to be a minimum with respect to other functions in the subspace W, as shown in Figure 5.32. But because integrands involving absolute value are often difficult to evaluate, it is more common to square the integrand to obtain

$$\int_a^b [f(x) - g(x)]^2\, dx.$$

With this criterion, the function g is called the **least squares approximation** of f with respect to the inner product space W.

Definition of Least Squares Approximation

Let f be continuous on $[a, b]$, and let W be a subspace of $C[a, b]$. A function g in W is called a **least squares approximation** of f with respect to W if the value of

$$I = \int_a^b [f(x) - g(x)]^2\, dx$$

is a minimum with respect to all other functions in W.

REMARK: If the subspace W in this definition is the entire space $C[a, b]$, then $g(x) = f(x)$, which implies that $I = 0$.

EXAMPLE 4 Finding a Least Squares Approximation

Find the least squares approximation $g(x) = a_0 + a_1x$ for

$$f(x) = e^x, \quad 0 \le x \le 1.$$

SOLUTION For this approximation you need to find the constants a_0 and a_1 that minimize the value of

$$I = \int_0^1 [f(x) - g(x)]^2\, dx$$

$$= \int_0^1 (e^x - a_0 - a_1x)^2\, dx.$$

Evaluating this integral, you have

$$
\begin{aligned}
I &= \int_0^1 (e^x - a_0 - a_1x)^2\,dx \\
&= \int_0^1 (e^{2x} - 2a_0e^x - 2a_1xe^x + a_0^2 + 2a_0a_1x + a_1^2x^2)\,dx \\
&= \left[\frac{1}{2}e^{2x} - 2a_0e^x - 2a_1e^x(x-1) + a_0^2x + a_0a_1x^2 + a_1^2\frac{x^3}{3}\right]_0^1 \\
&= \frac{1}{2}(e^2 - 1) - 2a_0(e - 1) - 2a_1 + a_0^2 + a_0a_1 + \frac{1}{3}a_1^2.
\end{aligned}
$$

Now, considering I to be a function of the variables a_0 and a_1, use calculus to determine the values of a_0 and a_1 that minimize I. Specifically, by setting the partial derivatives

$$\frac{\partial I}{\partial a_0} = 2a_0 - 2e + 2 + a_1$$

$$\frac{\partial I}{\partial a_1} = a_0 + \frac{2}{3}a_1 - 2$$

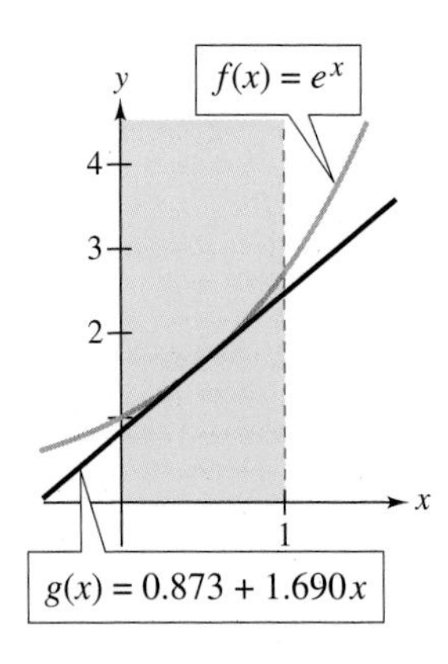

Figure 5.33

equal to zero, you obtain the following two linear equations in a_0 and a_1.

$$
\begin{aligned}
2a_0 + a_1 &= 2(e - 1) \\
3a_0 + 2a_1 &= 6
\end{aligned}
$$

The solution of this system is

$$a_0 = 4e - 10 \approx 0.873 \qquad \text{and} \qquad a_1 = 18 - 6e \approx 1.690.$$

So, the best *linear approximation* of $f(x) = e^x$ on the interval $[0, 1]$ is

$$
\begin{aligned}
g(x) &= 4e - 10 + (18 - 6e)x \\
&\approx 0.873 + 1.690x.
\end{aligned}
$$

Figure 5.33 shows the graphs of f and g on $[0, 1]$.

Of course, the approximation obtained in Example 4 depends on the definition of the best approximation. For instance, if the definition of "best" had been the *Taylor polynomial of degree 1* centered at 0.5, then the approximating function g would have been

$$
\begin{aligned}
g(x) &= f(0.5) + f'(0.5)(x - 0.5) \\
&= e^{0.5} + e^{0.5}(x - 0.5) \\
&\approx 0.824 + 1.649x.
\end{aligned}
$$

Moreover, the function g obtained in Example 4 is only the best *linear* approximation of f (according to the least squares criterion). In Example 5 you will find the best *quadratic* approximation.

EXAMPLE 5 **Finding a Least Squares Approximation**

Find the least squares approximation $g(x) = a_0 + a_1x + a_2x^2$ for

$$f(x) = e^x, \quad 0 \le x \le 1.$$

SOLUTION For this approximation you need to find the values of a_0, a_1, and a_2 that minimize the value of

$$\begin{aligned} I &= \int_0^1 [f(x) - g(x)]^2\, dx \\ &= \int_0^1 (e^x - a_0 - a_1x - a_2x^2)^2\, dx \\ &= \frac{1}{2}(e^2 - 1) + 2a_0(1 - e) + 2a_2(2 - e) \\ &\quad + a_0^2 + a_0a_1 + \frac{2}{3}a_0a_2 + \frac{1}{2}a_1a_2 + \frac{1}{3}a_1^2 + \frac{1}{5}a_2^2 - 2a_1. \end{aligned}$$

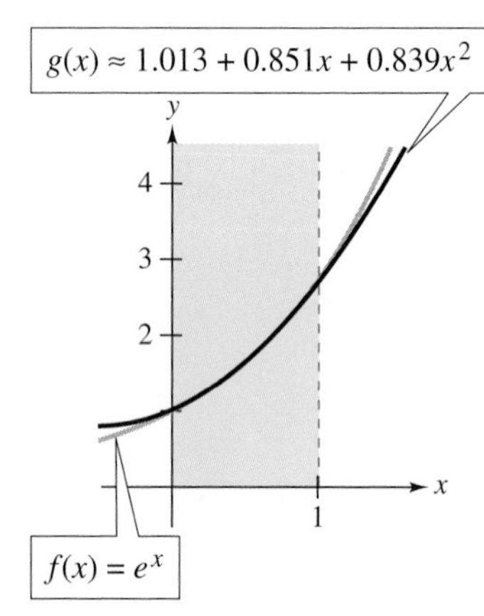

Figure 5.34

Integrating and then setting the partial derivatives of I (with respect to a_0, a_1, and a_2) equal to zero produces the following system of linear equations.

$$\begin{aligned} 6a_0 + 3a_1 + 2a_2 &= 6(e - 1) \\ 6a_0 + 4a_1 + 3a_2 &= 12 \\ 20a_0 + 15a_1 + 12a_2 &= 60(e - 2) \end{aligned}$$

The solution of this system is

$$\begin{aligned} a_0 &= -105 + 39e \approx 1.013 \\ a_1 &= 588 - 216e \approx 0.851 \\ a_2 &= -570 + 210e \approx 0.839. \end{aligned}$$

So, the approximating function g is

$$g(x) \approx 1.013 + 0.851x + 0.839x^2.$$

The graphs of f and g are shown in Figure 5.34.

The integral I (given in the definition of the least squares approximation) can be expressed in vector form. To do this, use the inner product defined in Example 5 in Section 5.2:

$$\langle f, g \rangle = \int_a^b f(x)g(x)\, dx.$$

With this inner product you have

$$I = \int_a^b [f(x) - g(x)]^2\, dx = \langle f - g, f - g \rangle = \|f - g\|^2.$$

This means that the least squares approximating function g is the function that minimizes $\|f - g\|^2$ or, equivalently, minimizes $\|f - g\|$. In other words, the least squares approximation of a function f is the function g (in the subspace W) closest to f in terms of the inner product $\langle f, g\rangle$. The next theorem gives you a way of determining the function g.

THEOREM 5.19 Least Squares Approximation

Let f be continuous on $[a, b]$, and let W be a finite-dimensional subspace of $C[a, b]$. The least squares approximating function of f with respect to W is given by

$$g = \langle f, \mathbf{w}_1\rangle \mathbf{w}_1 + \langle f, \mathbf{w}_2\rangle \mathbf{w}_2 + \cdot \cdot \cdot + \langle f, \mathbf{w}_n\rangle \mathbf{w}_n,$$

where $B = \{\mathbf{w}_1, \mathbf{w}_2, \ldots, \mathbf{w}_n\}$ is an orthonormal basis for W.

PROOF To show that g is the least squares approximating function of f, prove that the inequality

$$\|f - g\| \le \|f - \mathbf{w}\|$$

is true for any vector $\mathbf{w}$ in W. By writing $f - g$ as

$$f - g = f - \langle f, \mathbf{w}_1\rangle \mathbf{w}_1 - \langle f, \mathbf{w}_2\rangle \mathbf{w}_2 - \cdot \cdot \cdot - \langle f, \mathbf{w}_n\rangle \mathbf{w}_n,$$

you can see that $f - g$ is orthogonal to each $\mathbf{w}_i$, which in turn implies that it is orthogonal to each vector in W. In particular, $f - g$ is orthogonal to $g - \mathbf{w}$. This allows you to apply the Pythagorean Theorem to the vector sum

$$f - \mathbf{w} = (f - g) + (g - \mathbf{w})$$

to conclude that

$$\|f - \mathbf{w}\|^2 = \|f - g\|^2 + \|g - \mathbf{w}\|^2.$$

So, it follows that $\|f - g\|^2 \le \|f - \mathbf{w}\|^2$, which then implies that $\|f - g\| \le \|f - \mathbf{w}\|$.

Now observe how Theorem 5.19 can be used to produce the least squares approximation obtained in Example 4. First apply the Gram-Schmidt orthonormalization process to the standard basis $\{1, x\}$ to obtain the orthonormal basis $B = \{1, \sqrt{3}(2x - 1)\}$. Then, by Theorem 5.19, the least squares approximation for e^x in the subspace of all linear functions is

$$\begin{aligned} g(x) &= \langle e^x, 1\rangle(1) + \langle e^x, \sqrt{3}(2x - 1)\rangle \sqrt{3}(2x - 1) \\ &= \int_0^1 e^x\,dx + \sqrt{3}(2x - 1) \int_0^1 \sqrt{3}e^x(2x - 1)\,dx \\ &= \int_0^1 e^x\,dx + 3(2x - 1) \int_0^1 e^x(2x - 1)\,dx \\ &= (e - 1) + 3(2x - 1)(3 - e) \\ &= 4e - 10 + (18 - 6e)x, \end{aligned}$$

which agrees with the result obtained in Example 4.

EXAMPLE 6 **Finding a Least Squares Approximation**

Find the least squares approximation for $f(x) = \sin x$, $0 \le x \le \pi$, with respect to the subspace W of quadratic functions.

SOLUTION To use Theorem 5.19, apply the Gram-Schmidt orthonormalization process to the standard basis for W, $\{1, x, x^2\}$, to obtain the orthonormal basis

$$\begin{aligned} B &= \{\mathbf{w}_1, \mathbf{w}_2, \mathbf{w}_3\} \\ &= \left\{\frac{1}{\sqrt{\pi}}, \frac{\sqrt{3}}{\pi\sqrt{\pi}}(2x - \pi), \frac{\sqrt{5}}{\pi^2\sqrt{\pi}}(6x^2 - 6\pi x + \pi^2)\right\}. \end{aligned}$$

The least squares approximating function g is

$$g(x) = \langle f, \mathbf{w}_1\rangle\mathbf{w}_1 + \langle f, \mathbf{w}_2\rangle\mathbf{w}_2 + \langle f, \mathbf{w}_3\rangle\mathbf{w}_3,$$

and you have

$$\begin{aligned} \langle f, \mathbf{w}_1\rangle &= \frac{1}{\sqrt{\pi}}\int_0^{\pi} \sin x\, dx = \frac{2}{\sqrt{\pi}} \\ \langle f, \mathbf{w}_2\rangle &= \frac{\sqrt{3}}{\pi\sqrt{\pi}}\int_0^{\pi} \sin x(2x - \pi)\, dx = 0 \\ \langle f, \mathbf{w}_3\rangle &= \frac{\sqrt{5}}{\pi^2\sqrt{\pi}}\int_0^{\pi} \sin x(6x^2 - 6\pi x + \pi^2)\, dx \\ &= \frac{2\sqrt{5}}{\pi^2\sqrt{\pi}}(\pi^2 - 12). \end{aligned}$$

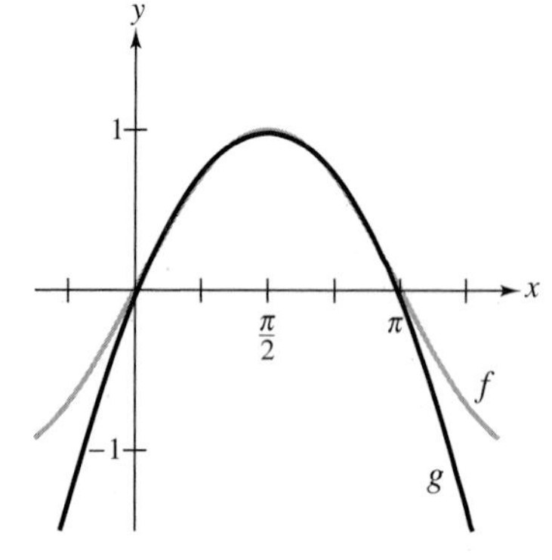

Figure 5.35

So, g is

$$\begin{aligned} g(x) &= \frac{2}{\pi} + \frac{10(\pi^2 - 12)}{\pi^5}(6x^2 - 6\pi x + \pi^2) \\ &\approx -0.4177x^2 + 1.3122x - 0.0505. \end{aligned}$$

The graphs of f and g are shown in Figure 5.35.

Fourier Approximations (Calculus)

You will now look at a special type of least squares approximation called a **Fourier approximation.** For this approximation, consider functions of the form

$$g(x) = \frac{a_0}{2} + a_1 \cos x + \cdot \cdot \cdot + a_n \cos nx + b_1 \sin x + \cdot \cdot \cdot + b_n \sin nx$$

in the subspace W of $C[0, 2\pi]$ spanned by the basis

$$S = \{1, \cos x, \cos 2x, \ldots, \cos nx, \sin x, \sin 2x, \ldots, \sin nx\}.$$

These $2n + 1$ vectors are *orthogonal* in the inner product space $C[0, 2\pi]$ because

$$\langle f, g \rangle = \int_0^{2\pi} f(x)g(x)\,dx = 0, \quad f \neq g,$$

as demonstrated in Example 3 in Section 5.3. Moreover, by normalizing each function in this basis, you obtain the orthonormal basis

$$\begin{aligned} B &= \{\mathbf{w}_0, \mathbf{w}_1, \ldots, \mathbf{w}_n, \mathbf{w}_{n+1}, \ldots, \mathbf{w}_{2n}\} \\ &= \left\{\frac{1}{\sqrt{2\pi}}, \frac{1}{\sqrt{\pi}}\cos x, \ldots, \frac{1}{\sqrt{\pi}}\cos nx, \frac{1}{\sqrt{\pi}}\sin x, \ldots, \frac{1}{\sqrt{\pi}}\sin nx\right\}. \end{aligned}$$

With this orthonormal basis, you can apply Theorem 5.19 to write

$$g(x) = \langle f, \mathbf{w}_0 \rangle \mathbf{w}_0 + \langle f, \mathbf{w}_1 \rangle \mathbf{w}_1 + \cdots + \langle f, \mathbf{w}_{2n} \rangle \mathbf{w}_{2n}.$$

The coefficients $a_0, a_1, \ldots, a_n, b_1, \ldots, b_n$ for $g(x)$ in the equation

$$g(x) = \frac{a_0}{2} + a_1 \cos x + \cdots + a_n \cos nx + b_1 \sin x + \cdots + b_n \sin nx$$

are shown by the following integrals.

$$a_0 = \langle f, \mathbf{w}_0 \rangle \frac{2}{\sqrt{2\pi}} = \frac{2}{\sqrt{2\pi}} \int_0^{2\pi} f(x) \frac{1}{\sqrt{2\pi}}\,dx = \frac{1}{\pi}\int_0^{2\pi} f(x)\,dx$$

$$a_1 = \langle f, \mathbf{w}_1 \rangle \frac{1}{\sqrt{\pi}} = \frac{1}{\sqrt{\pi}} \int_0^{2\pi} f(x) \frac{1}{\sqrt{\pi}}\cos x\,dx = \frac{1}{\pi}\int_0^{2\pi} f(x)\cos x\,dx$$

$$\vdots$$

$$a_n = \langle f, \mathbf{w}_n \rangle \frac{1}{\sqrt{\pi}} = \frac{1}{\sqrt{\pi}} \int_0^{2\pi} f(x) \frac{1}{\sqrt{\pi}}\cos nx\,dx = \frac{1}{\pi}\int_0^{2\pi} f(x)\cos nx\,dx$$

$$b_1 = \langle f, \mathbf{w}_{n+1} \rangle \frac{1}{\sqrt{\pi}} = \frac{1}{\sqrt{\pi}} \int_0^{2\pi} f(x) \frac{1}{\sqrt{\pi}}\sin x\,dx = \frac{1}{\pi}\int_0^{2\pi} f(x)\sin x\,dx$$

$$\vdots$$

$$b_n = \langle f, \mathbf{w}_{2n} \rangle \frac{1}{\sqrt{\pi}} = \frac{1}{\sqrt{\pi}} \int_0^{2\pi} f(x) \frac{1}{\sqrt{\pi}}\sin nx\,dx = \frac{1}{\pi}\int_0^{2\pi} f(x)\sin nx\,dx$$

The function $g(x)$ is called the ***n*th-order Fourier approximation** of f on the interval $[0, 2\pi]$. Like Fourier coefficients, this function is named after the French mathematician Jean-Baptiste Joseph Fourier (1768–1830). This brings you to Theorem 5.20.

THEOREM 5.20
Fourier Approximation

On the interval $[0, 2\pi]$, the least squares approximation of a continuous function f with respect to the vector space spanned by $\{1, \cos x, \ldots, \cos nx, \sin x, \ldots, \sin nx\}$ is given by

$$g(x) = \frac{a_0}{2} + a_1 \cos x + \cdots + a_n \cos nx + b_1 \sin x + \cdots + b_n \sin nx,$$

where the **Fourier coefficients** $a_0, a_1, \ldots, a_n, b_1, \ldots, b_n$ are

$$a_0 = \frac{1}{\pi}\int_0^{2\pi} f(x)\, dx$$

$$a_j = \frac{1}{\pi}\int_0^{2\pi} f(x) \cos jx\, dx, \quad j = 1, 2, \ldots, n$$

$$b_j = \frac{1}{\pi}\int_0^{2\pi} f(x) \sin jx\, dx, \quad j = 1, 2, \ldots, n.$$

EXAMPLE 7 Finding a Fourier Approximation

Find the third-order Fourier approximation of $f(x) = x$, $0 \le x \le 2\pi$.

SOLUTION Using Theorem 5.20, you have

$$g(x) = \frac{a_0}{2} + a_1 \cos x + a_2 \cos 2x + a_3 \cos 3x + b_1 \sin x + b_2 \sin 2x + b_3 \sin 3x,$$

where

$$a_0 = \frac{1}{\pi}\int_0^{2\pi} x\, dx = \frac{1}{\pi} 2\pi^2 = 2\pi$$

$$a_j = \frac{1}{\pi}\int_0^{2\pi} x \cos jx\, dx = \left[\frac{1}{\pi j^2}\cos jx + \frac{x}{\pi j}\sin jx\right]_0^{2\pi} = 0$$

$$b_j = \frac{1}{\pi}\int_0^{2\pi} x \sin jx\, dx = \left[\frac{1}{\pi j^2}\sin jx - \frac{x}{\pi j}\cos jx\right]_0^{2\pi} = -\frac{2}{j}.$$

This implies that $a_0 = 2\pi$, $a_1 = 0$, $a_2 = 0$, $a_3 = 0$, $b_1 = -2$, $b_2 = -\frac{2}{2} = -1$, and $b_3 = -\frac{2}{3}$. So, you have

$$g(x) = \frac{2\pi}{2} - 2 \sin x - \sin 2x - \frac{2}{3}\sin 3x$$

$$= \pi - 2 \sin x - \sin 2x - \frac{2}{3}\sin 3x.$$

The graphs of f and g are compared in Figure 5.36.

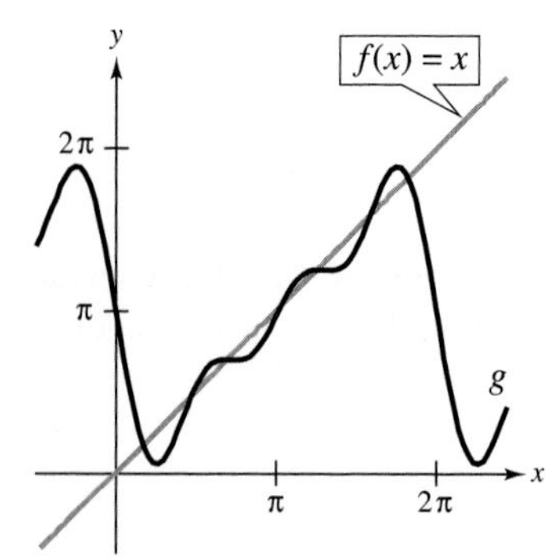

Third-Order Fourier Approximation

Figure 5.36

In Example 7 the general pattern for the Fourier coefficients appears to be $a_0 = 2\pi$, $a_1 = a_2 = \cdot \cdot \cdot = a_n = 0$, and

$$b_1 = -\frac{2}{1},\ b_2 = -\frac{2}{2},\ \ldots,\ b_n = -\frac{2}{n}.$$

The nth-order Fourier approximation of $f(x) = x$ is

$$g(x) = \pi - 2\left(\sin x + \frac{1}{2}\sin 2x + \frac{1}{3}\sin 3x + \cdot \cdot \cdot + \frac{1}{n}\sin nx\right).$$

As n increases, the Fourier approximation improves. For instance, Figure 5.37 shows the fourth- and fifth-order Fourier approximations of $f(x) = x,\ 0 \le x \le 2\pi$.

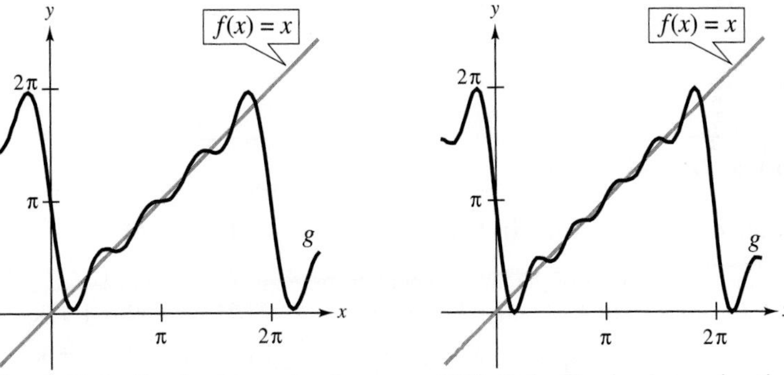

Figure 5.37

In advanced courses it is shown that as $n \to \infty$, the approximation error $\|f - g\|$ approaches zero for all x in the interval $(0, 2\pi)$. The infinite *series* for $g(x)$ is called a **Fourier series.**

EXAMPLE 8 Finding a Fourier Approximation

Find the fourth-order Fourier approximation of $f(x) = |x - \pi|,\ 0 \le x \le 2\pi$.

SOLUTION Using Theorem 5.20, find the Fourier coefficients as follows.

$$a_0 = \frac{1}{\pi}\int_0^{2\pi} |x - \pi|\, dx = \pi$$

$$\begin{aligned} a_j &= \frac{1}{\pi}\int_0^{2\pi} |x - \pi| \cos jx\, dx \\ &= \frac{2}{\pi}\int_0^{\pi} (\pi - x)\cos jx\, dx \\ &= \frac{2}{\pi j^2}(1 - \cos j\pi) \end{aligned}$$

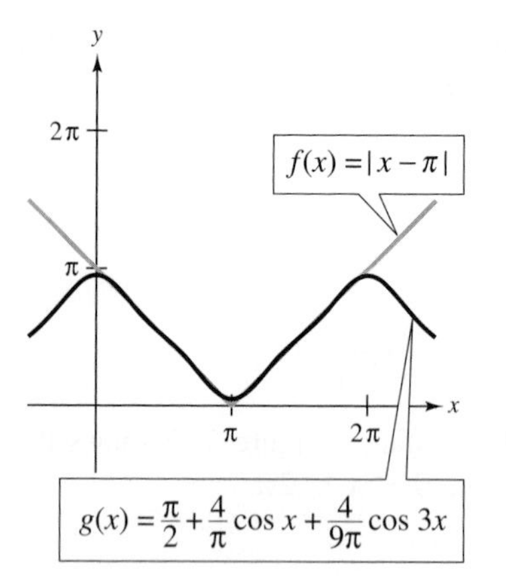

Figure 5.38

$$b_j = \frac{1}{\pi}\int_0^{2\pi} |x - \pi| \sin jx \, dx = 0$$

So, $a_0 = \pi$, $a_1 = 4/\pi$, $a_2 = 0$, $a_3 = 4/9\pi$, $a_4 = 0$, $b_1 = 0$, $b_2 = 0$, $b_3 = 0$, and $b_4 = 0$, which means that the fourth-order Fourier approximation of f is

$$g(x) = \frac{\pi}{2} + \frac{4}{\pi}\cos x + \frac{4}{9\pi}\cos 3x.$$

The graphs of f and g are compared in Figure 5.38.

SECTION 5.5 Exercises

The Cross Product of Two Vectors in Space

In Exercises 1–6, find the cross product of the unit vectors [where $\mathbf{i} = (1, 0, 0)$, $\mathbf{j} = (0, 1, 0)$, and $\mathbf{k} = (0, 0, 1)$]. Sketch your result.

1. $\mathbf{j} \times \mathbf{i}$

2. $\mathbf{i} \times \mathbf{j}$

3. $\mathbf{j} \times \mathbf{k}$

4. $\mathbf{k} \times \mathbf{j}$

5. $\mathbf{i} \times \mathbf{k}$

6. $\mathbf{k} \times \mathbf{i}$

In Exercises 7–16, find $\mathbf{u} \times \mathbf{v}$ and show that it is orthogonal to both $\mathbf{u}$ and $\mathbf{v}$.

7. $\mathbf{u} = (0, 1, -2)$, $\mathbf{v} = (1, -1, 0)$

8. $\mathbf{u} = (-1, 1, 2)$, $\mathbf{v} = (0, 1, -1)$

9. $\mathbf{u} = (12, -3, 1)$, $\mathbf{v} = (-2, 5, 1)$

10. $\mathbf{u} = (-2, 1, 1)$, $\mathbf{v} = (4, 2, 0)$

11. $\mathbf{u} = (2, -3, 1)$, $\mathbf{v} = (1, -2, 1)$

12. $\mathbf{u} = (4, 1, 0)$, $\mathbf{v} = (3, 2, -2)$

13. $\mathbf{u} = \mathbf{j} + 6\mathbf{k}$, $\mathbf{v} = 2\mathbf{i} - \mathbf{k}$

14. $\mathbf{u} = 2\mathbf{i} - \mathbf{j} + \mathbf{k}$, $\mathbf{v} = 3\mathbf{i} - \mathbf{j}$

15. $\mathbf{u} = \mathbf{i} + \mathbf{j} + \mathbf{k}$, $\mathbf{v} = 2\mathbf{i} + \mathbf{j} - \mathbf{k}$

16. $\mathbf{u} = \mathbf{i} - 2\mathbf{j} + \mathbf{k}$, $\mathbf{v} = -\mathbf{i} + 3\mathbf{j} - 2\mathbf{k}$

In Exercises 17–24, use a graphing utility with vector capabilities to find $\mathbf{u} \times \mathbf{v}$ and then show that it is orthogonal to both $\mathbf{u}$ and $\mathbf{v}$.

17. $\mathbf{u} = (1, 2, -1)$, $\mathbf{v} = (2, 1, 2)$

18. $\mathbf{u} = (1, 2, -3)$, $\mathbf{v} = (-1, 1, 2)$

19. $\mathbf{u} = (0, 1, -1)$, $\mathbf{v} = (1, 2, 0)$

20. $\mathbf{u} = (0, 1, -2)$, $\mathbf{v} = (0, 1, 4)$

21. $\mathbf{u} = 2\mathbf{i} - \mathbf{j} + \mathbf{k}$, $\mathbf{v} = \mathbf{i} - 2\mathbf{j} + \mathbf{k}$

22. $\mathbf{u} = 3\mathbf{i} - \mathbf{j} + \mathbf{k}$, $\mathbf{v} = 2\mathbf{i} + \mathbf{j} - \mathbf{k}$

23. $\mathbf{u} = 2\mathbf{i} + \mathbf{j} - \mathbf{k}$, $\mathbf{v} = \mathbf{i} - \mathbf{j} + 2\mathbf{k}$

24. $\mathbf{u} = 4\mathbf{i} + 2\mathbf{j}$, $\mathbf{v} = \mathbf{i} - 4\mathbf{k}$

In Exercises 25–28, find the area of the parallelogram that has the vectors as adjacent sides.

25. $\mathbf{u} = \mathbf{j}$, $\mathbf{v} = \mathbf{j} + \mathbf{k}$

26. $\mathbf{u} = \mathbf{i} + \mathbf{j} + \mathbf{k}$, $\mathbf{v} = \mathbf{j} + \mathbf{k}$

27. $\mathbf{u} = (3, 2, -1)$, $\mathbf{v} = (1, 2, 3)$

28. $\mathbf{u} = (2, -1, 0)$, $\mathbf{v} = (-1, 2, 0)$

In Exercises 29 and 30, verify that the points are the vertices of a parallelogram, then find its area.

29. $(1, 1, 1)$, $(2, 3, 4)$, $(6, 5, 2)$, $(7, 7, 5)$

30. $(2, -1, 1)$, $(5, 1, 4)$, $(0, 1, 1)$, $(3, 3, 4)$

In Exercises 31–34, find $\mathbf{u} \cdot (\mathbf{v} \times \mathbf{w})$. This quantity is called the **triple scalar product** of $\mathbf{u}$, $\mathbf{v}$, and $\mathbf{w}$.

31. $\mathbf{u} = \mathbf{i}$, $\mathbf{v} = \mathbf{j}$, $\mathbf{w} = \mathbf{k}$

32. $\mathbf{u} = -\mathbf{i}$, $\mathbf{v} = -\mathbf{j}$, $\mathbf{w} = \mathbf{k}$

33. $\mathbf{u} = (1, 1, 1)$, $\mathbf{v} = (2, 1, 0)$, $\mathbf{w} = (0, 0, 1)$

34. $\mathbf{u} = (2, 0, 1)$, $\mathbf{v} = (0, 3, 0)$, $\mathbf{w} = (0, 0, 1)$

35. Show that the volume of a parallelepiped having $\mathbf{u}$, $\mathbf{v}$, and $\mathbf{w}$ as adjacent sides is the triple scalar product $|\mathbf{u} \cdot (\mathbf{v} \times \mathbf{w})|$.

36. Use the result of Exercise 35 to find the volume of the parallelepiped with $\mathbf{u} = \mathbf{i} + \mathbf{j}$, $\mathbf{v} = \mathbf{j} + \mathbf{k}$, and $\mathbf{w} = \mathbf{i} + 2\mathbf{k}$ as adjacent edges. (See Figure 5.39.)

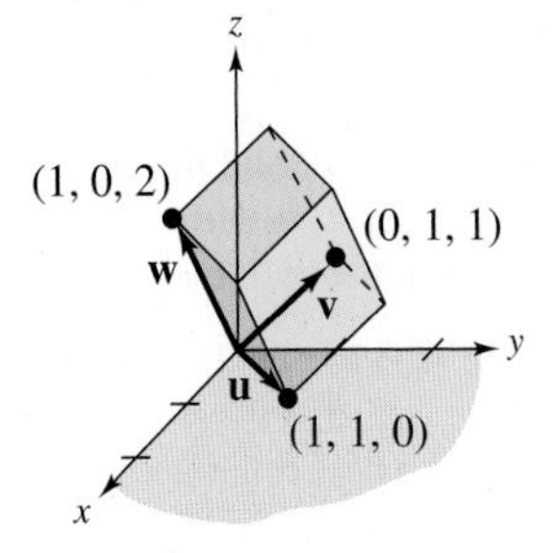

Figure 5.39

In Exercises 37 and 38, find the area of the triangle with the given vertices. Use the fact that the area of the triangle having $\mathbf{u}$ and $\mathbf{v}$ as adjacent sides is given by $A = \frac{1}{2}\|\mathbf{u} \times \mathbf{v}\|$.

37. $(1, 3, 5), (3, 3, 0), (-2, 0, 5)$

38. $(2, -3, 4), (0, 1, 2), (-1, 2, 0)$

39. Find the volume of the parallelepiped shown in Figure 5.40, with $\mathbf{u}$, $\mathbf{v}$, and $\mathbf{w}$ as adjacent sides.

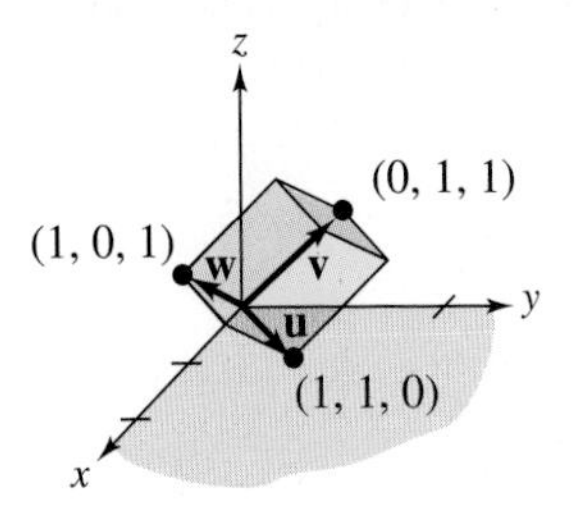

Figure 5.40

40. Prove that $c\mathbf{u} \times \mathbf{v} = c(\mathbf{u} \times \mathbf{v}) = \mathbf{u} \times c\mathbf{v}$.

41. Prove that $\mathbf{u} \times (\mathbf{v} + \mathbf{w}) = (\mathbf{u} \times \mathbf{v}) + (\mathbf{u} \times \mathbf{w})$.

42. Prove that $\mathbf{u} \times \mathbf{0} = \mathbf{0} \times \mathbf{u} = \mathbf{0}$.

43. Prove that $\mathbf{u} \times \mathbf{u} = \mathbf{0}$.

44. Prove that $\mathbf{u} \cdot (\mathbf{v} \times \mathbf{w}) = (\mathbf{u} \times \mathbf{v}) \cdot \mathbf{w}$.

45. Prove that $\mathbf{u} \times \mathbf{v}$ is orthogonal to both $\mathbf{u}$ and $\mathbf{v}$.

46. Prove that the angle θ between $\mathbf{u}$ and $\mathbf{v}$ is given by $\|\mathbf{u} \times \mathbf{v}\| = \|\mathbf{u}\|\,\|\mathbf{v}\| \sin \theta$.

47. Prove that $\mathbf{u} \times \mathbf{v} = \mathbf{0}$ if and only if $\mathbf{u}$ and $\mathbf{v}$ are parallel.

48. Prove **Lagrange's Identity:** $\|\mathbf{u} \times \mathbf{v}\|^2 = \|\mathbf{u}\|^2\|\mathbf{v}\|^2 - (\mathbf{u} \cdot \mathbf{v})^2$.

49. (a) Prove that $\mathbf{u} \times (\mathbf{v} \times \mathbf{w}) = (\mathbf{u} \cdot \mathbf{w})\mathbf{v} - (\mathbf{u} \cdot \mathbf{v})\mathbf{w}$.

(b) Find an example for which $\mathbf{u} \times (\mathbf{v} \times \mathbf{w}) \neq (\mathbf{u} \times \mathbf{v}) \times \mathbf{w}$.

Least Squares Approximations (Calculus)

In Exercises 50–56, (a) find the *linear* least squares approximating function g for the function f and (b) use a graphing utility to graph f and g.

50. $f(x) = x^2, \quad 0 \le x \le 1$

51. $f(x) = \sqrt{x}, \quad 1 \le x \le 4$

52. $f(x) = e^{2x}, \quad 0 \le x \le 1$

53. $f(x) = e^{-2x}, \quad 0 \le x \le 1$

54. $f(x) = \cos x, \quad 0 \le x \le \pi$

55. $f(x) = \sin x, \quad 0 \le x \le \pi/2$

56. $f(x) = \sin x, \quad -\pi/2 \le x \le \pi/2$

In Exercises 57–62, (a) find the *quadratic* least squares approximating function g for the function f and (b) graph f and g.

57. $f(x) = x^3, \quad 0 \le x \le 1$

58. $f(x) = \sqrt{x}, \quad 1 \le x \le 4$

59. $f(x) = \sin x, \quad 0 \le x \le \pi$

60. $f(x) = \sin x, \quad -\pi/2 \le x \le \pi/2$

61. $f(x) = \cos x, \quad -\pi/2 \le x \le \pi/2$

62. $f(x) = \cos x, \quad 0 \le x \le \pi$

Fourier Approximations (Calculus)

In Exercises 63–74, find the Fourier approximation of the specified order for the function on the interval $[0, 2\pi]$.

63. $f(x) = \pi - x$, third order

64. $f(x) = \pi - x$, fourth order

65. $f(x) = (x - \pi)^2$, third order

66. $f(x) = (x - \pi)^2$, fourth order

67. $f(x) = e^{-x}$, first order

68. $f(x) = e^{-x}$, second order

69. $f(x) = e^{-2x}$, first order

70. $f(x) = e^{-2x}$, second order

71. $f(x) = 1 + x$, third order

72. $f(x) = 1 + x$, fourth order

73. $f(x) = 2 \sin x \cos x$, fourth order

74. $f(x) = \sin^2 x$, fourth order

75. Use the results of Exercises 63 and 64 to find the nth-order Fourier approximation of $f(x) = \pi - x$ on the interval $[0, 2\pi]$.

76. Use the results of Exercises 65 and 66 to find the nth-order Fourier approximation of $f(x) = (x - \pi)^2$ on the interval $[0, 2\pi]$.

CHAPTER 5 Review Exercises

In Exercises 1–8, find (a) $\|\mathbf{u}\|$, (b) $\|\mathbf{v}\|$, (c) $\mathbf{u} \cdot \mathbf{v}$, and (d) $d(\mathbf{u}, \mathbf{v})$.

1. $\mathbf{u} = (1, 2),\ \ \mathbf{v} = (4, 1)$

2. $\mathbf{u} = (-1, 2),\ \ \mathbf{v} = (2, 3)$

3. $\mathbf{u} = (2, 1, 1),\ \ \mathbf{v} = (3, 2, -1)$

4. $\mathbf{u} = (1, -1, 2),\ \mathbf{v} = (2, 3, 1)$

5. $\mathbf{u} = (1, -2, 0, 1),\ \mathbf{v} = (1, 1, -1, 0)$

6. $\mathbf{u} = (1, -2, 2, 0),\ \ \mathbf{v} = (2, -1, 0, 2)$

7. $\mathbf{u} = (0, 1, -1, 1, 2),\ \ \mathbf{v} = (0, 1, -2, 1, 1)$

8. $\mathbf{u} = (1, -1, 0, 1, 1),\ \ \mathbf{v} = (0, 1, -2, 2, 1)$

In Exercises 9–12, find $\|\mathbf{v}\|$ and find a unit vector in the direction of $\mathbf{v}$.

9. $\mathbf{v} = (5, 3, -2)$

10. $\mathbf{v} = (1, -2, 1)$

11. $\mathbf{v} = (1, -1, 2)$

12. $\mathbf{v} = (0, 2, -1)$

In Exercises 13–18, find the angle between $\mathbf{u}$ and $\mathbf{v}$.

13. $\mathbf{u} = (2, 2),\ \ \mathbf{v} = (-3, 3)$

14. $\mathbf{u} = (1, -1),\ \ \mathbf{v} = (0, 1)$

15. $\mathbf{u} = \left(\cos \frac{3\pi}{4}, \sin \frac{3\pi}{4}\right),\ \ \mathbf{v} = \left(\cos \frac{2\pi}{3}, \sin \frac{2\pi}{3}\right)$

16. $\mathbf{u} = \left(\cos \frac{\pi}{6}, \sin \frac{\pi}{6}\right),\ \ \mathbf{v} = \left(\cos \frac{5\pi}{6}, \sin \frac{5\pi}{6}\right)$

17. $\mathbf{u} = (10, -5, 15),\ \ \mathbf{v} = (-2, 1, -3)$

18. $\mathbf{u} = (1, 0, -3, 0),\ \ \mathbf{v} = (2, -2, 1, 1)$

In Exercises 19–24, find $\text{proj}_{\mathbf{v}}\mathbf{u}$.

19. $\mathbf{u} = (2, 4),\ \ \mathbf{v} = (1, -5)$

20. $\mathbf{u} = (2, 3),\ \ \mathbf{v} = (0, 4)$

21. $\mathbf{u} = (1, 2),\ \ \mathbf{v} = (2, 5)$

22. $\mathbf{u} = (2, 5),\ \ \mathbf{v} = (0, 5)$

23. $\mathbf{u} = (0, -1, 2),\ \ \mathbf{v} = (3, 2, 4)$

24. $\mathbf{u} = (1, 2, -1),\ \ \mathbf{v} = (0, 2, 3)$

25. For $\mathbf{u} = \left(2, -\frac{1}{2}, 1\right)$ and $\mathbf{v} = \left(\frac{3}{2}, 2, -1\right)$, (a) find the inner product represented by $\langle \mathbf{u}, \mathbf{v} \rangle = u_1v_1 + 2u_2v_2 + 3u_3v_3$, and (b) use this inner product to find the distance between $\mathbf{u}$ and $\mathbf{v}$.

26. For $\mathbf{u} = \left(0, 3, \frac{1}{3}\right)$ and $\mathbf{v} = \left(\frac{4}{3}, 1, -3\right)$, (a) find the inner product represented by $\langle \mathbf{u}, \mathbf{v} \rangle = 2u_1v_1 + u_2v_2 + 2u_3v_3$, and (b) use this inner product to find the distance between $\mathbf{u}$ and $\mathbf{v}$.

27. Verify the Triangle Inequality and the Cauchy-Schwarz Inequality for $\mathbf{u}$ and $\mathbf{v}$ from Exercise 25. (Use the inner product from Exercise 25.)

28. Verify the Triangle Inequality and the Cauchy-Schwarz Inequality for $\mathbf{u}$ and $\mathbf{v}$ from Exercise 26. (Use the inner product given in Exercise 26.)

In Exercises 29–32, find all vectors orthogonal to $\mathbf{u}$.

29. $\mathbf{u} = (0, -4, 3)$

30. $\mathbf{u} = (1, -1, 2)$

31. $\mathbf{u} = (1, -2, 2, 1)$

32. $\mathbf{u} = (0, 1, 2, -1)$

In Exercises 33–36, use the Gram-Schmidt orthonormalization process to transform the basis into an orthonormal basis. (Use the Euclidean inner product.)

33. $B = \{(1, 1), (0, 1)\}$

34. $B = \{(3, 4), (1, 2)\}$

35. $B = \{(0, 3, 4), (1, 0, 0), (1, 1, 0)\}$

36. $B = \{(0, 0, 2), (0, 1, 1), (1, 1, 1)\}$

37. Let $B = \{(0, 2, -2), (1, 0, -2)\}$ be a basis for a subspace of R^3, and let $\mathbf{x} = (-1, 4, -2)$ be a vector in the subspace.

(a) Write $\mathbf{x}$ as a linear combination of the vectors in B. That is, find the coordinates of $\mathbf{x}$ relative to B.

(b) Use the Gram-Schmidt orthonormalization process to transform B into an orthonormal set B'.

(c) Write $\mathbf{x}$ as a linear combination of the vectors in B'. That is, find the coordinates of $\mathbf{x}$ relative to B'.

38. Repeat Exercise 37 for $B = \{(-1, 2, 2), (1, 0, 0)\}$ and $\mathbf{x} = (-3, 4, 4)$.

Calculus In Exercises 39–42, let f and g be functions in the vector space $C[a, b]$ with inner product

$$\langle f, g \rangle = \int_a^b f(x)g(x)\,dx.$$

39. Let $f(x) = x$ and $g(x) = x^2$ be vectors in $C[0, 1]$.
(a) Find $\langle f, g \rangle$.
(b) Find $\|g\|$.
(c) Find $d(f, g)$.
(d) Orthonormalize the set $B = \{f, g\}$.

40. Let $f(x) = x + 2$ and $g(x) = 15x - 8$ be vectors in $C[0, 1]$.
(a) Find $\langle f, g \rangle$.
(b) Find $\langle -4f, g \rangle$.
(c) Find $\|f\|$.
(d) Orthonormalize the set $B = \{f, g\}$.

41. Show that $f(x) = \sqrt{1 - x^2}$ and $g(x) = 2x\sqrt{1 - x^2}$ are orthogonal in $C[-1, 1]$.

42. Apply the Gram-Schmidt orthonormalization process to the set in $C[-\pi, \pi]$ shown below.

$$S = \{1, \cos x, \sin x, \cos 2x, \sin 2x, \ldots, \cos nx, \sin nx\}$$

43. Find an orthonormal basis for the following subspace of Euclidean 3-space.

$$W = \{(x_1, x_2, x_3): x_1 + x_2 + x_3 = 0\}$$

44. Find an orthonormal basis for the solution space of the homogeneous system of linear equations.

$$\begin{aligned} x + y - z + w &= 0 \\ 2x - y + z + 2w &= 0 \end{aligned}$$

Calculus In Exercises 45 and 46, (a) find the inner product, (b) determine whether the vectors are orthogonal, and (c) verify the Cauchy-Schwarz Inequality for the vectors.

45. $f(x) = x$, $g(x) = \dfrac{1}{x^2 + 1}$, $\langle f, g \rangle = \displaystyle\int_{-1}^{1} f(x)g(x)\,dx$

46. $f(x) = x$, $g(x) = 4x^2$, $\langle f, g \rangle = \displaystyle\int_{0}^{1} f(x)g(x)\,dx$

47. Prove that if $\mathbf{u}$ and $\mathbf{v}$ are vectors in an inner product space such that $\|\mathbf{u}\| \le 1$ and $\|\mathbf{v}\| \le 1$, then $|\langle \mathbf{u}, \mathbf{v} \rangle| \le 1$.

48. Prove that if $\mathbf{u}$ and $\mathbf{v}$ are vectors in an inner product space V, then

$$\big|\|\mathbf{u}\| - \|\mathbf{v}\|\big| \le \|\mathbf{u} \pm \mathbf{v}\|.$$

49. Let V be an m-dimensional subspace of R^n such that $m < n$. Prove that any vector $\mathbf{u}$ in R^n can be uniquely written in the form $\mathbf{u} = \mathbf{v} + \mathbf{w}$, where $\mathbf{v}$ is in V and $\mathbf{w}$ is orthogonal to every vector in V.

50. Let V be the two-dimensional subspace of R^4 spanned by $(0, 1, 0, 1)$ and $(0, 2, 0, 0)$. Write the vector $\mathbf{u} = (1, 1, 1, 1)$ in the form $\mathbf{u} = \mathbf{v} + \mathbf{w}$, where $\mathbf{v}$ is in V and $\mathbf{w}$ is orthogonal to every vector in V.

51. Let $\{\mathbf{u}_1, \mathbf{u}_2, \ldots, \mathbf{u}_m\}$ be an orthonormal subset of R^n, and let $\mathbf{v}$ be any vector in R^n. Prove that

$$\|\mathbf{v}\|^2 \ge \sum_{i=1}^{m} (\mathbf{v} \cdot \mathbf{u}_i)^2.$$

(This inequality is called **Bessel's Inequality.**)

52. Let $\{x_1, x_2, \ldots, x_n\}$ be a set of real numbers. Use the Cauchy-Schwarz Inequality to prove that

$$(x_1 + x_2 + \cdots + x_n)^2 \le n(x_1^2 + x_2^2 + \cdots + x_n^2).$$

53. Let $\mathbf{u}$ and $\mathbf{v}$ be vectors in an inner product space V. Prove that $\|\mathbf{u} + \mathbf{v}\| = \|\mathbf{u} - \mathbf{v}\|$ if and only if $\mathbf{u}$ and $\mathbf{v}$ are orthogonal.

54. Writing Let $\{\mathbf{u}_1, \mathbf{u}_2, \ldots, \mathbf{u}_n\}$ be a dependent set of vectors in an inner product space V. Describe the result of applying the Gram-Schmidt orthonormalization process to this set.

55. Find the orthogonal complement $S^\perp$ of the subspace S of R^3 spanned by the two column vectors of the matrix

$$A = \begin{bmatrix} 1 & 2 \\ 2 & 1 \\ 0 & -1 \end{bmatrix}.$$

56. Find the projection of the vector $\mathbf{v} = [1 \quad 0 \quad -2]^T$ onto the subspace

$$S = \operatorname{span}\left\{\begin{bmatrix} 0 \\ -1 \\ 1 \end{bmatrix}, \begin{bmatrix} 0 \\ 1 \\ 1 \end{bmatrix}\right\}.$$

57. Find bases for the four fundamental subspaces of the matrix

$$A = \begin{bmatrix} 0 & 1 & 0 \\ 0 & -3 & 0 \\ 1 & 0 & 1 \end{bmatrix}.$$

58. Find the least squares regression line for the set of data points

$\{(-2, 2), (-1, 1), (0, 1), (1, 3)\}$.

Graph the points and the line on the same set of axes.

Mathematical Modeling

59. The table shows the retail sales y (in millions of dollars) of running shoes in the United States during the years 1999 to 2005. Find the least squares regression line and least squares cubic regression polynomial for the data. Let t represent the year, with $t = -1$ corresponding to 1999. Which model is the better fit for the data? Use the model to predict the sales in the year 2010. (Source: National Sporting Goods Association)

Year	1999	2000	2001	2002
Sales, y	1502	1638	1670	1733

Year	2003	2004	2005
Sales, y	1802	1989	2049

60. The table shows the average salaries y (in thousands of dollars) for National Football League players during the years 2000 to 2005. Find the least squares regression line for the data. Let t represent the year, with $t = 0$ corresponding to 2000. (Source: National Football League)

Year	2000	2001	2002
Average Salary, y	787	986	1180

Year	2003	2004	2005
Average Salary, y	1259	1331	1440

61. The table shows the world energy consumption y (in quadrillions of Btu) during the years 1999 to 2004. Find the least squares regression line for the data. Let t represent the year, with $t = -1$ corresponding to 1999. (Source: U.S. Energy Information Administration)

Year	1999	2000	2001
Energy Consumption, y	389.1	399.5	403.5

Year	2002	2003	2004
Energy Consumption, y	409.7	425.7	446.4

62. The table shows the numbers of stores y for the Target Corporation during the years 1996 to 2007. Find the least squares regression quadratic polynomial that best fits the data. Let t represent the year, with $t = -4$ corresponding to 1996. Make separate models for the years 1996–2003 and 2004–2007. (Source: Target Corporation)

Year	1996	1997	1998	1999	2000	2001
Number of Stores, y	1101	1130	1182	1243	1307	1381

Year	2002	2003	2004	2005	2006	2007
Number of Stores, y	1475	1553	1308	1397	1495	1610

63. The table shows the revenues y (in millions of dollars) for eBay, Incorporated during the years 2000 to 2007. Find the least squares regression quadratic polynomial that best fits the data. Let t represent the year, with $t = 0$ corresponding to 2000. (Source: eBay, Incorporated)

Year	2000	2001	2002	2003
Revenue, y	431.4	748.8	1214.1	2165.1

Year	2004	2005	2006	2007
Revenue, y	3271.3	4552.4	5969.7	7150.0

64. The table shows the revenues y (in millions of dollars) for Google, Incorporated during the years 2002 to 2007. Find the least squares regression quadratic polynomial that best fits the data. Let t represent the year, with $t = 2$ corresponding to 2002. (Source: Google, Incorporated)

Year	2002	2003	2004
Revenue, y	439.5	1465.9	3189.2

Year	2005	2006	2007
Revenue, y	6138.6	10,604.9	16,000.0

65. The table shows the sales y (in millions of dollars) for Circuit City Stores during the years 2000 to 2007. Find the least squares regression quadratic polynomial that best fits the data. Let t represent the year, with $t = 0$ corresponding to 2000. (Source: Circuit City Stores)

Year	2000	2001	2002	2003
Sales, y	10,458.0	9589.8	9953.5	9745.4

Year	2004	2005	2006	2007
Sales, y	10,472.0	11,598.0	12,670.0	13,680.0

The Cross Product of Two Vectors in Space

In Exercises 66–69, find $\mathbf{u} \times \mathbf{v}$ and show that it is orthogonal to both $\mathbf{u}$ and $\mathbf{v}$.

66. $\mathbf{u} = (1, 1, 1)$, $\mathbf{v} = (1, 0, 0)$

67. $\mathbf{u} = (1, -1, 1)$, $\mathbf{v} = (0, 1, 1)$

68. $\mathbf{u} = \mathbf{j} + 6\mathbf{k}$, $\mathbf{v} = \mathbf{i} - 2\mathbf{j} + \mathbf{k}$

69. $\mathbf{u} = 2\mathbf{i} - \mathbf{k}$, $\mathbf{v} = \mathbf{i} + \mathbf{j} - \mathbf{k}$

70. Find the area of the parallelogram that has $\mathbf{u} = (1, 3, 0)$ and $\mathbf{v} = (-1, 0, 2)$ as adjacent sides.

71. Prove that $\|\mathbf{u} \times \mathbf{v}\| = \|\mathbf{u}\|\,\|\mathbf{v}\|$ if and only if $\mathbf{u}$ and $\mathbf{v}$ are orthogonal.

In Exercises 72 and 73, the volume of the parallelepiped having $\mathbf{u}$, $\mathbf{v}$, and $\mathbf{w}$ as adjacent sides is given by the **triple scalar product** $|\mathbf{u} \cdot (\mathbf{v} \times \mathbf{w})|$. Find the volume of the parallelepiped having the three vectors as adjacent sides.

72. $\mathbf{u} = (1, 0, 0)$, $\mathbf{v} = (0, 0, 1)$, $\mathbf{w} = (0, 1, 0)$

73. $\mathbf{u} = (1, 2, 1)$, $\mathbf{v} = (-1, -1, 0)$, $\mathbf{w} = (3, 4, -1)$

Least Squares Approximations (Calculus)

In Exercises 74–77, find the *linear* least squares approximating function g for the function f. Then sketch the graphs of f and g.

74. $f(x) = x^3$, $-1 \le x \le 1$

75. $f(x) = x^3$, $0 \le x \le 2$

76. $f(x) = \sin 2x$, $0 \le x \le \pi/2$

77. $f(x) = \sin x \cos x$, $0 \le x \le \pi$

In Exercises 78 and 79, find the *quadratic* least squares approximating function g for the function f. Then, using a graphing utility, graph f and g.

78. $f(x) = \sqrt{x}$, $0 \le x \le 1$

79. $f(x) = \dfrac{1}{x}$, $1 \le x \le 2$

Fourier Approximations (Calculus)

In Exercises 80 and 81, find the nth-order Fourier approximation of the function.

80. $f(x) = x^2$, $-\pi \le x \le \pi$, first order

81. $f(x) = x$, $-\pi \le x \le \pi$, second order

True or False? In Exercises 82 and 83, determine whether each statement is true or false. If a statement is true, give a reason or cite an appropriate statement from the text. If a statement is false, provide an example that shows the statement is not true in all cases or cite an appropriate statement from the text.

82. (a) The cross product of two nonzero vectors in R^3 yields a vector orthogonal to the two given vectors that produced it.

(b) The cross product of two nonzero vectors in R^3 is commutative.

(c) The least squares approximation of a function f is the function g (in the subspace W) closest to f in terms of the inner product $\langle f, g \rangle$.

83. (a) The vectors in R^3 $\mathbf{u} \times \mathbf{v}$ and $\mathbf{v} \times \mathbf{u}$ have equal lengths but opposite directions.

(b) If $\mathbf{u}$ and $\mathbf{v}$ are two nonzero vectors in R^3, then $\mathbf{u}$ and $\mathbf{v}$ are parallel if and only if $\mathbf{u} \times \mathbf{v} = \mathbf{0}$.

(c) A special type of least squares approximation, the Fourier approximation, is spanned by the basis $S = \{1, \cos x, \cos 2x, \ldots, \cos nx, \sin x, \sin 2x, \ldots, \sin nx\}$.

CHAPTER 5 Projects

1 The *QR*-Factorization

The Gram-Schmidt orthonormalization process leads to an important factorization of matrices called the ***QR*-factorization.** If A is an $m \times n$ matrix of rank n, then A can be expressed as the product $A = QR$ of an $m \times n$ matrix Q and an $n \times n$ matrix R, where Q has orthonormal columns and R is upper triangular.

The columns of A can be considered a basis for a subspace of R^m, and the columns of Q are the result of applying the Gram-Schmidt orthonormalization process to this set of column vectors.

Recall that in Example 7, Section 5.3, the Gram-Schmidt orthonormalization process was used on the column vectors of the matrix

$$A = \begin{bmatrix} 1 & 1 & 0 \\ 1 & 2 & 1 \\ 0 & 0 & 2 \end{bmatrix}.$$

An orthonormal basis for R^3 was obtained, which is labeled here as $\mathbf{q}_1, \mathbf{q}_2, \mathbf{q}_3$.

$$\mathbf{q}_1 = \left(\sqrt{2}/2, \sqrt{2}/2, 0\right)$$
$$\mathbf{q}_2 = \left(-\sqrt{2}/2, \sqrt{2}/2, 0\right)$$
$$\mathbf{q}_3 = (0, 0, 1)$$

These vectors form the columns of the matrix Q.

$$Q = \begin{bmatrix} \sqrt{2}/2 & -\sqrt{2}/2 & 0 \\ \sqrt{2}/2 & \sqrt{2}/2 & 0 \\ 0 & 0 & 1 \end{bmatrix}$$

The upper triangular matrix R consists of the following dot products.

$$\begin{bmatrix} \mathbf{v}_1 \cdot \mathbf{q}_1 & \mathbf{v}_2 \cdot \mathbf{q}_1 & \mathbf{v}_3 \cdot \mathbf{q}_1 \\ 0 & \mathbf{v}_2 \cdot \mathbf{q}_2 & \mathbf{v}_3 \cdot \mathbf{q}_2 \\ 0 & 0 & \mathbf{v}_3 \cdot \mathbf{q}_3 \end{bmatrix} = \begin{bmatrix} \sqrt{2} & 3\sqrt{2}/2 & \sqrt{2}/2 \\ 0 & \sqrt{2}/2 & \sqrt{2}/2 \\ 0 & 0 & 2 \end{bmatrix}$$

It is now an easy exercise to verify that $A = QR$.

In general, if A is an $m \times n$ matrix of rank n, then the QR-factorization of A can be constructed if you keep track of the dot products used in the Gram-Schmidt orthonormalization process as applied to the columns of A. The columns of the $m \times n$ matrix Q are the orthonormal vectors that result from the Gram-Schmidt orthonormalization process. The $n \times n$ upper triangular matrix R consists of certain dot products of the original column vectors $\mathbf{v}_i$ and the orthonormal column vectors $\mathbf{q}_i$. If the columns of the matrix A are denoted as $\mathbf{v}_1, \mathbf{v}_2, \ldots, \mathbf{v}_n$ and the columns of Q are denoted as $\mathbf{q}_1, \mathbf{q}_2, \ldots, \mathbf{q}_n$, then the QR-factorization of A is as follows.

$$A = QR$$

$$[\mathbf{v}_1 \ \ \mathbf{v}_2 \ \ . \ . \ . \ \ \mathbf{v}_n] = [\mathbf{q}_1 \ \ \mathbf{q}_2 \ \ . \ . \ . \ \ \mathbf{q}_n]\begin{bmatrix} \mathbf{v}_1 \cdot \mathbf{q}_1 & \mathbf{v}_2 \cdot \mathbf{q}_1 & \cdots & \mathbf{v}_n \cdot \mathbf{q}_1 \\ 0 & \mathbf{v}_2 \cdot \mathbf{q}_2 & \cdots & \mathbf{v}_n \cdot \mathbf{q}_2 \\ \vdots & \vdots & & \vdots \\ 0 & 0 & \cdots & \mathbf{v}_n \cdot \mathbf{q}_n \end{bmatrix}$$

1. Verify the matrix equation $A = QR$ for the preceding example.
2. Find the QR-factorization of each matrix.

(a) $A = \begin{bmatrix} 1 & 1 \\ 0 & 1 \\ 1 & 0 \end{bmatrix}$ (b) $A = \begin{bmatrix} 1 & 0 \\ 0 & 0 \\ 1 & 1 \\ 1 & 2 \end{bmatrix}$

(c) $A = \begin{bmatrix} 1 & 0 & 0 \\ 1 & 1 & 0 \\ 1 & 1 & 1 \end{bmatrix}$ (d) $A = \begin{bmatrix} 1 & 0 & -1 \\ 1 & 2 & 0 \\ 1 & 2 & 0 \\ 1 & 0 & 0 \end{bmatrix}$

3. Let $A = QR$ be the QR-factorization of the $m \times n$ matrix A of rank n. Show how the least squares problem can be solved using just matrix multiplication and back-substitution.
4. Use the result of part 3 to solve the least squares problem $A\mathbf{x} = \mathbf{b}$ if A is the matrix from part 2(a) and $\mathbf{b} = [-1 \quad 1 \quad -1]^T$.

The QR-factorization of a matrix forms the basis for many algorithms of linear algebra. Computer routines for the computation of eigenvalues (see Chapter 7) are based on this factorization, as are algorithms for computing the least squares regression line for a set of data points. It should also be mentioned that, in practice, techniques other than the Gram-Schmidt orthonormalization process are actually used to compute the QR-factorization of a matrix.

2 Orthogonal Matrices and Change of Basis

Let $B = \{\mathbf{v}_1, \mathbf{v}_2, . . . , \mathbf{v}_n\}$ be an ordered basis for the vector space V. Recall that the coordinate matrix of a vector $\mathbf{x} = c_1\mathbf{v}_1 + c_2\mathbf{v}_2 + \cdot \cdot \cdot + c_n\mathbf{v}_n$ in V is the column vector

$$[\mathbf{x}]_B = \begin{bmatrix} c_1 \\ c_2 \\ \vdots \\ c_n \end{bmatrix}.$$

If B' is another basis for V, then the transition matrix P from B' to B changes a coordinate matrix relative to B' into a coordinate matrix relative to B,

$$P[\mathbf{x}]_{B'} = [\mathbf{x}]_B.$$

The question you will explore now is whether there are transition matrices P that preserve the length of the coordinate matrix—that is, given $P[\mathbf{x}]_{B'} = [\mathbf{x}]_B$, does $\|[\mathbf{x}]_{B'}\| = \|[\mathbf{x}]_B\|$?

For example, consider the transition matrix from Example 5 in Section 4.7,

$$P = \begin{bmatrix} 3 & -2 \\ 2 & -1 \end{bmatrix}$$

relative to the bases for R^2,

$$B = \{(-3, 2), (4, -2)\} \quad \text{and} \quad B' = \{(-1, 2), (2, -2)\}.$$

If $\mathbf{x} = (-1, 2)$, then $[\mathbf{x}]_{B'} = [1 \quad 0]^T$ and $[\mathbf{x}]_B = P[\mathbf{x}]_{B'} = [3 \quad 2]^T$. So, using the Euclidean norm for R^2,

$$\|[\mathbf{x}]_{B'}\| = 1 \neq \sqrt{13} = \|[\mathbf{x}]_B\|.$$

You will see in this project that if the transition matrix P is **orthogonal,** then the norm of the coordinate vector will remain unchanged.

Definition of Orthogonal Matrix

The square matrix P is **orthogonal** if it is invertible and $P^{-1} = P^T$.

1. Show that the matrix P defined previously is *not* orthogonal.
2. Show that for any real number θ, the matrix $\begin{bmatrix} \cos\theta & -\sin\theta \\ \sin\theta & \cos\theta \end{bmatrix}$ is orthogonal.
3. Show that a matrix is orthogonal if and only if its columns are pairwise orthogonal.
4. Prove that the inverse of an orthogonal matrix is orthogonal.
5. Is the sum of orthogonal matrices orthogonal? Is the product of orthogonal matrices orthogonal? Illustrate your answers with appropriate examples.
6. What is the determinant of an orthogonal matrix?
7. Prove that if P is an $m \times n$ orthogonal matrix, then $\|P\mathbf{x}\| = \|\mathbf{x}\|$ for all vectors $\mathbf{x}$ in R^n.
8. Verify the result of part 7 using the bases $B = \{(1, 0), (0, 1)\}$ and

$$B' = \left\{\left(-\frac{2}{\sqrt{5}}, \frac{1}{\sqrt{5}}\right), \left(\frac{1}{\sqrt{5}}, \frac{2}{\sqrt{5}}\right)\right\}.$$

CHAPTERS 4 & 5 Cumulative Test

Take this test as you would take a test in class. After you are done, check your work against the answers in the back of the book.

1. Given the vectors $\mathbf{v} = (1, -2)$ and $\mathbf{w} = (2, -5)$, find and sketch each vector.

   (a) $\mathbf{v} + \mathbf{w}$ (b) $3\mathbf{v}$ (c) $2\mathbf{v} - 4\mathbf{w}$

2. If possible, write $\mathbf{w} = (2, 4, 1)$ as a linear combination of the vectors $\mathbf{v}_1$, $\mathbf{v}_2$, and $\mathbf{v}_3$.

   $\mathbf{v}_1 = (1, 2, 0), \quad \mathbf{v}_2 = (-1, 0, 1), \quad \mathbf{v}_3 = (0, 3, 0)$

3. Prove that the set of all singular 2×2 matrices is not a vector space.

4. Determine whether the set is a subspace of R^4.

   $\{(x, x + y, y, y): x, y \in R\}$

5. Determine whether the set is a subspace of R^3.

   $\{(x, xy, y): x, y \in R\}$

6. Determine whether the columns of matrix A span R^4.

$$A = \begin{bmatrix} 1 & 2 & -1 & 0 \\ 1 & 3 & 0 & 2 \\ 0 & 0 & 1 & -1 \\ 1 & 0 & 0 & 1 \end{bmatrix}$$

7. (a) Define what it means to say that a set of vectors is *linearly independent.*

   (b) Determine whether the set S is linearly dependent or independent.

   $S = \{(1, 0, 1, 0), (0, 3, 0, 1), (1, 1, 2, 2), (3, -4, 2, -3)\}$

8. Find the dimension of and a basis for the subspace of $M_{3,3}$ consisting of all the 3×3 symmetric matrices.

9. (a) Define *basis* of a vector space.

   (b) Determine if the set is a basis for R^3.

   $\{(1, 2, 1), (0, 1, 2), (2, 1, -3)\}$

10. Find a basis for the solution space of $A\mathbf{x} = \mathbf{0}$ if

$$A = \begin{bmatrix} 1 & 1 & 0 & 0 \\ -2 & -2 & 0 & 0 \\ 0 & 0 & 1 & 1 \\ 1 & 1 & 0 & 0 \end{bmatrix}.$$

11. Find the coordinates $[\mathbf{v}]_B$ of the vector $\mathbf{v} = (1, 2, -3)$ relative to the basis

    $B = \{(0, 1, 1), (1, 1, 1), (1, 0, 1)\}$.

12. Find the transition matrix from the basis $B = \{(2, 1, 0), (1, 0, 0), (0, 1, 1)\}$ to the basis

    $B' = \{(1, 1, 2), (1, 1, 1), (0, 1, 2)\}$.

13. Let $\mathbf{u} = (1, 0, 2)$ and $\mathbf{v} = (-2, 1, 3)$.

    (a) Find $\|\mathbf{u}\|$. (b) Find the distance between $\mathbf{u}$ and $\mathbf{v}$.

    (c) Find $\mathbf{u} \cdot \mathbf{v}$. (d) Find the angle θ between $\mathbf{u}$ and $\mathbf{v}$.

14. Find the inner product of $f(x) = x^2$ and $g(x) = x + 2$ from $C[0, 1]$ using the integral

$$\langle f, g \rangle = \int_0^1 f(x)g(x)\,dx.$$

15. Use the Gram-Schmidt orthonormalization process to transform the following set of vectors into an orthonormal basis for R^3.

$\{(2, 0, 0), (1, 1, 1), (0, 1, 2)\}$

16. Let $\mathbf{u} = (1, 2)$ and $\mathbf{v} = (-3, 2)$. Find $\text{proj}_{\mathbf{v}}\mathbf{u}$, and graph $\mathbf{u}$, $\mathbf{v}$, and $\text{proj}_{\mathbf{v}}\mathbf{u}$ on the same set of coordinate axes.

17. Find the four fundamental subspaces of the matrix

$$A = \begin{bmatrix} 0 & 1 & 1 & 0 \\ -1 & 0 & 0 & 1 \\ 1 & 1 & 1 & 1 \end{bmatrix}.$$

18. Find the orthogonal complement $S^\perp$ of the set

$$S = \text{span}\left(\begin{bmatrix} 1 \\ 0 \\ 1 \end{bmatrix}, \begin{bmatrix} -1 \\ 1 \\ 0 \end{bmatrix}\right).$$

19. Use the axioms for a vector space to prove that $0\mathbf{v} = \mathbf{0}$ for all vectors $\mathbf{v} \in V$.

20. Suppose that $\mathbf{x}_1, \ldots, \mathbf{x}_n$ are linearly independent vectors and $\mathbf{y}$ is a vector not in their span. Prove that the vectors $\mathbf{x}_1, \ldots, \mathbf{x}_n$ and $\mathbf{y}$ are linearly independent.

21. Let W be a subspace of an inner product space V. Prove that the set below is a subspace of V.

$W^\perp = \{\mathbf{v} \in V\colon \langle \mathbf{v}, \mathbf{w} \rangle = 0 \quad \text{for all} \quad \mathbf{w} \in W\}$

22. Find the least squares regression line for the points $\{(1, 1), (2, 0), (5, -5)\}$. Graph the points and the line.

23. The two matrices A and B are row-equivalent.

$$A = \begin{bmatrix} 2 & -4 & 0 & 1 & 7 & 11 \\ 1 & -2 & -1 & 1 & 9 & 12 \\ -1 & 2 & 1 & 3 & -5 & 16 \\ 4 & -8 & 1 & -1 & 6 & -2 \end{bmatrix} \qquad B = \begin{bmatrix} 1 & -2 & 0 & 0 & 3 & 2 \\ 0 & 0 & 1 & 0 & -5 & -3 \\ 0 & 0 & 0 & 1 & 1 & 7 \\ 0 & 0 & 0 & 0 & 0 & 0 \end{bmatrix}$$

(a) Find the rank of A.
(b) Find a basis for the row space of A.
(c) Find a basis for the column space of A.
(d) Find a basis for the nullspace of A.
(e) Is the last column of A in the span of the first three columns?
(f) Are the first three columns of A linearly independent?
(g) Is the last column of A in the span of columns 1, 3, and 4?
(h) Are columns 1, 3, and 4 linearly dependent?

24. Let $\mathbf{u}$ and $\mathbf{v}$ be vectors in an inner product space V. Prove that $\|\mathbf{u} + \mathbf{v}\| = \|\mathbf{u} - \mathbf{v}\|$ if and only if $\mathbf{u}$ and $\mathbf{v}$ are orthogonal.

Linear Transformations

CHAPTER OBJECTIVES

- Find the image and preimage of a function.
- Determine whether a function from one vector space to another is a linear transformation.
- Find the kernel, the range, and the bases for the kernel and range of a linear transformation T, and determine the nullity and rank of T.
- Determine whether a linear transformation is one-to-one or onto.
- Verify that a matrix defines a linear function that is one-to-one and onto.
- Determine whether two vector spaces are isomorphic.
- Find the standard matrix for a linear transformation and use this matrix to find the image of a vector and sketch the graph of the vector and its image.
- Find the standard matrix of the composition of a linear transformation.
- Determine whether a linear transformation is invertible and find its inverse, if it exists.
- Find the matrix of a linear transformation relative to a nonstandard basis.
- Know and use the definition and properties of similar matrices.
- Identify linear transformations defined by reflections, expansions, contractions, shears, and/or rotations.

6.1 Introduction to Linear Transformations

In this chapter you will learn about functions that **map** a vector space V into a vector space W. This type of function is denoted by

$$T: V \to W.$$

The standard function terminology is used for such functions. For instance, V is called the **domain** of T, and W is called the **codomain** of T. If $\mathbf{v}$ is in V and $\mathbf{w}$ is in W such that

$$T(\mathbf{v}) = \mathbf{w},$$

then $\mathbf{w}$ is called the **image** of $\mathbf{v}$ under T. The set of all images of vectors in V is called the **range** of T, and the set of all $\mathbf{v}$ in V such that $T(\mathbf{v}) = \mathbf{w}$ is called the **preimage** of $\mathbf{w}$. (See Figure 6.1 on the next page.)

REMARK: For a vector $\mathbf{v} = (v_1, v_2, \ldots, v_n)$ in R^n, it would be technically correct to use double parentheses to denote $T(\mathbf{v})$ as $T(\mathbf{v}) = T((v_1, v_2, \ldots, v_n))$. For convenience, however, one set of parentheses is dropped, producing

$$T(\mathbf{v}) = T(v_1, v_2, \ldots, v_n).$$

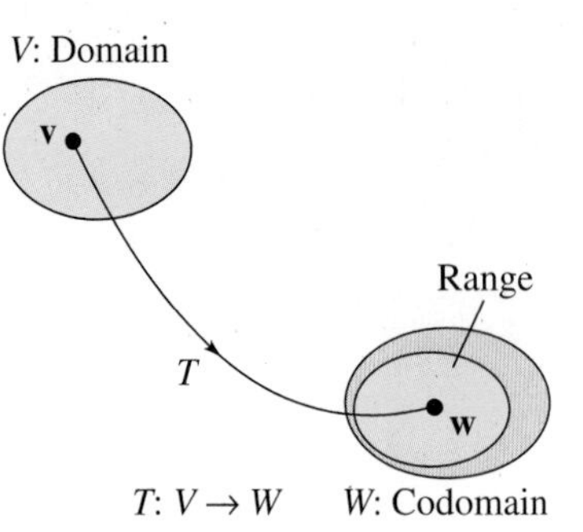

Figure 6.1

EXAMPLE 1 A Function from R^2 into R^2

For any vector $\mathbf{v} = (v_1, v_2)$ in R^2, let $T: R^2 \to R^2$ be defined by

$$T(v_1, v_2) = (v_1 - v_2, v_1 + 2v_2).$$

(a) Find the image of $\mathbf{v} = (-1, 2)$.

(b) Find the preimage of $\mathbf{w} = (-1, 11)$.

SOLUTION (a) For $\mathbf{v} = (-1, 2)$ you have

$$T(-1, 2) = (-1 - 2, -1 + 2(2)) = (-3, 3).$$

(b) If $T(\mathbf{v}) = (v_1 - v_2, v_1 + 2v_2) = (-1, 11)$, then

$$\begin{aligned} v_1 - v_2 &= -1 \\ v_1 + 2v_2 &= 11. \end{aligned}$$

This system of equations has the unique solution $v_1 = 3$ and $v_2 = 4$. So, the preimage of $(-1, 11)$ is the set in R^2 consisting of the single vector $(3, 4)$.

This chapter centers on functions (from one vector space to another) that preserve the operations of vector addition and scalar multiplication. Such functions are called **linear transformations.**

Definition of a Linear Transformation

Let V and W be vector spaces. The function $T: V \to W$ is called a **linear transformation** of V into W if the following two properties are true for all $\mathbf{u}$ and $\mathbf{v}$ in V and for any scalar c.

1. $T(\mathbf{u} + \mathbf{v}) = T(\mathbf{u}) + T(\mathbf{v})$
2. $T(c\mathbf{u}) = cT(\mathbf{u})$

A linear transformation is said to be *operation preserving,* because the same result occurs whether the operations of addition and scalar multiplication are performed before or after the linear transformation is applied. Although the same symbols are used to denote the vector operations in both V and W, you should note that the operations may be different, as indicated in the diagram below.

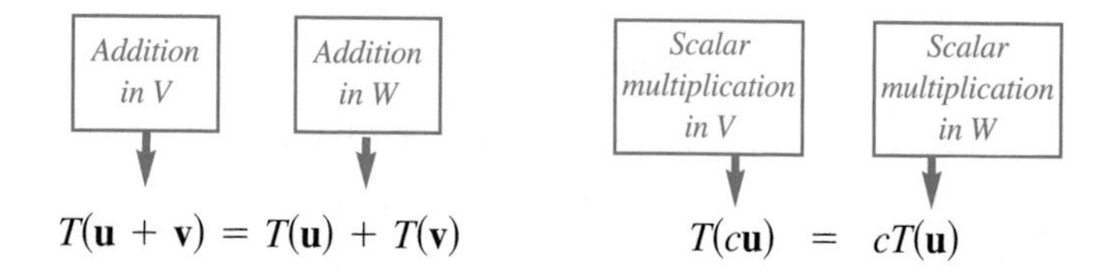

EXAMPLE 2 **Verifying a Linear Transformation from R^2 into R^2**

Show that the function given in Example 1 is a linear transformation from R^2 into R^2.

$$T(v_1, v_2) = (v_1 - v_2, v_1 + 2v_2)$$

SOLUTION To show that the function T is a linear transformation, you must show that it preserves vector addition and scalar multiplication. To do this, let $\mathbf{v} = (v_1, v_2)$ and $\mathbf{u} = (u_1, u_2)$ be vectors in R^2 and let c be any real number. Then, using the properties of vector addition and scalar multiplication, you have the two statements below.

1. Because $\mathbf{u} + \mathbf{v} = (u_1, u_2) + (v_1, v_2) = (u_1 + v_1, u_2 + v_2)$, you have

$$\begin{aligned} T(\mathbf{u} + \mathbf{v}) &= T(u_1 + v_1, u_2 + v_2) \\ &= ((u_1 + v_1) - (u_2 + v_2), (u_1 + v_1) + 2(u_2 + v_2)) \\ &= ((u_1 - u_2) + (v_1 - v_2), (u_1 + 2u_2) + (v_1 + 2v_2)) \\ &= (u_1 - u_2, u_1 + 2u_2) + (v_1 - v_2, v_1 + 2v_2) \\ &= T(\mathbf{u}) + T(\mathbf{v}). \end{aligned}$$

REMARK: A linear transformation $T: V \rightarrow V$ from a vector space into itself (as in Example 2) is called a **linear operator.**

2. Because $c\mathbf{u} = c(u_1, u_2) = (cu_1, cu_2)$, you have

$$\begin{aligned} T(c\mathbf{u}) &= T(cu_1, cu_2) \\ &= (cu_1 - cu_2, cu_1 + 2cu_2) \\ &= c(u_1 - u_2, u_1 + 2u_2) \\ &= cT(\mathbf{u}). \end{aligned}$$

So, T is a linear transformation.

Most of the common functions studied in calculus are not linear transformations.

EXAMPLE 3 **Some Functions That Are Not Linear Transformations**

(a) $f(x) = \sin x$ is not a linear transformation from R into R because, in general,

$$\sin(x_1 + x_2) \neq \sin x_1 + \sin x_2.$$

For instance, $\sin[(\pi/2) + (\pi/3)] \neq \sin(\pi/2) + \sin(\pi/3)$.

(b) $f(x) = x^2$ is not a linear transformation from R into R because, in general,

$$(x_1 + x_2)^2 \neq x_1^2 + x_2^2.$$

For instance, $(1 + 2)^2 \neq 1^2 + 2^2$.

(c) $f(x) = x + 1$ is not a linear transformation from R into R because

$$f(x_1 + x_2) = x_1 + x_2 + 1$$

whereas

$$f(x_1) + f(x_2) = (x_1 + 1) + (x_2 + 1) = x_1 + x_2 + 2.$$

So $f(x_1 + x_2) \neq f(x_1) + f(x_2)$.

REMARK: The function in Example 3(c) points out two uses of the term *linear.* In calculus, $f(x) = x + 1$ is called a linear function because its graph is a line. It is not a linear transformation from the vector space R into R, however, because it preserves neither vector addition nor scalar multiplication.

Two simple linear transformations are the **zero transformation** and the **identity transformation,** which are defined as follows.

1. $T(\mathbf{v}) = \mathbf{0}$, for all $\mathbf{v}$ — Zero transformation $(T: V \rightarrow W)$
2. $T(\mathbf{v}) = \mathbf{v}$, for all $\mathbf{v}$ — Identity transformation $(T: V \rightarrow V)$

The verifications of the linearity of these two transformations are left as exercises. (See Exercises 68 and 69.)

Note that the linear transformation in Example 2 has the property that the zero vector is mapped to itself. That is, $T(\mathbf{0}) = \mathbf{0}$. (Try checking this.) This property is true for all linear transformations, as stated in the next theorem.

THEOREM 6.1 Properties of Linear Transformations

Let T be a linear transformation from V into W, where $\mathbf{u}$ and $\mathbf{v}$ are in V. Then the following properties are true.

1. $T(\mathbf{0}) = \mathbf{0}$
2. $T(-\mathbf{v}) = -T(\mathbf{v})$
3. $T(\mathbf{u} - \mathbf{v}) = T(\mathbf{u}) - T(\mathbf{v})$
4. If $\mathbf{v} = c_1\mathbf{v}_1 + c_2\mathbf{v}_2 + \cdots + c_n\mathbf{v}_n$,

then

$$\begin{aligned} T(\mathbf{v}) &= T(c_1\mathbf{v}_1 + c_2\mathbf{v}_2 + \cdots + c_n\mathbf{v}_n) \\ &= c_1T(\mathbf{v}_1) + c_2T(\mathbf{v}_2) + \cdots + c_nT(\mathbf{v}_n). \end{aligned}$$

PROOF To prove the first property, note that $0\mathbf{v} = \mathbf{0}$. Then it follows that

$$T(\mathbf{0}) = T(0\mathbf{v}) = 0T(\mathbf{v}) = \mathbf{0}.$$

The second property follows from $-\mathbf{v} = (-1)\mathbf{v}$, which implies that

$$T(-\mathbf{v}) = T[(-1)\mathbf{v}] = (-1)T(\mathbf{v}) = -T(\mathbf{v}).$$

The third property follows from $\mathbf{u} - \mathbf{v} = \mathbf{u} + (-\mathbf{v})$, which implies that

$$T(\mathbf{u} - \mathbf{v}) = T[\mathbf{u} + (-1)\mathbf{v}] = T(\mathbf{u}) + (-1)T(\mathbf{v}) = T(\mathbf{u}) - T(\mathbf{v}).$$

The proof of the fourth property is left to you.

Property 4 of Theorem 6.1 tells you that a linear transformation $T: V \rightarrow W$ is determined completely by its action on a basis of V. In other words, if $\{\mathbf{v}_1, \mathbf{v}_2, \ldots, \mathbf{v}_n\}$ is a basis for the vector space V and if $T(\mathbf{v}_1), T(\mathbf{v}_2), \ldots, T(\mathbf{v}_n)$ are given, then $T(\mathbf{v})$ is determined for *any* $\mathbf{v}$ in V. The use of this property is demonstrated in Example 4.

EXAMPLE 4 Linear Transformations and Bases

Let $T: R^3 \rightarrow R^3$ be a linear transformation such that

$$T(1, 0, 0) = (2, -1, 4)$$
$$T(0, 1, 0) = (1, 5, -2)$$
$$T(0, 0, 1) = (0, 3, 1).$$

Find $T(2, 3, -2)$.

SOLUTION Because $(2, 3, -2)$ can be written as

$$(2, 3, -2) = 2(1, 0, 0) + 3(0, 1, 0) - 2(0, 0, 1),$$

you can use Property 4 of Theorem 6.1 to write

$$\begin{aligned} T(2, 3, -2) &= 2T(1, 0, 0) + 3T(0, 1, 0) - 2T(0, 0, 1) \\ &= 2(2, -1, 4) + 3(1, 5, -2) - 2(0, 3, 1) \\ &= (7, 7, 0). \end{aligned}$$

Another advantage of Theorem 6.1 is that it provides a quick way to spot functions that are not linear transformations. That is, because all four conditions of the theorem must be true of a linear transformation, it follows that if any one of the properties is not satisfied for a function T, then the function is not a linear transformation. For example, the function

$$T(x_1, x_2) = (x_1 + 1, x_2)$$

is not a linear transformation from R^2 to R^2 because $T(0, 0) \neq (0, 0)$.

In the next example, a matrix is used to define a linear transformation from R^2 into R^3. The vector $\mathbf{v} = (v_1, v_2)$ is written in the matrix form

$$\mathbf{v} = \begin{bmatrix} v_1 \\ v_2 \end{bmatrix},$$

so it can be multiplied *on the left* by a matrix of order 3×2.

EXAMPLE 5 A Linear Transformation Defined by a Matrix

The function $T: R^2 \rightarrow R^3$ is defined as follows.

$$T(\mathbf{v}) = A\mathbf{v} = \begin{bmatrix} 3 & 0 \\ 2 & 1 \\ -1 & -2 \end{bmatrix} \begin{bmatrix} v_1 \\ v_2 \end{bmatrix}$$

(a) Find $T(\mathbf{v})$, where $\mathbf{v} = (2, -1)$.

(b) Show that T is a linear transformation from R^2 into R^3.

SOLUTION (a) Because $\mathbf{v} = (2, -1)$, you have

$$T(\mathbf{v}) = A\mathbf{v} = \begin{bmatrix} 3 & 0 \\ 2 & 1 \\ -1 & -2 \end{bmatrix} \begin{bmatrix} 2 \\ -1 \end{bmatrix} = \begin{bmatrix} 6 \\ 3 \\ 0 \end{bmatrix}.$$

Vector in R^2 Vector in R^3

So, you have $T(2, -1) = (6, 3, 0)$.

(b) Begin by observing that T does map a vector in R^2 to a vector in R^3. To show that T is a linear transformation, use the properties of matrix multiplication, as shown in Theorem 2.3. For any vectors $\mathbf{u}$ and $\mathbf{v}$ in R^2, the distributive property of matrix multiplication over addition produces

$$T(\mathbf{u} + \mathbf{v}) = A(\mathbf{u} + \mathbf{v}) = A\mathbf{u} + A\mathbf{v} = T(\mathbf{u}) + T(\mathbf{v}).$$

Similarly, for any vector $\mathbf{u}$ in R^2 and any scalar c, the commutative property of scalar multiplication with matrix multiplication produces

$$T(c\mathbf{u}) = A(c\mathbf{u}) = c(A\mathbf{u}) = cT(\mathbf{u}).$$

Example 5 illustrates an important result regarding the representation of linear transformations from R^n into R^m. This result is presented in two stages. Theorem 6.2 on the next page states that every $m \times n$ matrix represents a linear transformation from R^n into R^m. Then, in Section 6.3, you will see the converse—that every linear transformation from R^n into R^m can be represented by an $m \times n$ matrix.

Note in part (b) of Example 5 that no reference is made to the specific matrix A. This verification serves as a general proof that the function defined by any $m \times n$ matrix is a linear transformation from R^n into R^m.

THEOREM 6.2
The Linear Transformation Given by a Matrix

Let A be an $m \times n$ matrix. The function T defined by

$$T(\mathbf{v}) = A\mathbf{v}$$

is a linear transformation from R^n into R^m. In order to conform to matrix multiplication with an $m \times n$ matrix, the vectors in R^n are represented by $n \times 1$ matrices and the vectors in R^m are represented by $m \times 1$ matrices.

REMARK: The $m \times n$ zero matrix corresponds to the zero transformation from R^n into R^m, and the $n \times n$ identity matrix I_n corresponds to the identity transformation from R^n into R^n.

Be sure you see that an $m \times n$ matrix A defines a linear transformation from R^n into R^m.

$$A\mathbf{v} = \begin{bmatrix} a_{11} & a_{12} & \cdots & a_{1n} \\ a_{21} & a_{22} & \cdots & a_{2n} \\ \vdots & \vdots & & \vdots \\ a_{m1} & a_{m2} & \cdots & a_{mn} \end{bmatrix} \begin{bmatrix} v_1 \\ v_2 \\ \vdots \\ v_n \end{bmatrix} = \begin{bmatrix} a_{11}v_1 + a_{12}v_2 + \cdots + a_{1n}v_n \\ a_{21}v_1 + a_{22}v_2 + \cdots + a_{2n}v_n \\ \vdots \\ a_{m1}v_1 + a_{m2}v_2 + \cdots + a_{mn}v_n \end{bmatrix}$$

Vector in R^n — Vector in R^m

EXAMPLE 6 **Linear Transformation Given by Matrices**

The linear transformation $T: R^n \rightarrow R^m$ is defined by $T(\mathbf{v}) = A\mathbf{v}$. Find the dimensions of R^n and R^m for the linear transformation represented by each matrix.

(a) $A = \begin{bmatrix} 0 & 1 & -1 \\ 2 & 3 & 0 \\ 4 & 2 & 1 \end{bmatrix}$

(b) $A = \begin{bmatrix} 2 & -3 \\ -5 & 0 \\ 0 & -2 \end{bmatrix}$

(c) $A = \begin{bmatrix} 1 & 0 & -1 & 2 \\ 3 & 1 & 0 & 0 \end{bmatrix}$

SOLUTION (a) Because the size of this matrix is 3×3, it defines a linear transformation from R^3 into R^3.

$$A\mathbf{v} = \begin{bmatrix} 0 & 1 & -1 \\ 2 & 3 & 0 \\ 4 & 2 & 1 \end{bmatrix} \begin{bmatrix} v_1 \\ v_2 \\ v_3 \end{bmatrix} = \begin{bmatrix} u_1 \\ u_2 \\ u_3 \end{bmatrix}$$

Vector in R^3 (under $\mathbf{v}$) — Vector in R^3 (under $\mathbf{u}$)

(b) Because the size of this matrix is 3×2, it defines a linear transformation from R^2 into R^3.

(c) Because the size of this matrix is 2×4, it defines a linear transformation from R^4 into R^2.

In the next example, a common type of linear transformation from R^2 into R^2 is discussed.

EXAMPLE 7 Rotation in the Plane

Show that the linear transformation $T: R^2 \rightarrow R^2$ represented by the matrix

$$A = \begin{bmatrix} \cos\theta & -\sin\theta \\ \sin\theta & \cos\theta \end{bmatrix}$$

has the property that it rotates every vector in R^2 counterclockwise about the origin through the angle θ.

SOLUTION From Theorem 6.2, you know that T is a linear transformation. To show that it rotates every vector in R^2 counterclockwise through the angle θ, let $\mathbf{v} = (x, y)$ be a vector in R^2. Using polar coordinates, you can write $\mathbf{v}$ as

$$\mathbf{v} = (x, y) = (r\cos\alpha, r\sin\alpha),$$

where r is the length of $\mathbf{v}$ and α is the angle from the positive x-axis counterclockwise to the vector $\mathbf{v}$. Now, applying the linear transformation T to $\mathbf{v}$ produces

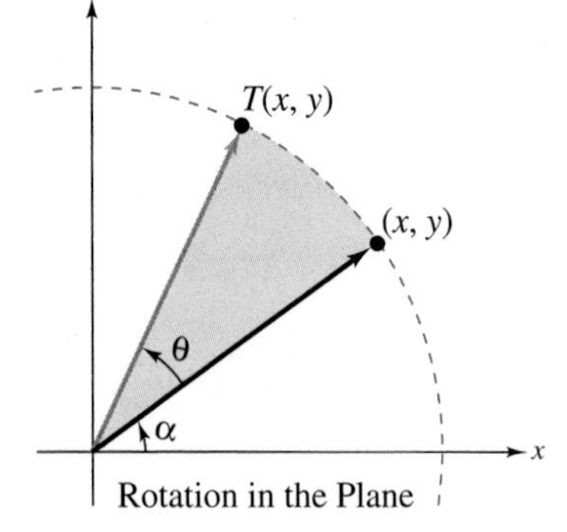

Figure 6.2

$$\begin{aligned} T(\mathbf{v}) = A\mathbf{v} &= \begin{bmatrix} \cos\theta & -\sin\theta \\ \sin\theta & \cos\theta \end{bmatrix} \begin{bmatrix} x \\ y \end{bmatrix} \\ &= \begin{bmatrix} \cos\theta & -\sin\theta \\ \sin\theta & \cos\theta \end{bmatrix} \begin{bmatrix} r\cos\alpha \\ r\sin\alpha \end{bmatrix} \\ &= \begin{bmatrix} r\cos\theta\cos\alpha - r\sin\theta\sin\alpha \\ r\sin\theta\cos\alpha + r\cos\theta\sin\alpha \end{bmatrix} \\ &= \begin{bmatrix} r\cos(\theta + \alpha) \\ r\sin(\theta + \alpha) \end{bmatrix}. \end{aligned}$$

So, the vector $T(\mathbf{v})$ has the same length as $\mathbf{v}$. Furthermore, because the angle from the positive x-axis to $T(\mathbf{v})$ is $\theta + \alpha$, $T(\mathbf{v})$ is the vector that results from rotating the vector $\mathbf{v}$ counterclockwise through the angle θ, as shown in Figure 6.2 on the previous page.

REMARK: The linear transformation in Example 7 is called a **rotation** in R^2. Rotations in R^2 preserve both vector length and the angle between two vectors. That is, the angle between $\mathbf{u}$ and $\mathbf{v}$ is equal to the angle between $T(\mathbf{u})$ and $T(\mathbf{v})$.

EXAMPLE 8 A Projection in R^3

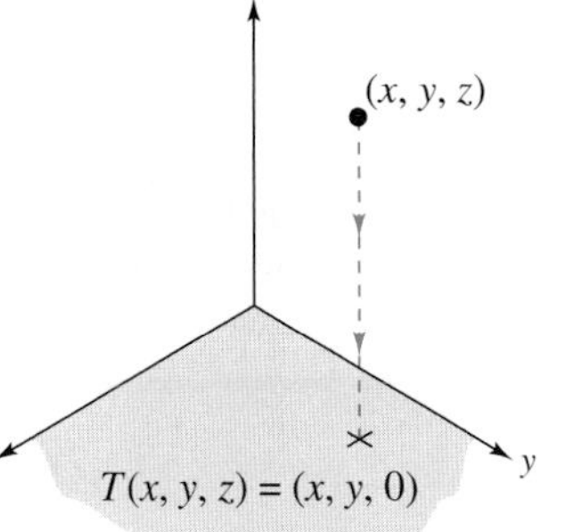

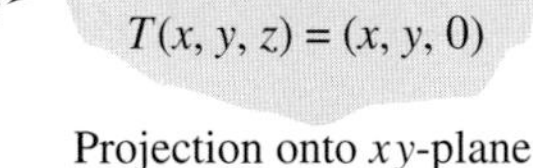

Projection onto xy-plane

Figure 6.3

The linear transformation $T: R^3 \to R^3$ represented by

$$A = \begin{bmatrix} 1 & 0 & 0 \\ 0 & 1 & 0 \\ 0 & 0 & 0 \end{bmatrix}$$

is called a **projection** in R^3. If $\mathbf{v} = (x, y, z)$ is a vector into R^3, then $T(\mathbf{v}) = (x, y, 0)$. In other words, T maps every vector in R^3 to its orthogonal projection in the xy-plane, as shown in Figure 6.3.

So far only linear transformations from R^n into R^m or from R^n into R^n have been discussed. In the remainder of this section, some linear transformations involving vector spaces other than R^n will be considered.

EXAMPLE 9 A Linear Transformation from $M_{m,n}$ into $M_{n,m}$

Let $T: M_{m,n} \to M_{n,m}$ be the function that maps an $m \times n$ matrix A to its transpose. That is,

$$T(A) = A^T.$$

Show that T is a linear transformation.

SOLUTION Let A and B be $m \times n$ matrices. From Theorem 2.6 you have

$$\begin{aligned} T(A + B) &= (A + B)^T \\ &= A^T + B^T \\ &= T(A) + T(B) \end{aligned}$$

and

$$\begin{aligned} T(cA) &= (cA)^T \\ &= c(A^T) \\ &= cT(A). \end{aligned}$$

So, T is a linear transformation from $M_{m,n}$ into $M_{n,m}$.

EXAMPLE 10 **The Differential Operator (Calculus)**

Let $C'[a, b]$ be the set of all functions whose derivatives are continuous on $[a, b]$. Show that the differential operator D_x defines a linear transformation from $C'[a, b]$ into $C[a, b]$.

SOLUTION Using operator notation, you can write

$$D_x(f) = \frac{d}{dx}[f],$$

where f is in $C'[a, b]$. To show that D_x is a linear transformation, you must use calculus. Specifically, because the derivative of the sum of two functions is equal to the sum of their derivatives and because the sum of two continuous functions is continuous, you have

$$\begin{aligned} D_x(f + g) &= \frac{d}{dx}[f + g] = \frac{d}{dx}[f] + \frac{d}{dx}[g] \\ &= D_x(f) + D_x(g). \end{aligned}$$

Similarly, because the derivative of a scalar multiple of a function is equal to the scalar multiple of the derivative and because the scalar multiple of a continuous function is continuous, you have

$$\begin{aligned} D_x(cf) &= \frac{d}{dx}[cf] = c\left(\frac{d}{dx}[f]\right) \\ &= cD_x(f). \end{aligned}$$

So, D_x is a linear transformation from $C'[a, b]$ into $C[a, b]$.

The linear transformation D_x in Example 10 is called the **differential operator.** For polynomials, the differential operator is a linear transformation from P_n into P_{n-1} because the derivative of a polynomial function of degree n is a polynomial function of degree $n - 1$ or less. That is,

$$D_x(a_nx^n + \cdot \cdot \cdot + a_1x + a_0) = na_nx^{n-1} + \cdot \cdot \cdot + a_1.$$

The next example describes a linear transformation from the vector space of polynomial functions P into the vector space of real numbers R.

EXAMPLE 11 **The Definite Integral as a Linear Transformation (Calculus)**

Let $T: P \rightarrow R$ be defined by

$$T(p) = \int_a^b p(x)\, dx.$$

Show that T is a linear transformation from P, the vector space of polynomial functions, into R, the vector space of real numbers.

SOLUTION Using properties of definite integrals, you can write

$$T(p+q) = \int_a^b [p(x)+q(x)]\,dx$$
$$= \int_a^b p(x)\,dx + \int_a^b q(x)\,dx$$
$$= T(p) + T(q)$$

and

$$T(cp) = \int_a^b c[p(x)]\,dx = c\int_a^b p(x)\,dx = cT(p).$$

So, T is a linear transformation.

SECTION 6.1 Exercises

In Exercises 1–8, use the function to find (a) the image of **v** and (b) the preimage of **w**.

1. $T(v_1, v_2) = (v_1 + v_2, v_1 - v_2)$, $\mathbf{v} = (3, -4)$, $\mathbf{w} = (3, 19)$

2. $T(v_1, v_2) = (2v_2 - v_1, v_1, v_2)$, $\mathbf{v} = (0, 6)$, $\mathbf{w} = (3, 1, 2)$

3. $T(v_1, v_2, v_3) = (v_2 - v_1, v_1 + v_2, 2v_1)$, $\mathbf{v} = (2, 3, 0)$, $\mathbf{w} = (-11, -1, 10)$

4. $T(v_1, v_2, v_3) = (2v_1 + v_2, 2v_2 - 3v_1, v_1 - v_3)$, $\mathbf{v} = (-4, 5, 1)$, $\mathbf{w} = (4, 1, -1)$

5. $T(v_1, v_2, v_3) = (4v_2 - v_1, 4v_1 + 5v_2)$, $\mathbf{v} = (2, -3, -1)$, $\mathbf{w} = (3, 9)$

6. $T(v_1, v_2, v_3) = (2v_1 + v_2, v_1 - v_2)$, $\mathbf{v} = (2, 1, 4)$, $\mathbf{w} = (-1, 2)$

7. $T(v_1, v_2) = \left(\frac{\sqrt{2}}{2}v_1 - \frac{\sqrt{2}}{2}v_2, v_1 + v_2, 2v_1 - v_2\right)$, $\mathbf{v} = (1, 1)$, $\mathbf{w} = \left(-5\sqrt{2}, -2, -16\right)$

8. $T(v_1, v_2) = \left(\frac{\sqrt{3}}{2}v_1 - \frac{1}{2}v_2, v_1 - v_2, v_2\right)$, $\mathbf{v} = (2, 4)$, $\mathbf{w} = \left(\sqrt{3}, 2, 0\right)$

In Exercises 9–22, determine whether the function is a linear transformation.

9. $T: R^2 \to R^2$, $T(x, y) = (x, 1)$

10. $T: R^2 \to R^2$, $T(x, y) = (x^2, y)$

11. $T: R^3 \to R^3$, $T(x, y, z) = (x + y, x - y, z)$

12. $T: R^3 \to R^3$, $T(x, y, z) = (x + 1, y + 1, z + 1)$

13. $T: R^2 \to R^3$, $T(x, y) = \left(\sqrt{x}, xy, \sqrt{y}\right)$

14. $T: R^2 \to R^3$, $T(x, y) = \left(x^2, xy, y^2\right)$

15. $T: M_{2,2} \to R$, $T(A) = |A|$

16. $T: M_{2,2} \to R$, $T(A) = a + b + c + d$, where $A = \begin{bmatrix} a & b \\ c & d \end{bmatrix}$.

17. $T: M_{3,3} \to M_{3,3}$, $T(A) = \begin{bmatrix} 0 & 0 & 1 \\ 0 & 1 & 0 \\ 1 & 0 & 0 \end{bmatrix} A$

18. $T: M_{3,3} \to M_{3,3}$, $T(A) = \begin{bmatrix} 1 & 0 & 0 \\ 0 & 1 & 0 \\ 0 & 0 & -1 \end{bmatrix} A$

19. $T: M_{2,2} \to M_{2,2}$, $T(A) = A^T$

20. $T: M_{2,2} \to M_{2,2}$, $T(A) = A^{-1}$

21. $T: P_2 \to P_2$, $T(a_0 + a_1x + a_2x^2) = (a_0 + a_1 + a_2) + (a_1 + a_2)x + a_2x^2$

22. $T: P_2 \to P_2$, $T(a_0 + a_1x + a_2x^2) = a_1 + 2a_2x$

In Exercises 23–26, let $T: R^3 \to R^3$ be a linear transformation such that $T(1, 0, 0) = (2, 4, -1)$, $T(0, 1, 0) = (1, 3, -2)$, and $T(0, 0, 1) = (0, -2, 2)$. Find

23. $T(0, 3, -1)$.

24. $T(2, -1, 0)$.

25. $T(2, -4, 1)$.

26. $T(-2, 4, -1)$.

In Exercises 27–30, let $T: R^3 \to R^3$ be a linear transformation such that $T(1, 1, 1) = (2, 0, -1)$, $T(0, -1, 2) = (-3, 2, -1)$, and $T(1, 0, 1) = (1, 1, 0)$. Find

27. $T(2, 1, 0)$.

28. $T(0, 2, -1)$.

29. $T(2, -1, 1)$.

30. $T(-2, 1, 0)$.

In Exercises 31–35, the linear transformation $T: R^n \to R^m$ is defined by $T(\mathbf{v}) = A\mathbf{v}$. Find the dimensions of R^n and R^m.

31. $A = \begin{bmatrix} 0 & 1 & -2 & 1 \\ -1 & 4 & 5 & 0 \\ 0 & 1 & 3 & 1 \end{bmatrix}$

32. $A = \begin{bmatrix} 1 & 2 \\ -2 & 4 \\ -2 & 2 \end{bmatrix}$

33. $A = \begin{bmatrix} -1 & 2 & 1 & 3 & 4 \\ 0 & 0 & 2 & -1 & 0 \end{bmatrix}$

34. $A = \begin{bmatrix} -1 & 0 & 0 & 0 \\ 0 & 1 & 0 & 0 \\ 0 & 0 & 2 & 0 \\ 0 & 0 & 0 & 1 \end{bmatrix}$

35. $A = \begin{bmatrix} 0 & -1 \\ -1 & 0 \end{bmatrix}$

36. For the linear transformation from Exercise 31, find (a) $T(1, 0, 2, 3)$ and (b) the preimage of $(0, 0, 0)$.

37. **Writing** For the linear transformation from Exercise 32, find (a) $T(2, 4)$ and (b) the preimage of $(-1, 2, 2)$. (c) Then explain why the vector $(1, 1, 1)$ has no preimage under this transformation.

38. For the linear transformation from Exercise 33, find (a) $T(1, 0, -1, 3, 0)$ and (b) the preimage of $(-1, 8)$.

39. For the linear transformation from Exercise 34, find (a) $T(1, 1, 1, 1)$ and (b) the preimage of $(1, 1, 1, 1)$.

40. For the linear transformation from Exercise 35, find (a) $T(1, 1)$, (b) the preimage of $(1, 1)$, and (c) the preimage of $(0, 0)$.

41. Let T be the linear transformation from R^2 into R^2 represented by $T(x, y) = (x \cos\theta - y \sin\theta, x \sin\theta + y \cos\theta)$.

Find (a) $T(4, 4)$ for $\theta = 45°$, (b) $T(4, 4)$ for $\theta = 30°$, and (c) $T(5, 0)$ for $\theta = 120°$.

42. For the linear transformation from Exercise 41, let $\theta = 45°$ and find the preimage of $\mathbf{v} = (1, 1)$.

In Exercises 43–46, let D_x be the linear transformation from $C'[a, b]$ into $C[a, b]$ from Example 10. Decide whether each statement is true or false. Explain your reasoning.

43. $D_x(e^{x^2} + 2x) = D_x(e^{x^2}) + 2D_x(x)$

44. $D_x(x^2 - \ln x) = D_x(x^2) - D_x(\ln x)$

45. $D_x(\sin 2x) = 2D_x(\sin x)$

46. $D_x\left(\cos \frac{x}{2}\right) = \frac{1}{2}D_x(\cos x)$

Calculus In Exercises 47–50, for the linear transformation from Example 10, find the preimage of each function.

47. $f(x) = 2x + 1$

48. $f(x) = e^x$

49. $f(x) = \sin x$

50. $f(x) = \frac{1}{x}$

51. **Calculus** Let T be the linear transformation from P into R shown by

$$T(p) = \int_0^1 p(x)\, dx.$$

Find (a) $T(3x^2 - 2)$, (b) $T(x^3 - x^5)$, and (c) $T(4x - 6)$.

52. **Calculus** Let T be the linear transformation from P_2 into R represented by the integral in Exercise 51. Find the preimage of 1. That is, find the polynomial function(s) of degree 2 or less such that $T(p) = 1$.

53. Let T be a linear transformation from R^2 into R^2 such that $T(1, 1) = (1, 0)$ and $T(1, -1) = (0, 1)$. Find $T(1, 0)$ and $T(0, 2)$.

54. Let T be a linear transformation from R^2 into R^2 such that $T(1, 0) = (1, 1)$ and $T(0, 1) = (-1, 1)$. Find $T(1, 4)$ and $T(-2, 1)$.

55. Let T be a linear transformation from P_2 into P_2 such that $T(1) = x$, $T(x) = 1 + x$, and $T(x^2) = 1 + x + x^2$. Find $T(2 - 6x + x^2)$.

56. Let T be a linear transformation from $M_{2,2}$ into $M_{2,2}$ such that

$$T\left(\begin{bmatrix} 1 & 0 \\ 0 & 0 \end{bmatrix}\right) = \begin{bmatrix} 1 & -1 \\ 0 & 2 \end{bmatrix}, \quad T\left(\begin{bmatrix} 0 & 1 \\ 0 & 0 \end{bmatrix}\right) = \begin{bmatrix} 0 & 2 \\ 1 & 1 \end{bmatrix},$$

$$T\left(\begin{bmatrix} 0 & 0 \\ 1 & 0 \end{bmatrix}\right) = \begin{bmatrix} 1 & 2 \\ 0 & 1 \end{bmatrix}, \quad T\left(\begin{bmatrix} 0 & 0 \\ 0 & 1 \end{bmatrix}\right) = \begin{bmatrix} 3 & -1 \\ 1 & 0 \end{bmatrix}.$$

Find $T\left(\begin{bmatrix} 1 & 3 \\ -1 & 4 \end{bmatrix}\right)$.

True or False? In Exercises 57 and 58, determine whether each statement is true or false. If a statement is true, give a reason or cite an appropriate statement from the text. If a statement is false, provide an example that shows the statement is not true in all cases or cite an appropriate statement from the text.

57. (a) Linear transformations are functions from one vector space to another that preserve the operations of vector addition and scalar multiplication.

(b) The function $f(x) = \cos x$ is a linear transformation from R into R.

(c) For polynomials, the differential operator D_x is a linear transformation from P_n into P_{n-1}.

58. (a) A linear transformation is operation preserving if the same result occurs whether the operations of addition and scalar multiplication are performed before or after the linear transformation is applied.

(b) The function $g(x) = x^3$ is a linear transformation from R into R.

(c) Any linear function of the form $f(x) = ax + b$ is a linear transformation from R into R.

59. Writing Suppose $T: R^2 \rightarrow R^2$ such that $T(1, 0) = (1, 0)$ and $T(0, 1) = (0, 0)$.

(a) Determine $T(x, y)$ for (x, y) in R^2.

(b) Give a geometric description of T.

60. Writing Suppose $T: R^2 \rightarrow R^2$ such that $T(1, 0) = (0, 1)$ and $T(0, 1) = (1, 0)$.

(a) Determine $T(x, y)$ for (x, y) in R^2.

(b) Give a geometric description of T.

61. Let T be the function from R^2 into R^2 such that $T(\mathbf{u}) = \text{proj}_{\mathbf{v}}\mathbf{u}$, where $\mathbf{v} = (1, 1)$.

(a) Find $T(x, y)$.   (b) Find $T(5, 0)$.

(c) Prove that $T(\mathbf{u} + \mathbf{w}) = T(\mathbf{u}) + T(\mathbf{w})$ for every $\mathbf{u}$ and $\mathbf{w}$ in R^2.

(d) Prove that $T(c\mathbf{u}) = cT(\mathbf{u})$ for every $\mathbf{u}$ in R^2. This result and the result in part (c) prove that T is a linear transformation from R^2 into R^2.

62. Writing Find $T(3, 4)$ and $T(T(3, 4))$ from Exercise 61 and give geometric descriptions of the results.

63. Show that T from Exercise 61 is represented by the matrix

$$A = \begin{bmatrix} \frac{1}{2} & \frac{1}{2} \\ \frac{1}{2} & \frac{1}{2} \end{bmatrix}.$$

64. Use the concept of a fixed point of a linear transformation $T: V \rightarrow V$. A vector $\mathbf{u}$ is a **fixed point** if $T(\mathbf{u}) = \mathbf{u}$.

(a) Prove that $\mathbf{0}$ is a fixed point of any linear transformation $T: V \rightarrow V$.

(b) Prove that the set of fixed points of a linear transformation $T: V \rightarrow V$ is a subspace of V.

(c) Determine all fixed points of the linear transformation $T: R^2 \rightarrow R^2$ represented by $T(x, y) = (x, 2y)$.

(d) Determine all fixed points of the linear transformation $T: R^2 \rightarrow R^2$ represented by $T(x, y) = (y, x)$.

65. A **translation** is a function of the form $T(x, y) = (x - h, y - k)$, where at least one of the constants h and k is nonzero.

(a) Show that a translation in the plane is not a linear transformation.

(b) For the translation $T(x, y) = (x - 2, y + 1)$, determine the images of $(0, 0)$, $(2, -1)$, and $(5, 4)$.

(c) Show that a translation in the plane has no fixed points.

66. Let $S = \{\mathbf{v}_1, \mathbf{v}_2, . . . , \mathbf{v}_n\}$ be a set of linearly dependent vectors in V, and let T be a linear transformation from V into V. Prove that the set

$$\{T(\mathbf{v}_1), T(\mathbf{v}_2), . . . , T(\mathbf{v}_n)\}$$

is linearly dependent.

67. Let $S = \{\mathbf{v}_1, \mathbf{v}_2, \mathbf{v}_3\}$ be a set of linearly independent vectors in R^3. Find a linear transformation T from R^3 into R^3 such that the set $\{T(\mathbf{v}_1), T(\mathbf{v}_2), T(\mathbf{v}_3)\}$ is linearly dependent.

68. Prove that the zero transformation $T: V \rightarrow W$ is a linear transformation.

69. Prove that the identity transformation $T: V \rightarrow V$ is a linear transformation.

70. Let V be an inner product space. For a fixed vector $\mathbf{v}_0$ in V, define $T: V \rightarrow R$ by $T(\mathbf{v}) = \langle \mathbf{v}, \mathbf{v}_0 \rangle$. Prove that T is a linear transformation.

71. Let $T: M_{n,n} \rightarrow R$ be defined by $T(A) = a_{11} + a_{22} + \cdot \cdot \cdot + a_{nn}$ (the trace of A). Prove that T is a linear transformation.

72. Let V be an inner product space with a subspace W having $B = \{\mathbf{w}_1, \mathbf{w}_2, . . . , \mathbf{w}_n\}$ as an orthonormal basis. Show that the function $T: V \rightarrow W$ represented by

$$T(\mathbf{v}) = \langle \mathbf{v}, \mathbf{w}_1 \rangle \mathbf{w}_1 + \langle \mathbf{v}, \mathbf{w}_2 \rangle \mathbf{w}_2 + \cdot \cdot \cdot + \langle \mathbf{v}, \mathbf{w}_n \rangle \mathbf{w}_n$$

is a linear transformation. T is called the **orthogonal projection of V onto W.**

73. Guided Proof Let $\{\mathbf{v}_1, \mathbf{v}_2, . . . , \mathbf{v}_n\}$ be a basis for a vector space V. Prove that if a linear transformation $T: V \rightarrow V$ satisfies $T(\mathbf{v}_i) = \mathbf{0}$ for $i = 1, 2, . . . , n$, then T is the zero transformation.

Getting Started: To prove that T is the zero transformation, you need to show that $T(\mathbf{v}) = \mathbf{0}$ for every vector $\mathbf{v}$ in V.

(i) Let $\mathbf{v}$ be an arbitrary vector in V such that

$\mathbf{v} = c_1\mathbf{v}_1 + c_2\mathbf{v}_2 + \cdot \cdot \cdot + c_n\mathbf{v}_n.$

(ii) Use the definition and properties of linear transformations to rewrite $T(\mathbf{v})$ as a linear combination of $T(\mathbf{v}_i)$.

(iii) Use the fact that $T(\mathbf{v}_i) = \mathbf{0}$ to conclude that $T(\mathbf{v}) = \mathbf{0}$, making T the zero transformation.

74. Guided Proof Prove that $T: V \rightarrow W$ is a linear transformation if and only if $T(a\mathbf{u} + b\mathbf{v}) = aT(\mathbf{u}) + bT(\mathbf{v})$ for all vectors $\mathbf{u}$ and $\mathbf{v}$ and all scalars a and b.

Getting Started: Because this is an "if and only if" statement, you need to prove the statement in both directions. To prove that T is a linear transformation, you need to show that the function satisfies the definition of a linear transformation. In the other direction, suppose T is a linear transformation. You can use the definition and properties of a linear transformation to prove that $T(a\mathbf{u} + b\mathbf{v}) = aT(\mathbf{u}) + bT(\mathbf{v})$.

(i) Suppose $T(a\mathbf{u} + b\mathbf{v}) = aT(\mathbf{u}) + bT(\mathbf{v})$. Show that T preserves the properties of vector addition and scalar multiplication by choosing appropriate values of a and b.

(ii) To prove the statement in the other direction, assume that T is a linear transformation. Use the properties and definition of a linear transformation to show that $T(a\mathbf{u} + b\mathbf{v}) = aT(\mathbf{u}) + bT(\mathbf{v})$.

6.2 The Kernel and Range of a Linear Transformation

You know from Theorem 6.1 that for any linear transformation $T: V \rightarrow W$, the zero vector in V is mapped to the zero vector in W. That is,

$$T(\mathbf{0}) = \mathbf{0}.$$

The first question you will consider in this section is whether there are *other* vectors $\mathbf{v}$ such that $T(\mathbf{v}) = \mathbf{0}$. The collection of all such elements is called the **kernel** of T. Note that the symbol $\mathbf{0}$ is used to represent the zero vector in both V and W, although these two zero vectors are often different.

Definition of Kernel of a Linear Transformation

Let $T: V \rightarrow W$ be a linear transformation. Then the set of all vectors $\mathbf{v}$ in V that satisfy $T(\mathbf{v}) = \mathbf{0}$ is called the **kernel** of T and is denoted by $\ker(T)$.

Sometimes the kernel of a transformation is obvious and can be found by inspection, as demonstrated in Examples 1, 2, and 3.

EXAMPLE 1 Finding the Kernel of a Linear Transformation

Let $T: M_{3,2} \rightarrow M_{2,3}$ be the linear transformation that maps a 3×2 matrix A to its transpose. That is,

$$T(A) = A^T.$$

Find the kernel of T.

SOLUTION For this linear transformation, the 3×2 zero matrix is clearly the only matrix in $M_{3,2}$ whose transpose is the zero matrix in $M_{2,3}$.

Zero Matrix in $M_{3,2}$ *Zero Matrix in* $M_{2,3}$

$$\mathbf{0} = \begin{bmatrix} 0 & 0 \\ 0 & 0 \\ 0 & 0 \end{bmatrix} \qquad \mathbf{0} = \begin{bmatrix} 0 & 0 & 0 \\ 0 & 0 & 0 \end{bmatrix}$$

So, the kernel of T consists of a single element: the zero matrix in $M_{3,2}$.

EXAMPLE 2 The Kernels of the Zero and Identity Transformations

(a) The kernel of the zero transformation $T: V \rightarrow W$ consists of all of V because $T(\mathbf{v}) = \mathbf{0}$ for every $\mathbf{v}$ in V. That is, $\ker(T) = V$.

(b) The kernel of the identity transformation $T: V \rightarrow V$ consists of the single element $\mathbf{0}$. That is, $\ker(T) = \{\mathbf{0}\}$.

EXAMPLE 3 Finding the Kernel of a Linear Transformation

Find the kernel of the projection $T: R^3 \rightarrow R^3$ represented by

$$T(x, y, z) = (x, y, 0).$$

SOLUTION This linear transformation projects the vector (x, y, z) in R^3 to the vector $(x, y, 0)$ in the xy-plane. The kernel consists of all vectors lying on the z-axis. That is,

$$\ker(T) = \{(0, 0, z): z \text{ is a real number}\}. \text{ (See Figure 6.4.)}$$

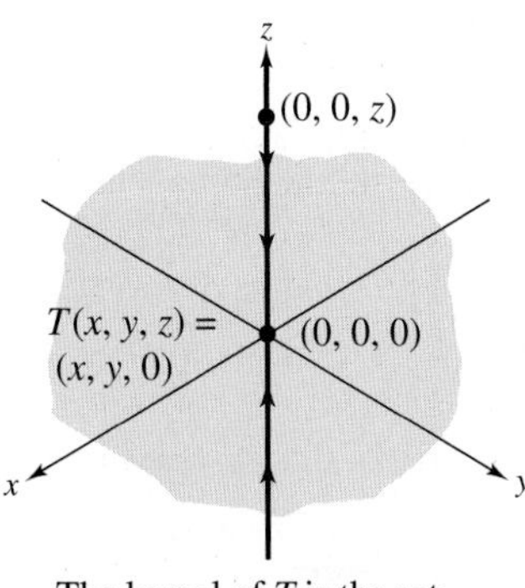

The kernel of T is the set of all vectors on the z-axis.

Figure 6.4

Finding the kernels of the linear transformations in Examples 1, 2, and 3 was fairly easy. Generally, the kernel of a linear transformation is not so obvious, and finding it requires a little work, as illustrated in the next two examples.

EXAMPLE 4 Finding the Kernel of a Linear Transformation

Find the kernel of the linear transformation $T: R^2 \to R^3$ represented by

$$T(x_1, x_2) = (x_1 - 2x_2, 0, -x_1).$$

SOLUTION To find ker(T), you need to find all $\mathbf{x} = (x_1, x_2)$ in R^2 such that

$$T(x_1, x_2) = (x_1 - 2x_2, 0, -x_1) = (0, 0, 0).$$

This leads to the homogeneous system

$$\begin{aligned} x_1 - 2x_2 &= 0 \\ 0 &= 0 \\ -x_1 \quad &= 0, \end{aligned}$$

which has only the trivial solution $(x_1, x_2) = (0, 0)$. So, you have

$$\ker(T) = \{(0, 0)\} = \{\mathbf{0}\}.$$

EXAMPLE 5 Finding the Kernel of a Linear Transformation

Find the kernel of the linear transformation $T: R^3 \to R^2$ defined by $T(\mathbf{x}) = A\mathbf{x}$, where

$$A = \begin{bmatrix} 1 & -1 & -2 \\ -1 & 2 & 3 \end{bmatrix}.$$

SOLUTION The kernel of T is the set of all $\mathbf{x} = (x_1, x_2, x_3)$ in R^3 such that

$$T(x_1, x_2, x_3) = (0, 0).$$

From this equation you can write the homogeneous system

$$\begin{bmatrix} 1 & -1 & -2 \\ -1 & 2 & 3 \end{bmatrix} \begin{bmatrix} x_1 \\ x_2 \\ x_3 \end{bmatrix} = \begin{bmatrix} 0 \\ 0 \end{bmatrix} \quad \longrightarrow \quad \begin{aligned} x_1 - x_2 - 2x_3 &= 0 \\ -x_1 + 2x_2 + 3x_3 &= 0. \end{aligned}$$

Writing the augmented matrix of this system in reduced row-echelon form produces

$$\begin{bmatrix} 1 & 0 & -1 & 0 \\ 0 & 1 & 1 & 0 \end{bmatrix} \quad \longrightarrow \quad \begin{aligned} x_1 &= x_3 \\ x_2 &= -x_3. \end{aligned}$$

Using the parameter $t = x_3$ produces the family of solutions

$$\begin{bmatrix} x_1 \\ x_2 \\ x_3 \end{bmatrix} = \begin{bmatrix} t \\ -t \\ t \end{bmatrix} = t \begin{bmatrix} 1 \\ -1 \\ 1 \end{bmatrix}.$$

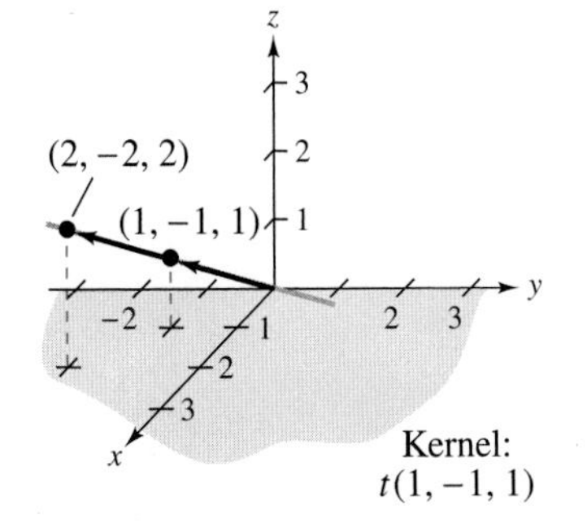

Figure 6.5

So, the kernel of T is represented by

$$\begin{aligned}\ker(T) &= \{t(1, -1, 1): t \text{ is a real number}\}\\ &= \operatorname{span}\{(1, -1, 1)\}.\end{aligned}$$

Note that in Example 5 the kernel of T contains an infinite number of vectors. Of course, the zero vector is in $\ker(T)$, but the kernel also contains such nonzero vectors as $(1, -1, 1)$ and $(2, -2, 2)$, as shown in Figure 6.5. Figure 6.5 shows that this particular kernel is a line passing through the origin, which implies that it is a subspace of R^3. In Theorem 6.3 you will now see that the kernel of every linear transformation $T: V \rightarrow W$ is a subspace of V.

THEOREM 6.3
The Kernel Is a Subspace of V

The kernel of a linear transformation $T: V \rightarrow W$ is a subspace of the domain V.

PROOF From Theorem 6.1 you know that $\ker(T)$ is a nonempty subset of V. So, by Theorem 4.5, you can show that $\ker(T)$ is a subspace of V by showing that it is closed under vector addition and scalar multiplication. To do so, let $\mathbf{u}$ and $\mathbf{v}$ be vectors in the kernel of T. Then

$$\begin{aligned}T(\mathbf{u} + \mathbf{v}) &= T(\mathbf{u}) + T(\mathbf{v})\\ &= \mathbf{0} + \mathbf{0}\\ &= \mathbf{0},\end{aligned}$$

which implies that $\mathbf{u} + \mathbf{v}$ is in the kernel. Moreover, if c is any scalar, then

$$\begin{aligned}T(c\mathbf{u}) &= cT(\mathbf{u})\\ &= c\mathbf{0}\\ &= \mathbf{0},\end{aligned}$$

which implies that $c\mathbf{u}$ is in the kernel.

REMARK: As a result of Theorem 6.3, the kernel of T is sometimes called the **nullspace** of T.

The next example shows how to find a basis for the kernel of a transformation defined by a matrix.

EXAMPLE 6 **Finding a Basis for the Kernel**

Let $T: R^5 \rightarrow R^4$ be defined by $T(\mathbf{x}) = A\mathbf{x}$, where $\mathbf{x}$ is in R^5 and

$$A = \begin{bmatrix} 1 & 2 & 0 & 1 & -1 \\ 2 & 1 & 3 & 1 & 0 \\ -1 & 0 & -2 & 0 & 1 \\ 0 & 0 & 0 & 2 & 8 \end{bmatrix}.$$

Find a basis for $\ker(T)$ as a subspace of R^5.

SOLUTION Using the procedure shown in Example 5, reduce the augmented matrix $[A \; \vdots \; \mathbf{0}]$ to echelon form as follows.

$$\begin{bmatrix} 1 & 0 & 2 & 0 & -1 & 0 \\ 0 & 1 & -1 & 0 & -2 & 0 \\ 0 & 0 & 0 & 1 & 4 & 0 \\ 0 & 0 & 0 & 0 & 0 & 0 \end{bmatrix} \longrightarrow \begin{aligned} x_1 &= -2x_3 + x_5 \\ x_2 &= x_3 + 2x_5 \\ x_4 &= -4x_5 \end{aligned}$$

Discovery

What is the rank of the matrix A in Example 6? Formulate a conjecture relating the dimension of the kernel, the rank, and the number of columns of A. Verify your conjecture for the matrix in Example 5.

Letting $x_3 = s$ and $x_5 = t$, you have

$$\mathbf{x} = \begin{bmatrix} x_1 \\ x_2 \\ x_3 \\ x_4 \\ x_5 \end{bmatrix} = \begin{bmatrix} -2s + t \\ s + 2t \\ s + 0t \\ 0s - 4t \\ 0s + t \end{bmatrix} = s\begin{bmatrix} -2 \\ 1 \\ 1 \\ 0 \\ 0 \end{bmatrix} + t\begin{bmatrix} 1 \\ 2 \\ 0 \\ -4 \\ 1 \end{bmatrix}.$$

So one basis for the kernel of T is

$$B = \{(-2, 1, 1, 0, 0), (1, 2, 0, -4, 1)\}.$$

In the solution of Example 6, a basis for the kernel of T was found by solving the homogeneous system represented by $A\mathbf{x} = \mathbf{0}$. This procedure is a familiar one—it is the same procedure used to find the *solution space* of $A\mathbf{x} = \mathbf{0}$. In other words, the kernel of T is the nullspace of the matrix A, as shown in the following corollary to Theorem 6.3.

COROLLARY TO THEOREM 6.3

Let $T: R^n \rightarrow R^m$ be the linear transformation given by $T(\mathbf{x}) = A\mathbf{x}$. Then the kernel of T is equal to the solution space of $A\mathbf{x} = \mathbf{0}$.

The Range of a Linear Transformation

The kernel is one of two critical subspaces associated with a linear transformation. The second is the **range** of T, denoted by range(T). Recall from Section 6.1 that the range of $T: V \rightarrow W$ is the set of all vectors $\mathbf{w}$ in W that are images of vectors in V. That is, range$(T) = \{T(\mathbf{v}): \mathbf{v} \text{ is in } V\}$.

THEOREM 6.4
The Range of T Is a Subspace of W

The range of a linear transformation $T: V \rightarrow W$ is a subspace of W.

PROOF The range of T is nonempty because $T(\mathbf{0}) = \mathbf{0}$ implies that the range contains the zero vector. To show that it is closed under vector addition, let $T(\mathbf{u})$ and $T(\mathbf{v})$ be vectors in the range of T. Because $\mathbf{u}$ and $\mathbf{v}$ are in V, it follows that $\mathbf{u} + \mathbf{v}$ is also in V. So, the sum $T(\mathbf{u}) + T(\mathbf{v}) = T(\mathbf{u} + \mathbf{v})$ is in the range of T.

To show closure under scalar multiplication, let $T(\mathbf{u})$ be a vector in the range of T and let c be a scalar. Because $\mathbf{u}$ is in V, it follows that $c\mathbf{u}$ is also in V. So, the scalar multiple $cT(\mathbf{u}) = T(c\mathbf{u})$ is in the range of T.

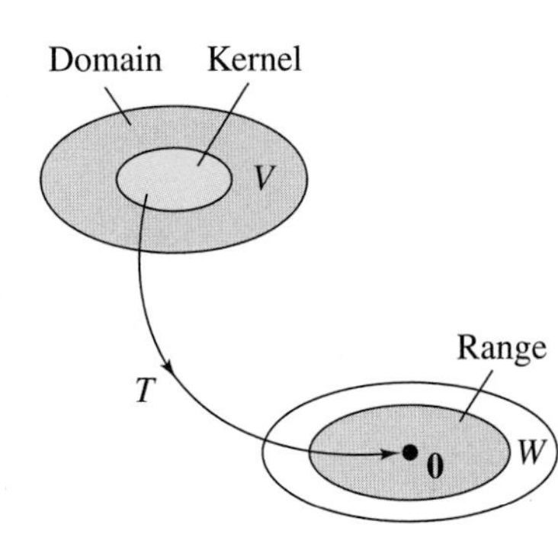

Figure 6.6

Note that the kernel and range of a linear transformation $T: V \rightarrow W$ are subspaces of V and W, respectively, as illustrated in Figure 6.6.

To find a basis for the range of a linear transformation defined by $T(\mathbf{x}) = A\mathbf{x}$, observe that the range consists of all vectors $\mathbf{b}$ such that the system $A\mathbf{x} = \mathbf{b}$ is consistent. By writing the system

$$\begin{bmatrix} a_{11} & a_{12} & \cdots & a_{1n} \\ a_{21} & a_{22} & \cdots & a_{2n} \\ \vdots & \vdots & & \vdots \\ a_{m1} & a_{m2} & \cdots & a_{mn} \end{bmatrix} \begin{bmatrix} x_1 \\ x_2 \\ \vdots \\ x_n \end{bmatrix} = \begin{bmatrix} b_1 \\ b_2 \\ \vdots \\ b_m \end{bmatrix}$$

in the form

$$A\mathbf{x} = x_1 \begin{bmatrix} a_{11} \\ a_{21} \\ \vdots \\ a_{m1} \end{bmatrix} + x_2 \begin{bmatrix} a_{12} \\ a_{22} \\ \vdots \\ a_{m2} \end{bmatrix} + \cdots + x_n \begin{bmatrix} a_{1n} \\ a_{2n} \\ \vdots \\ a_{mn} \end{bmatrix} = \begin{bmatrix} b_1 \\ b_2 \\ \vdots \\ b_m \end{bmatrix} = \mathbf{b}$$

you can see that $\mathbf{b}$ is in the range of T if and only if $\mathbf{b}$ is a linear combination of the column vectors of A. So *the column space of the matrix A is the same as the range of T.*

COROLLARY TO THEOREM 6.4

Let $T: R^n \rightarrow R^m$ be the linear transformation given by $T(\mathbf{x}) = A\mathbf{x}$. Then the column space of A is equal to the range of T.

In Example 4 in Section 4.6, you saw two procedures for finding a basis for the column space of a matrix. In the next example, the second procedure from Example 4 in Section 4.6 will be used to find a basis for the range of a linear transformation defined by a matrix.

EXAMPLE 7 Finding a Basis for the Range of a Linear Transformation

For the linear transformation $R^5 \rightarrow R^4$ from Example 6, find a basis for the range of T.

SOLUTION The echelon form of A was calculated in Example 6.

$$A = \begin{bmatrix} 1 & 2 & 0 & 1 & -1 \\ 2 & 1 & 3 & 1 & 0 \\ -1 & 0 & -2 & 0 & 1 \\ 0 & 0 & 0 & 2 & 8 \end{bmatrix} \Rightarrow \begin{bmatrix} 1 & 0 & 2 & 0 & -1 \\ 0 & 1 & -1 & 0 & -2 \\ 0 & 0 & 0 & 1 & 4 \\ 0 & 0 & 0 & 0 & 0 \end{bmatrix}$$

Because the leading 1's appear in columns 1, 2, and 4 of the reduced matrix on the right, the corresponding column vectors of A form a basis for the column space of A. One basis for the range of T is

$$B = \{(1, 2, -1, 0), (2, 1, 0, 0), (1, 1, 0, 2)\}.$$

The following definition gives the dimensions of the kernel and range of a linear transformation.

Definition of Rank and Nullity of a Linear Transformation

Let $T: V \rightarrow W$ be a linear transformation. The dimension of the kernel of T is called the **nullity** of T and is denoted by **nullity**(T). The dimension of the range of T is called the **rank** of T and is denoted by **rank**(T).

REMARK: If T is provided by a matrix A, then the rank of T is equal to the rank of A, as defined in Section 4.6.

In Examples 6 and 7, the nullity and rank of T are related to the dimension of the domain as follows.

$$\text{rank}(T) + \text{nullity}(T) = 3 + 2 = 5 = \text{dimension of domain}$$

This relationship is true for any linear transformation from a finite-dimensional vector space, as stated in the next theorem.

THEOREM 6.5 Sum of Rank and Nullity

Let $T: V \rightarrow W$ be a linear transformation from an n-dimensional vector space V into a vector space W. Then the sum of the dimensions of the range and kernel is equal to the dimension of the domain. That is,

$$\text{rank}(T) + \text{nullity}(T) = n$$

or

$$\dim(\text{range}) + \dim(\text{kernel}) = \dim(\text{domain}).$$

PROOF The proof provided here covers the case in which T is represented by an $m \times n$ matrix A. The general case will follow in the next section, where you will see that any linear transformation from an n-dimensional space to an m-dimensional space can be represented by a matrix. To prove this theorem, assume that the matrix A has a rank of r. Then you have

$$\text{rank}(T) = \dim(\text{range of } T) = \dim(\text{column space}) = \text{rank}(A) = r.$$

From Theorem 4.17, however, you know that

$$\text{nullity}(T) = \dim(\text{kernel of } T) = \dim(\text{solution space}) = n - r.$$

So, it follows that

$$\text{rank}(T) + \text{nullity}(T) = r + (n - r) = n.$$

EXAMPLE 8 **Finding the Rank and Nullity of a Linear Transformation**

Find the rank and nullity of the linear transformation $T: R^3 \rightarrow R^3$ defined by the matrix

$$A = \begin{bmatrix} 1 & 0 & -2 \\ 0 & 1 & 1 \\ 0 & 0 & 0 \end{bmatrix}.$$

SOLUTION Because A is in row-echelon form and has two nonzero rows, it has a rank of 2. So, the rank of T is 2, and the nullity is $\dim(\text{domain}) - \text{rank} = 3 - 2 = 1$.

REMARK: One way to visualize the relationship between the rank and the nullity of a linear transformation provided by a matrix is to observe that the rank is determined by the number of leading 1's, and the nullity by the number of free variables (columns without leading 1's). Their sum must be the total number of columns of the matrix, which is the dimension of the domain. In Example 8, the first two columns have leading 1's, indicating that the rank is 2. The third column corresponds to a free variable, indicating that the nullity is 1.

EXAMPLE 9 **Finding the Rank and Nullity of a Linear Transformation**

Let $T: R^5 \rightarrow R^7$ be a linear transformation.

(a) Find the dimension of the kernel of T if the dimension of the range is 2.
(b) Find the rank of T if the nullity of T is 4.
(c) Find the rank of T if $\ker(T) = \{\mathbf{0}\}$.

SOLUTION (a) By Theorem 6.5, with $n = 5$, you have

$$\dim(\text{kernel}) = n - \dim(\text{range}) = 5 - 2 = 3.$$

(b) Again by Theorem 6.5, you have

$$\text{rank}(T) = n - \text{nullity}(T) = 5 - 4 = 1.$$

(c) In this case, the nullity of T is 0. So

$$\text{rank}(T) = n - \text{nullity}(T) = 5 - 0 = 5.$$

One-to-One and Onto Linear Transformations

This section began with a question: How many vectors in the domain of a linear transformation are mapped to the zero vector? Theorem 6.6 (below) shows that if the zero vector is the only vector $\mathbf{v}$ such that $T(\mathbf{v}) = \mathbf{0}$, then T is one-to-one. A function $T: V \rightarrow W$ is called **one-to-one** if the preimage of every $\mathbf{w}$ in the range consists of a single vector, as shown in Figure 6.7. This is equivalent to saying that T is one-to-one if and only if, for all $\mathbf{u}$ and $\mathbf{v}$ in V, $T(\mathbf{u}) = T(\mathbf{v})$ implies $\mathbf{u} = \mathbf{v}$.

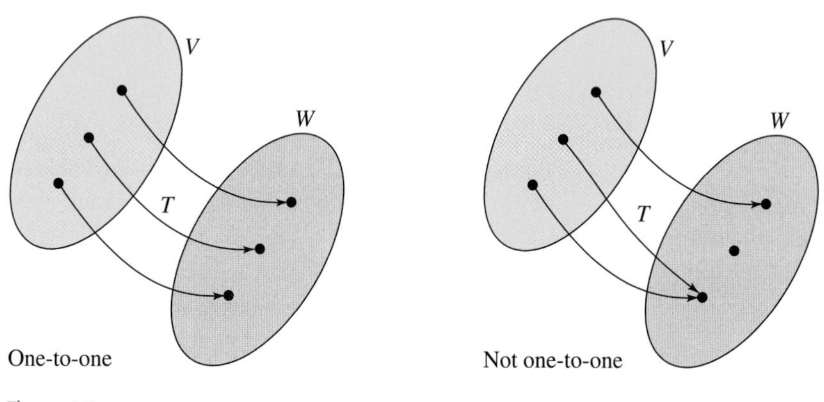

Figure 6.7

THEOREM 6.6 One-to-One Linear Transformations

Let $T: V \rightarrow W$ be a linear transformation. Then T is one-to-one if and only if $\ker(T) = \{\mathbf{0}\}$.

PROOF Suppose T is one-to-one. Then $T(\mathbf{v}) = \mathbf{0}$ can have only one solution: $\mathbf{v} = \mathbf{0}$. In that case, $\ker(T) = \{\mathbf{0}\}$. Conversely, suppose $\ker(T) = \{\mathbf{0}\}$ and $T(\mathbf{u}) = T(\mathbf{v})$. Because T is a linear transformation, it follows that

$$T(\mathbf{u} - \mathbf{v}) = T(\mathbf{u}) - T(\mathbf{v}) = \mathbf{0}.$$

This implies that the vector $\mathbf{u} - \mathbf{v}$ lies in the kernel of T and must equal 0. So, $\mathbf{u} - \mathbf{v} = \mathbf{0}$ and $\mathbf{u} = \mathbf{v}$, and you can conclude that T is one-to-one.

EXAMPLE 10 One-to-One and Not One-to-One Linear Transformations

(a) The linear transformation $T: M_{m,n} \rightarrow M_{n,m}$ represented by $T(A) = A^T$ is one-to-one because its kernel consists of only the $m \times n$ zero matrix.

(b) The zero transformation $T: R^3 \rightarrow R^3$ is not one-to-one because its kernel is all of R^3.

A function $T: V \rightarrow W$ is said to be **onto** if every element in W has a preimage in V. In other words, T is onto W when W is equal to the range of T.

THEOREM 6.7 Onto Linear Transformations

Let $T: V \rightarrow W$ be a linear transformation, where W is finite dimensional. Then T is onto if and only if the rank of T is equal to the dimension of W.

For vector spaces of equal dimensions, you can combine the results of Theorems 6.5, 6.6, and 6.7 to obtain the next theorem relating the concepts of one-to-one and onto.

THEOREM 6.8 One-to-One and Onto Linear Transformations

Let $T: V \rightarrow W$ be a linear transformation with vector spaces V and W *both* of dimension n. Then T is one-to-one if and only if it is onto.

PROOF If T is one-to-one, then by Theorem 6.6 $\ker(T) = \{\mathbf{0}\}$, and $\dim(\ker(T)) = 0$. In that case, Theorem 6.5 produces

$$\begin{aligned}\dim(\text{range of } T) &= n - \dim(\ker(T)) \\ &= n \\ &= \dim(W).\end{aligned}$$

Consequently, by Theorem 6.7, T is onto. Similarly, if T is onto, then

$$\dim(\text{range of } T) = \dim(W) = n,$$

which by Theorem 6.5 implies that $\dim(\ker(T)) = 0$. By Theorem 6.6, T is one-to-one.

The next example brings together several concepts related to the kernel and range of a linear transformation.

EXAMPLE 11 Summarizing Several Results

The linear transformation $T: R^n \rightarrow R^m$ is represented by $T(\mathbf{x}) = A\mathbf{x}$. Find the nullity and rank of T and determine whether T is one-to-one, onto, or neither.

(a) $A = \begin{bmatrix} 1 & 2 & 0 \\ 0 & 1 & 1 \\ 0 & 0 & 1 \end{bmatrix}$ (b) $A = \begin{bmatrix} 1 & 2 \\ 0 & 1 \\ 0 & 0 \end{bmatrix}$

(c) $A = \begin{bmatrix} 1 & 2 & 0 \\ 0 & 1 & -1 \end{bmatrix}$ (d) $A = \begin{bmatrix} 1 & 2 & 0 \\ 0 & 1 & 1 \\ 0 & 0 & 0 \end{bmatrix}$

SOLUTION Note that each matrix is already in echelon form, so that its rank can be determined by inspection.

$T: R^n \to R^m$	Dim(domain)	Dim(range) Rank(T)	Dim(kernel) Nullity(T)	One-to-One	Onto
(a) $T: R^3 \to R^3$	3	3	0	Yes	Yes
(b) $T: R^2 \to R^3$	2	2	0	Yes	No
(c) $T: R^3 \to R^2$	3	2	1	No	Yes
(d) $T: R^3 \to R^3$	3	2	1	No	No

Isomorphisms of Vector Spaces

This section ends with a very important concept that can be a great aid in your understanding of vector spaces. The concept provides a way to think of distinct vector spaces as being "essentially the same"—at least with respect to the operations of vector addition and scalar multiplication. For example, the vector spaces R^3 and $M_{3,1}$ are essentially the same with respect to their standard operations. Such spaces are said to be **isomorphic** to each other. (The Greek word *isos* means "equal.")

Definition of Isomorphism

A linear transformation $T: V \to W$ that is one-to-one and onto is called an **isomorphism.** Moreover, if V and W are vector spaces such that there exists an isomorphism from V to W, then V and W are said to be **isomorphic** to each other.

One way in which isomorphic spaces are "essentially the same" is that they have the same dimensions, as stated in the next theorem. In fact, the theorem goes even further, stating that if two vector spaces have the same finite dimension, then they must be isomorphic.

THEOREM 6.9 Isomorphic Spaces and Dimension

Two finite-dimensional vector spaces V and W are isomorphic if and only if they are of the same dimension.

PROOF Assume V is isomorphic to W, where V has dimension n. By the definition of isomorphic spaces, you know there exists a linear transformation $T: V \to W$ that is one-to-one and onto. Because T is one-to-one, it follows that dim(kernel) $= 0$, which also implies that

$$\dim(\text{range}) = \dim(\text{domain}) = n.$$

In addition, because T is onto, you can conclude that

$$\dim(\text{range}) = \dim(W) = n.$$

To prove the theorem in the other direction, assume V and W both have dimension n. Let $B = \{\mathbf{v}_1, \mathbf{v}_2, \ldots, \mathbf{v}_n\}$ be a basis of V, and let $B' = \{\mathbf{w}_1, \mathbf{w}_2, \ldots, \mathbf{w}_n\}$ be a basis of W. Then an arbitrary vector in V can be represented as

$$\mathbf{v} = c_1\mathbf{v}_1 + c_2\mathbf{v}_2 + \cdots + c_n\mathbf{v}_n,$$

and you can define a linear transformation $T: V \rightarrow W$ as follows.

$$T(\mathbf{v}) = c_1\mathbf{w}_1 + c_2\mathbf{w}_2 + \cdots + c_n\mathbf{w}_n$$

It can be shown that this linear transformation is both one-to-one and onto. So, V and W are isomorphic.

Our study of vector spaces has provided much greater coverage to R^n than to other vector spaces. This preference for R^n stems from its notational convenience and from the geometric models available for R^2 and R^3. Theorem 6.9 tells you that R^n is a perfect model for every n-dimensional vector space. Example 12 lists some vector spaces that are isomorphic to R^4.

EXAMPLE 12 Isomorphic Vector Spaces

The vector spaces listed below are isomorphic to each other.

(a) $R^4 =$ 4-space
(b) $M_{4,1} =$ space of all 4×1 matrices
(c) $M_{2,2} =$ space of all 2×2 matrices
(d) $P_3 =$ space of all polynomials of degree 3 or less
(e) $V = \{(x_1, x_2, x_3, x_4, 0): x_i \text{ is a real number}\}$ (subspace of R^5)

Example 12 tells you that the elements in these spaces behave the same way as *vectors* even though they are distinct mathematical entities. The convention of using the notation for an n-tuple and an $n \times 1$ matrix interchangeably is justified.

SECTION 6.2 Exercises

In Exercises 1–10, find the kernel of the linear transformation.

1. $T: R^3 \rightarrow R^3, T(x, y, z) = (0, 0, 0)$

2. $T: R^3 \rightarrow R^3, T(x, y, z) = (x, 0, z)$

3. $T: R^4 \rightarrow R^4, T(x, y, z, w) = (y, x, w, z)$

4. $T: R^3 \rightarrow R^3, T(x, y, z) = (z, y, x)$

5. $T: P_3 \rightarrow R, T(a_0 + a_1x + a_2x^2 + a_3x^3) = a_0$

6. $T: P_2 \rightarrow R, T(a_0 + a_1x + a_2x^2) = a_0$

7. $T: P_2 \rightarrow P_1, T(a_0 + a_1x + a_2x^2) = a_1 + 2a_2x$

8. $T: P_3 \rightarrow P_2,$
$T(a_0 + a_1x + a_2x^2 + a_3x^3) = a_1 + 2a_2x + 3a_3x^2$

9. $T: R^2 \rightarrow R^2, T(x, y) = (x + 2y, y - x)$

10. $T: R^2 \rightarrow R^2, T(x, y) = (x - y, y - x)$

In Exercises 11–18, the linear transformation T is represented by $T(\mathbf{v}) = A\mathbf{v}$. Find a basis for (a) the kernel of T and (b) the range of T.

11. $A = \begin{bmatrix} 1 & 2 \\ 3 & 4 \end{bmatrix}$

12. $A = \begin{bmatrix} 1 & 2 \\ -2 & -4 \end{bmatrix}$

13. $A = \begin{bmatrix} 1 & -1 & 2 \\ 0 & 1 & 2 \end{bmatrix}$

14. $A = \begin{bmatrix} 1 & -2 & 1 \\ 0 & 2 & 1 \end{bmatrix}$

15. $A = \begin{bmatrix} 1 & 2 \\ -1 & -2 \\ 1 & 1 \end{bmatrix}$

16. $A = \begin{bmatrix} 1 & 1 \\ -1 & 2 \\ 0 & 1 \end{bmatrix}$

17. $A = \begin{bmatrix} 1 & 2 & -1 & 4 \\ 3 & 1 & 2 & -1 \\ -4 & -3 & -1 & -3 \\ -1 & -2 & 1 & 1 \end{bmatrix}$

18. $A = \begin{bmatrix} -1 & 3 & 2 & 1 & 4 \\ 2 & 3 & 5 & 0 & 0 \\ 2 & 1 & 2 & 1 & 0 \end{bmatrix}$

In Exercises 19–30, the linear transformation T is defined by $T(\mathbf{x}) = A\mathbf{x}$. Find (a) $\ker(T)$, (b) nullity(T), (c) range(T), and (d) rank(T).

19. $A = \begin{bmatrix} -1 & 1 \\ 1 & 1 \end{bmatrix}$

20. $A = \begin{bmatrix} 3 & 2 \\ -9 & -6 \end{bmatrix}$

21. $A = \begin{bmatrix} 5 & -3 \\ 1 & 1 \\ 1 & -1 \end{bmatrix}$

22. $A = \begin{bmatrix} 4 & 1 \\ 0 & 0 \\ 2 & -3 \end{bmatrix}$

23. $A = \begin{bmatrix} 0 & -2 & 3 \\ 4 & 0 & 11 \end{bmatrix}$

24. $A = \begin{bmatrix} 1 & 1 & 0 & 0 \\ 0 & 0 & 1 & 1 \end{bmatrix}$

25. $A = \begin{bmatrix} \frac{9}{10} & \frac{3}{10} \\ \frac{3}{10} & \frac{1}{10} \end{bmatrix}$

26. $A = \begin{bmatrix} \frac{1}{26} & -\frac{5}{26} \\ -\frac{5}{26} & \frac{25}{26} \end{bmatrix}$

27. $A = \begin{bmatrix} \frac{4}{9} & -\frac{4}{9} & \frac{2}{9} \\ -\frac{4}{9} & \frac{4}{9} & -\frac{2}{9} \\ \frac{2}{9} & -\frac{2}{9} & \frac{1}{9} \end{bmatrix}$

28. $A = \begin{bmatrix} 1 & 0 & 0 \\ 0 & 0 & 0 \\ 0 & 0 & 1 \end{bmatrix}$

29. $A = \begin{bmatrix} 2 & 2 & -3 & 1 & 13 \\ 1 & 1 & 1 & 1 & -1 \\ 3 & 3 & -5 & 0 & 14 \\ 6 & 6 & -2 & 4 & 16 \end{bmatrix}$

30. $A = \begin{bmatrix} 3 & -2 & 6 & -1 & 15 \\ 4 & 3 & 8 & 10 & -14 \\ 2 & -3 & 4 & -4 & 20 \end{bmatrix}$

In Exercises 31–38, let $T: R^3 \to R^3$ be a linear transformation. Use the given information to find the nullity of T and give a geometric description of the kernel and range of T.

31. rank(T) = 2

32. rank(T) = 1

33. rank(T) = 0

34. rank(T) = 3

35. T is the counterclockwise rotation of 45° about the z-axis:

$$T(x, y, z) = \left(\frac{\sqrt{2}}{2}x - \frac{\sqrt{2}}{2}y, \frac{\sqrt{2}}{2}x + \frac{\sqrt{2}}{2}y, z\right)$$

36. T is the reflection through the yz-coordinate plane:

$$T(x, y, z) = (-x, y, z)$$

37. T is the projection onto the vector $\mathbf{v} = (1, 2, 2)$:

$$T(x, y, z) = \frac{x + 2y + 2z}{9}(1, 2, 2)$$

38. T is the projection onto the xy-coordinate plane:

$$T(x, y, z) = (x, y, 0)$$

In Exercises 39–42, find the nullity of T.

39. $T: R^4 \to R^2$, rank(T) = 2

40. $T: R^5 \to R^2$, rank(T) = 2

41. $T: R^4 \to R^4$, rank(T) = 0

42. $T: P_3 \to P_1$, rank(T) = 2

43. Identify the zero element and standard basis for each of the isomorphic vector spaces in Example 12.

44. Which vector spaces are isomorphic to R^6?

(a) $M_{2,3}$ (b) P_6 (c) $C[0, 6]$

(d) $M_{6,1}$ (e) P_5

(f) $\{(x_1, x_2, x_3, 0, x_5, x_6, x_7): x_i \text{ is a real number}\}$

45. **Calculus** Let $T: P_4 \to P_3$ be represented by $T(p) = p'$. What is the kernel of T?

46. **Calculus** Let $T: P_2 \to R$ be represented by

$$T(p) = \int_0^1 p(x)\, dx.$$

What is the kernel of T?

47. Let $T: R^3 \to R^3$ be the linear transformation that projects $\mathbf{u}$ onto $\mathbf{v} = (2, -1, 1)$.

(a) Find the rank and nullity of T.
(b) Find a basis for the kernel of T.

48. Repeat Exercise 47 for $\mathbf{v} = (3, 0, 4)$.

In Exercises 49–52, verify that the matrix defines a linear function T that is one-to-one and onto.

49. $A = \begin{bmatrix} -1 & 0 \\ 0 & 1 \end{bmatrix}$

50. $A = \begin{bmatrix} 1 & 0 \\ 0 & -1 \end{bmatrix}$

51. $A = \begin{bmatrix} 1 & 0 & 0 \\ 0 & 0 & 1 \\ 0 & 1 & 0 \end{bmatrix}$

52. $A = \begin{bmatrix} 1 & 2 & 3 \\ -1 & 2 & 4 \\ 0 & 4 & 1 \end{bmatrix}$

True or False? In Exercises 53 and 54, determine whether each statement is true or false. If a statement is true, give a reason or cite an appropriate statement from the text. If a statement is false, provide an example that shows the statement is not true in all cases or cite an appropriate statement from the text.

53. (a) The set of all vectors mapped from a vector space V to another vector space W by a linear transformation T is known as the kernel of T.

(b) The range of a linear transformation from a vector space V to a vector space W is a subspace of the vector space V.

(c) A linear transformation T from V to W is called one-to-one if and only if for all $\mathbf{u}$ and $\mathbf{v}$ in V, $T(\mathbf{u}) = T(\mathbf{v})$ implies that $\mathbf{u} = \mathbf{v}$.

(d) The vector spaces R^3 and $M_{3,1}$ are isomorphic to each other.

54. (a) The kernel of a linear transformation T from a vector space V to a vector space W is a subspace of the vector space V.

(b) The dimension of a linear transformation T from a vector space V to a vector space W is called the rank of T.

(c) A linear transformation T from V to W is one-to-one if the preimage of every $\mathbf{w}$ in the range consists of a single vector $\mathbf{v}$.

(d) The vector spaces R^2 and P_1 are isomorphic to each other.

55. For the transformation $T: R^n \rightarrow R^n$ represented by $T(\mathbf{v}) = A\mathbf{v}$, what can be said about the rank of T if (a) $\det(A) \neq 0$ and (b) $\det(A) = 0$?

56. Let $T: M_{n,n} \rightarrow M_{n,n}$ be represented by $T(A) = A - A^T$. Show that the kernel of T is the set of $n \times n$ symmetric matrices.

57. Determine a relationship among m, n, j, and k such that $M_{m,n}$ is isomorphic to $M_{j,k}$.

58. Guided Proof Let B be an invertible $n \times n$ matrix. Prove that the linear transformation $T: M_{n,n} \rightarrow M_{n,n}$ represented by $T(A) = AB$ is an isomorphism.

Getting Started: To show that the linear transformation is an isomorphism, you need to show that T is both onto and one-to-one.

(i) Because T is a linear transformation with vector spaces of equal dimension, then by Theorem 6.8, you only need to show that T is one-to-one.

(ii) To show that T is one-to-one, you need to determine the kernel of T and show that it is $\{\mathbf{0}\}$ (Theorem 6.6). Use the fact that B is an invertible $n \times n$ matrix and that $T(A) = AB$.

(iii) Conclude that T is an isomorphism.

59. Let $T: V \rightarrow W$ be a linear transformation. Prove that T is one-to-one if and only if the rank of T equals the dimension of V.

60. Let $T: V \rightarrow W$ be a linear transformation, and let U be a subspace of W. Prove that the set $T^{-1}(U) = \{\mathbf{v} \in V: T(\mathbf{v}) \in U\}$ is a subspace of V. What is $T^{-1}(U)$ if $U = \{\mathbf{0}\}$?

61. Writing Are the vector spaces R^4, $M_{2,2}$, and $M_{1,4}$ exactly the same? Describe their similarities and differences.

62. Writing Let $T: R^m \rightarrow R^n$ be a linear transformation. Explain the differences between the concepts of one-to-one and onto. What can you say about m and n if T is onto? What can you say about m and n if T is one-to-one?

6.3 Matrices for Linear Transformations

Which representation of $T: R^3 \rightarrow R^3$ is better,

$$T(x_1, x_2, x_3) = (2x_1 + x_2 - x_3, -x_1 + 3x_2 - 2x_3, 3x_2 + 4x_3)$$

or

$$T(\mathbf{x}) = A\mathbf{x} = \begin{bmatrix} 2 & 1 & -1 \\ -1 & 3 & -2 \\ 0 & 3 & 4 \end{bmatrix} \begin{bmatrix} x_1 \\ x_2 \\ x_3 \end{bmatrix}?$$

The second representation is better than the first for at least three reasons: it is simpler to write, simpler to read, and more easily adapted for computer use. Later you will see that matrix representation of linear transformations also has some theoretical advantages. In this section you will see that for linear transformations involving finite-dimensional vector spaces, matrix representation is always possible.

The key to representing a linear transformation $T: V \to W$ by a matrix is to determine how it acts on a basis of V. Once you know the image of every vector in the basis, you can use the properties of linear transformations to determine $T(\mathbf{v})$ for any $\mathbf{v}$ in V.

For convenience, the first three theorems in this section are stated in terms of linear transformations from R^n into R^m, relative to the standard bases in R^n and R^m. At the end of the section these results are generalized to include nonstandard bases and general vector spaces.

Recall that the standard basis for R^n, written in column vector notation, is represented by

$$B = \{\mathbf{e}_1, \mathbf{e}_2, \ldots, \mathbf{e}_n\} = \left\{ \begin{bmatrix} 1 \\ 0 \\ \vdots \\ 0 \end{bmatrix}, \begin{bmatrix} 0 \\ 1 \\ \vdots \\ 0 \end{bmatrix}, \ldots, \begin{bmatrix} 0 \\ 0 \\ \vdots \\ 1 \end{bmatrix} \right\}.$$

THEOREM 6.10

Standard Matrix for a Linear Transformation

Let $T: R^n \to R^m$ be a linear transformation such that

$$T(\mathbf{e}_1) = \begin{bmatrix} a_{11} \\ a_{21} \\ \vdots \\ a_{m1} \end{bmatrix}, T(\mathbf{e}_2) = \begin{bmatrix} a_{12} \\ a_{22} \\ \vdots \\ a_{m2} \end{bmatrix}, \ldots, T(\mathbf{e}_n) = \begin{bmatrix} a_{1n} \\ a_{2n} \\ \vdots \\ a_{mn} \end{bmatrix}.$$

Then the $m \times n$ matrix whose n columns correspond to $T(\mathbf{e}_i)$,

$$A = \begin{bmatrix} a_{11} & a_{12} & \cdots & a_{1n} \\ a_{21} & a_{22} & \cdots & a_{2n} \\ \vdots & \vdots & & \vdots \\ a_{m1} & a_{m2} & \cdots & a_{mn} \end{bmatrix},$$

is such that $T(\mathbf{v}) = A\mathbf{v}$ for every $\mathbf{v}$ in R^n. A is called the **standard matrix** for T.

PROOF To show that $T(\mathbf{v}) = A\mathbf{v}$ for any $\mathbf{v}$ in R^n, you can write

$$\mathbf{v} = \begin{bmatrix} v_1 \\ v_2 \\ \vdots \\ v_n \end{bmatrix} = v_1\mathbf{e}_1 + v_2\mathbf{e}_2 + \cdots + v_n\mathbf{e}_n.$$

Because T is a linear transformation, you have

$$\begin{aligned} T(\mathbf{v}) &= T(v_1\mathbf{e}_1 + v_2\mathbf{e}_2 + \cdots + v_n\mathbf{e}_n) \\ &= T(v_1\mathbf{e}_1) + T(v_2\mathbf{e}_2) + \cdots + T(v_n\mathbf{e}_n) \\ &= v_1T(\mathbf{e}_1) + v_2T(\mathbf{e}_2) + \cdots + v_nT(\mathbf{e}_n). \end{aligned}$$

On the other hand, the matrix product $A\mathbf{v}$ is represented by

$$A\mathbf{v} = \begin{bmatrix} a_{11} & a_{12} & \cdots & a_{1n} \\ a_{21} & a_{22} & \cdots & a_{2n} \\ \vdots & \vdots & & \vdots \\ a_{m1} & a_{m2} & \cdots & a_{mn} \end{bmatrix} \begin{bmatrix} v_1 \\ v_2 \\ \vdots \\ v_n \end{bmatrix} = \begin{bmatrix} a_{11}v_1 + a_{12}v_2 + \cdots + a_{1n}v_n \\ a_{21}v_1 + a_{22}v_2 + \cdots + a_{2n}v_n \\ \vdots \\ a_{m1}v_1 + a_{m2}v_2 + \cdots + a_{mn}v_n \end{bmatrix}$$

$$= v_1\begin{bmatrix} a_{11} \\ a_{21} \\ \vdots \\ a_{m1} \end{bmatrix} + v_2\begin{bmatrix} a_{12} \\ a_{22} \\ \vdots \\ a_{m2} \end{bmatrix} + \cdots + v_n\begin{bmatrix} a_{1n} \\ a_{2n} \\ \vdots \\ a_{mn} \end{bmatrix}$$

$$= v_1T(\mathbf{e}_1) + v_2T(\mathbf{e}_2) + \cdots + v_nT(\mathbf{e}_n).$$

So, $T(\mathbf{v}) = A\mathbf{v}$ for each $\mathbf{v}$ in R^n.

EXAMPLE 1 Finding the Standard Matrix for a Linear Transformation

Find the standard matrix for the linear transformation $T: R^3 \rightarrow R^2$ defined by

$$T(x, y, z) = (x - 2y, 2x + y).$$

SOLUTION Begin by finding the images of $\mathbf{e}_1$, $\mathbf{e}_2$, and $\mathbf{e}_3$.

Vector Notation	*Matrix Notation*
$T(\mathbf{e}_1) = T(1, 0, 0) = (1, 2)$	$T(\mathbf{e}_1) = T\left(\begin{bmatrix} 1 \\ 0 \\ 0 \end{bmatrix}\right) = \begin{bmatrix} 1 \\ 2 \end{bmatrix}$
$T(\mathbf{e}_2) = T(0, 1, 0) = (-2, 1)$	$T(\mathbf{e}_2) = T\left(\begin{bmatrix} 0 \\ 1 \\ 0 \end{bmatrix}\right) = \begin{bmatrix} -2 \\ 1 \end{bmatrix}$
$T(\mathbf{e}_3) = T(0, 0, 1) = (0, 0)$	$T(\mathbf{e}_3) = T\left(\begin{bmatrix} 0 \\ 0 \\ 1 \end{bmatrix}\right) = \begin{bmatrix} 0 \\ 0 \end{bmatrix}$

By Theorem 6.10, the columns of A consist of $T(\mathbf{e}_1)$, $T(\mathbf{e}_2)$, and $T(\mathbf{e}_3)$, and you have

$$A = [T(\mathbf{e}_1) \vdots T(\mathbf{e}_2) \vdots T(\mathbf{e}_3)] = \begin{bmatrix} 1 & -2 & 0 \\ 2 & 1 & 0 \end{bmatrix}.$$

As a check, note that

$$A\begin{bmatrix} x \\ y \\ z \end{bmatrix} = \begin{bmatrix} 1 & -2 & 0 \\ 2 & 1 & 0 \end{bmatrix}\begin{bmatrix} x \\ y \\ z \end{bmatrix} = \begin{bmatrix} x - 2y \\ 2x + y \end{bmatrix},$$

which is equivalent to $T(x, y, z) = (x - 2y, 2x + y)$.

A little practice will enable you to determine the standard matrix for a linear transformation, such as the one in Example 1, by inspection. For instance, the standard matrix for the linear transformation defined by

$$T(x_1, x_2, x_3) = (x_1 - 2x_2 + 5x_3, 2x_1 + 3x_3, 4x_1 + x_2 - 2x_3)$$

is found by using the coefficients of x_1, x_2, and x_3 to form the rows of A, as follows.

$$A = \begin{bmatrix} 1 & -2 & 5 \\ 2 & 0 & 3 \\ 4 & 1 & -2 \end{bmatrix} \begin{matrix} \longleftarrow & 1x_1 - 2x_2 + 5x_3 \\ \longleftarrow & 2x_1 + 0x_2 + 3x_3 \\ \longleftarrow & 4x_1 + 1x_2 - 2x_3 \end{matrix}$$

EXAMPLE 2 **Finding the Standard Matrix for a Linear Transformation**

The linear transformation $T: R^2 \rightarrow R^2$ is given by projecting each point in R^2 onto the x-axis, as shown in Figure 6.8. Find the standard matrix for T.

SOLUTION This linear transformation is represented by

$$T(x, y) = (x, 0).$$

So, the standard matrix for T is

$$A = [T(1, 0) \vdots T(0, 1)]$$

$$= \begin{bmatrix} 1 & 0 \\ 0 & 0 \end{bmatrix}.$$

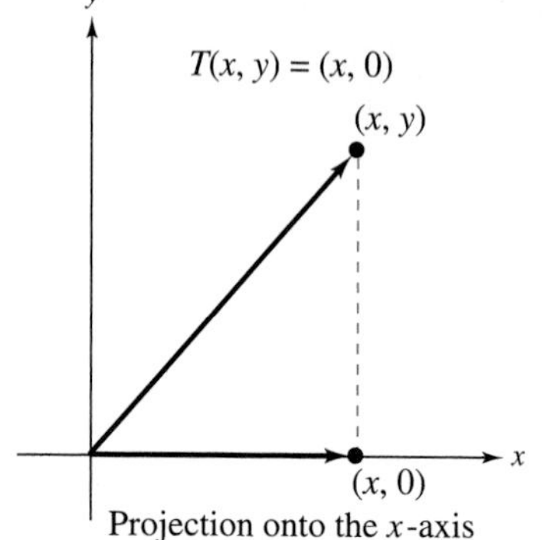

Figure 6.8

The standard matrix for the zero transformation from R^n into R^m is the $m \times n$ zero matrix, and the standard matrix for the identity transformation from R^n into R^n is I_n.

Composition of Linear Transformations

The **composition,** T, of $T_1: R^n \rightarrow R^m$ with $T_2: R^m \rightarrow R^p$ is defined by

$$T(\mathbf{v}) = T_2(T_1(\mathbf{v})),$$

where $\mathbf{v}$ is a vector in R^n. This composition is denoted by

$$T = T_2 \circ T_1.$$

The domain of T is defined as the domain of T_1. Moreover, the composition is not defined unless the range of T_1 lies within the domain of T_2, as shown in Figure 6.9.

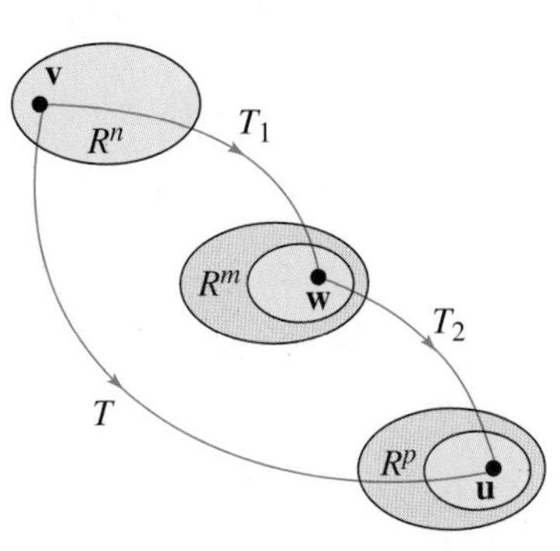

Composition of Transformations

Figure 6.9

The next theorem emphasizes the usefulness of matrices for representing linear transformations. This theorem not only states that the composition of two linear transformations is a linear transformation, but also says that the standard matrix for the composition is the product of the standard matrices for the two original linear transformations.

THEOREM 6.11 Composition of Linear Transformations

Let $T_1: R^n \rightarrow R^m$ and $T_2: R^m \rightarrow R^p$ be linear transformations with standard matrices A_1 and A_2. The **composition** $T: R^n \rightarrow R^p$, defined by $T(\mathbf{v}) = T_2(T_1(\mathbf{v}))$, is a linear transformation. Moreover, the standard matrix A for T is given by the matrix product

$$A = A_2A_1.$$

PROOF

To show that T is a linear transformation, let $\mathbf{u}$ and $\mathbf{v}$ be vectors in R^n and let c be any scalar. Then, because T_1 and T_2 are linear transformations, you can write

$$\begin{aligned} T(\mathbf{u} + \mathbf{v}) &= T_2(T_1(\mathbf{u} + \mathbf{v})) \\ &= T_2(T_1(\mathbf{u}) + T_1(\mathbf{v})) \\ &= T_2(T_1(\mathbf{u})) + T_2(T_1(\mathbf{v})) = T(\mathbf{u}) + T(\mathbf{v}) \end{aligned}$$

$$\begin{aligned} T(c\mathbf{v}) &= T_2(T_1(c\mathbf{v})) \\ &= T_2(cT_1(\mathbf{v})) \\ &= cT_2(T_1(\mathbf{v})) = cT(\mathbf{v}). \end{aligned}$$

Now, to show that A_2A_1 is the standard matrix for T, use the associative property of matrix multiplication to write

$$T(\mathbf{v}) = T_2(T_1(\mathbf{v})) = T_2(A_1\mathbf{v}) = A_2(A_1\mathbf{v}) = (A_2A_1)\mathbf{v}.$$

REMARK: Theorem 6.11 can be generalized to cover the composition of n linear transformations. That is, if the standard matrices of $T_1, T_2, \ldots, T_n$ are $A_1, A_2, \ldots, A_n$, then the standard matrix for the composition T is represented by

$$A = A_nA_{n-1} \cdot \cdot \cdot A_2A_1.$$

Because matrix multiplication is not commutative, order is important when the compositions of linear transformations are formed. In general, the composition $T_2 \circ T_1$ is not the same as $T_1 \circ T_2$, as demonstrated in the next example.

EXAMPLE 3 **The Standard Matrix for a Composition**

Let T_1 and T_2 be linear transformations from R^3 into R^3 such that

$$T_1(x, y, z) = (2x + y, 0, x + z) \quad \text{and} \quad T_2(x, y, z) = (x - y, z, y).$$

Find the standard matrices for the compositions $T = T_2 \circ T_1$ and $T' = T_1 \circ T_2$.

SOLUTION The standard matrices for T_1 and T_2 are

$$A_1 = \begin{bmatrix} 2 & 1 & 0 \\ 0 & 0 & 0 \\ 1 & 0 & 1 \end{bmatrix} \quad \text{and} \quad A_2 = \begin{bmatrix} 1 & -1 & 0 \\ 0 & 0 & 1 \\ 0 & 1 & 0 \end{bmatrix}.$$

By Theorem 6.11, the standard matrix for T is

$$A = A_2A_1 = \begin{bmatrix} 1 & -1 & 0 \\ 0 & 0 & 1 \\ 0 & 1 & 0 \end{bmatrix}\begin{bmatrix} 2 & 1 & 0 \\ 0 & 0 & 0 \\ 1 & 0 & 1 \end{bmatrix} = \begin{bmatrix} 2 & 1 & 0 \\ 1 & 0 & 1 \\ 0 & 0 & 0 \end{bmatrix},$$

and the standard matrix for T' is

$$A' = A_1A_2 = \begin{bmatrix} 2 & 1 & 0 \\ 0 & 0 & 0 \\ 1 & 0 & 1 \end{bmatrix}\begin{bmatrix} 1 & -1 & 0 \\ 0 & 0 & 1 \\ 0 & 1 & 0 \end{bmatrix} = \begin{bmatrix} 2 & -2 & 1 \\ 0 & 0 & 0 \\ 1 & 0 & 0 \end{bmatrix}.$$

Another benefit of matrix representation is that it can represent the **inverse** of a linear transformation. Before seeing how this works, consider the next definition.

Definition of Inverse Linear Transformation

If $T_1: R^n \to R^n$ and $T_2: R^n \to R^n$ are linear transformations such that for every $\mathbf{v}$ in R^n

$$T_2(T_1(\mathbf{v})) = \mathbf{v} \quad \text{and} \quad T_1(T_2(\mathbf{v})) = \mathbf{v},$$

then T_2 is called the **inverse** of T_1, and T_1 is said to be **invertible.**

Not every linear transformation has an inverse. If the transformation T_1 is invertible, however, then the inverse is unique and is denoted by T_1^{-1}.

Just as the inverse of a function of a real variable can be thought of as undoing what the function did, the inverse of a linear transformation T can be thought of as undoing the mapping done by T. For instance, if T is a linear transformation from R^3 onto R^3 such that

$$T(1, 4, -5) = (2, 3, 1)$$

and if T^{-1} exists, then T^{-1} maps (2, 3, 1) back to its preimage under T. That is,

$$T^{-1}(2, 3, 1) = (1, 4, -5).$$

The next theorem states that a linear transformation is invertible if and only if it is an isomorphism (one-to-one and onto). You are asked to prove this theorem in Exercise 78.

THEOREM 6.12 Existence of an Inverse Transformation

Let $T: R^n \rightarrow R^n$ be a linear transformation with standard matrix A. Then the following conditions are equivalent.

1. T is invertible.
2. T is an isomorphism.
3. A is invertible.

And, if T is invertible with standard matrix A, then the standard matrix for T^{-1} is A^{-1}.

REMARK: Several other conditions are equivalent to the three given in Theorem 6.12; see the summary of equivalent conditions from Section 4.6.

EXAMPLE 4 Finding the Inverse of a Linear Transformation

The linear transformation $T: R^3 \rightarrow R^3$ is defined by

$$T(x_1, x_2, x_3) = (2x_1 + 3x_2 + x_3, 3x_1 + 3x_2 + x_3, 2x_1 + 4x_2 + x_3).$$

Show that T is invertible, and find its inverse.

SOLUTION The standard matrix for T is

$$A = \begin{bmatrix} 2 & 3 & 1 \\ 3 & 3 & 1 \\ 2 & 4 & 1 \end{bmatrix}.$$

Using the techniques for matrix inversion (see Section 2.3), you can find that A is invertible and its inverse is

$$A^{-1} = \begin{bmatrix} -1 & 1 & 0 \\ -1 & 0 & 1 \\ 6 & -2 & -3 \end{bmatrix}.$$

So, T is invertible and its standard matrix is A^{-1}.

Using the standard matrix for the inverse, you can find the rule for T^{-1} by computing the image of an arbitrary vector $\mathbf{v} = (x_1, x_2, x_3)$.

$$A^{-1}\mathbf{v} = \begin{bmatrix} -1 & 1 & 0 \\ -1 & 0 & 1 \\ 6 & -2 & -3 \end{bmatrix} \begin{bmatrix} x_1 \\ x_2 \\ x_3 \end{bmatrix} = \begin{bmatrix} -x_1 + x_2 \\ -x_1 + x_3 \\ 6x_1 - 2x_2 - 3x_3 \end{bmatrix}$$

In other words,

$$T^{-1}(x_1, x_2, x_3) = (-x_1 + x_2, -x_1 + x_3, 6x_1 - 2x_2 - 3x_3).$$

Nonstandard Bases and General Vector Spaces

You will now consider the more general problem of finding a matrix for a linear transformation $T: V \rightarrow W$, where B and B' are ordered bases for V and W, respectively. Recall that the coordinate matrix of $\mathbf{v}$ relative to B is denoted by $[\mathbf{v}]_B$. In order to represent the linear transformation T, A must be multiplied by a *coordinate matrix relative to B*. The result of the multiplication will be a *coordinate matrix relative to B′*. That is,

$$[T(\mathbf{v})]_{B'} = A[\mathbf{v}]_B.$$

A is called the **matrix of T relative to the bases B and B'.**

To find the matrix A, you will use a procedure similar to the one used to find the standard matrix for T. That is, the images of the vectors in B are written as coordinate matrices relative to the basis B'. These coordinate matrices form the columns of A.

Transformation Matrix for Nonstandard Bases

Let V and W be finite-dimensional vector spaces with bases B and B', respectively, where

$$B = \{\mathbf{v}_1, \mathbf{v}_2, . . . , \mathbf{v}_n\}.$$

If $T: V \rightarrow W$ is a linear transformation such that

$$[T(\mathbf{v}_1)]_{B'} = \begin{bmatrix} a_{11} \\ a_{21} \\ \vdots \\ a_{m1} \end{bmatrix}, [T(\mathbf{v}_2)]_{B'} = \begin{bmatrix} a_{12} \\ a_{22} \\ \vdots \\ a_{m2} \end{bmatrix}, . . . , [T(\mathbf{v}_n)]_{B'} = \begin{bmatrix} a_{1n} \\ a_{2n} \\ \vdots \\ a_{mn} \end{bmatrix},$$

then the $m \times n$ matrix whose n columns correspond to $[T(\mathbf{v}_i)]_{B'}$,

$$A = \begin{bmatrix} a_{11} & a_{12} & . . . & a_{1n} \\ a_{21} & a_{22} & . . . & a_{2n} \\ \vdots & \vdots & & \vdots \\ a_{m1} & a_{m2} & . . . & a_{mn} \end{bmatrix},$$

is such that $[T(\mathbf{v})]_{B'} = A[\mathbf{v}]_B$ for every $\mathbf{v}$ in V.

EXAMPLE 5 **Finding a Matrix Relative to Nonstandard Bases**

Let $T: R^2 \rightarrow R^2$ be a linear transformation defined by

$$T(x_1, x_2) = (x_1 + x_2, 2x_1 - x_2).$$

Find the matrix for T relative to the bases

$$B = \{\overset{\mathbf{v}_1}{(1, 2)}, \overset{\mathbf{v}_2}{(-1, 1)}\} \quad \text{and} \quad B' = \{\overset{\mathbf{w}_1}{(1, 0)}, \overset{\mathbf{w}_2}{(0, 1)}\}.$$

SOLUTION By the definition of T, you have

$$T(\mathbf{v}_1) = T(1, 2) = (3, 0) = 3\mathbf{w}_1 + 0\mathbf{w}_2$$
$$T(\mathbf{v}_2) = T(-1, 1) = (0, -3) = 0\mathbf{w}_1 - 3\mathbf{w}_2.$$

The coordinate matrices for $T(\mathbf{v}_1)$ and $T(\mathbf{v}_2)$ relative to B' are

$$[T(\mathbf{v}_1)]_{B'} = \begin{bmatrix} 3 \\ 0 \end{bmatrix} \quad \text{and} \quad [T(\mathbf{v}_2)]_{B'} = \begin{bmatrix} 0 \\ -3 \end{bmatrix}.$$

The matrix for T relative to B and B' is formed by using these coordinate matrices as columns to produce

$$A = \begin{bmatrix} 3 & 0 \\ 0 & -3 \end{bmatrix}.$$

EXAMPLE 6 **Using a Matrix to Represent a Linear Transformation**

For the linear transformation $T: R^2 \rightarrow R^2$ from Example 5, use the matrix A to find $T(\mathbf{v})$, where $\mathbf{v} = (2, 1)$.

SOLUTION Using the basis $B = \{(1, 2), (-1, 1)\}$, you find

$$\mathbf{v} = (2, 1) = 1(1, 2) - 1(-1, 1),$$

which implies

$$[\mathbf{v}]_B = \begin{bmatrix} 1 \\ -1 \end{bmatrix}.$$

So, $[T(\mathbf{v})]_{B'}$ is

$$A[\mathbf{v}]_B = \begin{bmatrix} 3 & 0 \\ 0 & -3 \end{bmatrix} \begin{bmatrix} 1 \\ -1 \end{bmatrix} = \begin{bmatrix} 3 \\ 3 \end{bmatrix} = [T(\mathbf{v})]_{B'}.$$

Finally, because $B' = \{(1, 0), (0, 1)\}$, it follows that

$$T(\mathbf{v}) = 3(1, 0) + 3(0, 1) = (3, 3).$$

You can check this result by directly calculating $T(\mathbf{v})$ using the definition of T from Example 5:

$$T(2, 1) = (2 + 1, 2(2) - 1) = (3, 3).$$

In the special case where $V = W$ and $B = B'$, the matrix A is called the **matrix of T relative to the basis B.** In such cases the matrix of the identity transformation is simply I_n. To see this, let $B = \{\mathbf{v}_1, \mathbf{v}_2, \ldots, \mathbf{v}_n\}$. Because the identity transformation maps each $\mathbf{v}_i$ to itself, you have

$$[T(\mathbf{v}_1)]_B = \begin{bmatrix} 1 \\ 0 \\ \vdots \\ 0 \end{bmatrix}, [T(\mathbf{v}_2)]_B = \begin{bmatrix} 0 \\ 1 \\ \vdots \\ 0 \end{bmatrix}, \ldots, [T(\mathbf{v}_n)]_B = \begin{bmatrix} 0 \\ 0 \\ \vdots \\ 1 \end{bmatrix},$$

and it follows that $A = I_n$.

In the next example you will construct a matrix representing the differential operator discussed in Example 10 in Section 6.1.

EXAMPLE 7 A Matrix for the Differential Operator (Calculus)

Let $D_x: P_2 \rightarrow P_1$ be the differential operator that maps a quadratic polynomial p onto its derivative p'. Find the matrix for D_x using the bases

$$B = \{1, x, x^2\} \quad \text{and} \quad B' = \{1, x\}.$$

SOLUTION The derivatives of the basis vectors are

$$\begin{aligned} D_x(1) &= 0 = 0(1) + 0(x) \\ D_x(x) &= 1 = 1(1) + 0(x) \\ D_x(x^2) &= 2x = 0(1) + 2(x). \end{aligned}$$

So, the coordinate matrices relative to B' are

$$[D_x(1)]_{B'} = \begin{bmatrix} 0 \\ 0 \end{bmatrix}, \quad [D_x(x)]_{B'} = \begin{bmatrix} 1 \\ 0 \end{bmatrix}, \quad [D_x(x^2)]_{B'} = \begin{bmatrix} 0 \\ 2 \end{bmatrix},$$

and the matrix for D_x is

$$A = \begin{bmatrix} 0 & 1 & 0 \\ 0 & 0 & 2 \end{bmatrix}.$$

Note that this matrix *does* produce the derivative of a quadratic polynomial $p(x) = a + bx + cx^2$.

$$Ap = \begin{bmatrix} 0 & 1 & 0 \\ 0 & 0 & 2 \end{bmatrix} \begin{bmatrix} a \\ b \\ c \end{bmatrix} = \begin{bmatrix} b \\ 2c \end{bmatrix} \Rightarrow b + 2cx = D_x[a + bx + cx^2]$$

SECTION 6.3 Exercises

In Exercises 1–10, find the standard matrix for the linear transformation T.

1. $T(x, y) = (x + 2y, x - 2y)$
2. $T(x, y) = (3x + 2y, 2y - x)$
3. $T(x, y) = (2x - 3y, x - y, y - 4x)$
4. $T(x, y) = (4x + y, 0, 2x - 3y)$
5. $T(x, y, z) = (x + y, x - y, z - x)$
6. $T(x, y, z) = (5x - 3y + z, 2z + 4y, 5x + 3y)$
7. $T(x, y, z) = (3z - 2y, 4x + 11z)$
8. $T(x, y, z) = (3x - 2z, 2y - z)$
9. $T(x_1, x_2, x_3, x_4) = (0, 0, 0, 0)$
10. $T(x_1, x_2, x_3) = (0, 0, 0)$

In Exercises 11–16, use the standard matrix for the linear transformation T to find the image of the vector $\mathbf{v}$.

11. $T(x, y, z) = (13x - 9y + 4z, 6x + 5y - 3z)$, $\mathbf{v} = (1, -2, 1)$
12. $T(x, y, z) = (2x + y, 3y - z)$, $\mathbf{v} = (0, 1, -1)$
13. $T(x, y) = (x + y, x - y, 2x, 2y)$, $\mathbf{v} = (3, -3)$
14. $T(x, y) = (x - y, x + 2y, y)$, $\mathbf{v} = (2, -2)$
15. $T(x_1, x_2, x_3, x_4) = (x_1 + x_2, x_3 + x_4)$, $\mathbf{v} = (1, -1, 1, -1)$
16. $T(x_1, x_2, x_3, x_4) = (2x_1 - x_3, 3x_2 - 4x_4, 4x_3 - x_1, x_2 + x_4)$, $\mathbf{v} = (1, 2, 3, -2)$

In Exercises 17–34, (a) find the standard matrix A for the linear transformation T, (b) use A to find the image of the vector $\mathbf{v}$, and (c) sketch the graph of $\mathbf{v}$ and its image.

17. T is the reflection through the origin in R^2: $T(x, y) = (-x, -y)$, $\mathbf{v} = (3, 4)$.
18. T is the reflection in the line $y = x$ in R^2: $T(x, y) = (y, x)$, $\mathbf{v} = (3, 4)$.
19. T is the reflection in the y-axis in R^2: $T(x, y) = (-x, y)$, $\mathbf{v} = (2, -3)$.
20. T is the reflection in the x-axis in R^2: $T(x, y) = (x, -y)$, $\mathbf{v} = (4, -1)$.
21. T is the counterclockwise rotation of 135° in R^2, $\mathbf{v} = (4, 4)$.
22. T is the counterclockwise rotation of 120° in R^2, $\mathbf{v} = (2, 2)$.
23. T is the clockwise rotation (θ is negative) of 60° in R^2, $\mathbf{v} = (1, 2)$.
24. T is the clockwise rotation (θ is negative) of 30° in R^2, $\mathbf{v} = (2, 1)$.
25. T is the reflection through the xy-coordinate plane in R^3: $T(x, y, z) = (x, y, -z)$, $\mathbf{v} = (3, 2, 2)$.
26. T is the reflection through the yz-coordinate plane in R^3: $T(x, y, z) = (-x, y, z)$, $\mathbf{v} = (2, 3, 4)$.
27. T is the reflection through the xz-coordinate plane in R^3: $T(x, y, z) = (x, -y, z)$, $\mathbf{v} = (1, 2, -1)$.
28. T is the counterclockwise rotation of 45° in R^2, $\mathbf{v} = (2, 2)$.
29. T is the counterclockwise rotation of 30° in R^2, $\mathbf{v} = (1, 2)$.
30. T is the counterclockwise rotation of 180° in R^2, $\mathbf{v} = (1, 2)$.
31. T is the projection onto the vector $\mathbf{w} = (3, 1)$ in R^2: $T(\mathbf{v}) = \text{proj}_{\mathbf{w}}\mathbf{v}$, $\mathbf{v} = (1, 4)$.
32. T is the projection onto the vector $\mathbf{w} = (-1, 5)$ in R^2: $T(\mathbf{v}) = \text{proj}_{\mathbf{w}}\mathbf{v}$, $\mathbf{v} = (2, -3)$.
33. T is the reflection through the vector $\mathbf{w} = (3, 1)$ in R^2, $\mathbf{v} = (1, 4)$. [The reflection of a vector $\mathbf{v}$ through $\mathbf{w}$ is $T(\mathbf{v}) = 2\,\text{proj}_{\mathbf{w}}\mathbf{v} - \mathbf{v}$.]
34. Repeat Exercise 33 for $\mathbf{w} = (4, -2)$ and $\mathbf{v} = (5, 0)$.

In Exercises 35–38, (a) find the standard matrix A for the linear transformation T, (b) use A to find the image of the vector $\mathbf{v}$, and (c) use a graphing utility or computer software program and A to verify your result from part (b).

35. $T(x, y, z) = (2x + 3y - z, 3x - 2z, 2x - y + z)$, $\mathbf{v} = (1, 2, -1)$
36. $T(x, y, z) = (3x - 2y + z, 2x - 3y, y - 4z)$, $\mathbf{v} = (2, -1, -1)$
37. $T(x_1, x_2, x_3, x_4) = (x_1 - x_2, x_3, x_1 + 2x_2 - x_4, x_4)$, $\mathbf{v} = (1, 0, 1, -1)$
38. $T(x_1, x_2, x_3, x_4) = (x_1 + 2x_2, x_2 - x_1, 2x_3 - x_4, x_1)$, $\mathbf{v} = (0, 1, -1, 1)$

In Exercises 39–44, find the standard matrices for $T = T_2 \circ T_1$ and $T' = T_1 \circ T_2$.

39. $T_1: R^2 \rightarrow R^2$, $T_1(x, y) = (x - 2y, 2x + 3y)$
    $T_2: R^2 \rightarrow R^2$, $T_2(x, y) = (2x, x - y)$
40. $T_1: R^2 \rightarrow R^2$, $T_1(x, y) = (x - 2y, 2x + 3y)$
    $T_2: R^2 \rightarrow R^2$, $T_2(x, y) = (y, 0)$
41. $T_1: R^3 \rightarrow R^3$, $T_1(x, y, z) = (x, y, z)$
    $T_2: R^3 \rightarrow R^3$, $T_2(x, y, z) = (0, x, 0)$

42. $T_1: R^3 \to R^3, T_1(x, y, z) = (x + 2y, y - z, -2x + y + 2z)$
$T_2: R^3 \to R^3, T_2(x, y, z) = (y + z, x + z, 2y - 2z)$

43. $T_1: R^2 \to R^3, T_1(x, y) = (-x + 2y, x + y, x - y)$
$T_2: R^3 \to R^2, T_2(x, y, z) = (x - 3y, z + 3x)$

44. $T_1: R^2 \to R^3, T_1(x, y) = (x, y, y)$
$T_2: R^3 \to R^2, T_2(x, y, z) = (y, z)$

In Exercises 45–56, determine whether the linear transformation is invertible. If it is, find its inverse.

45. $T(x, y) = (x + y, x - y)$

46. $T(x, y) = (x + 2y, x - 2y)$

47. $T(x_1, x_2, x_3) = (x_1, x_1 + x_2, x_1 + x_2 + x_3)$

48. $T(x_1, x_2, x_3) = (x_1 + x_2, x_2 + x_3, x_1 + x_3)$

49. $T(x, y) = (2x, 0)$

50. $T(x, y) = (0, -y)$

51. $T(x, y) = (x + y, 3x + 3y)$

52. $T(x, y) = (x + 4y, x - 4y)$

53. $T(x, y) = (5x, 5y)$

54. $T(x, y) = (-2x, 2y)$

55. $T(x_1, x_2, x_3, x_4) = (x_1 - 2x_2, x_2, x_3 + x_4, x_3)$

56. $T(x_1, x_2, x_3, x_4) = (x_4, x_3, x_2, x_1)$

In Exercises 57–64, find $T(\mathbf{v})$ by using (a) the standard matrix and (b) the matrix relative to B and B'.

57. $T: R^2 \to R^3, T(x, y) = (x + y, x, y), \mathbf{v} = (5, 4)$,
$B = \{(1, -1), (0, 1)\}, B' = \{(1, 1, 0), (0, 1, 1), (1, 0, 1)\}$

58. $T: R^2 \to R^3, T(x, y) = (x - y, 0, x + y), \mathbf{v} = (-3, 2)$,
$B = \{(1, 2), (1, 1)\}, B' = \{(1, 1, 1), (1, 1, 0), (0, 1, 1)\}$

59. $T: R^3 \to R^2, T(x, y, z) = (x - y, y - z), \mathbf{v} = (1, 2, -3)$,
$B = \{(1, 1, 1), (1, 1, 0), (0, 1, 1)\}, B' = \{(1, 2), (1, 1)\}$

60. $T: R^3 \to R^2, T(x, y, z) = (2x - z, y - 2x), \mathbf{v} = (0, -5, 7)$,
$B = \{(2, 0, 1), (0, 2, 1), (1, 2, 1)\}, B' = \{(1, 1), (2, 0)\}$

61. $T: R^3 \to R^4, T(x, y, z) = (2x, x + y, y + z, x + z)$,
$\mathbf{v} = (1, -5, 2)$,
$B = \{(2, 0, 1), (0, 2, 1), (1, 2, 1)\}$,
$B' = \{(1, 0, 0, 1), (0, 1, 0, 1), (1, 0, 1, 0), (1, 1, 0, 0)\}$

62. $T: R^4 \to R^2, T(x_1, x_2, x_3, x_4) = (x_1 + x_2 + x_3 + x_4, x_4 - x_1)$,
$\mathbf{v} = (4, -3, 1, 1)$,
$B = \{(1, 0, 0, 1), (0, 1, 0, 1), (1, 0, 1, 0), (1, 1, 0, 0)\}$,
$B' = \{(1, 1), (2, 0)\}$

63. $T: R^3 \to R^3, T(x, y, z) = (x + y + z, 2z - x, 2y - z)$,
$\mathbf{v} = (4, -5, 10)$,
$B = \{(2, 0, 1), (0, 2, 1), (1, 2, 1)\}$,
$B' = \{(1, 1, 1), (1, 1, 0), (0, 1, 1)\}$

64. $T: R^2 \to R^2, T(x, y) = (2x - 12y, x - 5y), \mathbf{v} = (10, 5)$,
$B = B' = \{(4, 1), (3, 1)\}$

65. Let $T: P_2 \to P_3$ be given by $T(p) = xp$. Find the matrix for T relative to the bases $B = \{1, x, x^2\}$ and $B' = \{1, x, x^2, x^3\}$.

66. Let $T: P_2 \to P_4$ be given by $T(p) = x^2p$. Find the matrix for T relative to the bases $B = \{1, x, x^2\}$ and $B' = \{1, x, x^2, x^3, x^4\}$.

67. Calculus Let $B = \{1, x, e^x, xe^x\}$ be a basis of a subspace W of the space of continuous functions, and let D_x be the differential operator on W. Find the matrix for D_x relative to the basis B.

68. Calculus Repeat Exercise 67 for $B = \{e^{2x}, xe^{2x}, x^2e^{2x}\}$.

69. Calculus Use the matrix from Exercise 67 to evaluate $D_x[3x - 2xe^x]$.

70. Calculus Use the matrix from Exercise 68 to evaluate $D_x[5e^{2x} - 3xe^{2x} + x^2e^{2x}]$.

71. Calculus Let $B = \{1, x, x^2, x^3\}$ be a basis for P_3, and let $T: P_3 \to P_4$ be the linear transformation represented by

$$T(x^k) = \int_0^x t^k\,dt.$$

(a) Find the matrix A for T with respect to B and the standard basis for P_4.

(b) Use A to integrate $p(x) = 6 - 2x + 3x^3$.

True or False? In Exercises 72 and 73, determine whether each statement is true or false. If a statement is true, give a reason or cite an appropriate statement from the text. If a statement is false, provide an example that shows the statement is not true in all cases or cite an appropriate statement from the text.

72. (a) If T is a linear transformation R^n to R^m, the $m \times n$ matrix A is called the standard matrix such that $T(\mathbf{v}) = A\mathbf{v}$ for every $\mathbf{v}$ in R^n.

(b) The composition T of linear transformations T_1 and T_2, defined by $T(\mathbf{v}) = T_2(T_1(\mathbf{v}))$, has the standard matrix A represented by the matrix product $A = A_2A_1$.

(c) All linear transformations T have a unique inverse T^{-1}.

73. (a) The composition T of linear transformations T_1 and T_2, represented by $T(\mathbf{v}) = T_2(T_1(\mathbf{v}))$, is defined if the range of T_1 lies within the domain of T_2.

(b) In general, the compositions $T_2 \circ T_1$ and $T_1 \circ T_2$ have the same standard matrix A.

(c) If $T: R^n \to R^n$ is an invertible linear transformation with standard matrix A, then T^{-1} has the same standard matrix A.

74. Let T be a linear transformation such that $T(\mathbf{v}) = k\mathbf{v}$ for $\mathbf{v}$ in R^n. Find the standard matrix for T.

75. Let $T: M_{2,3} \to M_{3,2}$ be represented by $T(A) = A^T$. Find the matrix for T relative to the standard bases for $M_{2,3}$ and $M_{3,2}$.

76. Show that the linear transformation T given in Exercise 75 is an isomorphism, and find the matrix for the inverse of T.

77. Guided Proof Let $T_1: V \to V$ and $T_2: V \to V$ be one-to-one linear transformations. Prove that the composition $T = T_2 \circ T_1$ is one-to-one and that T^{-1} exists and is equal to $T_1^{-1} \circ T_2^{-1}$.

Getting Started: To show that T is one-to-one, you can use the definition of a one-to-one transformation and show that $T(\mathbf{u}) = T(\mathbf{v})$ implies $\mathbf{u} = \mathbf{v}$. For the second statement, you first need to use Theorems 6.8 and 6.12 to show that T is invertible, and then show that $T \circ (T_1^{-1} \circ T_2^{-1})$ and $(T_1^{-1} \circ T_2^{-1}) \circ T$ are identity transformations.

(i) Let $T(\mathbf{u}) = T(\mathbf{v})$. Recall that $(T_2 \circ T_1)(\mathbf{v}) = T_2(T_1(\mathbf{v}))$ for all vectors $\mathbf{v}$. Now use the fact that T_2 and T_1 are one-to-one to conclude that $\mathbf{u} = \mathbf{v}$.

(ii) Use Theorems 6.8 and 6.12 to show that T_1, T_2, and T are all invertible transformations. So T_1^{-1} and T_2^{-1} exist.

(iii) Form the composition $T' = T_1^{-1} \circ T_2^{-1}$. It is a linear transformation from V to V. To show that it is the inverse of T, you need to determine whether the composition of T with T' on both sides gives an identity transformation.

78. Prove Theorem 6.12.

79. Writing Is it always preferable to use the standard basis for R^n? Discuss the advantages and disadvantages of using different bases.

80. Writing Look back at Theorem 4.19 and rephrase it in terms of what you have learned in this chapter.

6.4 Transition Matrices and Similarity

In Section 6.3 you saw that the matrix for a linear transformation $T: V \to V$ depends on the basis of V. In other words, the matrix for T relative to a basis B is different from the matrix for T relative to another basis B'.

A classical problem in linear algebra is this: Is it possible to find a basis B such that the matrix for T relative to B is diagonal? The solution of this problem is discussed in Chapter 7. This section lays a foundation for solving the problem. You will see how the matrices for a linear transformation relative to two different bases are related. In this section, A, A', P, and P^{-1} represent the four square matrices listed below.

1. Matrix for T relative to B:	A
2. Matrix for T relative to B':	A'
3. Transition matrix from B' to B:	P
4. Transition matrix from B to B':	P^{-1}

Note that in Figure 6.10 there are two ways to get from the coordinate matrix $[\mathbf{v}]_{B'}$ to the coordinate matrix $[T(\mathbf{v})]_{B'}$. One way is direct, using the matrix A' to obtain

$$A'[\mathbf{v}]_{B'} = [T(\mathbf{v})]_{B'}.$$

The other way is indirect, using the matrices P, A, and P^{-1} to obtain

$$P^{-1}AP[\mathbf{v}]_{B'} = [T(\mathbf{v})]_{B'}.$$

But by the definition of the matrix of a linear transformation relative to a basis, this implies that

$$A' = P^{-1}AP.$$

This relationship is demonstrated in Example 1.

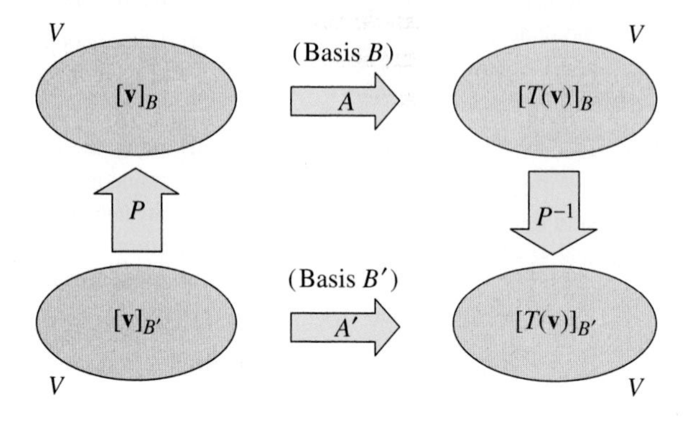

Figure 6.10

EXAMPLE 1 Finding a Matrix of a Linear Transformation

Find the matrix A' for $T: R^2 \rightarrow R^2$,

$$T(x_1, x_2) = (2x_1 - 2x_2, -x_1 + 3x_2),$$

relative to the basis $B' = \{(1, 0), (1, 1)\}$.

SOLUTION The standard matrix for T is

$$A = \begin{bmatrix} 2 & -2 \\ -1 & 3 \end{bmatrix}.$$

Furthermore, using the techniques of Section 4.7, you can find that the transition matrix from B' to the standard basis $B = \{(1, 0), (0, 1)\}$ is

$$P = \begin{bmatrix} 1 & 1 \\ 0 & 1 \end{bmatrix}.$$

The inverse of this matrix is the transition matrix from B to B',

$$P^{-1} = \begin{bmatrix} 1 & -1 \\ 0 & 1 \end{bmatrix}.$$

The matrix for T relative to B' is

$$A' = P^{-1}AP = \begin{bmatrix} 1 & -1 \\ 0 & 1 \end{bmatrix}\begin{bmatrix} 2 & -2 \\ -1 & 3 \end{bmatrix}\begin{bmatrix} 1 & 1 \\ 0 & 1 \end{bmatrix} = \begin{bmatrix} 3 & -2 \\ -1 & 2 \end{bmatrix}.$$

In Example 1, the basis B is the standard basis for R^2. In the next example, both B and B' are nonstandard bases.

EXAMPLE 2 **Finding a Matrix for a Linear Transformation**

Let

$$B = \{(-3, 2), (4, -2)\} \quad \text{and} \quad B' = \{(-1, 2), (2, -2)\}$$

be bases for R^2, and let

$$A = \begin{bmatrix} -2 & 7 \\ -3 & 7 \end{bmatrix}$$

be the matrix for $T: R^2 \rightarrow R^2$ relative to B. Find A', the matrix of T relative to B'.

SOLUTION In Example 5 in Section 4.7, you found that

$$P = \begin{bmatrix} 3 & -2 \\ 2 & -1 \end{bmatrix}$$

and

$$P^{-1} = \begin{bmatrix} -1 & 2 \\ -2 & 3 \end{bmatrix}.$$

So, the matrix of T relative to B' is

$$A' = P^{-1}AP = \begin{bmatrix} -1 & 2 \\ -2 & 3 \end{bmatrix}\begin{bmatrix} -2 & 7 \\ -3 & 7 \end{bmatrix}\begin{bmatrix} 3 & -2 \\ 2 & -1 \end{bmatrix} = \begin{bmatrix} 2 & 1 \\ -1 & 3 \end{bmatrix}.$$

The diagram in Figure 6.10 should help you to remember the roles of the matrices A, A', P, and P^{-1}.

EXAMPLE 3 **Using a Matrix for a Linear Transformation**

For the linear transformation $T: R^2 \rightarrow R^2$ from Example 2, find $[\mathbf{v}]_B$, $[T(\mathbf{v})]_B$, and $[T(\mathbf{v})]_{B'}$ for the vector $\mathbf{v}$ whose coordinate matrix is

$$[\mathbf{v}]_{B'} = \begin{bmatrix} -3 \\ -1 \end{bmatrix}.$$

SOLUTION To find $[\mathbf{v}]_B$, use the transition matrix P from B' to B.

$$[\mathbf{v}]_B = P[\mathbf{v}]_{B'} = \begin{bmatrix} 3 & -2 \\ 2 & -1 \end{bmatrix}\begin{bmatrix} -3 \\ -1 \end{bmatrix} = \begin{bmatrix} -7 \\ -5 \end{bmatrix}$$

To find $[T(\mathbf{v})]_B$, multiply $[\mathbf{v}]_B$ by the matrix A to obtain

$$[T(\mathbf{v})]_B = A[\mathbf{v}]_B = \begin{bmatrix} -2 & 7 \\ -3 & 7 \end{bmatrix}\begin{bmatrix} -7 \\ -5 \end{bmatrix} = \begin{bmatrix} -21 \\ -14 \end{bmatrix}.$$

To find $[T(\mathbf{v})]_{B'}$, multiply $[T(\mathbf{v})]_B$ by P^{-1} to obtain

$$[T(\mathbf{v})]_{B'} = P^{-1}[T(\mathbf{v})]_B = \begin{bmatrix} -1 & 2 \\ -2 & 3 \end{bmatrix}\begin{bmatrix} -21 \\ -14 \end{bmatrix} = \begin{bmatrix} -7 \\ 0 \end{bmatrix}$$

or multiply $[\mathbf{v}]_{B'}$ by A' to obtain

$$[T(\mathbf{v})]_{B'} = A'[\mathbf{v}]_{B'} = \begin{bmatrix} 2 & 1 \\ -1 & 3 \end{bmatrix}\begin{bmatrix} -3 \\ -1 \end{bmatrix} = \begin{bmatrix} -7 \\ 0 \end{bmatrix}.$$

REMARK: It is instructive to note that the transformation T in Examples 2 and 3 is represented by the rule $T(x, y) = \left(x - \frac{3}{2}y, 2x + 4y\right)$. Verify the results of Example 3 by showing that $\mathbf{v} = (1, -4)$ and $T(\mathbf{v}) = (7, -14)$.

Similar Matrices

Two square matrices A and A' that are related by an equation $A' = P^{-1}AP$ are called **similar** matrices, as indicated in the next definition.

Definition of Similar Matrices

For square matrices A and A' of order n, A' is said to be **similar** to A if there exists an invertible matrix P such that $A' = P^{-1}AP$.

If A' is similar to A, then it is also true that A is similar to A', as stated in the next theorem. So, it makes sense to say simply that **A and A' are similar.**

THEOREM 6.13 Properties of Similar Matrices

Let A, B, and C be square matrices of order n. Then the following properties are true.

1. A is similar to A.
2. If A is similar to B, then B is similar to A.
3. If A is similar to B and B is similar to C, then A is similar to C.

PROOF The first property follows from the fact that $A = I_nAI_n$. To prove the second property, write

$$\begin{aligned} A &= P^{-1}BP \\ PAP^{-1} &= P(P^{-1}BP)P^{-1} \\ PAP^{-1} &= B \\ Q^{-1}AQ &= B, \text{ where } Q = P^{-1}. \end{aligned}$$

The proof of the third property is left to you. (See Exercise 23.)

From the definition of similarity, it follows that any two matrices that represent the same linear transformation $T: V \to V$ with respect to different bases must be similar.

EXAMPLE 4 Similar Matrices

(a) From Example 1, the matrices

$$A = \begin{bmatrix} 2 & -2 \\ -1 & 3 \end{bmatrix} \quad \text{and} \quad A' = \begin{bmatrix} 3 & -2 \\ -1 & 2 \end{bmatrix}$$

are similar because $A' = P^{-1}AP$, where $P = \begin{bmatrix} 1 & 1 \\ 0 & 1 \end{bmatrix}$.

(b) From Example 2, the matrices

$$A = \begin{bmatrix} -2 & 7 \\ -3 & 7 \end{bmatrix} \quad \text{and} \quad A' = \begin{bmatrix} 2 & 1 \\ -1 & 3 \end{bmatrix}$$

are similar because $A' = P^{-1}AP$, where $P = \begin{bmatrix} 3 & -2 \\ 2 & -1 \end{bmatrix}$.

You have seen that the matrix for a linear transformation $T: V \to V$ depends on the basis used for V. This observation leads naturally to the question: What choice of basis will make the matrix for T as simple as possible? Is it always the *standard* basis? Not necessarily, as the next example demonstrates.

EXAMPLE 5 A Comparison of Two Matrices for a Linear Transformation

Suppose

$$A = \begin{bmatrix} 1 & 3 & 0 \\ 3 & 1 & 0 \\ 0 & 0 & -2 \end{bmatrix}$$

is the matrix for $T: R^3 \to R^3$ relative to the standard basis. Find the matrix for T relative to the basis

$$B' = \{(1, 1, 0), (1, -1, 0), (0, 0, 1)\}.$$

SOLUTION The transition matrix from B' to the standard matrix has columns consisting of the vectors in B',

$$P = \begin{bmatrix} 1 & 1 & 0 \\ 1 & -1 & 0 \\ 0 & 0 & 1 \end{bmatrix},$$

and it follows that

$$P^{-1} = \begin{bmatrix} \frac{1}{2} & \frac{1}{2} & 0 \\ \frac{1}{2} & -\frac{1}{2} & 0 \\ 0 & 0 & 1 \end{bmatrix}.$$

So, the matrix for T relative to B' is

$$\begin{aligned} A' &= P^{-1}AP \\ &= \begin{bmatrix} \frac{1}{2} & \frac{1}{2} & 0 \\ \frac{1}{2} & -\frac{1}{2} & 0 \\ 0 & 0 & 1 \end{bmatrix} \begin{bmatrix} 1 & 3 & 0 \\ 3 & 1 & 0 \\ 0 & 0 & -2 \end{bmatrix} \begin{bmatrix} 1 & 1 & 0 \\ 1 & -1 & 0 \\ 0 & 0 & 1 \end{bmatrix} \\ &= \begin{bmatrix} 4 & 0 & 0 \\ 0 & -2 & 0 \\ 0 & 0 & -2 \end{bmatrix}. \end{aligned}$$

Note that matrix A' is diagonal.

Diagonal matrices have many computational advantages over nondiagonal ones. For instance, for the diagonal matrix

$$D = \begin{bmatrix} d_1 & 0 & \cdots & 0 \\ 0 & d_2 & \cdots & 0 \\ \vdots & \vdots & & \vdots \\ 0 & 0 & \cdots & d_n \end{bmatrix},$$

the kth power is represented as follows.

$$D^k = \begin{bmatrix} d_1^k & 0 & \cdots & 0 \\ 0 & d_2^k & \cdots & 0 \\ \vdots & \vdots & & \vdots \\ 0 & 0 & \cdots & d_n^k \end{bmatrix}$$

A diagonal matrix is its own transpose. Moreover, if all the diagonal elements are nonzero, then the inverse of a diagonal matrix is the matrix whose main diagonal elements are the reciprocals of corresponding elements in the original matrix. With such computational advantages, it is important to find ways (if possible) to choose a basis for V such that the transformation matrix is diagonal, as it is in Example 5. You will pursue this problem in the next chapter.

SECTION 6.4 Exercises

In Exercises 1–8, (a) find the matrix A' for T relative to the basis B' and (b) show that A' is similar to A, the standard matrix for T.

1. $T: R^2 \to R^2$, $T(x, y) = (2x - y, y - x)$, $B' = \{(1, -2), (0, 3)\}$
2. $T: R^2 \to R^2$, $T(x, y) = (2x + y, x - 2y)$, $B' = \{(1, 2), (0, 4)\}$
3. $T: R^2 \to R^2$, $T(x, y) = (x + y, 4y)$, $B' = \{(-4, 1), (1, -1)\}$
4. $T: R^2 \to R^2$, $T(x, y) = (x - 2y, 4x)$, $B' = \{(-2, 1), (-1, 1)\}$
5. $T: R^3 \to R^3$, $T(x, y, z) = (x, y, z)$, $B' = \{(1, 1, 0), (1, 0, 1), (0, 1, 1)\}$
6. $T: R^3 \to R^3$, $T(x, y, z) = (0, 0, 0)$, $B' = \{(1, 1, 0), (1, 0, 1), (0, 1, 1)\}$
7. $T: R^3 \to R^3$, $T(x, y, z) = (x - y + 2z, 2x + y - z, x + 2y + z)$, $B' = \{(1, 0, 1), (0, 2, 2), (1, 2, 0)\}$
8. $T: R^3 \to R^3$, $T(x, y, z) = (x, x + 2y, x + y + 3z)$, $B' = \{(1, -1, 0), (0, 0, 1), (0, 1, -1)\}$

9. Let $B = \{(1, 3), (-2, -2)\}$ and $B' = \{(-12, 0), (-4, 4)\}$ be bases for R^2, and let

$$A = \begin{bmatrix} 3 & 2 \\ 0 & 4 \end{bmatrix}$$

be the matrix for $T: R^2 \to R^2$ relative to B.

(a) Find the transition matrix P from B' to B.

(b) Use the matrices A and P to find $[\mathbf{v}]_B$ and $[T(\mathbf{v})]_B$, where

$$[\mathbf{v}]_{B'} = \begin{bmatrix} -1 \\ 2 \end{bmatrix}.$$

(c) Find A' (the matrix for T relative to B') and P^{-1}.

(d) Find $[T(\mathbf{v})]_{B'}$ in two ways: first as $P^{-1}[T(\mathbf{v})]_B$ and then as $A'[\mathbf{v}]_{B'}$.

10. Repeat Exercise 9 for $B = \{(1, 1), (-2, 3)\}$, $B' = \{(1, -1), (0, 1)\}$, and

$$[\mathbf{v}]_{B'} = \begin{bmatrix} 1 \\ -3 \end{bmatrix}. \text{ (Use matrix } A \text{ provided in Exercise 9.)}$$

11. Let $B = \{(1, 1, 0), (1, 0, 1), (0, 1, 1)\}$ and $B' = \{(1, 0, 0), (0, 1, 0), (0, 0, 1)\}$ be bases for R^3, and let

$$A = \begin{bmatrix} \frac{3}{2} & -1 & -\frac{1}{2} \\ -\frac{1}{2} & 2 & \frac{1}{2} \\ \frac{1}{2} & 1 & \frac{5}{2} \end{bmatrix}$$

be the matrix for $T: R^3 \to R^3$ relative to B.

(a) Find the transition matrix P from B' to B.

(b) Use the matrices A and P to find $[\mathbf{v}]_B$ and $[T(\mathbf{v})]_B$, where

$$[\mathbf{v}]_{B'} = \begin{bmatrix} 1 \\ 0 \\ -1 \end{bmatrix}.$$

(c) Find A' (the matrix for T relative to B') and P^{-1}.

(d) Find $[T(\mathbf{v})]_{B'}$ in two ways: first as $P^{-1}[T(\mathbf{v})]_B$ and then as $A'[\mathbf{v}]_{B'}$.

12. Repeat Exercise 11 for $B = \{(1, 0, 0), (0, 1, 0), (0, 0, 1)\}$, $B' = \{(1, 1, -1), (1, -1, 1), (-1, 1, 1)\}$, and

$$[\mathbf{v}]_{B'} = \begin{bmatrix} 2 \\ 1 \\ 1 \end{bmatrix}. \text{ (Use matrix } A \text{ provided in Exercise 11.)}$$

13. Let $B = \{(1, 2), (-1, -1)\}$ and $B' = \{(-4, 1), (0, 2)\}$ be bases for R^2, and let $A = \begin{bmatrix} 2 & 1 \\ 0 & -1 \end{bmatrix}$ be the matrix for $T: R^2 \to R^2$ relative to B.

(a) Find the transition matrix P from B' to B.

(b) Use the matrices A and P to find $[\mathbf{v}]_B$ and $[T(\mathbf{v})]_B$, where

$$[\mathbf{v}]_{B'} = \begin{bmatrix} -1 \\ 4 \end{bmatrix}.$$

(c) Find A' (the matrix for T relative to B') and P^{-1}.

(d) Find $[T(\mathbf{v})]_{B'}$ in two ways: first as $P^{-1}[T(\mathbf{v})]_B$ and then as $A'[(\mathbf{v})]_{B'}$.

14. Repeat Exercise 13 for $B = \{(1, -1), (-2, 1)\}$, $B' = \{(-1, 1), (1, 2)\}$, and $[\mathbf{v}]_{B'} = \begin{bmatrix} 1 \\ -4 \end{bmatrix}$. (Use matrix A provided in Exercise 13.)

15. Prove that if A and B are similar, then $|A| = |B|$. Is the converse true?

16. Illustrate the result of Exercise 15 using the matrices

$$A = \begin{bmatrix} 1 & 0 & 0 \\ 0 & -2 & 0 \\ 0 & 0 & 3 \end{bmatrix}, \quad B = \begin{bmatrix} 11 & 7 & 10 \\ 10 & 8 & 10 \\ -18 & -12 & -17 \end{bmatrix},$$

$$P = \begin{bmatrix} -1 & 1 & 0 \\ 2 & 1 & 2 \\ 1 & 1 & 1 \end{bmatrix}, \quad P^{-1} = \begin{bmatrix} -1 & -1 & 2 \\ 0 & -1 & 2 \\ 1 & 2 & -3 \end{bmatrix},$$

where $B = P^{-1}AP$.

17. Let A and B be similar matrices.

(a) Prove that A^T and B^T are similar.

(b) Prove that if A is nonsingular, then B is also nonsingular and A^{-1} and B^{-1} are similar.

(c) Prove that there exists a matrix P such that $B^k = P^{-1}A^kP$.

18. Use the result of Exercise 17 to find B^4, where $B = P^{-1}AP$ for the matrices

$$A = \begin{bmatrix} 1 & 0 \\ 0 & 2 \end{bmatrix}, \quad B = \begin{bmatrix} -4 & -15 \\ 2 & 7 \end{bmatrix},$$

$$P = \begin{bmatrix} 2 & 5 \\ 1 & 3 \end{bmatrix}, \quad P^{-1} = \begin{bmatrix} 3 & -5 \\ -1 & 2 \end{bmatrix}.$$

19. Determine all $n \times n$ matrices that are similar to I_n.

20. Prove that if A is idempotent and B is similar to A, then B is idempotent. (An $n \times n$ matrix A is idempotent if $A = A^2$.)

21. Let A be an $n \times n$ matrix such that $A^2 = O$. Prove that if B is similar to A, then $B^2 = O$.

22. Let $B = P^{-1}AP$. Prove that if $A\mathbf{x} = \mathbf{x}$, then $PBP^{-1}\mathbf{x} = \mathbf{x}$.

23. Complete the proof of Theorem 6.13 by proving that if A is similar to B and B is similar to C, then A is similar to C.

24. **Writing** Suppose A and B are similar. Explain why they have the same rank.

25. Prove that if A and B are similar, then A^2 is similar to B^2.

26. Prove that if A and B are similar, then A^k is similar to B^k for any positive integer k.

27. Let $A = CD$, where C is an invertible $n \times n$ matrix. Prove that the matrix DC is similar to A.

28. Let $B = P^{-1}AP$, where B is a diagonal matrix with main diagonal entries $b_{11}, b_{22}, \ldots, b_{nn}$. Prove that

$$\begin{bmatrix} a_{11} & a_{12} & \cdots & a_{1n} \\ a_{21} & a_{22} & \cdots & a_{2n} \\ \vdots & \vdots & & \vdots \\ a_{n1} & a_{n2} & \cdots & a_{nn} \end{bmatrix} \begin{bmatrix} p_{1i} \\ p_{2i} \\ \vdots \\ p_{ni} \end{bmatrix} = b_{ii} \begin{bmatrix} p_{1i} \\ p_{2i} \\ \vdots \\ p_{ni} \end{bmatrix},$$

for $i = 1, 2, \ldots, n$.

29. **Writing** Let $B = \{\mathbf{v}_1, \mathbf{v}_2, \ldots, \mathbf{v}_n\}$ be a basis for the vector space V, let B' be the standard basis, and consider the identity transformation $I: V \to V$. What can you say about the matrix for I relative to both B and B'? What can you say about the matrix for I relative to B? Relative to B'?

True or False? In Exercises 30 and 31, determine whether each statement is true or false. If a statement is true, give a reason or cite an appropriate statement from the text. If a statement is false, provide an example that shows the statement is not true in all cases or cite an appropriate statement from the text.

30. (a) The matrix for a linear transformation A' relative to the basis B' is equal to the product $P^{-1}AP$, where P^{-1} is the transition matrix from B to B', A is the matrix for the linear transformation relative to basis B, and P is the transition matrix from B' to B.

(b) Two matrices that represent the same linear transformation $T: V \to V$ with respect to different bases are not necessarily similar.

31. (a) The matrix for a linear transformation A relative to the basis B is equal to the product $PA'P^{-1}$, where P is the transition matrix from B' to B, A' is the matrix for the linear transformation relative to basis B', and P^{-1} is the transition matrix from B to B'.

(b) The standard basis for R^n will always make the coordinate matrix for the linear transformation T the simplest matrix possible.

6.5 Applications of Linear Transformations

The Geometry of Linear Transformations in the Plane

This section gives geometric interpretations of linear transformations represented by 2×2 elementary matrices. A summary of the various types of 2×2 elementary matrices is followed by examples in which each type of matrix is examined in more detail.

Elementary Matrices for Linear Transformations in the Plane

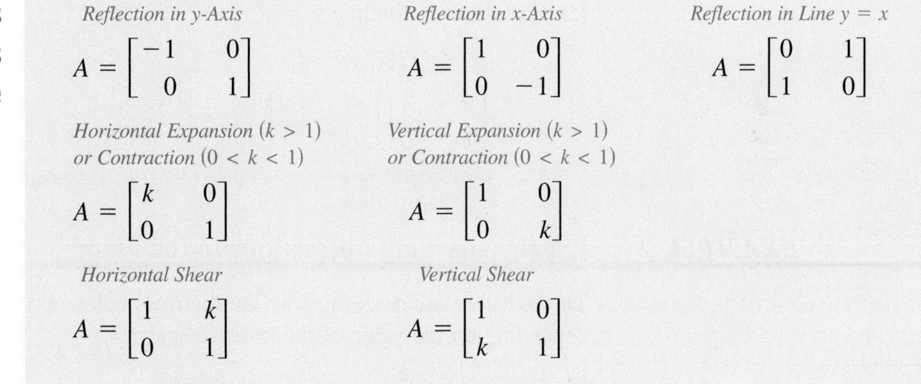

Reflection in y-Axis

$$A = \begin{bmatrix} -1 & 0 \\ 0 & 1 \end{bmatrix}$$

Reflection in x-Axis

$$A = \begin{bmatrix} 1 & 0 \\ 0 & -1 \end{bmatrix}$$

Reflection in Line $y = x$

$$A = \begin{bmatrix} 0 & 1 \\ 1 & 0 \end{bmatrix}$$

Horizontal Expansion $(k > 1)$ *or Contraction* $(0 < k < 1)$

$$A = \begin{bmatrix} k & 0 \\ 0 & 1 \end{bmatrix}$$

Vertical Expansion $(k > 1)$ *or Contraction* $(0 < k < 1)$

$$A = \begin{bmatrix} 1 & 0 \\ 0 & k \end{bmatrix}$$

Horizontal Shear

$$A = \begin{bmatrix} 1 & k \\ 0 & 1 \end{bmatrix}$$

Vertical Shear

$$A = \begin{bmatrix} 1 & 0 \\ k & 1 \end{bmatrix}$$

EXAMPLE 1 **Reflections in the Plane**

The transformations defined by the matrices listed below are called **reflections.** Reflections have the effect of mapping a point in the xy-plane to its “mirror image” with respect to one of the coordinate axes or the line $y = x$, as shown in Figure 6.11.

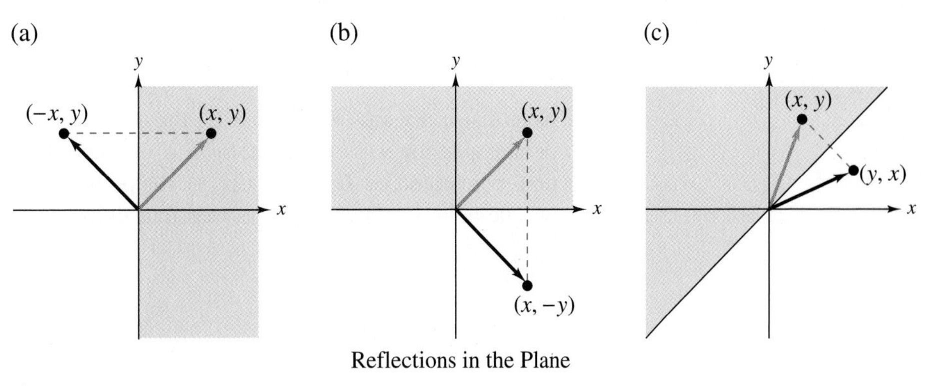

Reflections in the Plane

Figure 6.11

(a) Reflection in the y-axis:

$$T(x, y) = (-x, y)$$

$$\begin{bmatrix} -1 & 0 \\ 0 & 1 \end{bmatrix}\begin{bmatrix} x \\ y \end{bmatrix} = \begin{bmatrix} -x \\ y \end{bmatrix}$$

(b) Reflection in the x-axis:

$$T(x, y) = (x, -y)$$

$$\begin{bmatrix} 1 & 0 \\ 0 & -1 \end{bmatrix}\begin{bmatrix} x \\ y \end{bmatrix} = \begin{bmatrix} x \\ -y \end{bmatrix}$$

(c) Reflection in the line $y = x$:

$$T(x, y) = (y, x)$$

$$\begin{bmatrix} 0 & 1 \\ 1 & 0 \end{bmatrix}\begin{bmatrix} x \\ y \end{bmatrix} = \begin{bmatrix} y \\ x \end{bmatrix}$$

EXAMPLE 2 Expansions and Contractions in the Plane

The transformations defined by the matrices below are called **expansions** or **contractions,** depending on the value of the positive scalar k.

(a) Horizontal contractions and expansions:

$$T(x, y) = (kx, y)$$

$$\begin{bmatrix} k & 0 \\ 0 & 1 \end{bmatrix}\begin{bmatrix} x \\ y \end{bmatrix} = \begin{bmatrix} kx \\ y \end{bmatrix}$$

(b) Vertical contractions and expansions:

$$T(x, y) = (x, ky)$$

$$\begin{bmatrix} 1 & 0 \\ 0 & k \end{bmatrix}\begin{bmatrix} x \\ y \end{bmatrix} = \begin{bmatrix} x \\ ky \end{bmatrix}$$

Note that in Figures 6.12 and 6.13, the distance the point (x, y) is moved by a contraction or an expansion is proportional to its x- or y-coordinate. For instance, under the transformation represented by $T(x, y) = (2x, y)$, the point $(1, 3)$ would be moved one unit to the right, but the point $(4, 3)$ would be moved four units to the right.

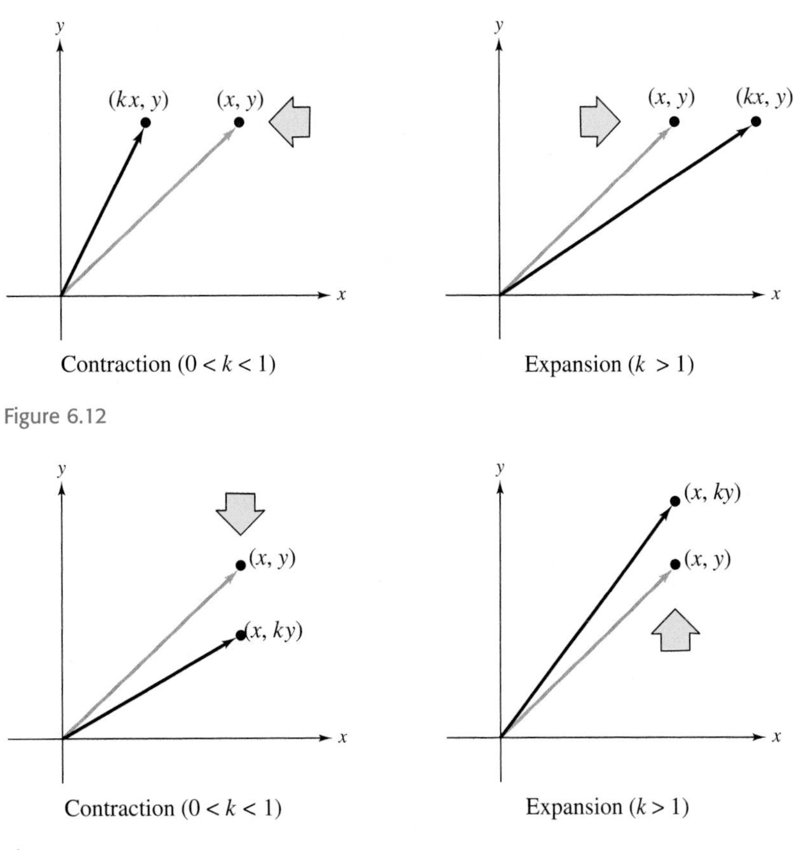

Figure 6.12

Figure 6.13

The third type of linear transformation in the plane corresponding to an elementary matrix is called a **shear,** as described in Example 3.

EXAMPLE 3 Shears in the Plane

The transformations defined by the following matrices are shears.

$$T(x, y) = (x + ky, y) \qquad\qquad T(x, y) = (x, y + kx)$$

$$\begin{bmatrix} 1 & k \\ 0 & 1 \end{bmatrix}\begin{bmatrix} x \\ y \end{bmatrix} = \begin{bmatrix} x + ky \\ y \end{bmatrix} \qquad\qquad \begin{bmatrix} 1 & 0 \\ k & 1 \end{bmatrix}\begin{bmatrix} x \\ y \end{bmatrix} = \begin{bmatrix} x \\ kx + y \end{bmatrix}$$

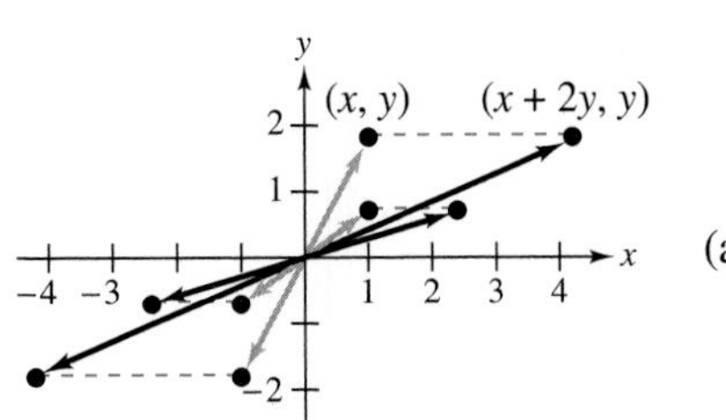

Figure 6.14

(a) The horizontal shear represented by $T(x, y) = (x + 2y, y)$ is shown in Figure 6.14. Under this transformation, points in the upper half-plane are "sheared" to the right by amounts proportional to their y-coordinates. Points in the lower half-plane are "sheared" to the left by amounts proportional to the absolute values of their y-coordinates. Points on the x-axis are unmoved by this transformation.

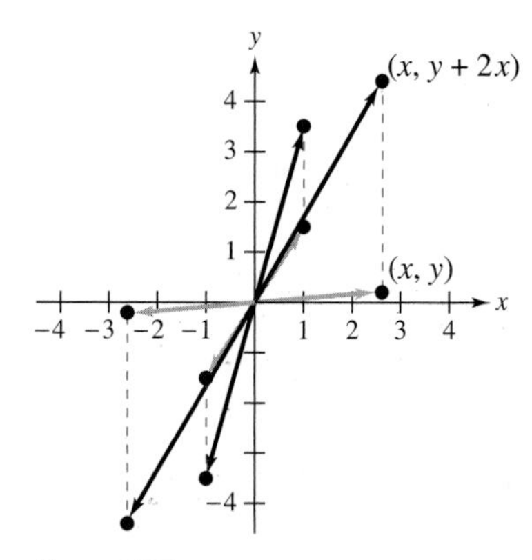

Figure 6.15

(b) The vertical shear represented by $T(x, y) = (x, y + 2x)$ is shown in Figure 6.15. Here, points in the right half-plane are "sheared" upward by amounts proportional to their x-coordinates. Points in the left half-plane are "sheared" downward by amounts proportional to the absolute values of their x-coordinates. Points on the y-axis are unmoved.

Computer Graphics

Linear transformations are useful in computer graphics. In Example 7 in Section 6.1, you saw how a linear transformation could be used to rotate figures in the plane. Here you will see how linear transformations can be used to rotate figures in three-dimensional space.

Suppose you want to rotate the point (x, y, z) counterclockwise about the z-axis through an angle θ, as shown in Figure 6.16. Letting the coordinates of the rotated point be (x', y', z'), you have

$$\begin{bmatrix} x' \\ y' \\ z' \end{bmatrix} = \begin{bmatrix} \cos\theta & -\sin\theta & 0 \\ \sin\theta & \cos\theta & 0 \\ 0 & 0 & 1 \end{bmatrix} \begin{bmatrix} x \\ y \\ z \end{bmatrix} = \begin{bmatrix} x\cos\theta - y\sin\theta \\ x\sin\theta + y\cos\theta \\ z \end{bmatrix}.$$

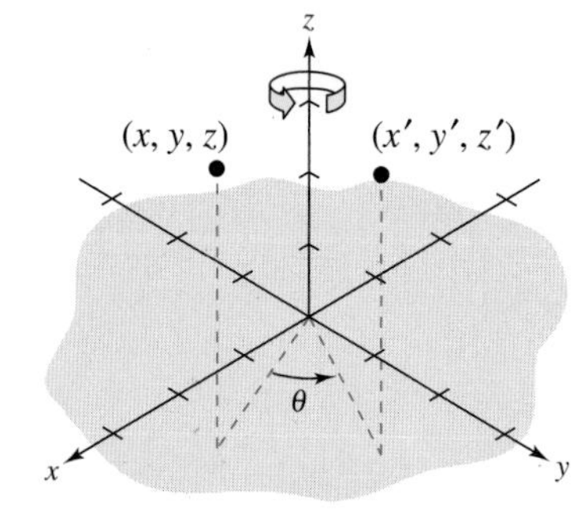

Figure 6.16

Example 4 shows how to use this matrix to rotate a figure in three-dimensional space.

EXAMPLE 4 **Rotation About the z-Axis**

The eight vertices of a rectangular box having sides of lengths 1, 2, and 3 are as follows.

$V_1 = (0, 0, 0)$, $V_2 = (1, 0, 0)$, $V_3 = (1, 2, 0)$, $V_4 = (0, 2, 0)$,
$V_5 = (0, 0, 3)$, $V_6 = (1, 0, 3)$, $V_7 = (1, 2, 3)$, $V_8 = (0, 2, 3)$

Find the coordinates of the box when it is rotated counterclockwise about the z-axis through each angle.

(a) $\theta = 60°$ (b) $\theta = 90°$ (c) $\theta = 120°$

SOLUTION The original box is shown in Figure 6.17.

(a) The matrix that yields a rotation of 60° is

$$A = \begin{bmatrix} \cos 60° & -\sin 60° & 0 \\ \sin 60° & \cos 60° & 0 \\ 0 & 0 & 1 \end{bmatrix} = \begin{bmatrix} 1/2 & -\sqrt{3}/2 & 0 \\ \sqrt{3}/2 & 1/2 & 0 \\ 0 & 0 & 1 \end{bmatrix}.$$

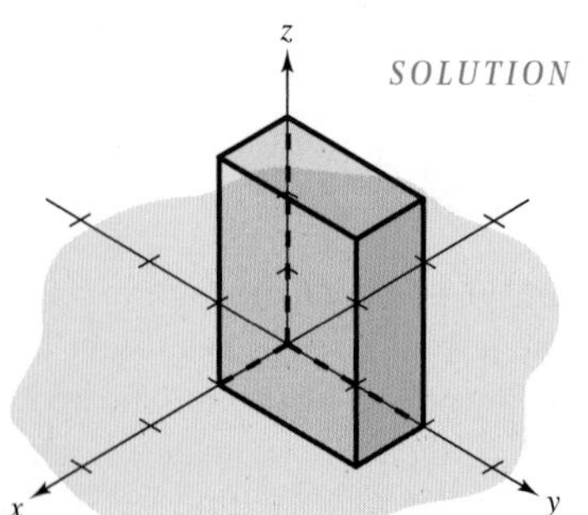

Figure 6.17

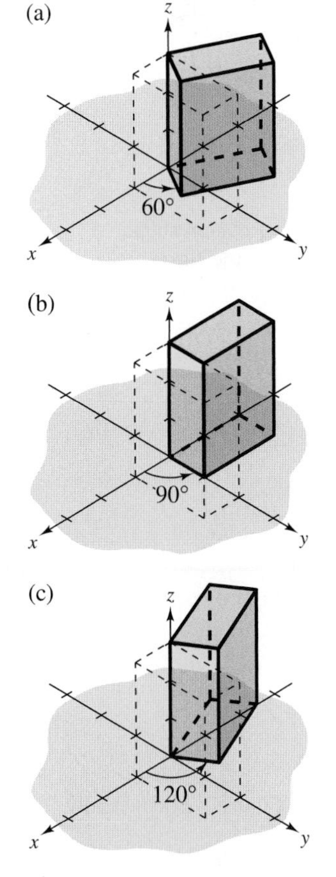

Figure 6.18

Multiplying this matrix by the eight vertices produces the rotated vertices listed below

Original Vertex	*Rotated Vertex*
$V_1 = (0, 0, 0)$	$(0, 0, 0)$
$V_2 = (1, 0, 0)$	$(0.5, 0.87, 0)$
$V_3 = (1, 2, 0)$	$(-1.23, 1.87, 0)$
$V_4 = (0, 2, 0)$	$(-1.73, 1, 0)$
$V_5 = (0, 0, 3)$	$(0, 0, 3)$
$V_6 = (1, 0, 3)$	$(0.5, 0.87, 3)$
$V_7 = (1, 2, 3)$	$(-1.23, 1.87, 3)$
$V_8 = (0, 2, 3)$	$(-1.73, 1, 3)$

A computer-generated graph of the rotated box is shown in Figure 6.18(a). Note that in this graph, line segments representing the sides of the box are drawn between images of pairs of vertices connected in the original box. For instance, because V_1 and V_2 are connected in the original box, the computer is told to connect the images of V_1 and V_2 in the rotated box.

(b) The matrix that yields a rotation of 90° is

$$A = \begin{bmatrix} \cos 90° & -\sin 90° & 0 \\ \sin 90° & \cos 90° & 0 \\ 0 & 0 & 1 \end{bmatrix} = \begin{bmatrix} 0 & -1 & 0 \\ 1 & 0 & 0 \\ 0 & 0 & 1 \end{bmatrix},$$

and the graph of the rotated box is shown in Figure 6.18(b).

(c) The matrix that yields a rotation of 120° is

$$A = \begin{bmatrix} \cos 120° & -\sin 120° & 0 \\ \sin 120° & \cos 120° & 0 \\ 0 & 0 & 1 \end{bmatrix} = \begin{bmatrix} -1/2 & -\sqrt{3}/2 & 0 \\ \sqrt{3}/2 & -1/2 & 0 \\ 0 & 0 & 1 \end{bmatrix},$$

and the graph of the rotated box is shown in Figure 6.18(c).

In Example 4, matrices were used to perform rotations about the z-axis. Similarly, you can use matrices to rotate figures about the x- or y-axis. All three types of rotations are summarized as follows.

Rotation About the x-Axis

$$\begin{bmatrix} 1 & 0 & 0 \\ 0 & \cos\theta & -\sin\theta \\ 0 & \sin\theta & \cos\theta \end{bmatrix}$$

Rotation About the y-Axis

$$\begin{bmatrix} \cos\theta & 0 & \sin\theta \\ 0 & 1 & 0 \\ -\sin\theta & 0 & \cos\theta \end{bmatrix}$$

Rotation About the z-Axis

$$\begin{bmatrix} \cos\theta & -\sin\theta & 0 \\ \sin\theta & \cos\theta & 0 \\ 0 & 0 & 1 \end{bmatrix}$$

In each case the rotation is oriented counterclockwise relative to a person facing the negative direction of the indicated axis, as shown in Figure 6.19.

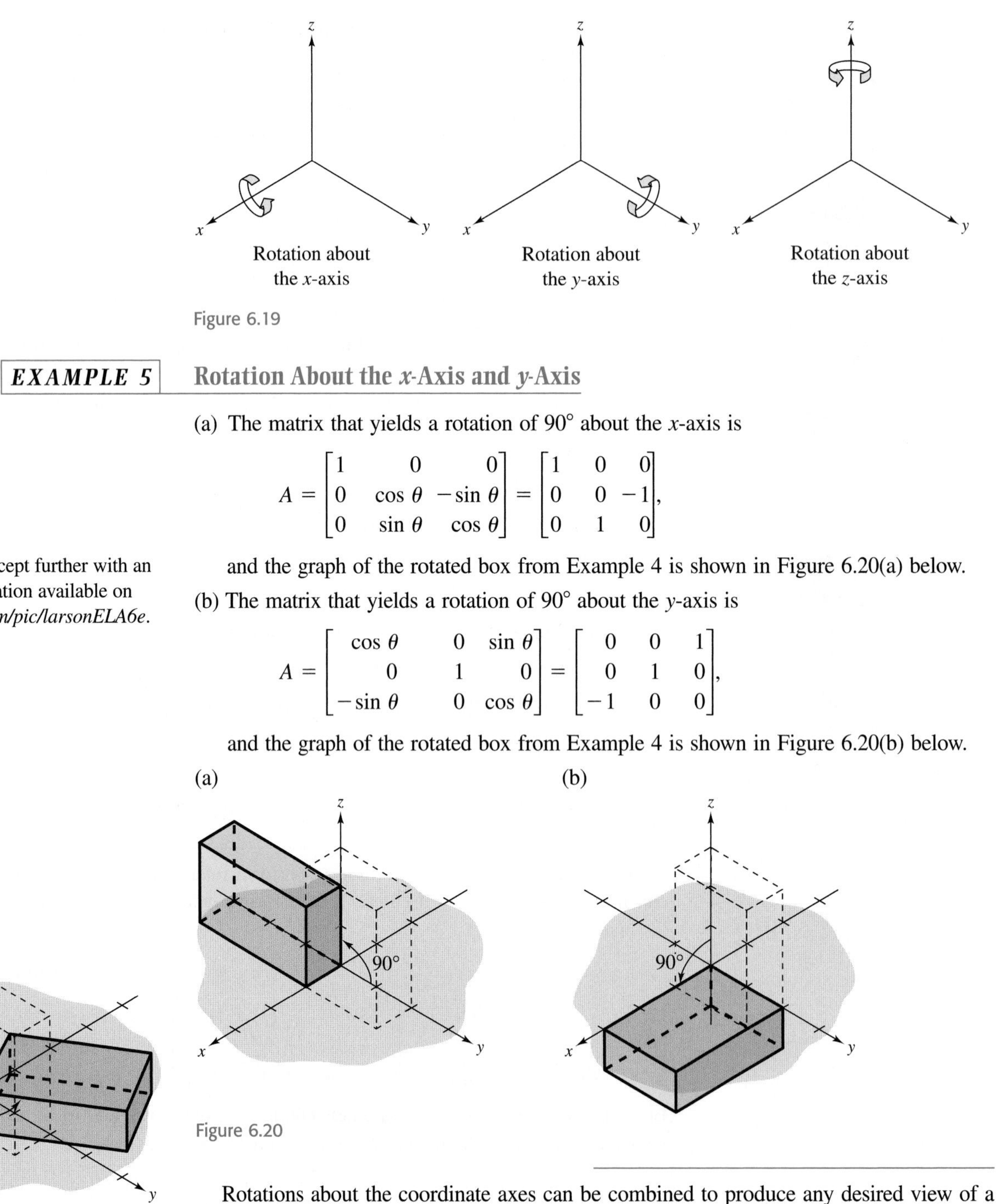

Figure 6.19

EXAMPLE 5 Rotation About the x-Axis and y-Axis

(a) The matrix that yields a rotation of 90° about the x-axis is

$$A = \begin{bmatrix} 1 & 0 & 0 \\ 0 & \cos\theta & -\sin\theta \\ 0 & \sin\theta & \cos\theta \end{bmatrix} = \begin{bmatrix} 1 & 0 & 0 \\ 0 & 0 & -1 \\ 0 & 1 & 0 \end{bmatrix},$$

and the graph of the rotated box from Example 4 is shown in Figure 6.20(a) below.

(b) The matrix that yields a rotation of 90° about the y-axis is

$$A = \begin{bmatrix} \cos\theta & 0 & \sin\theta \\ 0 & 1 & 0 \\ -\sin\theta & 0 & \cos\theta \end{bmatrix} = \begin{bmatrix} 0 & 0 & 1 \\ 0 & 1 & 0 \\ -1 & 0 & 0 \end{bmatrix},$$

and the graph of the rotated box from Example 4 is shown in Figure 6.20(b) below.

Simulation

Explore this concept further with an electronic simulation available on *college.hmco.com/pic/larsonELA6e*.

Figure 6.20

Rotations about the coordinate axes can be combined to produce any desired view of a figure. For instance, Figure 6.21 shows the rotation produced by first rotating the box (from Example 4) 90° about the y-axis, then further rotating the box 120° about the z-axis.

Figure 6.21

The use of computer graphics has become common among designers in many fields. By simply entering the coordinates that form the outline of an object into a computer, a designer can see the object before it is created. As a simple example, the images of the toy boat shown in Figure 6.22 were created using only 27 points in space. Once the points have been stored in the computer, the boat can be viewed from any perspective.

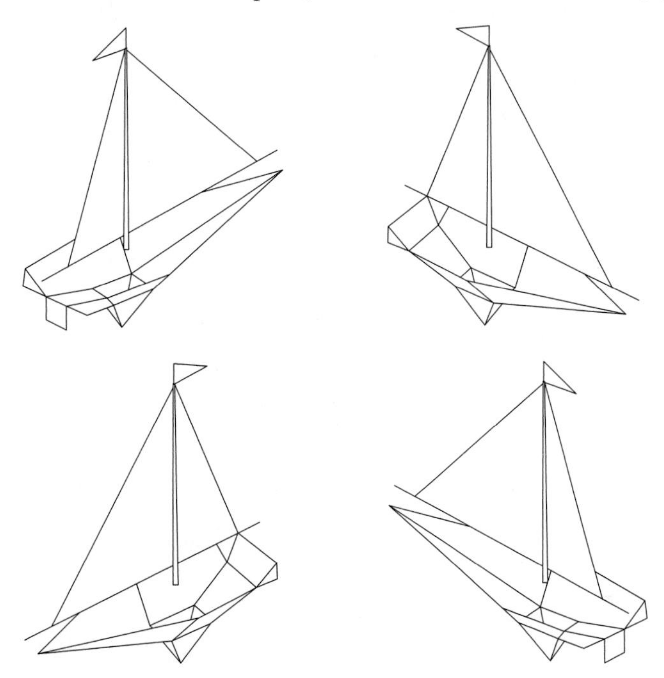

Figure 6.22

SECTION 6.5 Exercises

The Geometry of Linear Transformations in the Plane

1. Let $T: R^2 \to R^2$ be a reflection in the x-axis. Find the image of each vector.

(a) $(3, 5)$ (b) $(2, -1)$ (c) $(a, 0)$
(d) $(0, b)$ (e) $(-c, d)$ (f) $(f, -g)$

2. Let $T: R^2 \to R^2$ be a reflection in the y-axis. Find the image of each vector.

(a) $(2, 5)$ (b) $(-4, -1)$ (c) $(a, 0)$
(d) $(0, b)$ (e) $(c, -d)$ (f) (f, g)

3. Let $T: R^2 \to R^2$ be a reflection in the line $y = x$. Find the image of each vector.

(a) $(0, 1)$ (b) $(-1, 3)$ (c) $(a, 0)$
(d) $(0, b)$ (e) $(-c, d)$ (f) $(f, -g)$

4. Let $T: R^2 \to R^2$ be a reflection in the line $y = -x$. Find the image of each vector.

(a) $(-1, 2)$ (b) $(2, 3)$ (c) $(a, 0)$
(d) $(0, b)$ (e) $(e, -d)$ (f) $(-f, g)$

5. Let $T(1, 0) = (0, 1)$ and $T(0, 1) = (1, 0)$.

(a) Determine $T(x, y)$ for any (x, y).

(b) Give a geometric description of T.

6. Let $T(1, 0) = (2, 0)$ and $T(0, 1) = (0, 1)$.

(a) Determine $T(x, y)$ for any (x, y).

(b) Give a geometric description of T.

In Exercises 7–14, (a) identify the transformation and (b) graphically represent the transformation for an arbitrary vector in the plane.

7. $T(x, y) = (x, y/2)$

8. $T(x, y) = (x/4, y)$

9. $T(x, y) = (4x, y)$

10. $T(x, y) = (x, 2y)$

11. $T(x, y) = (x + 3y, y)$

12. $T(x, y) = (x + 4y, y)$

13. $T(x, y) = (x, 2x + y)$

14. $T(x, y) = (x, 4x + y)$

In Exercises 15–22, find all fixed points of the linear transformation. The vector $\mathbf{v}$ is a fixed point of T if $T(\mathbf{v}) = \mathbf{v}$.

15. A reflection in the y-axis

16. A reflection in the x-axis

17. A reflection in the line $y = x$

18. A reflection in the line $y = -x$

19. A vertical contraction

20. A horizontal expansion

21. A horizontal shear

22. A vertical shear

In Exercises 23–28, sketch the image of the unit square with vertices at $(0, 0)$, $(1, 0)$, $(1, 1)$, and $(0, 1)$ under the specified transformation.

23. T is a reflection in the x-axis.

24. T is a reflection in the line $y = x$.

25. T is the contraction given by $T(x, y) = (x/2, y)$.

26. T is the expansion given by $T(x, y) = (x, 3y)$.

27. T is the shear given by $T(x, y) = (x + 2y, y)$.

28. T is the shear given by $T(x, y) = (x, y + 3x)$.

In Exercises 29–34, sketch the image of the rectangle with vertices at $(0, 0)$, $(0, 2)$, $(1, 2)$, and $(1, 0)$ under the specified transformation.

29. T is a reflection in the y-axis.

30. T is a reflection in the line $y = x$.

31. T is the contraction represented by $T(x, y) = (x, y/2)$.

32. T is the expansion represented by $T(x, y) = (2x, y)$.

33. T is the shear represented by $T(x, y) = (x + y, y)$.

34. T is the shear represented by $T(x, y) = (x, y + 2x)$.

In Exercises 35–38, sketch each of the images with the given vertices under the specified transformations.

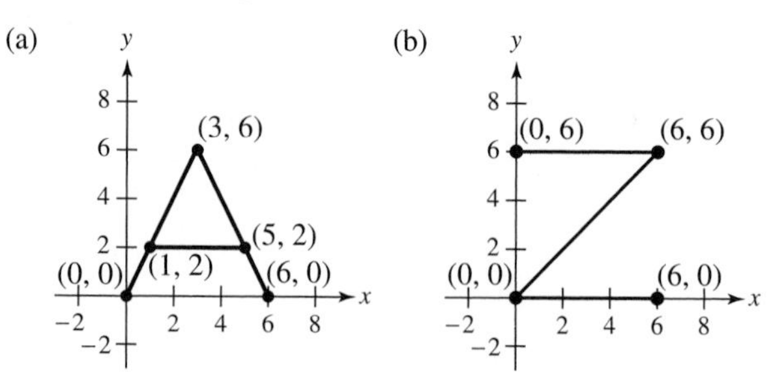

35. T is the shear represented by $T(x, y) = (x + y, y)$.

36. T is the shear represented by $T(x, y) = (x, x + y)$.

37. T is the expansion and contraction represented by $T(x, y) = \left(2x, \frac{1}{2}y\right)$.

38. T is the expansion and contraction represented by $T(x, y) = \left(\frac{1}{2}x, 2y\right)$.

39. The linear transformation defined by a diagonal matrix with positive main diagonal elements is called a **magnification.** Find the images of $(1, 0)$, $(0, 1)$, and $(2, 2)$ under the linear transformation A and graphically interpret your result.

$$A = \begin{bmatrix} 2 & 0 \\ 0 & 3 \end{bmatrix}$$

40. Repeat Exercise 39 for the linear transformation defined by

$$A = \begin{bmatrix} 3 & 0 \\ 0 & 3 \end{bmatrix}.$$

In Exercises 41–46, give a geometric description of the linear transformation defined by the elementary matrix.

41. $A = \begin{bmatrix} 2 & 0 \\ 0 & 1 \end{bmatrix}$

42. $A = \begin{bmatrix} 1 & 0 \\ 2 & 1 \end{bmatrix}$

43. $A = \begin{bmatrix} 0 & 1 \\ 1 & 0 \end{bmatrix}$

44. $A = \begin{bmatrix} 1 & 3 \\ 0 & 1 \end{bmatrix}$

45. $A = \begin{bmatrix} 1 & 0 \\ 0 & -2 \end{bmatrix}$

46. $A = \begin{bmatrix} -1 & 0 \\ 0 & 1 \end{bmatrix}$

In Exercises 47 and 48, give a geometric description of the linear transformation defined by the matrix product.

47. $A = \begin{bmatrix} 2 & 0 \\ 2 & 1 \end{bmatrix} = \begin{bmatrix} 2 & 0 \\ 0 & 1 \end{bmatrix} \begin{bmatrix} 1 & 0 \\ 2 & 1 \end{bmatrix}$

48. $A = \begin{bmatrix} 0 & 3 \\ 1 & 0 \end{bmatrix} = \begin{bmatrix} 0 & 1 \\ 1 & 0 \end{bmatrix} \begin{bmatrix} 1 & 0 \\ 0 & 3 \end{bmatrix}$

Computer Graphics

In Exercises 49–52, find the matrix that will produce the indicated rotation.

49. 30° about the z-axis

50. 60° about the x-axis

51. 60° about the y-axis

52. 120° about the x-axis

In Exercises 53–56, find the image of the vector $(1, 1, 1)$ for the indicated rotation.

53. 30° about the z-axis

54. 60° about the x-axis

55. 60° about the y-axis

56. 120° about the x-axis

In Exercises 57–62, determine which single counterclockwise rotation about the x-, y-, or z-axis will produce the indicated tetrahedron. The original tetrahedron position is illustrated in Figure 6.23.

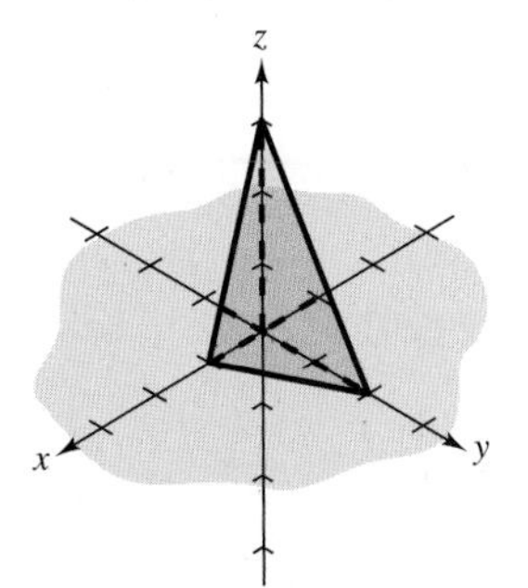

Figure 6.23

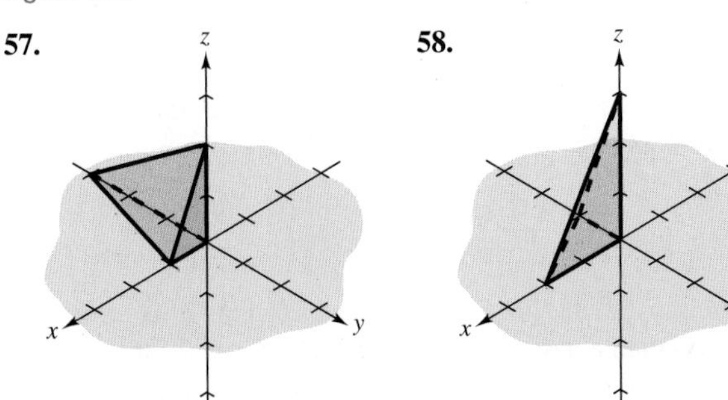

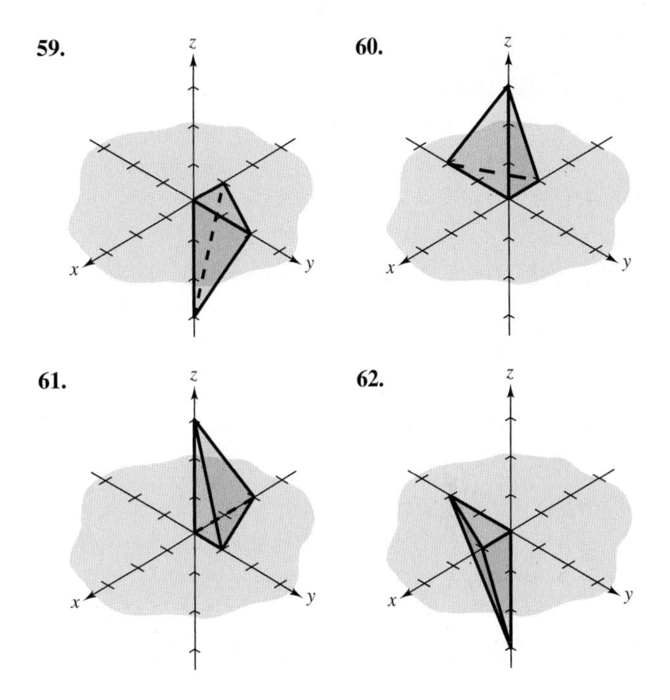

In Exercises 63–66, determine the matrix that will produce the indicated pair of rotations. Then find the image of the line segment from $(0, 0, 0)$ to $(1, 1, 1)$ under this composition.

63. 90° about the x-axis followed by 90° about the y-axis

64. 45° about the y-axis followed by 90° about the z-axis

65. 30° about the z-axis followed by 60° about the y-axis

66. 45° about the z-axis followed by 135° about the x-axis

CHAPTER 6 Review Exercises

In Exercises 1–4, find (a) the image of **v** and (b) the preimage of **w** for the linear transformation.

1. $T: R^2 \to R^2$, $T(v_1, v_2) = (v_1, v_1 + 2v_2)$, $\mathbf{v} = (2, -3)$, $\mathbf{w} = (4, 12)$
2. $T: R^2 \to R^2$, $T(v_1, v_2) = (v_1 + v_2, 2v_2)$, $\mathbf{v} = (4, -1)$, $\mathbf{w} = (8, 4)$
3. $T: R^3 \to R^3$, $T(v_1, v_2, v_3) = (0, v_1 + v_2, v_2 + v_3)$, $\mathbf{v} = (-3, 2, 5)$, $\mathbf{w} = (0, 2, 5)$
4. $T: R^3 \to R^3$, $T(v_1, v_2, v_3) = (v_1 + v_2, v_2 + v_3, v_3)$, $\mathbf{v} = (-2, 1, 2)$, $\mathbf{w} = (0, 1, 2)$

In Exercises 5–12, determine whether the function is a linear transformation. If it is, find its standard matrix A.

5. $T: R^2 \to R^2$, $T(x_1, x_2) = (x_1 + 2x_2, -x_1 - x_2)$
6. $T: R^2 \to R^2$, $T(x_1, x_2) = (x_1 + 3, x_2)$
7. $T: R^2 \to R^2$, $T(x, y) = (x - 2y, 2y - x)$
8. $T: R^2 \to R^2$, $T(x, y) = (x + y, y)$
9. $T: R^2 \to R^2$, $T(x, y) = (x + h, y + k)$, $h \neq 0$ or $k \neq 0$ (translation in the plane)
10. $T: R^2 \to R^2$, $T(x, y) = (|x|, |y|)$
11. $T: R^3 \to R^3$, $T(x_1, x_2, x_3) = (x_1 - x_2, x_2 - x_3, x_3 - x_1)$
12. $T: R^3 \to R^3$, $T(x, y, z) = (z, y, x)$
13. Let T be a linear transformation from R^2 to R^2 such that $T(2, 0) = (1, 1)$ and $T(0, 3) = (3, 3)$. Find $T(1, 1)$ and $T(0, 1)$.
14. Let T be a linear transformation from R^3 to R such that $T(1, 1, 1) = 1$, $T(1, 1, 0) = 2$, and $T(1, 0, 0) = 3$. Find $T(0, 1, 1)$.
15. Let T be a linear transformation from R^2 to R^2 such that $T(1, 1) = (2, 3)$ and $T(2, -1) = (1, 0)$. Find $T(0, -1)$.
16. Let T be a linear transformation from R^2 to R^2 such that $T(1, -1) = (2, -3)$ and $T(0, 2) = (0, 8)$. Find $T(2, 4)$.

In Exercises 17–20, find the indicated power of A, the standard matrix for T.

17. $T: R^3 \to R^3$, reflection in the xy-plane. Find A^2.
18. $T: R^3 \to R^3$, projection onto the xy-plane. Find A^2.
19. $T: R^2 \to R^2$, counterclockwise rotation through the angle θ. Find A^3.
20. **Calculus** $T: P_3 \to P_3$, differential operator. Find A^2.

In Exercises 21–28, the linear transformation $T: R^n \to R^m$ is defined by $T(\mathbf{v}) = A\mathbf{v}$. For each matrix A, (a) determine the dimensions of R^n and R^m, (b) find the image $T(\mathbf{v})$ of the given vector $\mathbf{v}$, and (c) find the preimage of the given vector $\mathbf{w}$.

21. $A = \begin{bmatrix} 0 & 1 & 2 \\ -2 & 0 & 0 \end{bmatrix}$, $\mathbf{v} = (6, 1, 1)$, $\mathbf{w} = (3, 5)$
22. $A = \begin{bmatrix} 1 & 2 & -1 \\ 1 & 0 & 1 \end{bmatrix}$, $\mathbf{v} = (5, 2, 2)$, $\mathbf{w} = (4, 2)$
23. $A = [1, 1]$, $\mathbf{v} = (2, 3)$, $\mathbf{w} = (4)$
24. $A = [2, -1]$, $\mathbf{v} = (1, 2)$, $\mathbf{w} = (-1)$
25. $A = \begin{bmatrix} 1 & 1 & 1 \\ 0 & 1 & 1 \\ 0 & 0 & 1 \end{bmatrix}$, $\mathbf{v} = (2, 1, -5)$, $\mathbf{w} = (6, 4, 2)$
26. $A = \begin{bmatrix} 2 & 1 \\ 0 & 1 \end{bmatrix}$, $\mathbf{v} = (8, 4)$, $\mathbf{w} = (5, 2)$
27. $A = \begin{bmatrix} 4 & 0 \\ 0 & 5 \\ 1 & 1 \end{bmatrix}$, $\mathbf{v} = (2, 2)$, $\mathbf{w} = (4, -5, 0)$
28. $A = \begin{bmatrix} 1 & 0 \\ 0 & -1 \\ 1 & 2 \end{bmatrix}$, $\mathbf{v} = (1, 2)$, $\mathbf{w} = (2, -5, 12)$

In Exercises 29–32, find a basis for (a) $\ker(T)$ and (b) range(T).

29. $T: R^4 \to R^3$, $T(w, x, y, z) = (2w + 4x + 6y + 5z, -w - 2x + 2y, 8y + 4z)$
30. $T: R^3 \to R^3$, $T(x, y, z) = (x + 2y, y + 2z, z + 2x)$
31. $T: R^3 \to R^3$, $T(x, y, z) = (x + y, y + z, x - z)$
32. $T: R^3 \to R^3$, $T(x, y, z) = (x, y + 2z, z)$

In Exercises 33–36, the linear transformation T is given by $T(\mathbf{v}) = A\mathbf{v}$. Find a basis for (a) the kernel of T and (b) the range of T, and then find (c) the rank of T and (d) the nullity of T.

33. $A = \begin{bmatrix} 1 & 2 \\ -1 & 0 \\ 1 & 1 \end{bmatrix}$
34. $A = \begin{bmatrix} 1 & 1 \\ 0 & -1 \\ 2 & 1 \end{bmatrix}$
35. $A = \begin{bmatrix} 2 & 1 & 3 \\ 1 & 1 & 0 \\ 0 & 1 & -3 \end{bmatrix}$
36. $A = \begin{bmatrix} 1 & 1 & -1 \\ 1 & 2 & 1 \\ 0 & 1 & 0 \end{bmatrix}$

37. Given $T: R^5 \to R^3$ and nullity$(T) = 2$, find rank(T).

38. Given $T: P_5 \to P_3$ and nullity$(T) = 4$, find rank(T).

39. Given $T: P_4 \to R^5$ and rank$(T) = 3$, find nullity(T).

40. Given $T: M_{2,2} \to M_{2,2}$ and rank$(T) = 3$, find nullity(T).

In Exercises 41–48, determine whether the transformation T has an inverse. If it does, find A and A^{-1}.

41. $T: R^2 \to R^2, T(x, y) = (2x, y)$

42. $T: R^2 \to R^2, T(x, y) = (0, y)$

43. $T: R^2 \to R^2, T(x, y) = (x \cos \theta - y \sin \theta, x \sin \theta + y \cos \theta)$

44. $T: R^2 \to R^2, T(x, y) = (x, -y)$

45. $T: R^3 \to R^3, T(x, y, z) = (x, y, 0)$

46. $T: R^3 \to R^3, T(x, y, z) = (x, -y, z)$

47. $T: R^3 \to R^2, T(x, y, z) = (x + y, y - z)$

48. $T: R^3 \to R^2, T(x, y, z) = (x + y + z, z)$

In Exercises 49 and 50, find the standard matrices for $T = T_1 \circ T_2$ and $T' = T_2 \circ T_1$.

49. $T_1: R^2 \to R^3, T_1(x, y) = (x, x + y, y)$
$T_2: R^3 \to R^2, T_2(x, y, z) = (0, y)$

50. $T_1: R \to R^2, T(x) = (x, 3x)$
$T_2: R^2 \to R, T(x, y) = (y + 2x)$

51. Use the standard matrix for counterclockwise rotation in R^2 to rotate the triangle with vertices $(3, 5)$, $(5, 3)$, and $(3, 0)$ counterclockwise $90°$ about the origin. Graph the triangles.

52. Rotate the triangle in Exercise 51 counterclockwise $90°$ about the point $(5, 3)$. Graph the triangles.

In Exercises 53–56, determine whether the linear transformation represented by the matrix A is (a) one-to-one, (b) onto, and (c) invertible.

53. $A = \begin{bmatrix} 2 & 0 \\ 0 & 3 \end{bmatrix}$

54. $A = \begin{bmatrix} 1 & 1 & 1 \\ 0 & 1 & 1 \end{bmatrix}$

55. $A = \begin{bmatrix} 1 & \frac{1}{4} \\ 0 & 1 \end{bmatrix}$

56. $A = \begin{bmatrix} 4 & 0 & 7 \\ 5 & 5 & 1 \\ 0 & 0 & 2 \end{bmatrix}$

In Exercises 57 and 58, find $T(\mathbf{v})$ by using (a) the standard matrix and (b) the matrix relative to B and B'.

57. $T: R^2 \to R^3, T(x, y) = (-x, y, x + y), \mathbf{v} = (0, 1)$,
$B = \{(1, 1), (1, -1)\}, B' = \{(0, 1, 0), (0, 0, 1), (1, 0, 0)\}$

58. $T: R^2 \to R^2, T(x, y) = (2y, 0), \mathbf{v} = (-1, 3)$,
$B = \{(2, 1), (-1, 0)\}, B' = \{(-1, 0), (2, 2)\}$

In Exercises 59 and 60, find the matrix A' for T relative to the basis B' and show that A' is similar to A, the standard matrix for T.

59. $T: R^2 \to R^2, T(x, y) = (x - 3y, y - x), B' = \{(1, -1), (1, 1)\}$

60. $T: R^3 \to R^3, T(x, y, z) = (x + 3y, 3x + y, -2z)$,
$B' = \{(1, 1, 0), (1, -1, 0), (0, 0, 1)\}$

61. Let $T: R^3 \to R^3$ be represented by $T(\mathbf{v}) = \text{proj}_{\mathbf{u}}\mathbf{v}$, where $\mathbf{u} = (0, 1, 2)$.
   (a) Find A, the standard matrix for T.
   (b) Let S be the linear transformation represented by $I - A$. Show that S is of the form $S(\mathbf{v}) = \text{proj}_{\mathbf{w}_1}\mathbf{v} + \text{proj}_{\mathbf{w}_2}\mathbf{v}$, where $\mathbf{w}_1$ and $\mathbf{w}_2$ are fixed vectors in R^3.
   (c) Show that the kernel of T is equal to the range of S.

62. Let $T: R^2 \to R^2$ be represented by $T(\mathbf{v}) = \text{proj}_{\mathbf{u}}\mathbf{v}$, where $\mathbf{u} = (4, 3)$.
   (a) Find A, the standard matrix for T, and show that $A^2 = A$.
   (b) Show that $(I - A)^2 = I - A$.
   (c) Find $A\mathbf{v}$ and $(I - A)\mathbf{v}$ for $\mathbf{v} = (5, 0)$.
   (d) Sketch the graph of $\mathbf{u}$, $\mathbf{v}$, $A\mathbf{v}$, and $(I - A)\mathbf{v}$.

63. Let S and T be linear transformations from V into W. Show that $S + T$ and kT are both linear transformations, where $(S + T)(\mathbf{v}) = S(\mathbf{v}) + T(\mathbf{v})$ and $(kT)(\mathbf{v}) = kT(\mathbf{v})$.

64. Suppose A and B are similar matrices and A is invertible.
   (a) Prove that B is invertible.
   (b) Prove that A^{-1} and B^{-1} are similar.

In Exercises 65 and 66, the sum $S + T$ of two linear transformations $S: V \to W$ and $T: V \to W$ is defined as $(S + T)(\mathbf{v}) = S(\mathbf{v}) + T(\mathbf{v})$.

65. Prove that rank$(S + T) \le$ rank(S) + rank(T).

66. Give an example for each.
   (a) Rank$(S + T)$ = rank(S) + rank(T)
   (b) Rank$(S + T)$ < rank(S) + rank(T)

67. Let $T: P_3 \to R$ such that $T(a_0 + a_1x + a_2x^2 + a_3x^3) = a_0 + a_1 + a_2 + a_3$.
   (a) Prove that T is linear.
   (b) Find the rank and nullity of T.
   (c) Find a basis for the kernel of T.

68. Let $T: V \to U$ and $S: U \to W$ be linear transformations.
   (a) Prove that if S and T are both one-to-one, then so is $S \circ T$.
   (b) Prove that the kernel of T is contained in the kernel of $S \circ T$.
   (c) Prove that if $S \circ T$ is onto, then so is S.

69. Let V be an inner product space. For a fixed nonzero vector $\mathbf{v}_0$ in V, let $T: V \rightarrow R$ be the linear transformation $T(\mathbf{v}) = \langle \mathbf{v}, \mathbf{v}_0 \rangle$. Find the kernel, range, rank, and nullity of T.

70. **Calculus** Let $B = \{1, x, \sin x, \cos x\}$ be a basis for a subspace W of the space of continuous functions, and let D_x be the differential operator on W. Find the matrix for D_x relative to the basis B. Find the range and kernel of D_x.

71. **Writing** Under what conditions are the spaces $M_{m,n}$ and $M_{p,q}$ isomorphic? Describe an isomorphism T in this case.

72. **Calculus** Let $T: P_3 \rightarrow P_3$ be represented by $T(p) = p(x) + p'(x)$. Find the rank and nullity of T.

The Geometry of Linear Transformations in the Plane

In Exercises 73–78, (a) identify the transformation and (b) graphically represent the transformation for an arbitrary vector in the plane.

73. $T(x, y) = (x, 2y)$

74. $T(x, y) = (x + y, y)$

75. $T(x, y) = (x, y + 3x)$

76. $T(x, y) = (5x, y)$

77. $T(x, y) = (x + 2y, y)$

78. $T(x, y) = (x, x + 2y)$

In Exercises 79–82, sketch the image of the triangle with vertices $(0, 0)$, $(1, 0)$, and $(0, 1)$ under the given transformation.

79. T is a reflection in the x-axis.

80. T is the expansion represented by $T(x, y) = (2x, y)$.

81. T is the shear represented by $T(x, y) = (x + 3y, y)$.

82. T is the shear represented by $T(x, y) = (x, y + 2x)$.

In Exercises 83 and 84, give a geometric description of the linear transformation defined by the matrix product.

83. $\begin{bmatrix} 0 & 2 \\ 1 & 0 \end{bmatrix} = \begin{bmatrix} 2 & 0 \\ 0 & 1 \end{bmatrix} \begin{bmatrix} 0 & 1 \\ 1 & 0 \end{bmatrix}$

84. $\begin{bmatrix} 1 & 0 \\ 6 & 2 \end{bmatrix} = \begin{bmatrix} 1 & 0 \\ 0 & 2 \end{bmatrix} \begin{bmatrix} 1 & 0 \\ 3 & 1 \end{bmatrix}$

Computer Graphics

In Exercises 85–88, find the matrix that will produce the indicated rotation and then find the image of the vector $(1, -1, 1)$.

85. 45° about the z-axis

86. 90° about the x-axis

87. 60° about the x-axis

88. 30° about the y-axis

In Exercises 89–92, determine the matrix that will produce the indicated pair of rotations.

89. 60° about the x-axis followed by 30° about the z-axis

90. 120° about the y-axis followed by 45° about the z-axis

91. 30° about the y-axis followed by 45° about the z-axis

92. 60° about the x-axis followed by 60° about the z-axis

In Exercises 93–96, find the image of the unit cube with vertices $(0, 0, 0)$, $(1, 0, 0)$, $(1, 1, 0)$, $(0, 1, 0)$, $(0, 0, 1)$, $(1, 0, 1)$, $(1, 1, 1)$, and $(0, 1, 1)$ when it is rotated by the given angle.

93. 45° about the z-axis

94. 90° about the x-axis

95. 30° about the x-axis

96. 120° about the z-axis

True or False? In Exercises 97–100, determine whether each statement is true or false. If a statement is true, give a reason or cite an appropriate statement from the text. If a statement is false, provide an example that shows the statement is not true in all cases or cite an appropriate statement from the text.

97. (a) Linear transformations called reflections that map a point in the xy-plane to its mirror image across the line $y = x$ are defined by the standard matrix $\begin{bmatrix} 1 & 0 \\ 0 & 1 \end{bmatrix}$ in $M_{2,2}$.

(b) The linear transformations called horizontal expansions or contractions are defined by the matrix $\begin{bmatrix} k & 0 \\ 0 & 1 \end{bmatrix}$ in $M_{2,2}$.

(c) The matrix $\begin{bmatrix} 1 & 0 & 0 \\ 0 & 1/2 & -\sqrt{3}/2 \\ 0 & \sqrt{3}/2 & 1/2 \end{bmatrix}$ would rotate a point 60° about the x-axis.

98. (a) Linear transformations called reflections that map a point in the xy-plane to its mirror image across the x-axis are defined by the matrix $\begin{bmatrix} 1 & 0 \\ 0 & -1 \end{bmatrix}$ in $M_{2,2}$.

(b) The linear transformations called vertical expansions or contractions are defined by the matrix $\begin{bmatrix} 1 & 0 \\ 0 & k \end{bmatrix}$ in $M_{2,2}$.

(c) The matrix $\begin{bmatrix} \sqrt{3}/2 & 0 & 1/2 \\ 0 & 1 & 0 \\ -1/2 & 0 & \sqrt{3}/2 \end{bmatrix}$ would rotate a point 30° about the y-axis.

99. (a) In calculus, any linear function is also a linear transformation from R^2 to R^2.

(b) A linear transformation is said to be onto if and only if for all $\mathbf{u}$ and $\mathbf{v}$ in V, $T(\mathbf{u}) = T(\mathbf{v})$ implies $\mathbf{u} = \mathbf{v}$.

(c) Because of the computational advantages, it is best to choose a basis for V such that the transformation matrix is diagonal.

100. (a) For polynomials, the differential operator D_x is a linear transformation from P_n to P_{n-1}.

(b) The set of all vectors $\mathbf{v}$ in V that satisfy $T(\mathbf{v}) = \mathbf{v}$ is called the kernel of T.

(c) The standard matrix A of the composition of two linear transformations $T(\mathbf{v}) = T_2(T_1(\mathbf{v}))$ is the product of the standard matrix for T_2 and the standard matrix for T_1.

CHAPTER 6 Projects

1 Reflections in the Plane (I)

Let ℓ be the line $ax + by = 0$ in the plane. The linear transformation $L: R^2 \rightarrow R^2$ that sends a point (x, y) to its mirror image in ℓ is called the **reflection** in ℓ. (See Figure 6.24.)

The goal of these two projects is to find the matrix for this reflection relative to the standard basis. The first project is based on transition matrices, and the second project uses projections.

1. Find the standard matrix for L for the line $x = 0$.
2. Find the standard matrix for L for the line $y = 0$.
3. Find the standard matrix for L for the line $x - y = 0$.
4. Consider the line ℓ represented by $x - 2y = 0$. Find a vector $\mathbf{v}$ parallel to ℓ and another vector $\mathbf{w}$ orthogonal to ℓ. Determine the matrix A for the reflection in ℓ relative to the ordered basis $\{\mathbf{v}, \mathbf{w}\}$. Finally, use the appropriate transition matrix to find the matrix for the reflection relative to the standard basis. Use this matrix to find $L(2, 1)$, $L(-1, 2)$, and $L(5, 0)$.
5. Consider the general line $ax + by = 0$. Let $\mathbf{v}$ be a vector parallel to ℓ, and let $\mathbf{w}$ be a vector orthogonal to ℓ. Determine the matrix A for the reflection in ℓ relative to the ordered basis $\{\mathbf{v}, \mathbf{w}\}$. Finally, use the appropriate transition matrix to find the matrix for L relative to the standard basis.
6. Find the standard matrix for the reflection in the line $3x + 4y = 0$. Use this matrix to find the images of the points $(3, 4)$, $(-4, 3)$, and $(0, 5)$.

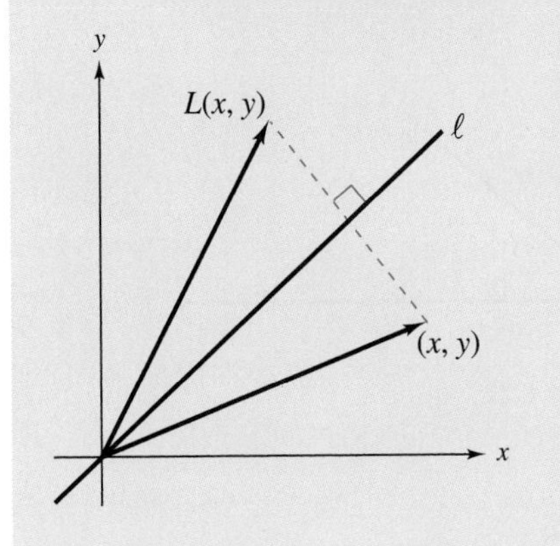

Figure 6.24

2 Reflections in the Plane (II)

In this second project, you will use projections to determine the standard matrix for the reflection L in the line $ax + by = 0$. (See Figure 6.25.) Recall that the projection of the vector $\mathbf{u}$ onto the vector $\mathbf{v}$ is represented by

$$\text{proj}_{\mathbf{v}}\mathbf{u} = \frac{\mathbf{u} \cdot \mathbf{v}}{\mathbf{v} \cdot \mathbf{v}}\mathbf{v}.$$

1. Find the standard matrix for the projection onto the y-axis. That is, find the standard matrix for $\text{proj}_{\mathbf{v}}\mathbf{u}$ if $\mathbf{v} = (0, 1)$.
2. Find the standard matrix for the projection onto the x-axis.
3. Consider the line ℓ represented by $x - 2y = 0$. Find a vector $\mathbf{v}$ parallel to ℓ and another vector $\mathbf{w}$ orthogonal to ℓ. Determine the matrix A for the projection onto ℓ relative to the ordered basis $\{\mathbf{v}, \mathbf{w}\}$. Finally, use the appropriate transition matrix to find the matrix for the projection relative to the standard basis. Use this matrix to find $\text{proj}_{\mathbf{v}}\mathbf{u}$ for the cases $\mathbf{u} = (2, 1)$, $\mathbf{u} = (-1, 2)$, and $\mathbf{u} = (5, 0)$.
4. Consider the general line $ax + by = 0$. Let $\mathbf{v}$ be a vector parallel to ℓ, and let $\mathbf{w}$ be a vector orthogonal to ℓ. Determine the matrix A for the projection onto ℓ relative to the ordered basis $\{\mathbf{v}, \mathbf{w}\}$. Finally, use the appropriate transition matrix to find the matrix for the projection relative to the standard basis.
5. Use Figure 6.26 to show that

   $$\text{proj}_{\mathbf{v}}\mathbf{u} = \tfrac{1}{2}(\mathbf{u} + L(\mathbf{u})),$$

   where L is the reflection in the line ℓ. Solve this equation for L and compare your answer with the formula from the first project.

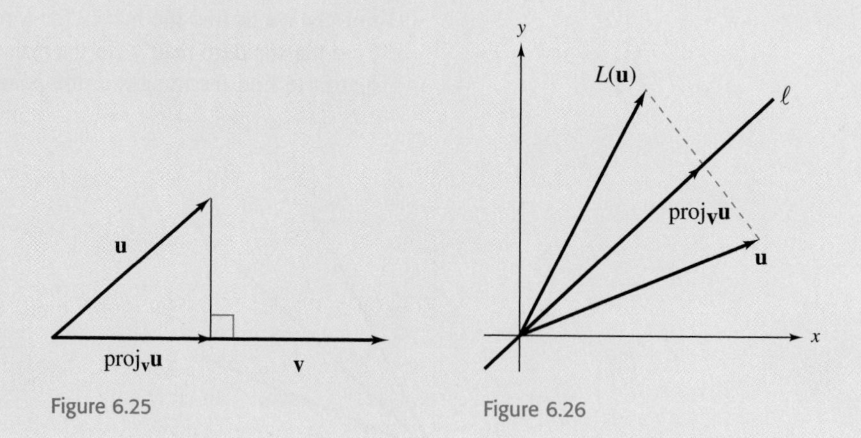

Figure 6.25

Figure 6.26

7 Eigenvalues and Eigenvectors

CHAPTER OBJECTIVES

- Find the eigenvalues and corresponding eigenvectors of a linear transformation, as well as the characteristic equation and the eigenvalues and corresponding eigenvectors of a matrix A.
- Demonstrate the Cayley-Hamilton Theorem for a matrix A.
- Find the eigenvalues of both an idempotent matrix and a nilpotent matrix.
- Determine whether a matrix is triangular, diagonalizable, symmetric, and/or orthogonal.
- Find (if possible) a nonsingular matrix P for a matrix A such that $P^{-1}AP$ is diagonal.
- Find a basis B (if possible) for the domain of a linear transformation T such that the matrix of T relative to B is diagonal.
- Find the eigenvalues of a symmetric matrix and determine the dimension of the corresponding eigenspace.
- Find an orthogonal matrix P that diagonalizes A.
- Find and use an age transition matrix and an age distribution vector to form a population model and find a stable age distribution for the population.
- Solve a system of first-order linear differential equations.
- Find a matrix of the quadratic form associated with a quadratic equation.
- Use the Principal Axes Theorem to perform a rotation of axes and eliminate the xy-, xz-, and yz-terms, and find the equation of the rotated quadratic surface.

7.1 Eigenvalues and Eigenvectors

This section presents one of the most important problems in linear algebra, the **eigenvalue problem.** Its central question can be stated as follows. If A is an $n \times n$ matrix, do nonzero vectors $\mathbf{x}$ in R^n exist such that $A\mathbf{x}$ is a scalar multiple of $\mathbf{x}$? The scalar, denoted by the Greek letter lambda (λ), is called an **eigenvalue** of the matrix A, and the nonzero vector $\mathbf{x}$ is called an **eigenvector** of A corresponding to λ. The terms *eigenvalue* and *eigenvector* are derived from the German word *Eigenwert*, meaning "proper value." So, you have

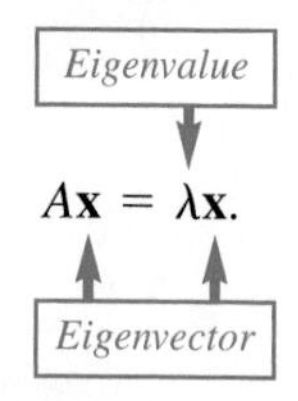

Although you looked at the eigenvalue problem briefly in Section 3.4, the approach in this chapter will not depend on that material.

Eigenvalues and eigenvectors have many important applications, some of which are discussed in Section 7.4. For now you will consider a geometric interpretation of the problem in R^2. If λ is an eigenvalue of a matrix A and $\mathbf{x}$ is an eigenvector of A corresponding to λ, then multiplication of $\mathbf{x}$ by the matrix A produces a vector $\lambda\mathbf{x}$ that is parallel to $\mathbf{x}$, as shown in Figure 7.1.

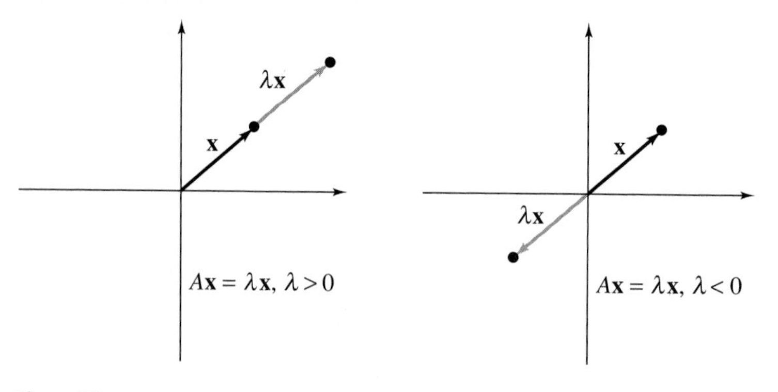

Figure 7.1

Definitions of Eigenvalue and Eigenvector

Let A be an $n \times n$ matrix. The scalar λ is called an **eigenvalue** of A if there is a *nonzero* vector $\mathbf{x}$ such that

$$A\mathbf{x} = \lambda\mathbf{x}.$$

The vector $\mathbf{x}$ is called an **eigenvector** of A corresponding to λ.

Only real eigenvalues are presented in this chapter.

REMARK: Note that an eigen*vector* cannot be zero. Allowing $\mathbf{x}$ to be the zero vector would render the definition meaningless, because $A\mathbf{0} = \lambda\mathbf{0}$ is true for all real values of λ. An eigen*value* of $\lambda = 0$, however, is possible. (See Example 2.)

A matrix can have more than one eigenvalue, as demonstrated in Examples 1 and 2.

EXAMPLE 1 Verifying Eigenvalues and Eigenvectors

For the matrix

$$A = \begin{bmatrix} 2 & 0 \\ 0 & -1 \end{bmatrix},$$

verify that $\mathbf{x}_1 = (1, 0)$ is an eigenvector of A corresponding to the eigenvalue $\lambda_1 = 2$, and that $\mathbf{x}_2 = (0, 1)$ is an eigenvector of A corresponding to the eigenvalue $\lambda_2 = -1$.

SOLUTION Multiplying $\mathbf{x}_1$ by A produces

$$A\mathbf{x}_1 = \begin{bmatrix} 2 & 0 \\ 0 & -1 \end{bmatrix}\begin{bmatrix} 1 \\ 0 \end{bmatrix} = \begin{bmatrix} 2 \\ 0 \end{bmatrix} = 2\begin{bmatrix} 1 \\ 0 \end{bmatrix}.$$

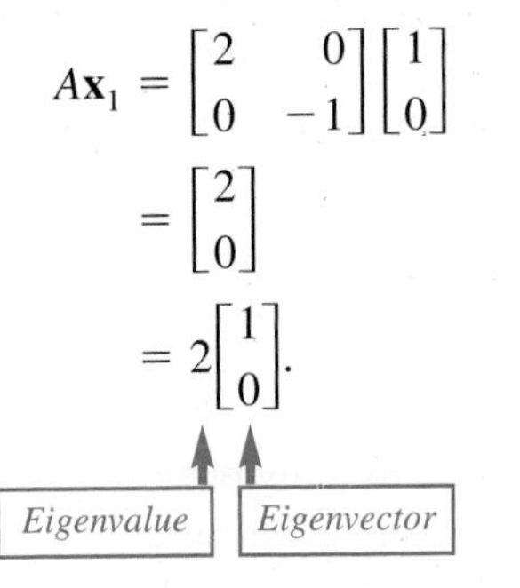

So, $\mathbf{x}_1 = (1, 0)$ is an eigenvector of A corresponding to the eigenvalue $\lambda_1 = 2$. Similarly, multiplying $\mathbf{x}_2$ by A produces

$$A\mathbf{x}_2 = \begin{bmatrix} 2 & 0 \\ 0 & -1 \end{bmatrix}\begin{bmatrix} 0 \\ 1 \end{bmatrix} = \begin{bmatrix} 0 \\ -1 \end{bmatrix} = -1\begin{bmatrix} 0 \\ 1 \end{bmatrix}.$$

So, $\mathbf{x}_2 = (0, 1)$ is an eigenvector of A corresponding to the eigenvalue $\lambda_2 = -1$.

EXAMPLE 2 Verifying Eigenvalues and Eigenvectors

For the matrix

$$A = \begin{bmatrix} 1 & -2 & 1 \\ 0 & 0 & 0 \\ 0 & 1 & 1 \end{bmatrix},$$

verify that

$$\mathbf{x}_1 = (-3, -1, 1) \quad \text{and} \quad \mathbf{x}_2 = (1, 0, 0)$$

are eigenvectors of A and find their corresponding eigenvalues.

SOLUTION Multiplying $\mathbf{x}_1$ by A produces

$$A\mathbf{x}_1 = \begin{bmatrix} 1 & -2 & 1 \\ 0 & 0 & 0 \\ 0 & 1 & 1 \end{bmatrix} \begin{bmatrix} -3 \\ -1 \\ 1 \end{bmatrix} = \begin{bmatrix} 0 \\ 0 \\ 0 \end{bmatrix} = 0 \begin{bmatrix} -3 \\ -1 \\ 1 \end{bmatrix}.$$

So, $\mathbf{x}_1 = (-3, -1, 1)$ is an eigenvector of A corresponding to the eigenvalue $\lambda_1 = 0$. Similarly, multiplying $\mathbf{x}_2$ by A produces

$$A\mathbf{x}_2 = \begin{bmatrix} 1 & -2 & 1 \\ 0 & 0 & 0 \\ 0 & 1 & 1 \end{bmatrix} \begin{bmatrix} 1 \\ 0 \\ 0 \end{bmatrix} = \begin{bmatrix} 1 \\ 0 \\ 0 \end{bmatrix} = 1 \begin{bmatrix} 1 \\ 0 \\ 0 \end{bmatrix}.$$

So, $\mathbf{x}_2 = (1, 0, 0)$ is an eigenvector of A corresponding to the eigenvalue $\lambda_2 = 1$.

Discovery

In Example 2, $\lambda_2 = 1$ is an eigenvalue of the matrix A. Calculate the determinant of the matrix $A - \lambda_2 I$, where I is the 3×3 identity matrix. Repeat this experiment for the other eigenvalue, $\lambda_1 = 0$. In general, if λ is an eigenvalue of the matrix A, what is the value of $|A - \lambda I|$?

Eigenspaces

Although Examples 1 and 2 list only one eigenvector for each eigenvalue, each of the four eigenvalues in Examples 1 and 2 has an infinite number of eigenvectors. For instance, in Example 1 the vectors $(2, 0)$ and $(-3, 0)$ are eigenvectors of A corresponding to the eigenvalue 2. In fact, if A is an $n \times n$ matrix with an eigenvalue λ and a corresponding eigenvector $\mathbf{x}$, then every nonzero scalar multiple of $\mathbf{x}$ is also an eigenvector of A. This may be seen by letting c be a nonzero scalar, which then produces

$$A(c\mathbf{x}) = c(A\mathbf{x}) = c(\lambda\mathbf{x}) = \lambda(c\mathbf{x}).$$

It is also true that if $\mathbf{x}_1$ and $\mathbf{x}_2$ are eigenvectors corresponding to the *same* eigenvalue λ, then their sum is also an eigenvector corresponding to λ, because

$$A(\mathbf{x}_1 + \mathbf{x}_2) = A\mathbf{x}_1 + A\mathbf{x}_2 = \lambda\mathbf{x}_1 + \lambda\mathbf{x}_2 = \lambda(\mathbf{x}_1 + \mathbf{x}_2).$$

In other words, the set of all eigenvectors of a given eigenvalue λ, together with the zero vector, is a subspace of R^n. This special subspace of R^n is called the **eigenspace** of λ.

THEOREM 7.1 Eigenvectors of λ Form a Subspace

If A is an $n \times n$ matrix with an eigenvalue λ, then the set of all eigenvectors of λ, together with the zero vector

$$\{\mathbf{0}\} \cup \{\mathbf{x}\colon \mathbf{x} \text{ is an eigenvector of } \lambda\},$$

is a subspace of R^n. This subspace is called the **eigenspace** of λ.

Determining the eigenvalues and corresponding eigenspaces of a matrix can be difficult. Occasionally, however, you can find eigenvalues and eigenspaces by simple inspection, as demonstrated in Example 3.

EXAMPLE 3 An Example of Eigenspaces in the Plane

Find the eigenvalues and corresponding eigenspaces of

$$A = \begin{bmatrix} -1 & 0 \\ 0 & 1 \end{bmatrix}.$$

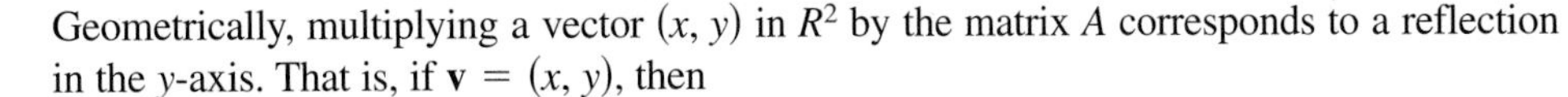

SOLUTION Geometrically, multiplying a vector (x, y) in R^2 by the matrix A corresponds to a reflection in the y-axis. That is, if $\mathbf{v} = (x, y)$, then

$$A\mathbf{v} = \begin{bmatrix} -1 & 0 \\ 0 & 1 \end{bmatrix}\begin{bmatrix} x \\ y \end{bmatrix} = \begin{bmatrix} -x \\ y \end{bmatrix}.$$

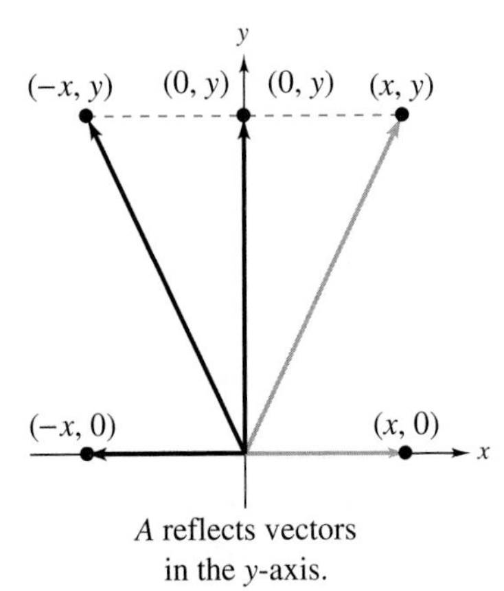

A reflects vectors in the y-axis.

Figure 7.2

Figure 7.2 illustrates that the only vectors reflected onto scalar multiples of themselves are those lying on either the x-axis or the y-axis.

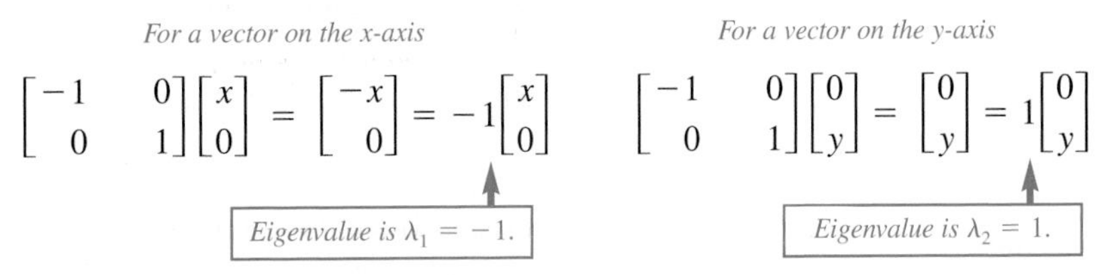

For a vector on the x-axis
For a vector on the y-axis
$\begin{bmatrix} -1 & 0 \\ 0 & 1 \end{bmatrix}\begin{bmatrix} x \\ 0 \end{bmatrix} = \begin{bmatrix} -x \\ 0 \end{bmatrix} = -1\begin{bmatrix} x \\ 0 \end{bmatrix}$
$\begin{bmatrix} -1 & 0 \\ 0 & 1 \end{bmatrix}\begin{bmatrix} 0 \\ y \end{bmatrix} = \begin{bmatrix} 0 \\ y \end{bmatrix} = 1\begin{bmatrix} 0 \\ y \end{bmatrix}$
Eigenvalue is $\lambda_1 = -1$.
Eigenvalue is $\lambda_2 = 1$.

So, the eigenvectors corresponding to $\lambda_1 = -1$ are the nonzero vectors on the x-axis, and the eigenvectors corresponding to $\lambda_2 = 1$ are the nonzero vectors on the y-axis. This implies that the eigenspace corresponding to $\lambda_1 = -1$ is the x-axis, and that the eigenspace corresponding to $\lambda_2 = 1$ is the y-axis.

Finding Eigenvalues and Eigenvectors

The geometric solution in Example 3 is not typical of the general eigenvalue problem. A general approach will now be described.

To find the eigenvalues and eigenvectors of an $n \times n$ matrix A, let I be the $n \times n$ identity matrix. Writing the equation $A\mathbf{x} = \lambda\mathbf{x}$ in the form $\lambda I\mathbf{x} = A\mathbf{x}$ then produces

$$(\lambda I - A)\mathbf{x} = \mathbf{0}.$$

This homogeneous system of equations has nonzero solutions if and only if the coefficient matrix $(\lambda I - A)$ is *not* invertible—that is, if and only if the determinant of $(\lambda I - A)$ is zero. This is formally stated in the next theorem.

THEOREM 7.2 Eigenvalues and Eigenvectors of a Matrix

Let A be an $n \times n$ matrix.

1. An eigenvalue of A is a scalar λ such that
$$\det(\lambda I - A) = 0.$$
2. The eigenvectors of A corresponding to λ are the nonzero solutions of
$$(\lambda I - A)\mathbf{x} = \mathbf{0}.$$

The equation $\det(\lambda I - A) = 0$ is called the **characteristic equation** of A. Moreover, when expanded to polynomial form, the polynomial

$$|\lambda I - A| = \lambda^n + c_{n-1}\lambda^{n-1} + \cdot \cdot \cdot + c_1\lambda + c_0$$

is called the **characteristic polynomial** of A. This definition tells you that the eigenvalues of an $n \times n$ matrix A correspond to the roots of the characteristic polynomial of A. Because the characteristic polynomial of A is of degree n, A can have at most n distinct eigenvalues.

REMARK: The Fundamental Theorem of Algebra states that an nth-degree polynomial has precisely n roots. These n roots, however, include both repeated and complex roots. In this chapter you will be concerned only with the real roots of characteristic polynomials—that is, real eigenvalues.

EXAMPLE 4 Finding Eigenvalues and Eigenvectors

Find the eigenvalues and corresponding eigenvectors of

$$A = \begin{bmatrix} 2 & -12 \\ 1 & -5 \end{bmatrix}.$$

SOLUTION The characteristic polynomial of A is

$$\begin{aligned} |\lambda I - A| &= \begin{vmatrix} \lambda - 2 & 12 \\ -1 & \lambda + 5 \end{vmatrix} \\ &= (\lambda - 2)(\lambda + 5) - (-12) \\ &= \lambda^2 + 3\lambda - 10 + 12 \\ &= \lambda^2 + 3\lambda + 2 \\ &= (\lambda + 1)(\lambda + 2). \end{aligned}$$

So, the characteristic equation is $(\lambda + 1)(\lambda + 2) = 0$, which gives $\lambda_1 = -1$ and $\lambda_2 = -2$ as the eigenvalues of A. To find the corresponding eigenvectors, use Gauss-Jordan elimination to solve the homogeneous linear system represented by $(\lambda I - A)\mathbf{x} = \mathbf{0}$ twice: first for $\lambda = \lambda_1 = -1$, and then for $\lambda = \lambda_2 = -2$. For $\lambda_1 = -1$, the coefficient matrix is

$$(-1)I - A = \begin{bmatrix} -1 - 2 & 12 \\ -1 & -1 + 5 \end{bmatrix} = \begin{bmatrix} -3 & 12 \\ -1 & 4 \end{bmatrix},$$

which row reduces to

$$\begin{bmatrix} 1 & -4 \\ 0 & 0 \end{bmatrix},$$

showing that $x_1 - 4x_2 = 0$. Letting $x_2 = t$, you can conclude that every eigenvector of λ_1 is of the form

$$\mathbf{x} = \begin{bmatrix} x_1 \\ x_2 \end{bmatrix} = \begin{bmatrix} 4t \\ t \end{bmatrix} = t\begin{bmatrix} 4 \\ 1 \end{bmatrix}, \quad t \neq 0.$$

For $\lambda_2 = -2$, you have

$$(-2)I - A = \begin{bmatrix} -2 - 2 & 12 \\ -1 & -2 + 5 \end{bmatrix} = \begin{bmatrix} -4 & 12 \\ -1 & 3 \end{bmatrix} \longrightarrow \begin{bmatrix} 1 & -3 \\ 0 & 0 \end{bmatrix}.$$

Letting $x_2 = t$, you can conclude that every eigenvector of λ_2 is of the form

$$\mathbf{x} = \begin{bmatrix} x_1 \\ x_2 \end{bmatrix} = \begin{bmatrix} 3t \\ t \end{bmatrix} = t\begin{bmatrix} 3 \\ 1 \end{bmatrix}, \quad t \neq 0.$$

Try checking $A\mathbf{x} = \lambda_i \mathbf{x}$ for the eigenvalues and eigenvectors in this example.

The homogeneous systems that arise when you are finding eigenvectors will always row reduce to a matrix having at least one row of zeros, because the systems must have nontrivial solutions. The steps used to find the eigenvalues and corresponding eigenvectors of a matrix are summarized as follows.

Finding Eigenvalues and Eigenvectors

Let A be an $n \times n$ matrix.

1. Form the characteristic equation $|\lambda I - A| = 0$. It will be a polynomial equation of degree n in the variable λ.
2. Find the real roots of the characteristic equation. These are the eigenvalues of A.
3. For each eigenvalue λ_i, find the eigenvectors corresponding to λ_i by solving the homogeneous system $(\lambda_i I - A)\mathbf{x} = \mathbf{0}$. This requires row reducing of an $n \times n$ matrix. The resulting reduced row-echelon form must have at least one row of zeros.

Finding the eigenvalues of an $n \times n$ matrix can be difficult because it involves the factorization of an nth-degree polynomial. Once an eigenvalue has been found, however, finding the corresponding eigenvectors is a straightforward application of Gauss-Jordan reduction.

EXAMPLE 5 **Finding Eigenvalues and Eigenvectors**

Find the eigenvalues and corresponding eigenvectors of

$$A = \begin{bmatrix} 2 & 1 & 0 \\ 0 & 2 & 0 \\ 0 & 0 & 2 \end{bmatrix}.$$

What is the dimension of the eigenspace of each eigenvalue?

SOLUTION The characteristic polynomial of A is

$$|\lambda I - A| = \begin{vmatrix} \lambda - 2 & -1 & 0 \\ 0 & \lambda - 2 & 0 \\ 0 & 0 & \lambda - 2 \end{vmatrix} = (\lambda - 2)^3.$$

So, the characteristic equation is $(\lambda - 2)^3 = 0$.

So, the only eigenvalue is $\lambda = 2$. To find the eigenvectors of $\lambda = 2$, solve the homogeneous linear system represented by $(2I - A)\mathbf{x} = \mathbf{0}$.

$$2I - A = \begin{bmatrix} 0 & -1 & 0 \\ 0 & 0 & 0 \\ 0 & 0 & 0 \end{bmatrix}$$

This implies that $x_2 = 0$. Using the parameters $s = x_1$ and $t = x_3$, you can find that the eigenvectors of $\lambda = 2$ are of the form

$$\mathbf{x} = \begin{bmatrix} x_1 \\ x_2 \\ x_3 \end{bmatrix} = \begin{bmatrix} s \\ 0 \\ t \end{bmatrix} = s\begin{bmatrix} 1 \\ 0 \\ 0 \end{bmatrix} + t\begin{bmatrix} 0 \\ 0 \\ 1 \end{bmatrix}, \quad s \text{ and } t \text{ not both zero.}$$

Because $\lambda = 2$ has two linearly independent eigenvectors, the dimension of its eigenspace is 2.

If an eigenvalue λ_1 occurs as a *multiple root* (k times) of the characteristic polynomial, then λ_1 has **multiplicity** k. This implies that $(\lambda - \lambda_1)^k$ is a factor of the characteristic polynomial and $(\lambda - \lambda_1)^{k+1}$ is not a factor of the characteristic polynomial. For instance, in Example 5 the eigenvalue $\lambda = 2$ has a multiplicity of 3.

Also note that in Example 5 the dimension of the eigenspace of $\lambda = 2$ is 2. In general, the multiplicity of an eigenvalue is greater than or equal to the dimension of its eigenspace.

EXAMPLE 6 Finding Eigenvalues and Eigenvectors

Find the eigenvalues of

$$A = \begin{bmatrix} 1 & 0 & 0 & 0 \\ 0 & 1 & 5 & -10 \\ 1 & 0 & 2 & 0 \\ 1 & 0 & 0 & 3 \end{bmatrix}$$

and find a basis for each of the corresponding eigenspaces.

SOLUTION The characteristic polynomial of A is

$$|\lambda I - A| = \begin{vmatrix} \lambda - 1 & 0 & 0 & 0 \\ 0 & \lambda - 1 & -5 & 10 \\ -1 & 0 & \lambda - 2 & 0 \\ -1 & 0 & 0 & \lambda - 3 \end{vmatrix}$$
$$= (\lambda - 1)^2(\lambda - 2)(\lambda - 3).$$

So, the characteristic equation is $(\lambda - 1)^2(\lambda - 2)(\lambda - 3) = 0$ and the eigenvalues are $\lambda_1 = 1$, $\lambda_2 = 2$, and $\lambda_3 = 3$. (Note that $\lambda_1 = 1$ has a multiplicity of 2.)

You can find a basis for the eigenspace of $\lambda_1 = 1$ as follows.

$$(1)I - A = \begin{bmatrix} 0 & 0 & 0 & 0 \\ 0 & 0 & -5 & 10 \\ -1 & 0 & -1 & 0 \\ -1 & 0 & 0 & -2 \end{bmatrix} \longrightarrow \begin{bmatrix} 1 & 0 & 0 & 2 \\ 0 & 0 & 1 & -2 \\ 0 & 0 & 0 & 0 \\ 0 & 0 & 0 & 0 \end{bmatrix}$$

Letting $s = x_2$ and $t = x_4$ produces

$$\mathbf{x} = \begin{bmatrix} x_1 \\ x_2 \\ x_3 \\ x_4 \end{bmatrix} = \begin{bmatrix} 0s - 2t \\ s + 0t \\ 0s + 2t \\ 0s + t \end{bmatrix} = s\begin{bmatrix} 0 \\ 1 \\ 0 \\ 0 \end{bmatrix} + t\begin{bmatrix} -2 \\ 0 \\ 2 \\ 1 \end{bmatrix}.$$

A basis for the eigenspace corresponding to $\lambda_1 = 1$ is

$B_1 = \{(0, 1, 0, 0), (-2, 0, 2, 1)\}$. **Basis for $\lambda_1 = 1$**

For $\lambda_2 = 2$ and $\lambda_3 = 3$, follow the same pattern to obtain the eigenspace bases

$B_2 = \{(0, 5, 1, 0)\}$ **Basis for $\lambda_2 = 2$**

$B_3 = \{(0, -5, 0, 1)\}$. **Basis for $\lambda_3 = 3$**

Finding eigenvalues and eigenvectors of matrices of order $n \geq 4$ can be tedious. Moreover, the procedure followed in Example 6 is generally inefficient when used on a computer, because finding roots on a computer is both time consuming and subject to roundoff error. Consequently, numerical methods of approximating the eigenvalues of large matrices are required. These numerical methods can be found in texts on advanced linear algebra and numerical analysis.

Technology Note

Many computer software programs and graphing utilities have built-in programs to approximate the eigenvalues and eigenvectors of an $n \times n$ matrix. If you enter the matrix A from Example 6, you should obtain the four eigenvalues

$$\{1\ 1\ 2\ 3\}.$$

Your computer software program or graphing utility should also be able to produce a matrix in which the columns are the corresponding eigenvectors, which are sometimes scalar multiples of those you would obtain by hand calculations. Keystrokes and programming syntax for these utilities/programs applicable to Example 6 are provided in the **Online Technology Guide,** available at *college.hmco.com/pic/larsonELA6e.*

There are a few types of matrices for which eigenvalues are easy to find. The next theorem states that the eigenvalues of an $n \times n$ triangular matrix are the entries on the main diagonal. Its proof follows from the fact that the determinant of a triangular matrix is the product of its diagonal elements.

THEOREM 7.3 Eigenvalues of Triangular Matrices

If A is an $n \times n$ triangular matrix, then its eigenvalues are the entries on its main diagonal.

EXAMPLE 7 Finding Eigenvalues of Diagonal and Triangular Matrices

Find the eigenvalues of each matrix.

(a) $A = \begin{bmatrix} 2 & 0 & 0 \\ -1 & 1 & 0 \\ 5 & 3 & -3 \end{bmatrix}$ (b) $A = \begin{bmatrix} -1 & 0 & 0 & 0 & 0 \\ 0 & 2 & 0 & 0 & 0 \\ 0 & 0 & 0 & 0 & 0 \\ 0 & 0 & 0 & -4 & 0 \\ 0 & 0 & 0 & 0 & 3 \end{bmatrix}$

SOLUTION (a) Without using Theorem 7.3, you can find that

$$|\lambda I - A| = \begin{vmatrix} \lambda - 2 & 0 & 0 \\ 1 & \lambda - 1 & 0 \\ -5 & -3 & \lambda + 3 \end{vmatrix}$$

$$= (\lambda - 2)(\lambda - 1)(\lambda + 3).$$

So, the eigenvalues are $\lambda_1 = 2$, $\lambda_2 = 1$, and $\lambda_3 = -3$, which are simply the main diagonal entries of A.

(b) In this case, use Theorem 7.3 to conclude that the eigenvalues are the main diagonal entries $\lambda_1 = -1$, $\lambda_2 = 2$, $\lambda_3 = 0$, $\lambda_4 = -4$, and $\lambda_5 = 3$.

Eigenvalues and Eigenvectors of Linear Transformations

This section began with definitions of eigenvalues and eigenvectors in terms of matrices. They can also be defined in terms of linear transformations. A number λ is called an **eigenvalue** of a linear transformation $T: V \to V$ if there is a nonzero vector $\mathbf{x}$ such that $T(\mathbf{x}) = \lambda\mathbf{x}$. The vector $\mathbf{x}$ is called an **eigenvector** of T corresponding to λ, and the set of all eigenvectors of λ (with the zero vector) is called the **eigenspace** of λ.

Consider the linear transformation $T: R^3 \to R^3$, whose matrix relative to the standard basis is

$$A = \begin{bmatrix} 1 & 3 & 0 \\ 3 & 1 & 0 \\ 0 & 0 & -2 \end{bmatrix}.$$

Standard basis:
$B = \{(1, 0, 0), (0, 1, 0), (0, 0, 1)\}$

In Example 5 of Section 6.4, you found that the matrix of T relative to the basis B' is the diagonal matrix

$$A' = \begin{bmatrix} 4 & 0 & 0 \\ 0 & -2 & 0 \\ 0 & 0 & -2 \end{bmatrix}.$$

Nonstandard basis:
$B' = \{(1, 1, 0), (1, -1, 0), (0, 0, 1)\}$

The question now is: "For a given transformation T, can you find a basis B' whose corresponding matrix is diagonal?" The next example gives an indication of the answer.

EXAMPLE 8 Finding Eigenvalues and Eigenspaces

Find the eigenvalues and corresponding eigenspaces of

$$A = \begin{bmatrix} 1 & 3 & 0 \\ 3 & 1 & 0 \\ 0 & 0 & -2 \end{bmatrix}.$$

SOLUTION Because

$$\begin{aligned} |\lambda I - A| &= \begin{vmatrix} \lambda - 1 & -3 & 0 \\ -3 & \lambda - 1 & 0 \\ 0 & 0 & \lambda + 2 \end{vmatrix} \\ &= (\lambda + 2)[(\lambda - 1)^2 - 9] \\ &= (\lambda + 2)(\lambda^2 - 2\lambda - 8) = (\lambda + 2)^2(\lambda - 4), \end{aligned}$$

the eigenvalues of A are $\lambda_1 = 4$ and $\lambda_2 = -2$. The eigenspaces for these two eigenvalues are as follows.

$B_1 = \{(1, 1, 0)\}$ Basis for $\lambda_1 = 4$

$B_2 = \{(1, -1, 0), (0, 0, 1)\}$ Basis for $\lambda_2 = -2$

Example 8 illustrates two important and perhaps surprising results.

1. Let $T: R^3 \rightarrow R^3$ be the linear transformation whose standard matrix is A, and let B' be the basis of R^3 made up of the three linearly independent eigenvectors found in Example 8. Then A', the matrix of T relative to the basis B', is diagonal.

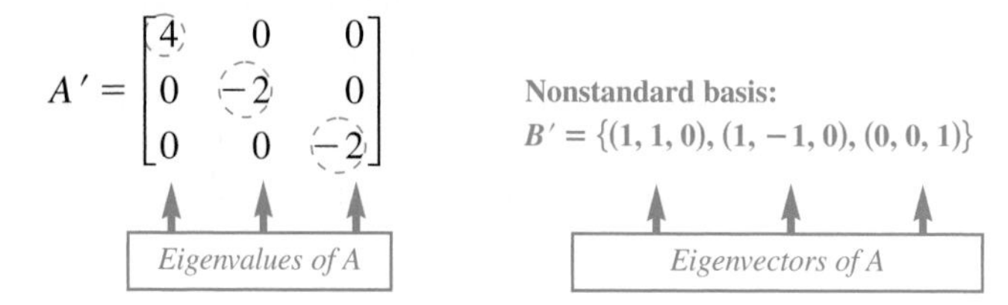

2. The main diagonal entries of the matrix A' are the eigenvalues of A.

The next section formalizes these two results and also characterizes linear transformations that can be represented by diagonal matrices.

SECTION 7.1 Exercises

In Exercises 1–8, verify that λ_i is an eigenvalue of A and that $\mathbf{x}_i$ is a corresponding eigenvector.

1. $A = \begin{bmatrix} 1 & 0 \\ 0 & -1 \end{bmatrix}$, $\lambda_1 = 1, \mathbf{x}_1 = (1, 0)$; $\lambda_2 = -1, \mathbf{x}_2 = (0, 1)$

2. $A = \begin{bmatrix} 4 & -5 \\ 2 & -3 \end{bmatrix}$, $\lambda_1 = -1, \mathbf{x}_1 = (1, 1)$; $\lambda_2 = 2, \mathbf{x}_2 = (5, 2)$

3. $A = \begin{bmatrix} 1 & 1 \\ 1 & 1 \end{bmatrix}$, $\lambda_1 = 0, \mathbf{x}_1 = (1, -1)$; $\lambda_2 = 2, \mathbf{x}_2 = (1, 1)$

4. $A = \begin{bmatrix} -2 & 4 \\ 1 & 1 \end{bmatrix}$, $\lambda_1 = 2, \mathbf{x}_1 = (1, 1)$; $\lambda_2 = -3, \mathbf{x}_2 = (-4, 1)$

5. $A = \begin{bmatrix} 2 & 3 & 1 \\ 0 & -1 & 2 \\ 0 & 0 & 3 \end{bmatrix}$, $\lambda_1 = 2, \mathbf{x}_1 = (1, 0, 0)$; $\lambda_2 = -1, \mathbf{x}_2 = (1, -1, 0)$; $\lambda_3 = 3, \mathbf{x}_3 = (5, 1, 2)$

6. $A = \begin{bmatrix} -2 & 2 & -3 \\ 2 & 1 & -6 \\ -1 & -2 & 0 \end{bmatrix}$, $\lambda_1 = 5, \mathbf{x}_1 = (1, 2, -1)$; $\lambda_2 = -3, \mathbf{x}_2 = (-2, 1, 0)$; $\lambda_3 = -3, \mathbf{x}_3 = (3, 0, 1)$

7. $A = \begin{bmatrix} 0 & 1 & 0 \\ 0 & 0 & 1 \\ 1 & 0 & 0 \end{bmatrix}$, $\lambda_1 = 1, \mathbf{x}_1 = (1, 1, 1)$

8. $A = \begin{bmatrix} 4 & -1 & 3 \\ 0 & 2 & 1 \\ 0 & 0 & 3 \end{bmatrix}$, $\lambda_1 = 4, \mathbf{x}_1 = (1, 0, 0)$; $\lambda_2 = 2, \mathbf{x}_2 = (1, 2, 0)$; $\lambda_3 = 3, \mathbf{x}_3 = (-2, 1, 1)$

9. Use A, λ_i, and $\mathbf{x}_i$ from Exercise 3 to show that
(a) $A(c\mathbf{x}_1) = 0(c\mathbf{x}_1)$ for any real number c.
(b) $A(c\mathbf{x}_2) = 2(c\mathbf{x}_2)$ for any real number c.

10. Use A, λ_i, and $\mathbf{x}_i$ from Exercise 5 to show that
(a) $A(c\mathbf{x}_1) = 2(c\mathbf{x}_1)$ for any real number c.
(b) $A(c\mathbf{x}_2) = -(c\mathbf{x}_2)$ for any real number c.
(c) $A(c\mathbf{x}_3) = 3(c\mathbf{x}_3)$ for any real number c.

In Exercises 11–14, determine whether $\mathbf{x}$ is an eigenvector of A.

11. $A = \begin{bmatrix} 7 & 2 \\ 2 & 4 \end{bmatrix}$

(a) $\mathbf{x} = (1, 2)$
(b) $\mathbf{x} = (2, 1)$
(c) $\mathbf{x} = (1, -2)$
(d) $\mathbf{x} = (-1, 0)$

12. $A = \begin{bmatrix} -3 & 10 \\ 5 & 2 \end{bmatrix}$

(a) $\mathbf{x} = (4, 4)$
(b) $\mathbf{x} = (-8, 4)$
(c) $\mathbf{x} = (-4, 8)$
(d) $\mathbf{x} = (5, -3)$

13. $A = \begin{bmatrix} -1 & -1 & 1 \\ -2 & 0 & -2 \\ 3 & -3 & 1 \end{bmatrix}$

(a) $\mathbf{x} = (2, -4, 6)$
(b) $\mathbf{x} = (2, 0, 6)$
(c) $\mathbf{x} = (2, 2, 0)$
(d) $\mathbf{x} = (-1, 0, 1)$

14. $A = \begin{bmatrix} 1 & 0 & 5 \\ 0 & -2 & 4 \\ 1 & -2 & 9 \end{bmatrix}$

(a) $\mathbf{x} = (1, 1, 0)$
(b) $\mathbf{x} = (-5, 2, 1)$
(c) $\mathbf{x} = (0, 0, 0)$
(d) $\mathbf{x} = \left(2\sqrt{6} - 3, -2\sqrt{6} + 6, 3\right)$

In Exercises 15–28, find (a) the characteristic equation and (b) the eigenvalues (and corresponding eigenvectors) of the matrix.

15. $\begin{bmatrix} 6 & -3 \\ -2 & 1 \end{bmatrix}$

16. $\begin{bmatrix} 1 & -4 \\ -2 & 8 \end{bmatrix}$

17. $\begin{bmatrix} 1 & -\frac{3}{2} \\ \frac{1}{2} & -1 \end{bmatrix}$

18. $\begin{bmatrix} \frac{1}{4} & \frac{1}{4} \\ \frac{1}{2} & 0 \end{bmatrix}$

19. $\begin{bmatrix} 2 & 0 & 1 \\ 0 & 3 & 4 \\ 0 & 0 & 1 \end{bmatrix}$

20. $\begin{bmatrix} -5 & 0 & 0 \\ 3 & 7 & 0 \\ 4 & -2 & 3 \end{bmatrix}$

21. $\begin{bmatrix} 2 & -2 & 3 \\ 0 & 3 & -2 \\ 0 & -1 & 2 \end{bmatrix}$

22. $\begin{bmatrix} 3 & 2 & 1 \\ 0 & 0 & 2 \\ 0 & 2 & 0 \end{bmatrix}$

23. $\begin{bmatrix} 1 & 2 & -2 \\ -2 & 5 & -2 \\ -6 & 6 & -3 \end{bmatrix}$

24. $\begin{bmatrix} 3 & 2 & -3 \\ -3 & -4 & 9 \\ -1 & -2 & 5 \end{bmatrix}$

25. $\begin{bmatrix} 0 & -3 & 5 \\ -4 & 4 & -10 \\ 0 & 0 & 4 \end{bmatrix}$

26. $\begin{bmatrix} 1 & -\frac{3}{2} & \frac{5}{2} \\ -2 & \frac{13}{2} & -10 \\ \frac{3}{2} & -\frac{9}{2} & 8 \end{bmatrix}$

27. $\begin{bmatrix} 2 & 0 & 0 & 0 \\ 0 & 2 & 0 & 0 \\ 0 & 0 & 3 & 0 \\ 0 & 0 & 4 & 0 \end{bmatrix}$

28. $\begin{bmatrix} 3 & 0 & 0 & 0 \\ 4 & 1 & 0 & 0 \\ 0 & 0 & 2 & 1 \\ 0 & 0 & 0 & 2 \end{bmatrix}$

In Exercises 29–40, use a graphing utility with matrix capabilities or a computer software program to find the eigenvalues of the matrix.

29. $\begin{bmatrix} -4 & 5 \\ -2 & 3 \end{bmatrix}$

30. $\begin{bmatrix} 2 & 3 \\ 3 & -6 \end{bmatrix}$

31. $\begin{bmatrix} 7 & 2 \\ -2 & 3 \end{bmatrix}$

32. $\begin{bmatrix} -6 & 2 \\ 3 & -1 \end{bmatrix}$

33. $\begin{bmatrix} 0 & -\frac{1}{2} & 5 \\ -\frac{1}{3} & -\frac{1}{6} & -\frac{1}{4} \\ 0 & 0 & 4 \end{bmatrix}$

34. $\begin{bmatrix} \frac{1}{2} & 0 & 5 \\ -2 & \frac{1}{5} & \frac{1}{4} \\ 0 & 0 & 3 \end{bmatrix}$

35. $\begin{bmatrix} 2 & 4 & 2 \\ 1 & 0 & 1 \\ 1 & -4 & 5 \end{bmatrix}$

36. $\begin{bmatrix} 1 & 2 & -1 \\ 1 & 0 & 1 \\ 1 & -1 & 2 \end{bmatrix}$

37. $\begin{bmatrix} 1 & 0 & -1 & 1 \\ 0 & 1 & 0 & 1 \\ -2 & 0 & 2 & -2 \\ 0 & 2 & 0 & 2 \end{bmatrix}$

38. $\begin{bmatrix} 1 & -3 & 3 & 3 \\ -1 & 4 & -3 & -3 \\ -2 & 0 & 1 & 1 \\ 1 & 0 & 0 & 0 \end{bmatrix}$

39. $\begin{bmatrix} 1 & 1 & 2 & 3 \\ 2 & 2 & 4 & 6 \\ 3 & 3 & 6 & 9 \\ 4 & 4 & 8 & 12 \end{bmatrix}$

40. $\begin{bmatrix} 1 & 1 & 0 & 0 \\ 4 & 4 & 0 & 0 \\ 0 & 0 & 1 & 1 \\ 0 & 0 & 2 & 2 \end{bmatrix}$

In Exercises 41–48, demonstrate the Cayley-Hamilton Theorem for the given matrix. The **Cayley-Hamilton Theorem** states that a matrix satisfies its characteristic equation. For example, the characteristic equation of

$$A = \begin{bmatrix} 1 & -3 \\ 2 & 5 \end{bmatrix}$$

is $\lambda^2 - 6\lambda + 11 = 0$, and by the theorem you have $A^2 - 6A + 11I_2 = O$.

41. $\begin{bmatrix} 4 & 0 \\ -3 & 2 \end{bmatrix}$

42. $\begin{bmatrix} 6 & -1 \\ 1 & 5 \end{bmatrix}$

43. $\begin{bmatrix} 2 & -2 \\ 1 & 5 \end{bmatrix}$

44. $\begin{bmatrix} 4 & 1 \\ -2 & 1 \end{bmatrix}$

45. $\begin{bmatrix} 0 & 2 & -1 \\ -1 & 3 & 1 \\ 0 & 0 & -1 \end{bmatrix}$

46. $\begin{bmatrix} 3 & 1 & 4 \\ 2 & 4 & 0 \\ 5 & 5 & 6 \end{bmatrix}$

47. $\begin{bmatrix} 1 & 0 & -4 \\ 0 & 3 & 1 \\ 2 & 0 & 1 \end{bmatrix}$

48. $\begin{bmatrix} -3 & 1 & 0 \\ -1 & 3 & 2 \\ 0 & 4 & 3 \end{bmatrix}$

49. Perform the computational checks listed below on the eigenvalues found in Exercises 15–27 odd.

(a) The sum of the n eigenvalues equals the sum of the diagonal entries of the matrix. (This sum is called the **trace** of A.)

(b) The product of the n eigenvalues equals $|A|$.

(If λ is an eigenvalue of multiplicity k, remember to enter it k times in the sum or product of these checks.)

50. Perform the computational checks listed below on the eigenvalues found in Exercises 16–28 even.

(a) The sum of the n eigenvalues equals the sum of the diagonal entries of the matrix. (This sum is called the **trace** of A.)

(b) The product of the n eigenvalues equals $|A|$.

(If λ is an eigenvalue of multiplicity k, remember to enter it k times in the sum or product of these checks.)

51. Show that if A is an $n \times n$ matrix whose ith row is identical to the ith row of I, then 1 is an eigenvalue of A.

52. Prove that $\lambda = 0$ is an eigenvalue of A if and only if A is singular.

53. Writing For an invertible matrix A, prove that A and A^{-1} have the same eigenvectors. How are the eigenvalues of A related to the eigenvalues of A^{-1}?

54. Writing Prove that A and A^T have the same eigenvalues. Are the eigenspaces the same?

55. Prove that the constant term of the characteristic polynomial is $\pm|A|$.

56. Let $T: R^2 \rightarrow R^2$ be represented by $T(\mathbf{v}) = \text{proj}_{\mathbf{u}}\mathbf{v}$, where $\mathbf{u}$ is a fixed vector in R^2. Show that the eigenvalues of A (the standard matrix of T) are 0 and 1.

57. Guided Proof Prove that a triangular matrix is nonsingular if and only if its eigenvalues are real and nonzero.

Getting Started: Because this is an "if and only if" statement, you must prove that the statement is true in both directions. Review Theorems 3.2 and 3.7.

(i) To prove the statement in one direction, assume that the triangular matrix A is nonsingular. Use your knowledge of nonsingular and triangular matrices and determinants to conclude that the entries on the main diagonal of A are nonzero.

(ii) Because A is triangular, you can use Theorem 7.3 and part (i) to conclude that the eigenvalues are real and nonzero.

(iii) To prove the statement in the other direction, assume that the eigenvalues of the triangular matrix A are real and nonzero. Repeat parts (i) and (ii) in reverse order to prove that A is nonsingular.

58. Guided Proof Prove that if $A^2 = O$, then 0 is the only eigenvalue of A.

Getting Started: You need to show that if there exists a nonzero vector $\mathbf{x}$ and a real number λ such that $A\mathbf{x} = \lambda\mathbf{x}$, then if $A^2 = O$, λ must be zero.

(i) Because $A^2 = A \cdot A$, you can write $A^2\mathbf{x}$ as $A(A\mathbf{x})$.

(ii) Use the fact that $A\mathbf{x} = \lambda\mathbf{x}$ and the properties of matrix multiplication to conclude that $A^2\mathbf{x} = \lambda^2\mathbf{x}$.

(iii) Because A^2 is a zero matrix, you can conclude that λ must be zero.

59. If the eigenvalues of

$$A = \begin{bmatrix} a & b \\ 0 & d \end{bmatrix}$$

are $\lambda_1 = 0$ and $\lambda_2 = 1$, what are the possible values of a and d?

60. Show that

$$A = \begin{bmatrix} 0 & 1 \\ -1 & 0 \end{bmatrix}$$

has no real eigenvalues.

True or False? In Exercises 61 and 62, determine whether each statement is true or false. If a statement is true, give a reason or cite an appropriate statement from the text. If a statement is false, provide an example that shows the statement is not true in all cases or cite an appropriate statement from the text.

61. (a) The scalar λ is an eigenvalue of an $n \times n$ matrix A if there exists a vector $\mathbf{x}$ such that $A\mathbf{x} = \lambda\mathbf{x}$.

(b) If A is an $n \times n$ matrix with eigenvalue λ and corresponding eigenvector $\mathbf{x}$, then every nonzero scalar multiple of $\mathbf{x}$ is also an eigenvector of A.

(c) To find the eigenvalue(s) of an $n \times n$ matrix A, you can solve the characteristic equation, $\det(\lambda I - A) = 0$.

62. (a) Geometrically, if λ is an eigenvalue of a matrix A and $\mathbf{x}$ is an eigenvector of A corresponding to λ, then multiplying $\mathbf{x}$ by A produces a vector $\lambda\mathbf{x}$ parallel to $\mathbf{x}$.

(b) An $n \times n$ matrix A can have only one eigenvalue.

(c) If A is an $n \times n$ matrix with an eigenvalue λ, then the set of all eigenvectors of λ is a subspace of R^n.

In Exercises 63–66, find the dimension of the eigenspace corresponding to the eigenvalue $\lambda = 3$.

63. $A = \begin{bmatrix} 3 & 0 & 0 \\ 0 & 3 & 0 \\ 0 & 0 & 3 \end{bmatrix}$ **64.** $A = \begin{bmatrix} 3 & 1 & 0 \\ 0 & 3 & 0 \\ 0 & 0 & 3 \end{bmatrix}$

65. $A = \begin{bmatrix} 3 & 1 & 0 \\ 0 & 3 & 1 \\ 0 & 0 & 3 \end{bmatrix}$ **66.** $A = \begin{bmatrix} 3 & 1 & 1 \\ 0 & 3 & 1 \\ 0 & 0 & 3 \end{bmatrix}$

67. Calculus Let $T: C'[0, 1] \rightarrow C[0, 1]$ be given by $T(f) = f'$. Show that $\lambda = 1$ is an eigenvalue of T with corresponding eigenvector $f(x) = e^x$.

68. Calculus For the linear transformation given in Exercise 67, find the eigenvalue corresponding to the eigenvector $f(x) = e^{-2x}$.

69. Let $T: P_2 \rightarrow P_2$ be represented by

$$T(a_0 + a_1x + a_2x^2) = (-3a_1 + 5a_2) + (-4a_0 + 4a_1 - 10a_2)x + 4a_2x^2.$$

Find the eigenvalues and the eigenvectors of T relative to the standard basis $\{1, x, x^2\}$.

70. Let $T: P_2 \rightarrow P_2$ be represented by

$$T(a_0 + a_1x + a_2x^2) = (2a_0 + a_1 - a_2) + (-a_1 + 2a_2)x - a_2x^2.$$

Find the eigenvalues and eigenvectors of T relative to the standard basis $\{1, x, x^2\}$.

71. Let $T: M_{2,2} \rightarrow M_{2,2}$ be represented by

$$T\left(\begin{bmatrix} a & b \\ c & d \end{bmatrix}\right) = \begin{bmatrix} a - c + d & b + d \\ -2a + 2c - 2d & 2b + 2d \end{bmatrix}.$$

Find the eigenvalues and eigenvectors of T relative to the standard basis

$$B = \left\{\begin{bmatrix} 1 & 0 \\ 0 & 0 \end{bmatrix}, \begin{bmatrix} 0 & 1 \\ 0 & 0 \end{bmatrix}, \begin{bmatrix} 0 & 0 \\ 1 & 0 \end{bmatrix}, \begin{bmatrix} 0 & 0 \\ 0 & 1 \end{bmatrix}\right\}.$$

72. A square matrix A is called **idempotent** if $A^2 = A$. What are the possible eigenvalues of an idempotent matrix?

73. A square matrix A is called **nilpotent** if there exists a positive integer k such that $A^k = 0$. What are the possible eigenvalues of a nilpotent matrix?

74. Find all values of the angle θ for which the matrix

$$A = \begin{bmatrix} \cos\theta & -\sin\theta \\ \sin\theta & \cos\theta \end{bmatrix}$$

has real eigenvalues. Interpret your answer geometrically.

75. Let A be an $n \times n$ matrix such that the sum of the entries in each row is a fixed constant r. Prove that r is an eigenvalue of A. Illustrate this result with a specific example.

7.2 Diagonalization

The preceding section discussed the eigenvalue problem. In this section, you will look at another classic problem in linear algebra called the **diagonalization problem.** Expressed in terms of matrices*, the problem is this: "For a square matrix A, does there exist an invertible matrix P such that $P^{-1}AP$ is diagonal?"

Recall from Section 6.4 that two square matrices A and B are called **similar** if there exists an invertible matrix P such that $B = P^{-1}AP$.

Matrices that are similar to diagonal matrices are called **diagonalizable.**

Definition of a Diagonalizable Matrix

An $n \times n$ matrix A is **diagonalizable** if A is similar to a diagonal matrix. That is, A is diagonalizable if there exists an invertible matrix P such that $P^{-1}AP$ is a diagonal matrix.

* At the end of this section, the diagonalization problem will be expressed in terms of linear transformations.

Provided with this definition, the diagonalization problem can be stated as follows: "Which square matrices are diagonalizable?" Clearly, every diagonal matrix D is diagonalizable, because the identity matrix I can play the role of P to yield $D = I^{-1}DI$. Example 1 shows another example of a diagonalizable matrix.

EXAMPLE 1 A Diagonalizable Matrix

The matrix from Example 5 in Section 6.4,

$$A = \begin{bmatrix} 1 & 3 & 0 \\ 3 & 1 & 0 \\ 0 & 0 & -2 \end{bmatrix},$$

is diagonalizable because

$$P = \begin{bmatrix} 1 & 1 & 0 \\ 1 & -1 & 0 \\ 0 & 0 & 1 \end{bmatrix}$$

has the property

$$P^{-1}AP = \begin{bmatrix} 4 & 0 & 0 \\ 0 & -2 & 0 \\ 0 & 0 & -2 \end{bmatrix}.$$

As indicated in Example 8 in the preceding section, the eigenvalue problem is related closely to the diagonalization problem. The next two theorems shed more light on this relationship. The first theorem tells you that similar matrices must have the same eigenvalues.

THEOREM 7.4
Similar Matrices Have the Same Eigenvalues

If A and B are similar $n \times n$ matrices, then they have the same eigenvalues.

PROOF Because A and B are similar, there exists an invertible matrix P such that $B = P^{-1}AP$. By the properties of determinants, it follows that

$$\begin{aligned} |\lambda I - B| = |\lambda I - P^{-1}AP| &= |P^{-1}\lambda IP - P^{-1}AP| \\ &= |P^{-1}(\lambda I - A)P| \\ &= |P^{-1}||\lambda I - A||P| \\ &= |P^{-1}||P||\lambda I - A| \\ &= |P^{-1}P||\lambda I - A| \\ &= |\lambda I - A|. \end{aligned}$$

But this means that A and B have the same characteristic polynomial. So, they must have the same eigenvalues.

EXAMPLE 2 Finding Eigenvalues of Similar Matrices

The matrices A and D are similar.

$$A = \begin{bmatrix} 1 & 0 & 0 \\ -1 & 1 & 1 \\ -1 & -2 & 4 \end{bmatrix} \quad \text{and} \quad D = \begin{bmatrix} 1 & 0 & 0 \\ 0 & 2 & 0 \\ 0 & 0 & 3 \end{bmatrix}$$

Use Theorem 7.4 to find the eigenvalues of A and D.

SOLUTION Because D is a diagonal matrix, its eigenvalues are simply the entries on its main diagonal—that is,

$$\lambda_1 = 1,$$
$$\lambda_2 = 2, \text{ and}$$
$$\lambda_3 = 3.$$

Moreover, because A is said to be similar to D, you know from Theorem 7.4 that A has the same eigenvalues. Check this by showing that the characteristic polynomial of A is

$$|\lambda I - A| = (\lambda - 1)(\lambda - 2)(\lambda - 3).$$

REMARK: Example 2 simply states that matrices A and D are similar. Try checking $D = P^{-1}AP$ using the matrices

$$P = \begin{bmatrix} 1 & 0 & 0 \\ 1 & 1 & 1 \\ 1 & 1 & 2 \end{bmatrix} \quad \text{and} \quad P^{-1} = \begin{bmatrix} 1 & 0 & 0 \\ -1 & 2 & -1 \\ 0 & -1 & 1 \end{bmatrix}.$$

In fact, the columns of P are precisely the eigenvectors of A corresponding to the eigenvalues 1, 2, and 3.

The two diagonalizable matrices in Examples 1 and 2 provide a clue to the diagonalization problem. Each of these matrices has a set of three linearly independent eigenvectors. (See Example 3.) This is characteristic of diagonalizable matrices, as stated in Theorem 7.5.

THEOREM 7.5 Condition for Diagonalization

An $n \times n$ matrix A is diagonalizable if and only if it has n linearly independent eigenvectors.

PROOF First, assume A is diagonalizable. Then there exists an invertible matrix P such that $P^{-1}AP = D$ is diagonal. Letting the main entries of D be $\lambda_1, \lambda_2, \ldots, \lambda_n$ and the column vectors of P be $\mathbf{p}_1, \mathbf{p}_2, \ldots, \mathbf{p}_n$ produces

$$PD = [\mathbf{p}_1 \vdots \mathbf{p}_2 \vdots \cdots \vdots \mathbf{p}_n]\begin{bmatrix} \lambda_1 & 0 & \cdots & 0 \\ 0 & \lambda_2 & \cdots & 0 \\ \vdots & \vdots & & \vdots \\ 0 & 0 & \cdots & \lambda_n \end{bmatrix}$$

$$= [\lambda_1\mathbf{p}_1 \vdots \lambda_2\mathbf{p}_2 \vdots \cdots \vdots \lambda_n\mathbf{p}_n].$$

Because $P^{-1}AP = D$, $AP = PD$, which implies

$$[A\mathbf{p}_1 \vdots A\mathbf{p}_2 \vdots \cdots \vdots A\mathbf{p}_n] = [\lambda_1\mathbf{p}_1 \vdots \lambda_2\mathbf{p}_2 \vdots \cdots \vdots \lambda_n\mathbf{p}_n].$$

In other words, $A\mathbf{p}_i = \lambda_i\mathbf{p}_i$ for each column vector $\mathbf{p}_i$. This means that the column vectors $\mathbf{p}_i$ of P are eigenvectors of A. Moreover, because P is invertible, its column vectors are linearly independent. So, A has n linearly independent eigenvectors.

Conversely, assume A has n linearly independent eigenvectors $\mathbf{p}_1, \mathbf{p}_2, \ldots, \mathbf{p}_n$ with corresponding eigenvalues $\lambda_1, \lambda_2, \ldots, \lambda_n$. Let P be the matrix whose columns are these n eigenvectors. That is, $P = [\mathbf{p}_1 \vdots \mathbf{p}_2 \vdots \cdots \vdots \mathbf{p}_n]$. Because each $\mathbf{p}_i$ is an eigenvector of A, you have $A\mathbf{p}_i = \lambda_i\mathbf{p}_i$ and

$$AP = A[\mathbf{p}_1 \vdots \mathbf{p}_2 \vdots \cdots \vdots \mathbf{p}_n] = [\lambda_1\mathbf{p}_1 \vdots \lambda_2\mathbf{p}_2 \vdots \cdots \vdots \lambda_n\mathbf{p}_n].$$

The right-hand matrix in this equation can be written as the matrix product below.

$$AP = [\mathbf{p}_1 \vdots \mathbf{p}_2 \vdots \cdots \vdots \mathbf{p}_n]\begin{bmatrix} \lambda_1 & 0 & \cdots & 0 \\ 0 & \lambda_2 & \cdots & 0 \\ \vdots & \vdots & & \vdots \\ 0 & 0 & \cdots & \lambda_n \end{bmatrix} = PD$$

Finally, because the vectors $\mathbf{p}_1, \mathbf{p}_2, \ldots, \mathbf{p}_n$ are linearly independent, P is invertible and you can write the equation $AP = PD$ as $P^{-1}AP = D$, which means that A is diagonalizable.

A key result of this proof is the fact that for diagonalizable matrices, *the columns of P consist of the n linearly independent eigenvectors.* Example 3 verifies this important property for the matrices in Examples 1 and 2.

EXAMPLE 3 Diagonalizable Matrices

(a) The matrix in Example 1 has the eigenvalues and corresponding eigenvectors listed below.

$$\lambda_1 = 4, \mathbf{p}_1 = \begin{bmatrix} 1 \\ 1 \\ 0 \end{bmatrix}; \quad \lambda_2 = -2, \mathbf{p}_2 = \begin{bmatrix} 1 \\ -1 \\ 0 \end{bmatrix}; \quad \lambda_3 = -2, \mathbf{p}_3 = \begin{bmatrix} 0 \\ 0 \\ 1 \end{bmatrix}$$

The matrix P whose columns correspond to these eigenvectors is

$$P = \begin{bmatrix} 1 & 1 & 0 \\ 1 & -1 & 0 \\ 0 & 0 & 1 \end{bmatrix}.$$

Moreover, because P is row-equivalent to the identity matrix, the eigenvectors $\mathbf{p}_1$, $\mathbf{p}_2$, and $\mathbf{p}_3$ are linearly independent.

(b) The matrix in Example 2 has the eigenvalues and corresponding eigenvectors listed below.

$$\lambda_1 = 1, \mathbf{p}_1 = \begin{bmatrix} 1 \\ 1 \\ 1 \end{bmatrix}; \quad \lambda_2 = 2, \mathbf{p}_2 = \begin{bmatrix} 0 \\ 1 \\ 1 \end{bmatrix}; \quad \lambda_3 = 3, \mathbf{p}_3 = \begin{bmatrix} 0 \\ 1 \\ 2 \end{bmatrix}$$

The matrix P whose columns correspond to these eigenvectors is

$$P = \begin{bmatrix} 1 & 0 & 0 \\ 1 & 1 & 1 \\ 1 & 1 & 2 \end{bmatrix}.$$

Again, because P is row-equivalent to the identity matrix, the eigenvectors $\mathbf{p}_1$, $\mathbf{p}_2$, and $\mathbf{p}_3$ are linearly independent.

The second part of the proof of Theorem 7.5 and Example 3 suggest the steps listed below for diagonalizing a matrix.

Steps for Diagonalizing an $n \times n$ Square Matrix

Let A be an $n \times n$ matrix.

1. Find n linearly independent eigenvectors $\mathbf{p}_1, \mathbf{p}_2, \ldots, \mathbf{p}_n$ for A with corresponding eigenvalues $\lambda_1, \lambda_2, \ldots, \lambda_n$. If n linearly independent eigenvectors do not exist, then A is not diagonalizable.
2. If A has n linearly independent eigenvectors, let P be the $n \times n$ matrix whose columns consist of these eigenvectors. That is,

$$P = [\mathbf{p}_1 \vdots \mathbf{p}_2 \vdots \cdots \vdots \mathbf{p}_n].$$

3. The diagonal matrix $D = P^{-1}AP$ will have the eigenvalues $\lambda_1, \lambda_2, \ldots, \lambda_n$ on its main diagonal (and zeros elsewhere). Note that the order of the eigenvectors used to form P will determine the order in which the eigenvalues appear on the main diagonal of D.

EXAMPLE 4 **A Matrix That Is Not Diagonalizable**

Show that the matrix A is not diagonalizable.

$$A = \begin{bmatrix} 1 & 2 \\ 0 & 1 \end{bmatrix}$$

SOLUTION Because A is triangular, the eigenvalues are simply the entries on the main diagonal. So, the only eigenvalue is $\lambda = 1$. The matrix $(I - A)$ has the reduced row-echelon form shown below.

$$I - A = \begin{bmatrix} 0 & -2 \\ 0 & 0 \end{bmatrix} \longrightarrow \begin{bmatrix} 0 & 1 \\ 0 & 0 \end{bmatrix}$$

This implies that $x_2 = 0$, and letting $x_1 = t$, you can find that every eigenvector of A has the form

$$\mathbf{x} = \begin{bmatrix} x_1 \\ x_2 \end{bmatrix} = \begin{bmatrix} t \\ 0 \end{bmatrix} = t\begin{bmatrix} 1 \\ 0 \end{bmatrix}.$$

So, A does not have two linearly independent eigenvectors, and you can conclude that A is not diagonalizable.

EXAMPLE 5 Diagonalizing a Matrix

Show that the matrix A is diagonalizable.

$$A = \begin{bmatrix} 1 & -1 & -1 \\ 1 & 3 & 1 \\ -3 & 1 & -1 \end{bmatrix}$$

Then find a matrix P such that $P^{-1}AP$ is diagonal.

SOLUTION The characteristic polynomial of A is

$$|\lambda I - A| = \begin{vmatrix} \lambda - 1 & 1 & 1 \\ -1 & \lambda - 3 & -1 \\ 3 & -1 & \lambda + 1 \end{vmatrix} = (\lambda - 2)(\lambda + 2)(\lambda - 3).$$

So, the eigenvalues of A are $\lambda_1 = 2$, $\lambda_2 = -2$, and $\lambda_3 = 3$. From these eigenvalues you obtain the reduced row-echelon forms and corresponding eigenvectors shown below.

Eigenvector

$$2I - A = \begin{bmatrix} 1 & 1 & 1 \\ -1 & -1 & -1 \\ 3 & -1 & 3 \end{bmatrix} \longrightarrow \begin{bmatrix} 1 & 0 & 1 \\ 0 & 1 & 0 \\ 0 & 0 & 0 \end{bmatrix} \qquad \begin{bmatrix} -1 \\ 0 \\ 1 \end{bmatrix}$$

$$-2I - A = \begin{bmatrix} -3 & 1 & 1 \\ -1 & -5 & -1 \\ 3 & -1 & -1 \end{bmatrix} \longrightarrow \begin{bmatrix} 1 & 0 & -\frac{1}{4} \\ 0 & 1 & \frac{1}{4} \\ 0 & 0 & 0 \end{bmatrix} \qquad \begin{bmatrix} 1 \\ -1 \\ 4 \end{bmatrix}$$

$$3I - A = \begin{bmatrix} 2 & 1 & 1 \\ -1 & 0 & -1 \\ 3 & -1 & 4 \end{bmatrix} \longrightarrow \begin{bmatrix} 1 & 0 & 1 \\ 0 & 1 & -1 \\ 0 & 0 & 0 \end{bmatrix} \qquad \begin{bmatrix} -1 \\ 1 \\ 1 \end{bmatrix}$$

Form the matrix P whose columns are the eigenvectors just obtained.

$$P = \begin{bmatrix} -1 & 1 & -1 \\ 0 & -1 & 1 \\ 1 & 4 & 1 \end{bmatrix}$$

This matrix is nonsingular, which implies that the eigenvectors are linearly independent and A is diagonalizable. The inverse of P is

$$P^{-1} = \begin{bmatrix} -1 & -1 & 0 \\ \frac{1}{5} & 0 & \frac{1}{5} \\ \frac{1}{5} & 1 & \frac{1}{5} \end{bmatrix},$$

and it follows that

$$P^{-1}AP = \begin{bmatrix} 2 & 0 & 0 \\ 0 & -2 & 0 \\ 0 & 0 & 3 \end{bmatrix}.$$

EXAMPLE 6 **Diagonalizing a Matrix**

Show that the matrix A is diagonalizable.

$$A = \begin{bmatrix} 1 & 0 & 0 & 0 \\ 0 & 1 & 5 & -10 \\ 1 & 0 & 2 & 0 \\ 1 & 0 & 0 & 3 \end{bmatrix}$$

Then find a matrix P such that $P^{-1}AP$ is diagonal.

SOLUTION In Example 6 in Section 7.1, you found that the three eigenvalues $\lambda_1 = 1$, $\lambda_2 = 2$, and $\lambda_3 = 3$ have the eigenvectors shown below.

$$\lambda_1: \begin{bmatrix} 0 \\ 1 \\ 0 \\ 0 \end{bmatrix}, \begin{bmatrix} -2 \\ 0 \\ 2 \\ 1 \end{bmatrix} \quad \lambda_2: \begin{bmatrix} 0 \\ 5 \\ 1 \\ 0 \end{bmatrix} \quad \lambda_3: \begin{bmatrix} 0 \\ -5 \\ 0 \\ 1 \end{bmatrix}$$

The matrix whose columns consist of these eigenvectors is

$$P = \begin{bmatrix} 0 & -2 & 0 & 0 \\ 1 & 0 & 5 & -5 \\ 0 & 2 & 1 & 0 \\ 0 & 1 & 0 & 1 \end{bmatrix}.$$

Because P is invertible (check this), its column vectors form a linearly independent set.

$$P^{-1} = \begin{bmatrix} -\frac{5}{2} & 1 & -5 & 5 \\ -\frac{1}{2} & 0 & 0 & 0 \\ 1 & 0 & 1 & 0 \\ \frac{1}{2} & 0 & 0 & 1 \end{bmatrix}$$

So, A is diagonalizable, and you have

$$P^{-1}AP = \begin{bmatrix} 1 & 0 & 0 & 0 \\ 0 & 1 & 0 & 0 \\ 0 & 0 & 2 & 0 \\ 0 & 0 & 0 & 3 \end{bmatrix}.$$

For a square matrix A of order n to be diagonalizable, the sum of the dimensions of the eigenspaces must be equal to n. One way this can happen is if A has n distinct eigenvalues. So, you have the next theorem.

THEOREM 7.6 Sufficient Condition for Diagonalization

If an $n \times n$ matrix A has n *distinct* eigenvalues, then the corresponding eigenvectors are linearly independent and A is diagonalizable.

PROOF Let $\lambda_1, \lambda_2, \ldots, \lambda_n$ be n distinct eigenvalues of A corresponding to the eigenvectors $\mathbf{x}_1, \mathbf{x}_2, \ldots, \mathbf{x}_n$. To begin, assume the set of eigenvectors is linearly dependent. Moreover, consider the eigenvectors to be ordered so that the first m eigenvectors are linearly independent, but the first $m + 1$ are dependent, where $m < n$. Then $\mathbf{x}_{m+1}$ can be written as a linear combination of the first m eigenvectors:

$$\mathbf{x}_{m+1} = c_1\mathbf{x}_1 + c_2\mathbf{x}_2 + \cdots + c_m\mathbf{x}_m, \qquad \text{Equation 1}$$

where the c_i's are not all zero. Multiplication of both sides of Equation 1 by A yields

$$\begin{aligned} A\mathbf{x}_{m+1} &= Ac_1\mathbf{x}_1 + Ac_2\mathbf{x}_2 + \cdots + Ac_m\mathbf{x}_m \\ \lambda_{m+1}\mathbf{x}_{m+1} &= c_1\lambda_1\mathbf{x}_1 + c_2\lambda_2\mathbf{x}_2 + \cdots + c_m\lambda_m\mathbf{x}_m, \end{aligned} \qquad \text{Equation 2}$$

whereas multiplication of Equation 1 by λ_{m+1} yields

$$\lambda_{m+1}\mathbf{x}_{m+1} = c_1\lambda_{m+1}\mathbf{x}_1 + c_2\lambda_{m+1}\mathbf{x}_2 + \cdots + c_m\lambda_{m+1}\mathbf{x}_m. \qquad \text{Equation 3}$$

Now, subtracting Equation 2 from Equation 3 produces

$$c_1(\lambda_{m+1} - \lambda_1)\mathbf{x}_1 + c_2(\lambda_{m+1} - \lambda_2)\mathbf{x}_2 + \cdots + c_m(\lambda_{m+1} - \lambda_m)\mathbf{x}_m = \mathbf{0},$$

and, using the fact that the first m eigenvectors are linearly independent, you can conclude that all coefficients of this equation must be zero. That is,

$$c_1(\lambda_{m+1} - \lambda_1) = c_2(\lambda_{m+1} - \lambda_2) = \cdots = c_m(\lambda_{m+1} - \lambda_m) = 0.$$

Because all the eigenvalues are distinct, it follows that $c_i = 0$, $i = 1, 2, \ldots, m$. But this result contradicts our assumption that $\mathbf{x}_{m+1}$ can be written as a linear combination of the first m eigenvectors. So, the set of eigenvectors is linearly independent, and from Theorem 7.5 you can conclude that A is diagonalizable.

EXAMPLE 7 **Determining Whether a Matrix Is Diagonalizable**

Determine whether the matrix A is diagonalizable.

$$A = \begin{bmatrix} 1 & -2 & 1 \\ 0 & 0 & 1 \\ 0 & 0 & -3 \end{bmatrix}$$

SOLUTION Because A is a triangular matrix, its eigenvalues are the main diagonal entries

$$\lambda_1 = 1, \qquad \lambda_2 = 0, \qquad \lambda_3 = -3.$$

Moreover, because these three values are distinct, you can conclude from Theorem 7.6 that A is diagonalizable.

REMARK: Remember that the condition in Theorem 7.6 is sufficient but not necessary for diagonalization, as demonstrated in Example 6. In other words, a diagonalizable matrix need not have distinct eigenvalues.

Diagonalization and Linear Transformations

So far in this section, the diagonalization problem has been considered in terms of matrices. In terms of linear transformations, the diagonalization problem can be stated as follows. For a linear transformation

$$T: V \rightarrow V,$$

does there exist a basis B for V such that the matrix for T relative to B is diagonal? The answer is "yes," provided the standard matrix for T is diagonalizable.

EXAMPLE 8 **Finding a Diagonal Matrix for a Linear Transformation**

Let $T: R^3 \rightarrow R^3$ be the linear transformation represented by

$$T(x_1, x_2, x_3) = (x_1 - x_2 - x_3, x_1 + 3x_2 + x_3, -3x_1 + x_2 - x_3).$$

If possible, find a basis B for R^3 such that the matrix for T relative to B is diagonal.

SOLUTION The standard matrix for T is represented by

$$A = \begin{bmatrix} 1 & -1 & -1 \\ 1 & 3 & 1 \\ -3 & 1 & -1 \end{bmatrix}.$$

From Example 5, you know that A is diagonalizable. So, the three linearly independent eigenvectors found in Example 5 can be used to form the basis B. That is,

$$B = \{(-1, 0, 1), (1, -1, 4), (-1, 1, 1)\}.$$

The matrix for T relative to this basis is

$$D = \begin{bmatrix} 2 & 0 & 0 \\ 0 & -2 & 0 \\ 0 & 0 & 3 \end{bmatrix}.$$

SECTION 7.2 Exercises

In Exercises 1–8, verify that A is diagonalizable by computing $P^{-1}AP$.

1. $A = \begin{bmatrix} -11 & 36 \\ -3 & 10 \end{bmatrix}$, $P = \begin{bmatrix} -3 & -4 \\ -1 & -1 \end{bmatrix}$

2. $A = \begin{bmatrix} 1 & 3 \\ -1 & 5 \end{bmatrix}$, $P = \begin{bmatrix} 3 & 1 \\ 1 & 1 \end{bmatrix}$

3. $A = \begin{bmatrix} -2 & 4 \\ 1 & 1 \end{bmatrix}$, $P = \begin{bmatrix} 1 & -4 \\ 1 & 1 \end{bmatrix}$

4. $A = \begin{bmatrix} 4 & -5 \\ 2 & -3 \end{bmatrix}$, $P = \begin{bmatrix} 1 & 5 \\ 1 & 2 \end{bmatrix}$

5. $A = \begin{bmatrix} -1 & 1 & 0 \\ 0 & 3 & 0 \\ 4 & -2 & 5 \end{bmatrix}$, $P = \begin{bmatrix} 0 & 1 & -3 \\ 0 & 4 & 0 \\ 1 & 2 & 2 \end{bmatrix}$

6. $A = \begin{bmatrix} 2 & 3 & 1 \\ 0 & -1 & 2 \\ 0 & 0 & 3 \end{bmatrix}$, $P = \begin{bmatrix} 1 & 1 & 5 \\ 0 & -1 & 1 \\ 0 & 0 & 2 \end{bmatrix}$

7. $A = \begin{bmatrix} 4 & -1 & 3 \\ 0 & 2 & 1 \\ 0 & 0 & 3 \end{bmatrix}$, $P = \begin{bmatrix} 1 & 1 & -2 \\ 0 & 2 & 1 \\ 0 & 0 & 1 \end{bmatrix}$

8. $A = \begin{bmatrix} 0.80 & 0.10 & 0.05 & 0.05 \\ 0.10 & 0.80 & 0.05 & 0.05 \\ 0.05 & 0.05 & 0.80 & 0.10 \\ 0.05 & 0.05 & 0.10 & 0.80 \end{bmatrix}$, $P = \begin{bmatrix} 1 & -1 & 0 & 1 \\ 1 & -1 & 0 & -1 \\ 1 & 1 & 1 & 0 \\ 1 & 1 & -1 & 0 \end{bmatrix}$

In Exercises 9–16, show that the matrix is not diagonalizable.

9. $\begin{bmatrix} 0 & 0 \\ 5 & 0 \end{bmatrix}$

10. $\begin{bmatrix} 1 & \frac{1}{2} \\ -2 & -1 \end{bmatrix}$

11. $\begin{bmatrix} 1 & 1 \\ 0 & 1 \end{bmatrix}$

12. $\begin{bmatrix} 1 & 0 \\ -2 & 1 \end{bmatrix}$

13. $\begin{bmatrix} 1 & -2 & 1 \\ 0 & 1 & 4 \\ 0 & 0 & 2 \end{bmatrix}$

14. $\begin{bmatrix} 2 & 1 & -1 \\ 0 & -1 & 2 \\ 0 & 0 & -1 \end{bmatrix}$

15. $\begin{bmatrix} 1 & 0 & -1 & 1 \\ 0 & 1 & 0 & 1 \\ -2 & 0 & 2 & -2 \\ 0 & 2 & 0 & 2 \end{bmatrix}$
(See Exercise 37, Section 7.1.)

16. $\begin{bmatrix} 1 & -3 & 3 & 3 \\ -1 & 4 & -3 & -3 \\ -2 & 0 & 1 & 1 \\ 1 & 0 & 0 & 0 \end{bmatrix}$
(See Exercise 38, Section 7.1.)

In Exercises 17–20, find the eigenvalues of the matrix and determine whether there is a sufficient number to guarantee that the matrix is diagonalizable. (Recall that the matrix may be diagonalizable even though it is not guaranteed to be diagonalizable by Theorem 7.6.)

17. $\begin{bmatrix} 1 & 1 \\ 1 & 1 \end{bmatrix}$

18. $\begin{bmatrix} 2 & 0 \\ 5 & 2 \end{bmatrix}$

19. $\begin{bmatrix} 3 & 2 & -3 \\ -3 & -4 & 9 \\ -1 & -2 & 5 \end{bmatrix}$

20. $\begin{bmatrix} 4 & 3 & -2 \\ 0 & 1 & 1 \\ 0 & 0 & -2 \end{bmatrix}$

In Exercises 21–34, for each matrix A, find (if possible) a nonsingular matrix P such that $P^{-1}AP$ is diagonal. Verify that $P^{-1}AP$ is a diagonal matrix with the eigenvalues on the diagonal.

21. $A = \begin{bmatrix} 6 & -3 \\ -2 & 1 \end{bmatrix}$

(See Exercise 15, Section 7.1.)

22. $A = \begin{bmatrix} 1 & -4 \\ -2 & 8 \end{bmatrix}$

(See Exercise 16, Section 7.1.)

23. $A = \begin{bmatrix} 1 & -\frac{3}{2} \\ \frac{1}{2} & -1 \end{bmatrix}$

(See Exercise 17, Section 7.1.)

24. $A = \begin{bmatrix} \frac{1}{4} & \frac{1}{4} \\ \frac{1}{2} & 0 \end{bmatrix}$

(See Exercise 18, Section 7.1.)

25. $A = \begin{bmatrix} 2 & -2 & 3 \\ 0 & 3 & -2 \\ 0 & -1 & 2 \end{bmatrix}$

(See Exercise 21, Section 7.1.)

26. $A = \begin{bmatrix} 3 & 2 & 1 \\ 0 & 0 & 2 \\ 0 & 2 & 0 \end{bmatrix}$

(See Exercise 22, Section 7.1.)

27. $A = \begin{bmatrix} 1 & 2 & -2 \\ -2 & 5 & -2 \\ -6 & 6 & -3 \end{bmatrix}$

(See Exercise 23, Section 7.1.)

28. $A = \begin{bmatrix} 3 & 2 & -3 \\ -3 & -4 & 9 \\ -1 & -2 & 5 \end{bmatrix}$

(See Exercise 24, Section 7.1.)

29. $A = \begin{bmatrix} 0 & -3 & 5 \\ -4 & 4 & -10 \\ 0 & 0 & 4 \end{bmatrix}$

(See Exercise 25, Section 7.1.)

30. $A = \begin{bmatrix} 1 & -\frac{3}{2} & \frac{5}{2} \\ -2 & \frac{13}{2} & -10 \\ \frac{3}{2} & -\frac{9}{2} & 8 \end{bmatrix}$

(See Exercise 26, Section 7.1.)

31. $A = \begin{bmatrix} 1 & 0 & 0 \\ 1 & 2 & 1 \\ 1 & 0 & 2 \end{bmatrix}$

32. $A = \begin{bmatrix} 4 & 0 & 0 \\ 2 & 2 & 0 \\ 0 & 2 & 2 \end{bmatrix}$

33. $A = \begin{bmatrix} 2 & 0 & 0 & 0 \\ 3 & -1 & 0 & 0 \\ 0 & 1 & 1 & 0 \\ 0 & 0 & 1 & -2 \end{bmatrix}$

34. $A = \begin{bmatrix} 1 & 0 & 0 & 0 \\ 1 & 0 & 1 & 0 \\ 0 & 0 & 1 & 0 \\ 0 & 1 & 0 & 1 \end{bmatrix}$

In Exercises 35–38, find a basis B for the domain of T such that the matrix of T relative to B is diagonal.

35. $T: R^2 \to R^2$: $T(x, y) = (x + y, x + y)$

36. $T: R^3 \to R^3$: $T(x, y, z) = (-2x + 2y - 3z, 2x + y - 6z, -x - 2y)$

37. $T: P_1 \to P_1$: $T(a + bx) = a + (a + 2b)x$

38. $T: P_2 \to P_2$: $T(a_0 + a_1x + a_2x^2) = (2a_0 + a_2) + (3a_1 + 4a_2)x + a_2x^2$

39. Let A be a diagonalizable $n \times n$ matrix and P an invertible $n \times n$ matrix such that $B = P^{-1}AP$ is the diagonal form of A. Prove that

(a) $B^k = P^{-1}A^kP$, where k is a positive integer.

(b) $A^k = PB^kP^{-1}$, where k is a positive integer.

40. Let $\lambda_1, \lambda_2, \ldots, \lambda_n$ be n distinct eigenvalues of the $n \times n$ matrix A. Use the result of Exercise 39 to find the eigenvalues of A^k.

In Exercises 41–44, use the result of Exercise 39 to find the indicated power of A.

41. $A = \begin{bmatrix} 10 & 18 \\ -6 & -11 \end{bmatrix}$, A^6

42. $A = \begin{bmatrix} 1 & 3 \\ 2 & 0 \end{bmatrix}$, A^7

43. $A = \begin{bmatrix} 3 & 2 & -3 \\ -3 & -4 & 9 \\ -1 & -2 & 5 \end{bmatrix}$, A^8

44. $A = \begin{bmatrix} 2 & 0 & -2 \\ 0 & 2 & -2 \\ 3 & 0 & -3 \end{bmatrix}$, A^5

True or False? In Exercises 45 and 46, determine whether each statement is true or false. If a statement is true, give a reason or cite an appropriate statement from the text. If a statement is false, provide an example that shows the statement is not true in all cases or cite an appropriate statement from the text.

45. (a) If A and B are similar $n \times n$ matrices, then they always have the same characteristic polynomial equation.

(b) The fact that an $n \times n$ matrix A has n distinct eigenvalues does not guarantee that A is diagonalizable.

46. (a) If A is a diagonalizable matrix, then it has n linearly independent eigenvectors.

(b) If an $n \times n$ matrix A is diagonalizable, then it must have n distinct eigenvalues.

47. **Writing** Can a matrix be similar to two different diagonal matrices? Explain your answer.

48. Are the two matrices similar? If so, find a matrix P such that $B = P^{-1}AP$.

$$A = \begin{bmatrix} 1 & 0 & 0 \\ 0 & 2 & 0 \\ 0 & 0 & 3 \end{bmatrix} \qquad B = \begin{bmatrix} 3 & 0 & 0 \\ 0 & 2 & 0 \\ 0 & 0 & 1 \end{bmatrix}$$

49. Prove that if A is diagonalizable, then A^T is diagonalizable.

50. Prove that the matrix

$$A = \begin{bmatrix} a & b \\ c & d \end{bmatrix}$$

is diagonalizable if $-4bc < (a - d)^2$ and is not diagonalizable if $-4bc > (a - d)^2$.

51. Prove that if A is diagonalizable with n real eigenvalues $\lambda_1, \lambda_2, \ldots, \lambda_n$, then $|A| = \lambda_1\lambda_2 \cdots \lambda_n$.

52. Calculus If x is a real number, then e^x can be defined by the series

$$e^x = 1 + x + \frac{x^2}{2!} + \frac{x^3}{3!} + \frac{x^4}{4!} + \cdots.$$

In a similar way, if X is a square matrix, you can define e^X by the series

$$e^X = I + X + \frac{1}{2!}X^2 + \frac{1}{3!}X^3 + \frac{1}{4!}X^4 + \cdots.$$

Evaluate e^X, where X is the indicated square matrix.

(a) $X = \begin{bmatrix} 1 & 0 \\ 0 & 1 \end{bmatrix}$ (b) $X = \begin{bmatrix} 0 & 0 \\ 0 & 0 \end{bmatrix}$

(c) $X = \begin{bmatrix} 1 & 0 \\ 1 & 0 \end{bmatrix}$ (d) $X = \begin{bmatrix} 0 & 1 \\ 1 & 0 \end{bmatrix}$

(e) $X = \begin{bmatrix} 2 & 0 \\ 0 & -2 \end{bmatrix}$

53. Guided Proof Prove that if the eigenvalues of a diagonalizable matrix A are all ± 1, then the matrix is equal to its inverse.

Getting Started: To show that the matrix is equal to its inverse, use the fact that there exists an invertible matrix P such that $D = P^{-1}AP$, where D is a diagonal matrix with ± 1 along its main diagonal.

(i) Let $D = P^{-1}AP$, where D is a diagonal matrix with ± 1 along its main diagonal.

(ii) Find A in terms of P, P^{-1}, and D.

(iii) Use the properties of the inverse of a product of matrices and the fact that D is diagonal to expand to find A^{-1}.

(iv) Conclude that $A^{-1} = A$.

54. Guided Proof Prove that nonzero nilpotent matrices are not diagonalizable.

Getting Started: From Exercise 73 in Section 7.1, you know that 0 is the only eigenvalue of the nilpotent matrix A. Show that it is impossible for A to be diagonalizable.

(i) Assume A is diagonalizable, so there exists an invertible matrix P such that $P^{-1}AP = D$, where D is the zero matrix.

(ii) Find A in terms of P, P^{-1}, and D.

(iii) Find a contradiction and conclude that nonzero nilpotent matrices are not diagonalizable.

55. Prove that if A is a nonsingular diagonalizable matrix, then A^{-1} is also diagonalizable.

In Exercises 56 and 57, show that the matrix is not diagonalizable. Then write a brief statement explaining your reasoning.

56. $\begin{bmatrix} 3 & k \\ 0 & 3 \end{bmatrix}, k \neq 0$ **57.** $\begin{bmatrix} k & 0 \\ 0 & k \end{bmatrix}$

7.3 Symmetric Matrices and Orthogonal Diagonalization

For most matrices you must go through much of the diagonalization process before you can finally determine whether diagonalization is possible. One exception is a triangular matrix with distinct entries on the main diagonal. Such a matrix can be recognized as diagonalizable by simple inspection. In this section you will study another type of matrix that is guaranteed to be diagonalizable: a **symmetric** matrix.

Definition of Symmetric Matrix

A square matrix A is **symmetric** if it is equal to its transpose:

$$A = A^T.$$

You can determine easily whether a matrix is symmetric by checking whether it is symmetric with respect to its main diagonal.

EXAMPLE 1 **Symmetric Matrices and Nonsymmetric Matrices**

The matrices A and B are symmetric, but the matrix C is not.

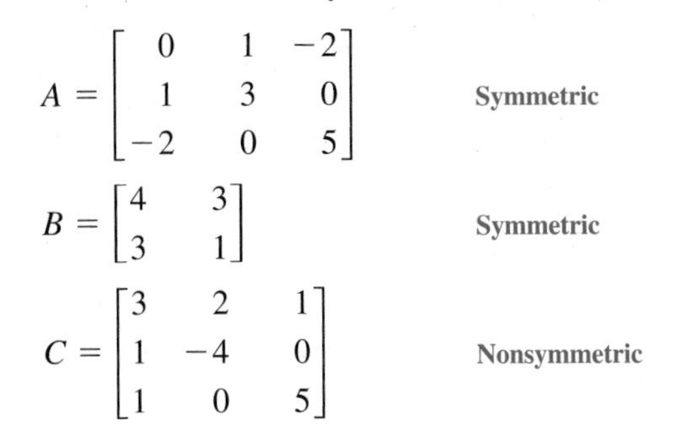

$$A = \begin{bmatrix} 0 & 1 & -2 \\ 1 & 3 & 0 \\ -2 & 0 & 5 \end{bmatrix} \qquad \text{Symmetric}$$

$$B = \begin{bmatrix} 4 & 3 \\ 3 & 1 \end{bmatrix} \qquad \text{Symmetric}$$

$$C = \begin{bmatrix} 3 & 2 & 1 \\ 1 & -4 & 0 \\ 1 & 0 & 5 \end{bmatrix} \qquad \text{Nonsymmetric}$$

Discovery

If you have access to a computer software program or a graphing utility that can find eigenvalues, try the following experiment. Pick an arbitrary square matrix and calculate its eigenvalues. Can you find a matrix for which the eigenvalues are not real? Now pick an arbitrary symmetric matrix and calculate its eigenvalues. Can you find a symmetric matrix for which the eigenvalues are not real? What can you conclude about the eigenvalues of a symmetric matrix?

Nonsymmetric matrices have the following special properties that are not exhibited by symmetric matrices.

1. A nonsymmetric matrix may not be diagonalizable.
2. A nonsymmetric matrix can have eigenvalues that are not real. For instance, the matrix

$$A = \begin{bmatrix} 0 & -1 \\ 1 & 0 \end{bmatrix}$$

   has a characteristic equation of $\lambda^2 + 1 = 0$. So, its eigenvalues are the imaginary numbers $\lambda_1 = i$ and $\lambda_2 = -i$.
3. For a nonsymmetric matrix, the number of linearly independent eigenvectors corresponding to an eigenvalue can be less than the multiplicity of the eigenvalue. (See Example 4, Section 7.2.)

None of these three properties is exhibited by symmetric matrices.

THEOREM 7.7 Eigenvalues of Symmetric Matrices

If A is an $n \times n$ symmetric matrix, then the following properties are true.

1. A is diagonalizable.
2. All eigenvalues of A are real.
3. If λ is an eigenvalue of A with multiplicity k, then λ has k linearly independent eigenvectors. That is, the eigenspace of λ has dimension k.

REMARK: Theorem 7.7 is called the **Real Spectral Theorem,** and the set of eigenvalues of A is called the **spectrum** of A.

A general proof of Theorem 7.7 is beyond the scope of this text. The next example verifies that every 2×2 symmetric matrix is diagonalizable.

EXAMPLE 2 **The Eigenvalues and Eigenvectors of a 2 × 2 Symmetric Matrix**

Prove that a symmetric matrix

$$A = \begin{bmatrix} a & c \\ c & b \end{bmatrix}$$

is diagonalizable.

SOLUTION The characteristic polynomial of A is

$$|\lambda I - A| = \begin{vmatrix} \lambda - a & -c \\ -c & \lambda - b \end{vmatrix} = \lambda^2 - (a + b)\lambda + ab - c^2.$$

As a quadratic in λ, this polynomial has a discriminant of

$$\begin{aligned} (a + b)^2 - 4(ab - c^2) &= a^2 + 2ab + b^2 - 4ab + 4c^2 \\ &= a^2 - 2ab + b^2 + 4c^2 \\ &= (a - b)^2 + 4c^2. \end{aligned}$$

Because this discriminant is the sum of two squares, it must be either zero or positive. If $(a - b)^2 + 4c^2 = 0$, then $a = b$ and $c = 0$, which implies that A is already diagonal. That is,

$$A = \begin{bmatrix} a & 0 \\ 0 & a \end{bmatrix}.$$

On the other hand, if $(a - b)^2 + 4c^2 > 0$, then by the Quadratic Formula the characteristic polynomial of A has two distinct real roots, which implies that A has two distinct real eigenvalues. So, A is diagonalizable in this case also.

EXAMPLE 3 **Dimensions of the Eigenspaces of a Symmetric Matrix**

Find the eigenvalues of the symmetric matrix

$$A = \begin{bmatrix} 1 & -2 & 0 & 0 \\ -2 & 1 & 0 & 0 \\ 0 & 0 & 1 & -2 \\ 0 & 0 & -2 & 1 \end{bmatrix}$$

and determine the dimensions of the corresponding eigenspaces.

SOLUTION The characteristic polynomial of A is represented by

$$|\lambda I - A| = \begin{vmatrix} \lambda - 1 & 2 & 0 & 0 \\ 2 & \lambda - 1 & 0 & 0 \\ 0 & 0 & \lambda - 1 & 2 \\ 0 & 0 & 2 & \lambda - 1 \end{vmatrix} = (\lambda + 1)^2(\lambda - 3)^2.$$

So, the eigenvalues of A are $\lambda_1 = -1$ and $\lambda_2 = 3$. Because each of these eigenvalues has a multiplicity of 2, you know from Theorem 7.7 that the corresponding eigenspaces also have dimension 2. Specifically, the eigenspace of $\lambda_1 = -1$ has a basis of $B_1 = \{(1, 1, 0, 0), (0, 0, 1, 1)\}$ and the eigenspace of $\lambda_2 = 3$ has a basis of $B_2 = \{(1, -1, 0, 0), (0, 0, 1, -1)\}$.

Orthogonal Matrices

To diagonalize a square matrix A, you need to find an *invertible* matrix P such that $P^{-1}AP$ is diagonal. For symmetric matrices, you will see that the matrix P can be chosen to have the special property that $P^{-1} = P^T$. This unusual matrix property is defined as follows.

Definition of an Orthogonal Matrix

A square matrix P is called **orthogonal** if it is invertible and if

$$P^{-1} = P^T.$$

EXAMPLE 4 Orthogonal Matrices

(a) The matrix

$$P = \begin{bmatrix} 0 & 1 \\ -1 & 0 \end{bmatrix}$$

is orthogonal because

$$P^{-1} = P^T = \begin{bmatrix} 0 & -1 \\ 1 & 0 \end{bmatrix}.$$

(b) The matrix

$$P = \begin{bmatrix} \frac{3}{5} & 0 & -\frac{4}{5} \\ 0 & 1 & 0 \\ \frac{4}{5} & 0 & \frac{3}{5} \end{bmatrix}$$

is orthogonal because

$$P^{-1} = P^T = \begin{bmatrix} \frac{3}{5} & 0 & \frac{4}{5} \\ 0 & 1 & 0 \\ -\frac{4}{5} & 0 & \frac{3}{5} \end{bmatrix}.$$

In parts (a) and (b) of Example 4, the columns of the matrices P form orthonormal sets in R^2 and R^3, respectively. This suggests the next theorem.

THEOREM 7.8
Property of Orthogonal Matrices

An $n \times n$ matrix P is orthogonal if and only if its column vectors form an orthonormal set.

PROOF Suppose the column vectors of P form an orthonormal set:

$$P = [\mathbf{p}_1 \vdots \mathbf{p}_2 \vdots \cdots \vdots \mathbf{p}_n] = \begin{bmatrix} p_{11} & p_{12} & \cdots & p_{1n} \\ p_{21} & p_{22} & \cdots & p_{2n} \\ \vdots & \vdots & & \vdots \\ p_{n1} & p_{n2} & \cdots & p_{nn} \end{bmatrix}.$$

Then the product P^TP has the form

$$P^TP = \begin{bmatrix} p_{11} & p_{21} & \cdots & p_{n1} \\ p_{12} & p_{22} & \cdots & p_{n2} \\ \vdots & \vdots & & \vdots \\ p_{1n} & p_{2n} & \cdots & p_{nn} \end{bmatrix} \begin{bmatrix} p_{11} & p_{12} & \cdots & p_{1n} \\ p_{21} & p_{22} & \cdots & p_{2n} \\ \vdots & \vdots & & \vdots \\ p_{n1} & p_{n2} & \cdots & p_{nn} \end{bmatrix}$$

$$P^TP = \begin{bmatrix} \mathbf{p}_1 \cdot \mathbf{p}_1 & \mathbf{p}_1 \cdot \mathbf{p}_2 & \cdots & \mathbf{p}_1 \cdot \mathbf{p}_n \\ \mathbf{p}_2 \cdot \mathbf{p}_1 & \mathbf{p}_2 \cdot \mathbf{p}_2 & \cdots & \mathbf{p}_2 \cdot \mathbf{p}_n \\ \vdots & \vdots & & \vdots \\ \mathbf{p}_n \cdot \mathbf{p}_1 & \mathbf{p}_n \cdot \mathbf{p}_2 & \cdots & \mathbf{p}_n \cdot \mathbf{p}_n \end{bmatrix}.$$

Because the set $\{\mathbf{p}_1, \mathbf{p}_2, \ldots, \mathbf{p}_n\}$ is orthonormal, you have

$$\mathbf{p}_i \cdot \mathbf{p}_j = 0,\ i \neq j \quad \text{and} \quad \mathbf{p}_i \cdot \mathbf{p}_i = \|\mathbf{p}_i\|^2 = 1.$$

So, the matrix composed of dot products has the form

$$P^TP = \begin{bmatrix} 1 & 0 & \cdots & 0 \\ 0 & 1 & \cdots & 0 \\ \vdots & \vdots & & \vdots \\ 0 & 0 & \cdots & 1 \end{bmatrix} = I_n.$$

This implies that $P^T = P^{-1}$, and you can conclude that P is orthogonal.

Conversely, if P is orthogonal, you can reverse the steps above to verify that the column vectors of P form an orthonormal set.

EXAMPLE 5 **An Orthogonal Matrix**

Show that

$$P = \begin{bmatrix} \frac{1}{3} & \frac{2}{3} & \frac{2}{3} \\ -\frac{2}{\sqrt{5}} & \frac{1}{\sqrt{5}} & 0 \\ -\frac{2}{3\sqrt{5}} & -\frac{4}{3\sqrt{5}} & \frac{5}{3\sqrt{5}} \end{bmatrix}$$

is orthogonal by showing that $PP^T = I$. Then show that the column vectors of P form an orthonormal set.

SOLUTION Because

$$PP^T = \begin{bmatrix} \frac{1}{3} & \frac{2}{3} & \frac{2}{3} \\ -\frac{2}{\sqrt{5}} & \frac{1}{\sqrt{5}} & 0 \\ -\frac{2}{3\sqrt{5}} & -\frac{4}{3\sqrt{5}} & \frac{5}{3\sqrt{5}} \end{bmatrix} \begin{bmatrix} \frac{1}{3} & -\frac{2}{\sqrt{5}} & -\frac{2}{3\sqrt{5}} \\ \frac{2}{3} & \frac{1}{\sqrt{5}} & -\frac{4}{3\sqrt{5}} \\ \frac{2}{3} & 0 & \frac{5}{3\sqrt{5}} \end{bmatrix}$$

$$= \begin{bmatrix} 1 & 0 & 0 \\ 0 & 1 & 0 \\ 0 & 0 & 1 \end{bmatrix} = I_3,$$

it follows that $P^T = P^{-1}$, and you can conclude that P is orthogonal. Moreover, letting

$$\mathbf{p}_1 = \begin{bmatrix} \frac{1}{3} \\ -\frac{2}{\sqrt{5}} \\ -\frac{2}{3\sqrt{5}} \end{bmatrix}, \mathbf{p}_2 = \begin{bmatrix} \frac{2}{3} \\ \frac{1}{\sqrt{5}} \\ -\frac{4}{3\sqrt{5}} \end{bmatrix}, \text{ and } \mathbf{p}_3 = \begin{bmatrix} \frac{2}{3} \\ 0 \\ \frac{5}{3\sqrt{5}} \end{bmatrix}$$

produces

$$\mathbf{p}_1 \cdot \mathbf{p}_2 = \mathbf{p}_1 \cdot \mathbf{p}_3 = \mathbf{p}_2 \cdot \mathbf{p}_3 = 0$$

and

$$\|\mathbf{p}_1\| = \|\mathbf{p}_2\| = \|\mathbf{p}_3\| = 1.$$

So, $\{\mathbf{p}_1, \mathbf{p}_2, \mathbf{p}_3\}$ is an orthonormal set, as guaranteed by Theorem 7.8.

It can be shown that for a symmetric matrix, the eigenvectors corresponding to distinct eigenvalues are orthogonal. This property is stated in the next theorem.

THEOREM 7.9
Property of Symmetric Matrices

Let A be an $n \times n$ symmetric matrix. If λ_1 and λ_2 are distinct eigenvalues of A, then their corresponding eigenvectors $\mathbf{x}_1$ and $\mathbf{x}_2$ are orthogonal.

PROOF Let λ_1 and λ_2 be distinct eigenvalues of A with corresponding eigenvectors $\mathbf{x}_1$ and $\mathbf{x}_2$. So,

$$A\mathbf{x}_1 = \lambda_1\mathbf{x}_1 \quad \text{and} \quad A\mathbf{x}_2 = \lambda_2\mathbf{x}_2.$$

To prove the theorem, use the matrix form of the dot product shown below.

$$\mathbf{x}_1 \cdot \mathbf{x}_2 = \begin{bmatrix} x_{11} & x_{12} & \cdots & x_{1n} \end{bmatrix} \begin{bmatrix} x_{21} \\ x_{22} \\ \vdots \\ x_{2n} \end{bmatrix} = \mathbf{x}_1^T\mathbf{x}_2$$

Now you can write

$$\begin{aligned} \lambda_1(\mathbf{x}_1 \cdot \mathbf{x}_2) &= (\lambda_1\mathbf{x}_1) \cdot \mathbf{x}_2 \\ &= (A\mathbf{x}_1) \cdot \mathbf{x}_2 \\ &= (A\mathbf{x}_1)^T\mathbf{x}_2 \\ &= (\mathbf{x}_1^TA^T)\mathbf{x}_2 \\ &= (\mathbf{x}_1^TA)\mathbf{x}_2 && \text{Because } A \text{ is symmetric, } A = A^T. \\ &= \mathbf{x}_1^T(A\mathbf{x}_2) \\ &= \mathbf{x}_1^T(\lambda_2\mathbf{x}_2) \\ &= \mathbf{x}_1 \cdot (\lambda_2\mathbf{x}_2) \\ &= \lambda_2(\mathbf{x}_1 \cdot \mathbf{x}_2). \end{aligned}$$

This implies that $(\lambda_1 - \lambda_2)(\mathbf{x}_1 \cdot \mathbf{x}_2) = 0$, and because $\lambda_1 \neq \lambda_2$ it follows that $\mathbf{x}_1 \cdot \mathbf{x}_2 = 0$. So, $\mathbf{x}_1$ and $\mathbf{x}_2$ are orthogonal.

EXAMPLE 6 Eigenvectors of a Symmetric Matrix

Show that any two eigenvectors of

$$A = \begin{bmatrix} 3 & 1 \\ 1 & 3 \end{bmatrix}$$

corresponding to distinct eigenvalues are orthogonal.

SOLUTION The characteristic polynomial of A is

$$|\lambda I - A| = \begin{vmatrix} \lambda - 3 & -1 \\ -1 & \lambda - 3 \end{vmatrix} = (\lambda - 2)(\lambda - 4),$$

which implies that the eigenvalues of A are $\lambda_1 = 2$ and $\lambda_2 = 4$. Every eigenvector corresponding to $\lambda_1 = 2$ is of the form

$$\mathbf{x}_1 = \begin{bmatrix} s \\ -s \end{bmatrix}, \quad s \neq 0$$

and every eigenvector corresponding to $\lambda_2 = 4$ is of the form

$$\mathbf{x}_2 = \begin{bmatrix} t \\ t \end{bmatrix}, \quad t \neq 0.$$

So,

$$\mathbf{x}_1 \cdot \mathbf{x}_2 = \begin{bmatrix} s \\ -s \end{bmatrix} \cdot \begin{bmatrix} t \\ t \end{bmatrix} = st - st = 0,$$

and you can conclude that $\mathbf{x}_1$ and $\mathbf{x}_2$ are orthogonal.

Orthogonal Diagonalization

A matrix A is **orthogonally diagonalizable** if there exists an orthogonal matrix P such that $P^{-1}AP = D$ is diagonal. The following important theorem states that the set of orthogonally diagonalizable matrices is precisely the set of symmetric matrices.

THEOREM 7.10
Fundamental Theorem of Symmetric Matrices

Let A be an $n \times n$ matrix. Then A is orthogonally diagonalizable and has real eigenvalues if and only if A is symmetric.

PROOF The proof of the theorem in one direction is fairly straightforward. That is, if you assume A is orthogonally diagonalizable, then there exists an orthogonal matrix P such that $D = P^{-1}AP$ is diagonal. Moreover, because $P^{-1} = P^T$, you have $A = PDP^{-1} = PDP^T$, which implies that $A^T = (PDP^T)^T = (P^T)^T D^T P^T = PDP^T = A$. So, A is symmetric.

The proof of the theorem in the other direction is more involved, but it is important because it is constructive. Assume A is symmetric. If A has an eigenvalue λ of multiplicity k, then by Theorem 7.7, λ has k linearly independent eigenvectors. Through the Gram-Schmidt orthonormalization process, this set of k vectors can be used to form an orthonormal basis of eigenvectors for the eigenspace corresponding to λ. This procedure is repeated for each eigenvalue of A. The collection of all resulting eigenvectors is orthogonal by Theorem 7.9, and you know from the normalization process that the collection is also orthonormal. Now let P be the matrix whose columns consist of these n orthonormal eigenvectors. By Theorem 7.8, P is an orthogonal matrix. Finally, by Theorem 7.5, you can conclude that $P^{-1}AP$ is diagonal. So, A is orthogonally diagonalizable.

EXAMPLE 7 **Determining Whether a Matrix Is Orthogonally Diagonalizable**

Which matrices are orthogonally diagonalizable?

$$A_1 = \begin{bmatrix} 1 & 1 & 1 \\ 1 & 0 & 1 \\ 1 & 1 & 1 \end{bmatrix} \qquad A_2 = \begin{bmatrix} 5 & 2 & 1 \\ 2 & 1 & 8 \\ -1 & 8 & 0 \end{bmatrix}$$

$$A_3 = \begin{bmatrix} 3 & 2 & 0 \\ 2 & 0 & 1 \end{bmatrix} \qquad A_4 = \begin{bmatrix} 0 & 0 \\ 0 & -2 \end{bmatrix}$$

SOLUTION By Theorem 7.10, the orthogonally diagonalizable matrices are the symmetric ones: A_1 and A_4.

It was mentioned that the second part of the proof of Theorem 7.10 is *constructive.* That is, it gives you steps to follow to diagonalize a symmetric matrix orthogonally. These steps are summarized as follows.

Orthogonal Diagonalization of a Symmetric Matrix

Let A be an $n \times n$ symmetric matrix.

1. Find all eigenvalues of A and determine the multiplicity of each.
2. For *each* eigenvalue of multiplicity 1, choose a unit eigenvector. (Choose any eigenvector and then normalize it.)
3. For each eigenvalue of multiplicity $k \geq 2$, find a set of k linearly independent eigenvectors. (You know from Theorem 7.7 that this is possible.) If this set is not orthonormal, apply the Gram-Schmidt orthonormalization process.
4. The composite of steps 2 and 3 produces an orthonormal set of n eigenvectors. Use these eigenvectors to form the columns of P. The matrix $P^{-1}AP = P^TAP = D$ will be diagonal. (The main diagonal entries of D are the eigenvalues of A.)

EXAMPLE 8 **Orthogonal Diagonalization**

Find an orthogonal matrix P that orthogonally diagonalizes

$$A = \begin{bmatrix} -2 & 2 \\ 2 & 1 \end{bmatrix}.$$

SOLUTION 1. The characteristic polynomial of A is

$$|\lambda I - A| = \begin{vmatrix} \lambda + 2 & -2 \\ -2 & \lambda - 1 \end{vmatrix} = (\lambda + 3)(\lambda - 2).$$

So the eigenvalues are $\lambda_1 = -3$ and $\lambda_2 = 2$.

2. For each eigenvalue, find an eigenvector by converting the matrix $\lambda I - A$ to reduced row-echelon form.

Eigenvector

$$-3I - A = \begin{bmatrix} -1 & -2 \\ -2 & -4 \end{bmatrix} \rightarrow \begin{bmatrix} 1 & 2 \\ 0 & 0 \end{bmatrix} \rightarrow \begin{bmatrix} -2 \\ 1 \end{bmatrix}$$

$$2I - A = \begin{bmatrix} 4 & -2 \\ -2 & 1 \end{bmatrix} \rightarrow \begin{bmatrix} 1 & -\frac{1}{2} \\ 0 & 0 \end{bmatrix} \rightarrow \begin{bmatrix} 1 \\ 2 \end{bmatrix}$$

The eigenvectors $(-2, 1)$ and $(1, 2)$ form an *orthogonal* basis for R^2. Each of these eigenvectors is normalized to produce an *orthonormal* basis.

$$\mathbf{p}_1 = \frac{(-2, 1)}{\|(-2, 1)\|} = \frac{1}{\sqrt{5}}(-2, 1) = \left(-\frac{2}{\sqrt{5}}, \frac{1}{\sqrt{5}}\right)$$

$$\mathbf{p}_2 = \frac{(1, 2)}{\|(1, 2)\|} = \frac{1}{\sqrt{5}}(1, 2) = \left(\frac{1}{\sqrt{5}}, \frac{2}{\sqrt{5}}\right)$$

3. Because each eigenvalue has a multiplicity of 1, go directly to step 4.
4. Using $\mathbf{p}_1$ and $\mathbf{p}_2$ as column vectors, construct the matrix P.

$$P = \begin{bmatrix} -\frac{2}{\sqrt{5}} & \frac{1}{\sqrt{5}} \\ \frac{1}{\sqrt{5}} & \frac{2}{\sqrt{5}} \end{bmatrix}$$

Verify that P is correct by computing $P^{-1}AP = P^TAP$.

$$P^TAP = \begin{bmatrix} -\frac{2}{\sqrt{5}} & \frac{1}{\sqrt{5}} \\ \frac{1}{\sqrt{5}} & \frac{2}{\sqrt{5}} \end{bmatrix} \begin{bmatrix} -2 & 2 \\ 2 & 1 \end{bmatrix} \begin{bmatrix} -\frac{2}{\sqrt{5}} & \frac{1}{\sqrt{5}} \\ \frac{1}{\sqrt{5}} & \frac{2}{\sqrt{5}} \end{bmatrix} = \begin{bmatrix} -3 & 0 \\ 0 & 2 \end{bmatrix}$$

EXAMPLE 9 Orthogonal Diagonalization

Find an orthogonal matrix P that diagonalizes

$$A = \begin{bmatrix} 2 & 2 & -2 \\ 2 & -1 & 4 \\ -2 & 4 & -1 \end{bmatrix}.$$

SOLUTION

1. The characteristic polynomial of A, $|\lambda I - A| = (\lambda - 3)^2(\lambda + 6)$, yields the eigenvalues $\lambda_1 = -6$ and $\lambda_2 = 3$. λ_1 has a multiplicity of 1 and λ_2 has a multiplicity of 2.
2. An eigenvector for λ_1 is $\mathbf{v}_1 = (1, -2, 2)$, which normalizes to

$$\mathbf{u}_1 = \frac{\mathbf{v}_1}{\|\mathbf{v}_1\|} = \left(\frac{1}{3}, -\frac{2}{3}, \frac{2}{3}\right).$$

3. Two eigenvectors for λ_2 are $\mathbf{v}_2 = (2, 1, 0)$ and $\mathbf{v}_3 = (-2, 0, 1)$. Note that $\mathbf{v}_1$ is orthogonal to $\mathbf{v}_2$ and $\mathbf{v}_3$, as guaranteed by Theorem 7.9. The eigenvectors $\mathbf{v}_2$ and $\mathbf{v}_3$, however, are not orthogonal to each other. To find two orthonormal eigenvectors for λ_2, use the Gram-Schmidt process as follows.

$$\mathbf{w}_2 = \mathbf{v}_2 = (2, 1, 0)$$

$$\mathbf{w}_3 = \mathbf{v}_3 - \left(\frac{\mathbf{v}_3 \cdot \mathbf{w}_2}{\mathbf{w}_2 \cdot \mathbf{w}_2}\right)\mathbf{w}_2 = \left(-\frac{2}{5}, \frac{4}{5}, 1\right)$$

These vectors normalize to

$$\mathbf{u}_2 = \frac{\mathbf{w}_2}{\|\mathbf{w}_2\|} = \left(\frac{2}{\sqrt{5}}, \frac{1}{\sqrt{5}}, 0\right)$$

$$\mathbf{u}_3 = \frac{\mathbf{w}_3}{\|\mathbf{w}_3\|} = \left(-\frac{2}{3\sqrt{5}}, \frac{4}{3\sqrt{5}}, \frac{5}{3\sqrt{5}}\right).$$

4. The matrix P has $\mathbf{u}_1$, $\mathbf{u}_2$, and $\mathbf{u}_3$ as its column vectors.

$$P = \begin{bmatrix} \frac{1}{3} & \frac{2}{\sqrt{5}} & -\frac{2}{3\sqrt{5}} \\ -\frac{2}{3} & \frac{1}{\sqrt{5}} & \frac{4}{3\sqrt{5}} \\ \frac{2}{3} & 0 & \frac{5}{3\sqrt{5}} \end{bmatrix}$$

A check shows that

$$P^{-1}AP = P^TAP = \begin{bmatrix} -6 & 0 & 0 \\ 0 & 3 & 0 \\ 0 & 0 & 3 \end{bmatrix}.$$

SECTION 7.3 Exercises

In Exercises 1–6, determine whether the matrix is symmetric.

1. $\begin{bmatrix} 1 & 3 \\ 3 & -1 \end{bmatrix}$

2. $\begin{bmatrix} 6 & -2 \\ -2 & 1 \end{bmatrix}$

3. $\begin{bmatrix} 4 & -2 & 1 \\ 3 & 1 & 2 \\ 1 & 2 & 1 \end{bmatrix}$

4. $\begin{bmatrix} 1 & -5 & 4 \\ -5 & 3 & 6 \\ -4 & 6 & 2 \end{bmatrix}$

5. $\begin{bmatrix} 0 & 1 & 2 & -1 \\ 1 & 0 & -3 & 2 \\ 2 & -3 & 0 & 1 \\ -1 & 2 & 1 & -2 \end{bmatrix}$

6. $\begin{bmatrix} 2 & 0 & 3 & 5 \\ 0 & 11 & 0 & -2 \\ 3 & 0 & 5 & 0 \\ 5 & -2 & 0 & 1 \end{bmatrix}$

In Exercises 7–14, find the eigenvalues of the symmetric matrix. For each eigenvalue, find the dimension of the corresponding eigenspace.

7. $\begin{bmatrix} 3 & 1 \\ 1 & 3 \end{bmatrix}$

8. $\begin{bmatrix} 2 & 0 \\ 0 & 2 \end{bmatrix}$

9. $\begin{bmatrix} 3 & 0 & 0 \\ 0 & 2 & 0 \\ 0 & 0 & 2 \end{bmatrix}$

10. $\begin{bmatrix} 2 & 1 & 1 \\ 1 & 2 & 1 \\ 1 & 1 & 2 \end{bmatrix}$

11. $\begin{bmatrix} 0 & 2 & 2 \\ 2 & 0 & 2 \\ 2 & 2 & 0 \end{bmatrix}$

12. $\begin{bmatrix} 0 & 4 & 4 \\ 4 & 2 & 0 \\ 4 & 0 & -2 \end{bmatrix}$

13. $\begin{bmatrix} 0 & 1 & 1 \\ 1 & 0 & 1 \\ 1 & 1 & 1 \end{bmatrix}$

14. $\begin{bmatrix} 2 & -1 & -1 \\ -1 & 2 & -1 \\ -1 & -1 & 2 \end{bmatrix}$

In Exercises 15–22, determine whether the matrix is orthogonal.

15. $\begin{bmatrix} \frac{\sqrt{2}}{2} & \frac{\sqrt{2}}{2} \\ -\frac{\sqrt{2}}{2} & \frac{\sqrt{2}}{2} \end{bmatrix}$

16. $\begin{bmatrix} \frac{2}{3} & -\frac{2}{3} \\ \frac{2}{3} & \frac{1}{3} \end{bmatrix}$

17. $\begin{bmatrix} -4 & 0 & 3 \\ 0 & 1 & 0 \\ 3 & 0 & 4 \end{bmatrix}$

18. $\begin{bmatrix} -\frac{4}{5} & 0 & \frac{3}{5} \\ 0 & 1 & 0 \\ \frac{3}{5} & 0 & \frac{4}{5} \end{bmatrix}$

19. $\begin{bmatrix} \frac{\sqrt{2}}{2} & -\frac{\sqrt{6}}{6} & \frac{\sqrt{3}}{3} \\ 0 & \frac{\sqrt{6}}{3} & \frac{\sqrt{3}}{3} \\ \frac{\sqrt{2}}{2} & \frac{\sqrt{6}}{6} & -\frac{\sqrt{3}}{3} \end{bmatrix}$

20. $\begin{bmatrix} \frac{\sqrt{2}}{3} & 0 & \frac{\sqrt{5}}{2} \\ 0 & \frac{2\sqrt{5}}{5} & 0 \\ -\frac{\sqrt{2}}{6} & -\frac{\sqrt{5}}{5} & \frac{1}{2} \end{bmatrix}$

21. $\begin{bmatrix} \frac{1}{10}\sqrt{10} & 0 & 0 & -\frac{3}{10}\sqrt{10} \\ 0 & 0 & 1 & 0 \\ 0 & 1 & 0 & 0 \\ \frac{3}{10}\sqrt{10} & 0 & 0 & \frac{1}{10}\sqrt{10} \end{bmatrix}$

22. $\begin{bmatrix} \frac{1}{8} & 0 & 0 & \frac{3\sqrt{7}}{8} \\ 0 & 1 & 0 & 0 \\ 0 & 0 & 1 & 0 \\ \frac{3\sqrt{7}}{8} & 0 & 0 & \frac{1}{8} \end{bmatrix}$

In Exercises 23–32, find an orthogonal matrix P such that P^TAP diagonalizes A. Verify that P^TAP gives the proper diagonal form.

23. $A = \begin{bmatrix} 1 & 1 \\ 1 & 1 \end{bmatrix}$

24. $A = \begin{bmatrix} 4 & 2 \\ 2 & 4 \end{bmatrix}$

25. $A = \begin{bmatrix} 2 & \sqrt{2} \\ \sqrt{2} & 1 \end{bmatrix}$

26. $A = \begin{bmatrix} 0 & 1 & 1 \\ 1 & 0 & 1 \\ 1 & 1 & 0 \end{bmatrix}$

27. $A = \begin{bmatrix} 0 & 10 & 10 \\ 10 & 5 & 0 \\ 10 & 0 & -5 \end{bmatrix}$

28. $A = \begin{bmatrix} 0 & 3 & 0 \\ 3 & 0 & 4 \\ 0 & 4 & 0 \end{bmatrix}$

29. $A = \begin{bmatrix} 1 & -1 & 2 \\ -1 & 1 & 2 \\ 2 & 2 & 2 \end{bmatrix}$

30. $A = \begin{bmatrix} -2 & 2 & 4 \\ 2 & -2 & 4 \\ 4 & 4 & 4 \end{bmatrix}$

31. $A = \begin{bmatrix} 4 & 2 & 0 & 0 \\ 2 & 4 & 0 & 0 \\ 0 & 0 & 4 & 2 \\ 0 & 0 & 2 & 4 \end{bmatrix}$

32. $A = \begin{bmatrix} 1 & 1 & 0 & 0 \\ 1 & 1 & 0 & 0 \\ 0 & 0 & 1 & 1 \\ 0 & 0 & 1 & 1 \end{bmatrix}$

True or False? In Exercises 33 and 34, determine whether each statement is true or false. If a statement is true, give a reason or cite an appropriate statement from the text. If a statement is false, provide an example that shows the statement is not true in all cases or cite an appropriate statement from the text.

33. (a) Let A be an $n \times n$ matrix. Then A is symmetric if and only if A is orthogonally diagonalizable.

(b) The eigenvectors corresponding to distinct eigenvalues are orthogonal for symmetric matrices.

34. (a) A square matrix P is orthogonal if it is invertible—that is, if $P^{-1} = P^T$.

(b) If A is an $n \times n$ symmetric matrix, then A is orthogonally diagonalizable and has real eigenvalues.

35. Prove that if A is an $m \times n$ matrix, then A^TA and AA^T are symmetric.

36. Find A^TA and AA^T for the matrix below.

$$A = \begin{bmatrix} 1 & -3 & 2 \\ 4 & -6 & 1 \end{bmatrix}$$

37. Prove that if A is an orthogonal matrix, then $|A| = \pm 1$.

38. Prove that if A and B are $n \times n$ orthogonal matrices, then AB and BA are orthogonal.

39. Show that the matrix below is orthogonal for any value of θ.

$$A = \begin{bmatrix} \cos\theta & -\sin\theta \\ \sin\theta & \cos\theta \end{bmatrix}$$

40. Prove that if a symmetric matrix A has only one eigenvalue λ, then $A = \lambda I$.

41. Prove that if A is an orthogonal matrix, then so are A^T and A^{-1}.

7.4 Applications of Eigenvalues and Eigenvectors

Population Growth

Matrices can be used to form models for population growth. The first step in this process is to group the population into age classes of equal duration. For instance, if the maximum life span of a member is L years, then the age classes are represented by the n intervals shown below.

First age class, *Second age class*, ..., *nth age class*

$$\left[0, \frac{L}{n}\right), \quad \left[\frac{L}{n}, \frac{2L}{n}\right), \; . \; . \; . \; , \left[\frac{(n-1)L}{n}, L\right]$$

The number of population members in each age class is then represented by the **age distribution vector**

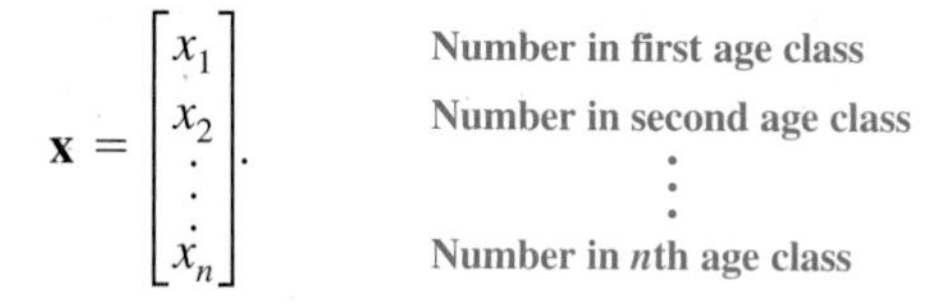

$$\mathbf{x} = \begin{bmatrix} x_1 \\ x_2 \\ \vdots \\ x_n \end{bmatrix}.$$

Over a period of L/n years, the *probability* that a member of the ith age class will survive to become a member of the $(i + 1)$th age class is given by p_i, where

$$0 \le p_i \le 1, \quad i = 1, 2, . . . , n - 1.$$

The *average number* of offspring produced by a member of the ith age class is given by b_i, where

$$0 \le b_i, \quad i = 1, 2, . . . , n.$$

These numbers can be written in matrix form, as shown below.

$$A = \begin{bmatrix} b_1 & b_2 & b_3 & \cdots & b_{n-1} & b_n \\ p_1 & 0 & 0 & \cdots & 0 & 0 \\ 0 & p_2 & 0 & \cdots & 0 & 0 \\ \vdots & \vdots & \vdots & & \vdots & \vdots \\ 0 & 0 & 0 & \cdots & p_{n-1} & 0 \end{bmatrix}$$

Multiplying this **age transition matrix** by the age distribution vector for a specific time period produces the age distribution vector for the next time period. That is,

$$A\mathbf{x}_i = \mathbf{x}_{i+1}.$$

This procedure is illustrated in Example 1.

EXAMPLE 1 A Population Growth Model

A population of rabbits raised in a research laboratory has the characteristics listed below.

(a) Half of the rabbits survive their first year. Of those, half survive their second year. The maximum life span is 3 years.

(b) During the first year, the rabbits produce no offspring. The average number of offspring is 6 during the second year and 8 during the third year.

The laboratory population now consists of 24 rabbits in the first age class, 24 in the second, and 20 in the third. How many rabbits will be in each age class in 1 year?

SOLUTION The current age distribution vector is

$$\mathbf{x}_1 = \begin{bmatrix} 24 \\ 24 \\ 20 \end{bmatrix} \begin{matrix} 0 \le \text{age} < 1 \\ 1 \le \text{age} < 2 \\ 2 \le \text{age} \le 3 \end{matrix}$$

and the age transition matrix is

$$A = \begin{bmatrix} 0 & 6 & 8 \\ 0.5 & 0 & 0 \\ 0 & 0.5 & 0 \end{bmatrix}.$$

After 1 year the age distribution vector will be

$$\mathbf{x}_2 = A\mathbf{x}_1 = \begin{bmatrix} 0 & 6 & 8 \\ 0.5 & 0 & 0 \\ 0 & 0.5 & 0 \end{bmatrix} \begin{bmatrix} 24 \\ 24 \\ 20 \end{bmatrix} = \begin{bmatrix} 304 \\ 12 \\ 12 \end{bmatrix}. \begin{matrix} 0 \le \text{age} < 1 \\ 1 \le \text{age} < 2 \\ 2 \le \text{age} \le 3 \end{matrix}$$

If the pattern of growth in Example 1 continued for another year, the rabbit population would be

$$\mathbf{x}_3 = A\mathbf{x}_2 = \begin{bmatrix} 0 & 6 & 8 \\ 0.5 & 0 & 0 \\ 0 & 0.5 & 0 \end{bmatrix} \begin{bmatrix} 304 \\ 12 \\ 12 \end{bmatrix} = \begin{bmatrix} 168 \\ 152 \\ 6 \end{bmatrix}. \quad \begin{matrix} 0 \le \text{age} < 1 \\ 1 \le \text{age} < 2 \\ 2 \le \text{age} \le 3 \end{matrix}$$

From the age distribution vectors $\mathbf{x}_1$, $\mathbf{x}_2$, and $\mathbf{x}_3$, you can see that the percent of rabbits in each of the three age classes changes each year. Suppose the laboratory prefers a stable growth pattern, one in which the percent in each age class remains the same each year. For this stable growth pattern to be achieved, the $(n + 1)$th age distribution vector must be a scalar multiple of the nth age distribution vector. That is, $A\mathbf{x}_n = \mathbf{x}_{n+1} = \lambda\mathbf{x}_n$. So, the laboratory can obtain a growth pattern in which the percent in each age class remains constant each year by finding an eigenvector of A. Example 2 shows how to solve this problem.

EXAMPLE 2 Finding a Stable Age Distribution Vector

Find a stable age distribution vector for the population in Example 1.

SOLUTION To solve this problem, find an eigenvalue λ and a corresponding eigenvector $\mathbf{x}$ such that

$$A\mathbf{x} = \lambda\mathbf{x}.$$

Simulation

Explore this concept further with an electronic simulation available at *college.hmco.com/pic/larsonELA6e.*

The characteristic polynomial of A is

$$|\lambda I - A| = \begin{vmatrix} \lambda & -6 & -8 \\ -0.5 & \lambda & 0 \\ 0 & -0.5 & \lambda \end{vmatrix} = \lambda^3 - 3\lambda - 2 = (\lambda + 1)^2(\lambda - 2),$$

which implies that the eigenvalues are -1 and 2. Choosing the positive value, let $\lambda = 2$. To find a corresponding eigenvector, row reduce the matrix $2I - A$ to obtain

$$\begin{bmatrix} 2 & -6 & -8 \\ -0.5 & 2 & 0 \\ 0 & -0.5 & 2 \end{bmatrix} \longrightarrow \begin{bmatrix} 1 & 0 & -16 \\ 0 & 1 & -4 \\ 0 & 0 & 0 \end{bmatrix}.$$

So, the eigenvectors of $\lambda = 2$ are of the form

$$\mathbf{x} = \begin{bmatrix} x_1 \\ x_2 \\ x_3 \end{bmatrix} = \begin{bmatrix} 16t \\ 4t \\ t \end{bmatrix} = t\begin{bmatrix} 16 \\ 4 \\ 1 \end{bmatrix}.$$

For instance, if $t = 2$, then the initial age distribution vector would be

$$\mathbf{x}_1 = \begin{bmatrix} 32 \\ 8 \\ 2 \end{bmatrix} \quad \begin{matrix} 0 \le \text{age} < 1 \\ 1 \le \text{age} < 2 \\ 2 \le \text{age} \le 3 \end{matrix}$$

and the age distribution vector for the next year would be

$$\mathbf{x}_2 = A\mathbf{x}_1 = \begin{bmatrix} 0 & 6 & 8 \\ 0.5 & 0 & 0 \\ 0 & 0.5 & 0 \end{bmatrix} \begin{bmatrix} 32 \\ 8 \\ 2 \end{bmatrix} = \begin{bmatrix} 64 \\ 16 \\ 4 \end{bmatrix}. \quad \begin{matrix} 0 \le \text{age} < 1 \\ 1 \le \text{age} < 2 \\ 2 \le \text{age} \le 3 \end{matrix}$$

Notice that the ratio of the three age classes is still 16 : 4 : 1, and so the percent of the population in each age class remains the same.

Systems of Linear Differential Equations (Calculus)

A **system of first-order linear differential equations** has the form

$$\begin{aligned} y_1' &= a_{11}y_1 + a_{12}y_2 + \cdot \cdot \cdot + a_{1n}y_n \\ y_2' &= a_{21}y_1 + a_{22}y_2 + \cdot \cdot \cdot + a_{2n}y_n \\ &\vdots \\ y_n' &= a_{n1}y_1 + a_{n2}y_2 + \cdot \cdot \cdot + a_{nn}y_n, \end{aligned}$$

where each y_i is a function of t and $y_i' = dy_i/dt$. If you let

$$\mathbf{y} = \begin{bmatrix} y_1 \\ y_2 \\ \vdots \\ y_n \end{bmatrix} \quad \text{and} \quad \mathbf{y}' = \begin{bmatrix} y_1' \\ y_2' \\ \vdots \\ y_n' \end{bmatrix},$$

then the system can be written in matrix form as $\mathbf{y}' = A\mathbf{y}$.

EXAMPLE 3 Solving a System of Linear Differential Equations

Solve the system of linear differential equations.

$$\begin{aligned} y_1' &= 4y_1 \\ y_2' &= -y_2 \\ y_3' &= 2y_3 \end{aligned}$$

SOLUTION From calculus you know that the solution of the differential equation $y' = ky$ is

$$y = Ce^{kt}.$$

So, the solution of the system is

$$\begin{aligned} y_1 &= C_1e^{4t} \\ y_2 &= C_2e^{-t} \\ y_3 &= C_3e^{2t}. \end{aligned}$$

The matrix form of the system of linear differential equations in Example 3 is $\mathbf{y}' = A\mathbf{y}$, or

$$\begin{bmatrix} y_1' \\ y_2' \\ y_3' \end{bmatrix} = \begin{bmatrix} 4 & 0 & 0 \\ 0 & -1 & 0 \\ 0 & 0 & 2 \end{bmatrix} \begin{bmatrix} y_1 \\ y_2 \\ y_3 \end{bmatrix}.$$

So, the coefficients of t in the solutions $y_i = C_i e^{\lambda_i t}$ are provided by the *eigenvalues* of the matrix A.

If A is a *diagonal* matrix, then the solution of

$$\mathbf{y}' = A\mathbf{y}$$

can be obtained immediately, as in Example 3. If A is *not* diagonal, then the solution requires a little more work. First attempt to find a matrix P that diagonalizes A. Then the change of variables represented by $\mathbf{y} = P\mathbf{w}$ and $\mathbf{y}' = P\mathbf{w}'$ produces

$$P\mathbf{w}' = \mathbf{y}' = A\mathbf{y} = AP\mathbf{w} \quad \longrightarrow \quad \mathbf{w}' = P^{-1}AP\mathbf{w},$$

where $P^{-1}AP$ is a diagonal matrix. This procedure is demonstrated in Example 4.

EXAMPLE 4 Solving a System of Linear Differential Equations

Solve the system of linear differential equations.

$$\begin{aligned} y_1' &= 3y_1 + 2y_2 \\ y_2' &= 6y_1 - y_2 \end{aligned}$$

SOLUTION First find a matrix P that diagonalizes $A = \begin{bmatrix} 3 & 2 \\ 6 & -1 \end{bmatrix}$. The eigenvalues of A are $\lambda_1 = -3$ and $\lambda_2 = 5$, with corresponding eigenvectors

$$\mathbf{p}_1 = \begin{bmatrix} 1 \\ -3 \end{bmatrix} \quad \text{and} \quad \mathbf{p}_2 = \begin{bmatrix} 1 \\ 1 \end{bmatrix}.$$

Diagonalize A using the matrix P whose columns consist of $\mathbf{p}_1$ and $\mathbf{p}_2$ to obtain

$$P = \begin{bmatrix} 1 & 1 \\ -3 & 1 \end{bmatrix}, \quad P^{-1} = \begin{bmatrix} \frac{1}{4} & -\frac{1}{4} \\ \frac{3}{4} & \frac{1}{4} \end{bmatrix}, \quad \text{and} \quad P^{-1}AP = \begin{bmatrix} -3 & 0 \\ 0 & 5 \end{bmatrix}.$$

The system represented by $\mathbf{w}' = P^{-1}AP\mathbf{w}$ has the following form.

$$\begin{bmatrix} w_1' \\ w_2' \end{bmatrix} = \begin{bmatrix} -3 & 0 \\ 0 & 5 \end{bmatrix} \begin{bmatrix} w_1 \\ w_2 \end{bmatrix} \quad \longrightarrow \quad \begin{aligned} w_1' &= -3w_1 \\ w_2' &= 5w_2 \end{aligned}$$

The solution of this system of equations is

$$\begin{aligned} w_1 &= C_1 e^{-3t} \\ w_2 &= C_2 e^{5t}. \end{aligned}$$

To return to the original variables y_1 and y_2, use the substitution $\mathbf{y} = P\mathbf{w}$ and write

$$\begin{bmatrix} y_1 \\ y_2 \end{bmatrix} = \begin{bmatrix} 1 & 1 \\ -3 & 1 \end{bmatrix} \begin{bmatrix} w_1 \\ w_2 \end{bmatrix},$$

which implies that the solution is

$$\begin{aligned} y_1 &= w_1 + w_2 = C_1 e^{-3t} + C_2 e^{5t} \\ y_2 &= -3w_1 + w_2 = -3C_1 e^{-3t} + C_2 e^{5t}. \end{aligned}$$

For the systems of linear differential equations in Examples 3 and 4, you found that each y_i can be written as a linear combination of $e^{\lambda_1 t}, e^{\lambda_2 t}, \ldots, e^{\lambda_n t}$, where $\lambda_1, \lambda_2, \ldots, \lambda_n$ are *distinct real eigenvalues* of the $n \times n$ matrix A. If A has eigenvalues with multiplicity greater than 1 or if A has complex eigenvalues, then the technique for solving the system must be modified. If you take a course on differential equations you will cover these two cases. For now, you can get an idea of the type of modification required from the next two systems of linear differential equations.

1. *Eigenvalues with multiplicity greater than* 1: The coefficient matrix of the system

$$\begin{aligned} y_1' &= y_2 \\ y_2' &= -4y_1 + 4y_2 \end{aligned} \quad \text{is} \quad A = \begin{bmatrix} 0 & 1 \\ -4 & 4 \end{bmatrix}.$$

The only eigenvalue of A is $\lambda = 2$, and the solution of the system of linear differential equations is

$$\begin{aligned} y_1 &= C_1 e^{2t} + C_2 t e^{2t} \\ y_2 &= (2C_1 + C_2)e^{2t} + 2C_2 t e^{2t}. \end{aligned}$$

2. *Complex eigenvalues:* The coefficient matrix of the system

$$\begin{aligned} y_1' &= -y_2 \\ y_2' &= y_1 \end{aligned} \quad \text{is} \quad A = \begin{bmatrix} 0 & -1 \\ 1 & 0 \end{bmatrix}.$$

The eigenvalues of A are $\lambda_1 = i$ and $\lambda_2 = -i$, and the solution of the system of linear differential equations is

$$\begin{aligned} y_1 &= C_1 \cos t + C_2 \sin t \\ y_2 &= -C_2 \cos t + C_1 \sin t. \end{aligned}$$

Try checking these solutions by differentiating and substituting into the original systems of equations.

Quadratic Forms

Eigenvalues and eigenvectors can be used to solve the rotation of axes problem introduced in Section 4.8. Recall that classifying the graph of the quadratic equation

$$ax^2 + bxy + cy^2 + dx + ey + f = 0 \qquad \textbf{Quadratic equation}$$

is fairly straightforward as long as the equation has no xy-term (that is, $b = 0$). If the equation has an xy-term, however, then the classification is accomplished most easily by first performing a rotation of axes that eliminates the xy-term. The resulting equation (relative to the new $x'y'$-axes) will then be of the form

$$a'(x')^2 + c'(y')^2 + d'x' + e'y' + f' = 0.$$

You will see that the coefficients a' and c' are eigenvalues of the matrix

$$A = \begin{bmatrix} a & b/2 \\ b/2 & c \end{bmatrix}.$$

The expression

$$ax^2 + bxy + cy^2$$ Quadratic form

is called the **quadratic form** associated with the quadratic equation $ax^2 + bxy + cy^2 + dx + ey + f = 0$, and the matrix A is called the **matrix of the quadratic form.** Note that the matrix A *is symmetric* by definition. Moreover, the matrix A will be diagonal if and only if its corresponding quadratic form has no xy-term, as illustrated in Example 5.

EXAMPLE 5 Finding the Matrix of a Quadratic Form

Find the matrix of the quadratic form associated with each quadratic equation.

(a) $4x^2 + 9y^2 - 36 = 0$ (b) $13x^2 - 10xy + 13y^2 - 72 = 0$

SOLUTION (a) Because $a = 4$, $b = 0$, and $c = 9$, the matrix is

$$A = \begin{bmatrix} 4 & 0 \\ 0 & 9 \end{bmatrix}.$$ Diagonal matrix (no xy-term)

(b) Because $a = 13$, $b = -10$, and $c = 13$, the matrix is

$$A = \begin{bmatrix} 13 & -5 \\ -5 & 13 \end{bmatrix}.$$ Nondiagonal matrix (xy-term)

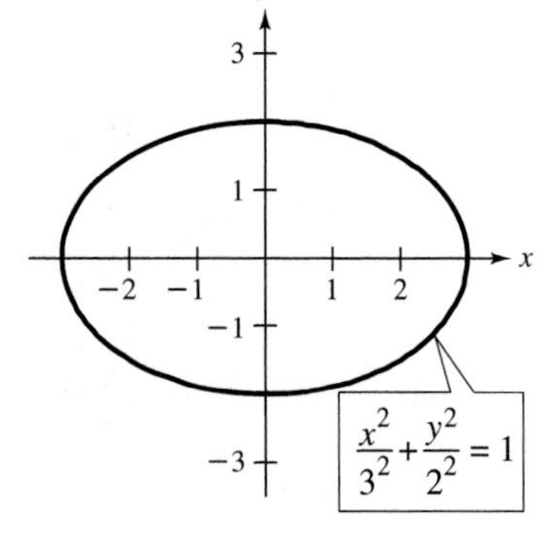

Figure 7.3

In standard form, the equation $4x^2 + 9y^2 - 36 = 0$ is

$$\frac{x^2}{3^2} + \frac{y^2}{2^2} = 1,$$

which is the equation of the ellipse shown in Figure 7.3. Although it is not apparent by simple inspection, the graph of the equation $13x^2 - 10xy + 13y^2 - 72 = 0$ is similar. In fact, if you rotate the x- and y-axes counterclockwise 45° to form a new $x'y'$-coordinate system, this equation takes the form

$$\frac{(x')^2}{3^2} + \frac{(y')^2}{2^2} = 1,$$

which is the equation of the ellipse shown in Figure 7.4.

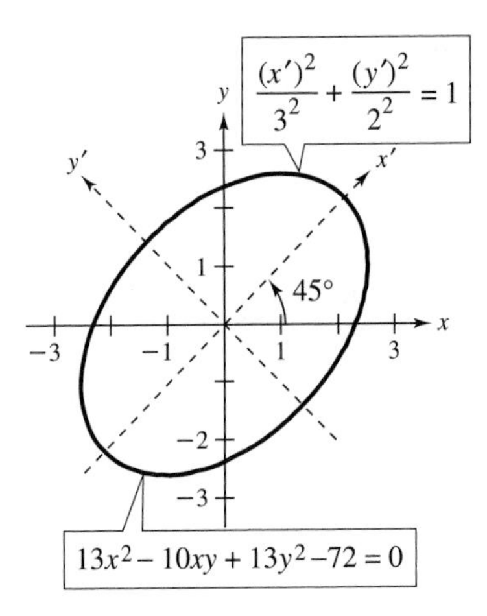

Figure 7.4

To see how the matrix of a quadratic form can be used to perform a rotation of axes, let

$$X = \begin{bmatrix} x \\ y \end{bmatrix}.$$

Then the quadratic expression $ax^2 + bxy + cy^2 + dx + ey + f$ can be written in matrix form as follows.

$$\begin{aligned} X^TAX + [d \quad e]X + f &= [x \quad y]\begin{bmatrix} a & b/2 \\ b/2 & c \end{bmatrix}\begin{bmatrix} x \\ y \end{bmatrix} + [d \quad e]\begin{bmatrix} x \\ y \end{bmatrix} + f \\ &= ax^2 + bxy + cy^2 + dx + ey + f \end{aligned}$$

If $b = 0$, no rotation is necessary. But if $b \neq 0$, then because A is symmetric, you can apply Theorem 7.10 to conclude that there exists an orthogonal matrix P such that $P^TAP = D$ is diagonal. So, if you let

$$P^TX = X' = \begin{bmatrix} x' \\ y' \end{bmatrix},$$

it follows that $X = PX'$, and you have

$$\begin{aligned} X^TAX &= (PX')^TA(PX') \\ &= (X')^TP^TAPX' \\ &= (X')^TDX'. \end{aligned}$$

The choice of the matrix P must be made with care. Because P is orthogonal, its determinant will be ± 1. It can be shown (see Exercise 55) that if P is chosen so $|P| = 1$, then P will be of the form

$$P = \begin{bmatrix} \cos\theta & -\sin\theta \\ \sin\theta & \cos\theta \end{bmatrix},$$

where θ gives the angle of rotation of the conic measured from the positive x-axis to the positive x'-axis. This brings you to the next theorem, the **Principal Axes Theorem.**

Principal Axes Theorem

For a conic whose equation is $ax^2 + bxy + cy^2 + dx + ey + f = 0$, the rotation given by $X = PX'$ eliminates the xy-term if P is an orthogonal matrix, with $|P| = 1$, that diagonalizes A. That is,

$$P^TAP = \begin{bmatrix} \lambda_1 & 0 \\ 0 & \lambda_2 \end{bmatrix},$$

where λ_1 and λ_2 are eigenvalues of A. The equation of the rotated conic is given by

$$\lambda_1(x')^2 + \lambda_2(y')^2 + [d \quad e]PX' + f = 0.$$

REMARK: Note that the matrix product $[d \quad e]PX'$ has the form

$$[d \quad e]PX' = (d\cos\theta + e\sin\theta)x' + (-d\sin\theta + e\cos\theta)y'.$$

EXAMPLE 6 **Rotation of a Conic**

Perform a rotation of axes to eliminate the xy-term in the quadratic equation

$$13x^2 - 10xy + 13y^2 - 72 = 0.$$

SOLUTION The matrix of the quadratic form associated with this equation is

$$A = \begin{bmatrix} 13 & -5 \\ -5 & 13 \end{bmatrix}.$$

Because the characteristic polynomial of A is

$$\begin{vmatrix} \lambda - 13 & 5 \\ 5 & \lambda - 13 \end{vmatrix} = (\lambda - 13)^2 - 25 = (\lambda - 8)(\lambda - 18),$$

it follows that the eigenvalues of A are $\lambda_1 = 8$ and $\lambda_2 = 18$. So, the equation of the rotated conic is

$$8(x')^2 + 18(y')^2 - 72 = 0,$$

which, when written in the standard form

$$\frac{(x')^2}{3^2} + \frac{(y')^2}{2^2} = 1,$$

is the equation of an ellipse. (See Figure 7.4.)

In Example 6, the eigenvectors of the matrix A are

$$\mathbf{x}_1 = \begin{bmatrix} 1 \\ 1 \end{bmatrix} \quad \text{and} \quad \mathbf{x}_2 = \begin{bmatrix} -1 \\ 1 \end{bmatrix},$$

which you can normalize to form the columns of P, as follows.

$$P = \begin{bmatrix} \frac{1}{\sqrt{2}} & -\frac{1}{\sqrt{2}} \\ \frac{1}{\sqrt{2}} & \frac{1}{\sqrt{2}} \end{bmatrix} = \begin{bmatrix} \cos\theta & -\sin\theta \\ \sin\theta & \cos\theta \end{bmatrix}$$

Note first that $|P| = 1$, which implies that P is a rotation. Moreover, because $\cos 45° = 1/\sqrt{2} = \sin 45°$, you can conclude that $\theta = 45°$, as shown in Figure 7.4.

The orthogonal matrix P specified in the Principal Axes Theorem is not unique. Its entries depend on the ordering of the eigenvalues λ_1 and λ_2 *and* on the subsequent choice of eigenvectors $\mathbf{x}_1$ and $\mathbf{x}_2$. For instance, in the solution of Example 6, any of the following choices of P (see the next page) would have worked.

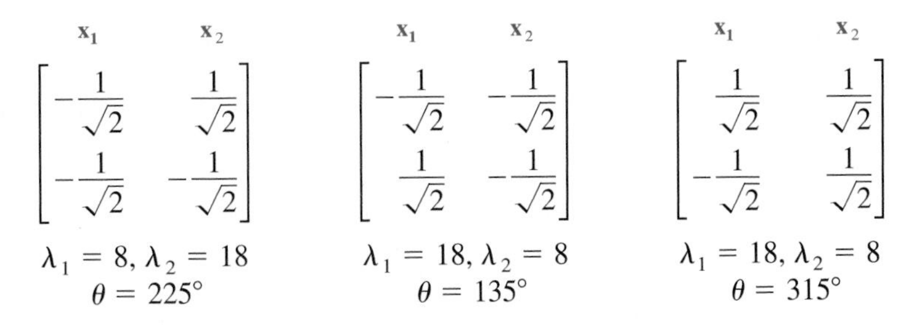

$$\begin{array}{ccc} \begin{bmatrix} -\frac{1}{\sqrt{2}} & \frac{1}{\sqrt{2}} \\ -\frac{1}{\sqrt{2}} & -\frac{1}{\sqrt{2}} \end{bmatrix} & \begin{bmatrix} -\frac{1}{\sqrt{2}} & -\frac{1}{\sqrt{2}} \\ \frac{1}{\sqrt{2}} & -\frac{1}{\sqrt{2}} \end{bmatrix} & \begin{bmatrix} \frac{1}{\sqrt{2}} & \frac{1}{\sqrt{2}} \\ -\frac{1}{\sqrt{2}} & \frac{1}{\sqrt{2}} \end{bmatrix} \\ \lambda_1 = 8, \lambda_2 = 18 & \lambda_1 = 18, \lambda_2 = 8 & \lambda_1 = 18, \lambda_2 = 8 \\ \theta = 225^\circ & \theta = 135^\circ & \theta = 315^\circ \end{array}$$

(In each matrix, the first column is labelled $\mathbf{x}_1$ and the second column $\mathbf{x}_2$.)

For any of these choices of P, the graph of the rotated conic will, of course, be the same. (See Figure 7.5.)

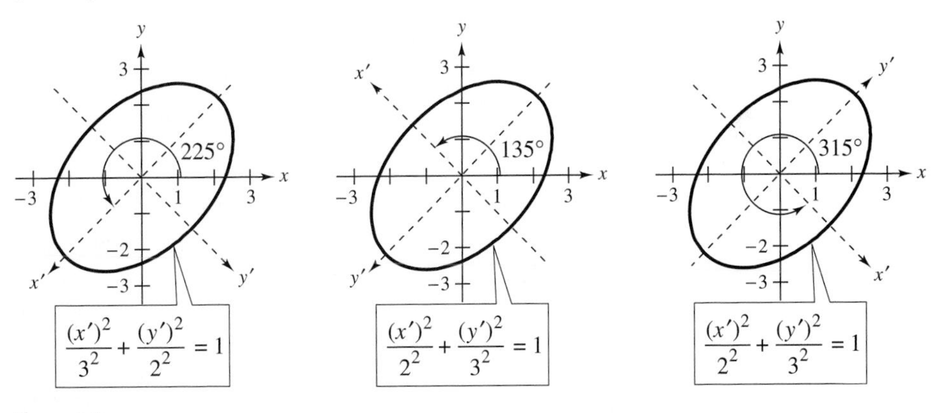

Figure 7.5

The steps used to apply the Principal Axes Theorem are summarized as follows.

1. Form the matrix A and find its eigenvalues λ_1 and λ_2.
2. Find eigenvectors corresponding to λ_1 and λ_2. Normalize these eigenvectors to form the columns of P.
3. If $|P| = -1$, then multiply one of the columns of P by -1 to obtain a matrix of the form

$$P = \begin{bmatrix} \cos\theta & -\sin\theta \\ \sin\theta & \cos\theta \end{bmatrix}.$$

4. The angle θ represents the angle of rotation of the conic.
5. The equation of the rotated conic is $\lambda_1(x')^2 + \lambda_2(y')^2 + [d \quad e]PX' + f = 0$.

Example 7 shows how to apply the Principal Axes Theorem to rotate a conic whose center has been translated away from the origin.

EXAMPLE 7 Rotation of a Conic

Perform a rotation of axes to eliminate the xy-term in the quadratic equation

$$3x^2 - 10xy + 3y^2 + 16\sqrt{2}x - 32 = 0.$$

SOLUTION The matrix of the quadratic form associated with this equation is

$$A = \begin{bmatrix} 3 & -5 \\ -5 & 3 \end{bmatrix}.$$

The eigenvalues of A are $\lambda_1 = 8$ and $\lambda_2 = -2$, with corresponding eigenvectors of $\mathbf{x}_1 = (-1, 1)$ and $\mathbf{x}_2 = (-1, -1)$. This implies that the matrix P is

$$P = \begin{bmatrix} -\frac{1}{\sqrt{2}} & -\frac{1}{\sqrt{2}} \\ \frac{1}{\sqrt{2}} & -\frac{1}{\sqrt{2}} \end{bmatrix} = \begin{bmatrix} \cos\theta & -\sin\theta \\ \sin\theta & \cos\theta \end{bmatrix}, \text{ where } |P| = 1.$$

Because $\cos 135° = -1/\sqrt{2}$ and $\sin 135° = 1/\sqrt{2}$, you can conclude that the angle of rotation is 135°. Finally, from the matrix product

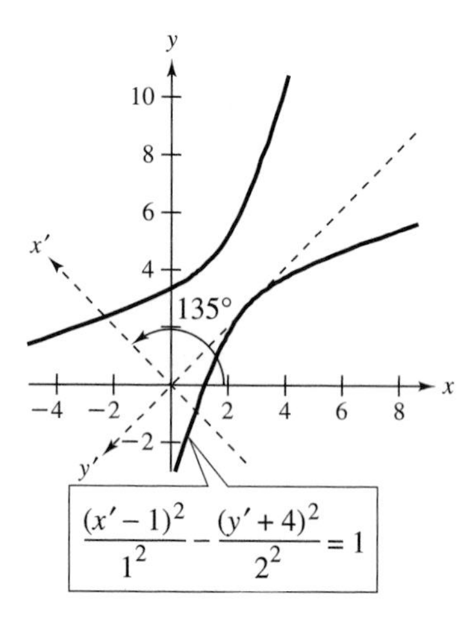

Figure 7.6

$$\begin{aligned}[d \quad e]PX' &= [16\sqrt{2} \quad 0]\begin{bmatrix} -\frac{1}{\sqrt{2}} & -\frac{1}{\sqrt{2}} \\ \frac{1}{\sqrt{2}} & -\frac{1}{\sqrt{2}} \end{bmatrix}\begin{bmatrix} x' \\ y' \end{bmatrix} \\ &= -16x' - 16y',\end{aligned}$$

you can conclude that the equation of the rotated conic is

$$8(x')^2 - 2(y')^2 - 16x' - 16y' - 32 = 0.$$

In standard form, the equation

$$\frac{(x'-1)^2}{1^2} - \frac{(y'+4)^2}{2^2} = 1$$

is the equation of a hyperbola. Its graph is shown in Figure 7.6.

Quadratic forms can also be used to analyze equations of quadric surfaces in space, which are the three-dimensional analogues of conic sections. The equation of a quadric surface in space is a second-degree polynomial of the form

$$ax^2 + by^2 + cz^2 + dxy + exz + fyz + gx + hy + iz + j = 0.$$

There are six basic types of quadric surfaces: ellipsoids, hyperboloids of one sheet, hyperboloids of two sheets, elliptic cones, elliptic paraboloids, and hyperbolic paraboloids. The intersection of a surface with a plane, called the **trace** of the surface in the plane, is useful to help visualize the graph of the surface in space. The six basic types of quadric surfaces, together with their traces, are shown on the next two pages.

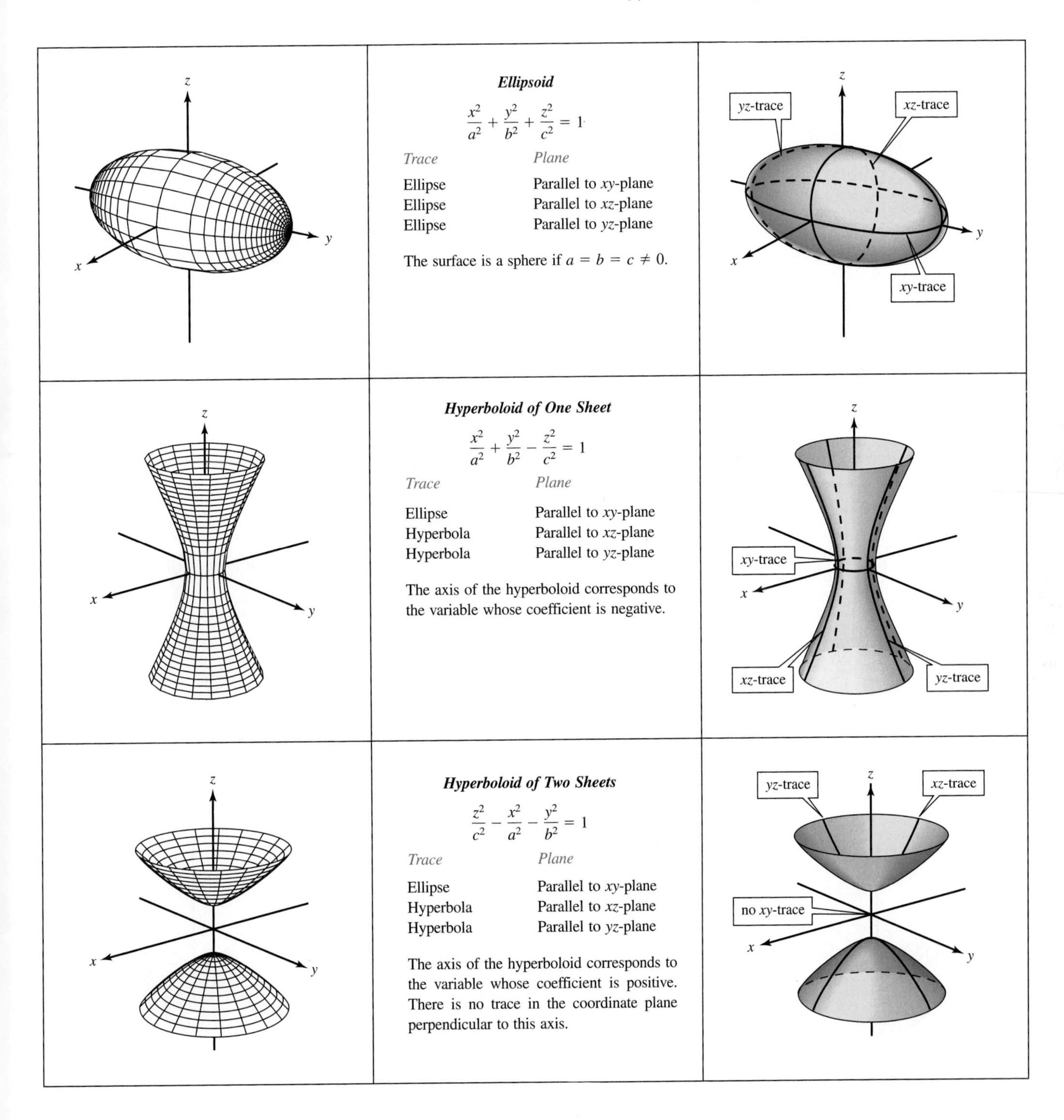

Ellipsoid

$$\frac{x^2}{a^2} + \frac{y^2}{b^2} + \frac{z^2}{c^2} = 1$$

Trace	*Plane*
Ellipse	Parallel to xy-plane
Ellipse	Parallel to xz-plane
Ellipse	Parallel to yz-plane

The surface is a sphere if $a = b = c \neq 0$.

Hyperboloid of One Sheet

$$\frac{x^2}{a^2} + \frac{y^2}{b^2} - \frac{z^2}{c^2} = 1$$

Trace	*Plane*
Ellipse	Parallel to xy-plane
Hyperbola	Parallel to xz-plane
Hyperbola	Parallel to yz-plane

The axis of the hyperboloid corresponds to the variable whose coefficient is negative.

Hyperboloid of Two Sheets

$$\frac{z^2}{c^2} - \frac{x^2}{a^2} - \frac{y^2}{b^2} = 1$$

Trace	*Plane*
Ellipse	Parallel to xy-plane
Hyperbola	Parallel to xz-plane
Hyperbola	Parallel to yz-plane

The axis of the hyperboloid corresponds to the variable whose coefficient is positive. There is no trace in the coordinate plane perpendicular to this axis.

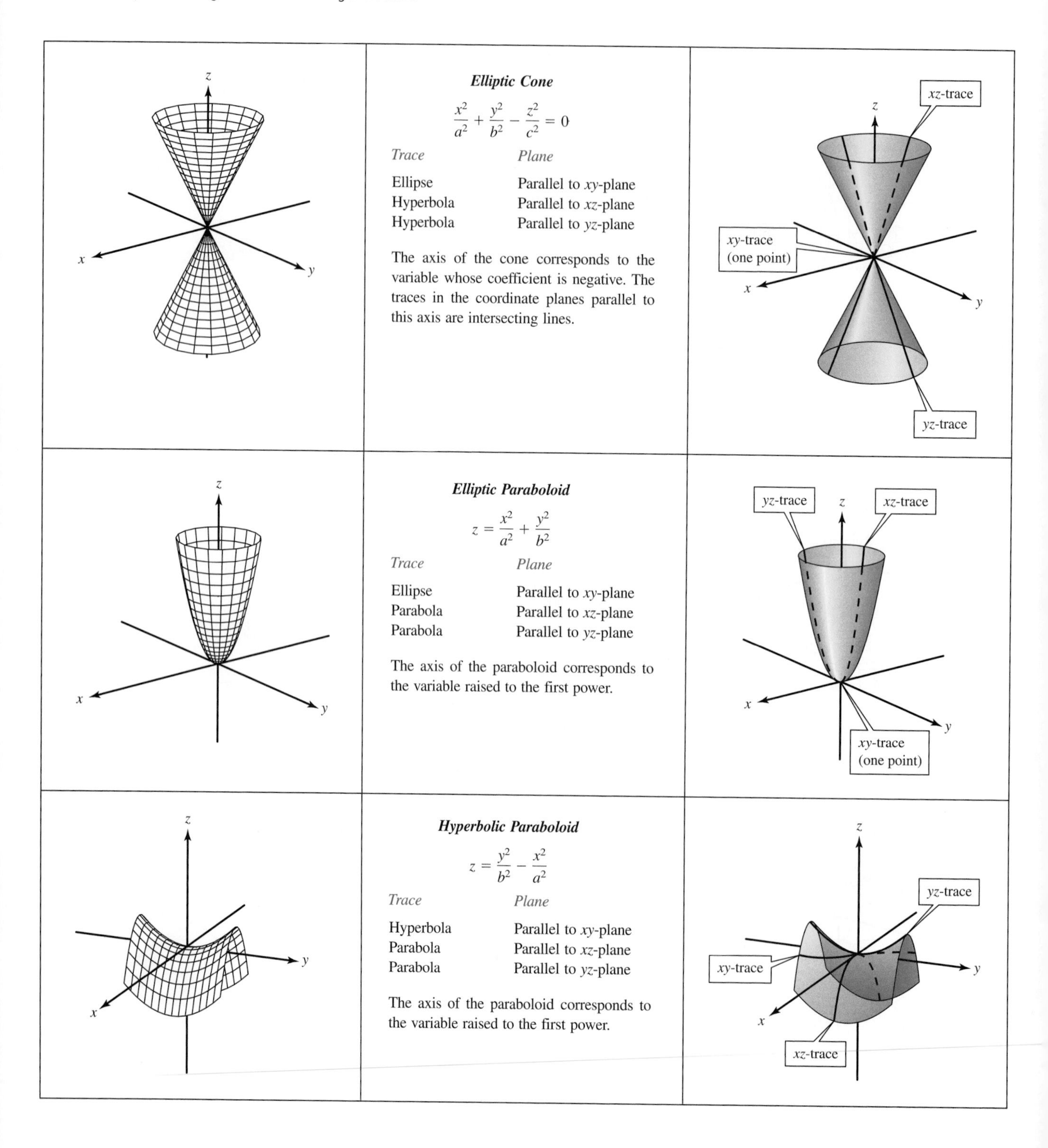
z
x
y
Elliptic Cone
$\frac{x^2}{a^2} + \frac{y^2}{b^2} - \frac{z^2}{c^2} = 0$
Trace Plane
Ellipse Parallel to xy-plane
Hyperbola Parallel to xz-plane
Hyperbola Parallel to yz-plane
The axis of the cone corresponds to the variable whose coefficient is negative. The traces in the coordinate planes parallel to this axis are intersecting lines.
xz-trace
xy-trace (one point)
yz-trace
Elliptic Paraboloid
$z = \frac{x^2}{a^2} + \frac{y^2}{b^2}$
Trace Plane
Ellipse Parallel to xy-plane
Parabola Parallel to xz-plane
Parabola Parallel to yz-plane
The axis of the paraboloid corresponds to the variable raised to the first power.
yz-trace
xz-trace
xy-trace (one point)
Hyperbolic Paraboloid
$z = \frac{y^2}{b^2} - \frac{x^2}{a^2}$
Trace Plane
Hyperbola Parallel to xy-plane
Parabola Parallel to xz-plane
Parabola Parallel to yz-plane
The axis of the paraboloid corresponds to the variable raised to the first power.
yz-trace
xy-trace
xz-trace

The quadratic form of the equation

$$ax^2 + by^2 + cz^2 + dxy + exz + fyz + gx + hy + iz + j = 0 \qquad \textbf{Quadric surface}$$

is defined as

$$ax^2 + by^2 + cz^2 + dxy + exz + fyz. \qquad \textbf{Quadratic form}$$

The corresponding matrix is

$$A = \begin{bmatrix} a & \frac{d}{2} & \frac{e}{2} \\ \frac{d}{2} & b & \frac{f}{2} \\ \frac{e}{2} & \frac{f}{2} & c \end{bmatrix}.$$

In its three-dimensional version, the Principal Axes Theorem relates the eigenvalues and eigenvectors of A to the equation of the rotated surface, as shown in Example 8.

EXAMPLE 8 Rotation of a Quadric Surface

Perform a rotation of axes to eliminate the xz-term in the quadratic equation

$$5x^2 + 4y^2 + 5z^2 + 8xz - 36 = 0.$$

SOLUTION The matrix A associated with this quadratic equation is

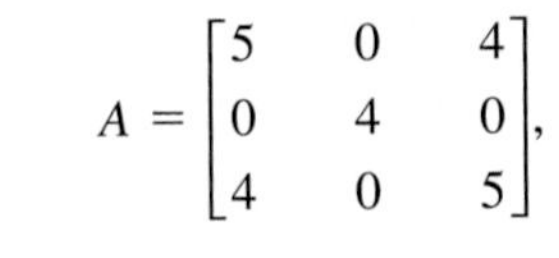

$$A = \begin{bmatrix} 5 & 0 & 4 \\ 0 & 4 & 0 \\ 4 & 0 & 5 \end{bmatrix},$$

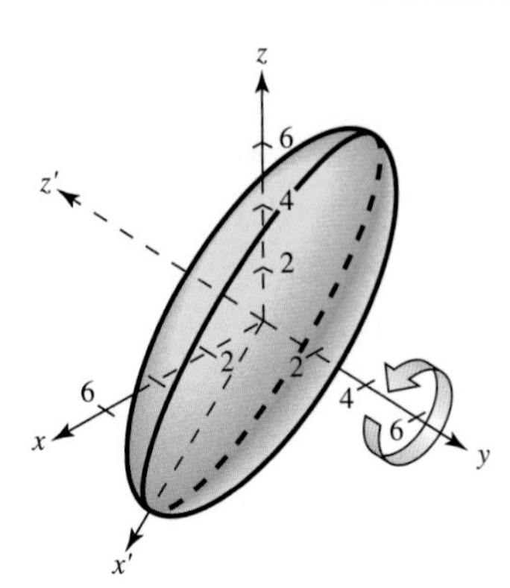

Figure 7.7

which has eigenvalues of $\lambda_1 = 1$, $\lambda_2 = 4$, and $\lambda_3 = 9$. So, in the rotated $x'y'z'$-system, the quadratic equation is $(x')^2 + 4(y')^2 + 9(z')^2 - 36 = 0$, which in standard form is

$$\frac{(x')^2}{6^2} + \frac{(y')^2}{3^2} + \frac{(z')^2}{2^2} = 1.$$

The graph of this equation is an ellipsoid. As shown in Figure 7.7, the $x'y'z'$-axes represent a counterclockwise rotation of 45° about the y-axis. Moreover, the orthogonal matrix

$$P = \begin{bmatrix} \frac{1}{\sqrt{2}} & 0 & \frac{1}{\sqrt{2}} \\ 0 & 1 & 0 \\ -\frac{1}{\sqrt{2}} & 0 & \frac{1}{\sqrt{2}} \end{bmatrix},$$

whose columns are the eigenvectors of A, has the property that P^TAP is diagonal.

SECTION 7.4 Exercises

Population Growth

In Exercises 1–4, use the age transition matrix A and age distribution vector $\mathbf{x}_1$ to find the age distribution vectors $\mathbf{x}_2$ and $\mathbf{x}_3$.

1. $A = \begin{bmatrix} 0 & 2 \\ \frac{1}{2} & 0 \end{bmatrix}, \mathbf{x}_1 = \begin{bmatrix} 10 \\ 10 \end{bmatrix}$

2. $A = \begin{bmatrix} 0 & 4 \\ \frac{1}{16} & 0 \end{bmatrix}, \mathbf{x}_1 = \begin{bmatrix} 160 \\ 160 \end{bmatrix}$

3. $A = \begin{bmatrix} 0 & 3 & 4 \\ 1 & 0 & 0 \\ 0 & \frac{1}{2} & 0 \end{bmatrix}, \mathbf{x}_1 = \begin{bmatrix} 12 \\ 12 \\ 12 \end{bmatrix}$

4. $A = \begin{bmatrix} 0 & 2 & 2 & 0 \\ \frac{1}{4} & 0 & 0 & 0 \\ 0 & 1 & 0 & 0 \\ 0 & 0 & \frac{1}{2} & 0 \end{bmatrix}, \mathbf{x}_1 = \begin{bmatrix} 100 \\ 100 \\ 100 \\ 100 \end{bmatrix}$

5. Find a stable age distribution vector for the age transition matrix in Exercise 1.

6. Find a stable age distribution vector for the age transition matrix in Exercise 2.

7. Find a stable age distribution vector for the age transition matrix in Exercise 3.

8. Find a stable age distribution vector for the age transition matrix in Exercise 4.

9. A population has the characteristics listed below.

(a) A total of 75% of the population survives its first year. Of that 75%, 25% survives the second year. The maximum life span is 3 years.

(b) The average number of offspring for each member of the population is 2 the first year, 4 the second year, and 2 the third year.

The population now consists of 120 members in each of the three age classes. How many members will there be in each age class in 1 year? In 2 years?

10. A population has the characteristics listed below.

(a) A total of 80% of the population survives its first year. Of that 80%, 25% survives the second year. The maximum life span is 3 years.

(b) The average number of offspring for each member of the population is 3 the first year, 6 the second year, and 3 the third year.

The population now consists of 150 members in each of the three age classes. How many members will there be in each age class in 1 year? In 2 years?

11. A population has the characteristics listed below.

(a) A total of 60% of the population survives its first year. Of that 60%, 50% survives the second year. The maximum life span is 3 years.

(b) The average number of offspring for each member of the population is 2 the first year, 5 the second year, and 2 the third year.

The population now consists of 100 members in each of the three age classes. How many members will there be in each age class in 1 year? In 2 years?

12. Find the limit (if it exists) of $A^n\mathbf{x}_1$ as n approaches infinity for the following matrices.

$$A = \begin{bmatrix} 0 & 2 \\ \frac{1}{2} & 0 \end{bmatrix} \text{ and } \mathbf{x}_1 = \begin{bmatrix} a \\ a \end{bmatrix}$$

Systems of Linear Differential Equations (Calculus)

In Exercises 13–18, solve the system of first-order linear differential equations.

13. $y_1' = 2y_1$
$y_2' = y_2$

14. $y_1' = -3y_1$
$y_2' = 4y_2$

15. $y_1' = -y_1$
$y_2' = 6y_2$
$y_3' = y_3$

16. $y_1' = 5y_1$
$y_2' = -2y_2$
$y_3' = -3y_3$

17. $y_1' = 2y_1$
$y_2' = -y_2$
$y_3' = y_3$

18. $y_1' = -y_1$
$y_2' = -2y_2$
$y_3' = y_3$

In Exercises 19–26, solve the system of first-order linear differential equations.

19. $y_1' = y_1 - 4y_2$
$y_2' = 2y_2$

20. $y_1' = y_1 - 4y_2$
$y_2' = -2y_1 + 8y_2$

21. $y_1' = y_1 + 2y_2$
$y_2' = 2y_1 + y_2$

22. $y_1' = y_1 - y_2$
$y_2' = 2y_1 + 4y_2$

23. $y_1' = -3y_2 + 5y_3$
$y_2' = -4y_1 + 4y_2 - 10y_3$
$y_3' = 4y_3$

24. $y_1' = -2y_1 + y_3$
$y_2' = 3y_2 + 4y_3$
$y_3' = y_3$

25. $y_1' = y_1 - 2y_2 + y_3$
$y_2' = 2y_2 + 4y_3$
$y_3' = 3y_3$

26. $y_1' = 2y_1 + y_2 + y_3$
$y_2' = y_1 + y_2$
$y_3' = y_1 + y_3$

In Exercises 27–30, write out the system of first-order linear differential equations represented by the matrix equation $\mathbf{y}' = A\mathbf{y}$. Then verify the indicated general solution.

27. $A = \begin{bmatrix} 1 & 1 \\ 0 & 1 \end{bmatrix}$, $\begin{aligned} y_1 &= C_1e^t + C_2te^t \\ y_2 &= C_2e^t \end{aligned}$

28. $A = \begin{bmatrix} 1 & -1 \\ 1 & 1 \end{bmatrix}$, $\begin{aligned} y_1 &= C_1e^t \cos t + C_2e^t \sin t \\ y_2 &= -C_2e^t \cos t + C_1e^t \sin t \end{aligned}$

29. $A = \begin{bmatrix} 0 & 1 & 0 \\ 0 & 0 & 1 \\ 0 & -4 & 0 \end{bmatrix}$, $\begin{aligned} y_1 &= C_1 + C_2 \cos 2t + C_3 \sin 2t \\ y_2 &= 2C_3 \cos 2t - 2C_2 \sin 2t \\ y_3 &= -4C_2 \cos 2t - 4C_3 \sin 2t \end{aligned}$

30. $A = \begin{bmatrix} 0 & 1 & 0 \\ 0 & 0 & 1 \\ 1 & -3 & 3 \end{bmatrix}$,

$$\begin{aligned} y_1 &= C_1e^t + C_2te^t + C_3t^2e^t \\ y_2 &= (C_1 + C_2)e^t + (C_2 + 2C_3)te^t + C_3t^2e^t \\ y_3 &= (C_1 + 2C_2 + 2C_3)e^t + (C_2 + 4C_3)te^t + C_3t^2e^t \end{aligned}$$

Quadratic Forms

In Exercises 31–36, find the matrix of the quadratic form associated with the equation.

31. $x^2 + y^2 - 4 = 0$

32. $x^2 - 4xy + y^2 - 4 = 0$

33. $9x^2 + 10xy - 4y^2 - 36 = 0$

34. $12x^2 - 5xy - x + 2y - 20 = 0$

35. $10xy - 10y^2 + 4x - 48 = 0$

36. $16x^2 - 4xy + 20y^2 - 72 = 0$

In Exercises 37–42, find the matrix A of the quadratic form associated with the equation. In each case, find the eigenvalues of A and an orthogonal matrix P such that P^TAP is diagonal.

37. $2x^2 - 3xy - 2y^2 + 10 = 0$

38. $5x^2 - 2xy + 5y^2 + 10x - 17 = 0$

39. $13x^2 + 6\sqrt{3}\,xy + 7y^2 - 16 = 0$

40. $3x^2 - 2\sqrt{3}xy + y^2 + 2x + 2\sqrt{3}y = 0$

41. $16x^2 - 24xy + 9y^2 - 60x - 80y + 100 = 0$

42. $17x^2 + 32xy - 7y^2 - 75 = 0$

In Exercises 43–50, use the Principal Axes Theorem to perform a rotation of axes to eliminate the xy-term in the quadratic equation. Identify the resulting rotated conic and give its equation in the new coordinate system.

43. $13x^2 - 8xy + 7y^2 - 45 = 0$

44. $x^2 + 4xy + y^2 - 9 = 0$

45. $7x^2 + 32xy - 17y^2 - 50 = 0$

46. $2x^2 - 4xy + 5y^2 - 36 = 0$

47. $2x^2 + 4xy + 2y^2 + 6\sqrt{2}x + 2\sqrt{2}y + 4 = 0$

48. $8x^2 + 8xy + 8y^2 + 10\sqrt{2}x + 26\sqrt{2}y + 31 = 0$

49. $xy + x - 2y + 3 = 0$

50. $5x^2 - 2xy + 5y^2 + 10\sqrt{2}x = 0$

In Exercises 51–54, find the matrix A of the quadratic form associated with the equation. Then find the equation of the rotated quadric surface in which the xy-, xz-, and yz-terms have been eliminated.

51. $3x^2 - 2xy + 3y^2 + 8z^2 - 16 = 0$

52. $2x^2 + 2y^2 + 2z^2 + 2xy + 2xz + 2yz - 1 = 0$

53. $x^2 + 2y^2 + 2z^2 + 2yz - 1 = 0$

54. $x^2 + y^2 + z^2 + 2xy - 8 = 0$

55. Let P be a 2×2 orthogonal matrix such that $|P| = 1$. Show that there exists a number θ, $0 \le \theta < 2\pi$, such that

$$P = \begin{bmatrix} \cos\theta & -\sin\theta \\ \sin\theta & \cos\theta \end{bmatrix}.$$

CHAPTER 7 Review Exercises

In Exercises 1–6, find (a) the characteristic equation of A, (b) the real eigenvalues of A, and (c) a basis for the eigenspace corresponding to each eigenvalue.

1. $A = \begin{bmatrix} 2 & 1 \\ 5 & -2 \end{bmatrix}$

2. $A = \begin{bmatrix} 2 & 1 \\ -4 & -2 \end{bmatrix}$

3. $A = \begin{bmatrix} 9 & 4 & -3 \\ -2 & 0 & 6 \\ -1 & -4 & 11 \end{bmatrix}$

4. $A = \begin{bmatrix} -4 & 1 & 2 \\ 0 & 1 & 1 \\ 0 & 0 & 3 \end{bmatrix}$

5. $A = \begin{bmatrix} 2 & 0 & 1 \\ 0 & 3 & 4 \\ 0 & 0 & 1 \end{bmatrix}$

6. $A = \begin{bmatrix} 1 & 0 & 4 \\ 0 & 1 & -2 \\ 1 & 0 & -2 \end{bmatrix}$

In Exercises 7 and 8, use a graphing utility or computer software program to find (a) the characteristic equation of A, (b) the real eigenvalues of A, and (c) a basis for the eigenspace corresponding to each eigenvalue.

7. $A = \begin{bmatrix} 2 & 1 & 0 & 0 \\ 1 & 2 & 0 & 0 \\ 0 & 0 & 2 & 1 \\ 0 & 0 & 1 & 2 \end{bmatrix}$

8. $A = \begin{bmatrix} 3 & 0 & 2 & 0 \\ 1 & 3 & 1 & 0 \\ 0 & 1 & 1 & 0 \\ 0 & 0 & 0 & 4 \end{bmatrix}$

In Exercises 9–12, determine whether A is diagonalizable. If it is, find a nonsingular matrix P such that $P^{-1}AP$ is diagonal.

9. $A = \begin{bmatrix} -2 & -1 & 3 \\ 0 & 1 & 2 \\ 0 & 0 & 1 \end{bmatrix}$

10. $A = \begin{bmatrix} 3 & -2 & 2 \\ -2 & 0 & -1 \\ 2 & -1 & 0 \end{bmatrix}$

11. $A = \begin{bmatrix} 1 & 0 & 2 \\ 0 & 1 & 0 \\ 2 & 0 & 1 \end{bmatrix}$

12. $A = \begin{bmatrix} 2 & -1 & 1 \\ -2 & 3 & -2 \\ -1 & 1 & 0 \end{bmatrix}$

13. Show that if $0 < \theta < \pi$, then the transformation for a counterclockwise rotation through an angle θ has no real eigenvalues.

14. For what value(s) of a does the matrix

$$A = \begin{bmatrix} 0 & 1 \\ a & 1 \end{bmatrix}$$

have the characteristics listed below?

(a) A has an eigenvalue of multiplicity 2.
(b) A has -1 and 2 as eigenvalues.
(c) A has real eigenvalues.

Writing In Exercises 15–18, explain why the matrix is not diagonalizable.

15. $A = \begin{bmatrix} 0 & 2 \\ 0 & 0 \end{bmatrix}$

16. $A = \begin{bmatrix} -1 & 2 \\ 0 & -1 \end{bmatrix}$

17. $A = \begin{bmatrix} 3 & 0 & 0 \\ 1 & 3 & 0 \\ 0 & 0 & 3 \end{bmatrix}$

18. $A = \begin{bmatrix} -2 & 3 & 1 \\ 0 & 4 & 3 \\ 0 & 0 & -2 \end{bmatrix}$

In Exercises 19–22, determine whether the matrices are similar. If they are, find a matrix P such that $A = P^{-1}BP$.

19. $A = \begin{bmatrix} 1 & 0 \\ 0 & 2 \end{bmatrix}, B = \begin{bmatrix} 2 & 0 \\ 0 & 1 \end{bmatrix}$

20. $A = \begin{bmatrix} 5 & 0 \\ 0 & 3 \end{bmatrix}, B = \begin{bmatrix} 7 & 2 \\ -4 & 1 \end{bmatrix}$

21. $A = \begin{bmatrix} 1 & 1 & 0 \\ 0 & 1 & 1 \\ 0 & 0 & 1 \end{bmatrix}, B = \begin{bmatrix} 1 & 1 & 0 \\ 0 & 1 & 0 \\ 0 & 0 & 1 \end{bmatrix}$

22. $A = \begin{bmatrix} 1 & 0 & 0 \\ 0 & -2 & 0 \\ 0 & 0 & -2 \end{bmatrix}, B = \begin{bmatrix} 1 & -3 & -3 \\ 3 & -5 & -3 \\ -3 & 3 & 1 \end{bmatrix}$

In Exercises 23–28, determine whether the matrix is symmetric, orthogonal, both, or neither.

23. $A = \begin{bmatrix} -\frac{\sqrt{2}}{2} & \frac{\sqrt{2}}{2} \\ \frac{\sqrt{2}}{2} & \frac{\sqrt{2}}{2} \end{bmatrix}$

24. $A = \begin{bmatrix} \frac{2\sqrt{5}}{5} & \frac{\sqrt{5}}{5} \\ \frac{\sqrt{5}}{5} & -\frac{2\sqrt{5}}{5} \end{bmatrix}$

25. $A = \begin{bmatrix} 0 & 0 & 1 \\ 0 & 1 & 0 \\ 1 & 0 & 1 \end{bmatrix}$

26. $A = \begin{bmatrix} \frac{\sqrt{3}}{3} & \frac{\sqrt{3}}{3} & \frac{\sqrt{3}}{3} \\ \frac{\sqrt{3}}{3} & \frac{2\sqrt{3}}{3} & 0 \\ \frac{\sqrt{3}}{3} & 0 & \frac{\sqrt{3}}{3} \end{bmatrix}$

27. $A = \begin{bmatrix} -\frac{2}{3} & \frac{1}{3} & -\frac{2}{3} \\ \frac{2}{3} & \frac{2}{3} & -\frac{1}{3} \\ \frac{1}{3} & -\frac{2}{3} & \frac{2}{3} \end{bmatrix}$

28. $A = \begin{bmatrix} \frac{4}{5} & 0 & \frac{3}{5} \\ 0 & 1 & 0 \\ -\frac{3}{5} & 0 & \frac{4}{5} \end{bmatrix}$

In Exercises 29–32, find an orthogonal matrix P that diagonalizes A.

29. $A = \begin{bmatrix} 3 & 4 \\ 4 & -3 \end{bmatrix}$ **30.** $A = \begin{bmatrix} 8 & 15 \\ 15 & -8 \end{bmatrix}$

31. $A = \begin{bmatrix} 2 & 0 & -1 \\ 0 & 1 & 0 \\ -1 & 0 & 2 \end{bmatrix}$ **32.** $A = \begin{bmatrix} 1 & 2 & 0 \\ 2 & 1 & 0 \\ 0 & 0 & 5 \end{bmatrix}$

In Exercises 33–40, find the steady state probability vector (if it exists) for the matrix. An eigenvector $\mathbf{v}$ of an $n \times n$ matrix A is called a **steady state probability vector** if $A\mathbf{v} = \mathbf{v}$ and the components of $\mathbf{v}$ add up to 1.

33. $A = \begin{bmatrix} \frac{2}{3} & \frac{1}{2} \\ \frac{1}{3} & \frac{1}{2} \end{bmatrix}$ **34.** $A = \begin{bmatrix} \frac{1}{2} & 1 \\ \frac{1}{2} & 0 \end{bmatrix}$

35. $A = \begin{bmatrix} 0.8 & 0.3 \\ 0.2 & 0.7 \end{bmatrix}$ **36.** $A = \begin{bmatrix} 0.4 & 0.2 \\ 0.6 & 0.8 \end{bmatrix}$

37. $A = \begin{bmatrix} \frac{1}{2} & \frac{1}{4} & 0 \\ \frac{1}{2} & \frac{1}{2} & \frac{1}{2} \\ 0 & \frac{1}{4} & \frac{1}{2} \end{bmatrix}$ **38.** $A = \begin{bmatrix} \frac{1}{3} & \frac{2}{3} & \frac{1}{3} \\ \frac{1}{3} & \frac{1}{3} & 0 \\ \frac{1}{3} & 0 & \frac{2}{3} \end{bmatrix}$

39. $A = \begin{bmatrix} 0.7 & 0.1 & 0.1 \\ 0.2 & 0.7 & 0.1 \\ 0.1 & 0.2 & 0.8 \end{bmatrix}$ **40.** $A = \begin{bmatrix} 0.3 & 0.1 & 0.4 \\ 0.2 & 0.4 & 0.0 \\ 0.5 & 0.5 & 0.6 \end{bmatrix}$

41. Prove that if A is an $n \times n$ symmetric matrix, then P^TAP is symmetric for any $n \times n$ matrix P.

42. Show that the characteristic equation of

$$A = \begin{bmatrix} 0 & 1 & 0 & 0 & \cdots & 0 \\ 0 & 0 & 1 & 0 & \cdots & 0 \\ \vdots & \vdots & \vdots & \vdots & & \vdots \\ 0 & 0 & 0 & 0 & \cdots & 1 \\ -a_0/a_n & -a_1/a_n & -a_2/a_n & -a_3/a_n & \cdots & -a_{n-1}/a_n \end{bmatrix},$$

$a_n \neq 0$, is $p(\lambda) = a_n\lambda^n + a_{n-1}\lambda^{n-1} + \cdots + a_3\lambda^3 + a_2\lambda^2 + a_1\lambda + a_0 = 0$. A is called the **companion matrix** of the polynomial p.

In Exercises 43 and 44, use the result of Exercise 42 to find the companion matrix A of the polynomial and find the eigenvalues of A.

43. $p(\lambda) = -9\lambda + 4\lambda^2$

44. $p(\lambda) = 189 - 120\lambda - 7\lambda^2 + 2\lambda^3$

45. The characteristic equation of the matrix

$$A = \begin{bmatrix} 8 & -4 \\ 2 & 2 \end{bmatrix}$$

is $\lambda^2 - 10\lambda + 24 = 0$. Because $A^2 - 10A + 24I_2 = O$, you can find powers of A by the process shown below.

$A^2 = 10A - 24I_2, \quad A^3 = 10A^2 - 24A,$
$A^4 = 10A^3 - 24A^2, \ldots$

Use this process to find the matrices A^2 and A^3.

46. Repeat Exercise 45 for the matrix

$$A = \begin{bmatrix} 9 & 4 & -3 \\ -2 & 0 & 6 \\ -1 & -4 & 11 \end{bmatrix}.$$

47. Let A be an $n \times n$ matrix.

(a) Prove or disprove that an eigenvector of A is also an eigenvector of A^2.

(b) Prove or disprove that an eigenvector of A^2 is also an eigenvector of A.

48. Let A be an $n \times n$ matrix. Prove that if $A\mathbf{x} = \lambda\mathbf{x}$, then $\mathbf{x}$ is an eigenvector of $(A + cI)$. What is the corresponding eigenvalue?

49. Let A and B be $n \times n$ matrices. Prove that if A is nonsingular, then AB is similar to BA.

50. (a) Find a symmetric matrix B such that $B^2 = A$ for the matrix

$$A = \begin{bmatrix} 2 & 1 \\ 1 & 2 \end{bmatrix}.$$

(b) Generalize the result of part (a) by proving that if A is an $n \times n$ symmetric matrix with positive eigenvalues, then there exists a symmetric matrix B such that $B^2 = A$.

51. Find an orthogonal matrix P such that $P^{-1}AP$ is diagonal for the matrix

$$A = \begin{bmatrix} a & b \\ b & a \end{bmatrix}.$$

52. Writing Let A be an $n \times n$ idempotent matrix (that is, $A^2 = A$). Describe the eigenvalues of A.

53. Writing The matrix below has an eigenvalue $\lambda = 2$ of multiplicity 4.

$$A = \begin{bmatrix} 2 & a & 0 & 0 \\ 0 & 2 & b & 0 \\ 0 & 0 & 2 & c \\ 0 & 0 & 0 & 2 \end{bmatrix}$$

(a) Under what conditions is A diagonalizable?

(b) Under what conditions does the eigenspace of $\lambda = 2$ have dimension 1? 2? 3?

54. Determine all $n \times n$ symmetric matrices that have 0 as their only eigenvalue.

True or False? In Exercises 55 and 56, determine whether each statement is true or false. If a statement is true, give a reason or cite an appropriate statement from the text. If a statement is false, provide an example that shows the statement is not true in all cases or cite an appropriate statement from the text.

55. (a) An eigenvector of an $n \times n$ matrix A is a nonzero vector $\mathbf{x}$ in R^n such that $A\mathbf{x}$ is a scalar multiple of $\mathbf{x}$.
(b) Similar matrices may or may not have the same eigenvalues.
(c) To diagonalize a square matrix A, you need to find an invertible matrix P such that $P^{-1}AP$ is diagonal.

56. (a) An eigenvalue of a matrix A is a scalar λ such that $\det(\lambda I - A) = 0$.
(b) An eigenvector may be the zero vector $\mathbf{0}$.
(c) A matrix A is orthogonally diagonalizable if there exists an orthogonal matrix P such that $P^{-1}AP = D$ is diagonal.

Population Growth

In Exercises 57–60, use the age transition matrix A and the age distribution vector $\mathbf{x}_1$ to find the age distribution vectors $\mathbf{x}_2$ and $\mathbf{x}_3$. Then find a stable age distribution vector for the population.

57. $A = \begin{bmatrix} 0 & 1 \\ \frac{1}{4} & 0 \end{bmatrix}, \mathbf{x}_1 = \begin{bmatrix} 100 \\ 100 \end{bmatrix}$

58. $A = \begin{bmatrix} 0 & 1 \\ \frac{3}{4} & 0 \end{bmatrix}, \mathbf{x}_1 = \begin{bmatrix} 32 \\ 32 \end{bmatrix}$

59. $A = \begin{bmatrix} 0 & 3 & 12 \\ 1 & 0 & 0 \\ 0 & \frac{1}{6} & 0 \end{bmatrix}, \mathbf{x}_1 = \begin{bmatrix} 300 \\ 300 \\ 300 \end{bmatrix}$

60. $A = \begin{bmatrix} 0 & 2 & 2 \\ \frac{1}{2} & 0 & 0 \\ 0 & 0 & 0 \end{bmatrix}, \mathbf{x}_1 = \begin{bmatrix} 240 \\ 240 \\ 240 \end{bmatrix}$

61. A population has the characteristics listed below.
(a) A total of 90% of the population survives its first year. Of that 90%, 75% survives the second year. The maximum life span is 3 years.
(b) The average number of offspring for each member of the population is 4 the first year, 6 the second year, and 2 the third year.

The population now consists of 120 members in each of the three age classes. How many members will there be in each age class in 1 year? In 2 years?

62. A population has the characteristics listed below.
(a) A total of 75% of the population survives its first year. Of that 75%, 60% survives the second year. The maximum life span is 3 years.
(b) The average number of offspring for each member of the population is 4 the first year, 8 the second year, and 2 the third year.

The population now consists of 120 members in each of the three age classes. How many members will there be in each age class in 1 year? In 2 years?

Systems of Linear Differential Equations (Calculus)

In Exercises 63–66, solve the system of first-order linear differential equations.

63. $y_1' = y_1 + 2y_2$
$y_2' = 0$

64. $y_1' = 3y_1$
$y_2' = y_1 - y_2$

65. $y_1' = y_2$
$y_2' = y_1$
$y_3' = 0$

66. $y_1' = 6y_1 - y_2 + 2y_3$
$y_2' = 3y_2 - y_3$
$y_3' = y_3$

Quadratic Forms

In Exercises 67–70, find the matrix A of the quadratic form associated with the equation. In each case, find an orthogonal matrix P such that P^TAP is diagonal. Sketch the graph of each equation.

67. $x^2 + 3xy + y^2 - 3 = 0$

68. $x^2 - \sqrt{3}xy + 2y^2 - 10 = 0$

69. $xy - 2 = 0$

70. $9x^2 - 24xy + 16y^2 - 400x - 300y = 0$

CHAPTER 7 Projects

1 Population Growth and Dynamical Systems (I)

Systems of differential equations often arise in biological applications of population growth of various species of animals. These equations are called **dynamical systems** because they describe the changes in a system as functions of time. Suppose that over time t you are studying the populations of predator sharks $y_1(t)$ and their small fish prey $y_2(t)$. One simple model for the relative growths of these populations is

$$y_1'(t) = ay_1(t) + by_2(t) \qquad \textbf{Predator}$$

$$y_2'(t) = cy_1(t) + dy_2(t) \qquad \textbf{Prey}$$

where a, b, c, and d are constants that depend on the particular species being studied and on factors such as environment, available food, other competing species, and so on. Generally, the constants a and d are positive, reflecting the growth rates of the individual species. If the species are in a predator–prey relationship, then $b > 0$ and $c < 0$ indicate that an increase in prey fish y_2 would cause an increase in y_1, whereas an increase in the predator sharks y_1 would cause a decrease in y_2.

Suppose the system of linear differential equations shown below models the populations of sharks $y_1(t)$ and prey fish $y_2(t)$ with the given initial populations at time $t = 0$.

$$y_1'(t) = 0.5y_1(t) + 0.6y_2(t) \qquad y_1(0) = 36$$

$$y_2'(t) = -0.4y_1(t) + 3.0y_2(t) \qquad y_2(0) = 121$$

1. Use the diagonalization techniques of this chapter to find the populations $y_1(t)$ and $y_2(t)$ at any time $t > 0$.
2. Interpret the solutions in terms of the long-term population trends for the two species. Does one species ultimately disappear? Why or why not?
3. If you have access to a computer software program or graphing utility, graph the solutions $y_1(t)$ and $y_2(t)$ over the domain $0 \le t \le 3$.
4. Use the explicit solution found in part 1 to explain why the quotient $y_2(t)/y_1(t)$ approaches a limit as t increases.
5. If you have access to a computer software program or graphing utility that can solve differential equations numerically, use it to graph the solution of the original system of equations. Does this numerical approximation appear to be accurate?

For Further Reference You can learn more about dynamical systems and population modeling in most books on differential equations. For example, *Differential Equations and Their Applications,* fourth edition, by Martin Braun, Springer-Verlag, 1993, discusses the theory and applications of systems of linear differential equations. An especially interesting application of *nonlinear* differential equations is given in Section 4.10: "Predator-prey problems; or why the percentage of sharks caught in the Mediterranean Sea rose dramatically during World War I."

2 The Fibonacci Sequence

The **Fibonacci sequence** is named after the Italian mathematician Leonard Fibonacci of Pisa (1170–1250). The simplest way to form this sequence is to define the first two terms as $x_1 = 1$ and $x_2 = 1$, and then define the nth term as the sum of its two immediate predecessors. That is,

$$x_n = x_{n-1} + x_{n-2}.$$

So, the third term is $2 = 1 + 1$, the fourth term is $3 = 2 + 1$, and so on. The formula $x_n = x_{n-1} + x_{n-2}$ is called *recursive* because the first $n - 1$ terms must be calculated before the nth term can be calculated. Is it possible to find an *explicit* formula for the nth term of the Fibonacci sequence? In this project, you will use eigenvalues and diagonalization to derive such a formula.

1. Use the formula $x_n = x_{n-1} + x_{n-2}$ to calculate the first 12 terms in the Fibonacci sequence.
2. Explain how the matrix identity

$$\begin{bmatrix} 1 & 1 \\ 1 & 0 \end{bmatrix} \begin{bmatrix} x_{n-1} \\ x_{n-2} \end{bmatrix} = \begin{bmatrix} x_{n-1} + x_{n-2} \\ x_{n-1} \end{bmatrix}$$

   can be used to generate recursively the Fibonacci sequence.
3. Starting with $\begin{bmatrix} x_1 \\ x_2 \end{bmatrix} = \begin{bmatrix} 1 \\ 1 \end{bmatrix}$, show that $A^{n-2}\begin{bmatrix} 1 \\ 1 \end{bmatrix} = \begin{bmatrix} x_n \\ x_{n-1} \end{bmatrix}$, where $A = \begin{bmatrix} 1 & 1 \\ 1 & 0 \end{bmatrix}$.
4. Find a matrix P that diagonalizes A.
5. Derive an explicit formula for the nth term of the Fibonacci sequence. Use this formula to calculate x_1, x_2, and x_3.
6. Use the explicit formula for the nth term of the Fibonacci sequence together with a computer or graphing utility to find x_{10} and x_{20}.
7. Calculate the quotient x_n/x_{n-1} for various large values of n. Does the quotient appear to be approaching a fixed number as n tends to infinity?
8. Determine the limit of x_n/x_{n-1} as n approaches infinity. Do you recognize this number?

For Further Reference You can learn more about Fibonacci numbers in most books on number theory. You might find it interesting to look at the *Fibonacci Quarterly,* the official journal of the Fibonacci Association.

CHAPTERS 6 & 7 Cumulative Test

Take this test as you would take a test in class. After you are done, check your work against the answers in the back of the book.

1. Determine whether the function $T: R^3 \rightarrow R^2$, $T(x, y, z) = (2x, x + y)$ is a linear transformation.
2. Determine whether the function $T: M_{2,2} \rightarrow R$, $T(A) = |A + A^T|$ is a linear transformation.
3. Let $T: R^2 \rightarrow R^3$ be the linear transformation defined by $T(\mathbf{v}) = A\mathbf{v}$, where

$$A = \begin{bmatrix} 1 & 0 \\ -1 & 0 \\ 0 & 0 \end{bmatrix}.$$

Find (a) $T(1, -2)$ and (b) the preimage of $(5, -5, 0)$.

4. Find the kernel of the linear transformation $T: R^4 \rightarrow R^4$, $T(x_1, x_2, x_3, x_4) = (x_1 - x_2, x_2 - x_1, 0, x_3 + x_4)$.
5. Let $T: R^4 \rightarrow R^2$ be the linear transformation represented by $T(\mathbf{v}) = A\mathbf{v}$, where

$$A = \begin{bmatrix} 1 & 0 & 1 & 0 \\ 0 & -1 & 0 & -1 \end{bmatrix}.$$

Find a basis for (a) the kernel of T and (b) the range of T. (c) Determine the rank and nullity of T.

6. Find the standard matrix for the linear transformation represented by

$$T(x, y, z) = (x + y, y + z, x - z).$$

7. Find the standard matrix A of the linear transformation $\text{proj}_{\mathbf{v}}\mathbf{u}: R^2 \rightarrow R^2$ that projects an arbitrary vector $\mathbf{u}$ onto the vector

$$\mathbf{v} = \begin{bmatrix} 1 \\ -1 \end{bmatrix}$$

as shown in Figure 7.8. Use this matrix to find the images of the vectors $(1, 1)$ and $(-2, 2)$.

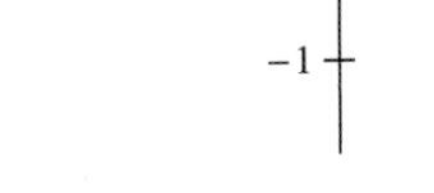

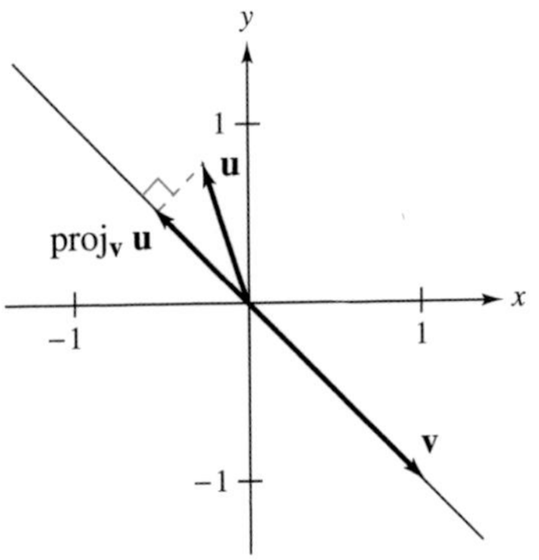

Figure 7.8

8. Find the inverse of the linear transformation $T: R^2 \rightarrow R^2$ represented by $T(x, y) = (x - y, 2x + y)$. Verify that $(T^{-1} \circ T)(3, -2) = (3, -2)$.
9. Find the matrix of the linear transformation $T(x, y) = (y, 2x, x + y)$ relative to the bases $B = \{(1, 1), (1, 0)\}$ for R^2 and $B' = \{(1, 0, 0), (1, 1, 0), (1, 1, 1)\}$ for R^3. Use this matrix to find the image of the vector $(0, 1)$.
10. Let $B = \{(1, 0), (0, 1)\}$ and $B' = \{(1, 1), (1, 2)\}$ be bases for R^2.
    (a) Find the matrix A of $T: R^2 \rightarrow R^2$, $T(x, y) = (x - 2y, x + 4y)$ relative to the basis B.
    (b) Find the transition matrix P from B' to B.
    (c) Find the matrix A' of T relative to the basis B'.
    (d) Find $[T(\mathbf{v})]_{B'}$ if $[\mathbf{v}]_{B'} = \begin{bmatrix} 3 \\ -2 \end{bmatrix}$.
    (e) Verify your answer in part (d) by finding $[\mathbf{v}]_B$ and $[T(\mathbf{v})]_B$.
11. Find the eigenvalues and the corresponding eigenvectors of the matrix

$$A = \begin{bmatrix} 1 & 2 & 1 \\ 0 & 3 & 1 \\ 0 & -3 & -1 \end{bmatrix}.$$

12. Find the eigenvalues and the corresponding eigenvectors of the matrix

$$A = \begin{bmatrix} 1 & -1 & 1 \\ 0 & 1 & 2 \\ 0 & 0 & 1 \end{bmatrix}.$$

13. Find a nonsingular matrix P such that $P^{-1}AP$ is diagonal if

$$A = \begin{bmatrix} 1 & 3 \\ -1 & 5 \end{bmatrix}.$$

14. Find a basis B for R^3 such that the matrix for $T: R^3 \rightarrow R^3$, $T(x, y, z) = (2x - 2z, 2y - 2z, 3x - 3z)$ relative to B is diagonal.
15. Find an orthogonal matrix P such that P^TAP diagonalizes the symmetric matrix

$$A = \begin{bmatrix} 1 & 3 \\ 3 & 1 \end{bmatrix}.$$

16. Use the Gram-Schmidt orthonormalization process to find an orthogonal matrix P such that P^TAP diagonalizes the symmetric matrix

$$A = \begin{bmatrix} 0 & 1 & 1 \\ 1 & 0 & 1 \\ 1 & 1 & 0 \end{bmatrix}.$$

17. Solve the system of differential equations.

$$y_1' = y_1$$
$$y_2' = 3y_2$$

18. Find the matrix of the quadratic form associated with the quadratic equation $4x^2 - 8xy + 4y^2 - 1 = 0$.
19. A population has the characteristics listed below.

    (a) A total of 80% of the population survives its first year. Of that 80%, 40% survives its second year. The maximum life span is 3 years.

    (b) The average number of offspring for each member of the population is 3 the first year, 6 the second year, and 3 the third year.

    The population now consists of 150 members in each of the three age classes. How many members will there be in each age class in 1 year? In 2 years?
20. Let A be an $n \times n$ matrix. Define the terms *eigenvalue* and *eigenvector* of A. How many eigenvalues can A have?
21. Define an *orthogonal matrix* and determine the possible values of its determinant.
22. Prove that if A is similar to B and A is diagonalizable, then B is diagonalizable.
23. Prove that if 0 is an eigenvalue of A, then A is singular.
24. Prove that the eigenvectors corresponding to distinct eigenvalues of a symmetric matrix are orthogonal.
25. Prove that the range of a linear transformation $T: V \rightarrow W$ is a subspace of W.
26. Prove that a linear transformation is one-to-one if and only if its kernel is $\{\mathbf{0}\}$.
27. Find all eigenvalues of A if $A^2 = O$.

APPENDIX Mathematical Induction and Other Forms of Proofs

In this appendix you will study some basic strategies for writing mathematical proofs—mathematical induction, proof by contradiction, and the use of counterexamples.

Mathematical Induction

To see the need for using mathematical induction, study the problem situation in the next example.

EXAMPLE 1 Sum of Odd Integers

Use the pattern to propose a formula for the sum of the first n odd integers.

$$1 = 1$$
$$1 + 3 = 4$$
$$1 + 3 + 5 = 9$$
$$1 + 3 + 5 + 7 = 16$$
$$1 + 3 + 5 + 7 + 9 = 25$$

SOLUTION Notice that the sums on the right are equal to the squares 1^2, 2^2, 3^2, 4^2, and 5^2. Judging from this pattern, you can *surmise* that the sum of the first n odd integers is n^2,

$$1 + 3 + 5 + 7 + \cdots + (2n - 1) = n^2.$$

Although this formula is valid, it is important to see that recognizing a pattern and then simply *jumping to the conclusion* that the pattern must be true for all values of n is not a logically valid method of proof. There are many examples in which a pattern appears to be developing for small values of n and then at some point the pattern fails. One of the most famous cases of this was the conjecture by the French mathematician Pierre de Fermat (1601–1665), who speculated that all numbers of the form

$$F_n = 2^{2^n} + 1, \quad n = 0, 1, 2, \ldots$$

are prime. For $n = 0, 1, 2, 3,$ and 4, the conjecture is true.

$$F_0 = 3$$
$$F_1 = 5$$
$$F_2 = 17$$
$$F_3 = 257$$
$$F_4 = 65{,}537$$

The size of the next Fermat number ($F_5 = 4{,}294{,}967{,}297$) is so great that it was difficult for Fermat to determine whether it was prime or not. Another well-known mathematician, Leonhard Euler (1707–1783), later found the factorization

$$F_5 = 4{,}294{,}967{,}297 = (641)(6{,}700{,}417),$$

which proved that F_5 is *not* prime and Fermat's conjecture was false.

Just because a rule, pattern, or formula seems to work for several values of n, you cannot simply decide that it is valid for all values of n without proving it to be so. One legitimate method of proof for such conjectures is the **Principle of Mathematical Induction.**

Principle of Mathematical Induction

Let P_n be a statement involving the positive integer n. If

1. P_1 is true, and
2. the truth of P_k implies the truth of P_{k+1} for every positive integer k, then P_n must be true for all positive integers n.

In the next example, the Principle of Mathematical Induction is used to prove the conjecture from Example 1.

EXAMPLE 2 Using Mathematical Induction

Use mathematical induction to prove that the following formula is valid for all positive integers n.

$$S_n = 1 + 3 + 5 + 7 + \cdots + (2n - 1) = n^2$$

SOLUTION Mathematical induction consists of two distinct parts. First, you must show that the formula is true when $n = 1$.

1. When $n = 1$, the formula is valid because $S_1 = 1 = 1^2$.

The second part of mathematical induction has two steps. The first step is to assume the formula is valid for some integer k (the **induction hypothesis**). The second step is to use this assumption to prove that the formula is valid for the next integer, $k + 1$.

2. Assuming the formula

$$S_k = 1 + 3 + 5 + 7 + \cdots + (2k - 1) = k^2$$

is true, you must show that the formula

$$S_{k+1} = (k + 1)^2$$

is true. Notice in the steps shown below that because $(2k - 1)$ is the kth term of the sum, $[2(k + 1) - 1]$ is the $(k + 1)$st term.

$$\begin{aligned} S_{k+1} &= 1 + 3 + 5 + 7 + \cdots + (2k - 1) + [2(k + 1) - 1] \\ &= [1 + 3 + 5 + 7 + \cdots + (2k - 1)] + [2k + 1] \\ &= S_k + (2k + 1) \\ &= k^2 + 2k + 1 && \textbf{$S_k = k^2$ from induction hypothesis.} \\ &= (k + 1)^2 \end{aligned}$$

Combining the results of parts (1) and (2), you can conclude by mathematical induction that the formula is valid for *all* positive integers n.

A well-known illustration used to explain why mathematical induction works is the unending line of dominoes shown in the figure at the left. If the line actually contains infinitely many dominoes, then it is clear that you could not knock the entire line down by knocking down only *one domino* at a time. Suppose, however, it were true that each domino would knock down the next one as it fell. Then you could knock them all down simply by pushing the first one and starting a chain reaction.

Mathematical induction works in the same way. If the truth of P_k implies the truth of P_{k+1} and if P_1 is true, then the chain reaction proceeds as follows:

P_1 implies P_2

P_2 implies P_3

P_3 implies P_4, and so on.

In the next example you will see the proof of a formula that is often used in calculus.

EXAMPLE 3 **Using Mathematical Induction**

Use mathematical induction to prove the formula for the sum of the first n squares.

$$S_n = 1^2 + 2^2 + 3^2 + 4^2 + \cdots + n^2 = \frac{n(n+1)(2n+1)}{6}$$

SOLUTION 1. The formula is valid when $n = 1$ because

$$S_1 = 1^2 = \frac{1(1+1)[2(1)+1]}{6} = \frac{1(2)(3)}{6} = 1.$$

2. Assuming the formula is true for k,

$$S_k = 1^2 + 2^2 + 3^2 + 4^2 + \cdots + k^2 = \frac{k(k+1)(2k+1)}{6},$$

you must show that it is true for $k + 1$,

$$S_{k+1} = \frac{(k+1)[(k+1)+1][2(k+1)+1]}{6} = \frac{(k+1)(k+2)(2k+3)}{6}.$$

To do this, write S_{k+1} as the sum of S_k and the $(k + 1)$st term, $(k + 1)^2$, as follows.

$$\begin{aligned} S_{k+1} &= 1^2 + 2^2 + 3^2 + 4^2 + \cdots + k^2 + (k+1)^2 \\ &= [1^2 + 2^2 + 3^2 + 4^2 + \cdots + k^2] + (k+1)^2 \\ &= \frac{k(k+1)(2k+1)}{6} + (k+1)^2 \qquad \textbf{Induction hypothesis} \\ &= \frac{k(k+1)(2k+1) + 6(k+1)^2}{6} \end{aligned}$$

$$= \frac{(k + 1)[k(2k + 1) + 6(k + 1)]}{6}$$

$$= \frac{(k + 1)[2k^2 + 7k + 6]}{6}$$

$$= \frac{(k + 1)(k + 2)(2k + 3)}{6}$$

Combining the results of parts (1) and (2), you can conclude by mathematical induction that the formula is valid for *all* positive integers n.

Many of the proofs in linear algebra use mathematical induction. Here is an example from Chapter 2.

EXAMPLE 4 Using Mathematical Induction in Linear Algebra

If $A_1, A_2, \ldots, A_n$ are invertible matrices, prove the generalization of Theorem 2.9.

$$(A_1A_2A_3 \cdots A_n)^{-1} = A_n^{-1} \cdots A_3^{-1}A_2^{-1}A_1^{-1}$$

SOLUTION

1. The formula is valid trivially when $n = 1$ because $A_1^{-1} = A_1^{-1}$.
2. Assuming the formula is valid for k,

$$(A_1A_2A_3 \cdots A_k)^{-1} = A_k^{-1} \cdots A_3^{-1}A_2^{-1}A_1^{-1},$$

you must show that it is valid for $k + 1$. To do this, use Theorem 2.9, which says that the inverse of a product of two invertible matrices is the product of their inverses in reverse order.

$$\begin{aligned}(A_1A_2A_3 \cdots A_kA_{k+1})^{-1} &= [(A_1A_2A_3 \cdots A_k)A_{k+1}]^{-1} \\ &= A_{k+1}^{-1}(A_1A_2A_3 \cdots A_k)^{-1} && \textbf{Theorem 2.9} \\ &= A_{k+1}^{-1}(A_k^{-1} \cdots A_3^{-1}A_2^{-1}A_1^{-1}) && \textbf{Induction hypothesis} \\ &= A_{k+1}^{-1}A_k^{-1} \cdots A_3^{-1}A_2^{-1}A_1^{-1}\end{aligned}$$

So, the formula is valid for $k + 1$.

Combining the results of parts (1) and (2), you can conclude by mathematical induction that the formula is valid for *all* positive integers n.

Proof by Contradiction

A second basic strategy for writing a proof is *proof by contradiction*. In mathematical logic, proof by contradiction is described by the following equivalence.

p implies q if and only if not q implies not p.

One way to prove that q is a true statement is to assume that q is not true. If this leads you to a statement that you know is false, then you have proved that q must be true.

The next example shows how to use proof by contradiction to prove that $\sqrt{2}$ is irrational.

EXAMPLE 5 **Using Proof by Contradiction**

Prove that $\sqrt{2}$ is an irrational number.

SOLUTION Begin by assuming that $\sqrt{2}$ is not an irrational number. $\sqrt{2}$ is rational and can be written as the quotient of two integers a and b that have no common factor.

$$\sqrt{2} = \frac{a}{b} \qquad \text{Assume that } \sqrt{2} \text{ is a rational number.}$$

$$2 = \frac{a^2}{b^2} \qquad \text{Square each side.}$$

$$2b^2 = a^2 \qquad \text{Multiply each side by } b^2.$$

This implies that 2 is a factor of a^2. So, 2 is also a factor of a, and a can be written as $2c$.

$$2b^2 = (2c)^2 \qquad \text{Substitute } 2c \text{ for } a.$$

$$2b^2 = 4c^2 \qquad \text{Simplify.}$$

$$b^2 = 2c^2 \qquad \text{Divide each side by 2.}$$

2 is a factor of b^2, and it is also a factor of b. So, 2 is a factor of both a and b. But this is impossible because a and b have no common factor. It must be impossible that $\sqrt{2}$ is a rational number. You can conclude that $\sqrt{2}$ must be an irrational number.

Proof by contradiction is not a new technique. The proof in the next example was provided by Euclid around 300 B.C.

EXAMPLE 6 **Using Proof by Contradiction**

A positive integer greater than 1 is a *prime* if its only positive factors are 1 and itself. Prove that there are infinitely many prime numbers.

SOLUTION Assume there are only finitely many primes, $p_1, p_2, \ldots, p_n$. Consider the number $N = p_1 p_2 \cdots p_n + 1$. This number is either prime or composite. If it is composite, then it can be factored as the product of primes. But, none of the primes $(p_1, p_2, \ldots, p_n)$ divide evenly into N. N is itself a prime, and you have found a new prime number, which contradicts the assumption that there are only n prime numbers.

It follows that no list of prime numbers is complete. There are infinitely many prime numbers.

You can use proof by contradiction to prove many theorems in linear algebra. On the next page is an example from Chapter 3.

EXAMPLE 7 Using Proof by Contradiction in Linear Algebra

Let A and B be $n \times n$ matrices such that AB is singular. Prove that either A or B is singular.

SOLUTION Assume that neither A nor B is singular. Because you know that a matrix is singular if and only if its determinant is zero, $\det(A)$ and $\det(B)$ are both nonzero real numbers. By Theorem 3.5, $\det(AB) = \det(A)\det(B)$. So, $\det(AB)$ is not zero because it is a product of two nonzero real numbers. But this contradicts that AB is a singular matrix. So, you can conclude that the assumption was wrong and that either A or B must be singular.

Using Counterexamples

Often you can disprove a statement using a *counterexample*. For instance, when Euler disproved Fermat's conjecture about prime numbers of the form $F_n = 2^{2^n} + 1$, $n = 0, 1, 2, \ldots$, he used the counterexample $F_5 = 4{,}294{,}967{,}297$, which is not prime.

EXAMPLE 8 Using a Counterexample

Use a counterexample to show that the statement is false.

Every odd number is a prime.

SOLUTION Certainly, you can list many odd numbers that are prime (3, 5, 7, 11), but the statement above is not true, because 9 and 15 are odd but they are not prime numbers. The numbers 9 and 15 are counterexamples.

Counterexamples can be used to disprove statements in linear algebra, as shown in the next example.

EXAMPLE 9 Using a Counterexample

Use a counterexample to show that the statement is false.

If A and B are square singular matrices of order n, then $A + B$ is a singular matrix of order n.

SOLUTION Let $A = \begin{bmatrix} 1 & 0 \\ 0 & 0 \end{bmatrix}$ and $B = \begin{bmatrix} 0 & 0 \\ 0 & 1 \end{bmatrix}$. Both A and B are singular of order 2, but

$$A + B = \begin{bmatrix} 1 & 0 \\ 0 & 1 \end{bmatrix}$$

is the identity matrix of order 2, which is not singular.

Exercises

In Exercises 1–4, use mathematical induction to prove that the formula is valid for all positive integers n.

1. $1 + 2 + 3 + \cdots + n = \dfrac{n(n+1)}{2}$

2. $1^3 + 2^3 + 3^3 + \cdots + n^3 = \dfrac{n^2(n+1)^2}{4}$

3. $3 + 7 + 11 + 15 + \cdots + (4n - 1) = n(2n + 1)$

4. $\left(1 + \dfrac{1}{1}\right)\left(1 + \dfrac{1}{2}\right)\left(1 + \dfrac{1}{3}\right) \cdots \left(1 + \dfrac{1}{n}\right) = n + 1$

In Exercises 5 and 6, propose a formula for the sum of the first n terms of the sequence. Then use mathematical induction to prove that the formula is valid.

5. $2^1, 2^2, 2^3, 2^4, \ldots$

6. $\dfrac{1}{1 \cdot 2}, \dfrac{1}{2 \cdot 3}, \dfrac{1}{3 \cdot 4}, \dfrac{1}{4 \cdot 5}, \ldots$

In Exercises 7 and 8, use mathematical induction to prove the inequality for the indicated integer values of n.

7. $n! > 2^n, \quad n \geq 4$

8. $\dfrac{1}{\sqrt{1}} + \dfrac{1}{\sqrt{2}} + \dfrac{1}{\sqrt{3}} + \cdots + \dfrac{1}{\sqrt{n}} > \sqrt{n}, \quad n \geq 2$

9. Prove that for all integers $n > 0$,

$$a^0 + a^1 + a^2 + a^3 + \cdots + a^n = \frac{1 - a^{n+1}}{1 - a}, \quad a \neq 1.$$

10. (From Chapter 2) Use mathematical induction to prove that

$$(A_1 A_2 A_3 \cdots A_n)^T = A_n^T \cdots A_3^T A_2^T A_1^T,$$

assuming that $A_1, A_2, A_3, \ldots, A_n$ are matrices with sizes such that the multiplications are defined.

11. (From Chapter 3) Use mathematical induction to prove that

$$|A_1 A_2 A_3 \cdots A_n| = |A_1||A_2||A_3| \cdots |A_n|,$$

where $A_1, A_2, A_3, \ldots, A_n$ are square matrices of the same size.

In Exercises 12–17, use proof by contradiction to prove the statement.

12. If p is an integer and p^2 is odd, then p is odd. (*Hint:* An odd number can be written as $2n + 1$, where n is an integer.)

13. If a and b are real numbers and $a \leq b$, then $a + c \leq b + c$.

14. If a, b, and c are real numbers such that $ac \geq bc$ and $c > 0$, then $a \geq b$.

15. If a and b are real numbers and $1 < a < b$, then $\dfrac{1}{a} > \dfrac{1}{b}$.

16. If a and b are real numbers and $(a + b)^2 = a^2 + b^2$, then $a = 0$ or $b = 0$ or $a = b = 0$.

17. If a is a real number and $0 < a < 1$, then $a^2 < a$.

18. Use proof by contradiction to prove that the sum of a rational number and an irrational number is irrational.

19. (From Chapter 4) Use proof by contradiction to prove that in a given vector space, the zero vector is unique.

20. (From Chapter 4) Let $S = \{\mathbf{u}, \mathbf{v}\}$ be a linearly independent set. Use proof by contradiction to prove that the set $\{\mathbf{u} - \mathbf{v}, \mathbf{u} + \mathbf{v}\}$ is linearly independent.

In Exercises 21–27, use a counterexample to show that the statement is false.

21. If a and b are real numbers and $a < b$, then $a^2 < b^2$.

22. The product of two irrational numbers is irrational.

23. If a and b are real numbers such that $a \neq 0$ and $b \neq 0$, then $(a + b)^3 = a^3 + b^3$.

24. If f is a polynomial function and $f(a) = f(b)$, then $a = b$.

25. If f and g are differentiable functions and $y = f(x)g(x)$, then $\frac{dy}{dx} = f'(x)g'(x)$.

26. (From Chapter 2) If A, B, and C are matrices and $AC = BC$, then $A = B$.

27. (From Chapter 3) If A is a matrix, then $\det(A^{-1}) = \frac{1}{\det A}$.

ANSWER KEY

Chapter 1

Section 1.1 *(page 11)*

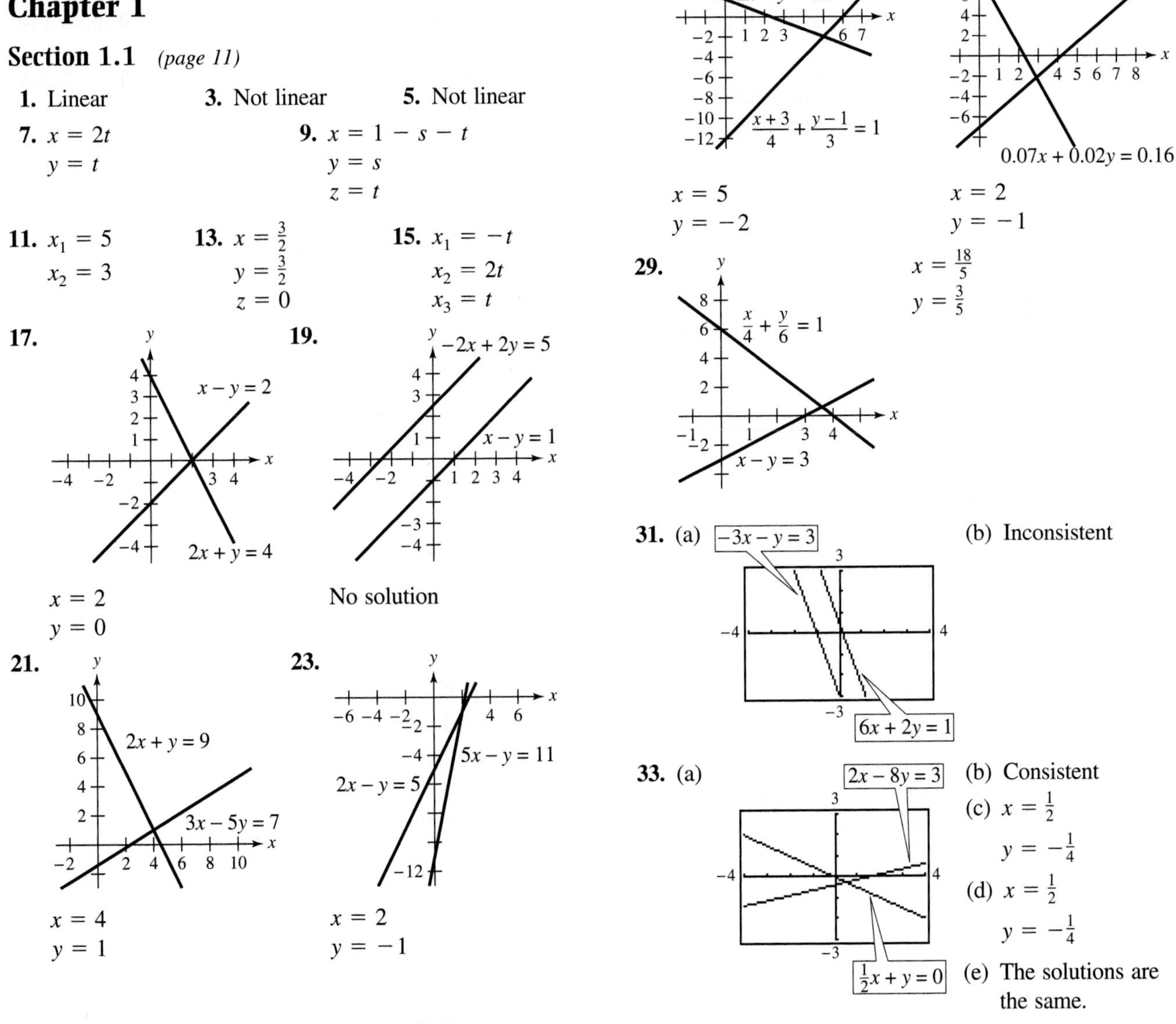

1. Linear **3.** Not linear **5.** Not linear

7. $x = 2t$, $y = t$

9. $x = 1 - s - t$, $y = s$, $z = t$

11. $x_1 = 5$, $x_2 = 3$

13. $x = \frac{3}{2}$, $y = \frac{3}{2}$, $z = 0$

15. $x_1 = -t$, $x_2 = 2t$, $x_3 = t$

17. $x = 2$, $y = 0$

19. No solution

21. $x = 4$, $y = 1$

23. $x = 2$, $y = -1$

25. $x = 5$, $y = -2$

27. $x = 2$, $y = -1$

29. $x = \frac{18}{5}$, $y = \frac{3}{5}$

31. (a) (b) Inconsistent

33. (a) (b) Consistent
(c) $x = \frac{1}{2}$, $y = -\frac{1}{4}$
(d) $x = \frac{1}{2}$, $y = -\frac{1}{4}$
(e) The solutions are the same.

35. (a)

3, −4, 4, −3

$4x - 8y = 9$

$0.8x - 1.6y = 1.8$

(b) Consistent

(c) There are infinite solutions.

(d) $x = \frac{9}{4} + 2t$

$y = t$

(e) The solutions are consistent.

37. $x_1 = -1$
$x_2 = -1$

39. $u = 40$
$v = 40$

41. $x = -\frac{1}{3}$
$y = -\frac{2}{3}$

43. $x = 7$
$y = 1$

45. $x_1 = 8$
$x_2 = 7$

47. $x = 1$
$y = 2$
$z = 3$

49. No solution

51. $x_1 = \frac{5}{2} - \frac{1}{2}t$
$x_2 = 4t - 1$
$x_3 = t$

53. No solution

55. $x = 1$
$y = 0$
$z = 3$
$w = 2$

57. $x_1 = -15$
$x_2 = 40$
$x_3 = 45$
$x_4 = -75$

59. $x = -1.2$
$y = -0.6$
$z = 2.4$

61. $x_1 = \frac{1}{5}$
$x_2 = -\frac{4}{5}$
$x_3 = \frac{1}{2}$

63. $x = 6.8813$
$y = -163.3111$
$z = -210.2915$
$w = -59.2913$

65. This system must have at least one solution because $x = y = z = 0$ is an obvious solution.

Solution: $x = 0$
$y = 0$
$z = 0$

This system has exactly one solution.

67. This system must have at least one solution because $x = y = z = 0$ is an obvious solution.

Solution: $x = -\frac{3}{5}t$
$y = \frac{4}{5}t$
$z = t$

This system has an infinite number of solutions.

69. (a) True. You can describe the entire solution set using parametric representation.
$ax + by = c$
Choosing $y = t$ as the free variable, the solution is
$x = \frac{c}{a} - \frac{b}{a}t$, $y = t$, where t is any real number.

(b) False. For example, consider the system
$x_1 + x_2 + x_3 = 1$
$x_1 + x_2 + x_3 = 2$
which is an inconsistent system.

(c) False. A consistent system may have only one solution.

71. $3x_1 - x_2 = 4$
$-3x_1 + x_2 = -4$
(The answer is not unique.)

73. $x = 3$
$y = -4$

75. $x = \dfrac{2}{5 - t}$
$y = \dfrac{1}{4t - 1}$
$z = \dfrac{1}{t}$ where $t \neq 5, \frac{1}{4}, 0$

77. $x = \cos\theta$
$y = \sin\theta$

79. $k = -2$

81. All $k \neq \pm 1$

83. $k = \frac{8}{3}$

85. $k = 1, -2$

87. (a) Three lines intersecting at one point
(b) Three coincident lines
(c) Three lines having no common point

89. Answers will vary. (*Hint:* Choose three different values of x and solve the resulting system of linear equations in the variables a, b, and c.)

91.

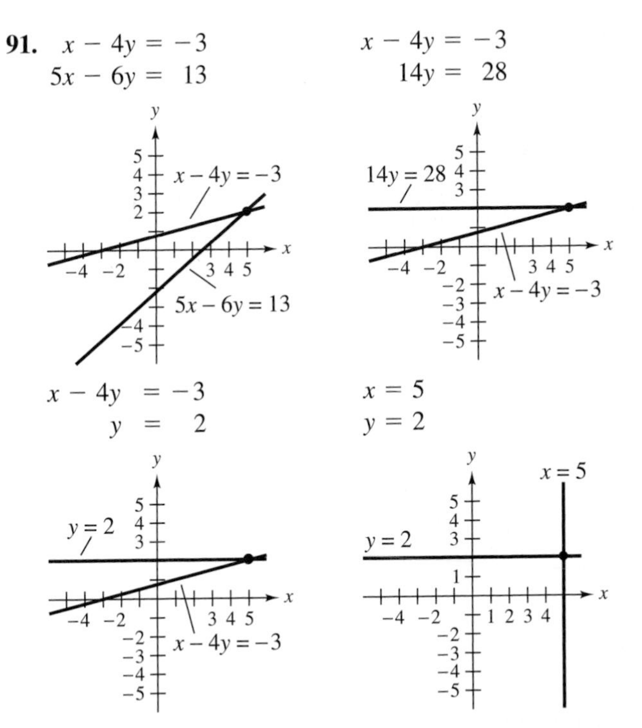

The intersection points are all the same.

93. $x = 39{,}600$

$y = 398$

The graphs are misleading because, while they appear parallel, when the equations are solved for y, they have slightly different slopes.

Section 1.2 *(page 26)*

1. 3×3 **3.** 2×4 **5.** 1×5 **7.** 4×5

9. Reduced row-echelon form

11. Not in row-echelon form

13. Not in row-echelon form

15. $x_1 = 0$, $x_2 = 2$

17. $x_1 = 2$, $x_2 = -1$, $x_3 = -1$

19. $x_1 = 1$, $x_2 = 1$, $x_3 = 0$

21. $x_1 = -26$, $x_2 = 13$, $x_3 = -7$, $x_4 = 4$

23. $x = 3$, $y = 2$

25. No solution

27. $x = 4$, $y = -2$

29. $x_1 = 4$, $x_2 = -3$, $x_3 = 2$

31. $x_1 = 1 + 2t$, $x_2 = 2 + 3t$, $x_3 = t$

33. $x = 100 + 96t - 3s$, $y = s$, $z = 54 + 52t$, $w = t$

35. $x = 0$, $y = 2 - 4t$, $z = t$

37. $x_1 = 23.5361 + 0.5278t$, $x_2 = 18.5444 + 4.1111t$, $x_3 = 7.4306 + 2.1389t$, $x_4 = t$

39. $x_1 = 2$, $x_2 = -2$, $x_3 = 3$, $x_4 = -5$, $x_5 = 1$

41. $x_1 = 1$, $x_2 = -2$, $x_3 = 6$, $x_4 = -3$, $x_5 = -4$, $x_6 = 3$

43. $x_1 = 0$, $x_2 = -t$, $x_3 = t$

45. $x_1 = -t$, $x_2 = s$, $x_3 = 0$, $x_4 = t$

47. (a) Two equations in two variables
(b) All real $k \neq -\frac{4}{3}$
(c) Two equations in three variables
(d) All real k

49. (a) $a + b + c = 0$
(b) $a + b + c \neq 0$
(c) Not possible

51. (a) $x = \frac{8}{3} - \frac{5}{6}t$, $y = -\frac{8}{3} + \frac{5}{6}t$, $z = t$
(b) $x = \frac{18}{7} - \frac{11}{14}t$, $y = -\frac{20}{7} + \frac{13}{14}t$, $z = t$
(c) $x = 3 - t$, $y = -3 + t$, $z = t$
(d) Each system has an infinite number of solutions.

53. $\begin{bmatrix} 1 & 0 \\ 0 & 1 \end{bmatrix}$

55. $\begin{bmatrix} 1 & 0 \\ 0 & 1 \end{bmatrix}, \begin{bmatrix} 1 & k \\ 0 & 0 \end{bmatrix}, \begin{bmatrix} 0 & 1 \\ 0 & 0 \end{bmatrix}, \begin{bmatrix} 0 & 0 \\ 0 & 0 \end{bmatrix}$

57. (a) True. In the notation $m \times n$, m is the number of rows of the matrix. So, a 6×3 matrix has six rows.

(b) True. At the top of page 19, the sentence reads, "It can be shown that every matrix is row-equivalent to a matrix in row-echelon form."

(c) False. Consider the row-echelon form

$$\begin{bmatrix} 1 & 0 & 0 & 0 & 0 \\ 0 & 1 & 0 & 0 & 1 \\ 0 & 0 & 1 & 0 & 2 \\ 0 & 0 & 0 & 1 & 3 \end{bmatrix}$$

which gives the solution $x_1 = 0$, $x_2 = 1$, $x_3 = 2$, and $x_4 = 3$.

(d) True. Theorem 1.1 states that if a homogeneous system has fewer equations than variables, then it must have an infinite number of solutions.

59. $ad - bc \neq 0$

61. $\lambda = 1, 3$

63. Yes, it is possible:

$$x_1 + x_2 + x_3 = 0$$
$$x_1 + x_2 + x_3 = 1$$

65. The rows have been interchanged. The first elementary row operation is redundant, so you can just use the second and third elementary row operations.

67. An inconsistent matrix in row-echelon form would have a row consisting of all zeros except for the last entry.

69. In the matrix in reduced row-echelon form, there would be zeros above any leading ones.

Section 1.3 *(page 38)*

1. (a) $p(x) = 29 - 18x + 3x^2$

(b)

3. (a) $p(x) = 2x$

(b)

5. (a) $p(z) = 7 + \frac{7}{2}z + \frac{3}{2}z^2$

$p(x) = 7 + \frac{7}{2}(x - 2007) + \frac{3}{2}(x - 2007)^2$

(b)

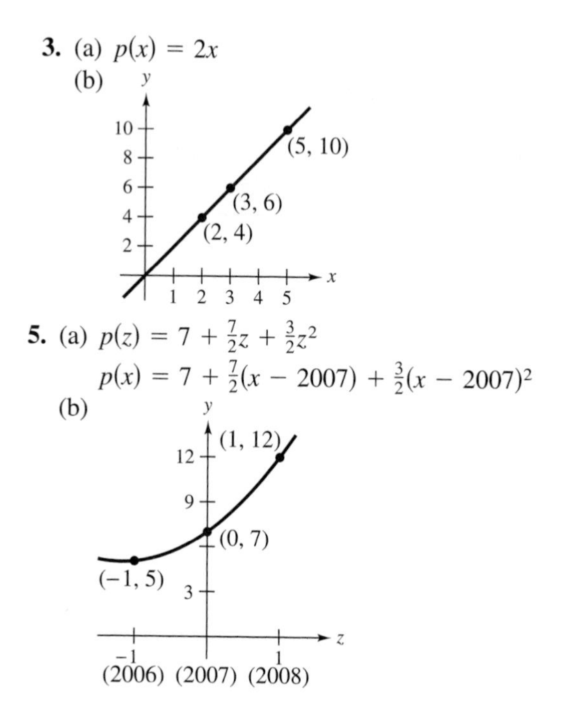

7. y is not a function of x because the x-value of 3 is repeated.

9. $p(x) = 1 + x$

11. $p(x) = -3x + x^3$

13. $p(z) = 249 + 2.7z + 0.05z^2$ (where $z = x - 1990$)
Year 2010: $p = 323$ million
Year 2020: $p = 375$ million

15. (a)

$$\begin{aligned} a_0 + a_1 + a_2 + a_3 &= 10{,}003 \\ a_0 + 3a_1 + 9a_2 + 27a_3 &= 10{,}526 \\ a_0 + 5a_1 + 25a_2 + 125a_3 &= 12{,}715 \\ a_0 + 7a_1 + 49a_2 + 343a_3 &= 14{,}410 \end{aligned}$$

(b) $p(z) = 11{,}041.25 - 1606.5z + 613.25z^2 - 45z^3$
(where $z = x - 2000$)
No, the profits have increased every year except 2006 and our model predicts a decrease in 2008. This is not a reasonable estimate.

17. $p(x) = -\frac{4}{\pi^2}x^2 + \frac{4}{\pi}x$

$\sin\frac{\pi}{3} \approx \frac{8}{9} \approx 0.889$

(Actual value is $\sqrt{3}/2 \approx 0.866$.)

19. Solve the system:
$p(-1) = a_0 - a_1 + a_2 = 0$
$p(0) = a_0 = 0$
$p(1) = a_0 + a_1 + a_2 = 0$
$a_0 = a_1 = a_2 = 0$

21. (a) $x_1 = s$, $x_2 = t$, $x_3 = 600 - s$, $x_4 = s - t$, $x_5 = 500 - t$, $x_6 = s$, $x_7 = t$
(b) $x_1 = 0$, $x_2 = 0$, $x_3 = 600$, $x_4 = 0$, $x_5 = 500$, $x_6 = 0$, $x_7 = 0$
(c) $x_1 = 0$, $x_2 = -500$, $x_3 = 600$, $x_4 = 500$, $x_5 = 1000$, $x_6 = 0$, $x_7 = -500$

23. (a) $x_1 = 100 + t$, $x_2 = -100 + t$, $x_3 = 200 + t$, $x_4 = t$
(b) $x_1 = 100$, $x_2 = -100$, $x_3 = 200$, $x_4 = 0$
(c) $x_1 = 200$, $x_2 = 0$, $x_3 = 300$, $x_4 = 100$

25. $I_1 = 0$, $I_2 = 1$, $I_3 = 1$

27. (a) $I_1 = 1$, $I_2 = 2$, $I_3 = 1$ (b) $I_1 = 0$, $I_2 = 1$, $I_3 = 1$

29. $\frac{1}{x-1} + \frac{3}{x+1} - \frac{2}{(x+1)^2}$

31. $\frac{1}{x+2} - \frac{2}{x-2} + \frac{4}{(x-2)^2}$

33. $x = 2$, $y = 2$, $\lambda = -4$

35. $x_1 + x_2 + x_3 = 13$
$6x_1 + x_2 + 3x_3 = 46$
$x_2 - x_3 = 0$
where x_1 = touchdowns, x_2 = extra points, and x_3 = field goals
$x_1 = 5$
$x_2 = 4$
$x_3 = 4$

Review Exercises – Chapter 1 *(page 41)*

1. Not linear **3.** Linear

5. Not linear **7.** Linear

9. $x = -\frac{1}{4} + \frac{1}{2}s - \frac{3}{2}t$, $y = s$, $z = t$

11. $x = \frac{1}{2}$, $y = \frac{3}{2}$ **13.** $x = -12$, $y = -8$ **15.** $x = 0$, $y = 0$

17. No solution

19. $x_1 = -\frac{1}{2}$, $x_2 = \frac{4}{5}$ **21.** $x = 0$, $y = 0$ **23.** 2×3

25. Row-echelon form (not reduced)

27. Not in row-echelon form

29. $x_1 = -2t$, $x_2 = t$, $x_3 = 0$ **31.** $x = 2$, $y = -3$, $z = 3$ **33.** $x = \frac{1}{2}$, $y = -\frac{1}{3}$, $z = 1$

35. $x = 4 + 3t$, $y = 5 + 2t$, $z = t$ **37.** $x = \frac{3}{2} - 2t$, $y = 1 + 2t$, $z = t$ **39.** $x_1 = 1$, $x_2 = 4$, $x_3 = -3$, $x_4 = -2$

41. $x = 0$, $y = 2 - 4t$, $z = t$ **43.** $x = 1$, $y = 0$, $z = 4$, $w = -2$ **45.** $x = -2t$, $y = t$, $z = t$, $w = 0$

47. $x_1 = 0$, $x_2 = 0$, $x_3 = 0$ **49.** $x_1 = -4t$, $x_2 = -\frac{1}{2}t$, $x_3 = t$ **51.** $k = \pm 1$

53. (a) $b = 2a$ and $a \neq -3$
(b) $b \neq 2a$
(c) $a = -3$ and $b = -6$

55. Use an elimination method to get both matrices in reduced row-echelon form. The two matrices are row-equivalent because each is row-equivalent to
$$\begin{bmatrix} 1 & 0 & 0 \\ 0 & 1 & 0 \\ 0 & 0 & 1 \end{bmatrix}.$$

57. $$\begin{bmatrix} 1 & 0 & -1 & -2 & \cdots & 2-n \\ 0 & 1 & 2 & 3 & \cdots & n-1 \\ 0 & 0 & 0 & 0 & \cdots & 0 \\ \vdots & & & & & \vdots \\ 0 & 0 & 0 & 0 & \cdots & 0 \end{bmatrix}$$

59. (a) False. See page 3, following Example 2.
(b) True. See page 5, Example 4(b).

61. (a) $$\begin{aligned} 3x_1 + 2x_2 + x_3 &= 59 \\ 3x_1 - x_2 &= 0 \\ x_2 - x_3 &= 1 \end{aligned}$$
where x_1 = number of three-point baskets,
x_2 = number of two-point baskets,
x_3 = number of one-point free throws
(b) $x_1 = 5$
$x_2 = 15$
$x_3 = 14$

63. $\dfrac{1}{x+2} + \dfrac{2}{x-2} + \dfrac{1}{(x-2)^2}$

65. (a) $p(x) = 90 - \frac{135}{2}x + \frac{25}{2}x^2$
(b)

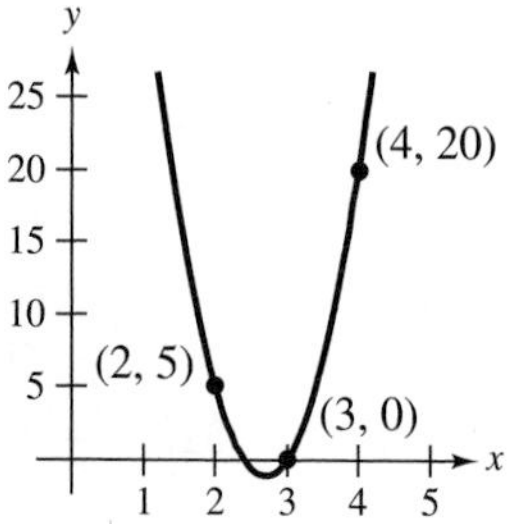

67. $p(x) = 50 + \frac{15}{2}x + \frac{5}{2}x^2$
(First year is represented by $x = 0$.)
Fourth-year sales: $p(3) = 95$

69. (a) $$\begin{aligned} a_0 &= 80 \\ a_0 + 4a_1 + 16a_2 &= 68 \\ a_0 + 80a_1 + 6400a_2 &= 30 \end{aligned}$$
(b) and (c) $a_0 = 80$
$a_1 = -\frac{25}{8}$
$a_2 = \frac{1}{32}$
So, $y = \frac{1}{32}x^2 - \frac{25}{8}x + 80$.
(d) The results to (b) and (c) are the same.
(e) There is precisely one polynomial function of degree $n - 1$ (or less) that fits n distinct points.

71. $I_1 = \frac{5}{13}$
$I_2 = \frac{6}{13}$
$I_3 = \frac{1}{13}$

Chapter 2

Section 2.1 *(page 56)*

1. (a) $\begin{bmatrix} 3 & -2 \\ 1 & 7 \end{bmatrix}$ (b) $\begin{bmatrix} -1 & 0 \\ 3 & -9 \end{bmatrix}$ (c) $\begin{bmatrix} 2 & -2 \\ 4 & -2 \end{bmatrix}$
(d) $\begin{bmatrix} 0 & -1 \\ 5 & -10 \end{bmatrix}$ (e) $\begin{bmatrix} \frac{5}{2} & -\frac{3}{2} \\ 0 & \frac{15}{2} \end{bmatrix}$

3. (a) $\begin{bmatrix} 7 & 3 \\ 1 & 9 \\ -2 & 15 \end{bmatrix}$ (b) $\begin{bmatrix} 5 & -5 \\ 3 & -1 \\ -4 & -5 \end{bmatrix}$ (c) $\begin{bmatrix} 12 & -2 \\ 4 & 8 \\ -6 & 10 \end{bmatrix}$
(d) $\begin{bmatrix} 11 & -6 \\ 5 & 3 \\ -7 & 0 \end{bmatrix}$ (e) $\begin{bmatrix} 4 & \frac{7}{2} \\ 0 & 7 \\ -\frac{1}{2} & \frac{25}{2} \end{bmatrix}$

5. (a) $\begin{bmatrix} 3 & 4 & 0 \\ 7 & 8 & 7 \\ 2 & 2 & 2 \end{bmatrix}$ (b) $\begin{bmatrix} 3 & 0 & -2 \\ -3 & 0 & 3 \\ -2 & 0 & 2 \end{bmatrix}$
(c) $\begin{bmatrix} 6 & 4 & -2 \\ 4 & 8 & 10 \\ 0 & 2 & 4 \end{bmatrix}$ (d) $\begin{bmatrix} 6 & 2 & -3 \\ -1 & 4 & 8 \\ -2 & 1 & 4 \end{bmatrix}$

(e) $\begin{bmatrix} \frac{3}{2} & 3 & \frac{1}{2} \\ 6 & 6 & \frac{9}{2} \\ 2 & \frac{3}{2} & 1 \end{bmatrix}$

7. (a) $c_{21} = -6$ (b) $c_{13} = 29$

9. $x = 3,\ y = 2,\ z = 1$

11. (a) $\begin{bmatrix} 0 & 15 \\ 6 & 12 \end{bmatrix}$ (b) $\begin{bmatrix} -2 & 2 \\ 31 & 14 \end{bmatrix}$

13. (a) Not defined (b) $\begin{bmatrix} 3 & -4 \\ 10 & 16 \\ 26 & 46 \end{bmatrix}$

15. (a) $\begin{bmatrix} -1 & 19 \\ 4 & -27 \\ 0 & 14 \end{bmatrix}$ (b) Not defined

17. (a) $\begin{bmatrix} 60 & 72 \\ -20 & -24 \\ 10 & 12 \\ 60 & 72 \end{bmatrix}$ (b) Not defined

19. (a) $\begin{bmatrix} 5 & -2 & 5 & 6 & 1 & 8 \\ 0 & 5 & 5 & -1 & -3 & 8 \\ 6 & -8 & -1 & 4 & 7 & -9 \\ 5 & 0 & 9 & 2 & 3 & 3 \\ 12 & 1 & -7 & -8 & 6 & 8 \\ 5 & 2 & 10 & -10 & -6 & -5 \end{bmatrix}$

(b) $\begin{bmatrix} 1 & 8 & -13 & 11 & 3 & 3 \\ 7 & -13 & 1 & 11 & -2 & 3 \\ -3 & -3 & -10 & -2 & 0 & 1 \\ 1 & 7 & 6 & 6 & 2 & -5 \\ 1 & -4 & -7 & 4 & 11 & 3 \\ 1 & -8 & 9 & -2 & -11 & -1 \end{bmatrix}$

(c) $\begin{bmatrix} 2 & -2 & -5 & -2 & -3 & -8 \\ 6 & -27 & 10 & -2 & -21 & 1 \\ 1 & 22 & -34 & 15 & 37 & 7 \\ 4 & -4 & -11 & 1 & 2 & -1 \\ 8 & -4 & -11 & 9 & -10 & 17 \\ -2 & -11 & -30 & 10 & 3 & 9 \end{bmatrix}$

(d) $\begin{bmatrix} 8 & 16 & 21 & -21 & -8 & 28 \\ 15 & -19 & 20 & 6 & 6 & -8 \\ -4 & 0 & -12 & -2 & -5 & 11 \\ 16 & -6 & 12 & 3 & 11 & 6 \\ 20 & 9 & 1 & -26 & -6 & 23 \\ -10 & -26 & 3 & 33 & 9 & -33 \end{bmatrix}$

21. 3×4 **23.** 4×2 **25.** 3×2

27. Not defined, sizes do not match.

29. $\begin{bmatrix} -1 & 1 \\ -2 & 1 \end{bmatrix}\begin{bmatrix} x_1 \\ x_2 \end{bmatrix} = \begin{bmatrix} 4 \\ 0 \end{bmatrix}$

$\begin{bmatrix} x_1 \\ x_2 \end{bmatrix} = \begin{bmatrix} 4 \\ 8 \end{bmatrix}$

31. $\begin{bmatrix} -2 & -3 \\ 6 & 1 \end{bmatrix}\begin{bmatrix} x_1 \\ x_2 \end{bmatrix} = \begin{bmatrix} -4 \\ -36 \end{bmatrix}$

$\begin{bmatrix} x_1 \\ x_2 \end{bmatrix} = \begin{bmatrix} -7 \\ 6 \end{bmatrix}$

33. $\begin{bmatrix} 1 & -2 & 3 \\ -1 & 3 & -1 \\ 2 & -5 & 5 \end{bmatrix}\begin{bmatrix} x_1 \\ x_2 \\ x_3 \end{bmatrix} = \begin{bmatrix} 9 \\ -6 \\ 17 \end{bmatrix}$

$\begin{bmatrix} x_1 \\ x_2 \\ x_3 \end{bmatrix} = \begin{bmatrix} 1 \\ -1 \\ 2 \end{bmatrix}$

35. $\begin{bmatrix} 1 & -5 & 2 \\ -3 & 1 & -1 \\ 0 & -2 & 5 \end{bmatrix}\begin{bmatrix} x_1 \\ x_2 \\ x_3 \end{bmatrix} = \begin{bmatrix} -20 \\ 8 \\ -16 \end{bmatrix}$

$\begin{bmatrix} x_1 \\ x_2 \\ x_3 \end{bmatrix} = \begin{bmatrix} -1 \\ 3 \\ -2 \end{bmatrix}$

37. $\begin{bmatrix} -5 & 2 \\ 3 & -1 \end{bmatrix}$

39. $a = 7, b = -4, c = -\frac{1}{2}, d = \frac{7}{2}$

41. $w = z, x = -y$

43. $\begin{bmatrix} 1 & 0 & 0 \\ 0 & 4 & 0 \\ 0 & 0 & 9 \end{bmatrix}$

45. $AB = \begin{bmatrix} -10 & 0 \\ 0 & -12 \end{bmatrix}$

$BA = \begin{bmatrix} -10 & 0 \\ 0 & -12 \end{bmatrix}$

47. Proof **49.** 2 **51.** 4 **53.** Proof

55. Let $A = \begin{bmatrix} a_{11} & a_{12} \\ a_{21} & a_{22} \end{bmatrix}$.

Then the given matrix equation expands to

$$\begin{bmatrix} a_{11} + a_{21} & a_{12} + a_{22} \\ a_{11} + a_{21} & a_{12} + a_{22} \end{bmatrix} = \begin{bmatrix} 1 & 0 \\ 0 & 1 \end{bmatrix}.$$

Because $a_{11} + a_{21} = 1$ and $a_{11} + a_{21} = 0$ cannot both be true, you can conclude that there is no solution.

57. (a) $A^2 = \begin{bmatrix} i^2 & 0 \\ 0 & i^2 \end{bmatrix} = \begin{bmatrix} -1 & 0 \\ 0 & -1 \end{bmatrix}$

$A^3 = \begin{bmatrix} i^3 & 0 \\ 0 & i^3 \end{bmatrix} = \begin{bmatrix} -i & 0 \\ 0 & -i \end{bmatrix}$

$A^4 = \begin{bmatrix} i^4 & 0 \\ 0 & i^4 \end{bmatrix} = \begin{bmatrix} 1 & 0 \\ 0 & 1 \end{bmatrix}$

(b) $B^2 = \begin{bmatrix} -i^2 & 0 \\ 0 & -i^2 \end{bmatrix} = \begin{bmatrix} 1 & 0 \\ 0 & 1 \end{bmatrix} = I$

59. Proof **61.** Proof

63. $\begin{bmatrix} 84 & 60 & 30 \\ 42 & 120 & 84 \end{bmatrix}$

65. [$1037.50 $1400.00 $1012.50]
Each entry represents the total profit at each outlet.

67. (a) True. On page 51, ". . . for the product of two matrices to be defined, the number of columns of the first matrix must equal the number of rows of the second matrix."

(b) True. On page 55, ". . . the system $A\mathbf{x} = \mathbf{b}$ is consistent if and only if $\mathbf{b}$ can be expressed as . . . a linear combination, where the coefficients of the linear combination are a solution of the system."

69. $PP = \begin{bmatrix} 0.75 & 0.15 & 0.10 \\ 0.20 & 0.60 & 0.20 \\ 0.30 & 0.40 & 0.30 \end{bmatrix}\begin{bmatrix} 0.75 & 0.15 & 0.10 \\ 0.20 & 0.60 & 0.20 \\ 0.30 & 0.40 & 0.30 \end{bmatrix}$

$= \begin{bmatrix} 0.6225 & 0.2425 & 0.135 \\ 0.33 & 0.47 & 0.20 \\ 0.395 & 0.405 & 0.20 \end{bmatrix}$

This product represents the changes in party affiliation after *two* elections.

71. $\left[\begin{array}{cc|c} -1 & 4 & 0 \\ -1 & 1 & 0 \\ \hline 0 & 0 & 5 \end{array}\right]$

73. $\mathbf{b} = 3\begin{bmatrix} 1 \\ 3 \end{bmatrix} + 0\begin{bmatrix} -1 \\ -3 \end{bmatrix} - 2\begin{bmatrix} 2 \\ 1 \end{bmatrix} = \begin{bmatrix} -1 \\ 7 \end{bmatrix}$
(The answer is not unique.)

75. $\mathbf{b} = 1\begin{bmatrix} 1 \\ 1 \\ 2 \end{bmatrix} + 2\begin{bmatrix} 1 \\ 0 \\ -1 \end{bmatrix} + 0\begin{bmatrix} -5 \\ -1 \\ -1 \end{bmatrix} = \begin{bmatrix} 3 \\ 1 \\ 0 \end{bmatrix}$

Section 2.2 *(page 70)*

1. $\begin{bmatrix} 3 & 2 \\ 13 & 4 \end{bmatrix}$ **3.** $\begin{bmatrix} 0 & -12 \\ 12 & -24 \end{bmatrix}$ **5.** $\begin{bmatrix} 7 & 7 \\ 28 & 14 \end{bmatrix}$

7. (a) $\begin{bmatrix} 3 & \frac{2}{3} \\ -\frac{4}{3} & \frac{11}{3} \\ \frac{10}{3} & 0 \end{bmatrix}$ (b) $\begin{bmatrix} -\frac{13}{3} & -\frac{10}{3} \\ 4 & -5 \\ -\frac{26}{3} & -\frac{16}{3} \end{bmatrix}$

(c) $\begin{bmatrix} -14 & -4 \\ 7 & -17 \\ -17 & -2 \end{bmatrix}$ (d) $\begin{bmatrix} -\frac{13}{6} & 1 \\ -\frac{1}{3} & -\frac{17}{6} \\ 0 & \frac{10}{3} \end{bmatrix}$

9. $\begin{bmatrix} -3 & -5 & -10 \\ -2 & -5 & -5 \end{bmatrix}$ **11.** $\begin{bmatrix} 1 & 6 & -1 \\ -2 & -2 & -8 \end{bmatrix}$

13. $\begin{bmatrix} 12 & -4 \\ 8 & 4 \end{bmatrix}$ **15.** $AC = BC = \begin{bmatrix} 2 & 3 \\ 2 & 3 \end{bmatrix}$

17. Proof **19.** $\begin{bmatrix} 1 & 0 \\ 0 & 1 \end{bmatrix}$ **21.** $\begin{bmatrix} 2 & 2 \\ 0 & 0 \end{bmatrix}$

23. (a) $\begin{bmatrix} 4 & 0 \\ 2 & 2 \\ 1 & -1 \end{bmatrix}$ (b) $\begin{bmatrix} 16 & 8 & 4 \\ 8 & 8 & 0 \\ 4 & 0 & 2 \end{bmatrix}$

(c) $\begin{bmatrix} 21 & 3 \\ 3 & 5 \end{bmatrix}$

25. (a) $\begin{bmatrix} 2 & 1 & 0 \\ 1 & 4 & 2 \\ -3 & 1 & 1 \end{bmatrix}$ (b) $\begin{bmatrix} 5 & 6 & -5 \\ 6 & 21 & 3 \\ -5 & 3 & 11 \end{bmatrix}$

(c) $\begin{bmatrix} 14 & 3 & -1 \\ 3 & 18 & 9 \\ -1 & 9 & 5 \end{bmatrix}$

27. (a) $\begin{bmatrix} 0 & 8 & -2 & 0 \\ -4 & 4 & 3 & 0 \\ 3 & 0 & 5 & -3 \\ 2 & 1 & 1 & 2 \end{bmatrix}$

(b) $\begin{bmatrix} 68 & 26 & -10 & 6 \\ 26 & 41 & 3 & -1 \\ -10 & 3 & 43 & 5 \\ 6 & -1 & 5 & 10 \end{bmatrix}$

(c) $\begin{bmatrix} 29 & -14 & 5 & -5 \\ -14 & 81 & -3 & 2 \\ 5 & -3 & 39 & -13 \\ -5 & 2 & -13 & 13 \end{bmatrix}$

29. $(A + B)(A - B) = A^2 + BA - AB - B^2$, which is not necessarily equal to $A^2 - B^2$ because AB is not necessarily equal to BA.

31. $(AB)^T = B^TA^T = \begin{bmatrix} 2 & -5 \\ 4 & -1 \end{bmatrix}$

33. $(AB)^T = B^TA^T = \begin{bmatrix} 4 & 0 & -4 \\ 10 & 4 & -2 \\ 1 & -1 & -3 \end{bmatrix}$

35. (a) True. See Theorem 2.1, part 1.
(b) True. See Theorem 2.3, part 1.
(c) False. See Theorem 2.6, part 4, or Example 9.
(d) True. See Example 10.

37. (a) $a = 3$ and $b = -1$
(b) $\begin{aligned} a + b &= 1 \\ b &= 1 \\ a \quad &= 1 \end{aligned}$
No solution
(c) $\begin{aligned} a + b + c &= 0 \\ b + c &= 0 \\ a \quad + c &= 0 \end{aligned}$; $a = -c \to b = 0 \to c = 0 \to a = 0$
(d) $a = -3t$; let $t = 1$: $a = -3, b = 1, c = 1$
$b = t$
$c = t$

39. $\begin{bmatrix} 1 & 0 & 0 \\ 0 & -1 & 0 \\ 0 & 0 & 1 \end{bmatrix}$

41. $\begin{bmatrix} \pm 3 & 0 \\ 0 & \pm 2 \end{bmatrix}$

43. $\begin{bmatrix} -4 & 0 \\ 8 & 2 \end{bmatrix}$

45. $\begin{bmatrix} -1 & -1 \\ 1 & 1 \end{bmatrix}$

47–55. Proof

57. Skew-symmetric

59. Symmetric

61. Proof

63. (a) $\frac{1}{2}(A + A^T)$

$$= \frac{1}{2}\left(\begin{bmatrix} a_{11} & a_{12} & \cdots & a_{1n} \\ a_{21} & a_{22} & \cdots & a_{2n} \\ \vdots & \vdots & & \vdots \\ a_{n1} & a_{n2} & \cdots & a_{nn} \end{bmatrix} + \begin{bmatrix} a_{11} & a_{21} & \cdots & a_{n1} \\ a_{12} & a_{22} & \cdots & a_{n2} \\ \vdots & \vdots & & \vdots \\ a_{1n} & a_{2n} & \cdots & a_{nn} \end{bmatrix}\right)$$

$$= \frac{1}{2}\begin{bmatrix} 2a_{11} & a_{12} + a_{21} & \cdots & a_{1n} + a_{n1} \\ a_{21} + a_{12} & 2a_{22} & \cdots & a_{2n} + a_{n2} \\ \vdots & \vdots & & \vdots \\ a_{n1} + a_{1n} & a_{n2} + a_{2n} & \cdots & 2a_{nn} \end{bmatrix}$$

(b) $\frac{1}{2}(A - A^T)$

$$= \frac{1}{2}\left(\begin{bmatrix} a_{11} & a_{12} & \cdots & a_{1n} \\ a_{21} & a_{22} & \cdots & a_{2n} \\ \vdots & \vdots & & \vdots \\ a_{n1} & a_{n2} & \cdots & a_{nn} \end{bmatrix} - \begin{bmatrix} a_{11} & a_{21} & \cdots & a_{n1} \\ a_{12} & a_{22} & \cdots & a_{n2} \\ \vdots & \vdots & & \vdots \\ a_{1n} & a_{2n} & \cdots & a_{nn} \end{bmatrix}\right)$$

$$= \frac{1}{2}\begin{bmatrix} 0 & a_{12} - a_{21} & \cdots & a_{1n} - a_{n1} \\ a_{21} - a_{12} & 0 & \cdots & a_{2n} - a_{n2} \\ \vdots & \vdots & & \vdots \\ a_{n1} - a_{1n} & a_{n2} - a_{2n} & \cdots & 0 \end{bmatrix}$$

(c) Proof

(d) $A = \frac{1}{2}(A - A^T) + \frac{1}{2}(A + A^T) =$

$$\begin{bmatrix} 0 & 4 & -\frac{1}{2} \\ -4 & 0 & -\frac{1}{2} \\ \frac{1}{2} & \frac{1}{2} & 0 \end{bmatrix} + \begin{bmatrix} 2 & 1 & \frac{7}{2} \\ 1 & 6 & \frac{1}{2} \\ \frac{7}{2} & \frac{1}{2} & 1 \end{bmatrix}$$

Skew-symmetric Symmetric

65. (a) $A = \begin{bmatrix} 0 & 1 \\ 1 & 0 \end{bmatrix}$, $B = \begin{bmatrix} -1 & 1 \\ 1 & 0 \end{bmatrix}$
(The answer is not unique.)

(b) Proof

Section 2.3 *(page 84)*

1. $AB = \begin{bmatrix} 1 & 0 \\ 0 & 1 \end{bmatrix} = BA$

3. $AB = \begin{bmatrix} 1 & 0 & 0 \\ 0 & 1 & 0 \\ 0 & 0 & 1 \end{bmatrix} = BA$

5. $\begin{bmatrix} 7 & -2 \\ -3 & 1 \end{bmatrix}$

7. $\begin{bmatrix} -19 & -33 \\ -4 & -7 \end{bmatrix}$

9. $\begin{bmatrix} 1 & 1 & -1 \\ -3 & 2 & -1 \\ 3 & -3 & 2 \end{bmatrix}$

11. Singular

13. $\begin{bmatrix} -\frac{3}{2} & \frac{3}{2} & 1 \\ \frac{9}{2} & -\frac{7}{2} & -3 \\ -1 & 1 & 1 \end{bmatrix}$

15. $\begin{bmatrix} 0 & -2 & 0.8 \\ -10 & 4 & 4.4 \\ 10 & -2 & -3.2 \end{bmatrix}$

17. $\begin{bmatrix} 1 & 0 & 0 \\ -\frac{3}{4} & \frac{1}{4} & 0 \\ \frac{7}{20} & -\frac{1}{4} & \frac{1}{5} \end{bmatrix}$

19. Singular

21. $\begin{bmatrix} -24 & 7 & 1 & -2 \\ -10 & 3 & 0 & -1 \\ -29 & 7 & 3 & -2 \\ 12 & -3 & -1 & 1 \end{bmatrix}$

23. Singular

25. (a) $x = 1$, $y = -1$ (b) $x = 2$, $y = 4$ (c) $x = -\frac{3}{2}$, $y = -\frac{3}{4}$

27. (a) $x_1 = 1$, $x_2 = 1$, $x_3 = -1$ (b) $x_1 = 0$, $x_2 = 1$, $x_3 = -1$

29. $x_1 = 0$, $x_2 = 1$, $x_3 = 2$, $x_4 = -1$, $x_5 = 0$

31. $x_1 = 1$, $x_2 = -2$, $x_3 = 3$, $x_4 = 0$, $x_5 = 1$, $x_6 = -2$

33. (a) $\begin{bmatrix} 35 & 17 \\ 4 & 10 \end{bmatrix}$ (b) $\begin{bmatrix} 2 & -7 \\ 5 & 6 \end{bmatrix}$

(c) $\begin{bmatrix} -31 & 40 \\ -56 & 1 \end{bmatrix}$ (d) $\begin{bmatrix} 1 & \frac{5}{2} \\ -\frac{7}{2} & 3 \end{bmatrix}$

35. (a) $\frac{1}{16}\begin{bmatrix} 138 & 56 & -84 \\ 37 & 26 & -71 \\ 24 & 34 & 3 \end{bmatrix}$ (b) $\frac{1}{4}\begin{bmatrix} 4 & 6 & 1 \\ -2 & 2 & 4 \\ 3 & -8 & 2 \end{bmatrix}$

(c) $\frac{1}{16}\begin{bmatrix} 7 & 0 & 34 \\ 28 & -40 & -14 \\ 30 & 14 & -25 \end{bmatrix}$ (d) $\frac{1}{8}\begin{bmatrix} 4 & -2 & 3 \\ 6 & 2 & -8 \\ 1 & 4 & 2 \end{bmatrix}$

37. $x = 4$

39. $x = 6$

41. $\begin{bmatrix} -1 & \frac{1}{2} \\ \frac{3}{4} & -\frac{1}{4} \end{bmatrix}$

43. Proof

45. (a) True. See Theorem 2.7.
(b) True. See Theorem 2.10, part 1.
(c) False. See Theorem 2.9.
(d) True. See "Finding the Inverse of a Matrix by Gauss-Jordan Elimination," part 2, page 76.

47–53. Proof

55. The sum of two invertible matrices is not necessarily invertible. For example, let

$$A = \begin{bmatrix} 1 & 0 \\ 0 & 1 \end{bmatrix} \text{ and } B = \begin{bmatrix} -1 & 0 \\ 0 & -1 \end{bmatrix}.$$

57. (a) $\begin{bmatrix} -1 & 0 & 0 \\ 0 & \frac{1}{3} & 0 \\ 0 & 0 & \frac{1}{2} \end{bmatrix}$ (b) $\begin{bmatrix} 2 & 0 & 0 \\ 0 & 3 & 0 \\ 0 & 0 & 4 \end{bmatrix}$

59. (a) Proof (b) $H = \begin{bmatrix} 0 & -1 & 0 \\ -1 & 0 & 0 \\ 0 & 0 & 1 \end{bmatrix}$

61. $A = PDP^{-1}$
No, A is not necessarily equal to D.

63. $\begin{bmatrix} 1 & 0 \\ -1 & 0 \end{bmatrix}$

Section 2.4 *(page 96)*

1. Elementary, multiply Row 2 by 2.

3. Elementary, add 2 times Row 1 to Row 2.

5. Not elementary

7. Elementary, add -5 times Row 2 to Row 3.

9. $\begin{bmatrix} 0 & 0 & 1 \\ 0 & 1 & 0 \\ 1 & 0 & 0 \end{bmatrix}$

11. $\begin{bmatrix} 0 & 0 & 1 \\ 0 & 1 & 0 \\ 1 & 0 & 0 \end{bmatrix}$

13. $\begin{bmatrix} 0 & 1 \\ 1 & 0 \end{bmatrix}$ **15.** $\begin{bmatrix} 0 & 0 & 1 \\ 0 & 1 & 0 \\ 1 & 0 & 0 \end{bmatrix}$

17. $\begin{bmatrix} \frac{1}{k} & 0 \\ 0 & 1 \end{bmatrix}$ **19.** $\begin{bmatrix} 1 & 0 & 0 \\ 0 & 0 & 1 \\ 0 & 1 & 0 \end{bmatrix}$

21. $\begin{bmatrix} 0 & 1 \\ -\frac{1}{2} & \frac{3}{2} \end{bmatrix}$ **23.** $\begin{bmatrix} 1 & 0 & \frac{1}{4} \\ 0 & \frac{1}{6} & \frac{1}{24} \\ 0 & 0 & \frac{1}{4} \end{bmatrix}$

25. $\begin{bmatrix} 1 & 0 \\ 1 & 1 \end{bmatrix}\begin{bmatrix} 1 & -1 \\ 0 & 1 \end{bmatrix}\begin{bmatrix} 1 & 0 \\ 0 & -2 \end{bmatrix}$
(The answer is not unique.)

27. $\begin{bmatrix} 1 & 1 \\ 0 & 1 \end{bmatrix}\begin{bmatrix} 1 & 0 \\ 3 & 1 \end{bmatrix}\begin{bmatrix} 1 & 0 \\ 0 & -1 \end{bmatrix}$
(The answer is not unique.)

29. $\begin{bmatrix} 1 & 0 & 0 \\ -1 & 1 & 0 \\ 0 & 0 & 1 \end{bmatrix}\begin{bmatrix} 1 & -2 & 0 \\ 0 & 1 & 0 \\ 0 & 0 & 1 \end{bmatrix}$
(The answer is not unique.)

31. $\begin{bmatrix} 1 & 0 & 0 & 0 \\ 0 & -1 & 0 & 0 \\ 0 & 0 & 1 & 0 \\ 0 & 0 & 0 & 1 \end{bmatrix}\begin{bmatrix} 1 & 0 & 0 & 0 \\ 0 & 1 & 0 & 0 \\ 0 & 0 & 2 & 0 \\ 0 & 0 & 0 & 1 \end{bmatrix}$

$\begin{bmatrix} 1 & 0 & 0 & 0 \\ 0 & 1 & 0 & 0 \\ 0 & 0 & 1 & 0 \\ 0 & 0 & 0 & -1 \end{bmatrix}\begin{bmatrix} 1 & 0 & 0 & 0 \\ 0 & 1 & 0 & 0 \\ 0 & 0 & 1 & 0 \\ 0 & 0 & -1 & 1 \end{bmatrix}$

$\begin{bmatrix} 1 & 0 & 0 & 1 \\ 0 & 1 & 0 & 0 \\ 0 & 0 & 1 & 0 \\ 0 & 0 & 0 & 1 \end{bmatrix}\begin{bmatrix} 1 & 0 & 0 & 0 \\ 0 & 1 & -3 & 0 \\ 0 & 0 & 1 & 0 \\ 0 & 0 & 0 & 1 \end{bmatrix}$
(The answer is not unique.)

33. (a) True. See "Remark" following "Definition of Elementary Matrix," page 87.
(b) False. Multiplication of a matrix by a scalar is not a single elementary row operation so it cannot be represented by a corresponding elementary matrix.
(c) True. See "Definition of Row Equivalence," page 90.
(d) True. See Theorem 2.13.

35. (a) EA will have two rows interchanged. (The same rows are interchanged in E.)
(b) $E^2 = I_n$

37. $A^{-1} = \begin{bmatrix} 1 & -a & 0 \\ -b & ab+1 & 0 \\ 0 & 0 & \frac{1}{c} \end{bmatrix}$

39. No. For example, $\begin{bmatrix} 1 & 0 \\ 2 & 1 \end{bmatrix}\begin{bmatrix} 1 & 1 \\ 0 & 1 \end{bmatrix} = \begin{bmatrix} 1 & 1 \\ 2 & 3 \end{bmatrix}$.

41. $\begin{bmatrix} 1 & 0 \\ -2 & 1 \end{bmatrix}\begin{bmatrix} 1 & 0 \\ 0 & 1 \end{bmatrix}$
(The answer is not unique.)

43. $\begin{bmatrix} 1 & 0 & 0 \\ 2 & 1 & 0 \\ -1 & 1 & 1 \end{bmatrix}\begin{bmatrix} 3 & 0 & 1 \\ 0 & 1 & -1 \\ 0 & 0 & 2 \end{bmatrix}$
(The answer is not unique.)

45. (a) $\begin{bmatrix} 1 & 0 & 0 \\ 0 & 1 & 0 \\ -1 & 2 & 1 \end{bmatrix}\begin{bmatrix} 2 & 1 & 0 \\ 0 & 1 & -1 \\ 0 & 0 & 3 \end{bmatrix}$
(The answer is not unique.)
(b) $\mathbf{y} = \begin{bmatrix} 1 \\ 2 \\ -5 \end{bmatrix}$ (c) $\mathbf{x} = \begin{bmatrix} \frac{1}{3} \\ \frac{1}{3} \\ -\frac{5}{3} \end{bmatrix}$

47. First, factor the matrix $A = LU$. Then, for each right-hand side $\mathbf{b}_i$, solve $L\mathbf{y} = \mathbf{b}_i$ and $U\mathbf{x} = \mathbf{y}$.

49. Idempotent **51.** Not idempotent

53. Not idempotent

55. *Case 1*: $b = 1, a = 0$
Case 2: $b = 0, a =$ any real number

57–61. Proofs

Section 2.5 *(page 112)*

1. Not stochastic **3.** Stochastic

5. Stochastic **7.** Next month: 350 people
In 2 months: 475 people

9.

	In 1 month	In 2 months
Nonsmokers	5025	5047
Smokers of less than 1 pack/day	2500	2499
Smokers of more than 1 pack/day	2475	2454

11. Tomorrow: 25 students
In 2 days: 44 students
In 30 days: 40 students

13. Proof

15. Uncoded: $[19\ \ 5\ \ 12], [12\ \ 0\ \ 3], [15\ \ 14\ \ 19], [15\ \ 12\ \ 9], [4\ \ 1\ \ 20], [5\ \ 4\ \ 0]$
Encoded: $-48, 5, 31, -6, -6, 9, -85, 23, 43, -27, 3, 15, -115, 36, 59, 9, -5, -4$

17. Uncoded: $[3\ \ 15], [13\ \ 5], [0\ \ 8], [15\ \ 13], [5\ \ 0], [19\ \ 15], [15\ \ 14]$
Encoded: 48, 81, 28, 51, 24, 40, 54, 95, 5, 10, 64, 113, 57, 100

19. HAPPY_NEW_YEAR

21. ICEBERG_DEAD_AHEAD

23. MEET_ME_TONIGHT_RON

25. $A^{-1} = \begin{bmatrix} -3 & 2 & 2 \\ -4 & 2 & 3 \\ 2 & -1 & -1 \end{bmatrix}$;

_SEPTEMBER_THE_ELEVENTH_WE_WILL
_ALWAYS_REMEMBER_

27. $D = \begin{bmatrix} 0.1 & 0.2 \\ 0.8 & 0.1 \end{bmatrix}$ (columns: Coal, Steel; rows: Coal, Steel) $\quad X = \begin{bmatrix} 20{,}000 \\ 40{,}000 \end{bmatrix}$ Coal, Steel

29. $X = \begin{bmatrix} 8622.0 \\ 4685.0 \\ 3661.4 \end{bmatrix}$ Farmer, Baker, Grocer

31. (a)

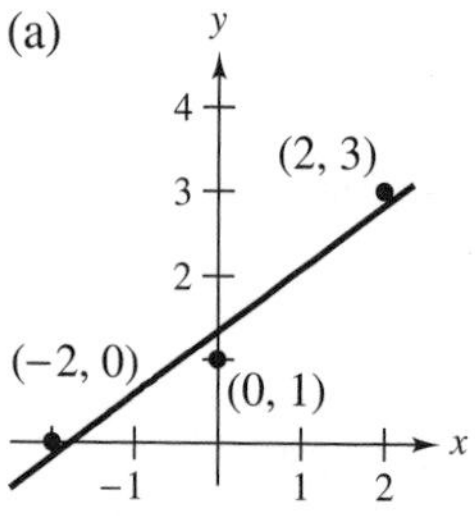

(b) $y = \frac{4}{3} + \frac{3}{4}x$
(c) $\frac{1}{6}$

33. (a)

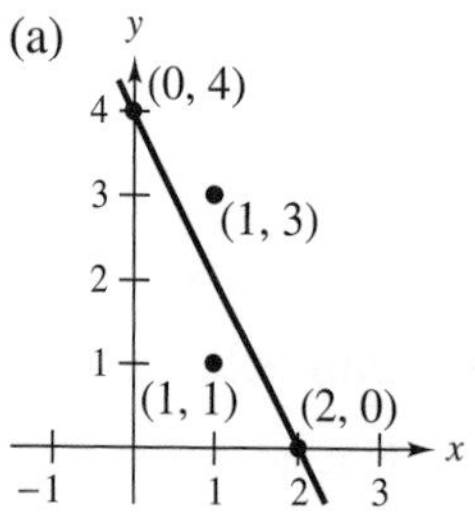

(b) $y = 4 - 2x$
(c) 2

35. $y = -\frac{1}{3} + 2x$

37. $y = 1.3 + 0.6x$

39. $y = 0.412x + 3$

41. $y = -0.5x + 7.5$

43. (a) $y = 11{,}650 - 2400x$
(b) 3490 gallons

45. (a) $y = 3.24t + 223.5$
(b) $y = 3.24t + 223.5$

Review Exercises – Chapter 2 *(page 115)*

1. $\begin{bmatrix} -13 & -8 & 18 \\ 0 & 11 & -19 \end{bmatrix}$

3. $\begin{bmatrix} 14 & -2 & 8 \\ 14 & -10 & 40 \\ 36 & -12 & 48 \end{bmatrix}$

5. $\begin{bmatrix} 4 & 6 & 3 \\ 0 & 6 & -10 \\ 0 & 0 & 6 \end{bmatrix}$

7. $\begin{aligned} 5x + 4y &= 2 \\ -x + y &= -22 \end{aligned}$

9. $\begin{aligned} x_2 - 2x_3 &= -1 \\ -x_1 + 3x_2 + x_3 &= 0 \\ 2x_1 - 2x_2 + 4x_3 &= 2 \end{aligned}$

11. $\begin{bmatrix} 2 & -1 \\ 3 & 2 \end{bmatrix}\begin{bmatrix} x \\ y \end{bmatrix} = \begin{bmatrix} 5 \\ -4 \end{bmatrix}$

13. $\begin{bmatrix} 2 & 3 & 1 \\ 2 & -3 & -3 \\ 4 & -2 & 3 \end{bmatrix}\begin{bmatrix} x_1 \\ x_2 \\ x_3 \end{bmatrix} = \begin{bmatrix} 10 \\ 22 \\ -2 \end{bmatrix}$

15. $A^T = \begin{bmatrix} 1 & 0 \\ 2 & 1 \\ -3 & 2 \end{bmatrix}, A^TA = \begin{bmatrix} 1 & 2 & -3 \\ 2 & 5 & -4 \\ -3 & -4 & 13 \end{bmatrix},$

$AA^T = \begin{bmatrix} 14 & -4 \\ -4 & 5 \end{bmatrix}$

17. $A^T = [1\ \ \ 3\ \ -1], A^TA = [11]$

$AA^T = \begin{bmatrix} 1 & 3 & -1 \\ 3 & 9 & -3 \\ -1 & -3 & 1 \end{bmatrix}$

19. $\begin{bmatrix} 1 & -1 \\ 2 & -3 \end{bmatrix}$

21. $\begin{bmatrix} \frac{3}{20} & \frac{3}{20} & \frac{1}{10} \\ \frac{3}{10} & -\frac{1}{30} & -\frac{2}{15} \\ -\frac{1}{5} & -\frac{1}{5} & \frac{1}{5} \end{bmatrix}$

23. $\begin{bmatrix} 5 & 4 \\ -1 & 1 \end{bmatrix}\begin{bmatrix} x_1 \\ x_2 \end{bmatrix} = \begin{bmatrix} 2 \\ -22 \end{bmatrix}$

$A^{-1} = \begin{bmatrix} \frac{1}{9} & -\frac{4}{9} \\ \frac{1}{9} & \frac{5}{9} \end{bmatrix}$

$\begin{bmatrix} x_1 \\ x_2 \end{bmatrix} = \begin{bmatrix} 10 \\ -12 \end{bmatrix}$

25. $\begin{bmatrix} -1 & 1 & 2 \\ 2 & 3 & 1 \\ 5 & 4 & 2 \end{bmatrix}\begin{bmatrix} x_1 \\ x_2 \\ x_3 \end{bmatrix} = \begin{bmatrix} 1 \\ -2 \\ 4 \end{bmatrix}$

$A^{-1} = \begin{bmatrix} -\frac{2}{15} & -\frac{2}{5} & \frac{1}{3} \\ -\frac{1}{15} & \frac{4}{5} & -\frac{1}{3} \\ \frac{7}{15} & -\frac{3}{5} & \frac{1}{3} \end{bmatrix}$

$\begin{bmatrix} x_1 \\ x_2 \\ x_3 \end{bmatrix} = \begin{bmatrix} 2 \\ -3 \\ 3 \end{bmatrix}$

27. $\begin{bmatrix} \frac{1}{14} & \frac{1}{42} \\ -\frac{1}{21} & \frac{2}{21} \end{bmatrix}$

29. $x \neq 3$

31. $\begin{bmatrix} 1 & 0 & -4 \\ 0 & 1 & 0 \\ 0 & 0 & 1 \end{bmatrix}$

33. $\begin{bmatrix} 1 & 3 \\ 0 & 1 \end{bmatrix}\begin{bmatrix} 2 & 0 \\ 0 & 1 \end{bmatrix}$
(The answer is not unique.)

35. $\begin{bmatrix} 1 & 0 & 0 \\ 0 & 1 & 0 \\ 0 & 0 & 4 \end{bmatrix}\begin{bmatrix} 1 & 0 & 0 \\ 0 & 1 & -2 \\ 0 & 0 & 1 \end{bmatrix}\begin{bmatrix} 1 & 0 & 1 \\ 0 & 1 & 0 \\ 0 & 0 & 1 \end{bmatrix}$
(The answer is not unique.)

37. $\begin{bmatrix} -1 & 0 \\ 0 & -1 \end{bmatrix}$ and $\begin{bmatrix} 1 & 0 \\ 0 & 1 \end{bmatrix}$
(The answer is not unique.)

39. $\begin{bmatrix} 0 & 0 \\ 0 & 0 \end{bmatrix}, \begin{bmatrix} 1 & 0 \\ 0 & 1 \end{bmatrix}$, and $\begin{bmatrix} 1 & 0 \\ 0 & 0 \end{bmatrix}$
(The answer is not unique.)

41. (a) $a = -1$
$b = -1$
$c = 1$
(b) Proof

43. Proof

45. $\begin{bmatrix} 1 & 0 \\ 3 & 1 \end{bmatrix}\begin{bmatrix} 2 & 5 \\ 0 & -1 \end{bmatrix}$
(The answer is not unique.)

47. $x = 4,\ y = 1, z = -1$

49. (a) False. See Theorem 2.1, part 1, page 61.
(b) True. See Theorem 2.6, part 2, page 68.

51. (a) False. The matrix $\begin{bmatrix} 1 & 0 \\ 0 & 0 \end{bmatrix}$ is not invertible.
(b) False. See Exercise 55, page 72.

53. (a) $\begin{bmatrix} 5455 & 128.2 \\ 3551 & 77.6 \\ 7591 & 178.6 \end{bmatrix}$

The first column of the matrix gives the total sales for each type of gas and the second column gives the profit for each type of gas.

(b) \$384.40

55. (a) $B = [2 \quad \frac{1}{2} \quad 3]$ (b) $BA = [473.5 \quad 588.5]$
(c) The matrix BA represents the numbers of calories burned by the 120-pound person and the 150-pound person.

57. Not stochastic

59. $PX = \begin{bmatrix} 80 \\ 112 \end{bmatrix}, P^2X = \begin{bmatrix} 68 \\ 124 \end{bmatrix}, P^3X = \begin{bmatrix} 65 \\ 127 \end{bmatrix}$

61. (a) $\begin{bmatrix} 110{,}000 \\ 100{,}000 \\ 90{,}000 \end{bmatrix}$ Region 1, Region 2, Region 3 (b) $\begin{bmatrix} 123{,}125 \\ 100{,}000 \\ 76{,}875 \end{bmatrix}$ Region 1, Region 2, Region 3

63. Uncoded: [15 14], [5 0], [9 6], [0 2], [25 0], [12 1], [14 4]
Encoded: 103, 44, 25, 10, 57, 24, 4, 2, 125, 50, 62, 25, 78, 32

65. $A^{-1} = \begin{bmatrix} 3 & 2 \\ 4 & 3 \end{bmatrix}$; ALL_SYSTEMS_GO

67. $A^{-1} = \begin{bmatrix} -2 & -1 & 0 \\ 0 & 1 & 1 \\ -5 & -3 & -3 \end{bmatrix}$; INVASION_AT_DAWN

69. _CAN_YOU_HEAR_ME_NOW

71. $D = \begin{bmatrix} 0.20 & 0.50 \\ 0.30 & 0.10 \end{bmatrix}, X \approx \begin{bmatrix} 133{,}333 \\ 133{,}333 \end{bmatrix}$

73. $y = \frac{20}{3} - \frac{3}{2}x$

75. $y = \frac{2}{5} - \frac{9}{5}x$

77. (a) $y = 19 + 14x$
(b) 41.4 kilograms per square kilometer

79. (a) $1.828x + 30.81$ (b) $1.828x + 30.81$
The models are the same.

(c)

x	Actual	Estimated
0	30.37	30.81
1	32.87	32.64
2	34.71	34.47
3	36.59	36.29
4	38.14	38.12
5	39.63	39.95

The estimated values are very close to the actual values.
(d) 49.09 (e) 2011

81. (a) $0.12x + 1.9$ (b) $0.12x + 1.9$
The models are the same.

(c)

x	Actual	Estimated
0	1.8	1.9
1	2.1	2.0
2	2.3	2.1
3	2.4	2.3
4	2.3	2.4
5	2.5	2.5

The estimated values are close to the actual values.
(d) 3.1 million (e) 2015

Chapter 3

Section 3.1 *(page 130)*

1. 1 **3.** 5 **5.** 27 **7.** -24

9. 6 **11.** $\lambda^2 - 4\lambda - 5$

13. (a) $M_{11} = 4$, $M_{12} = 3$, $M_{21} = 2$, $M_{22} = 1$
(b) $C_{11} = 4$, $C_{12} = -3$, $C_{21} = -2$, $C_{22} = 1$

15. (a) $M_{11} = 23$ $M_{12} = -8$ $M_{13} = -22$
$M_{21} = 5$ $M_{22} = -5$ $M_{23} = 5$
$M_{31} = 7$ $M_{32} = -22$ $M_{33} = -23$

(b) $C_{11} = 23$ $C_{12} = 8$ $C_{13} = -22$
$C_{21} = -5$ $C_{22} = -5$ $C_{23} = -5$
$C_{31} = 7$ $C_{32} = 22$ $C_{33} = -23$

17. (a) $4(-5) + 5(-5) + 6(-5) = -75$
(b) $2(8) + 5(-5) - 3(22) = -75$

19. -58 **21.** -30 **23.** -0.022

25. $4x - 2y - 2$ **27.** -168 **29.** 0

31. $65{,}644w + 62{,}256x + 12{,}294y - 24{,}672z$

33. -100 **35.** -43.5 **37.** -1098

39. 329 **41.** -24 **43.** 0 **45.** -30

47. (a) False. See "Definition of the Determinant of a 2×2 Matrix," page 123.
(b) True. See the first line after "Remark," page 124.
(c) False. See "Definitions of Minors and Cofactors of a Matrix," page 124.

49. $x = -1, -4$ **51.** $x = 0, 1$

53. $x = -1, 4$ **55.** $\lambda = -1 \pm \sqrt{3}$

57. $\lambda = -2, 0, \text{ or } 1$ **59.** $8uv - 1$

61. e^{5x} **63.** $1 - \ln x$

65. Expanding along the first row, the determinant of a 4×4 matrix involves four 3×3 determinants. Each of these 3×3 determinants requires six triple products. So, there are $4(6) = 24$ quadruple products.

67. $wz - xy$ **69.** $wz - xy$

71. $xy^2 - xz^2 + yz^2 - x^2y + x^2z - y^2z$

73. $bc^2 + ca^2 + ab^2 - ba^2 - ac^2 - cb^2$

75. (a) Proof

(b) $$\begin{vmatrix} x & 0 & 0 & d \\ -1 & x & 0 & c \\ 0 & -1 & x & b \\ 0 & 0 & -1 & a \end{vmatrix}$$

Section 3.2 *(page 140)*

1. The first row is 2 times the second row. If one row of a matrix is a multiple of another row, then the determinant of the matrix is zero.

3. The second row consists entirely of zeros. If one row of a matrix consists entirely of zeros, then the determinant of the matrix is zero.

5. The second and third columns are interchanged. If two columns of a matrix are interchanged, then the determinant of the matrix changes sign.
7. The first row of the matrix is multiplied by 5. If a row in a matrix is multiplied by a scalar, then the determinant of the matrix is multiplied by that scalar.
9. A 4 is factored out of the second column and a 3 is factored out of the third column. If a column of a matrix is multiplied by a scalar, then the determinant of the matrix is multiplied by that scalar.
11. The matrix is multiplied by 5. If an $n \times n$ matrix is multiplied by a scalar c, then the determinant of the matrix is multiplied by c^n.
13. -4 times the first row is added to the second row. If a scalar multiple of one row of a matrix is added to another row, then the determinant of the matrix is unchanged.
15. A multiple of the first row is added to the second row. If a scalar multiple of one row is added to another row, then the determinants are equal.
17. The second row of the matrix is multiplied by -1. If a row of a matrix is multiplied by a scalar, then the determinant is multiplied by that scalar.
19. The sixth column is 2 times the first column. If one column of a matrix is a multiple of another column, then the determinant of the matrix is zero.

21. -1 23. 19 25. 28
27. 17 29. -60 31. 223
33. -1344 35. 136 37. -1100

39. (a) True. See Theorem 3.3, part 1, page 134.
(b) True. See Theorem 3.3, part 3, page 134.
(c) True. See Theorem 3.4, part 2, page 136.

41. k 43. -1 45. 1 47. Proof

49. $\cos^2\theta + \sin^2\theta = 1$ 51. $\sin^2\theta - 1 = -\cos^2\theta$

53. Not possible. The determinant is equal to $\cos^2 x + \sin^2 x$, which cannot equal zero because $\cos^2 x + \sin^2 x = 1$.

55. Proof

Section 3.3 *(page 149)*

1. (a) 0 (b) -1
(c) $\begin{bmatrix} -2 & -3 \\ 4 & 6 \end{bmatrix}$
(d) 0

3. (a) 2 (b) -6
(c) $\begin{bmatrix} 1 & 4 & 3 \\ -1 & 0 & 3 \\ 0 & 2 & 0 \end{bmatrix}$
(d) -12

5. (a) 3 (b) 6
(c) $\begin{bmatrix} 6 & 3 & -2 & 2 \\ 2 & 1 & 0 & -1 \\ 9 & 4 & -3 & 8 \\ 8 & 5 & -4 & 5 \end{bmatrix}$
(d) 18

7. -44 9. 54

11. (a) -2 (b) -2 (c) 0

13. (a) 0 (b) -1 (c) -15

15. (a) 14 (b) 196 (c) 196 (d) 56 (e) $\frac{1}{14}$

17. (a) 29 (b) 841 (c) 841 (d) 232 (e) $\frac{1}{29}$

19. (a) 22 (b) 22 (c) 484 (d) 88 (e) $\frac{1}{22}$

21. (a) -115 (b) -115 (c) 13,225 (d) -1840 (e) $-\frac{1}{115}$

23. (a) -15 (b) -125 (c) 243 (d) -15 (e) $-\frac{1}{5}$

25. (a) 8 (b) 4 (c) 64 (d) 8 (e) $\frac{1}{2}$

27. Singular 29. Nonsingular
31. Nonsingular 33. Singular

35. $\frac{1}{5}$ 37. $-\frac{1}{2}$ 39. $\frac{1}{24}$

41. The solution is not unique because the determinant of the coefficient matrix is zero.

43. The solution is unique because the determinant of the coefficient matrix is nonzero.

45. $k = -1, 4$ 47. $k = 24$ 49. Proof

51. $\begin{bmatrix} 0 & 1 \\ 0 & 0 \end{bmatrix}$ and $\begin{bmatrix} 1 & 0 \\ 0 & 0 \end{bmatrix}$
(The answer is not unique.)

53. 0

55. Proof

57. No; in general, $P^{-1}AP \neq A$. For example, let

$$P = \begin{bmatrix} 1 & 2 \\ 3 & 5 \end{bmatrix},\ P^{-1} = \begin{bmatrix} -5 & 2 \\ 3 & -1 \end{bmatrix},$$

and $A = \begin{bmatrix} 2 & 1 \\ -1 & 0 \end{bmatrix}$.

Then you have

$$P^{-1}AP = \begin{bmatrix} -27 & -49 \\ 16 & 29 \end{bmatrix} \neq A.$$

The equation $|P^{-1}AP| = |A|$ is true in general because

$$|P^{-1}AP| = |P^{-1}||A||P| = |P^{-1}||P||A| = \frac{1}{|P|}|P||A| = |A|.$$

59. (a) False. See Theorem 3.6, page 144.
(b) True. See Theorem 3.8, page 146.
(c) True. See "Equivalent Conditions for a Nonsingular Matrix," parts 1 and 2, page 147.

61. Proof **63.** Orthogonal **65.** Not orthogonal

67. Orthogonal **69.** Proof

71. (a) $\begin{bmatrix} \frac{2}{3} & \frac{2}{3} & \frac{1}{3} \\ -\frac{2}{3} & \frac{1}{3} & \frac{2}{3} \\ \frac{1}{3} & -\frac{2}{3} & \frac{2}{3} \end{bmatrix}$ (b) $\begin{bmatrix} \frac{2}{3} & \frac{2}{3} & \frac{1}{3} \\ -\frac{2}{3} & \frac{1}{3} & \frac{2}{3} \\ \frac{1}{3} & -\frac{2}{3} & \frac{2}{3} \end{bmatrix}$ (c) 1

A is orthogonal.

73. Proof

Section 3.4 *(page 157)*

1. $\begin{bmatrix} 1 & 2 \\ 0 & -3 \end{bmatrix}\begin{bmatrix} 1 \\ 0 \end{bmatrix} = 1\begin{bmatrix} 1 \\ 0 \end{bmatrix}$;

$\begin{bmatrix} 1 & 2 \\ 0 & -3 \end{bmatrix}\begin{bmatrix} -1 \\ 2 \end{bmatrix} = \begin{bmatrix} 3 \\ -6 \end{bmatrix} = -3\begin{bmatrix} -1 \\ 2 \end{bmatrix}$

3. $\begin{bmatrix} 1 & 1 & 1 \\ 0 & 1 & 0 \\ 1 & 1 & 1 \end{bmatrix}\begin{bmatrix} 1 \\ 0 \\ 1 \end{bmatrix} = \begin{bmatrix} 2 \\ 0 \\ 2 \end{bmatrix} = 2\begin{bmatrix} 1 \\ 0 \\ 1 \end{bmatrix}$

$\begin{bmatrix} 1 & 1 & 1 \\ 0 & 1 & 0 \\ 1 & 1 & 1 \end{bmatrix}\begin{bmatrix} -1 \\ 0 \\ 1 \end{bmatrix} = \begin{bmatrix} 0 \\ 0 \\ 0 \end{bmatrix} = 0\begin{bmatrix} -1 \\ 0 \\ 1 \end{bmatrix}$

$\begin{bmatrix} 1 & 1 & 1 \\ 0 & 1 & 0 \\ 1 & 1 & 1 \end{bmatrix}\begin{bmatrix} -1 \\ 1 \\ -1 \end{bmatrix} = \begin{bmatrix} -1 \\ 1 \\ -1 \end{bmatrix} = 1\begin{bmatrix} -1 \\ 1 \\ -1 \end{bmatrix}$

5. (a) $\lambda^2 - \lambda - 2 = 0$ (b) $2, -1$
(c) $\lambda = 2$: $\mathbf{x} = \begin{bmatrix} 5 \\ 2 \end{bmatrix}$; $\lambda = -1$: $\mathbf{x} = \begin{bmatrix} 1 \\ 1 \end{bmatrix}$

7. (a) $\lambda^2 - 2\lambda - 3 = 0$ (b) $3, -1$
(c) $\lambda = 3$: $\mathbf{x} = \begin{bmatrix} 1 \\ 1 \end{bmatrix}$; $\lambda = -1$: $\mathbf{x} = \begin{bmatrix} -1 \\ 3 \end{bmatrix}$

9. (a) $\lambda^2 - 3\lambda - 18 = 0$ (b) $6, -3$
(c) $\lambda = 6$: $\mathbf{x} = \begin{bmatrix} 1 \\ 2 \end{bmatrix}$; $\lambda = -3$: $\mathbf{x} = \begin{bmatrix} -4 \\ 1 \end{bmatrix}$

11. (a) $\lambda^3 - 3\lambda^2 - 4\lambda + 12 = 0$ (b) $2, 3, -2$
(c) $\lambda = 2$: $\mathbf{x} = \begin{bmatrix} -1 \\ 0 \\ 1 \end{bmatrix}$; $\lambda = 3$: $\mathbf{x} = \begin{bmatrix} -1 \\ 1 \\ 1 \end{bmatrix}$;
$\lambda = -2$: $\mathbf{x} = \begin{bmatrix} 1 \\ -1 \\ 4 \end{bmatrix}$

13. (a) $\lambda^3 - 3\lambda^2 - \lambda + 3 = 0$ (b) $-1, 1, 3$
(c) $\lambda = 1$: $\mathbf{x} = \begin{bmatrix} 0 \\ -1 \\ 2 \end{bmatrix}$; $\lambda = -1$: $\mathbf{x} = \begin{bmatrix} -1 \\ 0 \\ 2 \end{bmatrix}$
$\lambda = 3$: $\mathbf{x} = \begin{bmatrix} 1 \\ 0 \\ 2 \end{bmatrix}$

15. Eigenvalues: $\lambda = -3, 1$

Eigenvectors: $\lambda = -3$: $\mathbf{x} = \begin{bmatrix} -1 \\ 1 \end{bmatrix}$

$\lambda = 1$: $\mathbf{x} = \begin{bmatrix} -5 \\ 1 \end{bmatrix}$

17. Eigenvalues: $\lambda = 1, 2, 3$
Eigenvectors:

$\lambda = 1$: $\mathbf{x} = \begin{bmatrix} 0 \\ -1 \\ 1 \end{bmatrix}$; $\lambda = 2$: $\mathbf{x} = \begin{bmatrix} 1 \\ 0 \\ 1 \end{bmatrix}$; $\lambda = 3$: $\mathbf{x} = \begin{bmatrix} 2 \\ 0 \\ 1 \end{bmatrix}$

19. Eigenvalues: $\lambda = 1, -2$

Eigenvectors:

$$\lambda = 1: \mathbf{x} = \begin{bmatrix}1\\0\\0\end{bmatrix}; \lambda = -2: \mathbf{x} = \begin{bmatrix}1\\0\\3\end{bmatrix}$$

21. Eigenvalues: $\lambda = -3, -1, 3, 5$

Eigenvectors:

$$\lambda = -3: \mathbf{x} = \begin{bmatrix}0\\0\\-1\\1\end{bmatrix}; \lambda = -1: \mathbf{x} = \begin{bmatrix}0\\1\\0\\0\end{bmatrix};$$

$$\lambda = 3: \mathbf{x} = \begin{bmatrix}1\\0\\0\\0\end{bmatrix}; \lambda = 5: \mathbf{x} = \begin{bmatrix}0\\0\\5\\3\end{bmatrix}$$

23. Eigenvalues: $\lambda = -2, -1, 1, 3$

Eigenvectors:

$$\lambda = -2: \mathbf{x} = \begin{bmatrix}0\\1\\0\\0\end{bmatrix}; \lambda = -1: \mathbf{x} = \begin{bmatrix}1\\0\\-2\\6\end{bmatrix}$$

$$\lambda = 1: \mathbf{x} = \begin{bmatrix}1\\0\\0\\0\end{bmatrix}; \lambda = 3: \mathbf{x} = \begin{bmatrix}1\\0\\2\\2\end{bmatrix}$$

25. (a) False. If $\mathbf{x}$ is an eigenvector corresponding to λ, then any *nonzero* multiple of $\mathbf{x}$ is also an eigenvector corresponding to λ. See page 153, first paragraph.

(b) False. If $\lambda = a$ is an eigenvalue of the matrix A, then $\lambda = a$ is a solution of the characteristic equation $|\lambda I - A| = 0$. See page 153, second paragraph.

Section 3.5 *(page 168)*

1. $\text{adj}(A) = \begin{bmatrix}4 & -2\\-3 & 1\end{bmatrix}$, $A^{-1} = \begin{bmatrix}-2 & 1\\\frac{3}{2} & -\frac{1}{2}\end{bmatrix}$

3. $\text{adj}(A) = \begin{bmatrix}0 & 0 & 0\\0 & -12 & -6\\0 & 4 & 2\end{bmatrix}$, A^{-1} does not exist.

5. $\text{adj}(A) = \begin{bmatrix}-7 & -12 & 13\\2 & 3 & -5\\2 & 3 & -2\end{bmatrix}$, $A^{-1} = \begin{bmatrix}\frac{7}{3} & 4 & -\frac{13}{3}\\-\frac{2}{3} & -1 & \frac{5}{3}\\-\frac{2}{3} & -1 & \frac{2}{3}\end{bmatrix}$

7. $\text{adj}(A) = \begin{bmatrix}7 & 1 & 9 & -13\\7 & 1 & 0 & -4\\-4 & 2 & -9 & 10\\2 & -1 & 9 & -5\end{bmatrix}$,

$$A^{-1} = \begin{bmatrix}\frac{7}{9} & \frac{1}{9} & 1 & -\frac{13}{9}\\\frac{7}{9} & \frac{1}{9} & 0 & -\frac{4}{9}\\-\frac{4}{9} & \frac{2}{9} & -1 & \frac{10}{9}\\\frac{2}{9} & -\frac{1}{9} & 1 & -\frac{5}{9}\end{bmatrix}$$

9. Proof

11. Proof

13. $|\text{adj}(A)| = \begin{vmatrix}-2 & 0\\-1 & 1\end{vmatrix} = -2,$

$|A|^{2-1} = \begin{vmatrix}1 & 0\\1 & -2\end{vmatrix}^{2-1} = -2$

15. Proof

17. $x_1 = 1$, $x_2 = 2$

19. $x_1 = 2$, $x_2 = -2$

21. $x_1 = \frac{3}{4}$, $x_2 = -\frac{1}{2}$

23. Cramer's Rule does not apply because the coefficient matrix has a determinant of zero.

25. Cramer's Rule does not apply because the coefficient matrix has a determinant of zero.

27. $x_1 = 1$, $x_2 = 1$, $x_3 = 2$

29. $x_1 = 1$, $x_2 = \frac{1}{2}$, $x_3 = \frac{3}{2}$

31. $x_1 = 0$, $x_2 = -\frac{1}{2}$, $x_3 = \frac{1}{2}$

33. Cramer's Rule does not apply because the coefficient matrix has a determinant of zero.

35. $x_1 = -4$

37. $x_1 = -1$

39. $x_1 = -7$

41. $x_1 = 5$

43. $x = \dfrac{4k-3}{2k-1}$, $y = \dfrac{4k-1}{2k-1}$

The system will be inconsistent if $k = \frac{1}{2}$.

45. 3

47. 3

49. Collinear

51. Not collinear

53. $3y - 4x = 0$

55. $x = -2$

57. $\frac{1}{3}$

59. 2

61. Not coplanar

63. Coplanar

65. $4x - 10y + 3z = 27$

67. $x + y + z = 0$

69. Incorrect. The numerator and denominator should be interchanged.

71. Correct

73. (a) $49a + 7b + c = 4380$

$64a + 8b + c = 4439$

$81a + 9b + c = 4524$

(b) $a = 13, b = -136, c = 4695$

(c) 4600

0 10

4300

(d) The polynomial fits the data exactly.

Review Exercises – Chapter 3 *(page 171)*

1. 10

3. 0

5. 0

7. -6

9. 1620

11. 82

13. -64

15. -1

17. -1

19. Because the second row is a multiple of the first row, the determinant is zero.

21. A -4 has been factored out of the second column and a 3 has been factored out of the third column. If a column of a matrix is multiplied by a scalar, then the determinant of the matrix is also multiplied by that scalar.

23. (a) -1 (b) -5 (c) $\begin{bmatrix} 1 & -2 \\ 2 & 1 \end{bmatrix}$ (d) 5

25. (a) -12 (b) -1728 (c) 144 (d) -300

27. (a) -20 (b) $-\frac{1}{20}$

29. $\frac{1}{6}$

31. $-\frac{1}{10}$

33. $x_1 = 0$
$x_2 = -\frac{1}{2}$
$x_3 = \frac{1}{2}$

35. $x_1 = -3$
$x_2 = -1$
$x_3 = 2$

37. Unique solution

39. Unique solution

41. Not a unique solution

43. (a) False. See "Definitions of Minors and Cofactors of a Matrix," page 124.

(b) False. See Theorem 3.3, part 1, page 134.

(c) False. See Theorem 3.9, page 148.

45. 128

47. Proof

49. 0

51. $\lambda = 7$: $\mathbf{x} = \begin{bmatrix} 1 \\ 1 \end{bmatrix}$ and $\lambda = -8$: $\mathbf{x} = \begin{bmatrix} -2 \\ 1 \end{bmatrix}$

53. $\lambda = 1$: $\mathbf{x} = \begin{bmatrix} 1 \\ 1 \\ 0 \end{bmatrix}$; $\lambda = 3$: $\mathbf{x} = \begin{bmatrix} 0 \\ 1 \\ 0 \end{bmatrix}$;

and $\lambda = 4$: $\mathbf{x} = \begin{bmatrix} 0 \\ 0 \\ 1 \end{bmatrix}$

55. $-\frac{1}{2}$

57. $-uv$

59. Row reduction is generally preferred for matrices with few zeros. For a matrix with many zeros, it is often easier to expand along a row or column having many zeros.

61. $\begin{bmatrix} 1 & -1 \\ 2 & 0 \end{bmatrix}$

63. Unique solution: $x = 0.6$
$y = 0.5$

65. Unique solution: $x_1 = \frac{1}{2}$
$x_2 = -\frac{1}{3}$
$x_3 = 1$

67. (a) $100a + 10b + c = 308.9$

$400a + 20b + c = 335.8$

$900a + 30b + c = 363.6$

(b) $a = 0.0045$
$b = 2.555$
$c = 282.9$

(c)

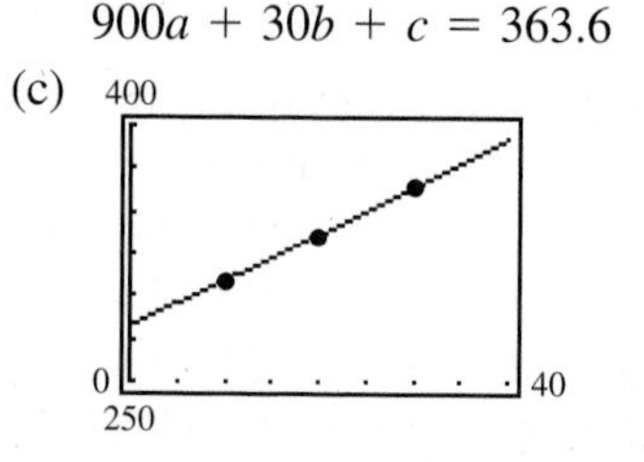

(d) The polynomial fits the data exactly.

69. 16

71. $x - 2y = -4$

73. $9x + 4y - 3z = 0$

75. (a) False. See Theorem 3.11, page 163.

(b) False. See "Test for Collinear Points in the xy-Plane," page 165.

Cumulative Test Chapters 1–3 *(page 177)*

1. $x_1 = 2,\ x_2 = -3,\ x_3 = -2$

2. $x_1 = s - 2t,\ x_2 = 2 + t,\ x_3 = t,\ x_4 = s$

3. $x_1 = -2s,\ x_2 = s,\ x_3 = 2t,\ x_4 = t$ **4.** $k = 12$

5. $BA = [13{,}275.00 \quad 15{,}500.00]$
The entries represent the total values (in dollars) of the products sent to the two warehouses.

6. $x = -3,\ y = 4$

7. $A^TA = \begin{bmatrix} 17 & 22 & 27 \\ 22 & 29 & 36 \\ 27 & 36 & 45 \end{bmatrix}$

8. (a) $\begin{bmatrix} -\frac{1}{4} & \frac{1}{8} \\ \frac{1}{6} & \frac{1}{12} \end{bmatrix}$ (b) $\begin{bmatrix} -\frac{2}{7} & \frac{1}{7} \\ \frac{1}{7} & \frac{2}{21} \end{bmatrix}$

9. $\begin{bmatrix} 1 & 0 & -1 \\ 0 & 0 & 1 \\ \frac{3}{5} & \frac{1}{5} & -\frac{9}{5} \end{bmatrix}$

10. $\begin{bmatrix} 0 & 1 \\ 1 & 0 \end{bmatrix}\begin{bmatrix} 1 & 0 \\ 2 & 1 \end{bmatrix}\begin{bmatrix} 1 & 0 \\ 0 & -4 \end{bmatrix}$ **11.** -34
(The answer is not unique.)

12. (a) 14 (b) -10 (c) -140 (d) $\frac{1}{14}$

13. (a) 567 (b) 7 (c) $\frac{1}{7}$ (d) 343

14. $\begin{bmatrix} \frac{4}{11} & -\frac{10}{11} & \frac{7}{11} \\ -\frac{1}{11} & -\frac{3}{11} & \frac{1}{11} \\ -\frac{2}{11} & \frac{5}{11} & \frac{2}{11} \end{bmatrix}$ **15.** $a = 1,\ b = 0,\ c = 2$
(The answer is not unique.)

16. $y = \frac{7}{6}x^2 + \frac{1}{6}x + 1$

17. $3x + 2y = 11$ **18.** 16

19. $\lambda = -1, \begin{bmatrix} -5 \\ 1 \\ 1 \end{bmatrix}$; $\lambda = -2, \begin{bmatrix} -2 \\ 0 \\ 1 \end{bmatrix}$; $\lambda = 2, \begin{bmatrix} -2 \\ -2 \\ 1 \end{bmatrix}$

20. No; proof **21.** Proof **22.** Proof

23. (a) A is row-equivalent to B if there exist elementary matrices, $E_1, \ldots, E_k$, such that $A = E_k \ldots E_1B$.
(b) Proof

Chapter 4

Section 4.1 *(page 188)*

1. $\mathbf{v} = (4, 5)$

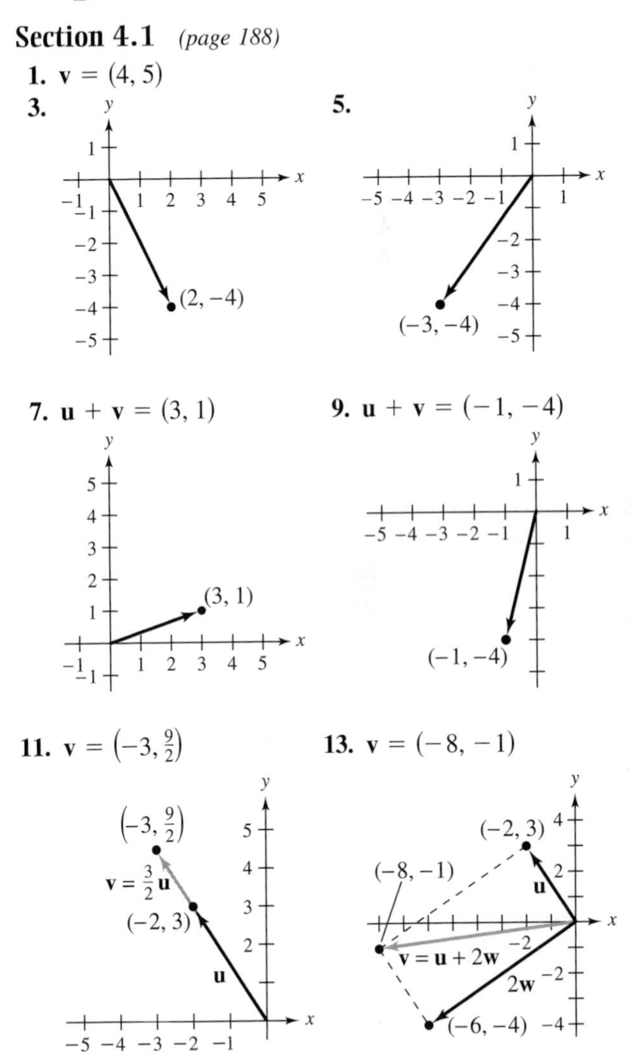

15. $\mathbf{v} = \left(-\frac{9}{2}, \frac{7}{2}\right)$

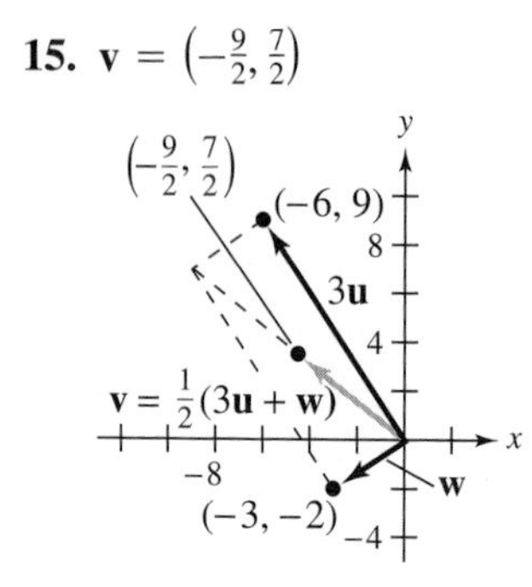

17. (a) (b)

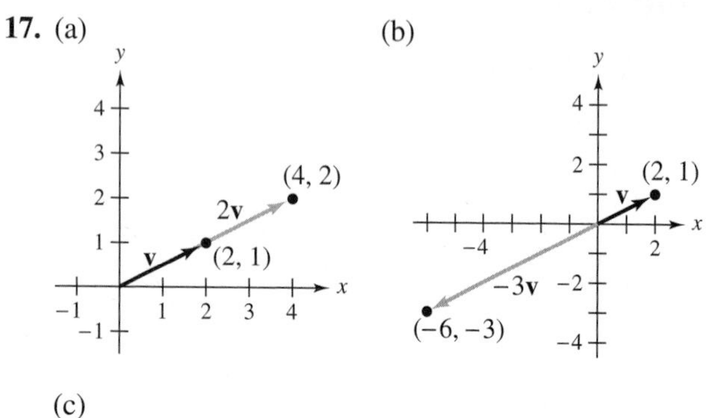

(c)

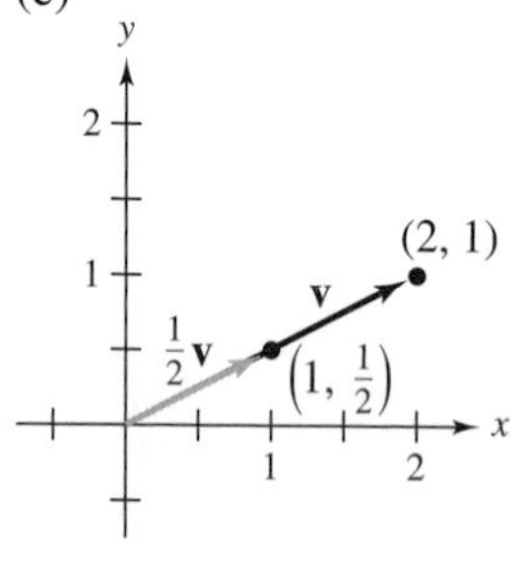

19. $\mathbf{u} - \mathbf{v} = (-1, 0, 4)$
$\mathbf{v} - \mathbf{u} = (1, 0, -4)$

21. $(6, 12, 6)$

23. $\left(\frac{7}{2}, 3, \frac{5}{2}\right)$

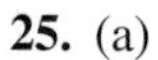

25. (a) (b)

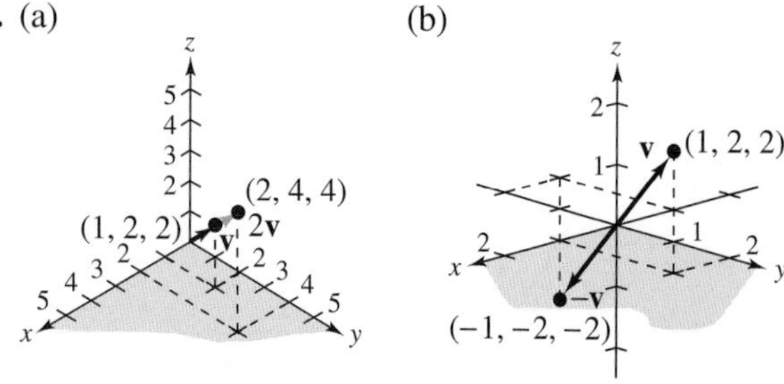

(c)

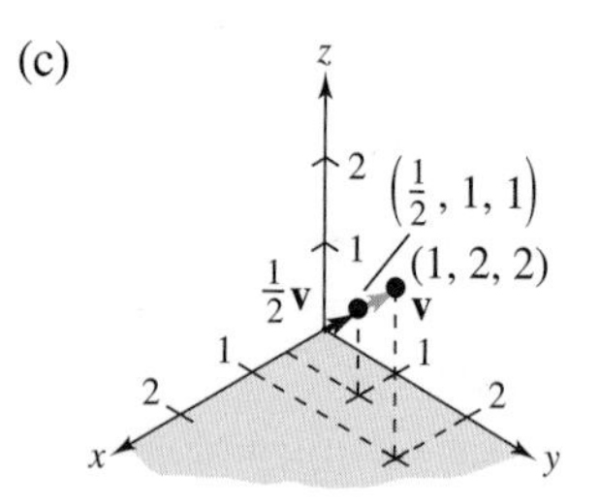

27. (a) and (b)

29. (a) $(4, -2, -8, 1)$
(b) $(8, 12, 24, 34)$
(c) $(-4, 4, 13, 3)$

31. (a) $(-9, 3, 2, -3, 6)$
(b) $(-2, -18, -12, 18, 36)$
(c) $(11, -6, -4, 6, -3)$

33. (a) $(1, 6, -5, -3)$ (b) $(-1, -8, 10, 0)$
(c) $\left(-\frac{3}{2}, 11, -\frac{13}{2}, -\frac{21}{2}\right)$ (d) $\left(\frac{1}{4}, 3, -3, -1\right)$

35. $\left(\frac{1}{2}, -\frac{7}{2}, -\frac{9}{2}, 2\right)$

37. $(4, 8, 18, -2)$

39. $\mathbf{v} = \mathbf{u} + \mathbf{w}$

41. $\mathbf{v} = \mathbf{u} + 2\mathbf{w}$

43. $\mathbf{v} = -\mathbf{u}$

45. $\left(-1, \frac{5}{3}, 6, \frac{2}{3}\right)$

47. $\mathbf{v} = \mathbf{u}_1 + 2\mathbf{u}_2 - 3\mathbf{u}_3$

49. It is not possible to write $\mathbf{v}$ as a linear combination of $\mathbf{u}_1$, $\mathbf{u}_2$, and $\mathbf{u}_3$.

51. $\mathbf{v} = 2\mathbf{u}_1 + \mathbf{u}_2 - 2\mathbf{u}_3 + \mathbf{u}_4 - \mathbf{u}_5$

53. $\mathbf{v} = 5\mathbf{u}_1 - \mathbf{u}_2 + \mathbf{u}_3 + 2\mathbf{u}_4 - 5\mathbf{u}_5 + 3\mathbf{u}_6$

55. (a) True. Two vectors in R^n are equal if and only if their corresponding components are equal, that is, $\mathbf{u} = \mathbf{v}$ if and only if $\mathbf{u}_1 = \mathbf{v}_1, \mathbf{u}_2 = \mathbf{v}_2, \ldots, \mathbf{u}_u = \mathbf{v}_u$.
(b) False. The vector $c\mathbf{v}$ is $|c|$ times as long as $\mathbf{v}$ and has the same direction as $\mathbf{v}$ if c is positive and the opposite direction if c is negative.

57. No

59. Answers will vary.

61. Proof

63. (a) Add $-\mathbf{v}$ to both sides.
(b) Associative property and additive identity
(c) Additive inverse
(d) Commutative property
(e) Additive identity

65. (a) Additive identity
(b) Distributive property
(c) Add $-c\mathbf{0}$ to both sides.
(d) Additive inverse and associative property
(e) Additive inverse
(f) Additive identity

67. (a) Additive inverse
(b) Transitive property
(c) Add $\mathbf{v}$ to both sides.
(d) Associative property
(e) Additive inverse
(f) Additive identity

69. No

71. You could describe vector subtraction as follows:

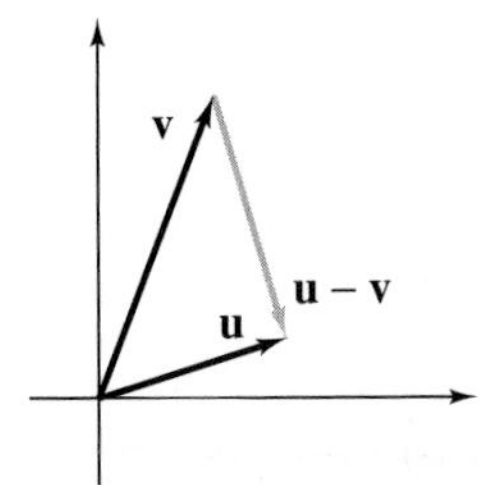

or write subtraction in terms of addition, $\mathbf{u} - \mathbf{v} = \mathbf{u} + (-1)\mathbf{v}$.

Section 4.2 *(page 197)*

1. $(0, 0, 0, 0)$

3. $\begin{bmatrix} 0 & 0 & 0 \\ 0 & 0 & 0 \end{bmatrix}$

5. $0 + 0x + 0x^2 + 0x^3$

7. $-(v_1, v_2, v_3, v_4) = (-v_1, -v_2, -v_3, -v_4)$

9. $-\begin{bmatrix} a_{11} & a_{12} & a_{13} \\ a_{21} & a_{22} & a_{23} \end{bmatrix} = \begin{bmatrix} -a_{11} & -a_{12} & -a_{13} \\ -a_{21} & -a_{22} & -a_{23} \end{bmatrix}$

11. $-(a_0 + a_1x + a_2x^2 + a_3x^3) = -a_0 - a_1x - a_2x^2 - a_3x^3$

13. The set is a vector space.

15. The set is not a vector space. Axiom 1 fails because $x^3 + (-x^3 + 1) = 1$, which is not a third-degree polynomial. (Axioms 4, 5, and 6 also fail.)

17. The set is not a vector space. Axiom 4 fails.

19. The set is not a vector space. Axiom 6 fails because $(-1)(x, y) = (-x, -y)$, which is not in the set when $x \neq 0$.

21. The set is a vector space.

23. The set is a vector space.

25. The set is not a vector space. Axiom 1 fails because

$$\begin{bmatrix} 1 & 0 \\ 0 & 0 \end{bmatrix} + \begin{bmatrix} 0 & 0 \\ 0 & 1 \end{bmatrix} = \begin{bmatrix} 1 & 0 \\ 0 & 1 \end{bmatrix},$$

which is not singular.

27. The set is a vector space.

29. (a) The set is not a vector space. Axiom 8 fails because

$$(1 + 2)(1, 1) = 3(1, 1) = (3, 1)$$
$$1(1, 1) + 2(1, 1) = (1, 1) + (2, 1) = (3, 2).$$

(b) The set is not a vector space. Axiom 2 fails because

$$(1, 2) + (2, 1) = (1, 0)$$
$$(2, 1) + (1, 2) = (2, 0).$$

(Axioms 4, 5, and 8 also fail.)

(c) The set is not a vector space. Axiom 6 fails because $(-1)(1, 1) = (\sqrt{-1}, \sqrt{-1})$, which is not in R^2. (Axioms 8 and 9 also fail.)

31. Proof

33. The set is not a vector space. Axiom 5 fails because $(1, 1)$ is the additive identity so $(0, 0)$ has no additive inverse. (Axioms 7 and 8 also fail.)

35. (a) True. See page 191.
(b) False. See Example 6, page 195.
(c) False. With standard operations on R^2, the additive inverse axiom is not satisfied.

37. (a) Add $-\mathbf{w}$ to both sides.
(b) Associative property
(c) Additive inverse
(d) Additive identity

39. Proof

41. Proof

Section 4.3 *(page 205)*

1. Because W is nonempty and $W \subset R^4$, you need only check that W is closed under addition and scalar multiplication. Given

$(x_1, x_2, x_3, 0) \in W$ and $(y_1, y_2, y_3, 0) \in W$,

it follows that

$$(x_1, x_2, x_3, 0) + (y_1, y_2, y_3, 0) = (x_1 + y_1, x_2 + y_2, x_3 + y_3, 0) \in W.$$

So, for any real number c and $(x_1, x_2, x_3, 0) \in W$, it follows that

$$c(x_1, x_2, x_3, 0) = (cx_1, cx_2, cx_3, 0) \in W.$$

3. Because W is nonempty and $W \subset M_{2,2}$, you need only check that W is closed under addition and scalar multiplication. Given

$$\begin{bmatrix} 0 & a_1 \\ b_1 & 0 \end{bmatrix} \in W \quad \text{and} \quad \begin{bmatrix} 0 & a_2 \\ b_2 & 0 \end{bmatrix} \in W,$$

it follows that

$$\begin{bmatrix} 0 & a_1 \\ b_1 & 0 \end{bmatrix} + \begin{bmatrix} 0 & a_2 \\ b_2 & 0 \end{bmatrix} = \begin{bmatrix} 0 & a_1 + a_2 \\ b_1 + b_2 & 0 \end{bmatrix} \in W.$$

So, for any real number c and

$\begin{bmatrix} 0 & a \\ b & 0 \end{bmatrix} \in W$, it follows that

$$c\begin{bmatrix} 0 & a \\ b & 0 \end{bmatrix} = \begin{bmatrix} 0 & ca \\ cb & 0 \end{bmatrix} \in W.$$

5. Recall from calculus that continuity implies integrability; $W \subset V$. So, because W is nonempty, you need only check that W is closed under addition and scalar multiplication. Given continuous functions $f, g \in W$, it follows that $f + g$ is continuous and $f + g \in W$. Also, for any real number c and for a continuous function $f \in W$, cf is continuous. So, $cf \in W$.

7. Not closed under addition:
$(0, 0, -1) + (0, 0, -1) = (0, 0, -2)$
Not closed under scalar multiplication:
$2(0, 0, -1) = (0, 0, -2)$

9. Not closed under scalar multiplication:
$\sqrt{2}(1, 1) = (\sqrt{2}, \sqrt{2})$

11. Not closed under scalar multiplication: $(-1)e^x = -e^x$

13. Not closed under scalar multiplication:
$(-2)(1, 1, 1) = (-2, -2, -2)$

15. Not closed under addition:

$$\begin{bmatrix} 1 & 0 \\ 0 & 0 \end{bmatrix} + \begin{bmatrix} 0 & 0 \\ 0 & 1 \end{bmatrix} = \begin{bmatrix} 1 & 0 \\ 0 & 1 \end{bmatrix}$$

17. Not closed under addition:
$(2, 8) + (3, 27) = (5, 35)$
Not closed under scalar multiplication:
$2(3, 27) = (6, 54)$

19. Not a subspace **21.** Subspace **23.** Subspace

25. Subspace **27.** Subspace **29.** Not a subspace

31. W is a subspace of R^3. (W is nonempty and closed under addition and scalar multiplication.)

33. W is a subspace of R^3. (W is nonempty and closed under addition and scalar multiplication.)

35. W is not a subspace of R^3.
Not closed under addition:
$(1, 1, 1) + (1, 1, 1) = (2, 2, 2)$
Not closed under scalar multiplication:
$2(1, 1, 1) = (2, 2, 2)$

37. (a) True. See "Remark," page 199.
(b) True. See Theorem 4.6, page 202.
(c) False. There may be elements of W which are not elements of U, or vice-versa.

39–47. Proof

Section 4.4 *(page 219)*

1. (a) **u** cannot be written as a linear combination of the given vectors.
(b) $\mathbf{v} = \frac{1}{4}(2, -1, 3) + \frac{3}{2}(5, 0, 4)$
(c) $\mathbf{w} = 8(2, -1, 3) - 3(5, 0, 4)$
(d) $\mathbf{z} = 2(2, -1, 3) - (5, 0, 4)$

3. (a) $\mathbf{u} = -\frac{7}{4}(2, 0, 7) + \frac{5}{4}(2, 4, 5) + 0(2, -12, 13)$
(b) **v** cannot be written as a linear combination of the given vectors.
(c) $\mathbf{w} = -\frac{1}{6}(2, 0, 7) + \frac{1}{3}(2, 4, 5) + 0(2, -12, 13)$
(d) $\mathbf{z} = -4(2, 0, 7) + 5(2, 4, 5) + 0(2, -12, 13)$

5. S spans R^2. **7.** S spans R^2.

9. S does not span R^2. (It spans a line in R^2.)

11. S does not span R^2. (It spans a line in R^2.)

13. S does not span R^2. (It spans a line in R^2.)

15. S spans R^2. **17.** S spans R^3.

19. S does not span R^3. (It spans a plane in R^3.)

21. S does not span R^3. (It spans a plane in R^3.)

23. Linearly independent **25.** Linearly dependent

27. Linearly independent **29.** Linearly dependent

31. Linearly independent **33.** Linearly independent

35. $(3, 4) - 4(-1, 1) - \frac{7}{2}(2, 0) = (0, 0)$,
$(3, 4) = 4(-1, 1) + \frac{7}{2}(2, 0)$
(The answer is not unique.)

37. $(1, 1, 1) - (1, 1, 0) - (0, 0, 1) - 0(0, 1, 1) = (0, 0, 0)$
$(1, 1, 1) = (1, 1, 0) + (0, 0, 1) - 0(0, 1, 1)$
(The answer is not unique.)

39. (a) All $t \neq 1, -2$ (b) All $t \neq \frac{1}{2}$

41. (a) $\begin{bmatrix} 6 & -19 \\ 10 & 7 \end{bmatrix} = 3A - 2B$
(b) Not a linear combination of A and B
(c) $\begin{bmatrix} -2 & 28 \\ 1 & -11 \end{bmatrix} = -A + 5B$
(d) $\begin{bmatrix} 0 & 0 \\ 0 & 0 \end{bmatrix} = 0A + 0B$

43. Linearly dependent **45.** Linearly independent

47. S does not span P_2.

49. (a) Any set of three vectors in R^2 must be linearly dependent.
(b) The second vector is a scalar multiple of the first vector.
(c) The first vector is the zero vector.

51. S_1 and S_2 span the same subspace.

53. (a) False. See "Definition of Linear Dependence and Linear Independence," page 213.
(b) True. See corollary to Theorem 4.8, page 218.

55–61. Proof

63. The theorem requires that only one of the vectors be a linear combination of the others. Because $(-1, 0, 2) = 0(1, 2, 3) - (1, 0, -2)$, there is no contradiction.

65. Proof

67. On $[0, 1]$,
$f_2(x) = |x| = x = \frac{1}{3}(3x)$
$= \frac{1}{3}f_1(x) \Rightarrow \{f_1, f_2\}$ dependent

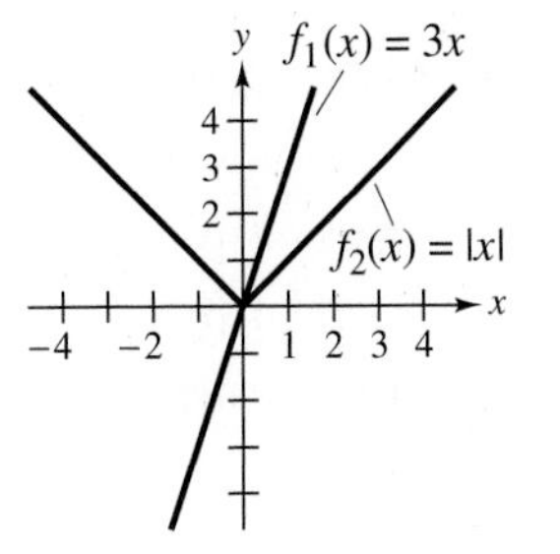

On $[-1, 1]$, f_1 and f_2 are not multiples of each other. For if they were, $cf_1(x) = f_2(x)$, then $c(3x) = |x|$. But if $x = 1$, $c = \frac{1}{3}$, whereas if $x = -1$, $c = -\frac{1}{3}$.

69. Proof

Section 4.5 *(page 230)*

1. R^6: $\{(1, 0, 0, 0, 0, 0), (0, 1, 0, 0, 0, 0), (0, 0, 1, 0, 0, 0), (0, 0, 0, 1, 0, 0), (0, 0, 0, 0, 1, 0), (0, 0, 0, 0, 0, 1)\}$

3. $M_{2,4}$: $\left\{\begin{bmatrix} 1 & 0 & 0 & 0 \\ 0 & 0 & 0 & 0 \end{bmatrix}, \begin{bmatrix} 0 & 1 & 0 & 0 \\ 0 & 0 & 0 & 0 \end{bmatrix}, \begin{bmatrix} 0 & 0 & 1 & 0 \\ 0 & 0 & 0 & 0 \end{bmatrix}, \begin{bmatrix} 0 & 0 & 0 & 1 \\ 0 & 0 & 0 & 0 \end{bmatrix}, \begin{bmatrix} 0 & 0 & 0 & 0 \\ 1 & 0 & 0 & 0 \end{bmatrix}, \begin{bmatrix} 0 & 0 & 0 & 0 \\ 0 & 1 & 0 & 0 \end{bmatrix}, \begin{bmatrix} 0 & 0 & 0 & 0 \\ 0 & 0 & 1 & 0 \end{bmatrix}, \begin{bmatrix} 0 & 0 & 0 & 0 \\ 0 & 0 & 0 & 1 \end{bmatrix}\right\}$

5. P_4: $\{1, x, x^2, x^3, x^4\}$ **7.** S is linearly dependent.

9. S is linearly dependent and does not span R^2.

11. S is linearly dependent and does not span R^2.

13. S does not span R^2.

15. S is linearly dependent and does not span R^3.

17. S does not span R^3.

19. S is linearly dependent and does not span R^3.

21. S is linearly dependent. **23.** S is linearly dependent.

25. S does not span $M_{2,2}$.

27. S is linearly dependent and does not span $M_{2,2}$.

29. The set is a basis for R^2.

31. The set is not a basis for R^2.

33. The set is not a basis for R^2.

35. S is a basis for R^2. **37.** S is a basis for R^3.

39. S is not a basis for R^3.

41. S is a basis for R^4. **43.** S is a basis for $M_{2,2}$.

45. S is a basis for P_3. **47.** S is not a basis for P_3.

49. S is a basis for R^3.
$(8, 3, 8) = 2(4, 3, 2) - (0, 3, 2) + 3(0, 0, 2)$

51. S is not a basis for R^3. **53.** S is not a basis for R^3.

55. 6 **57.** 1 **59.** 8 **61.** 6

63. $\begin{bmatrix}1&0&0\\0&0&0\\0&0&0\end{bmatrix}, \begin{bmatrix}0&0&0\\0&1&0\\0&0&0\end{bmatrix}, \begin{bmatrix}0&0&0\\0&0&0\\0&0&1\end{bmatrix},$
$\dim(D_{3,3}) = 3$

65. $\{(1, 0), (0, 1)\}, \{(1, 0), (1, 1)\}, \{(0, 1), (1, 1)\}$

67. $\{(1, 1), (1, 0)\}$ (The answer is not unique.)

69. (a) Line through the origin
(b) $\{(2, 1)\}$ (c) 1

71. (a) Line through the origin
(b) $\{(2, 1, -1)\}$ (c) 1

73. (a) $\{(2, 1, 0, 1), (-1, 0, 1, 0)\}$ (b) 2

75. (a) $\{(0, 6, 1, -1)\}$ (b) 1

77. (a) False. If the dimension of V is n, then every spanning set of V must have at least n vectors.
(b) True. Find a set of n basis vectors in V that will span V and add any other vector.

79. Proof **81.** Proof

83. (a) Basis for S_1: $\{(1, 0, 0), (1, 1, 0)\}$, dimension $= 2$
Basis for S_2: $\{(0, 0, 1), (0, 1, 0)\}$, dimension $= 2$
Basis for $S_1 \cap S_2$: $\{(0, 1, 0)\}$, dimension $= 1$
Basis for $S_1 + S_2$: $\{(1,0,0), (0,1,0), (0,0,1)\}$, dimension $= 3$
(b) No, it is not possible.

85. Proof

Section 4.6 *(page 246)*

1. (a) 2 (b) $\{(1, 0), (0, 1)\}$
(c) $\left\{\begin{bmatrix}1\\0\end{bmatrix}, \begin{bmatrix}0\\1\end{bmatrix}\right\}$

3. (a) 1 (b) $\{(1, 2, 3)\}$ (c) $\{[1]\}$

5. (a) 2 (b) $\{(1, 0, \frac{1}{2}), (0, 1, -\frac{1}{2})\}$
(c) $\left\{\begin{bmatrix}1\\0\end{bmatrix}, \begin{bmatrix}0\\1\end{bmatrix}\right\}$

7. (a) 2 (b) $\{(1, 0, \frac{1}{4}), (0, 1, \frac{3}{2})\}$
(c) $\left\{\begin{bmatrix}1\\0\\-\frac{2}{5}\end{bmatrix}, \begin{bmatrix}0\\1\\\frac{3}{5}\end{bmatrix}\right\}$

9. (a) 2 (b) $\{(1, 2, -2, 0), (0, 0, 0, 1)\}$
(c) $\left\{\begin{bmatrix}1\\0\\\frac{19}{7}\end{bmatrix}, \begin{bmatrix}0\\1\\\frac{8}{7}\end{bmatrix}\right\}$

11. (a) 5
(b) $\{(1, 0, 0, 0, 0), (0, 1, 0, 0, 0), (0, 0, 1, 0, 0), (0, 0, 0, 1, 0), (0, 0, 0, 0, 1)\}$
(c) $\left\{\begin{bmatrix}1\\0\\0\\0\\0\end{bmatrix}, \begin{bmatrix}0\\1\\0\\0\\0\end{bmatrix}, \begin{bmatrix}0\\0\\1\\0\\0\end{bmatrix}, \begin{bmatrix}0\\0\\0\\1\\0\end{bmatrix}, \begin{bmatrix}1\\0\\0\\0\\1\end{bmatrix}\right\}$

13. $\{(1, 0, 0), (0, 1, 0), (0, 0, 1)\}$

15. $\{(1, 1, 0), (0, 0, 1)\}$

17. $\{(1, 0, -1, 0), (0, 1, 0, 0), (0, 0, 0, 1)\}$

19. $\{(1, 0, 0, 0), (0, 1, 0, 0), (0, 0, 1, 0), (0, 0, 0, 1)\}$

21. $\{(0, 0)\}$, dim $= 0$

23. $\{(-2, 1, 0), (-3, 0, 1)\}$, dim $= 2$

25. $\{(-3, 0, 1)\}$, dim $= 1$

27. $\{(-1, 2, 1)\}$, dim $= 1$

29. $\{(2, -2, 0, 1), (-1, 1, 1, 0)\}$, dim $= 2$

31. $\{(0, 0, 0, 0)\}$, dim $= 0$

33. (a) $\{(-1, -3, 2)\}$ (b) 1

35. (a) $\{(-3, 0, 1), (2, 1, 0)\}$ (b) 2

37. (a) $\{(-4, -1, 1, 0), (-3, -\frac{2}{3}, 0, 1)\}$ (b) 2

39. (a) $\{(8, -9, -6, 6)\}$ (b) 1

41. (a) Consistent
(b) $\mathbf{x} = t(2, -4, 1) + (3, 5, 0)$

43. (a) Inconsistent (b) Not applicable

45. (a) Consistent
(b) $\mathbf{x} = t(5, 0, -6, -4, 1) + s(-2, 1, 0, 0, 0) + (1, 0, 2, -3, 0)$

47. $\begin{bmatrix}-1\\4\end{bmatrix} + 2\begin{bmatrix}2\\0\end{bmatrix} = \begin{bmatrix}3\\4\end{bmatrix}$

49. $-\frac{5}{4}\begin{bmatrix}1\\-1\\2\end{bmatrix} + \frac{3}{4}\begin{bmatrix}3\\1\\0\end{bmatrix} - \frac{1}{2}\begin{bmatrix}0\\0\\1\end{bmatrix} = \begin{bmatrix}1\\2\\-3\end{bmatrix}$

51. Four vectors in R^3 must be linearly dependent.

53. Proof

55. (a) $\begin{bmatrix}1 & 0\\0 & 1\end{bmatrix}, \begin{bmatrix}0 & 1\\1 & 0\end{bmatrix}$

(b) $\begin{bmatrix}1 & 0\\0 & 0\end{bmatrix}, \begin{bmatrix}0 & 1\\0 & 0\end{bmatrix}$

(c) $\begin{bmatrix}1 & 0\\0 & 0\end{bmatrix}, \begin{bmatrix}0 & 0\\0 & 1\end{bmatrix}$

57. (a) m (b) r (c) r (d) R^n (e) R^m

59. Answers will vary. **61.** Proof

63. (a) True. See Theorem 4.13, page 233.

(b) False. See Theorem 4.17, page 241.

65. (a) True. the columns of A become the rows of A^T, so the columns of A span the same space as the rows of A^T.

(b) False. The elementary row operations on A do not change linear dependency relationships of the columns of A but may change the column space of A.

67. (a) $\text{rank}(A) = \text{rank}(B) = 3$

$\text{nullity}(A) = n - r = 5 - 3 = 2$

(b) Choosing x_3 and x_5 as the free variables,

$x_1 = -s - t$

$x_2 = 2s - 3t$

$x_3 = s$

$x_4 = 5t$

$x_5 = t.$

A basis for the nullspace is

$$\left\{\begin{bmatrix}-1\\2\\1\\0\\0\end{bmatrix}, \begin{bmatrix}-1\\-3\\0\\5\\1\end{bmatrix}\right\}.$$

(c) $\{(1, 0, 1, 0, 1), (0, 1, -2, 0, 3), (0, 0, 0, 1, -5)\}$

(d) $\{(-2, 1, 3, 1), (-5, 3, 11, 7), (0, 1, 7, 5)\}$

(e) Linearly dependent

(f) (i) and (iii)

69. Proof **71.** Proof

Section 4.7 *(page 260)*

1. $\begin{bmatrix}8\\-3\end{bmatrix}$ **3.** $\begin{bmatrix}5\\4\\3\end{bmatrix}$ **5.** $\begin{bmatrix}-1\\2\\0\\1\end{bmatrix}$ **7.** $\begin{bmatrix}3\\2\end{bmatrix}$

9. $\begin{bmatrix}1\\-1\\2\end{bmatrix}$ **11.** $\begin{bmatrix}0\\-1\\2\end{bmatrix}$ **13.** $\begin{bmatrix}\frac{3}{2} & -\frac{1}{2}\\-2 & 1\end{bmatrix}$

15. $\begin{bmatrix}2 & -1\\4 & 3\end{bmatrix}$ **17.** $\begin{bmatrix}1 & 2 & -\frac{1}{2}\\0 & \frac{1}{2} & 0\\0 & -\frac{1}{3} & \frac{1}{12}\end{bmatrix}$

19. $\begin{bmatrix}\frac{9}{5} & \frac{4}{5}\\\frac{8}{5} & \frac{3}{5}\end{bmatrix}$

21. $\begin{bmatrix}1 & 1 & 1\\3 & 5 & 4\\3 & 6 & 5\end{bmatrix}$ **23.** $\begin{bmatrix}-7 & 3 & 10\\5 & -1 & -6\\11 & -3 & -10\end{bmatrix}$

25. $\begin{bmatrix}-24 & 7 & 1 & -2\\-10 & 3 & 0 & -1\\-29 & 7 & 3 & -2\\12 & -3 & -1 & 1\end{bmatrix}$

27. $\begin{bmatrix}1 & -\frac{3}{11} & \frac{5}{11} & 0 & -\frac{7}{11}\\0 & -\frac{2}{11} & \frac{3}{22} & 0 & -\frac{1}{11}\\-\frac{5}{4} & \frac{9}{22} & -\frac{19}{44} & -\frac{1}{4} & \frac{21}{22}\\-\frac{3}{4} & \frac{1}{2} & -\frac{1}{4} & \frac{1}{4} & \frac{1}{2}\\0 & -\frac{1}{11} & -\frac{2}{11} & 0 & \frac{5}{11}\end{bmatrix}$

29. (a) $\begin{bmatrix}-\frac{1}{3} & \frac{1}{3}\\\frac{3}{4} & -\frac{1}{2}\end{bmatrix}$ (b) $\begin{bmatrix}6 & 4\\9 & 4\end{bmatrix}$

(c) Verify. (d) $\begin{bmatrix}6\\3\end{bmatrix}$

31. (a) $\begin{bmatrix}4 & 5 & 1\\-7 & -10 & -1\\-2 & -2 & 0\end{bmatrix}$ (b) $\begin{bmatrix}\frac{1}{2} & \frac{1}{2} & -\frac{5}{4}\\-\frac{1}{2} & -\frac{1}{2} & \frac{3}{4}\\\frac{3}{2} & \frac{1}{2} & \frac{5}{4}\end{bmatrix}$

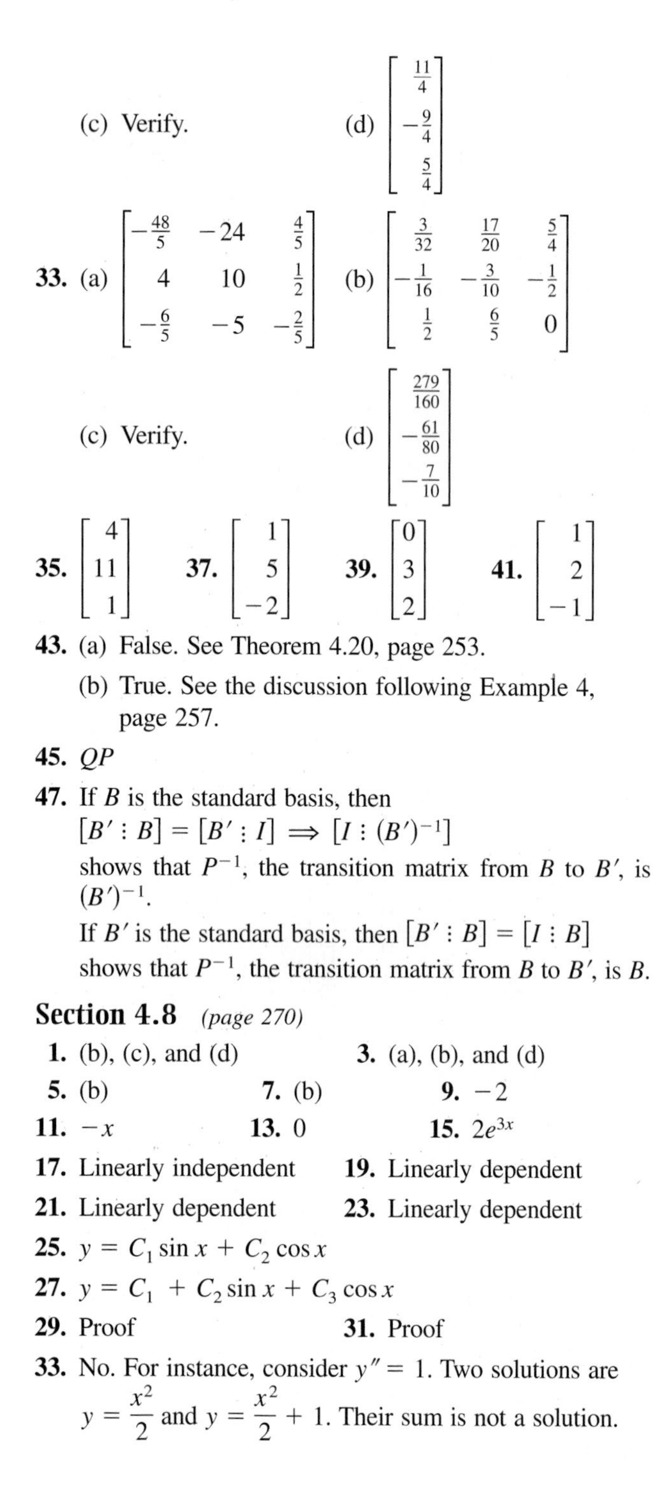

(c) Verify. (d) $\begin{bmatrix} \frac{11}{4} \\ -\frac{9}{4} \\ \frac{5}{4} \end{bmatrix}$

33. (a) $\begin{bmatrix} -\frac{48}{5} & -24 & \frac{4}{5} \\ 4 & 10 & \frac{1}{2} \\ -\frac{6}{5} & -5 & -\frac{2}{5} \end{bmatrix}$ (b) $\begin{bmatrix} \frac{3}{32} & \frac{17}{20} & \frac{5}{4} \\ -\frac{1}{16} & -\frac{3}{10} & -\frac{1}{2} \\ \frac{1}{2} & \frac{6}{5} & 0 \end{bmatrix}$

(c) Verify. (d) $\begin{bmatrix} \frac{279}{160} \\ -\frac{61}{80} \\ -\frac{7}{10} \end{bmatrix}$

35. $\begin{bmatrix} 4 \\ 11 \\ 1 \end{bmatrix}$ **37.** $\begin{bmatrix} 1 \\ 5 \\ -2 \end{bmatrix}$ **39.** $\begin{bmatrix} 0 \\ 3 \\ 2 \end{bmatrix}$ **41.** $\begin{bmatrix} 1 \\ 2 \\ -1 \end{bmatrix}$

43. (a) False. See Theorem 4.20, page 253.
(b) True. See the discussion following Example 4, page 257.

45. QP

47. If B is the standard basis, then
$[B' \vdots B] = [B' \vdots I] \Rightarrow [I \vdots (B')^{-1}]$
shows that P^{-1}, the transition matrix from B to B', is $(B')^{-1}$.
If B' is the standard basis, then $[B' \vdots B] = [I \vdots B]$ shows that P^{-1}, the transition matrix from B to B', is B.

Section 4.8 *(page 270)*

1. (b), (c), and (d) **3.** (a), (b), and (d)
5. (b) **7.** (b) **9.** -2
11. $-x$ **13.** 0 **15.** $2e^{3x}$
17. Linearly independent **19.** Linearly dependent
21. Linearly dependent **23.** Linearly dependent
25. $y = C_1 \sin x + C_2 \cos x$
27. $y = C_1 + C_2 \sin x + C_3 \cos x$
29. Proof **31.** Proof
33. No. For instance, consider $y'' = 1$. Two solutions are $y = \frac{x^2}{2}$ and $y = \frac{x^2}{2} + 1$. Their sum is not a solution.

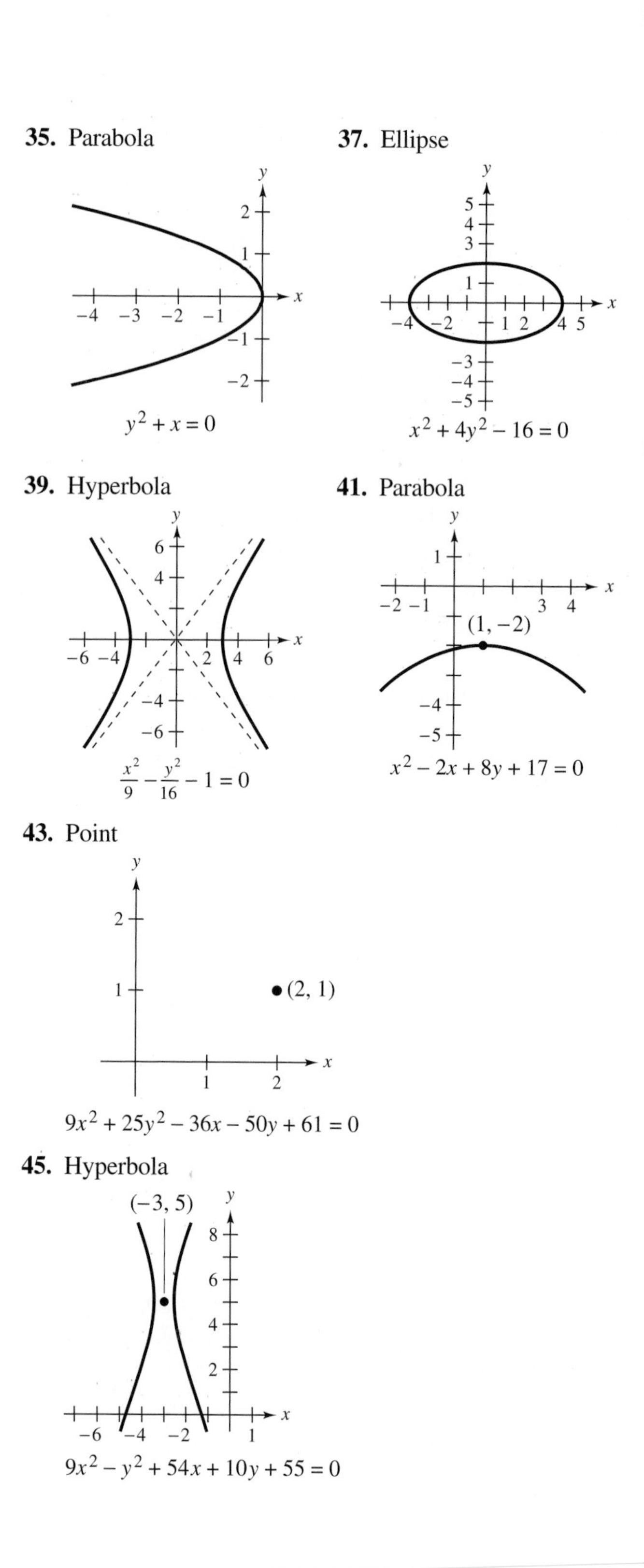

47. Ellipse

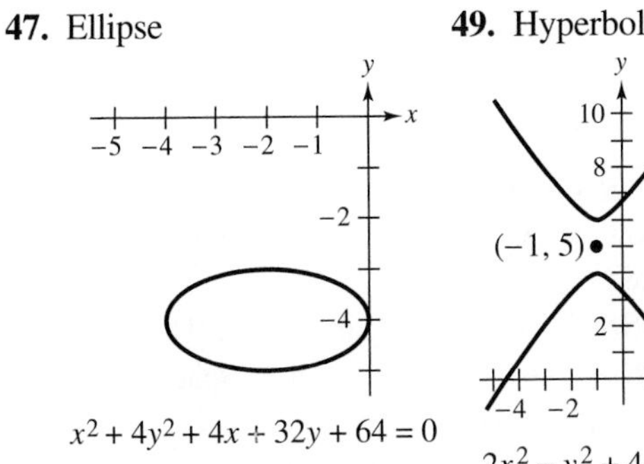

$x^2 + 4y^2 + 4x + 32y + 64 = 0$

49. Hyperbola

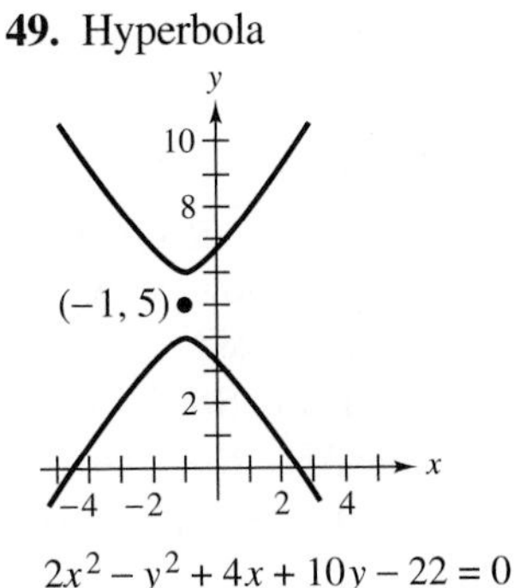

$2x^2 - y^2 + 4x + 10y - 22 = 0$

51. Parabola

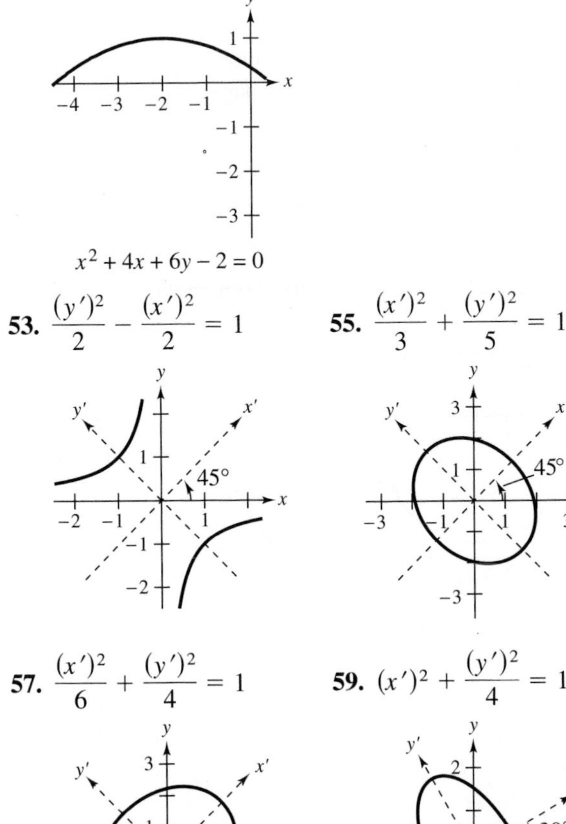

$x^2 + 4x + 6y - 2 = 0$

53. $\dfrac{(y')^2}{2} - \dfrac{(x')^2}{2} = 1$

55. $\dfrac{(x')^2}{3} + \dfrac{(y')^2}{5} = 1$

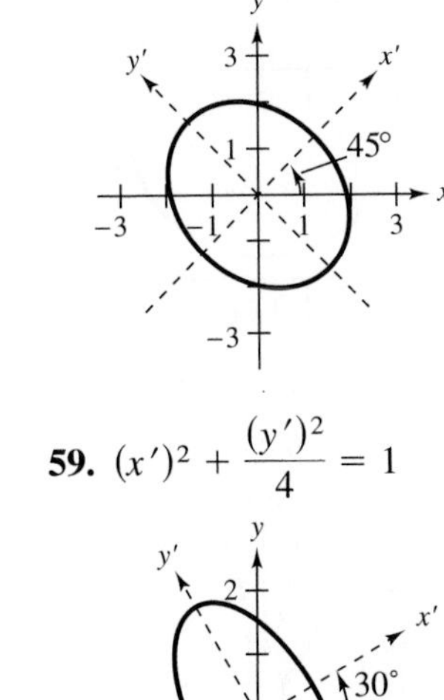

57. $\dfrac{(x')^2}{6} + \dfrac{(y')^2}{4} = 1$

59. $(x')^2 + \dfrac{(y')^2}{4} = 1$

61. $y' + 4 = -(x')^2$

63. $y' = 0$

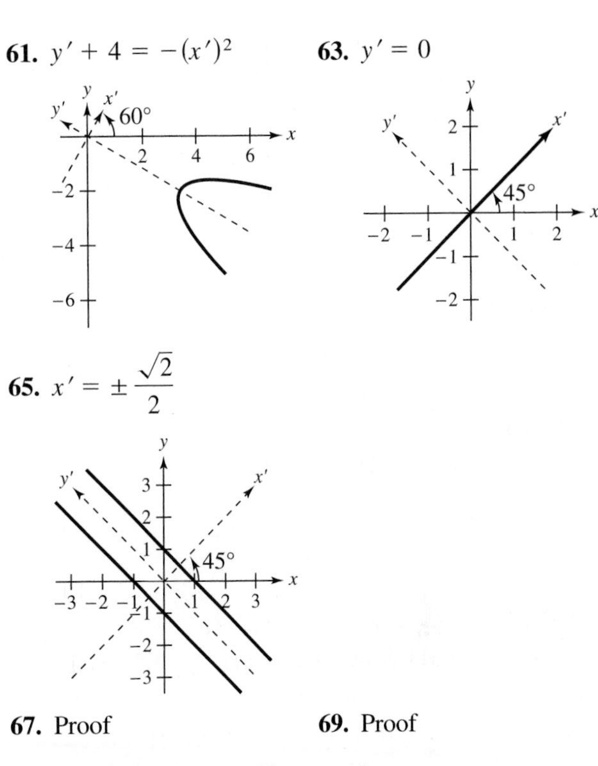

65. $x' = \pm\dfrac{\sqrt{2}}{2}$

67. Proof

69. Proof

Review Exercises – Chapter 4 *(page 272)*

1. (a) $(0, 2, 5)$
(b) $(2, 0, 4)$
(c) $(-2, 2, 1)$
(d) $(-5, 6, 5)$

3. (a) $(3, 1, 4, 4)$
(b) $(0, 4, 4, 2)$
(c) $(3, -3, 0, 2)$
(d) $(9, -7, 2, 7)$

5. $\left(\frac{1}{2}, -4, -4\right)$

7. $\left(\frac{5}{2}, -6, 0\right)$

9. $\mathbf{v} = 2\mathbf{u}_1 - \mathbf{u}_2 + 3\mathbf{u}_3$

11. $\mathbf{v} = \frac{9}{8}\mathbf{u}_1 + \frac{1}{8}\mathbf{u}_2 + 0\mathbf{u}_3$

13. $O_{3,4} = \begin{bmatrix} 0 & 0 & 0 & 0 \\ 0 & 0 & 0 & 0 \\ 0 & 0 & 0 & 0 \end{bmatrix},$

$-A = \begin{bmatrix} -a_{11} & -a_{12} & -a_{13} & -a_{14} \\ -a_{21} & -a_{22} & -a_{23} & -a_{24} \\ -a_{31} & -a_{32} & -a_{33} & -a_{34} \end{bmatrix}$

15. $O = (0, 0, 0)$
$-A = (-a_1, -a_2, -a_3)$

17. W is a subspace of R^2.

19. W is not a subspace of R^2.

21. W is a subspace of R^3.

23. W is not a subspace of $C[-1, 1]$.

25. (a) W is a subspace of R^3.
(b) W is not a subspace of R^3.

27. (a) Yes (b) Yes (c) Yes

29. (a) No (b) No (c) No

31. (a) Yes (b) No (c) No

33. S is a basis for P_3.

35. The set is not a basis for $M_{2,2}$.

37. (a) $\{(-3, 0, 4, 1), (-2, 1, 0, 0)\}$ (b) 2

39. (a) $\{(2, 3, 7, 0), (-1, 0, 0, 1)\}$ (b) 2

41. $\{(8, 5)\}$
$\text{Rank}(A) + \text{nullity}(A) = 1 + 1 = 2$

43. $\{(3, 0, 1, 0), (-1, -2, 0, 1)\}$
$\text{Rank}(A) + \text{nullity}(A) = 2 + 2 = 4$

45. $\{(4, -2, 1)\}$
$\text{Rank}(A) + \text{nullity}(A) = 2 + 1 = 3$

47. (a) 2 (b) $\{(1, 0), (0, 1)\}$

49. (a) 1 (b) $\{(1, -4, 0, 4)\}$

51. (a) 3 (b) $\{(1, 0, 0), (0, 1, 0), (0, 0, 1)\}$

53. $\begin{bmatrix} -2 \\ 8 \end{bmatrix}$ **55.** $\begin{bmatrix} \frac{3}{4} \\ \frac{1}{4} \end{bmatrix}$ **57.** $\begin{bmatrix} 2 \\ -1 \\ -1 \end{bmatrix}$

59. $\begin{bmatrix} \frac{2}{5} \\ -\frac{1}{4} \end{bmatrix}$ **61.** $\begin{bmatrix} -1 \\ 4 \\ \frac{3}{2} \end{bmatrix}$ **63.** $\begin{bmatrix} 3 \\ 1 \\ 0 \\ 1 \end{bmatrix}$

65. $\begin{bmatrix} -12 \\ 6 \end{bmatrix}$ **67.** $\begin{bmatrix} -2 \\ 1 \\ -2 \end{bmatrix}$ **69.** $\begin{bmatrix} 1 & 3 \\ -1 & 1 \end{bmatrix}$

71. $\begin{bmatrix} 0 & 0 & 1 \\ 0 & 1 & 0 \\ 1 & 0 & 0 \end{bmatrix}$

73. Basis for W: $\{x, x^2, x^3\}$
Basis for U: $\{(x - 1), x(x - 1), x^2(x - 1)\}$
Basis for $W \cap U$: $\{x(x - 1), x^2(x - 1)\}$

75. No. For example, the set
$\{x^2 + x, x^2 - x, 1\}$
is a basis for P_2.

77. Yes, W is a subspace of V.

79. Proof

81. Answers will vary.

83. (a) True. See discussion above "Definitions of Vector Addition and Scalar Multiplication in R^n," page 183.
(b) False. See Theorem 4.3, part 2, page 186.
(c) True. See "Definition of Vector Space" and the discussion following, page 191.

85. (a) True. See discussion under "Vectors in R^n," page 183.
(b) False. See "Definition of Vector Space," part 4, page 191.
(c) True. See discussion following "Summary of Important Vector Spaces," page 194.

87. (a) and (d) are solutions.

89. (a) is a solution.

91. e^x

93. -8

95. Linearly independent

97. Linearly dependent

99. Circle

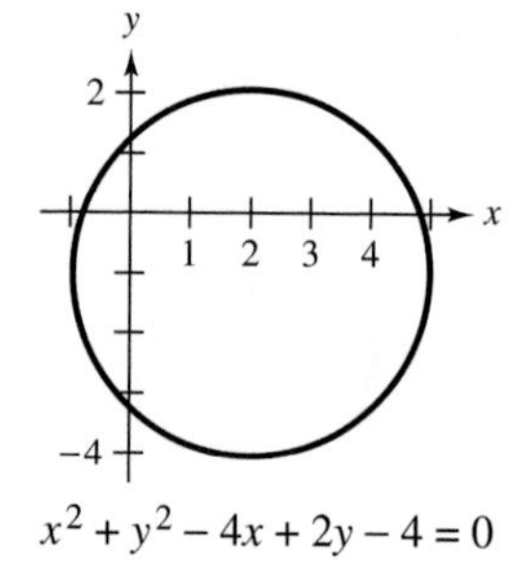

$x^2 + y^2 - 4x + 2y - 4 = 0$

101. Hyperbola

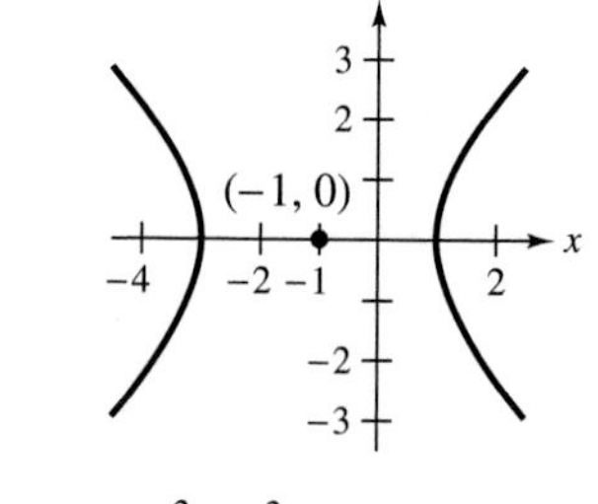

$x^2 - y^2 + 2x - 3 = 0$

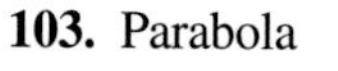

103. Parabola

105. Ellipse

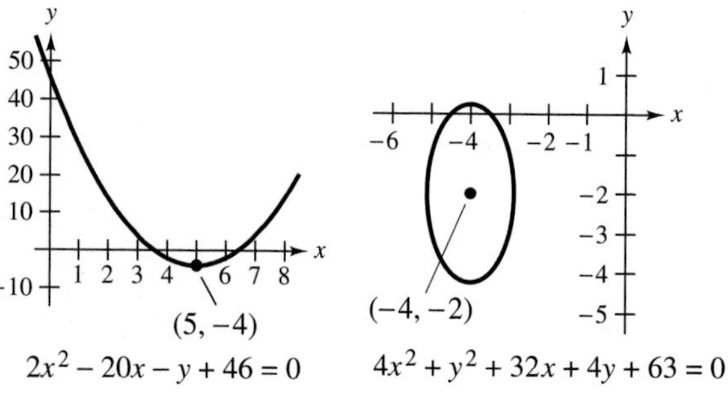

$2x^2 - 20x - y + 46 = 0$

$4x^2 + y^2 + 32x + 4y + 63 = 0$

107. $\dfrac{(x')^2}{6} - \dfrac{(y')^2}{6} = 1$ **109.** $(x')^2 = 4(y' - 1)$

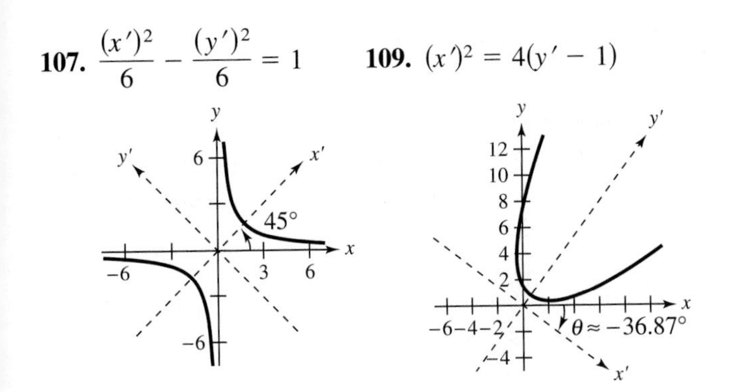

Chapter 5

Section 5.1 *(page 290)*

1. 5 **3.** 3 **5.** $3\sqrt{6}$

7. (a) $\dfrac{\sqrt{17}}{4}$ (b) $\dfrac{5\sqrt{41}}{8}$ (c) $\dfrac{\sqrt{577}}{8}$

9. (a) 5 (b) $\sqrt{6}$ (c) $\sqrt{21}$

11. (a) $\sqrt{6}$ (b) $\sqrt{11}$ (c) $\sqrt{13}$

13. (a) $\left(-\frac{5}{13}, \frac{12}{13}\right)$ (b) $\left(\frac{5}{13}, -\frac{12}{13}\right)$

15. (a) $\left(\dfrac{3}{\sqrt{38}}, \dfrac{2}{\sqrt{38}}, -\dfrac{5}{\sqrt{38}}\right)$
(b) $\left(-\dfrac{3}{\sqrt{38}}, -\dfrac{2}{\sqrt{38}}, \dfrac{5}{\sqrt{38}}\right)$

17. (a) $\left(\frac{1}{3}, 0, \frac{2}{3}, \frac{2}{3}\right)$ (b) $\left(-\frac{1}{3}, 0, -\frac{2}{3}, -\frac{2}{3}\right)$

19. $\pm\dfrac{1}{\sqrt{14}}$ **21.** $\left(2\sqrt{2}, 2\sqrt{2}\right)$

23. $\left(1, \sqrt{3}, 0\right)$ **25.** $\left(0, \dfrac{6}{\sqrt{6}}, \dfrac{3}{\sqrt{6}}, -\dfrac{3}{\sqrt{6}}\right)$

27. (a) $(4, 4, 3)$ (b) $\left(-2, -2, -\frac{3}{2}\right)$
(c) $(-16, -16, -12)$

29. $2\sqrt{2}$ **31.** $2\sqrt{3}$ **33.** $\sqrt{22}$

35. (a) -6 (b) 25 (c) 25 (d) $(-12, 18)$ (e) -30

37. (a) 0 (b) 6 (c) 6 (d) **0** (e) 0

39. (a) 5 (b) 50 (c) 50 (d) $(0, 10, 25, 20)$ (e) 25

41. -7

43. (a) $\|\mathbf{u}\| = 13$, $\|\mathbf{v}\| = 17$
(b) $\left(-\frac{8}{17}, -\frac{15}{17}\right)$
(c) $\left(-\frac{5}{13}, \frac{12}{13}\right)$
(d) 140
(e) 169
(f) 289

45. (a) $\|\mathbf{u}\| = 13$, $\|\mathbf{v}\| = 13$
(b) $\left(-\frac{12}{13}, \frac{5}{13}\right)$
(c) $\left(-\frac{5}{13}, -\frac{12}{13}\right)$
(d) 0
(e) 169
(f) 169

47. (a) $\|\mathbf{u}\| = 26$, $\|\mathbf{v}\| = 13$ (b) $\left(-\frac{5}{13}, -\frac{12}{13}\right)$
(c) $\left(-\frac{10}{26}, \frac{24}{26}\right) = \left(-\frac{5}{13}, \frac{12}{13}\right)$ (d) 238
(e) 676 (f) 169

49. (a) $\|\mathbf{u}\| = 13$, $\|\mathbf{v}\| = 13$ (b) $\left(0, -\frac{5}{13}, -\frac{12}{13}\right)$
(c) $\left(0, -\frac{5}{13}, -\frac{12}{13}\right)$ (d) -169
(e) 169 (f) 169

51. (a) $\|\mathbf{u}\| = 1.0843$, $\|\mathbf{v}\| = 0.3202$
(b) $(0, 0.7809, 0.6247)$
(c) $(-0.9223, -0.1153, -0.3689)$
(d) 0.1113 (e) 1.1756 (f) 0.1025

53. (a) $\|\mathbf{u}\| = 1.7321$, $\|\mathbf{v}\| = 2$
(b) $(-0.5, 0.7071, -0.5)$
(c) $(0, -0.5774, -0.8165)$
(d) 0 (e) 3 (f) 4

55. (a) $\|\mathbf{u}\| = 3.7417$, $\|\mathbf{v}\| = 3.7417$
(b) $(0.5345, 0, 0.2673, 0.2673, 0.5345, -0.5345)$
(c) $(0, -0.5345, -0.5345, 0.2673, -0.2673, 0.5345)$
(d) 7 (e) 14 (f) 14

57. (a) $\|\mathbf{u}\| = 3.7417$, $\|\mathbf{v}\| = 4$
(b) $(-0.25, 0, 0.25, 0.5, -0.5, 0.25, 0.25, -0.5)$
(c) $(0.2673, -0.2673, -0.5345, 0.2673, -0.2673, -0.2673, 0.5345, -0.2673)$
(d) -4 (e) 14 (f) 16

59. $|(3, 4) \cdot (2, -3)| \le \|(3, 4)\|\,\|(2, -3)\|$
$6 \le 5\sqrt{13}$

61. $|(1, 1, -2) \cdot (1, -3, -2)|$
$\le \|(1, 1, -2)\|\,\|(1, -3, -2)\|$
$2 \le 2\sqrt{21}$

63. 1.713 radians; (98.13°) **65.** $\dfrac{7\pi}{12}$; (105°)

67. 1.080 radians; (61.87°)

69. $\frac{\pi}{4}$ **71.** $\frac{\pi}{2}$ **73.** $\mathbf{v} = (t, 0)$

75. $\mathbf{v} = (2t, 3t)$ **77.** $\mathbf{v} = (t, 4t, s)$

79. $\mathbf{v} = (r, 0, s, t, w)$

81. Orthogonal **83.** Parallel **85.** Neither

87. Neither **89.** Orthogonal **91.** Neither

93. Orthogonal; $\mathbf{u} \cdot \mathbf{v} = 0$

95. (a) False. See "Definition of Length of a Vector in R^n," page 278.

(b) False. See "Definition of Dot Product in R^n," page 282.

97. (a) $\|\mathbf{u} \cdot \mathbf{v}\|$ is meaningless because $\mathbf{u} \cdot \mathbf{v}$ is a scalar.

(b) $\mathbf{u} + (\mathbf{u} \cdot \mathbf{v})$ is meaningless because $\mathbf{u}$ is a vector and $\mathbf{u} \cdot \mathbf{v}$ is a scalar.

99. $\|(5, 1)\| \le \|(4, 0)\| + \|(1, 1)\|$

$\sqrt{26} \le 4 + \sqrt{2}$

101. $\|(1, 1)\| \le \|(-1, 1)\| + \|(2, 0)\|$

$\sqrt{2} \le \sqrt{2} + 2$

103. $\|(2, 0)\|^2 = \|(1, -1)\|^2 + \|(1, 1)\|^2$

$4 = (\sqrt{2})^2 + (\sqrt{2})^2$

105. $\|(7, 1, -2)\|^2 = \|(3, 4, -2)\|^2 + \|(4, -3, 0)\|^2$

$54 = (\sqrt{29})^2 + 5^2$

107. (a) $\theta = \frac{\pi}{2}$ provided $\mathbf{u} \ne \mathbf{0}$ and $\mathbf{v} \ne \mathbf{0}$

(b) $0 \le \theta < \frac{\pi}{2}$ (c) $\frac{\pi}{2} < \theta \le \pi$

109. $(15, -8)$ and $(-15, 8)$, and $\left(\frac{15}{17}, -\frac{8}{17}\right)$ and $\left(-\frac{15}{17}, \frac{8}{17}\right)$ are orthogonal to $(8, 15)$. The answer is not unique.

111. $\theta = \cos^{-1}\left(\frac{\sqrt{6}}{3}\right) \approx 35.26°$

113–117. Proof

119. $A\mathbf{x} = \mathbf{0}$ means that the dot product of each row of A with the column vector $\mathbf{x}$ is zero. So, $\mathbf{x}$ is orthogonal to the row vectors of A.

121. Proof

Section 5.2 *(page 303)*

1. (a) -33 (b) 5 (c) 13 (d) $2\sqrt{65}$

3. (a) 15 (b) $\sqrt{57}$ (c) 5 (d) $2\sqrt{13}$

5. (a) -34 (b) $\sqrt{97}$ (c) $\sqrt{101}$ (d) $\sqrt{266}$

7. (a) 0 (b) $8\sqrt{3}$ (c) $\sqrt{411}$ (d) $3\sqrt{67}$

9. (a) 3 (b) $\sqrt{6}$ (c) 3 (d) 3

11. (a) $\frac{16}{15}$ (b) $\frac{\sqrt{10}}{5}$ (c) $\frac{2\sqrt{210}}{15}$ (d) $\sqrt{2}$

13. (a) $\frac{2}{e} \approx 0.736$ (b) $\frac{\sqrt{6}}{3} \approx 0.816$

(c) $\sqrt{\frac{e^2}{2} - \frac{1}{2e^2}} \approx 1.904$

(d) $\sqrt{\frac{e^2}{2} + \frac{2}{3} - \frac{1}{2e^2} - \frac{4}{e}} \approx 1.680$

15. (a) 0 (b) $\sqrt{2}$ (c) $\frac{2\sqrt{2}}{\sqrt{5}}$ (d) $\frac{3\sqrt{2}}{\sqrt{5}}$

17. (a) -6 (b) $\sqrt{35}$ (c) $\sqrt{7}$ (d) $3\sqrt{6}$

19. (a) -5 (b) $\sqrt{39}$ (c) $\sqrt{5}$ (d) $3\sqrt{6}$

21. (a) -4 (b) $\sqrt{11}$ (c) $\sqrt{2}$ (d) $\sqrt{21}$

23. (a) 0 (b) $\sqrt{2}$ (c) $\sqrt{2}$ (d) 2

25. Proof **27.** Proof

29. Axiom 4 fails, page 293.
$\langle (0, 1), (0, 1) \rangle = 0$, but $(0, 1) \ne \mathbf{0}$.

31. Axiom 4 fails, page 293.
$\langle (1, 1), (1, 1) \rangle = 0$, but $(1, 1) \ne \mathbf{0}$.

33. Axiom 2 fails, page 293.

$\langle (1, 0), (1, 0) + (1, 0) \rangle = \langle (1, 0), (2, 0) \rangle = 4$

$\langle (1, 0), (1, 0) \rangle + \langle (1, 0), (1, 0) \rangle = 1 + 1 = 2$

35. Axiom 1 fails, page 293. If $\mathbf{u} = (1, 2)$ and $\mathbf{v} = (2, 3)$,
$\langle \mathbf{u}, \mathbf{v} \rangle = 3(1)(3) - 2(2) = 5$ and
$\langle \mathbf{v}, \mathbf{u} \rangle = 3(2)(2) - 3(1) = 9$.

37. 2.103 radians (120.5°) **39.** 1.16 radians (66.59°)

41. $\frac{\pi}{2}$ **43.** 1.23 radians (70.53°)

45. $\frac{\pi}{2}$

47. (a) $|\langle(5, 12), (3, 4)\rangle| \leq \|(5, 12)\|\,\|(3, 4)\|$

$$63 \leq (13)(5)$$

(b) $\|(5, 12) + (3, 4)\| \leq \|(5, 12)\| + \|(3, 4)\|$

$$8\sqrt{5} \leq 13 + 5$$

49. (a) $|(1, 0, 4)| \cdot (-5, 4, 1)| \leq \sqrt{17}\sqrt{42}$

$$1 \leq \sqrt{714}$$

(b) $\|(-4, 4, 5)\| \leq \sqrt{17} + \sqrt{42}$

$$\sqrt{57} \leq \sqrt{17} + \sqrt{42}$$

51. (a) $|\langle 2x, 3x^2 + 1\rangle| \leq \|2x\|\,\|3x^2 + 1\|$

$$0 \leq (2)(\sqrt{10})$$

(b) $\|2x + 3x^2 + 1\| \leq \|2x\| + \|3x^2 + 1\|$

$$\sqrt{14} \leq 2 + \sqrt{10}$$

53. (a) $|0(-3) + 3(1) + 2(4) + 1(3)| \leq \sqrt{14}\sqrt{35}$

$$14 \leq \sqrt{14}\sqrt{35}$$

(b) $\left\|\begin{bmatrix} -3 & 4 \\ 6 & 4 \end{bmatrix}\right\| \leq \sqrt{14} + \sqrt{35}$

$$\sqrt{77} \leq \sqrt{14} + \sqrt{35}$$

55. (a) $|\langle \sin x, \cos x\rangle| \leq \|\sin x\|\,\|\cos x\|$

$$0 \leq (\sqrt{\pi})(\sqrt{\pi})$$

(b) $\|\sin x + \cos x\| \leq \|\sin x\| + \|\cos x\|$

$$\sqrt{2\pi} \leq \sqrt{\pi} + \sqrt{\pi}$$

57. (a) $|x, e^x| \leq \|x\|\,\|e^x\|$

$$1 \leq \sqrt{\tfrac{1}{3}} \cdot \sqrt{\tfrac{1}{2}e^2 - \tfrac{1}{2}}$$

(b) $\|x + e^x\| \leq \|x\| + \|e^x\|$

$$\sqrt{\tfrac{11}{6} + \tfrac{1}{2}e^2} \leq \sqrt{\tfrac{1}{3}} + \sqrt{\tfrac{1}{2}e^2 - \tfrac{1}{2}}$$

59. Because

$$\langle f, g\rangle = \int_{-\pi}^{\pi} \cos x \sin x \, dx = \frac{1}{2}\sin^2 x\Big]_{-\pi}^{\pi} = 0,$$

f and g are orthogonal.

61. The functions $f(x) = x$ and $g(x) = \frac{1}{2}(5x^3 - 3x)$ are orthogonal because

$$\langle f, g\rangle = \int_{-1}^{1} x\,\frac{1}{2}(5x^3 - 3x)\,dx = \frac{1}{2}\int_{-1}^{1}(5x^4 - 3x^2)\,dx = \frac{1}{2}(x^5 - x^3)\Big]_{-1}^{1} = 0.$$

63. (a) $\left(\frac{8}{5}, \frac{4}{5}\right)$ (b) $\left(\frac{4}{5}, \frac{8}{5}\right)$

(c)

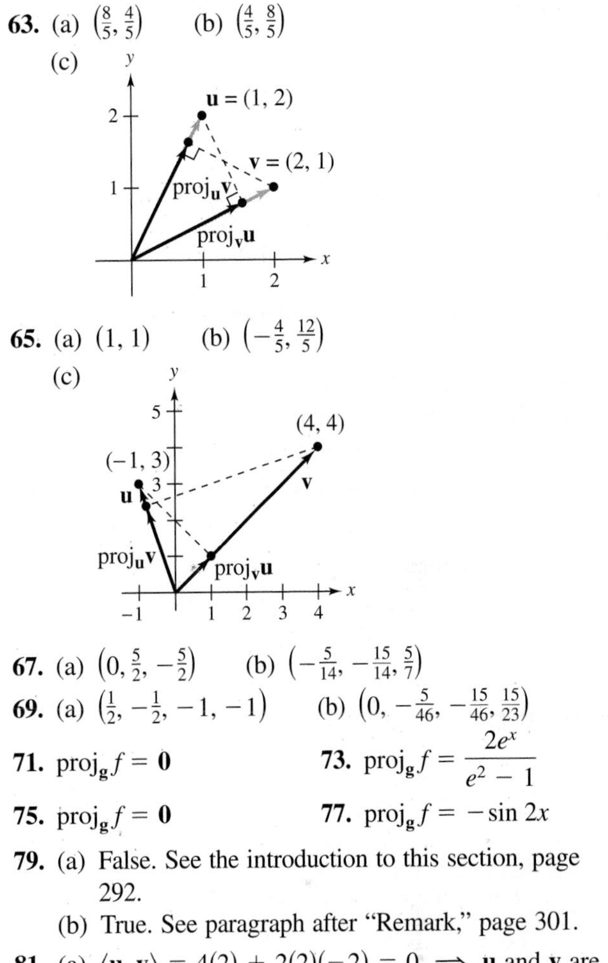

65. (a) $(1, 1)$ (b) $\left(-\frac{4}{5}, \frac{12}{5}\right)$

(c)

67. (a) $\left(0, \frac{5}{2}, -\frac{5}{2}\right)$ (b) $\left(-\frac{5}{14}, -\frac{15}{14}, \frac{5}{7}\right)$

69. (a) $\left(\frac{1}{2}, -\frac{1}{2}, -1, -1\right)$ (b) $\left(0, -\frac{5}{46}, -\frac{15}{46}, \frac{15}{23}\right)$

71. $\text{proj}_{\mathbf{g}} f = \mathbf{0}$

73. $\text{proj}_{\mathbf{g}} f = \dfrac{2e^x}{e^2 - 1}$

75. $\text{proj}_{\mathbf{g}} f = \mathbf{0}$

77. $\text{proj}_{\mathbf{g}} f = -\sin 2x$

79. (a) False. See the introduction to this section, page 292.

(b) True. See paragraph after "Remark," page 301.

81. (a) $\langle \mathbf{u}, \mathbf{v}\rangle = 4(2) + 2(2)(-2) = 0 \Rightarrow$ **u** and **v** are orthogonal.

(b)

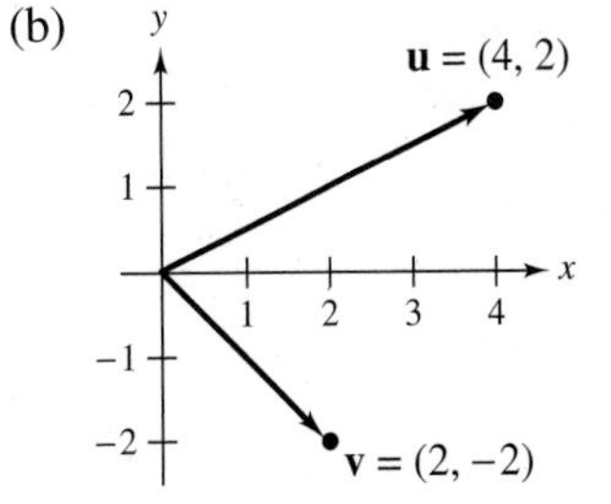

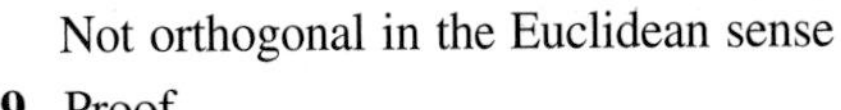

Not orthogonal in the Euclidean sense

83–89. Proof

Section 5.3 (page 318)

1. Orthogonal **3.** Neither **5.** Orthonormal
7. Orthogonal **9.** Orthonormal **11.** Orthogonal
13. Orthonormal

15. Orthogonal; $\left(-\frac{1}{\sqrt{17}}, \frac{4}{\sqrt{17}}\right), \left(\frac{4}{\sqrt{17}}, \frac{1}{\sqrt{17}}\right)$

17. Orthogonal; $\left(\frac{\sqrt{3}}{3}, \frac{\sqrt{3}}{3}, \frac{\sqrt{3}}{3}\right), \left(-\frac{\sqrt{2}}{2}, 0, \frac{\sqrt{2}}{2}\right)$

19. The set $\{1, x, x^2, x^3\}$ is orthogonal because
$\langle 1, x\rangle = 0, \langle 1, x^2\rangle = 0, \langle 1, x^3\rangle = 0,$
$\langle x, x^2\rangle = 0, \langle x, x^3\rangle = 0, \langle x^2, x^3\rangle = 0.$
Furthermore, the set is orthonormal because
$\|1\| = 1, \|x\| = 1, \|x^2\| = 1,$ and $\|x^3\| = 1.$
So, $\{1, x, x^2, x^3\}$ is an orthonormal basis for P_3.

21. $\begin{bmatrix} \frac{4\sqrt{13}}{13} \\ \frac{7\sqrt{13}}{13} \end{bmatrix}$ **23.** $\begin{bmatrix} \frac{\sqrt{10}}{2} \\ -2 \\ -\frac{\sqrt{10}}{2} \end{bmatrix}$ **25.** $\begin{bmatrix} 11 \\ 2 \\ 15 \end{bmatrix}$

27. $\left\{\left(\frac{3}{5}, \frac{4}{5}\right), \left(\frac{4}{5}, -\frac{3}{5}\right)\right\}$ **29.** $\{(0, 1), (1, 0)\}$

31. $\left\{\left(\frac{1}{3}, -\frac{2}{3}, \frac{2}{3}\right), \left(\frac{2}{3}, \frac{2}{3}, \frac{1}{3}\right), \left(\frac{2}{3}, -\frac{1}{3}, -\frac{2}{3}\right)\right\}$

33. $\left\{\left(\frac{4}{5}, -\frac{3}{5}, 0\right), \left(\frac{3}{5}, \frac{4}{5}, 0\right), (0, 0, 1)\right\}$

35. $\left\{\left(0, \frac{\sqrt{2}}{2}, \frac{\sqrt{2}}{2}\right), \left(\frac{\sqrt{6}}{3}, \frac{\sqrt{6}}{6}, -\frac{\sqrt{6}}{6}\right), \left(\frac{\sqrt{3}}{3}, -\frac{\sqrt{3}}{3}, \frac{\sqrt{3}}{3}\right)\right\}$

37. $\left\{\left(-\frac{4\sqrt{2}}{7}, \frac{3\sqrt{2}}{14}, \frac{5\sqrt{2}}{14}\right)\right\}$

39. $\left\{\left(\frac{3}{5}, \frac{4}{5}, 0\right), \left(\frac{4}{5}, -\frac{3}{5}, 0\right)\right\}$

41. $\left\{\left(\frac{\sqrt{6}}{6}, \frac{\sqrt{6}}{3}, -\frac{\sqrt{6}}{6}, 0\right), \left(\frac{\sqrt{3}}{3}, 0, \frac{\sqrt{3}}{3}, \frac{\sqrt{3}}{3}\right),\right.$
$\left.\left(\frac{\sqrt{3}}{3}, -\frac{\sqrt{3}}{3}, -\frac{\sqrt{3}}{3}, 0\right)\right\}$

43. $\langle x, 1\rangle = \int_{-1}^{1} x\, dx = \frac{x^2}{2}\Big]_{-1}^{1} = 0$

45. $\langle x^2, 1\rangle = \int_{-1}^{1} x^2\, dx = \frac{x^3}{3}\Big]_{-1}^{1} = \frac{2}{3}$

47. (a) True. See "Definitions of Orthogonal and Orthonormal Sets," page 306.
(b) True. See corollary to Theorem 5.10, page 310.
(c) False. See "Remark," page 314.

49. $\left\{\left(\frac{3\sqrt{10}}{10}, 0, \frac{\sqrt{10}}{10}, 0\right), \left(0, -\frac{2\sqrt{5}}{5}, 0, \frac{\sqrt{5}}{5}\right)\right\}$

51. $\left\{\left(\frac{\sqrt{2}}{2}, 0, \frac{\sqrt{2}}{2}, 0\right), \left(\frac{\sqrt{6}}{6}, 0, -\frac{\sqrt{6}}{6}, \frac{\sqrt{6}}{3}\right)\right\}$

53. $\left\{\left(-\frac{3\sqrt{10}}{10}, \frac{\sqrt{10}}{10}, 0\right), \left(\frac{3\sqrt{190}}{190}, \frac{9\sqrt{190}}{190}, \frac{\sqrt{190}}{19}\right)\right\}$

55. Orthonormal **57.** $\{x^2, x, 1\}$

59. $\left\{\frac{1}{\sqrt{2}}(x^2 - 1), -\frac{1}{\sqrt{6}}(x^2 - 2x + 1)\right\}$

61. $\left\{\left(\frac{2}{3}, -\frac{1}{3}\right), \left(\frac{\sqrt{2}}{6}, \frac{2\sqrt{2}}{3}\right)\right\}$

63. Proof **65.** Proof

67. $\left\{\left(\frac{1}{\sqrt{2}}, 0, \frac{1}{\sqrt{2}}, 0\right), \left(0, -\frac{1}{\sqrt{2}}, 0, \frac{1}{\sqrt{2}}\right),\right.$
$\left.\left(\frac{1}{\sqrt{2}}, 0, -\frac{1}{\sqrt{2}}, 0\right), \left(0, \frac{1}{\sqrt{2}}, 0, \frac{1}{\sqrt{2}}\right)\right\}$

69. $N(A)$ basis: $\{(3, -1, 2)\}$
$N(A^T)$ basis: $\{(-1, -1, 1)\}$
$R(A)$ basis: $\{(1, 0, 1), (1, 2, 3)\}$
$R(A^T)$ basis: $\{(1, 1, -1), (0, 2, 1)\}$

71. $N(A)$ basis: $\{(1, 0, -1)\}$
$N(A^T) = \{(0, 0)\}$
$R(A)$ basis: $\{(1, 1), (0, 1)\} = \mathbf{R}^2$
$R(A^T)$ basis: $\{(1, 0, 1), (1, 1, 1)\}$

73. Proof

Section 5.4 (page 333)

1. Not orthogonal **3.** Orthogonal

5. span $\left\{\begin{bmatrix} 0 \\ 1 \\ 0 \end{bmatrix}\right\}$ **7.** span $\left\{\begin{bmatrix} 0 \\ 0 \\ 1 \\ 0 \end{bmatrix}, \begin{bmatrix} 2 \\ -1 \\ 0 \\ 1 \end{bmatrix}\right\}$

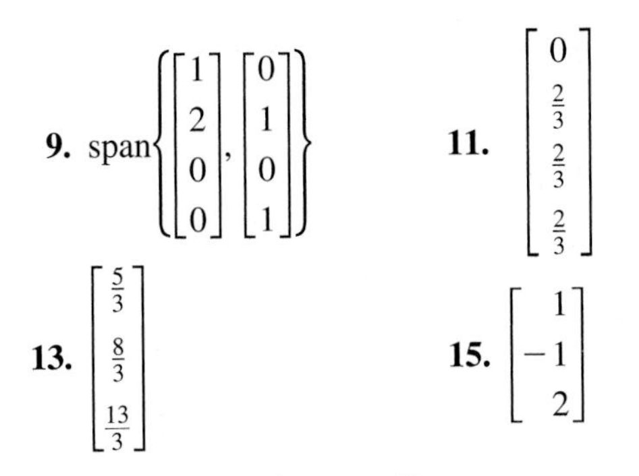

9. $\text{span}\left\{\begin{bmatrix}1\\2\\0\\0\end{bmatrix}, \begin{bmatrix}0\\1\\0\\1\end{bmatrix}\right\}$

11. $\begin{bmatrix}0\\\frac{2}{3}\\\frac{2}{3}\\\frac{2}{3}\end{bmatrix}$

13. $\begin{bmatrix}\frac{5}{3}\\\frac{8}{3}\\\frac{13}{3}\end{bmatrix}$

15. $\begin{bmatrix}1\\-1\\2\end{bmatrix}$

17. $N(A)$ basis: $\{(-3, 0, 1)\}$
$N(A^T) = \{(0, 0)\}$
$R(A)$ basis: $\{(1, 0), (2, 1)\} = \mathbf{R}^2$
$R(A^T)$ basis: $\{(1, 2, 3), (0, 1, 0)\}$

19. $N(A)$ basis: $\{(-1, -1, 0, 1), (0, -1, 1, 0)\}$
$N(A^T)$ basis: $\{(-1, -1, 1, 0), (-1, -2, 0, 1)\}$
$R(A)$ basis: $\{(1, 0, 1, 1), (0, 1, 1, 2)\}$
$R(A^T)$ basis: $\{(1, 0, 0, 1), (0, 1, 1, 1)\}$

21. $\mathbf{x} = \begin{bmatrix}1\\-1\end{bmatrix}$ **23.** $\mathbf{x} = \begin{bmatrix}2\\-2\\1\end{bmatrix}$ **25.** $\mathbf{x} = \begin{bmatrix}\frac{1}{3}\\0\\\frac{1}{3}\end{bmatrix}$

27. **29.**

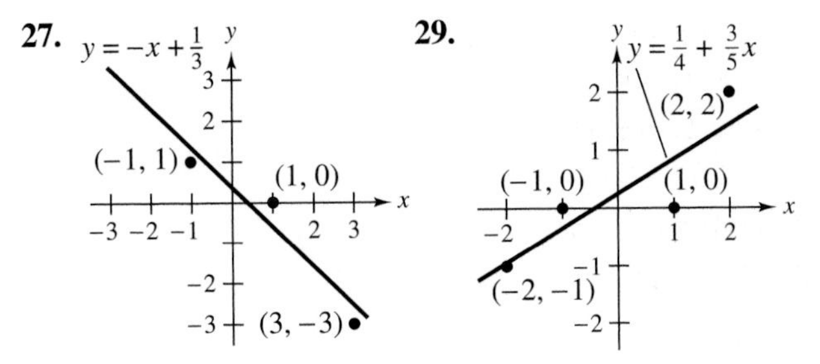

31.

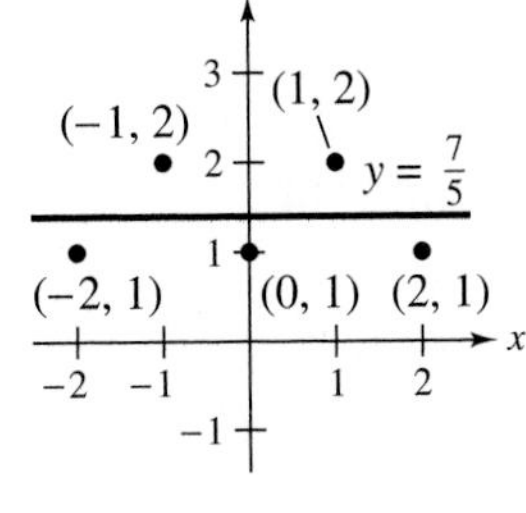

33. $y = x^2 - x$ **35.** $y = \frac{3}{7}x^2 + \frac{6}{5}x + \frac{26}{35}$

37. Advanced Auto Parts:
$S = 2.859t^3 - 32.81t^2 + 492.0t + 2234$
2010: $S \approx \$6732$ million
Auto Zone:
$S = 8.444t^3 - 105.48t^2 + 578.6t + 4444$
2010: $S \approx \$8126$ million

39. $y = 36.02t^2 + 87.0t + 6425$

41. $y = 2.416t^3 - 36.74t^2 + 4989.3t + 28{,}549$
Explanations will vary.

43. (a) False. See discussion after Example 2, page 322.
(b) True. See "Definition of Direct Sum," page 323.
(c) True. See discussion preceding Example 7, page 328.

45. Proof **47.** Proof

49. If A has orthonormal columns, then $A^TA = I$ and the normal equations become
$A^TA\mathbf{x} = A^T\mathbf{b}$
$\mathbf{x} = A^T\mathbf{b}$.

Section 5.5 *(page 350)*

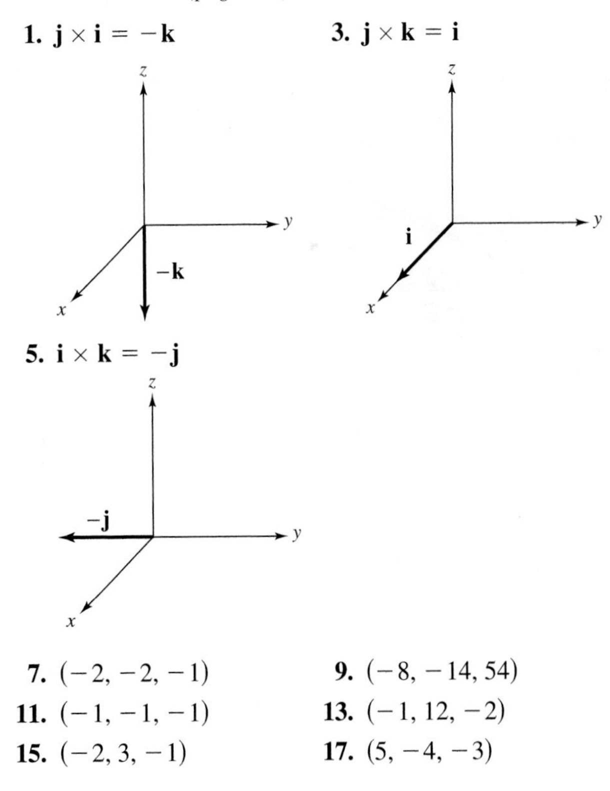

1. $\mathbf{j} \times \mathbf{i} = -\mathbf{k}$ **3.** $\mathbf{j} \times \mathbf{k} = \mathbf{i}$

5. $\mathbf{i} \times \mathbf{k} = -\mathbf{j}$

7. $(-2, -2, -1)$ **9.** $(-8, -14, 54)$

11. $(-1, -1, -1)$ **13.** $(-1, 12, -2)$

15. $(-2, 3, -1)$ **17.** $(5, -4, -3)$

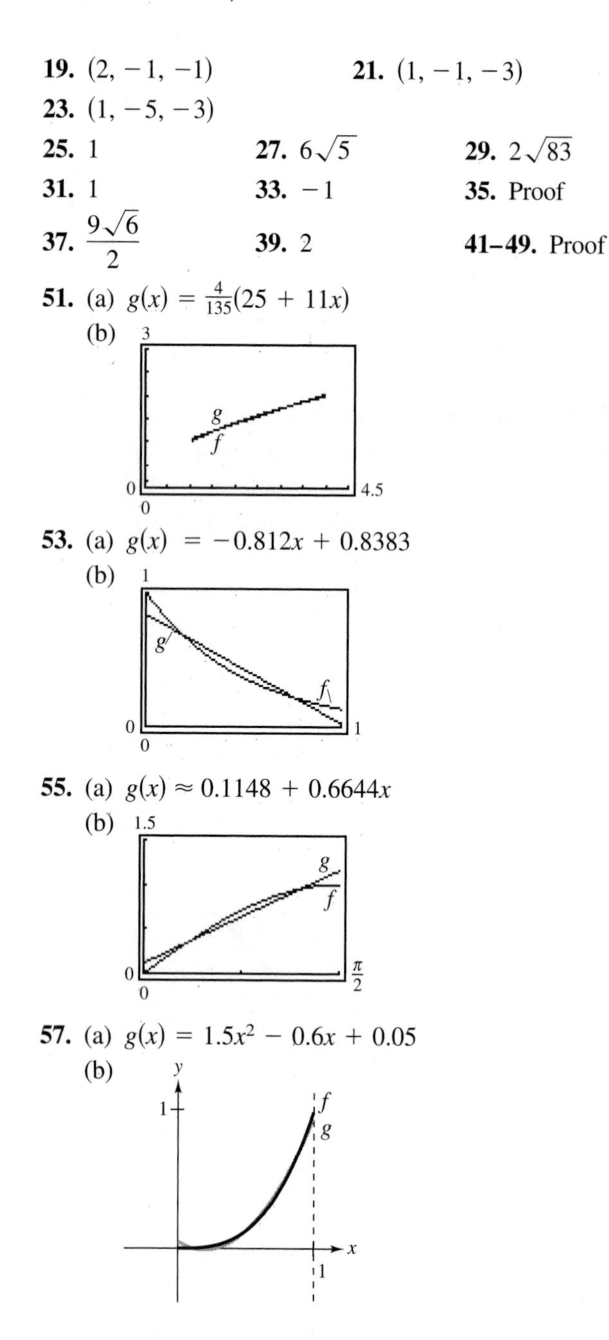

19. $(2, -1, -1)$ **21.** $(1, -1, -3)$

23. $(1, -5, -3)$

25. 1 **27.** $6\sqrt{5}$ **29.** $2\sqrt{83}$

31. 1 **33.** -1 **35.** Proof

37. $\dfrac{9\sqrt{6}}{2}$ **39.** 2 **41–49.** Proof

51. (a) $g(x) = \frac{4}{135}(25 + 11x)$

(b)

53. (a) $g(x) = -0.812x + 0.8383$

(b)

55. (a) $g(x) \approx 0.1148 + 0.6644x$

(b)

57. (a) $g(x) = 1.5x^2 - 0.6x + 0.05$

(b)

59. (a) $g(x) \approx -0.0505 + 1.3122x - 0.4177x^2$

(b)

61. (a) $g(x) \approx -0.4177x^2 + 0.9802$

(b)

63. $g(x) = 2 \sin x + \sin 2x + \frac{2}{3} \sin 3x$

65. $g(x) = \dfrac{\pi^2}{3} + 4 \cos x + \cos 2x + \dfrac{4}{9} \cos 3x$

67. $g(x) = \dfrac{1}{2\pi}(1 - e^{-2\pi})(1 + \cos x + \sin x)$

69. $g(x) = \dfrac{1 - e^{-4\pi}}{20\pi}(5 + 8 \cos x + 4 \sin x)$

71. $g(x) = (1 + \pi) - 2 \sin x - \sin 2x - \frac{2}{3} \sin 3x$

73. $g(x) = \sin 2x$

75. $g(x) = 2\left(\sin x + \dfrac{\sin 2x}{2} + \dfrac{\sin 3x}{3} + \cdots + \dfrac{\sin nx}{n}\right)$

Review Exercises – Chapter 5 *(page 352)*

1. (a) $\sqrt{5}$
(b) $\sqrt{17}$
(c) 6
(d) $\sqrt{10}$

3. (a) $\sqrt{6}$
(b) $\sqrt{14}$
(c) 7
(d) $\sqrt{6}$

5. (a) $\sqrt{6}$
(b) $\sqrt{3}$
(c) -1
(d) $\sqrt{11}$

7. (a) $\sqrt{7}$
(b) $\sqrt{7}$
(c) 6
(d) $\sqrt{2}$

9. $\|\mathbf{v}\| = \sqrt{38}$; $\mathbf{u} = \left(\frac{5}{\sqrt{38}}, \frac{3}{\sqrt{38}}, -\frac{2}{\sqrt{38}}\right)$

11. $\|\mathbf{v}\| = \sqrt{6}$; $\mathbf{u} = \left(\frac{1}{\sqrt{6}}, -\frac{1}{\sqrt{6}}, \frac{2}{\sqrt{6}}\right)$

13. $\frac{\pi}{2}$ **15.** $\frac{\pi}{12}$ **17.** π

19. $\left(-\frac{9}{13}, \frac{45}{13}\right)$ **21.** $\left(\frac{24}{29}, \frac{60}{29}\right)$ **23.** $\left(\frac{18}{29}, \frac{12}{29}, \frac{24}{29}\right)$

25. (a) -2 (b) $\frac{3\sqrt{11}}{2}$

27. Triangle Inequality:

$$\left\|\left(2, -\tfrac{1}{2}, 1\right) + \left(\tfrac{3}{2}, 2, -1\right)\right\| \le \left\|\left(2, -\tfrac{1}{2}, 1\right)\right\| + \left\|\left(\tfrac{3}{2}, 2, -1\right)\right\|$$

$$\frac{\sqrt{67}}{2} \le \sqrt{\frac{15}{2}} + \sqrt{\frac{53}{4}}$$

Cauchy-Schwarz Inequality:

$$\left|\left\langle\left(2, -\tfrac{1}{2}, 1\right), \left(\tfrac{3}{2}, 2, -1\right)\right\rangle\right| \le \left\|\left(2, -\tfrac{1}{2}, 1\right)\right\| \left\|\left(\tfrac{3}{2}, 2, -1\right)\right\|$$

$$2 \le \sqrt{\tfrac{15}{2}}\sqrt{\tfrac{53}{4}} \approx 9.969$$

29. $(s, 3t, 4t)$ **31.** $(2r - 2s - t, r, s, t)$

33. $\left\{\left(\frac{1}{\sqrt{2}}, \frac{1}{\sqrt{2}}\right), \left(-\frac{1}{\sqrt{2}}, \frac{1}{\sqrt{2}}\right)\right\}$

35. $\left\{\left(0, \frac{3}{5}, \frac{4}{5}\right), (1, 0, 0), \left(0, \frac{4}{5}, -\frac{3}{5}\right)\right\}$

37. (a) $(-1, 4, -2) = 2(0, 2, -2) - (1, 0, -2)$

(b) $\left\{\left(0, \frac{1}{\sqrt{2}}, -\frac{1}{\sqrt{2}}\right), \left(\frac{1}{\sqrt{3}}, -\frac{1}{\sqrt{3}}, -\frac{1}{\sqrt{3}}\right)\right\}$

(c) $(-1, 4, -2) = 3\sqrt{2}\left(0, \frac{1}{\sqrt{2}}, -\frac{1}{\sqrt{2}}\right) - \sqrt{3}\left(\frac{1}{\sqrt{3}}, -\frac{1}{\sqrt{3}}, -\frac{1}{\sqrt{3}}\right)$

39. (a) $\frac{1}{4}$ (b) $\frac{1}{\sqrt{5}}$ (c) $\frac{1}{\sqrt{30}}$

(d) $\left\{\sqrt{3}x, \sqrt{5}(4x^2 - 3x)\right\}$

41.
$$\langle f, g\rangle = \int_{-1}^{1} \sqrt{1 - x^2}\, 2x\sqrt{1 - x^2}\, dx = \int_{-1}^{1} 2x(1 - x^2)\, dx = 2\int_{-1}^{1} (x - x^3)\, dx = 2\left[\frac{x^2}{2} - \frac{x^4}{4}\right]_{-1}^{1} = 0$$

43. $\left\{\left(-\frac{1}{\sqrt{2}}, 0, \frac{1}{\sqrt{2}}\right), \left(-\frac{1}{\sqrt{6}}, \frac{2}{\sqrt{6}}, -\frac{1}{\sqrt{6}}\right)\right\}$

(The answer is not unique.)

45. (a) 0 (b) Orthogonal

(c) Because $\langle f, g\rangle = 0$, it follows that $|\langle f, g\rangle| \le \|f\|\|g\|$.

47–53. Proof

55. $\text{span}\left\{\begin{bmatrix} 2 \\ -1 \\ 3 \end{bmatrix}\right\}$

57. $N(A)$ basis: $\{(1, 0, -1)\}$

$N(A^T) = \{(3, 1, 0)\}$

$R(A)$ basis: $\{(0, 0, 1), (1, -3, 0)\}$

$R(A^T)$ basis: $\{(0, 1, 0), (1, 0, 1)\}$

59. $y = 1.778t^3 - 5.82t^2 + 77.9t + 1603$

2010: $y \approx \$3578$ million

61. $y = 10.61x + 396.4$

63. $y = 91.112x^2 + 365.26x + 315.0$

65. $y = 151.692x^2 - 542.62x + 10{,}265.4$

67. $(-2, -1, 1)$ **69.** $\mathbf{i} + \mathbf{j} + 2\mathbf{k}$

71. Proof **73.** 2

75. $g(x) = \frac{18}{5}x - \frac{8}{5}$ **77.** $g(x) = -\frac{3x}{\pi^2} + \frac{3}{2\pi}$

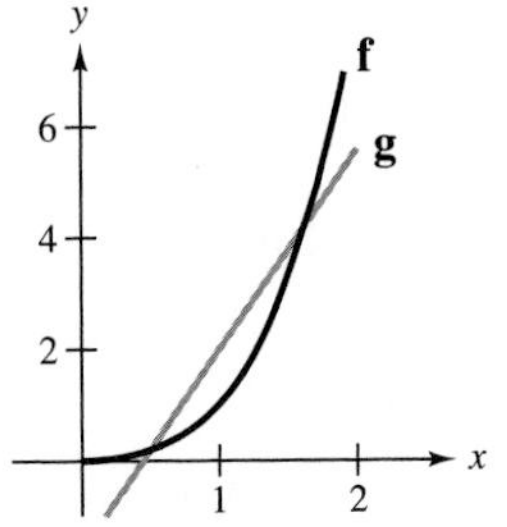

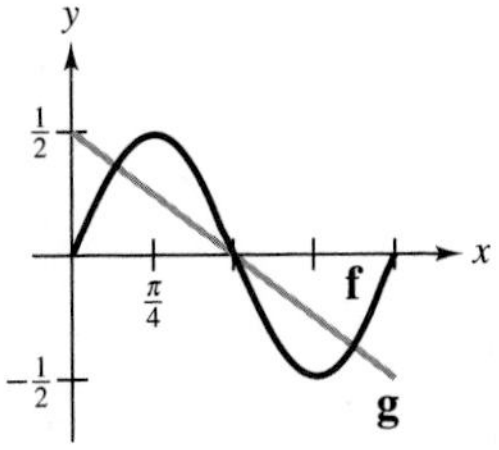

79. $y = 0.3274x^2 - 1.459x + 2.12$

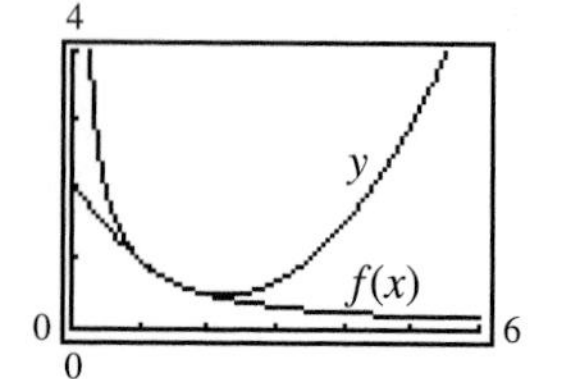

81. $g(x) = 2 \sin x - \sin 2x$

83. (a) True. See note following Theorem 5.17, page 338.
(b) True. See Theorem 3.18, page 339.
(c) True. See discussion on pages 346 and 347.

Cumulative Test—Chapters 4 and 5 *(page 359)*

1. (a) $(3, -7)$
(b) $(3, -6)$
(c) $(-6, 16)$

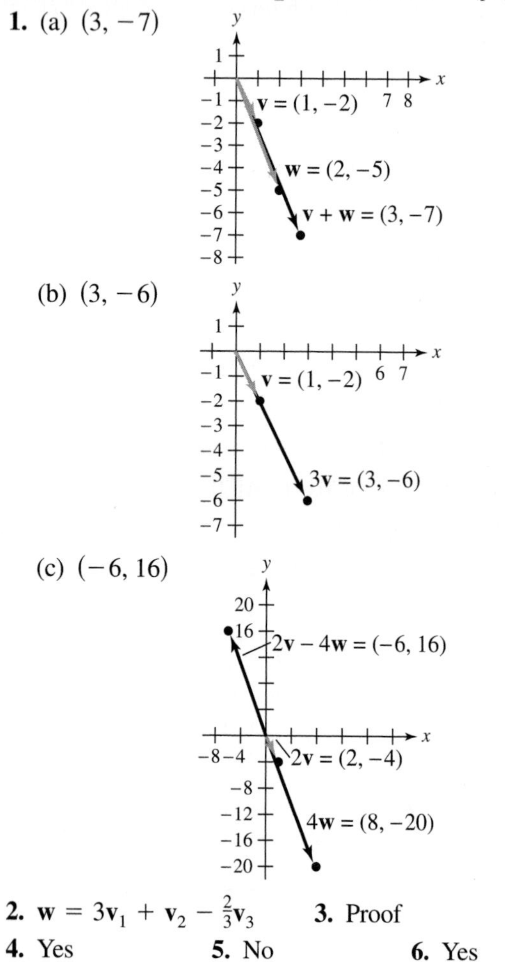

2. $\mathbf{w} = 3\mathbf{v}_1 + \mathbf{v}_2 - \frac{2}{3}\mathbf{v}_3$ **3.** Proof

4. Yes **5.** No **6.** Yes

7. (a) A set of vectors $\{\mathbf{v}_1, \ldots, \mathbf{v}_n\}$ is linearly independent if the vector equation $c_1\mathbf{v}_1 + \cdots + c_n\mathbf{v}_n = \mathbf{0}$ has only the trivial solution.
(b) Linearly dependent

8. The dimension is 6. One basis is

$$\left\{\begin{bmatrix}1&0&0\\0&0&0\\0&0&0\end{bmatrix}, \begin{bmatrix}0&1&0\\1&0&0\\0&0&0\end{bmatrix}, \begin{bmatrix}0&0&1\\0&0&0\\1&0&0\end{bmatrix}, \begin{bmatrix}0&0&0\\0&1&0\\0&0&0\end{bmatrix}, \begin{bmatrix}0&0&0\\0&0&1\\0&1&0\end{bmatrix}, \begin{bmatrix}0&0&0\\0&0&0\\0&0&1\end{bmatrix}\right\}.$$

9. (a) A set of vectors $\{\mathbf{v}_1, \ldots, \mathbf{v}_n\}$ in a vector space V is a basis for V if the set is linearly independent and spans V.
(b) Yes

10. $\left\{\begin{bmatrix}1\\-1\\0\\0\end{bmatrix}, \begin{bmatrix}0\\0\\1\\-1\end{bmatrix}\right\}$ **11.** $\begin{bmatrix}-4\\6\\-5\end{bmatrix}$

12. $\begin{bmatrix}0&1&-1\\2&0&1\\-1&-1&1\end{bmatrix}$

13. (a) $\sqrt{5}$ (b) $\sqrt{11}$ (c) 4
(d) 1.0723 radians; (61.45°)

14. $\frac{11}{12}$

15. $\left\{(1, 0, 0), \left(0, \frac{\sqrt{2}}{2}, \frac{\sqrt{2}}{2}\right), \left(0, -\frac{\sqrt{2}}{2}, \frac{\sqrt{2}}{2}\right)\right\}$

16. $\frac{1}{13}(-3, 2)$

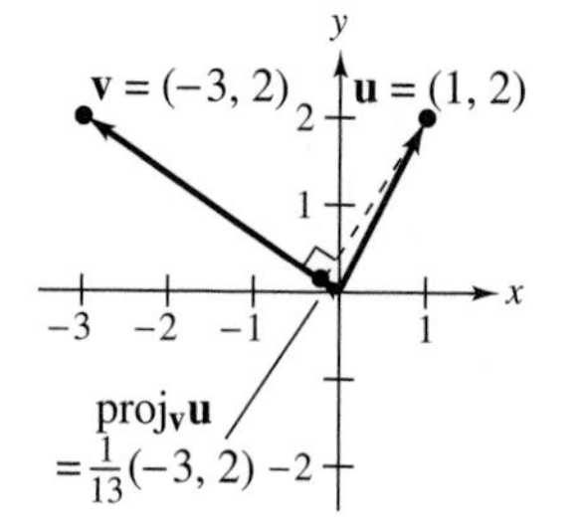

17. $N(A)$ basis: $\{(0, 1, -1, 0)\}$
$N(A^T) = \{(0, 0, 0)\}$
$R(A) = \mathbf{R}^3$
$R(A^T)$ basis: $\{(0, 1, 1, 0), (-1, 0, 0, 1), (1, 1, 1, 1)\}$

18. $\text{span}\left\{\begin{bmatrix} -1 \\ -1 \\ 1 \end{bmatrix}\right\}$

19–21. Proof

22. $y = \frac{36}{13} - \frac{20}{13}x$

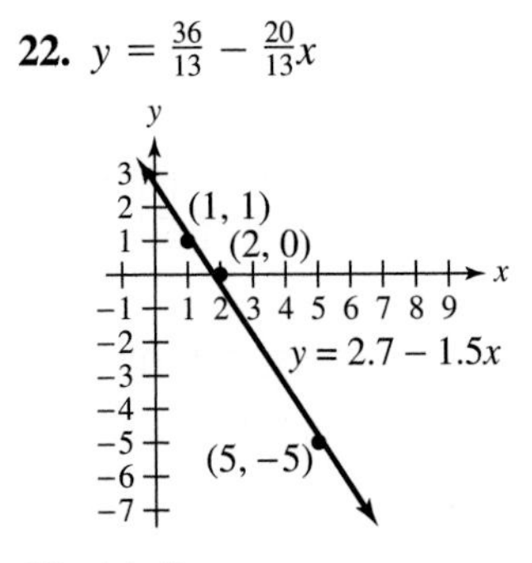

23. (a) 3
(b) One basis consists of the first three rows of A.
(c) One basis consists of columns 1, 3, and 4 of A.
(d) $\left\{\begin{bmatrix} 2 \\ 1 \\ 0 \\ 0 \\ 0 \\ 0 \end{bmatrix}, \begin{bmatrix} -3 \\ 0 \\ 5 \\ -1 \\ 1 \\ 0 \end{bmatrix}, \begin{bmatrix} -2 \\ 0 \\ 3 \\ -7 \\ 0 \\ 1 \end{bmatrix}\right\}$ (e) No
(f) No (g) Yes (h) No

24. Proof

Chapter 6

Section 6.1 *(page 371)*

1. (a) $(-1, 7)$ (b) $(11, -8)$
3. (a) $(1, 5, 4)$ (b) $(5, -6, t)$
5. (a) $(-14, -7)$ (b) $(1, 1, t)$
7. (a) $(0, 2, 1)$ (b) $(-6, 4)$
9. Not linear **11.** Linear **13.** Not linear
15. Not linear **17.** Linear **19.** Linear
21. Linear **23.** $(3, 11, -8)$ **25.** $(0, -6, 8)$
27. $(5, 0, 1)$ **29.** $\left(2, \frac{5}{2}, 2\right)$ **31.** $T: R^4 \rightarrow R^3$
33. $T: R^5 \rightarrow R^2$ **35.** $T: R^2 \rightarrow R^2$

37. (a) $(10, 12, 4)$ (b) $(-1, 0)$
(c) The system represented by
$$\begin{bmatrix} 1 & 2 \\ -2 & 4 \\ -2 & 2 \end{bmatrix}\begin{bmatrix} v_1 \\ v_2 \end{bmatrix} = \begin{bmatrix} 1 \\ 1 \\ 1 \end{bmatrix}$$
is inconsistent.

39. (a) $(-1, 1, 2, 1)$ (b) $\left(-1, 1, \frac{1}{2}, 1\right)$

41. (a) $\left(0, 4\sqrt{2}\right)$ (b) $\left(2\sqrt{3} - 2, 2\sqrt{3} + 2\right)$
(c) $\left(-\frac{5}{2}, \frac{5\sqrt{3}}{2}\right)$

43. True. D_x is a linear transformation and preserves addition and scalar multiplication.

45. False, because $\sin 2x \neq 2 \sin x$ for all x.

47. $g(x) = x^2 + x + C$ **49.** $g(x) = -\cos x + C$

51. (a) -1 (b) $\frac{1}{12}$ (c) -4

53. $T(1, 0) = \left(\frac{1}{2}, \frac{1}{2}\right)$ **55.** $x^2 - 3x - 5$
$T(0, 2) = (1, -1)$

57. (a) True. See discussion before "Definition of a Linear Transformation," page 362.
(b) False, because $\cos(x_1 + x_2) \neq \cos x_1 + \cos x_2$.
(c) True. See discussion following Example 10, page 370.

59. (a) $(x, 0)$ (b) Projection onto the x-axis

61. (a) $\left(\frac{1}{2}(x + y), \frac{1}{2}(x + y)\right)$ (b) $\left(\frac{5}{2}, \frac{5}{2}\right)$
(c) Proof (d) Proof

63. $A\mathbf{u} = \begin{bmatrix} \frac{1}{2} & \frac{1}{2} \\ \frac{1}{2} & \frac{1}{2} \end{bmatrix}\begin{bmatrix} x \\ y \end{bmatrix} = \begin{bmatrix} \frac{1}{2}x + \frac{1}{2}y \\ \frac{1}{2}x + \frac{1}{2}y \end{bmatrix} = T(\mathbf{u})$

65. (a) Assume that h and k are not both zero. Because $T(0, 0) = (0 - h, 0 - k) = (-h, -k) \neq (0, 0)$, it follows that T is not a linear transformation.
(b) $T(0, 0) = (-2, 1)$
$T(2, -1) = (0, 0)$
$T(5, 4) = (3, 5)$

(c) If $T(x, y) = (x - h, y - k) = (x, y)$, then $h = 0$ and $k = 0$, which contradicts the assumption that h and k are not both zero. Therefore, T has no fixed points.

67. Let $T: R^3 \to R^3$ be given by $T(x, y, z) = (0, 0, 0)$. Then, if $\{\mathbf{v}_1, \mathbf{v}_2, \mathbf{v}_3\}$ is any set of vectors in R^3, the set $\{T(\mathbf{v}_1), T(\mathbf{v}_2), T(\mathbf{v}_3)\} = \{\mathbf{0}, \mathbf{0}, \mathbf{0}\}$ is linearly dependent.

69–73. Proof

Section 6.2 *(page 385)*

1. R^3

3. $\{(0, 0, 0, 0)\}$

5. $\{a_1x + a_2x^2 + a_3x^3: a_1, a_2, a_3 \text{ are real}\}$

7. $\{a_0: a_0 \text{ is real}\}$

9. $\{(0, 0)\}$

11. (a) $\{(0, 0)\}$ (b) $\{(1, 0), (0, 1)\}$

13. (a) $\{(-4, -2, 1)\}$ (b) $\{(1, 0), (0, 1)\}$

15. (a) $\{(0, 0)\}$ (b) $\{(1, -1, 0), (0, 0, 1)\}$

17. (a) $\{(-1, 1, 1, 0)\}$
(b) $\{(1, 0, -1, 0), (0, 1, -1, 0), (0, 0, 0, 1)\}$

19. (a) $\{(0, 0)\}$ (b) 0 (c) R^2 (d) 2

21. (a) $\{(0, 0)\}$ (b) 0
(c) $\{(4s, 4t, s - t): s \text{ and } t \text{ are real}\}$
(d) 2

23. (a) $\{(-11t, 6t, 4t): t \text{ is real}\}$
(b) 1 (c) R^2 (d) 2

25. (a) $\{(t, -3t): t \text{ is real}\}$ (b) 1
(c) $\{(3t, t): t \text{ is real}\}$ (d) 1

27. (a) $\{(s + t, s, -2t): s \text{ and } t \text{ are real}\}$
(b) 2
(c) $\{(2t, -2t, t): t \text{ is real}\}$
(d) 1

29. (a) $\{(2s - t, t, 4s, -5s, s): s \text{ and } t \text{ are real}\}$
(b) 2
(c) $\{(7r, 7s, 7t, 8r + 20s + 2t): r, s, \text{ and } t \text{ are real}\}$
(d) 3

31. Nullity $= 1$
Kernel: a line
Range: a plane

33. Nullity $= 3$
Kernel: R^3
Range: $\{(0, 0, 0)\}$

35. Nullity $= 0$
Kernel: $\{(0, 0, 0)\}$
Range: R^3

37. Nullity $= 2$
Kernel: $\{(x, y, z): x + 2y + 2z = 0\}$ (plane)
Range: $\{(t, 2t, 2t), t \text{ is real}\}$ (line)

39. 2

41. 4

43.

	Zero	Standard Basis
(a)	$(0, 0, 0, 0)$	$\{(1, 0, 0, 0), (0, 1, 0, 0), (0, 0, 1, 0), (0, 0, 0, 1)\}$
(b)	$\begin{bmatrix}0\\0\\0\\0\end{bmatrix}$	$\left\{\begin{bmatrix}1\\0\\0\\0\end{bmatrix}, \begin{bmatrix}0\\1\\0\\0\end{bmatrix}, \begin{bmatrix}0\\0\\1\\0\end{bmatrix}, \begin{bmatrix}0\\0\\0\\1\end{bmatrix}\right\}$
(c)	$\begin{bmatrix}0 & 0\\0 & 0\end{bmatrix}$	$\left\{\begin{bmatrix}1 & 0\\0 & 0\end{bmatrix}, \begin{bmatrix}0 & 1\\0 & 0\end{bmatrix}, \begin{bmatrix}0 & 0\\1 & 0\end{bmatrix}, \begin{bmatrix}0 & 0\\0 & 1\end{bmatrix}\right\}$
(d)	$p(x) = 0$	$\{1, x, x^2, x^3\}$
(e)	$(0, 0, 0, 0, 0)$	$\{(1, 0, 0, 0, 0), (0, 1, 0, 0, 0), (0, 0, 1, 0, 0), (0, 0, 0, 1, 0)\}$

45. The set of constant functions: $p(x) = a_0$

47. (a) Rank $= 1$, nullity $= 2$
(b) $\{(1, 0, -2), (1, 2, 0)\}$

49. Because $|A| = -1 \neq 0$, the homogeneous equation $A\mathbf{x} = \mathbf{0}$ has only the trivial solution. So, $\ker(T) = \{(0, 0)\}$ and T is one-to-one (by Theorem 6.6). Furthermore, because

$$\begin{aligned}\text{rank}(T) &= \dim(R^2) - \text{nullity}(T)\\ &= 2 - 0 = 2 = \dim(R^2)\end{aligned}$$

T is onto (by Theorem 6.7).

51. Because $|A| = -1 \neq 0$, the homogeneous equation $A\mathbf{x} = \mathbf{0}$ has only the trivial solution. So, $\ker(T) = \{(0, 0, 0)\}$ and T is one-to-one (by Theorem 6.6). Furthermore, because

$$\begin{aligned}\text{rank}(T) &= \dim(R^3) - \text{nullity}(T)\\ &= 3 - 0 = 3 = \dim(R^3)\end{aligned}$$

T is onto (by Theorem 6.7).

53. (a) False. See "Definition of Kernel of a Linear Transformation," page 374.
(b) False. See Theorem 6.4, page 378.
(c) True. See discussion before Theorem 6.6, page 382.
(d) True. See discussion before "Definition of Isomor-phism," page 384.

55. (a) Rank $= n$ (b) Rank $< n$

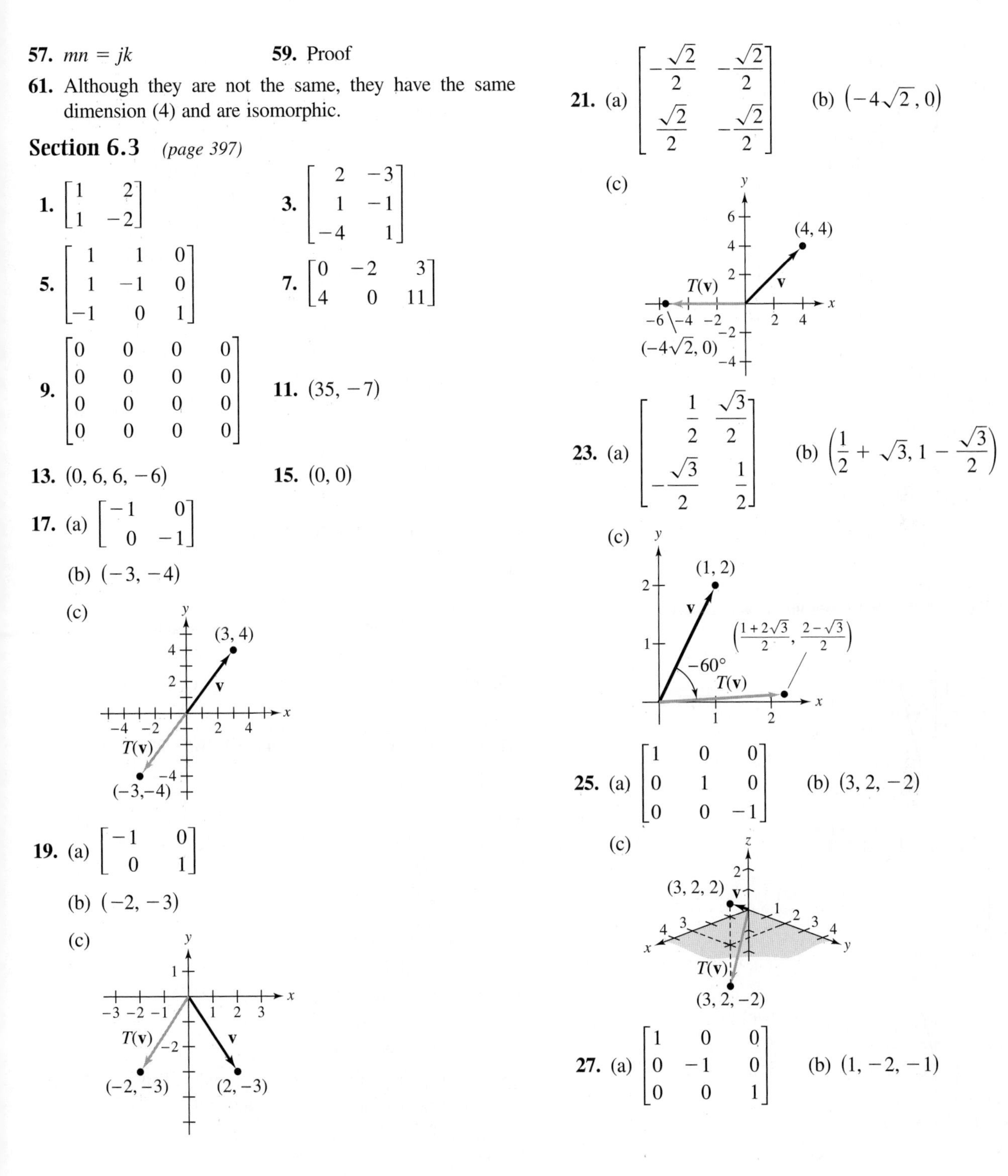

57. $mn = jk$ **59.** Proof

61. Although they are not the same, they have the same dimension (4) and are isomorphic.

Section 6.3 *(page 397)*

1. $\begin{bmatrix} 1 & 2 \\ 1 & -2 \end{bmatrix}$ **3.** $\begin{bmatrix} 2 & -3 \\ 1 & -1 \\ -4 & 1 \end{bmatrix}$

5. $\begin{bmatrix} 1 & 1 & 0 \\ 1 & -1 & 0 \\ -1 & 0 & 1 \end{bmatrix}$ **7.** $\begin{bmatrix} 0 & -2 & 3 \\ 4 & 0 & 11 \end{bmatrix}$

9. $\begin{bmatrix} 0 & 0 & 0 & 0 \\ 0 & 0 & 0 & 0 \\ 0 & 0 & 0 & 0 \\ 0 & 0 & 0 & 0 \end{bmatrix}$ **11.** $(35, -7)$

13. $(0, 6, 6, -6)$ **15.** $(0, 0)$

17. (a) $\begin{bmatrix} -1 & 0 \\ 0 & -1 \end{bmatrix}$

(b) $(-3, -4)$

(c)

19. (a) $\begin{bmatrix} -1 & 0 \\ 0 & 1 \end{bmatrix}$

(b) $(-2, -3)$

(c)

21. (a) $\begin{bmatrix} -\frac{\sqrt{2}}{2} & -\frac{\sqrt{2}}{2} \\ \frac{\sqrt{2}}{2} & -\frac{\sqrt{2}}{2} \end{bmatrix}$ (b) $(-4\sqrt{2}, 0)$

(c)

23. (a) $\begin{bmatrix} \frac{1}{2} & \frac{\sqrt{3}}{2} \\ -\frac{\sqrt{3}}{2} & \frac{1}{2} \end{bmatrix}$ (b) $\left(\frac{1}{2} + \sqrt{3}, 1 - \frac{\sqrt{3}}{2}\right)$

(c)

25. (a) $\begin{bmatrix} 1 & 0 & 0 \\ 0 & 1 & 0 \\ 0 & 0 & -1 \end{bmatrix}$ (b) $(3, 2, -2)$

(c)

27. (a) $\begin{bmatrix} 1 & 0 & 0 \\ 0 & -1 & 0 \\ 0 & 0 & 1 \end{bmatrix}$ (b) $(1, -2, -1)$

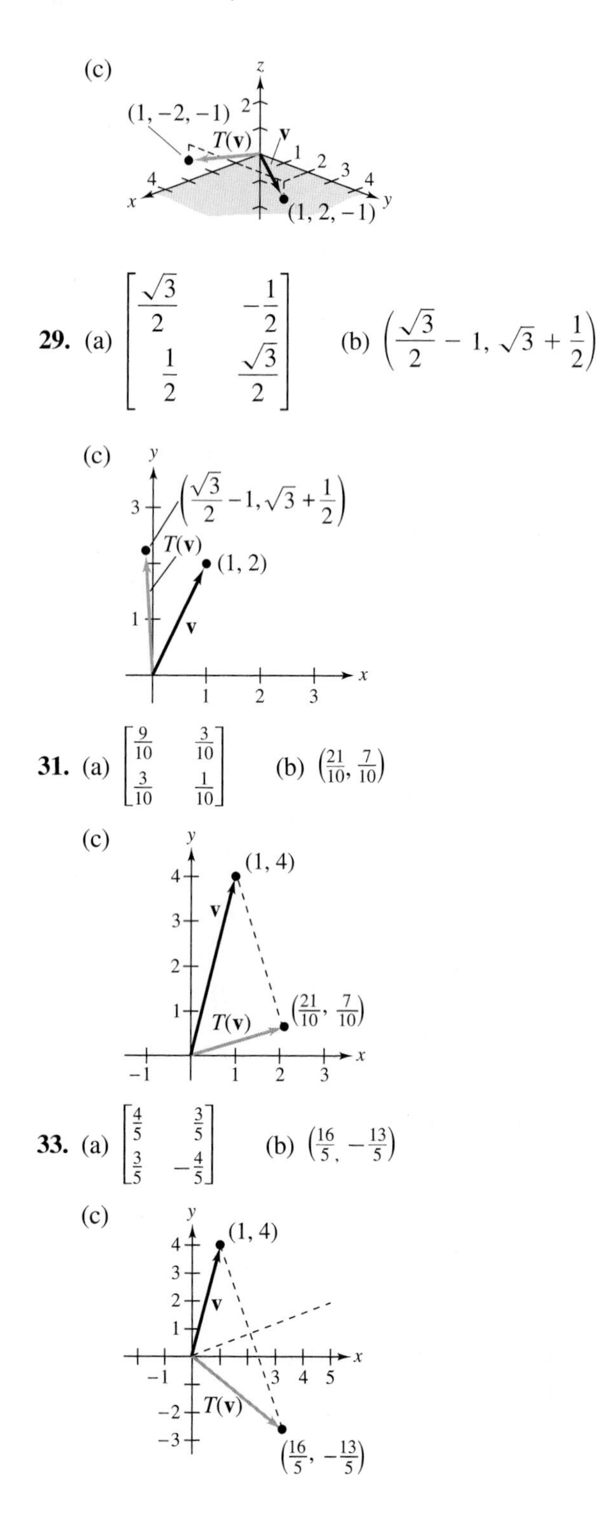

29. (a) $\begin{bmatrix} \frac{\sqrt{3}}{2} & -\frac{1}{2} \\ \frac{1}{2} & \frac{\sqrt{3}}{2} \end{bmatrix}$ (b) $\left(\frac{\sqrt{3}}{2} - 1, \sqrt{3} + \frac{1}{2}\right)$

31. (a) $\begin{bmatrix} \frac{9}{10} & \frac{3}{10} \\ \frac{3}{10} & \frac{1}{10} \end{bmatrix}$ (b) $\left(\frac{21}{10}, \frac{7}{10}\right)$

33. (a) $\begin{bmatrix} \frac{4}{5} & \frac{3}{5} \\ \frac{3}{5} & -\frac{4}{5} \end{bmatrix}$ (b) $\left(\frac{16}{5}, -\frac{13}{5}\right)$

35. (a) $\begin{bmatrix} 2 & 3 & -1 \\ 3 & 0 & -2 \\ 2 & -1 & 1 \end{bmatrix}$ (b) $\begin{bmatrix} 9 \\ 5 \\ -1 \end{bmatrix}$

37. (a) $\begin{bmatrix} 1 & -1 & 0 & 0 \\ 0 & 0 & 1 & 0 \\ 1 & 2 & 0 & -1 \\ 0 & 0 & 0 & 1 \end{bmatrix}$ (b) $\begin{bmatrix} 1 \\ 1 \\ 2 \\ -1 \end{bmatrix}$

39. $A = \begin{bmatrix} 2 & -4 \\ -1 & -5 \end{bmatrix}, A' = \begin{bmatrix} 0 & 2 \\ 7 & -3 \end{bmatrix}$

41. $A = \begin{bmatrix} 0 & 0 & 0 \\ 1 & 0 & 0 \\ 0 & 0 & 0 \end{bmatrix}, A' = \begin{bmatrix} 0 & 0 & 0 \\ 1 & 0 & 0 \\ 0 & 0 & 0 \end{bmatrix}$

43. $A = \begin{bmatrix} -4 & -1 \\ -2 & 5 \end{bmatrix}, A' = \begin{bmatrix} 5 & 3 & 2 \\ 4 & -3 & 1 \\ -2 & -3 & -1 \end{bmatrix}$

45. $T^{-1}(x, y) = \left(\frac{x + y}{2}, \frac{x - y}{2}\right)$

47. $T^{-1}(x_1, x_2, x_3) = (x_1, -x_1 + x_2, -x_2 + x_3)$

49. T is not invertible.

51. T is not invertible.

53. $T^{-1}(x, y) = \left(\frac{x}{5}, \frac{y}{5}\right)$

55. $T^{-1}(x_1, x_2, x_3, x_4) = (x_1 + 2x_2, x_2, x_4, x_3 - x_4)$

57. $(9, 5, 4)$

59. $(-1, 5)$

61. $(2, -4, -3, 3)$

63. $(9, 16, -20)$

65. $\begin{bmatrix} 0 & 0 & 0 \\ 1 & 0 & 0 \\ 0 & 1 & 0 \\ 0 & 0 & 1 \end{bmatrix}$

67. $\begin{bmatrix} 0 & 1 & 0 & 0 \\ 0 & 0 & 0 & 0 \\ 0 & 0 & 1 & 1 \\ 0 & 0 & 0 & 1 \end{bmatrix}$

69. $3 - 2e^x - 2xe^x$

71. (a) $\begin{bmatrix} 0 & 0 & 0 & 0 \\ 1 & 0 & 0 & 0 \\ 0 & \frac{1}{2} & 0 & 0 \\ 0 & 0 & \frac{1}{3} & 0 \\ 0 & 0 & 0 & \frac{1}{4} \end{bmatrix}$ (b) $6x - x^2 + \frac{3}{4}x^4$

73. (a) True. See discussion under "Composition of Linear Transformations," pages 390–391.

(b) False. See Example 3, page 392.

(c) False. See Theorem 6.12, page 393.

75. $\begin{bmatrix} 1 & 0 & 0 & 0 & 0 & 0 \\ 0 & 0 & 0 & 1 & 0 & 0 \\ 0 & 1 & 0 & 0 & 0 & 0 \\ 0 & 0 & 0 & 0 & 1 & 0 \\ 0 & 0 & 1 & 0 & 0 & 0 \\ 0 & 0 & 0 & 0 & 0 & 1 \end{bmatrix}$

77. Proof

79. Sometimes it is preferable to use a nonstandard basis. For example, some linear transformations have diagonal matrix representations relative to a nonstandard basis.

Section 6.4 *(page 405)*

1. (a) $A' = \begin{bmatrix} 4 & -3 \\ \frac{5}{3} & -1 \end{bmatrix}$

(b) $A' = \begin{bmatrix} 1 & 0 \\ \frac{2}{3} & \frac{1}{3} \end{bmatrix}\begin{bmatrix} 2 & -1 \\ -1 & 1 \end{bmatrix}\begin{bmatrix} 1 & 0 \\ -2 & 3 \end{bmatrix}$

3. (a) $A' = \begin{bmatrix} -\frac{1}{3} & \frac{4}{3} \\ -\frac{13}{3} & \frac{16}{3} \end{bmatrix}$

(b) $A' = \begin{bmatrix} -\frac{1}{3} & -\frac{1}{3} \\ -\frac{1}{3} & -\frac{4}{3} \end{bmatrix}\begin{bmatrix} 1 & 1 \\ 0 & 4 \end{bmatrix}\begin{bmatrix} -4 & 1 \\ 1 & -1 \end{bmatrix}$

5. (a) $A' = \begin{bmatrix} 1 & 0 & 0 \\ 0 & 1 & 0 \\ 0 & 0 & 1 \end{bmatrix}$

(b) $A' = \begin{bmatrix} \frac{1}{2} & \frac{1}{2} & -\frac{1}{2} \\ \frac{1}{2} & -\frac{1}{2} & \frac{1}{2} \\ -\frac{1}{2} & \frac{1}{2} & \frac{1}{2} \end{bmatrix}\begin{bmatrix} 1 & 0 & 0 \\ 0 & 1 & 0 \\ 0 & 0 & 1 \end{bmatrix}\begin{bmatrix} 1 & 1 & 0 \\ 1 & 0 & 1 \\ 0 & 1 & 1 \end{bmatrix}$

7. (a) $A' = \begin{bmatrix} \frac{7}{3} & \frac{10}{3} & -\frac{1}{3} \\ -\frac{1}{6} & \frac{4}{3} & \frac{8}{3} \\ \frac{2}{3} & -\frac{4}{3} & -\frac{2}{3} \end{bmatrix}$

(b) $A' = \begin{bmatrix} \frac{2}{3} & -\frac{1}{3} & \frac{1}{3} \\ -\frac{1}{3} & \frac{1}{6} & \frac{1}{3} \\ \frac{1}{3} & \frac{1}{3} & -\frac{1}{3} \end{bmatrix}\begin{bmatrix} 1 & -1 & 2 \\ 2 & 1 & -1 \\ 1 & 2 & 1 \end{bmatrix}\begin{bmatrix} 1 & 0 & 1 \\ 0 & 2 & 2 \\ 1 & 2 & 0 \end{bmatrix}$

9. (a) $\begin{bmatrix} 6 & 4 \\ 9 & 4 \end{bmatrix}$ (b) $[\mathbf{v}]_B = \begin{bmatrix} 2 \\ -1 \end{bmatrix}$, $[T(\mathbf{v})]_B = \begin{bmatrix} 4 \\ -4 \end{bmatrix}$

(c) $A' = \begin{bmatrix} 0 & -\frac{4}{3} \\ 9 & 7 \end{bmatrix}$, $P^{-1} = \begin{bmatrix} -\frac{1}{3} & \frac{1}{3} \\ \frac{3}{4} & -\frac{1}{2} \end{bmatrix}$

(d) $\begin{bmatrix} -\frac{8}{3} \\ 5 \end{bmatrix}$

11. (a) $\begin{bmatrix} \frac{1}{2} & \frac{1}{2} & -\frac{1}{2} \\ \frac{1}{2} & -\frac{1}{2} & \frac{1}{2} \\ -\frac{1}{2} & \frac{1}{2} & \frac{1}{2} \end{bmatrix}$

(b) $[\mathbf{v}]_B = \begin{bmatrix} 1 \\ 0 \\ -1 \end{bmatrix}$, $[T(\mathbf{v})]_B = \begin{bmatrix} 2 \\ -1 \\ -2 \end{bmatrix}$

(c) $A' = \begin{bmatrix} 1 & 0 & 0 \\ 0 & 2 & 0 \\ 0 & 0 & 3 \end{bmatrix}$, $P^{-1} = \begin{bmatrix} 1 & 1 & 0 \\ 1 & 0 & 1 \\ 0 & 1 & 1 \end{bmatrix}$

(d) $\begin{bmatrix} 1 \\ 0 \\ -3 \end{bmatrix}$

13. (a) $\begin{bmatrix} 5 & 2 \\ 9 & 2 \end{bmatrix}$

(b) $[\mathbf{v}]_B = \begin{bmatrix} 3 \\ -1 \end{bmatrix}$, $[T(\mathbf{v})]_B = \begin{bmatrix} 5 \\ 1 \end{bmatrix}$

(c) $A' = \begin{bmatrix} -7 & -2 \\ 27 & 8 \end{bmatrix}$, $P^{-1} = \begin{bmatrix} -\frac{1}{4} & \frac{1}{4} \\ \frac{9}{8} & -\frac{5}{8} \end{bmatrix}$

(d) $\begin{bmatrix} -1 \\ 5 \end{bmatrix}$

15. Proof **17.** Proof **19.** I_n

21–27. Proof

29. The matrix for I relative to B and B' is the square matrix whose columns are the coordinates of $\mathbf{v}_1, \ldots, \mathbf{v}_n$ relative to the standard basis. The matrix for I relative to B, or relative to B', is the identity matrix.

31. (a) True. See discussion, page 399–400, and note that $A' = P^{-1}AP \Rightarrow PA'P^{-1} = PP^{-1}APP^{-1} = A$.

(b) False. Unless it is a diagonal matrix; see Example 5, pages 403–404.

Section 6.5 *(page 413)*

1. (a) $(3, -5)$ (b) $(2, 1)$ (c) $(a, 0)$
(d) $(0, -b)$ (e) $(-c, -d)$ (f) (f, g)

3. (a) $(1, 0)$ (b) $(3, -1)$ (c) $(0, a)$
(d) $(b, 0)$ (e) $(d, -c)$ (f) $(-g, f)$

5. (a) (y, x) (b) Reflection in the line $y = x$

7. (a) Vertical contraction
(b)

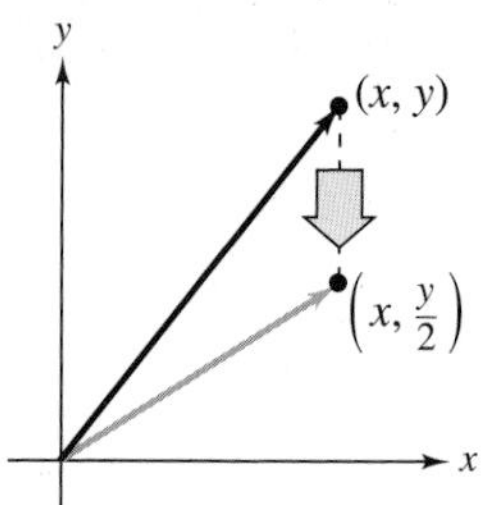

9. (a) Horizontal expansion
(b)

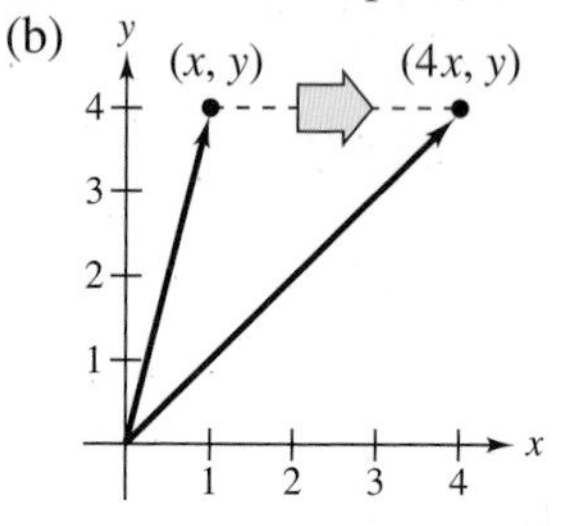

11. (a) Horizontal shear
(b)

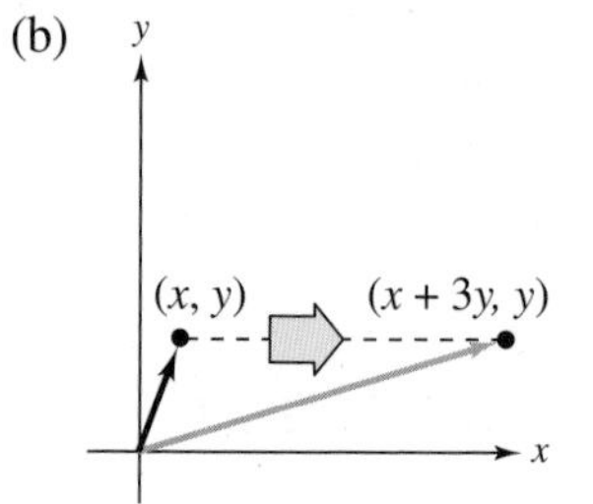

13. (a) Vertical shear
(b)

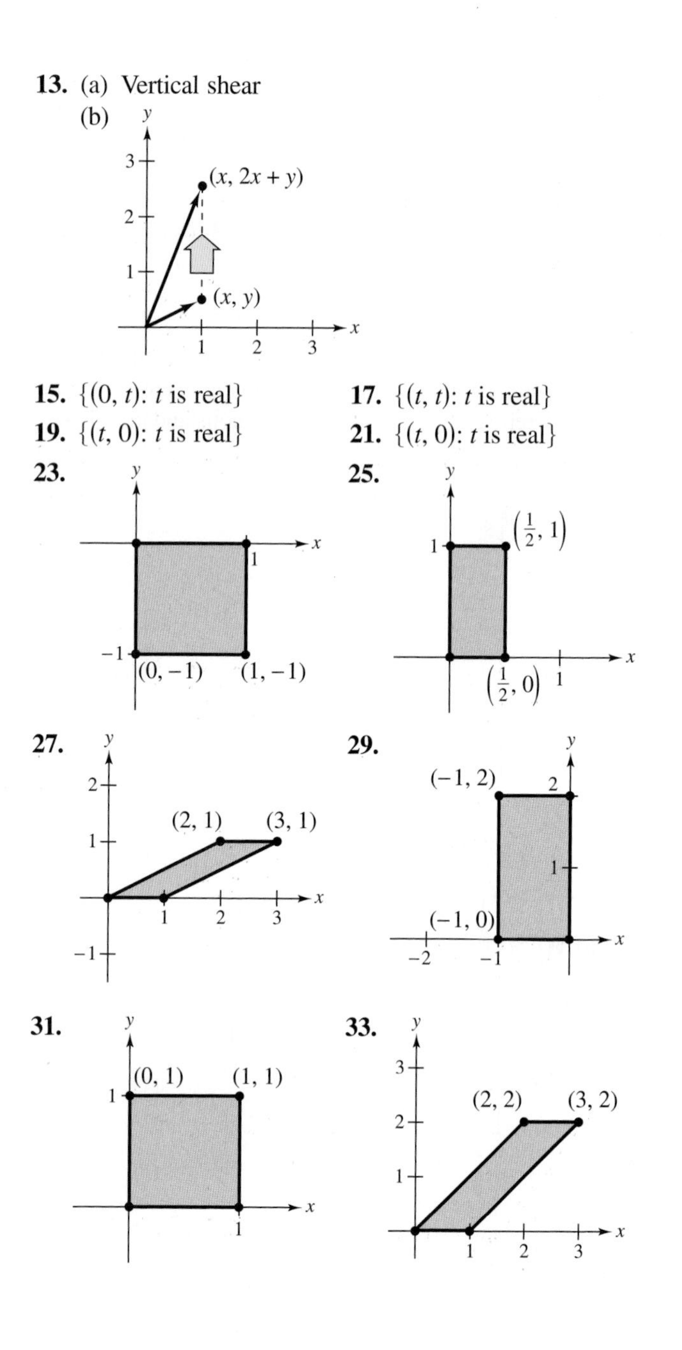

35. (a) (b)

37. (a) (b)

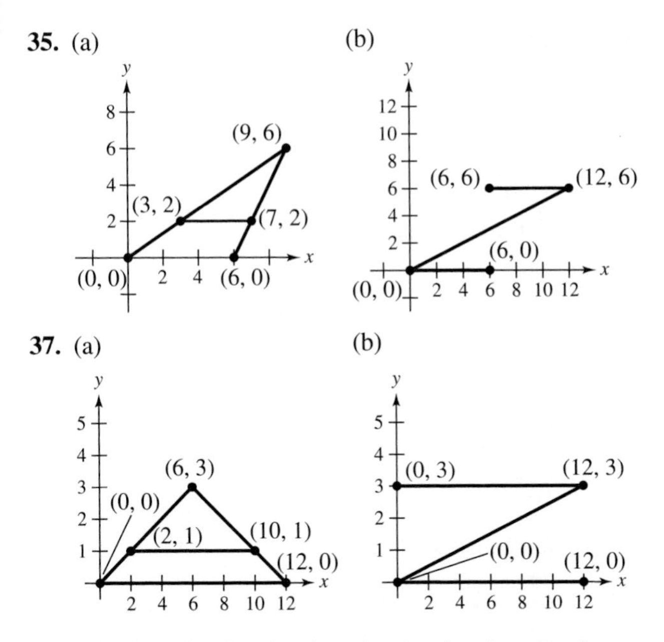

39. $T(1, 0) = (2, 0)$, $T(0, 1) = (0, 3)$, $T(2, 2) = (4, 6)$

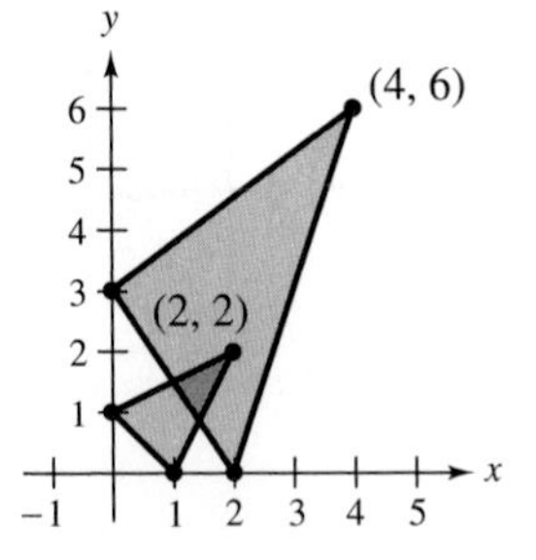

41. Horizontal expansion

43. Reflection in the line $y = x$

45. Reflection in the x-axis followed by a vertical expansion

47. Vertical shear followed by a horizontal expansion

49. $\begin{bmatrix} \frac{\sqrt{3}}{2} & -\frac{1}{2} & 0 \\ \frac{1}{2} & \frac{\sqrt{3}}{2} & 0 \\ 0 & 0 & 1 \end{bmatrix}$ **51.** $\begin{bmatrix} \frac{1}{2} & 0 & \frac{\sqrt{3}}{2} \\ 0 & 1 & 0 \\ -\frac{\sqrt{3}}{2} & 0 & \frac{1}{2} \end{bmatrix}$

53. $\left(\left(\sqrt{3} - 1\right)/2, \left(\sqrt{3} + 1\right)/2, 1\right)$

55. $\left(\left(1 + \sqrt{3}\right)/2, 1, \left(1 - \sqrt{3}\right)/2\right)$

57. 90° about the x-axis **59.** 180° about the y-axis

61. 90° about the z-axis

63. $\begin{bmatrix} 0 & 1 & 0 \\ 0 & 0 & -1 \\ -1 & 0 & 0 \end{bmatrix}$

Line segment from $(0, 0, 0)$ to $(1, -1, -1)$

65. $\begin{bmatrix} \frac{\sqrt{3}}{4} & -\frac{1}{4} & \frac{\sqrt{3}}{2} \\ \frac{1}{2} & \frac{\sqrt{3}}{2} & 0 \\ -\frac{3}{4} & \frac{\sqrt{3}}{4} & \frac{1}{2} \end{bmatrix}$

Line segment from $(0, 0, 0)$ to $\left(\left(3\sqrt{3} - 1\right)/4, \left(1 + \sqrt{3}\right)/2, \left(\sqrt{3} - 1\right)/4\right)$

Review Exercises – Chapter 6 *(page 416)*

1. (a) $(2, -4)$ (b) $(4, 4)$

3. (a) $(0, -1, 7)$ (b) $\{(t - 3, 5 - t, t)\text{: } t \text{ is real.})\}$

5. Linear, $\begin{bmatrix} 1 & 2 \\ -1 & -1 \end{bmatrix}$ **7.** Linear, $\begin{bmatrix} 1 & -2 \\ -1 & 2 \end{bmatrix}$

9. Not linear

11. Linear, $\begin{bmatrix} 1 & -1 & 0 \\ 0 & 1 & -1 \\ -1 & 0 & 1 \end{bmatrix}$

13. $T(1, 1) = \left(\frac{3}{2}, \frac{3}{2}\right)$, $T(0, 1) = (1, 1)$

15. $T(0, -1) = (-1, -2)$

17. $A^2 = I$

19. $A^3 = \begin{bmatrix} \cos 3\theta & -\sin 3\theta \\ \sin 3\theta & \cos 3\theta \end{bmatrix}$

21. (a) $T\text{: } R^3 \rightarrow R^2$ (b) $(3, -12)$
(c) $\left\{\left(-\frac{5}{2}, 3 - 2t, t\right)\text{: } t \text{ is real}\right\}$

23. (a) $T\text{: } R^2 \rightarrow R^1$ (b) 5
(c) $\{(4 - t, t)\text{: } t \text{ is real}\}$

25. (a) $T\text{: } R^3 \rightarrow R^3$
(b) $(-2, -4, -5)$
(c) $(2, 2, 2)$

27. (a) $T\text{: } R^2 \rightarrow R^3$
(b) $(8, 10, 4)$
(c) $(1, -1)$

29. (a) $\{(-2, 1, 0, 0), (2, 0, 1, -2)\}$
(b) $\{(5, 0, 4), (0, 5, 8)\}$

31. (a) $\{(1, -1, 1)\}$ (b) $\{(1, 0, 1), (0, 1, -1)\}$

33. (a) $\{(0, 0)\}$ (b) $\{(1, 0, \frac{1}{2}), (0, 1, -\frac{1}{2})\}$
(c) Rank $= 2$ (d) Nullity $= 0$

35. (a) $\{(-3, 3, 1)\}$ (b) $\{(1, 0, -1), (0, 1, 2)\}$
(c) Rank $= 2$ (d) Nullity $= 1$

37. 3 **39.** 2

41. $A = \begin{bmatrix} 2 & 0 \\ 0 & 1 \end{bmatrix}, A^{-1} = \begin{bmatrix} \frac{1}{2} & 0 \\ 0 & 1 \end{bmatrix}$

43. $A = \begin{bmatrix} \cos\theta & -\sin\theta \\ \sin\theta & \cos\theta \end{bmatrix}, A^{-1} = \begin{bmatrix} \cos\theta & \sin\theta \\ -\sin\theta & \cos\theta \end{bmatrix}$

45. T has no inverse. **47.** T has no inverse.

49. $A = \begin{bmatrix} 0 & 0 & 0 \\ 0 & 1 & 0 \\ 0 & 1 & 0 \end{bmatrix}, A' = \begin{bmatrix} 0 & 0 \\ 1 & 1 \end{bmatrix}$

51.

53. (a) One-to-one (b) Onto (c) Invertible

55. (a) One-to-one (b) Onto (c) Invertible

57. $(0, 1, 1)$

59 $A' = \begin{bmatrix} 3 & -1 \\ 1 & -1 \end{bmatrix}$,

$$A' = P^{-1}AP = \begin{bmatrix} \frac{1}{2} & -\frac{1}{2} \\ \frac{1}{2} & \frac{1}{2} \end{bmatrix} \begin{bmatrix} 1 & -3 \\ -1 & 1 \end{bmatrix} \begin{bmatrix} 1 & 1 \\ -1 & 1 \end{bmatrix}$$

61. (a) $\begin{bmatrix} 0 & 0 & 0 \\ 0 & \frac{1}{5} & \frac{2}{5} \\ 0 & \frac{2}{5} & \frac{4}{5} \end{bmatrix}$ (b) Answers will vary.
(c) Answers will vary.

63. Answers will vary. **65.** Proof

67. (a) Proof (b) Rank $= 1$, nullity $= 3$
(c) $\{1 - x, 1 - x^2, 1 - x^3\}$

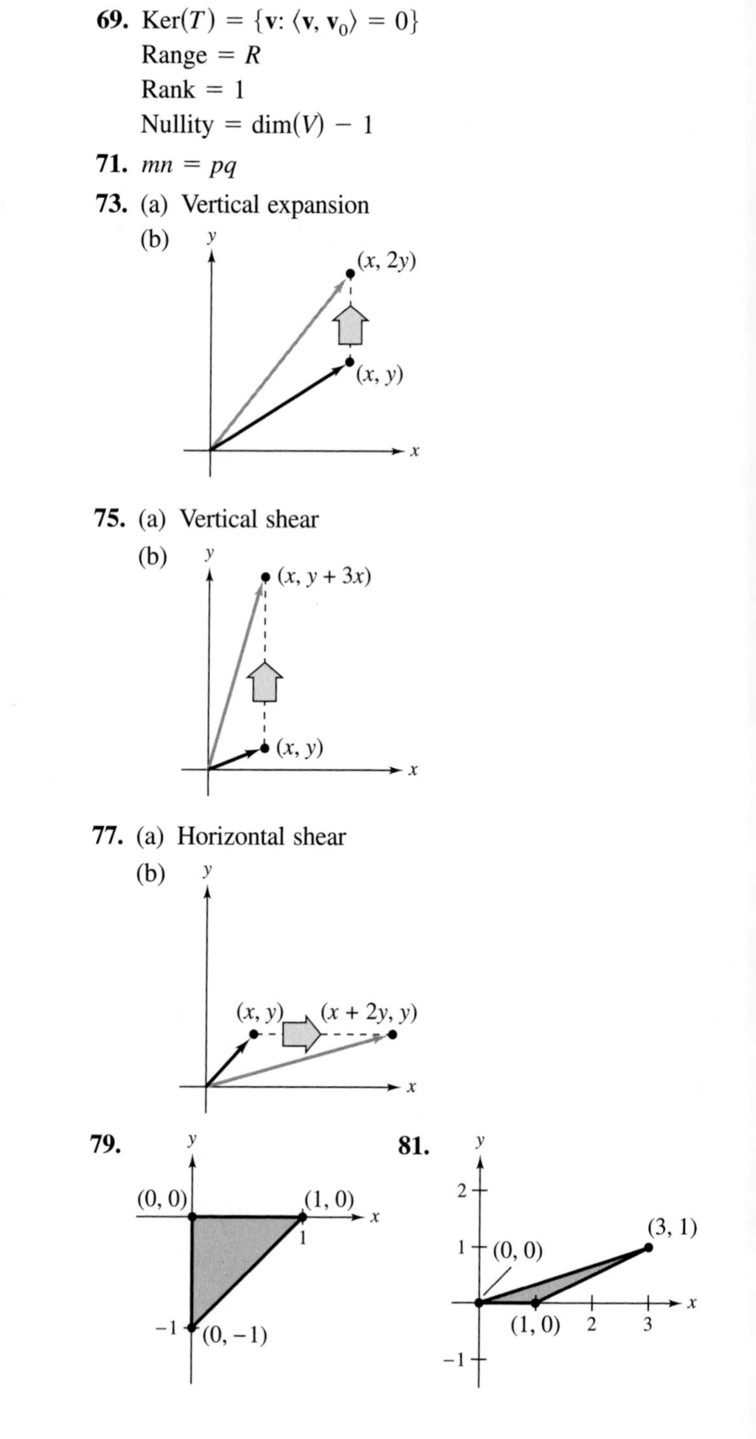

69. $\text{Ker}(T) = \{\mathbf{v}: \langle \mathbf{v}, \mathbf{v}_0 \rangle = 0\}$
Range $= R$
Rank $= 1$
Nullity $= \dim(V) - 1$

71. $mn = pq$

73. (a) Vertical expansion
(b)

75. (a) Vertical shear
(b)

77. (a) Horizontal shear
(b)

79.

81.

83. Reflection in the line $y = x$ followed by a horizontal expansion

85. $\begin{bmatrix} \sqrt{2}/2 & -\sqrt{2}/2 & 0 \\ \sqrt{2}/2 & \sqrt{2}/2 & 0 \\ 0 & 0 & 1 \end{bmatrix} (\sqrt{2}, 0, 1)$

87. $\begin{bmatrix} 1 & 0 & 0 \\ 0 & 1/2 & -\sqrt{3}/2 \\ 0 & \sqrt{3}/2 & 1/2 \end{bmatrix}$,

$(1, (-1 - \sqrt{3})/2, (-\sqrt{3} + 1)/2)$

89. $\begin{bmatrix} \sqrt{3}/2 & -1/4 & \sqrt{3}/4 \\ 1/2 & \sqrt{3}/4 & -3/4 \\ 0 & \sqrt{3}/2 & 1/2 \end{bmatrix}$

91. $\begin{bmatrix} \sqrt{6}/4 & -\sqrt{2}/2 & \sqrt{2}/4 \\ \sqrt{6}/4 & \sqrt{2}/2 & \sqrt{2}/4 \\ -1/2 & 0 & \sqrt{3}/2 \end{bmatrix}$

93. $(0, 0, 0), \left(\frac{\sqrt{2}}{2}, \frac{\sqrt{2}}{2}, 0\right), (0, \sqrt{2}, 0),$

$\left(-\frac{\sqrt{2}}{2}, \frac{\sqrt{2}}{2}, 0\right), (0, 0, 1), \left(\frac{\sqrt{2}}{2}, \frac{\sqrt{2}}{2}, 1\right),$

$(0, \sqrt{2}, 1), \left(-\frac{\sqrt{2}}{2}, \frac{\sqrt{2}}{2}, 1\right)$

95. $(0, 0, 0), (1, 0, 0), \left(1, \frac{\sqrt{3}}{2}, \frac{1}{2}\right), \left(0, \frac{\sqrt{3}}{2}, \frac{1}{2}\right),$

$\left(0, -\frac{1}{2}, \frac{\sqrt{3}}{2}\right), \left(1, -\frac{1}{2}, \frac{\sqrt{3}}{2}\right),$

$\left(1, \frac{-1 + \sqrt{3}}{2}, \frac{1 + \sqrt{3}}{2}\right), \left(0, \frac{-1 + \sqrt{3}}{2}, \frac{1 + \sqrt{3}}{2}\right)$

97. (a) False. See "Elementary Matrices for Linear Transformations in the Plane," page 407.
(b) True. See "Elementary Matrices for Linear Transformations in the Plane," page 407.
(c) True. See discussion following Example 4, page 411.

99. (a) False. See "Remark," page 364.
(b) False. See Theorem 6.7, page 383.
(c) True. See discussion following Example 5, page 404.

Chapter 7

Section 7.1 *(page 432)*

1. $\begin{bmatrix} 1 & 0 \\ 0 & -1 \end{bmatrix}\begin{bmatrix} 1 \\ 0 \end{bmatrix} = 1\begin{bmatrix} 1 \\ 0 \end{bmatrix}, \begin{bmatrix} 1 & 0 \\ 0 & -1 \end{bmatrix}\begin{bmatrix} 0 \\ 1 \end{bmatrix} = -1\begin{bmatrix} 0 \\ 1 \end{bmatrix}$

3. $\begin{bmatrix} 1 & 1 \\ 1 & 1 \end{bmatrix}\begin{bmatrix} 1 \\ -1 \end{bmatrix} = 0\begin{bmatrix} 1 \\ -1 \end{bmatrix}, \begin{bmatrix} 1 & 1 \\ 1 & 1 \end{bmatrix}\begin{bmatrix} 1 \\ 1 \end{bmatrix} = 2\begin{bmatrix} 1 \\ 1 \end{bmatrix}$

5. $\begin{bmatrix} 2 & 3 & 1 \\ 0 & -1 & 2 \\ 0 & 0 & 3 \end{bmatrix}\begin{bmatrix} 1 \\ 0 \\ 0 \end{bmatrix} = 2\begin{bmatrix} 1 \\ 0 \\ 0 \end{bmatrix},$

$\begin{bmatrix} 2 & 3 & 1 \\ 0 & -1 & 2 \\ 0 & 0 & 3 \end{bmatrix}\begin{bmatrix} 1 \\ -1 \\ 0 \end{bmatrix} = -1\begin{bmatrix} 1 \\ -1 \\ 0 \end{bmatrix},$

$\begin{bmatrix} 2 & 3 & 1 \\ 0 & -1 & 2 \\ 0 & 0 & 3 \end{bmatrix}\begin{bmatrix} 5 \\ 1 \\ 2 \end{bmatrix} = 3\begin{bmatrix} 5 \\ 1 \\ 2 \end{bmatrix}$

7. $\begin{bmatrix} 0 & 1 & 0 \\ 0 & 0 & 1 \\ 1 & 0 & 0 \end{bmatrix}\begin{bmatrix} 1 \\ 1 \\ 1 \end{bmatrix} = 1\begin{bmatrix} 1 \\ 1 \\ 1 \end{bmatrix}$

9. (a) $\begin{bmatrix} 1 & 1 \\ 1 & 1 \end{bmatrix}\begin{bmatrix} c \\ -c \end{bmatrix} = 0\begin{bmatrix} c \\ -c \end{bmatrix}$
(b) $\begin{bmatrix} 1 & 1 \\ 1 & 1 \end{bmatrix}\begin{bmatrix} c \\ c \end{bmatrix} = \begin{bmatrix} 2c \\ 2c \end{bmatrix} = 2\begin{bmatrix} c \\ c \end{bmatrix}$

11. (a) No (b) Yes (c) Yes (d) No

13. (a) Yes (b) No (c) Yes (d) Yes

15. (a) $\lambda(\lambda - 7) = 0$
(b) $\lambda = 0, (1, 2)$
$\lambda = 7, (3, -1)$

17. (a) $\lambda^2 - \frac{1}{4} = 0$
(b) $\lambda = -\frac{1}{2}, (1, 1)$
$\lambda = \frac{1}{2}, (3, 1)$

19. (a) $(\lambda - 2)(\lambda - 3)(\lambda - 1) = 0$
(b) $\lambda = 1, (1, 2, -1); \lambda = 2, (1, 0, 0); \lambda = 3, (0, 1, 0)$

21. (a) $(\lambda - 2)(\lambda - 4)(\lambda - 1) = 0$
(b) $\lambda = 4, (7, -4, 2); \lambda = 2, (1, 0, 0);$
$\lambda = 1, (-1, 1, 1)$

23. (a) $(\lambda + 3)(\lambda - 3)^2 = 0$
(b) $\lambda = -3, (1, 1, 3); \lambda = 3, (1, 0, -1), (1, 1, 0)$

25. (a) $(\lambda - 4)(\lambda - 6)(\lambda + 2) = 0$
(b) $\lambda = -2, (3, 2, 0); \lambda = 4, (5, -10, -2)$
$\lambda = 6, (1, -2, 0)$

27. (a) $\lambda(\lambda - 3)(\lambda - 2)^2 = 0$
(b) $\lambda = 2, (1, 0, 0, 0), (0, 1, 0, 0);$
$\lambda = 3, (0, 0, 3, 4); \lambda = 0, (0, 0, 0, 1)$

29. $\lambda = -2, 1$ **31.** $\lambda = 5, 5$ **33.** $\lambda = 4, -\frac{1}{2}, \frac{1}{3}$

35. $\lambda = -1, 4, 4$ **37.** $\lambda = 0, 3$ **39.** $\lambda = 0, 0, 0, 21$

41. $\lambda^2 - 6\lambda + 8$ **43.** $\lambda^2 - 7\lambda + 12$

45. $\lambda^3 - 2\lambda^2 - \lambda + 2$ **47.** $\lambda^3 - 5\lambda^2 + 15\lambda - 27$

49.

Exercise	Trace of A	Determinant of A
15	7	0
17	0	$-\frac{1}{4}$
19	6	6
21	7	8
23	3	-27
25	8	-48
27	7	0

51. Proof

53. Assume that λ is an eigenvalue of A, with corresponding eigenvector $\mathbf{x}$. Because A is invertible (from Exercise 52), $\lambda \neq 0$. Then, $A\mathbf{x} = \lambda\mathbf{x}$ implies that $\mathbf{x} = A^{-1}A\mathbf{x} = A^{-1}\lambda\mathbf{x} = \lambda A^{-1}\mathbf{x}$, which in turn implies that $(1/\lambda)\mathbf{x} = A^{-1}\mathbf{x}$. So, $\mathbf{x}$ is an eigenvector of A^{-1}, and its corresponding eigenvalue is $1/\lambda$.

55. Proof **57.** Proof

59. $a = 0,\ d = 1$ or $a = 1,\ d = 0$

61. (a) False.
(b) True. See discussion before Theorem 7.1, page 424.
(c) True. See Theorem 7.2, page 426.

63. Dim = 3 **65.** Dim = 1

67. $T(e^x) = \dfrac{d}{dx}[e^x] = e^x = 1(e^x)$

69. $\lambda = -2, 3 + 2x; \lambda = 4, -5 + 10x + 2x^2;$
$\lambda = 6, -1 + 2x$

71. $\lambda = 0, \begin{bmatrix} 1 & 0 \\ 1 & 0 \end{bmatrix}, \begin{bmatrix} 1 & 1 \\ 0 & -1 \end{bmatrix}; \lambda = 3, \begin{bmatrix} 1 & 0 \\ -2 & 0 \end{bmatrix}$

73. The only possible eigenvalue is 0.

75. Proof

Section 7.2 *(page 444)*

1. $P^{-1} = \begin{bmatrix} 1 & -4 \\ -1 & 3 \end{bmatrix}, P^{-1}AP = \begin{bmatrix} 1 & 0 \\ 0 & -2 \end{bmatrix}$

3. $P^{-1} = \begin{bmatrix} \frac{1}{5} & \frac{4}{5} \\ -\frac{1}{5} & \frac{1}{5} \end{bmatrix}, P^{-1}AP = \begin{bmatrix} 2 & 0 \\ 0 & -3 \end{bmatrix}$

5. $P^{-1} = \begin{bmatrix} \frac{2}{3} & -\frac{2}{3} & 1 \\ 0 & \frac{1}{4} & 0 \\ -\frac{1}{3} & \frac{1}{12} & 0 \end{bmatrix}, P^{-1}AP = \begin{bmatrix} 5 & 0 & 0 \\ 0 & 3 & 0 \\ 0 & 0 & -1 \end{bmatrix}$

7. $P^{-1} = \begin{bmatrix} 1 & -\frac{1}{2} & \frac{5}{2} \\ 0 & \frac{1}{2} & -\frac{1}{2} \\ 0 & 0 & 1 \end{bmatrix}, P^{-1}AP = \begin{bmatrix} 4 & 0 & 0 \\ 0 & 2 & 0 \\ 0 & 0 & 3 \end{bmatrix}$

9. There is only one eigenvalue, $\lambda = 0$, and the dimension of its eigenspace is 1. The matrix is not diagonalizable.

11. There is only one eigenvalue, $\lambda = 1$, and the dimension of its eigenspace is 1. The matrix is not diagonalizable.

13. There are two eigenvalues, 1 and 2. The dimension of the eigenspace for the repeated eigenvalue 1 is 1. The matrix is not diagonalizable.

15. There are two repeated eigenvalues, 0 and 3. The eigenspace associated with 3 is of dimension 1. The matrix is not diagonalizable.

17. $\lambda = 0, 2$ The matrix is diagonalizable.

19. $\lambda = 0, 2$ Insufficient number of eigenvalues to guarantee diagonalizability

21. $P = \begin{bmatrix} 1 & 3 \\ 2 & -1 \end{bmatrix}$ (The answer is not unique.)

23. $P = \begin{bmatrix} 3 & 1 \\ 1 & 1 \end{bmatrix}$ (The answer is not unique.)

25. $P = \begin{bmatrix} 7 & 1 & -1 \\ -4 & 0 & 1 \\ 2 & 0 & 1 \end{bmatrix}$ (The answer is not unique.)

27. $P = \begin{bmatrix} 1 & -1 & 1 \\ 1 & 0 & 1 \\ 3 & 1 & 0 \end{bmatrix}$ (The answer is not unique.)

29. $P = \begin{bmatrix} 3 & -1 & -5 \\ 2 & 2 & 10 \\ 0 & 0 & 2 \end{bmatrix}$ (The answer is not unique.)

31. A is not diagonalizable.

33. $P = \begin{bmatrix} 4 & 0 & 0 & 0 \\ 4 & 0 & -2 & 0 \\ 4 & 3 & 1 & 0 \\ 1 & 1 & 1 & 1 \end{bmatrix}$ (The answer is not unique.)

35. $\{(1, -1), (1, 1)\}$

37. $\{(-1 + x), x\}$

39. (a) and (b) Proof

41. $\begin{bmatrix} -188 & -378 \\ 126 & 253 \end{bmatrix}$

43. $\begin{bmatrix} 384 & 256 & -384 \\ -384 & -512 & 1152 \\ -128 & -256 & 640 \end{bmatrix}$

45. (a) True. See the proof of Theorem 7.4, pages 436–437.

(b) False. See Theorem 7.6, page 442.

47. Yes, the order of elements on the main diagonal may change.

49–55. Proof

57. The eigenvector for the eigenvalue $\lambda = k$ is $(0, 0)$. By Theorem 7.5, the matrix is not diagonalizable because it does not have two linearly independent vectors.

Section 7.3 *(page 456)*

1. Symmetric

3. Not symmetric

5. Symmetric

7. $\lambda = 2$, dim $= 1$
$\lambda = 4$, dim $= 1$

9. $\lambda = 2$, dim $= 2$
$\lambda = 3$, dim $= 1$

11. $\lambda = -2$, dim $= 2$
$\lambda = 4$, dim $= 1$

13. $\lambda = -1$, dim $= 1$
$\lambda = 1 + \sqrt{2}$, dim $= 1$
$\lambda = 1 - \sqrt{2}$, dim $= 1$

15. Orthogonal

17. Not orthogonal

19. Orthogonal

21. Orthogonal

23. $P = \begin{bmatrix} \sqrt{2}/2 & \sqrt{2}/2 \\ -\sqrt{2}/2 & \sqrt{2}/2 \end{bmatrix}$

(The answer is not unique.)

25. $P = \begin{bmatrix} \sqrt{3}/3 & \sqrt{6}/3 \\ -\sqrt{6}/3 & \sqrt{3}/3 \end{bmatrix}$

(The answer is not unique.)

27. $P = \begin{bmatrix} -\frac{2}{3} & -\frac{1}{3} & \frac{2}{3} \\ \frac{1}{3} & \frac{2}{3} & \frac{2}{3} \\ \frac{2}{3} & -\frac{2}{3} & \frac{1}{3} \end{bmatrix}$ (The answer is not unique.)

29. $\begin{bmatrix} -\sqrt{3}/3 & -\sqrt{2}/2 & \sqrt{6}/6 \\ -\sqrt{3}/3 & \sqrt{2}/2 & \sqrt{6}/6 \\ \sqrt{3}/3 & 0 & \sqrt{6}/3 \end{bmatrix}$

(The answer is not unique.)

31. $P = \begin{bmatrix} \sqrt{2}/2 & 0 & \sqrt{2}/2 & 0 \\ -\sqrt{2}/2 & 0 & \sqrt{2}/2 & 0 \\ 0 & \sqrt{2}/2 & 0 & \sqrt{2}/2 \\ 0 & -\sqrt{2}/2 & 0 & \sqrt{2}/2 \end{bmatrix}$

(The answer is not unique.)

33. (a) True. See Theorem 7.10, page 453.

(b) True. See Theorem 7.9, page 452.

35. Proof

37. Proof

39. $A^{-1} = \left(\dfrac{1}{\cos^2 \theta + \sin^2 \theta}\right)\begin{bmatrix} \cos \theta & \sin \theta \\ -\sin \theta & \cos \theta \end{bmatrix}$
$= \begin{bmatrix} \cos \theta & \sin \theta \\ -\sin \theta & \cos \theta \end{bmatrix} = A^T$

41. Proof

Section 7.4 *(page 472)*

1. $\mathbf{x}_2 = \begin{bmatrix} 20 \\ 5 \end{bmatrix}, \mathbf{x}_3 = \begin{bmatrix} 10 \\ 10 \end{bmatrix}$

3. $\mathbf{x}_2 = \begin{bmatrix} 84 \\ 12 \\ 6 \end{bmatrix}, \mathbf{x}_3 = \begin{bmatrix} 60 \\ 84 \\ 6 \end{bmatrix}$

5. $\mathbf{x} = t\begin{bmatrix} 2 \\ 1 \end{bmatrix}$

7. $\mathbf{x} = t\begin{bmatrix} 8 \\ 4 \\ 1 \end{bmatrix}$

9. $\mathbf{x}_2 = \begin{bmatrix} 960 \\ 90 \\ 30 \end{bmatrix}, \mathbf{x}_3 = \begin{bmatrix} 2340 \\ 720 \\ \frac{45}{2} \end{bmatrix}$

11. $\mathbf{x}_2 = \begin{bmatrix} 900 \\ 60 \\ 50 \end{bmatrix}, \mathbf{x}_3 = \begin{bmatrix} 2200 \\ 540 \\ 30 \end{bmatrix}$

13. $y_1 = C_1e^{2t}$
$y_2 = C_2e^{t}$

15. $y_1 = C_1e^{-t}$
$y_2 = C_2e^{6t}$
$y_3 = C_3e^{t}$

17. $y_1 = C_1e^{2t}$
$y_2 = C_2e^{-t}$
$y_3 = C_3e^{t}$

19. $y_1 = C_1e^{t} - 4C_2e^{2t}$
$y_2 = C_2e^{2t}$

21. $y_1 = C_1e^{-t} + C_2e^{3t}$
$y_2 = -C_1e^{-t} + C_2e^{3t}$

23. $y_1 = 3C_1e^{-2t} - 5C_2e^{4t} - C_3e^{6t}$
$y_2 = 2C_1e^{-2t} + 10C_2e^{4t} + 2C_3e^{6t}$
$y_3 = 2C_2e^{4t}$

25. $y_1 = C_1e^{t} - 2C_2e^{2t} - 7C_3e^{3t}$
$y_2 = C_2e^{2t} + 8C_3e^{3t}$
$y_3 = 2C_3e^{3t}$

27. $y_1{}' = y_1 + y_2$
$y_2{}' = y_2$

29. $y_1{}' = y_2$
$y_2{}' = y_3$
$y_3{}' = -4y_2$

31. $\begin{bmatrix} 1 & 0 \\ 0 & 1 \end{bmatrix}$ **33.** $\begin{bmatrix} 9 & 5 \\ 5 & -4 \end{bmatrix}$ **35.** $\begin{bmatrix} 0 & 5 \\ 5 & -10 \end{bmatrix}$

37. $A = \begin{bmatrix} 2 & -\frac{3}{2} \\ -\frac{3}{2} & -2 \end{bmatrix}$, $\lambda_1 = -\frac{5}{2}$, $\lambda_2 = \frac{5}{2}$,

$P = \begin{bmatrix} \frac{1}{\sqrt{10}} & -\frac{3}{\sqrt{10}} \\ \frac{3}{\sqrt{10}} & \frac{1}{\sqrt{10}} \end{bmatrix}$

39. $A = \begin{bmatrix} 13 & 3\sqrt{3} \\ 3\sqrt{3} & 7 \end{bmatrix}$, $\lambda_1 = 4$, $\lambda_2 = 16$,

$P = \begin{bmatrix} \frac{1}{2} & \frac{\sqrt{3}}{2} \\ -\frac{\sqrt{3}}{2} & \frac{1}{2} \end{bmatrix}$

41. $A = \begin{bmatrix} 16 & -12 \\ -12 & 9 \end{bmatrix}$, $\lambda_1 = 0$, $\lambda_2 = 25$, $P = \begin{bmatrix} \frac{3}{5} & -\frac{4}{5} \\ \frac{4}{5} & \frac{3}{5} \end{bmatrix}$

43. Ellipse, $5(x')^2 + 15(y')^2 - 45 = 0$

45. Hyperbola, $-25(x')^2 + 15(y')^2 - 50 = 0$

47. Parabola, $4(y')^2 + 4x' + 8y' + 4 = 0$

49. Hyperbola,
$\frac{1}{2}\left[-(x')^2 + (y')^2 - 3\sqrt{2}x' - \sqrt{2}y' + 6\right] = 0$

51. $A = \begin{bmatrix} 3 & -1 & 0 \\ -1 & 3 & 0 \\ 0 & 0 & 8 \end{bmatrix}$,
$2(x')^2 + 4(y')^2 + 8(z')^2 - 16 = 0$

53. $A = \begin{bmatrix} 1 & 0 & 0 \\ 0 & 2 & 1 \\ 0 & 1 & 2 \end{bmatrix}$,
$(x')^2 + (y')^2 + 3(z')^2 - 1 = 0$

55. Let $P = \begin{bmatrix} a & b \\ c & d \end{bmatrix}$ be a 2×2 orthogonal matrix such that $|P| = 1$. Define $\theta \in (0, 2\pi)$ as follows.

(i) If $a = 1$, then $c = 0$, $b = 0$, and $d = 1$, so let $\theta = 0$.
(ii) If $a = -1$, then $c = 0$, $b = 0$, and $d = -1$, so let $\theta = \pi$.
(iii) If $a \geq 0$ and $c > 0$, let $\theta = \arccos(a)$, $0 < \theta \leq \pi/2$.
(iv) If $a \geq 0$ and $c < 0$, let $\theta = 2\pi - \arccos(a)$, $3\pi/2 \leq \theta < 2\pi$.
(v) If $a \leq 0$ and $c > 0$, let $\theta = \arccos(a)$, $\pi/2 \leq \theta < \pi$.
(vi) If $a \leq 0$ and $c < 0$, let $\theta = 2\pi - \arccos(a)$, $\pi < \theta \leq 3\pi/2$.

In each of these cases, confirm that

$$P = \begin{bmatrix} a & b \\ c & d \end{bmatrix} = \begin{bmatrix} \cos\theta & -\sin\theta \\ \sin\theta & \cos\theta \end{bmatrix}.$$

Review Exercises – Chapter 7 *(page 474)*

1. (a) $\lambda^2 - 9 = 0$ (b) $\lambda = -3$, $\lambda = 3$
(c) A basis for $\lambda = -3$ is $(1, -5)$ and a basis for $\lambda = 3$ is $(1, 1)$.

3. (a) $(\lambda - 4)(\lambda - 8)^2 = 0$ (b) $\lambda = 4$, $\lambda = 8$
(c) A basis for $\lambda = 4$ is $(1, -2, -1)$ and a basis for $\lambda = 8$ is $(4, -1, 0)$, $(3, 0, 1)$.

5. (a) $(\lambda - 2)(\lambda - 3)(\lambda - 1) = 0$
(b) $\lambda = 1$, $\lambda = 2$, $\lambda = 3$
(c) A basis for $\lambda = 1$ is $(1, 2, -1)$, a basis for $\lambda = 2$ is $(1, 0, 0)$, and a basis for $\lambda = 3$ is $(0, 1, 0)$.

7. (a) $(\lambda - 1)^2(\lambda - 3)^2 = 0$ (b) $\lambda = 1$, $\lambda = 3$
(c) A basis for $\lambda = 1$ is $\{(1, -1, 0, 0), (0, 0, 1, -1)\}$ and a basis for $\lambda = 3$ is $\{(1, 1, 0, 0), (0, 0, 1, 1)\}$.

9. Not diagonalizable

11. $P = \begin{bmatrix} 1 & 0 & 1 \\ 0 & 1 & 0 \\ 1 & 0 & -1 \end{bmatrix}$ (The answer is not unique.)

13. The characteristic equation of

$$A = \begin{bmatrix} \cos\theta & -\sin\theta \\ \sin\theta & \cos\theta \end{bmatrix}$$

is $\lambda^2 - (2\cos\theta)\lambda + 1 = 0$. The roots of this equation are $\lambda = \cos\theta \pm \sqrt{\cos^2\theta - 1}$. If $0 < \theta < \pi$, then $-1 < \cos\theta < 1$, which implies that $\sqrt{\cos^2\theta - 1}$ is imaginary.

15. A has only one eigenvalue, $\lambda = 0$, and the dimension of its eigenspace is 1. So, the matrix is not diagonalizable.

17. A has only one eigenvalue, $\lambda = 3$, and the dimension of its eigenspace is 2. So, the matrix is not diagonalizable.

19. $P = \begin{bmatrix} 0 & 1 \\ 1 & 0 \end{bmatrix}$

21. Because the eigenspace corresponding to $\lambda = 1$ of matrix A has dimension 1, while that of matrix B has dimension 2, the matrices are not similar.

23. Both orthogonal and symmetric

25. Symmetric **27.** Neither

29. $P = \begin{bmatrix} \frac{2}{\sqrt{5}} & -\frac{1}{\sqrt{5}} \\ \frac{1}{\sqrt{5}} & \frac{2}{\sqrt{5}} \end{bmatrix}$ (The answer is not unique.)

31. $P = \begin{bmatrix} \frac{1}{\sqrt{2}} & 0 & \frac{1}{\sqrt{2}} \\ 0 & 1 & 0 \\ -\frac{1}{\sqrt{2}} & 0 & \frac{1}{\sqrt{2}} \end{bmatrix}$

(The answer is not unique.)

33. $\left(\frac{3}{5}, \frac{2}{5}\right)$ **35.** $\left(\frac{3}{5}, \frac{2}{5}\right)$ **37.** $\left(\frac{1}{4}, \frac{1}{2}, \frac{1}{4}\right)$

39. $\left(\frac{4}{16}, \frac{5}{16}, \frac{7}{16}\right)$ **41.** Proof

43. $A = \begin{bmatrix} 0 & 1 \\ 0 & \frac{9}{4} \end{bmatrix}$, $\lambda_1 = 0$, $\lambda_2 = \frac{9}{4}$

45. $A^2 = \begin{bmatrix} 56 & -40 \\ 20 & -4 \end{bmatrix}$, $A^3 = \begin{bmatrix} 368 & -304 \\ 152 & -88 \end{bmatrix}$

47. (a) and (b) Proof **49.** Proof

51. $P = \begin{bmatrix} \frac{1}{\sqrt{2}} & -\frac{1}{\sqrt{2}} \\ \frac{1}{\sqrt{2}} & \frac{1}{\sqrt{2}} \end{bmatrix}$

53. (a) $a = b = c = 0$

(b) Dim $= 1$ if $a \neq 0,\ b \neq 0,\ c \neq 0$.
Dim $= 2$ if exactly one is 0.
Dim $= 3$ if exactly two are 0.

55. (a) True. See "Definitions of Eigenvalue and Eigenvector," page 422.

(b) False. See Theorem 7.4, page 436.

(c) True. See "Definition of a Diagonalizable Matrix," page 435.

57. $\mathbf{x}_2 = \begin{bmatrix} 100 \\ 25 \end{bmatrix}$, $\mathbf{x}_3 = \begin{bmatrix} 25 \\ 25 \end{bmatrix}$, $\mathbf{x} = t\begin{bmatrix} 2 \\ 1 \end{bmatrix}$

59. $\mathbf{x}_2 = \begin{bmatrix} 4500 \\ 300 \\ 50 \end{bmatrix}$, $\mathbf{x}_3 = \begin{bmatrix} 1500 \\ 4500 \\ 50 \end{bmatrix}$, $\mathbf{x} = t\begin{bmatrix} 24 \\ 12 \\ 1 \end{bmatrix}$

61. $\mathbf{x}_2 = \begin{bmatrix} 1440 \\ 108 \\ 90 \end{bmatrix}$, $\mathbf{x}_3 = \begin{bmatrix} 6588 \\ 1296 \\ 81 \end{bmatrix}$

63. $y_1 = -2C_1 + C_2e^t$
$y_2 = C_1$

65. $y_1 = C_1e^t + C_2e^{-t}$
$y_2 = C_1e^t - C_2e^{-t}$
$y_3 = C_3$

67. $A = \begin{bmatrix} 1 & \frac{3}{2} \\ \frac{3}{2} & 1 \end{bmatrix}$

$P = \begin{bmatrix} \frac{1}{\sqrt{2}} & -\frac{1}{\sqrt{2}} \\ \frac{1}{\sqrt{2}} & \frac{1}{\sqrt{2}} \end{bmatrix}$

$5(x')^2 - (y')^2 = 6$

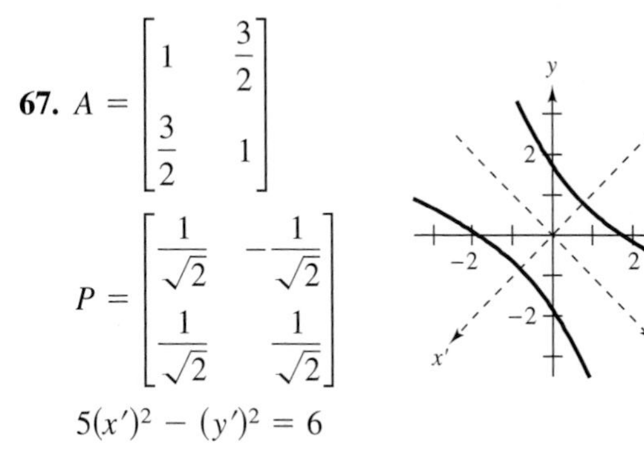

69. $A = \begin{bmatrix} 0 & \frac{1}{2} \\ \frac{1}{2} & 0 \end{bmatrix}$

$P = \begin{bmatrix} \frac{1}{\sqrt{2}} & -\frac{1}{\sqrt{2}} \\ \frac{1}{\sqrt{2}} & \frac{1}{\sqrt{2}} \end{bmatrix}$

$(x')^2 - (y')^2 = 4$

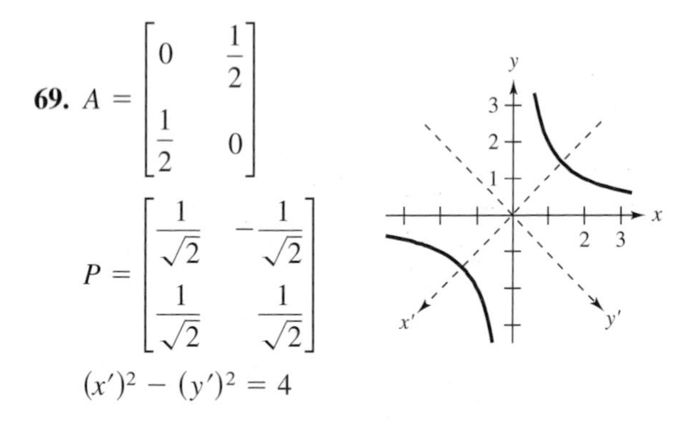

Cumulative Test – Chapters 6 and 7 *(page 479)*

1. Yes, T is a linear transformation.

2. No, T is not a linear transformation.

3. (a) $(1, -1, 0)$ (b) $(5, t)$

4. $\{(s, s, -t, t)\colon s, t \text{ are real}\}$

5. (a) $\text{Span}\{(0, -1, 0, 1), (1, 0, -1, 0)\}$
(b) $\text{Span}\{(1, 0), (0, 1)\}$
(c) $\text{Rank} = 2$, nullity $= 2$

6. $\begin{bmatrix} 1 & 1 & 0 \\ 0 & 1 & 1 \\ 1 & 0 & -1 \end{bmatrix}$

7. $\begin{bmatrix} \frac{1}{2} & -\frac{1}{2} \\ -\frac{1}{2} & \frac{1}{2} \end{bmatrix}$, $T(1, 1) = (0, 0)$, $T(-2, 2) = (-2, 2)$

8. $T^{-1}(x, y) = \left(\frac{1}{3}x + \frac{1}{3}y, -\frac{2}{3}x + \frac{1}{3}y\right)$

9. $\begin{bmatrix} -1 & -2 \\ 0 & 1 \\ 2 & 1 \end{bmatrix}$, $T(0, 1) = (1, 0, 1)$

10. (a) $A = \begin{bmatrix} 1 & -2 \\ 1 & 4 \end{bmatrix}$ (b) $P = \begin{bmatrix} 1 & 1 \\ 1 & 2 \end{bmatrix}$

(c) $A' = \begin{bmatrix} -7 & -15 \\ 6 & 12 \end{bmatrix}$ (d) $\begin{bmatrix} 9 \\ -6 \end{bmatrix}$

(e) $[\mathbf{v}]_B = \begin{bmatrix} 1 \\ -1 \end{bmatrix}$, $[T(\mathbf{v})]_B = \begin{bmatrix} 3 \\ -3 \end{bmatrix}$

11. $\lambda = 1, \begin{bmatrix} 1 \\ 0 \\ 0 \end{bmatrix}$; $\lambda = 0, \begin{bmatrix} -1 \\ -1 \\ 3 \end{bmatrix}$; $\lambda = 2, \begin{bmatrix} 1 \\ 1 \\ -1 \end{bmatrix}$

12. $\lambda = 1$ (three times), $\begin{bmatrix} 1 \\ 0 \\ 0 \end{bmatrix}$

13. $P = \begin{bmatrix} 3 & 1 \\ 1 & 1 \end{bmatrix}$ (The answer is not unique.)

14. $\{(0, 1, 0), (1, 1, 1), (2, 2, 3)\}$

15. $\begin{bmatrix} \frac{1}{\sqrt{2}} & \frac{1}{\sqrt{2}} \\ -\frac{1}{\sqrt{2}} & \frac{1}{\sqrt{2}} \end{bmatrix}$

16. $\begin{bmatrix} \frac{1}{\sqrt{3}} & \frac{1}{\sqrt{2}} & \frac{1}{\sqrt{6}} \\ \frac{1}{\sqrt{3}} & 0 & -\frac{2}{\sqrt{6}} \\ \frac{1}{\sqrt{3}} & -\frac{1}{\sqrt{2}} & \frac{1}{\sqrt{6}} \end{bmatrix}$

17. $y_1 = C_1e^t$
$y_2 = C_2e^{3t}$

18. $\begin{bmatrix} 4 & -4 \\ -4 & 4 \end{bmatrix}$

19. $\mathbf{x}_2 = \begin{bmatrix} 1800 \\ 120 \\ 60 \end{bmatrix}$, $\mathbf{x}_3 = \begin{bmatrix} 6300 \\ 1440 \\ 48 \end{bmatrix}$

20. λ is an eigenvalue of A if there exists a nonzero vector $\mathbf{x}$ such that $A\mathbf{x} = \lambda\mathbf{x}$. $\mathbf{x}$ is called an eigenvector of A. If A is an $n \times n$ matrix, then A can have n eigenvalues, possibly complex and possibly repeated.

21. P is orthogonal if $P^{-1} = P^T$. The possible eigenvalues of the determinant of an orthogonal matrix are 1 and -1.

22–26. Proof

27. 0 is the only eigenvalue.

INDEX